Earth Science for Civil and Environmental Engineers

This carefully targeted and rigorous new textbook introduces engineering students to the fundamental principles of applied earth science, highlighting how modern soil and rock mechanics, geomorphology, hydrogeology, seismology and environmental geochemistry affect geotechnical and environmental practice. Key geological topics of engineering relevance, including soils and sediments, rocks, groundwater and geologic hazards, are presented in an accessible and engaging way. A broad range of international case studies add real-world context and demonstrate practical applications in field and laboratory settings to guide site characterization. End-of-chapter problems are included for self-study and evaluation, and supplementary online materials include electronic figures, additional examples, solutions, and guidance on useful software.

Featuring a detailed glossary introducing key terminology, this text requires no prior geological training and is essential reading for senior undergraduate or graduate students in civil, geological, geotechnical and geoenvironmental engineering. It is also a useful reference and bridge for earth science graduates embarking on engineering geology courses.

Richard E. Jackson is a Fellow with Geofirma Engineering Ltd and Adjunct Professor of Earth and Environmental Sciences at the University of Waterloo, Ontario, Canada. He has practiced in the USA and Canada for 40 years and was responsible for establishing the Groundwater Contamination Program for the Canadian Department of the Environment. He is the recipient of several awards, including the 2008 Geoenvironmental Award from the Canadian Geotechnical Society, and the 2013 Robert N. Farvolden Award from the Canadian Geotechnical Society and the International Association of Hydrogeologists. He is an elected Honorary Fellow of Geoscientists Canada.

Earth Science for Civil and Environmental Engineers

Richard E. Jackson
Geofirma Engineering Ltd

CAMBRIDGE
UNIVERSITY PRESS

CAMBRIDGE
UNIVERSITY PRESS

University Printing House, Cambridge CB2 8BS, United Kingdom

One Liberty Plaza, 20th Floor, New York, NY 10006, USA

477 Williamstown Road, Port Melbourne, VIC 3207, Australia

314–321, 3rd Floor, Plot 3, Splendor Forum, Jasola District Centre, New Delhi – 110025, India

79 Anson Road, #06–04/06, Singapore 079906

Cambridge University Press is part of the University of Cambridge.

It furthers the University's mission by disseminating knowledge in the pursuit of education, learning, and research at the highest international levels of excellence.

www.cambridge.org
Information on this title: www.cambridge.org/9780521847254
DOI: 10.1017/9781139046336

First published 2019

Printed in the United Kingdom by TJ international Ltd, Padstow, Cornwall

A catalogue record for this publication is available from the British Library.

ISBN 978-0-521-84725-4 Hardback

Additional resources for this publication at www.cambridge.org/Jackson2019

CONTENTS

Preface   page xi
Acknowledgements   xv
List of Symbols Used in Earth Science   xvi
List of Greek Symbols   xix

1

Introduction   1

1.1   An Introduction for Geotechnical Practice   1
1.2   An Introduction to the Applied Earth Sciences   3
      1.2.1   Basic Geological Terms   3
      1.2.2   Heterogeneous Materials and the
              Representative Elementary Volume   7
      1.2.3   Rates of Geologic Processes   9
1.3   Elementary Principles   9
      1.3.1   Amonton's Law of Friction   10
      1.3.2   Mohr–Coulomb Criterion   10
      1.3.3   Elasticity and Hooke's Law   13
      1.3.4   Terzaghi's Principle of Effective Stress   14
      1.3.5   Porous Media and Darcy's Law   15
1.4   Summary   17
1.5   Further Reading   17
1.6   Questions   18

PART I   ROCKS AS ENGINEERING MATERIALS   19

2

The Structure and Composition of the Earth   21

2.1   Structure and Tectonics   21
2.2   Earth History   28
2.3   Rock-Forming Minerals   30
      2.3.1   Silicate Minerals   31
      2.3.2   Carbonate Minerals   33
      2.3.3   Evaporite Minerals   33
      2.3.4   Oxide Minerals   33
      2.3.5   Sulfide Minerals   33

      2.3.6   Mineral Identification   33
2.4   Rocks as Mineral Assemblages   38
      2.4.1   Igneous Rocks   40
      2.4.2   Sedimentary Rocks   42
      2.4.3   Metamorphic Rocks   48
      2.4.4   Ore Deposits   50
2.5   Summary   51
2.6   Further Reading   52
2.7   Questions   52

3

Geological Structures and Maps   53

3.1   Introduction   53
3.2   Basic Structural Measurements   55
      3.2.1   Strike and Dip in Folded Rocks   55
      3.2.2   The Plunge of a Fold   56
      3.2.3   Cross-Sections   57
      3.2.4   Field Measurements   57
3.3   Faults and Shear Zones   60
3.4   Joints   64
3.5   Unconformities   65
3.6   Introduction to Geological and Engineering
      Geology Maps   67
3.7   Stereographic Projection   72
3.8   Summary   73
3.9   Further Reading   73
3.10 Questions   73

4

Rock Mechanics   75

4.1   Introduction   75
4.2   Stress   75
      4.2.1   Origin of Stress Regimes   75
      4.2.2   Stress Regimes and Indicators   76
      4.2.3   Variation of Stress with Depth in
              Hard Rocks   79

4.2.4   Variation of Stress with Depth in Sedimentary Basins   80

4.2.5   Erosional Unloading   82

4.3   Strain and Deformation   82

4.3.1   The Load–Deformation Curve   82

4.3.2   Brittle and Ductile Deformation   83

4.3.3   Discontinuities and Fractures   86

4.4   Strength   87

4.4.1   Design of Underground Excavations   89

4.4.2   Assessment of Dam Foundations   90

4.5   Flow in Fractured Rocks   91

4.5.1   The Cubic Law   91

4.5.2   Groundwater Inflow into Tunnels   92

4.6   Rock-Slope Failure   93

4.7   Summary   95

4.8   Further Reading   96

4.9   Questions   96

5

Characterization of Rocks and Rock Masses   97

5.1   The Elements of Site Characterization   97

5.1.1   The Use of Geophysics   97

5.1.2   Drilling, Cores and Core Logging   100

5.2   Characterization of an Aggregate Quarry   101

5.3   Field and Laboratory Measurements   106

5.3.1   Index Tests   106

5.3.2   Strength   107

5.3.3   Porosity and Permeability   107

5.3.4   Discontinuities   108

5.4   Rock-Mass Classification   109

5.4.1   Rock-Mass Rating (RMR)   109

5.4.2   The Q system   109

5.4.3   Geological Strength Index (GSI)   110

5.4.4   Weak Rocks   112

5.5   Hydraulic and Geomechanical Testing in Boreholes   113

5.5.1   Hydraulic Testing   113

5.5.2   Geomechanical Testing   115

5.6   The Engineering Stratigraphic Column   116

5.7   Digital Photogrammetry and Monitoring   116

5.8   Summary   119

5.9   Further Reading   119

5.10   Questions   120

PART II   SOILS AND SEDIMENTS   121

6

Terrain Evolution and Analysis   123

6.1   Climate and the Hydrologic Cycle   123

6.1.1   Energy Budget of Solar Radiation   123

6.1.2   The Hydrologic Cycle   124

6.1.3   Streamflow Generation   126

6.2   Weathering of Rock   127

6.2.1   The Weathered Profile   127

6.2.2   The Effects of Climate   128

6.2.3   Mechanical Properties of Weathered Rocks   131

6.3   Weathering and Slope Movement   132

6.3.1   The Failure of Slopes   132

6.3.2   Clays and Shales   134

6.3.3   Igneous Rocks   136

6.3.4   Volcanic Rocks   136

6.4   Alluvial Channels   136

6.4.1   Stream Power   137

6.4.2   Hydraulic Geometry of Channels   138

6.4.3   Channel Morphology   139

6.4.4   Avulsion   140

6.4.5   Paleoflood Hydrology   142

6.5   Terrain Analysis   143

6.5.1   Air-Photo Analysis   144

6.5.2   LiDAR   145

6.5.3   Softcopy Mapping   146

6.5.4   Satellite Imagery   147

6.6   Summary   148

6.7   Further Reading   148

6.8   Questions   149

7

Environmental Geochemistry and Mineralogy   150

7.1   Introduction   150

7.2   Chemical Weathering   150

7.3   Ionic Strength and Activity   154

7.4   Solubility of Minerals   155

7.5   Alkalinity and Carbonate Mineral Dissolution   157

7.6   Redox Processes   158

7.7   Sorption and Sorbents   161

7.8	Acid-Rock Drainage	164
7.8.1	Metal-Sulfide Mineral Oxidation	164
7.8.2	Mineral Deposits	164
7.8.3	Reactions in Mine-Waste Tailings	166
7.9	Clay Minerals	167
7.9.1	Heave in Colorado Claystones	168
7.9.2	The Hong Kong Landslides	170
7.9.3	Clogging of Tunnel Boring Machines in "Sticky" Clay Soils	173
7.10	Summary	173
7.11	Further Reading	174
7.12	Questions	174

8

Glacial Sediments and Permafrost 175

8.1	Glaciation during the Quaternary	175
8.2	Ice Flow and Glaciotectonics	178
8.3	Glacial Erosion and Landforms	181
8.4	Glacial Sediments	184
8.4.1	Till	184
8.4.2	Glaciomarine Clays	186
8.4.3	Glaciofluvial Sediments	187
8.5	Permafrost	190
8.5.1	Thermal Regime	190
8.5.2	Periglacial Environments	191
8.5.3	Geotechnical Issues	193
8.6	Summary	194
8.7	Further Reading	195
8.8	Questions	195

9

Fluvial Processes and Sediments 196

9.1	Sediment Erosion and Transport	196
9.1.1	Entrainment of Sediment	196
9.1.2	Estimating Erosion Potential	199
9.1.3	Sedimentation in Reservoirs	202
9.2	Fluvial Sedimentary Regimes	204
9.2.1	Alluvial Deposits	204
9.2.2	Alluvial Fans	210
9.2.3	Deltaic Deposits	212
9.3	Diagenesis of Clastic Sediments	214
9.4	Summary	216
9.5	Further Reading	217
9.6	Questions	217

10

Characterization of Soils and Sediments 219

10.1	Elements of Site Characterization	220
10.1.1	Drilling Methods	222
10.1.2	Core Acquisition	224
10.1.3	Core Logs	226
10.1.4	Cone Penetration Testing	227
10.1.5	Geophysical Surveys	228
10.2	Textural Analysis of Soils and Sediments	231
10.2.1	Udden–Wentworth Scale	232
10.2.2	Inferences from Textural Analysis	232
10.3	Laboratory Testing of Soils and Sediments	236
10.3.1	Hydraulic Conductivity	238
10.3.2	Shear Strength	239
10.3.3	Consolidation of Soils and Sediments	240
10.4	Characterization of Aggregate Sources	241
10.5	Summary	242
10.6	Further Reading	243
10.7	Questions	243

PART III   GROUNDWATER 245

11

Hydrogeology 247

11.1	Flow through Porous Media	247
11.1.1	Porosity	247
11.1.2	Velocity, Permeability and Hydraulic Conductivity	247
11.1.3	Factors Affecting Hydraulic Conductivity	250
11.1.4	Fluid Potential	250
11.2	Compressibility, Storage and the Flow Equation	251
11.2.1	Compressibility	251
11.2.2	Storage and Transmissivity	252
11.2.3	The Groundwater Flow Equation	254
11.3	Wells and Hydraulic Testing	255
11.3.1	Wells	255
11.3.2	Hydraulic Testing	256
11.4	Groundwater Flow Systems	258
11.4.1	Idealized Flow Patterns in Groundwater Basins	259
11.4.2	Shallow Flow Systems	261
11.4.3	Transient-State Flow Systems	266

11.5 Aquifers and Aquitards 266
  11.5.1 Alluvial Aquifers 269
  11.5.2 Bedrock Aquifers 269
  11.5.3 Aquitards 271
11.6 Summary 272
11.7 Further Reading 273
11.8 Questions 273

12

Groundwater Quality and Contamination 275
12.1 Groundwater Quality 275
12.2 Environmental Isotopes 277
  12.2.1 Applications of Environmental Isotopes 277
  12.2.2 Radioisotopes 278
12.3 Groundwater Contamination 280
  12.3.1 The Geoenvironmental Perspective 280
  12.3.2 Solute Transport 281
12.4 Immiscible Contaminants 282
  12.4.1 Fluid Saturations and Interfacial Behaviour 283
  12.4.2 Migration of NAPLs 284
  12.4.3 Plume Generation by NAPLs 288
  12.4.4 Fugitive Gas Migration 290
12.5 Fate of Dissolved Contaminants 290
  12.5.1 Contaminant Sorption 292
  12.5.2 Contaminant Biodegradation 292
  12.5.3 Landfill Leachate 293
12.6 Characterization of Contaminated Sites 294
12.7 Seawater Intrusion 294
12.8 Wellhead Protection 297
12.9 Summary 298
12.10 Further Reading 298
12.11 Questions 298

PART IV   GEOLOGICAL HAZARDS 301

13

Land Subsidence and Karst 303
13.1 Groundwater-Extraction-Induced Subsidence 303
  13.1.1 Compaction 304
  13.1.2 Aquitard Drainage 305
  13.1.3 Fissure Development and Growth 308
  13.1.4 Measuring Subsidence 309
13.2 Mining-Induced Subsidence 311

13.3 Karst 316
  13.3.1 Dissolution of Soluble Rocks 316
  13.3.2 Groundwater Flow in Carbonate Aquifers 317
  13.3.3 Evolution of Karst Terrain 319
  13.3.4 Engineering Problems Associated with Karst 322
  13.3.5 Site Investigation in Karst Terrain 324
13.4 Summary 325
13.5 Further Reading 325
13.6 Questions 326

14

Seismicity and Earthquakes 327
14.1 Introduction to Seismic Faulting 327
14.2 Seismic Waves and Seismometry 331
14.3 Friction and Faults 335
14.4 Detection of Active Faults 337
14.5 Seismic Hazard Analysis 340
  14.5.1 Ground Motion 341
  14.5.2 Seismic Hazard Maps 344
  14.5.3 Liquefaction 348
14.6 Summary 352
14.7 Further Reading 352
14.8 Questions 353

15

Landslides 354
15.1 Definitions, Types and Processes 354
15.2 Landslide Triggering Mechanisms 356
  15.2.1 Hydrologic Triggering 356
  15.2.2 Seismic Triggering 359
  15.2.3 Volcanic Triggering 361
15.3 Characterization and Monitoring of Landslides 362
15.4 Slope Failure in Soils 366
  15.4.1 Measurement of the Critical Parameters 366
  15.4.2 Soil Slope Stability Analysis 366
  15.4.3 Unsaturated Soils 368
15.5 Slope Failure in Rock Masses 370
  15.5.1 Characterizing the Strength of Rock Slopes 370
  15.5.2 The Shear Strength of Discontinuities 372
  15.5.3 The Shear Strength of Rock Masses 373
  15.5.4 Rock-Slope Stability Analysis 373

15.6 Case History: The Vaiont Reservoir
      Disaster   374
15.7 Summary   378
15.8 Further Reading   378
15.9 Questions   379

16
Coastal Hazards   381
16.1 Coastal Landforms   383
16.2 Waves: Form and Energy   384
16.3 Sea-Level Change   388
  16.3.1 Tides   388
  16.3.2 The Rise in Mean Sea Level   389
  16.3.3 Storm Surges   392
  16.3.4 Tsunamis   397
  16.3.5 Coastal Subsidence   398

16.4 Stability of Coastal Cliffs and Bluffs   399
  16.4.1 Lake Erie Bluffs   399
  16.4.2 Californian Cliffs   401
  16.4.3 The English Chalk Cliffs   401
16.5 Nearshore Hazards   403
  16.5.1 Iceberg Scouring   403
  16.5.2 Gassy Sediments   406
  16.5.3 Submarine Landslides   406
16.6 Summary   408
16.7 Further Reading   408
16.8 Questions   408

Glossary   410
References   430
Index   455
The plate section can be found between pages 236 and 237

PREFACE

Motivation and Objectives

The purpose of this book is to introduce civil and environmental engineering students – and those active in engineering – to the fundamental principles of the earth sciences as they affect their practice. In particular, this book is intended for the use of geotechnical and geoenvironmental students and those in that practice. Hopefully, the reader will come to appreciate the complex problems that exist in engineering practice in a geological environment and find them intellectually rewarding. It also seeks to persuade the engineer that he or she should not try to solve all their problems without the advice of geoscientists. Rather, it is written to give the engineer an appreciation of the fundamentals of geological processes in order to work more effectively in a team comprising engineers and geoscientists.

This book is divided into four parts: Rocks as engineering materials; Soils and sediments; Groundwater; and Geological hazards. Additionally, the book contains supplementary material on a website, where additional case histories may be studied and where useful software is discussed. In many cases, this software is available free to the user; in other cases, the software is available at reasonable prices from software houses that cater to engineers. Chapters end with a set of questions to test understanding of the topics covered. Answers to the questions are to be found online in the material reserved for instructors.

The objectives of the four parts of this book may be summarized as follows:

Part I: Rocks as engineering materials. To understand the structure and properties of rocks as engineering materials in foundations, mineral extraction and waste disposal. This part necessarily begins with a discussion of minerals and the composition of rocks composed of these minerals before introducing the reader to geological structures and maps. The next chapter introduces the science of rock mechanics to allow the reader to appreciate its role in slope stability, tunnelling, waste disposal and seismicity, among other areas of engineering importance. Finally,

Part I ends with a discussion of how engineers and geoscientists characterize the properties of rocks important in their work.

Part II: Soils and sediments. To appreciate how soils and sediments are produced, transported and obtain their physical-chemical properties. This part introduces the engineer to weathering, glacial and fluvial processes and the characterization of soils and sediments. Furthermore, Part II devotes a chapter to assist the geoenvironmental engineer in understanding how geochemical processes and mineralogy can affect his or her work.

Part III: Groundwater. To understand groundwater flow and the cause and development of groundwater contamination. Here we consider the nature of hydraulic conductivity, groundwater flow systems, aquifers and aquitards. These discussions are followed by a review of the processes that generate groundwater quality, whether natural or contaminated.

Part IV: Geological hazards. To recognize the conditions under which geological hazards exist and may threaten lives and infrastructure. The hazards considered include land subsidence due to extraction of groundwater and minerals and to karst development, earthquakes and active faults, landslides in their many forms and coastal hazards, such as storm surges and beach erosion in an era of climate change.

Teaching Earth Sciences to Engineers

Geotechnical engineers face problems of enormous complexity. When the late John Harvey wrote the predecessor of this book – *Geology for Geotechnical Engineers* – in the early 1980s, the tunnel in the chalk beneath the English Channel linking France and England had not yet been started, and it was not completed until 1990. Vastly more difficult tunnels have been completed since, such as those in faulted rock in Taiwan, the Austrian Alps and Greece. Today, geotechnical engineers sometimes operate the tunnel boring machines, so complex has their task become. Geotechnical engineers have also learned much about protecting pipelines and transportation corridors from landslides and from seismic shock. Even if we are not quite ready to pronounce on the magnitude and timing of the next major displacement of the San Andreas Fault, when a magnitude 7.9 earthquake ruptured the Denali Fault in Alaska in 2002 and moved 4 m laterally and 0.5 m vertically, the Trans Alaska Oil Pipeline did not rupture. In fact, the pipeline had been designed to withstand 6.1 m of horizontal displacement and 1.5 m of vertical displacement at the location where the fault crossed the pipeline.

Similarly, today, geoenvironmental engineers design landfills, deep geological repositories and mine-tailings sites for the safe storage of hazardous wastes. Before geoenvironmental engineering developed, it was common for hazardous-waste sites to leak contaminants to groundwater and cause the closure of nearby public water-supply wells or to streams and cause fish kills in rivers.

Nevertheless, even today, many civil engineers graduate and enter practice without any academic introduction to the applied earth sciences irrespective of the onus on the engineer to protect public safety. Despite the advice of some of the greatest geotechnical engineers and engineering geologists of the twentieth century, there is no requirement for civil engineers to take a course in geology as applied to engineering in the USA; the situation is better in the UK and Canada, and perhaps elsewhere.

Karl Terzaghi, the father of geotechnical engineering, recommended "a two-semester course combined with field trips" taught by "a geologist who appreciates the requirements of engineers and an engineer who has learned from personal experience that geology is indispensable in the practice of his profession" (cited by Proctor, 1981). While acknowledging the need for natural science courses in the undergraduate civil engineering curriculum, the American Society of Civil Engineers (ASCE, 2008) merely suggests that geotechnical and environmental engineers would be well served by taking an introductory course in geology and geomorphology.

However, a typical elective course in physical geology is not necessarily helpful in that the lecturer is unlikely to have experience of geologic practice in engineering projects, because such lecturers are rare in earth science or civil engineering departments. Courses in introductory physical geology should be considered part of the liberal education of the engineer, not as preparation for engineering practice where public safety is the responsibility of the engineer.

This book subscribes to Terzaghi's opinion that geological training for civil engineers is essential to a civil engineer's education – in particular for those entering geotechnical and geoenvironmental practice – and therefore is not marginal to the engineer's future career. It has developed from John Harvey's earlier textbook (Harvey, 1982), which was based upon his lectures given to undergraduate civil engineers at the former Plymouth Polytechnic, now the University of Plymouth, in England. It has been written on the basis of the author's experience in practice in North America and through his links to European workers in environmental and engineering geoscience. Some of Harvey's original text remains in Chapter 2 and is acknowledged here to honour his role in teaching engineers.

One implicit goal of the book is to develop some preliminary judgement in evaluating geological phenomena. Here we will follow the advice of Terzaghi's colleague, Ralph Peck, late Professor of Geotechnical Engineering at the University of Illinois, who stressed the need for empiricism and theory in the development of engineering judgement (Peck, 1991). The former tells us what works and what does not, while the latter provides guidance when the engineer must project designs into unknown empirical territory. Judgement in decision making is not something that comes with an engineering degree; it is progressively learned. John Burland (2007), Emeritus Professor of Soil Mechanics at Imperial College, London, refers to this as "well-winnowed experience" (see Figure 1.1).

Richard Goodman of the University of California wrote (Goodman, 1993):

> No doubt, mastering advanced engineering mathematics or thermodynamics is "harder" for some students than understanding the principles of engineering geology. But in the practice of engineering, geology may prove to be the harder subject. The penalties for geologic mistakes can be severe, whereas the confidence that comes from having made the right choice cannot be obtained from a formula or theory. In my experience, most engineering students are more at home with formulas and analysis than with colors and grades of truth.

A second implicit goal of this book is to address the need to weave hydrogeological principles into geotechnical and geoenvironmental practice. In his brief history of geotechnical engineering, Burland (2012a) cited Terzaghi's complaint of 1939 that "in engineering practice difficulties with soils are almost exclusively due, not to the soils themselves, but to the water contained in their voids. On a planet without any water there would be no need for soil mechanics". Thus, Burland (2007) had earlier noted of geotechnical failures associated with site investigations that "nine failures out of ten result from a lack of knowledge about the ground profile – often the groundwater conditions".

Here I try to provide a physical context to interpret variations in hydraulic head and hydraulic gradient that might assist geotechnical engineers in site investigations. Splendid examples of such practice have been published recently for landslide sites in Western Canada (Eshraghian et al., 2008) and Northern Ireland (Hughes et al., 2016) and by Wyllie (2018) in his revision of Hoek and Bray's *Rock Slope Engineering*. Burland (2007) cites other examples from Terzaghi where hydrogeological phenomena – high heads and piping – complicated geotechnical practice; Burland (2012b) describes similar issues in his own work in guiding the design of the underground car park beneath the Houses of Parliament in London. Furthermore, hydrogeology has become central both to the practice of geoenvironmental engineering, where landfills and mine-tailings facilities must be designed and plumes of contaminated groundwater controlled and remediated, and to the control of land subsidence from over-extraction of groundwater and enhanced dissolution of karst rock.

In the pages that follow, we shall bear in mind ten areas of competency identified by Professor Allen Hatheway (2005), formerly of the Missouri School of Mines, as being required of the young engineer about to enter practice:

- an ability to define the physical properties and characteristics of soils, rock – especially weak rock – and groundwater;
- an appreciation of the manner in which these materials are found in nature;
- an appreciation of the regional geomorphology as the expression of the combined effects of climate, weathering and the sum history of all geologic forces and phenomena over history on the geology of the region;
- an understanding of how geological field data are collected, tested, evaluated, interpreted and then converted to specifications concerning the properties of the site;
- a sense of how anomalies occur routinely in geological materials and how such features can alter, disturb or remove what is most predictable about subsurface interpretations and projections;
- a realization of how geologic discontinuities can alter the properties of geological materials;
- a realization of how the presence of water in geological materials can effectively influence the nature of a site in terms of the construction and performance of engineered works;
- an appreciation of how dynamic earth processes are continually bringing change to the landscape and to the subsurface;
- a sense of the nature of risk as it relates to the potential for the presence and potential impact of undetected geologic features, or the absence thereof; and
- an appreciation of how to prepare a scope of work to seek geological specification of the nature of the proposed construction location, i.e., site characterization.

Like Terzaghi, the geotechnical engineer is urged to make geology an abiding interest. Ruth Doggett Terzaghi, his spouse, used her skills in the petrography of concrete and soils to advise the founder of soil mechanics. Perhaps this book will suffice as an introduction to the applied earth sciences for young engineers – and marriage to a geologist will not be necessary!

Readership

This book is suitable for (i) undergraduates in their final years of their degree course, (ii) graduate students entering geotechnical and/or geoenvironmental engineering courses and (iii) engineers in training and those beginning geotechnical and geoenvironmental practice. It is my belief that earth science is best introduced to engineering students following (i) their introduction to practice through work terms or summer jobs and (ii) their education in fluid, solid and soil mechanics. It is then that they can see that geological processes are cut from the same cloth as taught in engineering mechanics.

Earth Science for Civil and Environmental Engineers is intended to be suitable for those entering an M.Sc. or M.S. degree program – referred to elsewhere as MSCE, M.Eng. or MASc – in these engineering subdisciplines and those entering practice who have not had the benefit of earlier training in the applied earth sciences. Because licensing of US engineers may in future require an M.S. degree or equivalent, this book may become useful as the civil engineering profession in the USA develops the necessary "Body of Knowledge for Professional Practice" (ASCE, 2008) and engineering students in Europe enter new courses – often in the English language – guided by the Bologna process.

Terminology

The terminology used in this book is broadly North American, although some specific British usage is acknowledged, e.g., *superficial* rather than *surficial* deposits. Common geoscientific terms that are printed in **bold** in the text should be memorized: many are defined in the glossary at the end of the book; the definitions of others may be found in Bates and Jackson (1984). The term *geoenvironmental* is used to identify those issues that concern environmental engineers but are confined to the subsurface. Its use by the ASCE, as in their *Journal of Geotechnical and Geoenvironmental Engineering*, indicates that it is well established at least in the English-speaking world. Some terms in this book may strike American readers as somewhat unusual but it is important that they learn to understand documents written in English by the international engineering community because so much important work originates outside America.

ACKNOWLEDGEMENTS

I am most grateful to the late John Harvey, who wrote the predecessor book to this – *Geology for Geotechnical Engineers* (Cambridge University Press, 1982) – for encouraging me to write a sequel. Also, I am indebted to the many colleagues who helped me with text and illustrations: Jeff Keaton, of AMEC (now Wood) in Los Angeles, in particular, for all his help; Maurice Dusseault here at the University of Waterloo for his extensive reviews; Tony Philpotts of the University of Connecticut for his review of Chapter 2; and Bob Anderson of the State Seismic Commission in Sacramento for identifying errors and omissions in Chapter 14. I am also indebted to Chris Neville with Papadopulos, to Rob Sengebush and my former colleagues with INTERA in Texas, and to Ken Raven, John Avis and Robert Walsh with Geofirma Engineering for many helpful discussions over the years. All remaining errors and omissions in the book are mine alone.

Many others helped with comments on preliminary text or illustrations, including: Greg Brooks, Steve Grasby, Alfonso Rivera, David Sharpe and Baolin Wang of the Geological Survey of Canada; David Boore, Jeff Coe, Brian Collins, Devin Galloway, Cheryl Hapke, Ralph Haugerud, Lynn Highland, Tom Holzer, Steve Ingebritsen and Valerie Sahakian of the US Geological Survey; and, elsewhere, Philippa Black, Rob Blair, Terry Blair, Jean-Louis Briaud, Mike Church, John Clague, Ian Clark, John Dunnicliff, Grant Ferguson, Emil Frind, Martin Geertsema, Monica Ghirotti, Bob Graham, Bill Haneberg, Mike Hart, Stuart Haszeldine, Jurgen Heinz, Neal Iverson, Jean-Michel Lemieux, Jacques Locat, Jim Lolcama, Derek Martin, Piotr Migon, David Noe, Chuck O'Dale, Pete Pehme, Lynden Penner, George Priest, Pat Pringle, Peter Robertson, Eva Schandl, Paolo Semenza, Martin Shepley, Roy Shlemon, Norm Smith, Kevin Trenberth, Sai Vanapalli, Steve Worthington and Duncan Wyllie.

Susan Francis at Cambridge University Press and her merry band of assistants have been most encouraging and helpful throughout my writing – despite the fact I missed my original deadline by ten years! Carley Crann provided splendid figures and advice on the illustrations. Above all, I am indebted to my wife, Mary Sinclair, for her common sense, encouragement and forbearance in this task that I took on many years ago.

LIST OF SYMBOLS USED IN EARTH SCIENCE

a	annum or year; annual rate as in mm/a or m/a
a, a_c, a_{max}	ground acceleration (seismic) (m^2/s)
$a(f)$	acceleration spectral density function for stochastic model of ground motion, ground acceleration as a function of frequency of waveform
a_i	activity coefficient of ion i
b	aquifer thickness (m)
C	concentration of an analyte, in Chapter 12
C, C_0, C_s	celerity of sequence of wave crests in deep (0) or shallow (s) water, in Chapter 16 (m/s)
C	clay in the USCS of soil textures, e.g., GC, CH and CL for clayey gravel, high-plasticity clay and lean clay, respectively
C_D	damping parameter in slug testing (dimensionless)
$C_{e,i}$	effective solubility of an NAPL
C_H	Hazen coefficient (dimensionless)
$C_{s,i}$	aqueous solubility of an NAPL
C_u	uniformity coefficient (dimensionless)
c'	effective cohesion in a Mohr–Coulomb strength analysis (kPa)
cP	centipoise (unit of viscosity)
c_v	coefficient of consolidation (m^2/s)
D	darcy (unit of permeability in petroleum engineering, in Chapter 11)
D	deuterium (^{2}H)
D_H	hydraulic diameter of stream channel (m)
$D_{i,j}$	dispersion tensor, in Chapter 12 (m^2/s)
D_m	coefficient of molecular diffusion (m^2/s)
D_N	Newmark displacement (cm)
d	grain-size diameter (m)
d	slip distance in an earthquake (m)
d_{min}	minimum grain-size diameter in sieving (m)
d_{10}	grain-size diameter of which 10% of the grains are finer than (m)
E	Young's modulus of elasticity (Pa)
E	wave energy per unit surface area, in Chapter 16 (J/m^2)
E_H	redox potential, in Chapter 7 (V)
E_m	rock mass deformation modulus (Pa)
E_s	strain energy released by an earthquake (J or N m)
E_w	bulk modulus of water (Pa^{-1} or m^2/N)
e	void ratio
e_{max}	maximum void ratio (dimensionless)
F	force, as in F_T or F_N, in Part I (N)
F_D, F_L	drag and lift coefficients, in Chapter 9 (N)
F	fine facies, i.e., silts and clays, in Parts II and III, see Table 11.5
FS	factor of safety
f_c	seismic-source corner frequency, number of wave cycles per second (Hz)
f_s	sleeve resistance in cone penetration testing
G	gravel facies, in Chapter 9; and in USCS terminology as GW, GP, GM and GC for well-graded, poorly graded, silty and clayey gravel, respectively
G_s	specific gravity of soil grains (dimensionless)
g	acceleration of gravity (m/s^2)
H	hydrogen, as in pH or H_2O
H_0	initial water-level change in a slug test (m)
H_b	breaking wave height, in Chapter 16 (m)
H_c	minimum thickness for the dense flow phase of a turbidity current to begin migration (m)
H_{max}	maximum wave height, in Chapter 16 (m)
h	hydraulic head (m)
h_e	environmental water head (m)
h_f	equivalent fresh-water hydraulic head (m)
h	water depth, in Chapter 16 (m)
I	inflow or recharge rate in Figure 11.9 (m/s)
I	major textural component of hydrofacies, in Chapter 9
I_a	Arias intensity, in Chapter 15 (m/s)
I_L	index of liquidity in Atterberg limits (%)
I_P	index of plasticity in Atterberg limits (%)
i	well-loss exponent (dimensionless)
i	angle of inclination of rough surface in Patton's principle, in Chapter 4

K	hydraulic conductivity (m/s)	m	mass of object (kg)
K_0	saturated hydraulic conductivity; a term used only in the context of unsaturated soils (m/s)	mD	millidarcy (unit of permeability in petroleum engineering)
K_h	hydraulic conductivity in the horizontal direction (m/s)	m_i	molarity, in Chapter 7
K_v	hydraulic conductivity in the vertical direction (m/s)	Myr	duration in million years of a geologic event
$K_{x,y,z}$	hydraulic conductivity in x, y and z directions (m/s)	n	porosity
		n_e	effective porosity
K_f	hydraulic conductivity of a fracture, in Chapter 4 (m/s)	n_f	fracture porosity, in equation (5.2)
K	average ratio of horizontal to vertical total stresses, in Chapter 4	O	organic matter, in Chapter 10; in USCS as OL and OH for organic silt and organic clay, respectively
K_D	distribution coefficient of a contaminant (m^3/kg)	O	oxygen, as in dissolved oxygen, in Chapters 7 and 12; or oxygen-18, in Chapters 8 and 12
K_{eq}	equilibrium constant, in Chapter 7	P	poise (unit of viscosity, in Chapter 11)
K_{s0}	solubility product at zero ionic strength	P	partial pressure, as in P_{CO_2}, in Chapter 7
K_{AD}	conditional equilibrium sorption constant	$P^{\star}$	period of oscillation of sinusoid in temperature, in Chapter 8
K_D	sorption distribution coefficient		
K'	average ratio of horizontal to vertical effective stresses, in Chapter 4	P_b	wave power in watts per metre of shoreline (W/m)
K_f	hydraulic conductivity of uniformly fractured rocks, in Chapter 4 (m/s)	P_D, P_d	percentage of soil particles retained by consecutive sieves of sizes D and d, in Chapter 10
K_{oc}	organic carbon partition coefficient (mL/g)	PHA	peak horizontal acceleration (g)
k	specific or intrinsic permeability (m^2; darcy in petroleum engineering)	Pt	peat in USCS of soil textures, in Chapter 10
ka	thousand years ago	p	fluid pressure (Pa)
k_f	permeability of uniformly fractured rocks, in Chapter 4	p_w	pore pressure (Pa)
		p_c	capillary pressure (Pa)
k_{rw},	relative permeabilities of groundwater and NAPL	$p(d)$	Tóth's pore pressure profile (Pa)
k_{rd}	(dimensionless)	p_e	entry pressure (Pa)
L	original thickness of soil or rock sample under compression (m)	p_f	fracture pressure (Pa)
		p_r	reopening pressure (Pa)
L	fault length, in Chapter 14	per mil, $^0/_{00}$	parts per thousand
L_{SR}	length of surface rupture along fault (km)	Q	total fluid discharge or flow rate (m^3/t)
L, L_0	wavelength, in Chapter 16	q	specific discharge, also known as Darcy flux, i.e., Q/A (m^3/m^2 t)
L_0	the deep-water wavelength (m)		
L_e	effective length of the water column in slug testing (m)	q_c	sleeve resistance in cone penetration testing
L_{SR}	surface rupture length of a fault (km)	q_d, q$_w$	flow rates of DNAPL and groundwater in a porous medium (L^3/t)
M	molality, in Chapter 7		
M	silt, in Chapter 10, as in MH and GM, silt with high elasticity and silty gravel, respectively	R	recession rate of coastal cliffs or bluffs, in Chapter 16 (m/a)
MAW	multiple aquifer well	R_f	retardation factor of a contaminant versus groundwater (dimensionless)
M_0	seismic moment (N m, i.e., J)	$R(t)$	residual component of the measured sea level (m)
M_W	moment magnitude; Richter magnitude		
M_L	moment magnitude	REV	representative elementary volume
M_W	surface-wave magnitude	RQD	rock quality designation (%)
M_S	empirical measure (dimensionless)	Sal	salinity of pore water, in Chapter 8 (g/L)
		SSA	specific surface area of a soil sample (m^2/kg)

S	sand facies, in Table 11.5	UCS	unconfined or uniaxial compressive strength, often denoted as q_u, C_0 or σ_c (MPa)
S	shear force applied to the fracture, in Chapter 4 (N)	$\dot{u}$	long-term slip rate of an active fault (mm/a)
S	fluid saturation, fractional or percentage of pore volume	V	mean streamflow velocity, in Chapter 9
S_r	specific retention, fractional or percentage of total soil or rock volume	V_p	pore volume (m^3)
		V_{pm}	volume of porous medium, pore volume, etc. (m^3)
S_y	specific yield, fractional or percentage of total soil or rock volume	V_s	volume of solids (m^3)
S_w	water saturation, fractional or percentage of pore volume, in Part III	V_w	volume of groundwater (m^3)
		v	velocity, average linear velocity of groundwater, in Part III (m/s)
S	shear waves, in Chapter 14	v_S	S-wave velocity (km/s)
S_H	horizontal component of shear wave	v_P	P-wave velocity (km/s)
S_V	vertical component of shear wave	vol	volume of water entraining a particle (m^3)
S_t	sensitivity of clay sample, in Chapter 8 (dimensionless)	w	water content
		w_L, w_P	liquid limit and plastic limit, in Atterberg limits, in Part II (%)
S	storativity, in Part III (dimensionless)	w_0	settling velocity of a stream particle (mm/s)
SI	saturation index of a mineral in aqueous solution, in Chapter 7	$w(t)/H_0$	normalized head change in slug test, in Chapter 11
s_s	specific storage, in Part III (m^{-1})	X_i	mole fraction of compound i, in Chapter 12
s, s_u, s_{ur}	shear strength terms, in Parts II and IV (kPa)	$X(t)$	measured sea level (m)
T	transmissivity, in Part III (m^2/s)	y	displacement of a sinusoidal wave (m)
T, $\overline{T}_s$, T_{amp}	temperature terms, in Chapter 8	$\ddot{y}$	acceleration of a point by seismic vibration (m/s^2)
T	wave period, in Chapter 16 (s)		
$T(t)$	component of sea level associated with astronomical tide (m)	$Z_0(t)$	mean sea level over time (m)
		$\dot{Z}$	erosion rate (m/s)
T_D	DNAPL thickness causing penetration (m)	z	elevation head, in Part III (m)
t	time	z, z_{active}, $z_\star$	depth terms, in Chapter 8
U	uniformity coefficient (dimensionless)		

LIST OF GREEK SYMBOLS

α	compressibility of porous medium (Pa^{-1} or m^2/N)
α_L, α_T	longitudinal and transverse dispersivities (m)
β	angle of the failure plane in Mohr–Coulomb criterion (°)
β	compressibility of water (Pa^{-1} or m^2/N)
γ, γ_w	specific or unit weight of soil or fluid (kN/m^3)
Δp_w	excess pore pressure induced by seismic shaking (Pa)
ϵ_a	axial strain (dimensionless)
ϵ_l	lateral strain (dimensionless)
$\eta(x,t)$	displacement of water surface from mean sea level, in Chapter 16 (m)
θ	fractional moisture content of soil, referenced to porosity or angle of slope
κ	compressibility bulk modulus (Pa)
μ	dynamic viscosity (Pa s)
μ	coefficient of friction in Mohr–Coulomb criterion
μ	shear or rigidity modulus, in Chapter 14 (Pa)
μ_p	plastic viscosity of a non-Newtonian fluid (Pa s)
ν	kinematic viscosity, in Chapter 10 (m^2/s)
ν	Poisson's ratio, in Chapters 1 and 4 (dimensionless)
ρ	fluid density (kg/m^3)
ρ_f, ρ_s	density of fresh water and of salt water (kg/m^3)
ρ_d	dry bulk density (kg/m^3)
ρ_s	particle density (kg/m^3)
ρ_{avg}	average density of water between two depths (kg/m^3)

σ	stress
σ_a	axial stress
σ_e	effective stress
σ_n	normal stress
σ_v	vertical stress
$\sigma_{Hmax}, \sigma_{Hmin}$	maximum and minimum horizontal stresses (Pa)
σ_c	compressive strength of rock
σ_{ci}	laboratory intact uniaxial compressive strength
σ_{cm}	rock mass uniaxial compressive strength
σ	angular frequency of waves (1/s, Hz)
τ	shear stress (Pa)
τ_h	horizontal cyclic shear stress (Pa)
τ_0	cohesion, in Mohr–Coulomb criterion or average bed shear stress, in Chapter 9 (Pa)
τ_{fs}	frictional force per unit area (Pa)
τ_*	Shields parameter (Pa)
τ'	time required for $>90\%$ of ultimate compaction
τ_y	yield strength of a non-Newtonian fluid (Pa)
Φ	fluid potential (m^2/s^2)
ϕ	friction angle, in Chapters 1, 4 and 15 (°)
ϕ'	effective friction angle in shear-strength tests (°)
ϕ'_r	residual strength friction angle for drained samples, in Part II (°)
ψ	soil–water tension or suction (Pa)
ω	stream power per unit bed area (W/m^2)
ω	wave power or energy flux in watts per meter of wave, in Chapter 16 (W/m)

1 Introduction

This introductory chapter considers the nature of the collaborative work of geotechnical engineers and geoscientists in project teams. This requires defining the various disciplines involved and the competencies expected of engineers and geoscientists working on geotechnical problems. Secondly, we introduce the terminology of the geosciences and those concepts that are foreign to most engineering thinking, e.g., endless time, heterogeneity of materials and the importance of discontinuities in those materials. We conclude this chapter with a review of certain elementary ideas and principles of analysis that will be found throughout the book, e.g., Amonton's law, Mohr's analysis of stresses, elasticity and Hooke's law, Terzaghi's principle of effective stress and Darcy's law. Terms appearing in **bold** should be memorized.

1.1 An Introduction for Geotechnical Practice

The relationship between geotechnical engineers and engineering geologists, who are often their team partners, has been a matter of concern in the USA (Sitar, 1985), Canada (Hungr, 2001), Europe (Bock, 2006) and in particular the UK (Knill, 2003). This concern has often to do with the roles that geotechnical engineers and engineering geologists play in particular projects and how they act collaboratively. Schematics that show the roles played by geotechnical engineers – those practicing rock mechanics and soil mechanics – and engineering geologists usually involve a triangular relationship of mutual interdependence. But in the UK, John Knill (2003), late Professor of Engineering Geology at Imperial College in London, wrote of his regret that engineering geology was not considered by geotechnical engineers as being of equal importance to them as were soil and rock mechanics. If Knill was correct – and correct perhaps not just with respect to the UK – then it is likely that geotechnical project managers have failed to heed the wisdom of the founders of geotechnical engineering. For example, Karl Terzaghi (1961) wrote of himself that "as his experience in the practical application of soil mechanics broadened, he realized more and more the uncertainties associated with the results of even the most conscientious subsurface explorations. The nature and importance of these uncertainties depend entirely on the geological characteristics of the sites". Here then Terzaghi points to one cause of failure in the design of infrastructure, i.e., the complexity of geological phenomena even at the site scale.

Geotechnical teams must be carefully balanced and skillfully led while maintaining the flow of information between members during site characterization, experimental testing and numerically based design. Box 1.1 presents some definitions of the various professions that work in the geotechnical field. As a geotechnical or geoenvironmental engineer, it is most likely that your work will involve project teamwork with one or more of these professional disciplines throughout your career. A former aerospace engineer, James Adams of Stanford University, notes this about design teams: "The team must be large enough to include the necessary knowledge and skills but small enough to take advantage of the high quality of communication, creativity, and motivation found in small work groups ... there is no more rewarding job than being part of a motivated multidisciplinary design team working on a challenging and important product" (Adams, 1991).

According to John Burland (2012b), Emeritus Professor of Geotechnical Engineering at Imperial College, "four distinct but interlinked aspects", presented in Figure 1.1, define geotechnical practice:

BOX 1.1 | PROFESSIONAL DISCIPLINES PRESENT IN GEOTECHNICAL PROJECT TEAMS

Engineering geology: Application of geological principles to obtain information and understanding of geological structures, materials and processes, as needed for engineering analysis and design.

Environmental geology: Application of geology to obtain information and understanding of geological structures, materials and processes, as needed for the solution of environmental problems. There is much overlap with engineering geology.

Geoenvironmental engineering: Application of geological and engineering sciences to the solution of environmental problems, such as the design of sanitary landfills, industrial waste disposal facilities and other infrastructure associated with public health and environmental protection.

Geological engineering: Application of geological and engineering sciences to the analysis and design of soil, rock and groundwater resources. In North America, geological engineers act as a bridge between geotechnical engineers and geoscientists and often are licensed as both engineers and geoscientists.

Geomorphology: Application of geological and engineering principles to understand erosion and sedimentation leading to the creation of distinct landforms and to the description of rivers and floodplains. Engineering geomorphology is a field that has developed in the UK and is in its infancy in North America.

Geotechnical engineering: Application of the science of soil mechanics, rock mechanics, engineering geology and other related disciplines to engineering and environmental projects (after Morgenstern, 2000). The word *geotechnical* is often used to indicate all applied earth science, including engineering or geoscience, e.g., the Canadian Geotechnical Society. The title of this box shows an example of this broad use of the word.

Hydrogeology: Application of geological and engineering principles to the characterization of the hydraulic properties of geologic materials, the assessment of groundwater resources and their protection from contamination.

Source: Modified from Hungr (2001).

---

1. the ground profile, including groundwater conditions;
2. the observed or measured behaviour of the ground;
3. prediction using appropriate models; and
4. empirical procedures, judgement based on precedent and "well-winnowed" experience.

The ground profile, sometimes referred to as the "ground model" or geological model by engineering geologists and hydrogeologists, is the site description in terms of the soil and/or rock layers and the groundwater conditions as they vary across the site with careful acknowledgement of the boundary conditions. Burland notes that "nine failures out of ten result from a lack of knowledge about the ground profile – often the groundwater conditions". A Joint European Working Group (Bock, 2004) of geotechnical engineers and geologists defined the ground model as "the project specific idealisation of the ground incorporating the principal elements of the geological model and the relevant engineering parameters and material properties of the ground".

The observed or measured behaviour includes both laboratory testing of samples of soil and rock, and field testing to determine their hydraulic and mechanical properties. It also encompasses field observations of ground motion, groundwater flow or contaminant migration. Most recently it has come to include measurements of ground movement or topographic elevation by Earth-orbiting satellites.

The use of appropriate analytical and numerical models to simulate the distribution of stresses in rocks, the stability of slopes along transportation corridors, groundwater flow and contaminant transport near waste-disposal facilities, etc. has become central to geotechnical and geoenvironmental engineering. These models idealize the properties of the geological materials to allow the engineer to assess the likely behaviour of rock masses, soil slopes and pollutant plumes. It is vital that the engineer understands the limits of reliability of these models, which are usually determined by our imperfect knowledge of these properties and their spatial variability. Predictions must be validated and assessed by the engineer through other means, e.g., comparison with simpler solutions, laboratory experiments or previous experiences. In fact, before the engineer undertakes simulations of any great complexity, it is essential to have developed numerous "back-of-the-envelope"

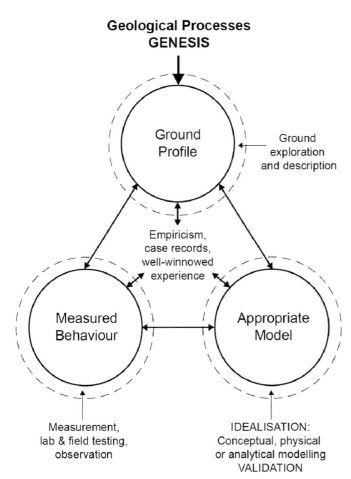

**Geological Processes
GENESIS**

Ground Profile

Ground exploration and description

Empiricism, case records, well-winnowed experience

Measured Behaviour

Appropriate Model

Measurement, lab & field testing, observation

IDEALISATION: Conceptual, physical or analytical modelling VALIDATION

Figure 1.1 Burland's geotechnical triangle. Each activity has its own distinct methodology and rigour. Good geotechnical engineering requires that the triangle is always kept in balance. The "model" is an idealization that is validated by consideration of the observed behaviour of materials at the site and by the ground profile, including the groundwater conditions and their variation across the site. In developing each node of the geotechnical triangle, the engineer should consider precedent, empiricism and "well-winnowed" experience. In Burland's words: "to analyze is to first idealize". (Reproduced from Burland (2012b) with the permission of the author.)

estimates of the problem under consideration so that the general domain of reasonable results is understood.

Burland's (2012b) final recommendation was to use "well-winnowed experience", i.e., that which has been continually refined by trial and error with respect to a particular procedure, or to employ an appropriate case history. One particular procedure – the observational method – was described by the late Ralph Peck, formerly of the University of Illinois (Peck, 1969). It stresses the critical need to continue the design process during construction so that modifications can be made to ensure satisfactory performance. This procedure involves careful examination of the soil and rock conditions that are actually encountered as well as monitoring ground

movements or water levels and comparing both with those anticipated during the design process. Thus, precedent, empiricism, experience and risk management are at the centre of what is referred to as Burland's geotechnical triangle.

Box 1.2 defines the professional competencies of geotechnical engineers as defined by the same Joint European Working Group mentioned above. The geotechnical engineer is assumed to understand the application of specialized testing methods to estimate parameters as well as the constitutive laws governing the behaviour of materials. He or she must be able to design geotechnical structures on the basis of a ground or site characterization model using appropriate numerical methods and tools. It is the engineering geoscientist's role (see Box 1.3) to specify the nature of site conditions and combine them into a geologic model that complements the ground model. This geologic model will employ a full range of modern mapping techniques, such as geographical information systems and digital photogrammetry.

1.2 An Introduction to the Applied Earth Sciences

There are not many geotechnical engineers who have earned PhDs in both geotechnical engineering and geology. One such person is Roland Pusch, formerly a Professor of Soil and Rock Mechanics in Sweden. In his introduction to *Rock Mechanics on a Geological Base*, Pusch (1995) notes "that many geologists have an education that does not make them able to fully understand technico-geological issues. The engineer should therefore have [his/her] own insight in geology that is sufficient to take geological features into consideration and to interpret the rock description given by geologists". We may translate "technico-geological" to read "engineering geological" issues. Unfortunately for the engineer, geology has acquired a huge vocabulary of terms that is utterly daunting at first sight. The handy paperback *Dictionary of Geological Terms* has 8000 terms that are in relatively common use and is a valuable reference for any engineer's bookshelf. Like it or not, if we are to heed Pusch's recommendation, we must plunge into this massive vocabulary in order to gain our own insight.

1.2.1 Basic Geological Terms

Atoms, which are just a few tenths of a nanometre in diameter, form mineral crystals that may grow to 1 mm or longer. These crystals form the basic building blocks of soil grains and rocks.

BOX 1.2 | PROFESSIONAL COMPETENCIES OF GEOTECHNICAL ENGINEERS

Specialised testing methods: The engineer is expected to understand the use and reliability of (1) such test instruments as the oedometer to measure soil consolidation, and triaxial testing machines for measurement of compressive strength and elastic moduli of rocks, and (2) field procedures such as slug tests to measure hydraulic conductivity.

Constitutive laws: These laws are invoked in the idealization of all aspects of material behaviour, including soils, rocks and fluids (groundwater, soil gas, oil). The engineer must understand the nature of each to identify the appropriate constitutive law. Complications arise due to the heterogeneity of geological materials, e.g., bedding and schistosity cause scale and orientation dependencies of the geological materials.

Numerical modelling of complex geotechnical structures: Such models must accommodate complex constitutive laws and the large spatial variability of the critical parameters in geological materials.

Size of the ground model and boundary conditions: Each geotechnical model requires that the boundaries be fully defined between the part of the ground affected by the engineering structure ("near field") and those parts unaffected by it ("far field") where the natural geological conditions prevail. This will require liaison between the engineer and engineering geologist so that the size of the ground model and the nature of the boundaries are well defined in terms of geomechanical and hydraulic properties and geological variability.

Uncertainty: Because of the above, uncertainties will always exist in geotechnical and geoenvironmental projects. Engineers cope with these uncertainties by specially adjusted site characterization, design, construction and contractual procedures. For example, the "observational method" involves the collection of observation and performance monitoring data during construction to allow the implementation of preconceived geotechnical design alternatives. Consequently, the design process continues throughout the construction period because the properties of the site are uncertain within bounds.

Source: After Bock (2004).

Rocks are therefore aggregates of minerals, such as silicates, sulfates, carbonates and oxides, the study of which is known as **mineralogy**. Why should a geotechnical engineer concern him- or herself with the mineral composition of rock? Pusch indicates that there are five principal reasons:

1. for estimating the wearing of drilling bits due to high quartz content;
2. for structural characterization of rock material with respect to the presence of zones of weak minerals, such as mica and chlorite bands;
3. for judging the sensitivity of a rock to chemical degradation by heat or dissolution (sulfates, chlorides and feldspars), or to mechanical degradation upon compaction of rockfill (richness in mica);
4. for identification of weathering that causes slaking or expansion on exposure to water, e.g., clay minerals; and
5. for estimating the mechanical properties of discontinuities (rock fractures) that are commonly coated or filled with minerals with special properties, e.g., chlorite, graphite and clay minerals.

The presence of quartz in rock indicates costly drilling operations due to prematurely worn drill bits. But quartz in rock aggregate makes asphalt more slip-resistant. Several of the minerals mentioned by Pusch are considered layer silicate minerals, e.g., mica, chlorite and clay minerals, which are insoluble but lack the strength of quartz. Clay minerals are perhaps the most important class of minerals in geotechnical and geoenvironmental engineering in that their presence is associated with heaving soils, ruptured pavements and slope instabilities. The mineral structure of others, e.g., sulfates and chlorides, is such that they weather readily when exposed at the surface; that is, they tend to break down and lose strength. To Pusch's list we might add metal sulfides, e.g., pyrite, whose weathering is the cause of **acid-rock drainage**, which occurs when pyrite-bearing shales and mudrocks are exposed to oxidation at the ground surface.

The description and classification of rocks by microscopic examination of thin sections of rock is known as **petrography**, while **petrology** deals with the origin, occurrence, structure and history of rocks. Thus a geotechnical engineer might request a petrographic analysis of a sample of rock that is

---

BOX 1.3 | PROFESSIONAL COMPETENCIES OF ENGINEERING GEOLOGISTS

Synthesis of fragmentary information into models: Engineering geologists bring an understanding of geological processes to bear in interpreting landforms, structures and geological materials in the site context. This allows them to develop geological models that synthesize diverse and fragmentary data from geological, geomorphological, hydrological, geotechnical and geophysical site data.

Training for site-related work: Engineering geologists are trained in fieldwork and the acquisition of information that such work produces. Thus engineering geologists can discern features in a landscape that are important in geotechnical engineering and may be unseen by the engineer.

Versatility in handling maps, aerial photos and GIS data: An engineering geologist is capable of employing and interpreting three-dimensional and time-variable data using classical geological maps as well as modern information technologies such as geographical information systems (GIS), digital photogrammetry and remote sensing systems, e.g., LiDAR.

Observation and analysis of geological data to minimize contractual disputes: Contractual disputes leading to litigation, excess costs and project delay have become familiar aspects of geotechnical projects. The training of engineering geologists permits them to observe, identify, describe and classify geological and technical phenomena in the field and on the construction site and then to provide objective interpretations of the data collected.

Familiarity with fractured and weathered materials: Rocks and overconsolidated soils are geological materials that are typically fractured by joints or faults or show signs of weakness due to long-term weathering. Engineering geologists have developed methods for characterizing these materials and features that serve the needs of geotechnical engineers.

Source: After Bock (2004).

---

to form the foundation of a structure after having referred to reports on the structural geology and petrology of this rock formation. We discuss mineralogy and petrology in Chapter 2.

We may define three principal rock types. **Igneous rocks** cool and crystallize from molten magma originating in the Earth's interior, e.g., granites and basalt. As magma intrudes into the Earth's surface, it metamorphoses those rocks already present into **metamorphic rocks**, producing gneiss and marble. Figure 1.2 shows an igneous intrusion into what was once a sedimentary rock that has been metamorphosed by the heat and stresses associated with the intrusion. Sediments originate as grains of rocks that have been weathered and have undergone erosion and transport of the eroded grains. Subsequent deposition by water, ice and wind yields **sedimentary rocks**, e.g., the sandstone, limestone and shale formations of the Grand Canyon of the Colorado River shown in Figure 1.3. Bedrock of any of these three types is often the foundation in and over which buildings, highways, pipelines, tunnels and other infrastructure are built.

A geologic structure is described by the geometry of the rocks composing it; **structural geology** deals with the geometry, spatial distribution and formation of geologic structures,

Figure 1.2 Granitic dike and sill complex in a road cut, Serpent River, Ontario. The 2 m thick sub-vertical dike has released thin (1–5 cm) horizontal sills that have penetrated along the bedding planes of the former sedimentary rock. The dike is part of the Cutler pluton that has been dated at approximately 1740 Ma (million years before present). Rock hammer placed for scale in the fracture in the centre of the dike. (Information courtesy of R.M. Easton, Ontario Geological Survey.)

Figure 1.3 The Grand Canyon of the Colorado River, Arizona, USA. Limestones and sandstones form the cliffs while the slopes are formed from shales, which are more readily weathered.

Figure 1.5 Fractured Precambrian siltstone, northern Ontario. Rock hammer for scale, length = 28 cm.

Figure 1.4 Faulting in sandstone and mudstone, Torrey Pines State Beach, San Diego, California.

in particular faults and folds in rocks. **Faults** are fracture zones along which there has been displacement of one side relative to the other parallel to the fracture, as shown in Figure 1.4, while **folds** are the deformed curved features created from an otherwise planar structure. These geological structures are the result of **tectonic forces**, which are those external and usually regional processes that create a set of structures within a particular region. Their importance will be examined in Chapter 3.

We must differentiate between **intact rock**, the **fractures** or **discontinuities** that may be contained within it and the **rock mass** as a whole. Figure 1.5 shows a fractured rock mass of siltstone that forms an **outcrop**, i.e., it is visible at the

Earth's surface. Intact rock is that which has no through-going fractures, which effectively reduce the tensile strength of the rock mass to zero. At a very small scale, approximately half the length of the rock hammer shown or about 15 cm, there are blocks of intact rock, which may be individually characterized on the basis of their density, deformability and strength. But quite clearly the outcrop shown is heavily fractured and the sum of intact rock plus the fractures constitutes the rock mass. The difference between the two terms – fractures and discontinuities – is a subtle one and their influence is discussed in Chapters 4 and 11.

Where, then, is the division between soil and rock? We shall follow Morgenstern's (2000) advice in which soil differs from rock by disintegrating when submerged in water. The term **soil** is one which requires some careful consideration because of its different meaning to geologists, soil scientists and engineers. The *Dictionary of Geological Terms* gives two definitions for soil: "1: the natural medium for growth of land plants; 2: in engineering geology, all unconsolidated material above bedrock". The first definition is essentially that of soil scientists, while the second is that of geotechnical engineers. In soil science, the soil profile is a characteristic sequence of unconsolidated materials each of which was generated by specific and well-understood soil development processes, which we shall discuss in Chapters 6 and 7. Residual soils are those that have developed in place from **weathering** (i.e., the physical and chemical degradation of rocks and minerals), whereas transported soils are those that have been transported by gravity, water, wind or ice. The geologist more precisely refers to the former as "soil" and the latter as "sediment".

Figure 1.7 Wind-blown silt, an eolian sediment, photographed near Walla Walla, Washington state, USA.

Figure 1.6 Fluvial gravels: the ingredients of a future conglomerate, Uranium Road, western Colorado. Hammer for scale. Some of the gravel pore spaces have been infilled by sand and mud, yielding a lower porosity and permeability but providing the cementing agents to convert the gravel to a conglomerate. The gravel is well rounded, as in a typical conglomerate, due to extensive abrasion in the stream that transported and then deposited it. Needless to say, this material is difficult to drill into and even more difficult to collect a sample of.

How did these differences in use develop? Perhaps it goes back to 1926 when many of the principles in Terzaghi's *Erdbaumechanik auf Bodenphysikalischer Grundlage*, which Goodman (1998) translates as *Earthwork Mechanics based on the Physics of Soils*, were translated in the USA in abbreviated form to be published as *Principles of Soil Mechanics, a summary of experimental studies of clay and sand*. We shall adhere to the meaning given to "soil" by engineering geologists and geotechnical engineers, i.e., *all unconsolidated material above bedrock*, but use **sediment** in the geological sense in conjunction with a genetic term, e.g., fluvial sediments, glacial sediments, so that the sense of solid grains and particles transported and deposited by water, wind, ice or by chemical precipitation is maintained because the mode of genesis is vitally important to the material properties. Rather

than merely describing the materials as sediments, we often define them by their texture, therefore Figure 1.6 shows some fluvial gravels in western Colorado (USA), while Figure 1.7 illustrates free-standing eolian, or wind-transported, silts in eastern Washington (USA).

This introduction to basic geological terms will suffice for the rest of this chapter. We will deepen our understanding in Part I as we consider mineralogy and petrology (Chapter 2), geological structures and maps (Chapter 3), elementary rock mechanics (Chapter 4) and the characterization of rock masses (Chapter 5). However, first we need to consider issues concerning geological space and time and the understanding of geological processes.

1.2.2 Heterogeneous Materials and the Representative Elementary Volume

Consider the following observation of Roland Pusch (1995): "the entire physical behavior of rock masses depend on their heterogeneity and it is becoming increasingly clear that the various discontinuities are responsible for the scale-dependence of all practically important bulk properties". The **discontinuities** referred to here are those associated with joints, fractures and faults, i.e., any discontinuous feature in an otherwise continuous rock mass that has no tensile strength, for example, the sets of fractures in the siltstone shown in Figure 1.5.

Heterogeneity in geological materials is a fact of life that makes them quite different from other materials that other engineers deal with, e.g., concrete, steel, silicon, polymer. This issue of heterogeneity, and in particular the expense of characterizing the heterogeneous properties of soil and rock, means that geotechnical and geoenvironmental engineers operate under very different constraints than do structural, mechanical or chemical engineers. It is typically the role of engineering geologists, hydrogeologists and geological engineers to specify the nature of this heterogeneity so that engineers might proceed with their designs. In so doing, we must introduce the concept of the **representative elementary volume** (REV).

The determination of an REV of soil or rock is required to analyze any problem involving stress and strain in geomechanics and fluid flow and pore volume in hydrogeology. For example, if we are concerned with defining the minimum volume of soil or rock that will provide a reliable parameter estimate of porosity, then sample size is important. Below a certain sample size, the variations in the measured porosity estimate fluctuate considerably – see Figure 1.8. The smallest volume sampled may give a porosity of 1 if the point sampled is within a pore. However, if the sampling point is part of a sand grain bounding the pore, then the porosity value is 0. Thus any domain that constitutes an REV must contain both a solid portion and a void space that may be filled by some water or gas.

Bear (2018) provides an extensive discussion of the REV concept and its role in quantifying fluid transport in continuum (fluid and solid) mechanics. He indicates that the size of the REV that is selected for analysis must meet the condition that "the average value at a point should remain, more or less, constant over a range of REV volumes that corresponds to the range of variation in the sample size, or in the instrument that monitors that average".

As shown in Figure 1.8, with the increasing volume of porous media sampled, the measured value of porosity begins to stabilize towards a representative or mean value. For sandy soils, a volume with a radius of 10 to 20 grain diameters may define the average volume, while petroleum engineers think of 100–1000 grain diameters for sandstones. However, if the sample size becomes too large, it is possible that the porosity estimate may again begin to show variability because of a textural change, i.e., the sample volume is so large that an adjacent soil or rock mass has become part of the sample, and the sample becomes inhomogeneous or heterogeneous. Quite obviously the REV is also strongly influenced by the presence of discontinuities as well as textural changes. Ultimately the use of the parameter must constrain how the REV is determined.

Other examples of geomechanical and hydrogeological parameters that require consideration of the REV are rock strength, stress and hydraulic conductivity. For structural geologists, a critical consideration is the representative elementary *area* (REA) that adequately describes the distribution of forces acting on any surface within a rock mass. Thus an REA will provide a meaningful average of the surface forces in the rock mass such that heterogeneities in stiffness and width of mineral grains do not unduly bias the measurement (Pollard and Fletcher, 2005).

We conclude that sample size is always an important consideration in testing specimens during laboratory analysis or *in situ* testing in the field. For example, the uniaxial strength of rock pillars in a mine is known to be a function of the sample

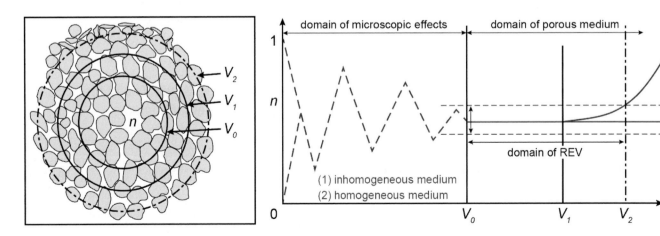

Figure 1.8 The REV of porosity (n) in a soil defined as existing only between V_0 and V_2. (Figure based on Bear (1972); reproduced with permission of Jacob Bear, Elsevier and Dover Books.)

size, such that the strength may decrease to 20–30% of its maximum over a distance of 1 m in sample size (Wyllie, 1999).

1.2.3 Rates of Geologic Processes

Geologic processes have occurred over a vast sea of time, which will be discussed in Chapter 2. However, this immense time scale of over four billion (i.e., 4 000 000 000) years means that what appear to be slow rates of change accumulate over time to result in impressive changes in the Earth's surface. Consider the following:

- The San Andreas Fault Zone in central California appears to be slipping at an average rate of about 35 mm/yr (Hough, 2002), less so in southern California. At that rate, about 16 million years ago the block containing the Pacific coast of San Francisco would have been at the latitude of Los Angeles.
- During the last Ice Age, about 40 000 years ago, Scandinavia was ice-covered to a depth of 2–3 km. When the present Interglacial period began about 10 000 years before present and the ice retreated, the surface of the Earth rebounded approximately 500 m. This is the elevation of raised beaches that were once at sea level. This isostatic rebound of Scandinavia due to a post-glacial uplift is approximately 20 mm/yr over the last 10 000 years whereas in Scotland it has been only 1 mm/yr during this period. Rates in eastern Canada have varied from 1.5–3.5 mm/yr during the Last Glacial Maximum 18 000 years before present to 0.5–1.3 mm/yr at present (Anderson et al., 2007).
- The Grand Canyon of the Colorado River in Arizona is interpreted to have eroded to its present geometry over a period of 4–5 million years. Its maximum depth is 1829 m and thus the rate of erosion of the Canyon is about 400 mm/yr or 400 m every million years. This is much more rapid than elsewhere in the world because of the steep hydraulic gradient and the large volumes of flow entering the Canyon following snowmelt in the Rocky Mountains. The Mississippi River Basin has an annual erosion rate of only 10 m per million years (Mathez and Webster, 2004).

These rates are averages determined from processes operating over a known time period with known displacements. They might not be representative of instantaneous rates. For example, the rise in sea level during the twenty-first century appears to be much higher than during the previous century because of global warming. Thus, a long-term average is inappropriate to estimate the future rise in sea level, which appears to be of the order of 3 mm/yr.

These are examples of geological processes with implications for infrastructure construction and damage. Such rate estimates allow us to consider the consequences and plan accordingly. For example, it is anticipated that every centimetre of sea-level rise will cause the loss of approximately 1 m of sandy beach. Given the hectic pace of beach-front construction in the USA over the past few decades, the consequence of this process is severe and indicates that coastline protection will become a major issue along the US East and Gulf coasts this century.

1.3 Elementary Principles

How is an engineer to develop a deep appreciation for geological materials and insight into geophysical processes? Here we use "geophysical" in the broadest sense of the term, i.e., the physical study of the Earth rather than the specific and more limited use of the term as in seismic exploration. There is no better point of departure than that presented by Norbert Morgenstern (2000) of the University of Alberta, an eminent geotechnical engineer with a profound appreciation of geology. Morgenstern has pointed out that the interactive aspects of the geotechnical method set out in Figure 1.1 can be illustrated by an origin–composition–consistency matrix of soil and rock that is presented in Table 1.1. Soil is differentiated from rock on the basis of whether it disintegrates when submerged in water, while the boundary between weak and hard rock is arbitrarily set at the compressive strength of concrete. This boundary is identified as one in which rocks will slake, i.e., disintegrate, or soften; such *weak rocks* are of great geotechnical concern and are discussed in Chapter 5.

Soils are classified as either cohesionless or cohesive. Cohesionless soils include alluvium deposited by rivers, e.g., sand and gravel, as well as calcareous and gypsiferous sands, topsoil and talus, which form slope debris. Cohesive soils include clays and glacial tills, clay-rich calcareous mud known as marl, peat and tropical laterite soils. These typically contain clay minerals and therefore exhibit plasticity. Rock flour simply comprises silt- and clay-sized particles without clay minerals and lacks plasticity. The Alberta oil sands are a particular case of cohesive sandy soils with the cohesion provided by the bitumen. Soils and sediments are the subject of Part II of this book.

If we define the compressive strength of soft rocks as that being 500 kPa to 1 MPa, then friable sandstones, mudstones, chalk and gypsum are typical soft rocks. Hard rocks generally have a compressive strength of >50 MPa, and these include sandstones, limestones, shales and igneous and metamorphic

Table 1.1 Geotechnical classifications of geological materials (after Morgenstern, 2000).

Rock type ⇒	Sedimentary					Igneous
Composition ⇒	Sandy	Clayey	Carbonate	Evaporite	Organic	Igneous
Consistency ⇓			Example			
Cohesionless soils	Alluvium	Rock flour	Calcareous	Gypsiferous	Topsoil	Talus
Cohesive soils	Oil sand	Clay + till	Marl		Peat	Laterite
Soft rock	Friable sst	Mudstones	Chalk	Gypsum	Lignite	Weathered granite
Hard rock	Sandstone	Shale	Limestone	Potash	Coal	Granite

Notes: Soft rock is defined by compressive strength of 0.5–1 MPa. Sst = sandstone.

rocks such as granites and gneisses. A distinguishing feature of soft rocks is their tendency to slake, i.e., break apart, and soften. In the past, it has often been thought that very hard rock was not of any geotechnical concern. However, the investigation of granites as potential nuclear waste repositories in Canada and Scandinavia has shown that even such strong rocks undergo progressive failure due to high internal stresses. We will return to this matter in Chapter 4.

Before proceeding to a discussion of geological materials and their dependence on structure, we will review some basic terms – porosity, permeability, stress, strength and stiffness – and their appearance in some fundamental laws and principles of the applied earth sciences.

1.3.1 Amonton's Law of Friction

Forces and stresses in soils and rocks arise from the effects of thermal and gravitational processes. The point of departure for this discussion are the forces on an inclined plane, perhaps an unstable slope or a fracture within a rock mass, and their vectorial representation.

Figure 1.9 shows the components of force on a block of soil or rock that has just started to slide down a plane inclined at an angle θ. The two components of the force (F) caused by this mass (m) are, firstly, in the downslope direction relative to the inclined plane,

$$F_t = mg \cdot \sin\theta, \quad (1.1)$$

where g is the acceleration due to gravity, and secondly, acting normal to the inclined plane,

$$F_n = mg \cdot \cos\theta. \quad (1.2)$$

The resistance experienced by the block to sliding is another surface force, friction. As the inclination of the plane is

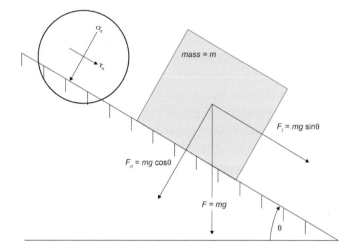

Figure 1.9 Amonton's law: forces on an inclined plane. The inset shows the equivalent stresses. (After Turcotte and Schubert (2002); reproduced with the permission of Cambridge University Press.)

increased, the frictional force per unit area (τ_{fs}) at which the block begins to slide is

$$\tau_{fs} = \frac{f_s \cdot F_n}{A} = f_s \cdot \sigma_n, \quad (1.3)$$

where f_s is the (dimensionless) coefficient of static friction, A is the contact area and σ_n is the normal stress exerted by the block on the inclined plane. Turcotte and Schubert (2002) refer to this as Amonton's law. They showed that $f_s \approx 0.85$ for a variety of very different rock types under drained conditions, i.e., the fluid pressure is zero. This discussion will be pursued in Chapters 14 and 15.

1.3.2 Mohr–Coulomb Criterion

In their introduction to the *Analysis of Geological Structures*, Price and Cosgrove (2005) present a compelling example of the importance of applied forces and stresses in rocks:

If a cube of granite with sides of 25 cm is submitted to an evenly distributed, compressive force of 10 tonnes (i.e., 10 000 kg) only infinitesimal deformation (strain) would be observed in the cube. If, however, the same load were applied to a cube of the same material with sides of one twentieth the length of the larger cube, the smaller granite cube would be pulverized by the action of the force.

Despite the fact that the force in both cases is the same, the stress ($= F \div A$) is obviously very different. In the case of the 25 cm cube, the stress is $\sim$1.6 MPa, while in the case of the smaller cube it is $\sim$630 MPa. Typical granite **strengths**, i.e., the stress necessary to cause the sample to fail, are of the order of 150–300 MPa; thus the small cube failed while the larger cube did not.

Note that the SI unit of stress (or pressure) is the pascal (Pa), which is the stress produced by a force of one newton (1 N) acting on an area of one square metre (1 m^2). The atmosphere at sea level exerts a pressure of 100 kPa. One megapascal (1 MPa) is the hydrostatic pressure at 102 m depth in fresh water and the lithostatic stress at 30–40 m depth in rock. In keeping with geological practice, we will typically refer to pressures when we are dealing with water and aqueous fluids, which lack any resistance to shear forces, and to stress when we are dealing with soils and rocks. Some fluids, of course, are viscous, e.g., lava, creosote.

Returning to the previous example of the forces affecting a block that is slipping on an inclined plane, we may use this definition of stress to define the tangential or frictional shear stress of the block on area A of the inclined plane as

$$\tau_f = \frac{F_t}{A} = \frac{mg}{A} \cdot \sin\theta \qquad (1.4)$$

and the normal stress as

$$\sigma_n = \frac{F_n}{A} = \frac{mg}{A} \cdot \cos\theta. \qquad (1.5)$$

It is common to identify shear stresses by the symbol τ and normal stresses by the symbol σ. Amonton's law (1.3) defines a particular case in which the shear stress on a sliding block (τ_{fs}) is a function of the frictional force per unit area and the coefficient of static friction and applies to pre-existing fractures.

In the case of the 10 tonne force applied to the granite cube, we assumed that the force was applied perpendicularly to the face of the cube so that there was no component of force acting parallel to the surface undergoing loading, i.e., a shear stress. Such a normal force would be considered a **principal stress**. If, and only if, three mutually orthogonal planes have shear stresses equal to zero acting along their surfaces, then their normal stresses (σ_1, σ_2, σ_3) are defined as principal stresses

and the major, intermediate and minor principal stresses are such that $\sigma_1 > \sigma_2 > \sigma_3$. Shear stresses will be identified by the use of two subscripts, e.g., τ_{xy} being the shear stress acting on the x plane in the y direction. Note that in systems under uniaxial stress, such as the granite cube, only one principal stress is nonzero; all three principal stresses are nonzero in the triaxial case. For the case of "hydrostatic" stress, $\sigma_1 = \sigma_2 = \sigma_3$, as in a fluid at rest.

The principal stress acting on the cube of granite discussed above was applying uniaxial compression. Unlike engineering mechanics in which extension is considered the positive direction, the geotechnical and geological convention is to identify compressive forces as operating in the positive direction and tensional forces to be in the negative direction. The reason for this departure from standard engineering convention is quite simple: in the earth sciences we deal predominantly with compressive forces and wish to avoid the frequent use of negative signs with all the potential for error and confusion that the standard engineering convention would produce. Figure 1.10 shows the normal and shear stresses acting in biaxial compression relative to Cartesian coordinate axes. Generally,

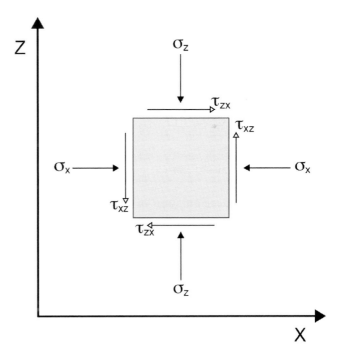

Figure 1.10 Terms defining the states of stress in biaxial compression with normal stresses indicated by σ and shear stresses by τ. Thus σ_z is the normal force in the z direction and τ_{zx} is the shear stress normal to it in the x direction. Compressive normal stresses are considered positive, while shear stress may be negative. Principal stresses are indicated by $\sigma_1, \sigma_2, \sigma_3$. (Modified from Middleton and Wilcock (1994); reproduced with the permission of Cambridge University Press.)

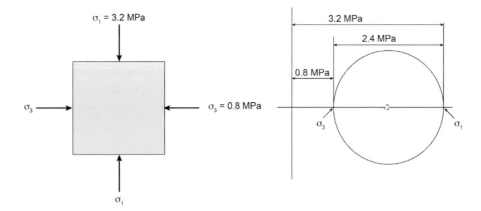

Figure 1.11 Principal stresses acting on a cubic element (left) and the resulting Mohr's circle diagram (right). Geological representations usually show the full Mohr's circle as drawn here, whereas geotechnical engineers typically show the upper quadrant of the circle that represents actual measurements, which are positive values.

geotechnical analysis concerns only σ_1 and σ_3; these are the maximum and minimum normal stresses.

Figure 1.10 depicts forces that are oriented with respect to horizontal and vertical planes. However, if an engineer is concerned with the two-dimensional stress field due to biaxial compression on a randomly inclined plane, the normal and shear stresses for that plane are resolved by the graphical technique known as Mohr's circle. The inclined plane may be a geological contact such as a fault or bedding plane. It can be shown that the normal and shear stresses on a plane inclined by the angle θ are, respectively,

$$\sigma_n = \sigma_1 \cos^2\theta + \sigma_3 \sin^2\theta = \frac{\sigma_1 + \sigma_3}{2} + \frac{\sigma_1 - \sigma_3}{2}\cos 2\theta, \tag{1.6}$$

$$\tau_n = (\sigma_1 - \sigma_3)\sin\theta\cos\theta = \frac{\sigma_1 - \sigma_3}{2}\sin 2\theta. \tag{1.7}$$

When $\theta = 45°$, the shear stress in (1.7) is at a maximum relative to the major principal plane, i.e., $\sin 2\theta = 1$ and $\cos 2\theta = 0$. Therefore, (1.6) and (1.7) simplify for this condition to $\tau_n = \frac{1}{2}(\sigma_1 - \sigma_3)$ and $\sigma_n = \frac{1}{2}(\sigma_1 + \sigma_3)$, that is, on the plane of maximum shear stress, the normal stress is one-half the sum of the major and minor principal stresses.

The Mohr circle of stress presents graphically solutions to (1.6) and (1.7) for various values of θ. The abscissa of the graph is σ, while the ordinate is τ. When each pair of stress values are plotted for a particular value of θ, the locus of points describes a circle. Figure 1.11 shows the construction of a Mohr's circle diagram for a maximum normal stress of $\sigma_1 = 3.2$ MPa and a minimum normal stress of $\sigma_3 = 0.8$ MPa. In conventional geotechnical analysis, counterclockwise shear stresses are assumed to be positive and the vertical normal stress is greater than the horizontal normal stress. The reader

is referred to the text of Hudson and Harrison (1997) for a review of Mohr's stress circle and its use in strain analysis in the context of rock mechanics where the vertical normal stress may be less than one of the horizontal stresses. They summarize several important points about Mohr's circles when used for stress analysis given the compression-positive convention that we are employing:

1. the rotation on the Mohr's circle is twice the real rotation;
2. the principal stresses are the maximum and minimum values of normal stress in the body;
3. two planes that are perpendicular in an outcrop of rock are represented on a Mohr's circle by points at the opposite ends of a diameter; and
4. the planes of maximum shear stress are oriented at an angle of 45° to the principal stress planes.

The failure of a rock sample would logically occur along planes exhibiting maximum shear stress, i.e., maximum and minimum τ. Because the normal stresses increase the rock sample strength, triaxial testing of rock samples indicates that the angle of inclination of the plane that fails is larger than 45° to σ_1. By repeating triaxial tests at various values of σ_1, we can plot a **failure or strength envelope** as a set of overlapping Mohr's circles as is shown in Figure 1.12. This analysis employs two concepts that are important properties of both soils and rocks: (1) the **cohesion** (τ_0) of the rock sample, which is the shear strength of the sample in the absence of either compressive or tensile normal stresses, and (2) the **angle of internal friction** (ϕ), which is equivalent to the angle of inclination of a surface within a block of rock at which sliding of the overlying part of the block occurs. Thus Mohr's circles are used to identify both the failure envelope and the rock-strength parameters of friction and cohesion.

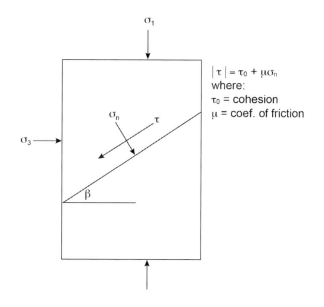

$$|\tau| = \tau_0 + \mu\sigma_n$$
where:
τ_0 = cohesion
μ = coef. of friction

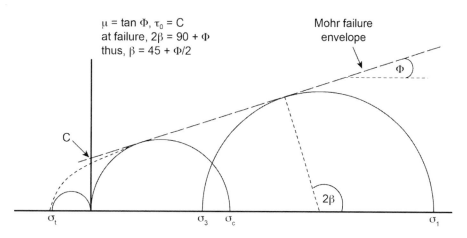

$\mu = \tan \Phi$, $\tau_0 = C$
at failure, $2\beta = 90 + \Phi$
thus, $\beta = 45 + \Phi/2$

Mohr failure
envelope

Figure 1.12 Mohr–Coulomb failure criterion. Note curvilinear failure envelope at low tensile strength. (From Hudson and Harrison (1997); reproduced with the permission of Elsevier.)

Rock failure occurs at some value of normal and shear stress indicated by the **Mohr–Coulomb strength criterion**,

$$|\tau| = \tau_0 + \mu\sigma_n, \qquad (1.8)$$

where τ_0 is the cohesion, μ is the coefficient of friction and σ_n is the normal stress shown in Figure 1.12. The shear stress at failure – the shear strength – is indicated as an absolute value. At very low confining stresses, rocks exhibit a curvilinear failure envelope but follow the Mohr–Coulomb criterion at higher confining stresses where $\beta = 45° + \phi/2$, where ϕ represents the friction angle.

1.3.3 Elasticity and Hooke's Law

Elastic materials, including rocks, display a reversible behaviour such that, if the applied stress is released, then the strain or deformation disappears. This principle linking stress and strain in uniaxial compression or tension is known as Hooke's law, in which

$$\sigma_a = E\epsilon_a, \qquad (1.9)$$

where σ_a is the axial stress, E is **Young's modulus** and ϵ_a is the strain (ΔL) in the axis parallel to the applied stress (dimensionless), where $\epsilon_a = \Delta L/L$, where L is length. Young's modulus is therefore a measure of the **stiffness** of the rock, that is, higher values of E indicate increasingly stiff rocks. A fresh sample of chalk, coal or claystone will likely have a Young's modulus of less than 10 GPa, while fresh samples of granites and similar hard rocks will have values in excess of 50 GPa. In laboratory tests, such rocks exhibit elastic behaviour. However, in the field, where rocks may contain fractures, bedding planes and clay seams that deform as plastic materials, such rocks will not behave in a perfectly elastic manner. We shall return to this issue in Chapter 4.

Because axial strain is a dimensionless length, we note that Young's modulus has the units of stress. A second parameter that is related to elasticity is **Poisson's ratio**, which is usually written as

$$\nu = \frac{\text{lateral strain}}{\text{axial strain}} = \frac{\epsilon_l}{\epsilon_a}. \tag{1.10}$$

The value ν varies between zero, for which the imposition of a vertical stress creates no horizontal strain (assuming no applied horizontal stress), and $\nu = 0.5$, which is that of a fluid indicating a constant volume elastic response. Most rocks have values of Poisson's ratio of $\nu = 0.25 \pm 0.10$, while soils have a range of $\nu = 0.10$–0.30 for loose sand in a drained condition to $\nu = 0.40$–0.50 for saturated (undrained) clays.

1.3.4 Terzaghi's Principle of Effective Stress

In the 1920s, Karl Terzaghi developed the principle of effective stress that is fundamental to many geomechanical phenomena in soil mechanics and, to a lesser extent, in rock mechanics. Terzaghi was concerned with understanding the processes involved in the consolidation of a water-saturated soil, which we consider in Chapter 10. The principle incorporates the effect of buoyancy imparted to the soil or rock mass causing the pore volume to share the overburden load with the soil or rock skeleton:

$$\sigma_T = \sigma_e + p_w, \tag{1.11}$$

where σ_T is the total stress, σ_e is the effective stress and p_w is the pore pressure. Thus the hydrostatic stress exerted in all directions by the pore water has a buoyant effect on the soil or rock skeleton that might otherwise deform. In the words of Bishop (1959): "The strength and deformability of a saturated soil are uniquely dependent on the difference between total stress and pore water pressure."

Although the principle of effective stress was introduced to explain soil mechanical phenomena rather than describing rock-mass behaviour, it can be incorporated into Amonton's law and represented in the Mohr circle diagram. Amonton's law is written for the effective normal stress acting on an inclined fault that is water-saturated as

$$\tau = f_s(\sigma_n - p_w). \tag{1.12}$$

Furthermore, we may write the representation of the effective stress on a Mohr's diagram as presented in Figure 1.13 for a fractured rock mass. The hydrostatic pore-water pressure acts in all directions in porous media, as shown in Figure 1.13(a) and may be represented by case 1 in Figure 1.13(b). This would also be the case for the interior of a fully saturated rock that lacks fractures or other discontinuities. If we consider the fracture shown in Figure 1.13(c), which we will initially assume is not saturated with groundwater, then case 1 shown in Figure 1.13(b) also applies.

But if this fracture becomes saturated with groundwater, it will exert a pore pressure within the fracture and the effective stress is thereby decreased because the vertical stress is

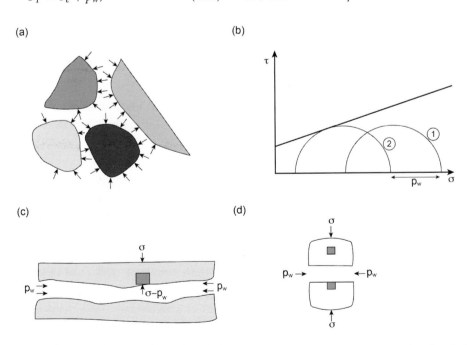

Figure 1.13 Effective stresses for intact rock and discontinuities. (From Hudson and Harrison (1997); reproduced with the permission of Elsevier.)

reduced by the same amount. The effective stress is indicated by the distance on the abscissa (σ) equal to the pore pressure, p_w, as shown in Figure 1.13(b). While the total stress circle (1) is safely removed from the failure plane, the effective stress circle (2) defines a condition of failure of the rock mass. The issue is more complex with fractured rock because the fracture and the intact rock form a dual-porosity system in which an element of the rock adjoining the fracture (i.e., Figure 1.13c) may fully respond immediately to any change in p_w while an element in the interior of the intact rock (i.e., Figure 1.13d) may not.

This of course leads to a critical difference between intact and fractured rock masses. Thus faults, which are large-scale discontinuities, can be made to slip by injection of fluids into them, thus causing earthquakes, which is what occurred when wastes were injected into deep faults in the Precambrian bedrock beneath Denver, Colorado, in the 1960s (Hsieh and Bredehoeft, 1981). This process has obvious consequences for the initiation of earthquakes (Chapter 14) and landslides (Chapter 15).

Returning to the context of soil and sediment mechanics (Head and Epps, 2011; Briaud, 2013), in using the Mohr–Coulomb criterion we will assume that the shear strength is given by the Coulomb equation,

$$s = c' + \sigma_e \tan \phi', \qquad (1.13)$$

where s is the shear strength of the sample in kPa, c' is the effective cohesion, i.e., either cementing of particles or particle interlocking, σ_e is the effective stress normal to the plane of failure, and ϕ' is the effective friction angle. We consider the measurement of the shear strength of soils and sediments in Chapter 10.

1.3.5 Porous Media and Darcy's Law

The movement of fluids in the Earth is a function of four properties: two of the fluid and two of the porous media. The fluid properties are those of density and viscosity; porosity and permeability are the properties of the soils and rocks. These constitute the subject of hydrogeology, petrophysics and other earth sciences because the presence and movement of fluids is critical to the resolution of many practical problems.

Porosity (n) is the fraction of void space within a porous medium (rock, sediment or soil) per unit volume and is therefore dimensionless or given as a percentage. Thus, for a total volume V, a pore volume V_p and a volume of solids V_s, one has

$$n = \frac{V - V_s}{V} = \frac{V_p}{V} = \frac{\text{pore volume}}{\text{total volume}}, \qquad (1.14)$$

and the *void ratio*, a term commonly used in soil mechanics, is

$$e = \frac{V_p}{V_s} = \frac{n}{1-n} = \frac{\text{pore volume}}{\text{solid volume}}. \qquad (1.15)$$

Porosities of sand deposits vary with the degree of **sorting**, which defines the homogeneity of grain sizes. A well-sorted or equigranular sand can have a porosity of 39–42%, while a poorly sorted sand (i.e., well graded in the language of soil mechanics) may be only 28–31% (Beard and Weyl, 1973). Consolidated rocks have a widely divergent set of porosities (0.5–46%) that depends on cementation, sorting and consolidation; we will discuss this later in Chapter 11. If the pores are interconnected, thereby allowing fluid flow, the rock or soil has the second critical property of earth materials: **permeability**, which is normally measured in units of length squared [L^2]. Figure 1.14 shows typical porosity variations in sediments and rocks in which the porosity in schematic A is somewhat less than that due to cubic packing while that in schematic C is close to rhombohedral packing. Sorting (see Chapter 9) is best in schematics A and C and is poor in schematics B and D. Sedimentary rocks may be fractured particularly along bedding planes, shown dipping in schematic F, or contain solution cavities in the case of carbonate and sulfate formations (schematic E), in which groundwater has dissolved away some of the soluble rock.

Permeability is the property of a porous medium that expresses its ability to transport fluids. The concept arose out of an experiment by the French engineer Henri Darcy, who was engaged in a study of the flow of the fountains of Dijon (Darcy, 1856). Darcy's experiment into the flow of water through a sand bed was described by de Marsily (1986) and used what we now call a constant-head permeameter. Darcy did not define the permeability as we now use it; instead he identified a term now known as the **hydraulic conductivity**, which is dependent upon the properties of both the fluid (i.e., water) and the porous medium. Furthermore, Darcy's experiment, like all profound experiments, identified just what additional knowledge needed to be created to produce a fundamental physical theory, in this case of groundwater flow, which we address in Chapter 11.

Hubbert (1940) considered a more general version of Darcy's experiment that is shown in Figure 1.15, and it is his analysis that is the basis for the interpretation of Darcy's experiment commonly followed in modern texts (e.g., Freeze and Cherry, 1979; Middleton and Wilcock, 1994). The soil column shown in Figure 1.15 would today be made of glass or a strong transparent plastic cylinder in which

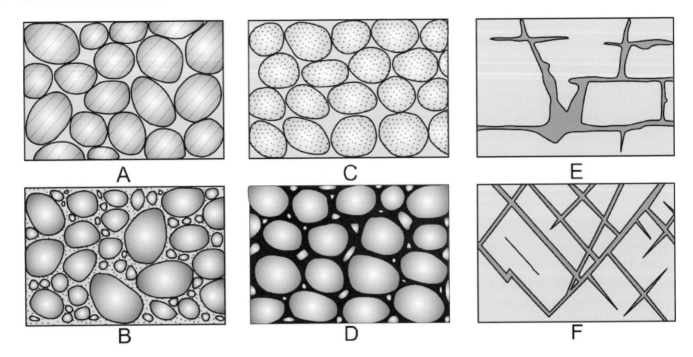

Figure 1.14 Porous, fractured and solution media, as visualized by Meinzer (1922). Meinzer was Chief of the USGS's Ground-Water Division for 34 years, during which time hydrogeology became a quantitative science. (Courtesy of the US Geological Survey.)

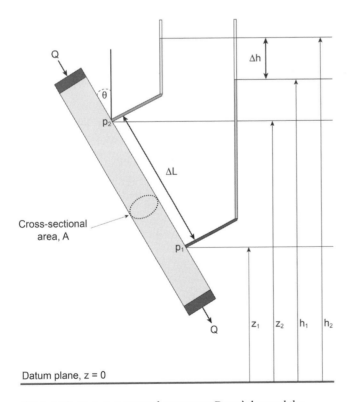

Figure 1.15 Permeameter to demonstrate Darcy's law and the meaning of hydraulic head. (After Tóth (2009); reproduced with the permission of Cambridge University Press.)

sands of various grain sizes can be placed and is the control volume for this discussion. The sands are retained in place by permeable filter plates at each end of the column that remain fully saturated during the experiment. The ends of the column are attached to fluid reservoirs whose elevations can be adjusted to vary the water levels across the column. In the soil column itself a pair of manometers are inserted to measure the water levels at their point of contact with the sand bed.

The reference datum is that fixed at an elevation $z = 0$ beneath the column so that all fluid elevations are measured relative to that datum. The sum of the pore pressure measured, p_w, plus the elevation of the datum give the **hydraulic head** (h). Let us also define the **specific discharge** (q) through the soil column shown in Figure 1.15 as being $q = Q/A$, where Q is the rate of fluid flow into and from the column that has a cross-sectional area A. We will also define the elevation of fluid rise in the manometers as the hydraulic head, h.

Darcy's experiment leads to three observations. Firstly, Q is proportional to the difference in hydraulic heads (h). Thus $Q \propto -(h_2 - h_1) \propto \Delta h$, where the negative sign indicates the convention that the direction of flow is one of higher head (h_2) to one of lower head (h_1). Secondly, changing the angle of tilt (α) while keeping Q constant does not affect our first observation regarding Q and Δh. Thirdly, changing the distance ΔL between the manometers while keeping Δh

constant shows that $Q \propto 1/\Delta L$. These observations can be related by

$$q = \frac{Q}{A} = -K \cdot \frac{\Delta h}{\Delta L} = -K \cdot \frac{dh}{dL}, \qquad (1.16)$$

where K is the constant of proportionality known as the **hydraulic conductivity**, which has units of length per unit time [L/T], and dh/dL is the **hydraulic gradient**, which is dimensionless. Equation (1.16) is known in the engineering and earth sciences as Darcy's law. The hydraulic conductivity is often referred to by engineers as the "coefficient of permeability" or, worse, as "permeability", which term is independent of the fluid in the pores, e.g., gas, water, oil. Darcy's law and values of hydraulic conductivity are discussed in more detail in Chapter 11.

1.4 Summary

In this chapter you have been introduced to geotechnical practice involving teams of engineers and geoscientists, who must work together with a common language to solve problems of construction, infrastructure remediation and the development and protection of natural resources, such as minerals or water. We have defined the professional competencies of both geotechnical engineers and engineering geologists so that areas of expertise are clearly identified.

We then developed a basic vocabulary of geoscientific terms that will be augmented as we work through the book. These terms are highlighted in **bold** to encourage the reader to develop a mental glossary of terms. We have identified rock types genetically – igneous, metamorphic and sedimentary rocks – and observed that we must differentiate between intact rock and the rock mass that contains discontinuities – joints, fractures and faults – that likely control the physical properties of the rock mass. We differentiate between soils and sediments by associating sediments with their genesis, e.g., fluvial sediments, although these too are soils in geotechnical parlance.

Lastly, we have defined a number of elementary principles that will guide us through the book. These are based on standard Newtonian mechanics – Amonton's law of friction, Hooke's law of elasticity and the Mohr–Coulomb criterion. To these we add the principle of effective stress established by Karl Terzaghi. We complement the principle of effective stress with Darcy's law, which describes fluid flow in porous media.

Morgenstern (2000) identified "unifying concepts" that are central to the geological appreciation of geotechnical and geoenvironmental engineering:

1. All geological materials identified in Table 1.1 are porous (and permeable) to varying degrees, and the concept of effective stress provides the fundamental basis for quantitative characterization.
2. All of these materials are stress-dependent to varying degrees; strength increases with normal stress, stiffness increases with normal stress and permeability generally decreases with normal stress.
3. All of these materials are structure-dependent to varying degrees; for some, such as homogeneous uniform clays, the structure is at a scale that can be characterized by the process of sampling and testing; for others, like a jointed, hard rock mass, the discontinuity (i.e., fracture) fabric dominates behaviour and scale effects affect the role of sampling and testing.

Therefore, with a basic geological vocabulary, some unifying concepts and a few laws and principles, we are now ready to consider how the applied earth sciences affect engineering practices.

1.5 FURTHER READING

Adams, J.L., 1991, *Flying Buttresses, Entropy, and O-Rings: The World of an Engineer*. Cambridge, MA: Harvard University Press. — Advice to the young engineer from a very experienced one who taught in the Science, Technology and Society Program at Stanford University.

Bates, R.L. and Jackson, J.A. (eds), 1984, *Dictionary of Geological Terms*, 3rd edn. New York, NY: Anchor Books. — An inexpensive glossary of geological terms.

Goodman, R.E., 1998, *Karl Terzaghi: The Engineer as Artist*. Reston, VA: ASCE Press. — A biography of the father of soil mechanics and a great engineering geologist who practiced in the first half of the twentieth century.

Hancock, P.L. and Skinner, B.J. (eds), 2000, *The Oxford Companion to the Earth*. Oxford, UK: Oxford University Press. — An encyclopedic guide to the earth sciences written in plain English that is a most helpful reference book for the library of any geotechnical practice.

Middleton, G.V. and Wilcock, P.R., 1994, *Mechanics in the Earth and Environmental Sciences*. Cambridge, UK: Cambridge University Press. — The role of elementary mechanics in solving problems in the earth sciences that makes use of principles familiar to the engineer.

1.6 Questions

Question 1.1. As a geotechnical engineer, you are appointed by the head of the civil engineering firm where you work to lead a project that will undertake a forensic analysis of why a highway bridge was destroyed by a landslide that also destroyed part of the highway leading to the bridge. You have funds to choose a colleague to work with you on this project so that a new bridge might be quickly constructed to replace the destroyed bridge. Whom would you choose?

Question 1.2. You established ACME Geoenvironmental Engineers to design sanitary landfills for cities and towns in the region where you are located. A nearby city has acquired land near to a municipal wellfield and requests your opinion on the suitability of this land as a landfill. In providing a cost bid for the work, identify the professionals that you plan to use in the site characterization.

Question 1.3. Why would two fracture surfaces that are perpendicular to one another in a rock outcrop (i.e., $\theta = 90°$) be represented on a Mohr's circle by points at the opposite ends of a diameter?

Question 1.4. The Mohr–Coulomb criterion contains a term for cohesion. What might cause cohesion in soils and sediments subject to shear stress?

Question 1.5. You have conducted uniaxial compression tests in the lab on three samples of intact rock: (1) the metamorphic rock shown adjacent to the granitic dike in Figure 1.2, (2) a sample of the shale shown in Figure 1.3, and (3) the siltstone in Figure 1.5. List in decreasing order the values of Young's modulus (in GPa) that would you expect to measure.

Question 1.6. A warehouse was built alongside a river; however, it was later discovered that the water-saturated soil beneath the warehouse was heavily contaminated with oil from an earlier industrial operation. The oil is discovered to be migrating with contaminated groundwater into the river. The owner is ordered to prevent the contamination of the river by the oil contamination by pumping the oil and groundwater from beneath the warehouse. After a few weeks, the walls of the warehouse started to crack. How is this explained by the principle of effective stress?

Question 1.7. If a head drop of 20 cm is recorded for a discharge of 10 mL/min from a permeameter of the kind shown in Figure 1.15, which has a diameter of 50 mm, and the distance between manometers is 50 cm, what is the hydraulic conductivity of the soil?

Rocks as Engineering Materials

Rocks are aggregates of minerals, which are composed of ordered arrangements of elements that regularly repeat themselves. Thus, atoms are formed into crystal structures, such as silicon–oxygen tetrahedra or sodium chloride cubes, which in turn are the basis for crystals of individual minerals. The atoms are typically less than a nanometre in radius, e.g., the ionic radius of oxygen is 0.14 nm and that of silicon is 0.04 nm. Minerals combine crystal structures to create grains of approximately 1 mm.

Rock-forming minerals and their composition in typical rocks are discussed in Chapter 2, which is devoted to mineralogy and petrology. It follows that we are most interested in the rock-forming minerals such as quartz (SiO_2), calcite ($CaCO_3$) and various complex silicate and oxide minerals that constitute the majority of rocks. Although mineralogy and petrology are the principal subjects of Chapter 2, we begin by describing the structure of the Earth as a planet and with the concept of plate tectonics that has been developed, principally since the 1960s, to explain how the continents have moved over geologic time and how continued movement creates geologic hazards with which the geotechnical engineer is concerned.

The stresses produced by plate tectonics are responsible for the development of structures in rock. Therefore, they also impart geotechnical and hydrogeological properties to these formations. Thus plate tectonics created the three-dimensional structures in rocks that are the domain of the engineering geologist and the subject of Chapter 3. Of particular concern in foundation engineering, engineering geology and hydrogeology are other three-dimensional (3D) features referred to as joints, fractures or discontinuities, such as are shown in Figure 1.5. The various kinds of geological maps that are prepared for different purposes are described, including engineering geology maps, which the engineer needs to be able to read just as he or she must be able to read and interpret a topographic or a highway map. We complete Chapter 3 with the principles of stereographic projection to reduce 3D field features to two-dimensional (2D) representations.

Chapter 4 is an introduction to rock mechanics – the theoretical basis for the design of built structures with foundations in rock and of structures that are wholly enclosed by rock under stress, e.g., tunnels and mines. We discriminate between the rock – a geological material defined by its mineralogy, colour, grain size, etc. – and the rock mass, which is the volume of rock that will affect the built structure and the properties of which are determined by its weathering, stresses and the discontinuities present in it. Then we address the stresses in rock masses from tectonic activity, the inherent strength of each rock mass and indicators of volumetric strain. We consider how fluid flow occurs in the discontinuities of rock masses as these can cause inflows into tunnels and mines, the undermining of dam foundations and the seepage of contaminants from hazardous-waste repositories created in mined cavities. We begin a discussion of rock-slope failure that is continued in Chapter 15.

Part I ends with a discussion in Chapter 5 of the engineering characterization of rock specimens and rock masses. We outline the work flow involved in site characterization, giving an example of defining the thickness of rock aggregate in a quarry suitable for the construction industry. Measurements of rock-mass properties are made at construction sites – *in situ* tests – and on rock samples recovered from the site and sent to geotechnical testing laboratories so that test results of the mechanical and hydraulic behaviour of the rock can be used to predict the performance of the rock foundation. Classification procedures, e.g., the geological strength index, are described to characterize rock masses. We describe hydraulic and geomechanical testing in boreholes, the use of engineering stratigraphic columns and digital photogrammetry to display rock-mass information.

2 The Structure and Composition of the Earth

The foundation of the applied earth sciences, the subject of this book, is over 200 years of scientific measurement and reasoning. This foundation comprises two scales of basic scientific investigation: (a) the structure, tectonic activity and history of the Earth's surface; and (b) the properties of minerals and rocks that constitute the Earth's crust. These topics are discussed in this chapter.

We begin with a discussion of the general structure of the Earth and the role of tectonics in forming the Earth's crust, which is the material with which the engineer must work. We then discuss the history of the Earth and the four general laws that govern the interpretation of geological formations. The third section addresses the nature of the minerals that form rocks, in particular silicate, carbonate, oxide and sulfide minerals. Finally, we consider how these minerals form assemblages known as rocks.

To illustrate how such fundamental geological science professionally affects geotechnical engineering, text boxes are included in the chapter relating to engineering activities in which the mineralogy of rock masses has caused geotechnical difficulties. One of these concerns the erosion of tunnel walls in the tailrace of a hydroelectric dam in Canada; the other concerns the complications induced by swelling clay minerals in the drilling of tunnels in New Zealand.

The reader will note terms throughout this book that are printed in **bold** typeface. These are the basic geological and engineering terms that the reader is encouraged to commit to memory; some of them are defined in the glossary at the end of the book; others may be found in Bates and Jackson (1984).

2.1 Structure and Tectonics

The Earth is not a rigid and static body but is in a continuous state of change, both internally and on its surface. Internal processes act to create new rock material within the Earth, while surficial processes – "superficial" processes in British terminology – act to break down rocks formed in the past. The product of this destruction is soil (or stream sediment), itself a new form of the material. Consequently the destructive forces – typically known as **weathering** – may also be thought of as constructive forces.

The age of the Earth is of the order of 4.5 billion years old – i.e., 4.5×10^9 yr. The age of our Solar System has been computed by isotopic dating of meteorites to be 4.566 billion years. The Earth is believed to have been formed by the collision and accretion of small rocks of up to kilometre size that orbited the Sun and which were known as planetesimals. The **Earth's core** formed some 30 million years after accretion; the potential energy it released by the sinking of iron and nickel to the core was sufficient to melt a large fraction of the Earth. The Earth's oldest rocks – the Acasta Gneiss in the Canadian north – have been dated at about 4 billion years old. All dates were determined by **radiometric methods**, i.e., those based on the radioactive decay of natural isotopes found in certain minerals that comprise these rocks and which allow dating through the growth of daughter isotopes, in particular the decay of uranium isotopes to stable lead isotopes. Box 2.1 describes how the age of the Earth was a subject of much discussion one hundred years ago because physical theory produced conclusions that were far removed from geological reasoning.

The interior of the Earth is built of concentric shells of rock material, named the **crust** and the **mantle**, which surround a central **core**, as shown in Figure 2.1. The thin outer shell is the crust, which is thicker beneath the continents than beneath the oceans for reasons that shall become evident. The upper 100 km of crust and the upper part of the underlying mantle form the **lithosphere**, which is much more rigid than the

BOX 2.1 | LORD KELVIN AND THE AGE OF THE EARTH

The age of the Earth has been a topic of great interest to many civilizations. In 1650 Archbishop Ussher in Ireland estimated that the Earth was created on the evening of October 22, 4004 BC or 6000 ka (6000 years ago) and this date was duly recorded in most Bibles of the nineteenth century. During the late 1800s there was a fierce debate between Lord Kelvin, Professor of Physics at the University of Glasgow in Scotland, and the early geological community in England over the age of the Earth. Based upon observations of geological processes, such as sedimentation and lava flow thicknesses from volcanoes, the geologists were of the opinion that hundreds of millions of years were required to account for the geological features that they were mapping and describing in print. Lord Kelvin disagreed and used his status and the principles of mathematical physics to refute the geologists. In 1897 he refined his estimates of the age of the Earth from at most 100 Ma (100 million years), which many geologists could live with, to between 20 to 40 Ma. This was greeted with dismay by geologists and by Charles Darwin, who agreed with the geological community because he felt much longer was needed for natural selection to have occurred.

Kelvin's physical model of the infant Earth was based upon the conductive cooling of a simple molten solid having an initial temperature of 1300°C and a surface temperature of 0°C. Kelvin's model Earth was a rigid, homogeneous, semi-infinite half-space with no source of heat energy other than its initial thermal energy. However, Kelvin's assistant, the young engineer John Perry, suggested that convection in the Earth's interior would transfer heat to the Earth's surface much more effectively than would conduction. Perry proposed that convection heated a surface layer the thickness of which is given by an expression that implies the duration of cooling, i.e., the age of the Earth. This surface layer would cool while the deep interior transferred a steady heat flux to its base, thus the surface heat flux would be sustained. The consequence of this two-layer model was that Kelvin was underestimating the age of the Earth.

Although he was unable to calculate the role of convection in a mathematically complete way, Perry's model of 1895 included an outer conducting surface of 50–100 km thickness with an interior having perfect thermal conductivity. The computed surface heat flux for this model is roughly consistent with measured heat fluxes and gives an age of up to 2 Ga or 2×10^9 years, about half of the present estimate. However, Perry failed to convince Kelvin of the merits of his model.

A myth grew in the twentieth century that Kelvin was wrong because of the subsequent discovery of radioactivity as a second source of Earth's internal heat, one unknown until 1903. However, England et al. (2007) have shown that this is not the case. Perry's model roughly accounts for the surface heat flux now measured if the thermal conductivity of the interior is 100 times that of a 50 km surface layer. Conduction of heat from radioactive decay together with the residual heat from the Earth's formation is insufficient to explain today's heat flux (80 mW/m²). Ironically, the development of radiometric dating in the twentieth century would show that Kelvin's model greatly underestimated the age of the Earth because of his assumption of simple conduction rather than coupled interior convection and surface-layer conduction.

We might conclude that simple physical models are most helpful as a tool in geological reasoning as long as they are correctly formulated. Unfortunately, Perry's account was considered of no interest to the geological community, which paid for its negligence for 60 years by failing to realize the role that interior convection might play in explaining continental drift, another controversial idea.

Source: After England et al. (2007).

underlying **asthenosphere**. The average specific gravity of the Earth as determined by the mathematics of planetary motion is 5.5; however, the average specific gravity of the rocks found on the Earth's surface is just half this value, i.e., 2.7. Consequently, the interior of the Earth must be much denser.

Therefore the structure of the planet is like an egg in that it has a thin crust (the shell) in which occur events and processes that are the subject of this book. The interior contains a mantle and a core – egg white and yolk – in which geothermal energy from the original heat of accretion of planetesimals

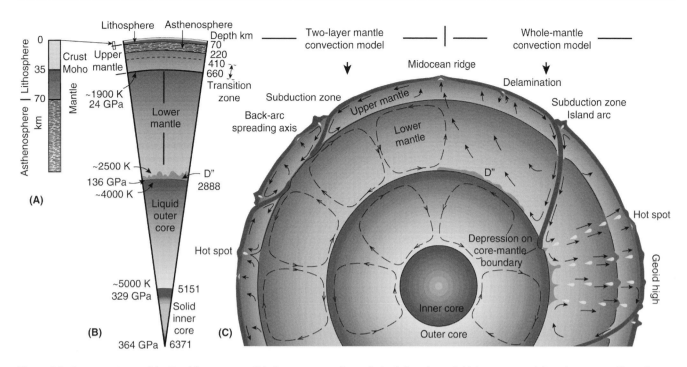

Figure 2.1 Cross-sections of the Earth's structure. (A) Cross-section through the lithosphere. (B) Temperature (K) and stress profile with depth at major seismic discontinuities. (C) Cross-section illustrating convection in mantle and subduction zones. This structure is based on the interpretation of seismic wave paths and velocities. The core is occupied by iron, nickel and some lighter elements. The mantle behaves as a plastic material exhibiting slow convection. (From Klein and Philpotts (2016); reproduced with permission of Cambridge University Press.)

(and core formation) and from radioactive decay drive convection within the mantle, which in turn drives plate tectonics.

The evidence for the existence of a structure containing distinct shells of different densities as shown in Figure 2.1 has come from measurements of the rates at which seismic waves from earthquakes travel through the Earth. The properties of the different types of seismic waves are considered in Chapter 14; P and S waves are discussed in Chapter 5, as they are used as an engineering measure of rock properties. Results obtained from the study of these waves – the science of **seismology** – have shown that there are relatively rapid changes in the specific gravity of the Earth at depths of 35, 70 and 2900 km, which are referred to by geophysicists and seismologists as discontinuities. These relatively rapid changes in specific gravity cause seismic waves from earthquakes (or nuclear weapons tests) to be reflected and refracted; the subsequent paths followed by the waves can be detected, and their study forms the basis of our knowledge of the Earth's interior. It should be noted that engineering geologists and geotechnical engineers attach a specific meaning to discontinuities (see Chapters 3 and 4), indicating fractures or open joints that interrupt the continuity of the properties of the rock and which are zones of weakness.

The concept that the Earth's continents migrated apart – as is suggested by the fit of Africa and South America – goes back to a map by the Flemish cartographer Ortelius in

1596 and was remarked upon by several later scientists such as Francis Bacon in England and Benjamin Franklin in the USA. However, it was the German scientist Alfred Wegener (1880–1930) who, in 1912, argued that the continents are light and mobile barges that plough through the denser and static oceanic crust. Wegener's theory was debated during the first half of the twentieth century without resolution, but it was the deployment of technological advances introduced in the 1950s that prompted the new conceptual model of plate tectonics to be developed from Wegener's idea of continental drift and which made sense of the many valid geological observations of Wegener that supported his hypothesis.

Plate tectonics is the well-established theory of global tectonics (i.e., global crustal evolution) based upon the observation that the lithosphere is divided into a number of rigid plates that move and interact with one another at their boundaries, along which most volcanic, seismic and tectonic activity occurs. Figure 2.2 shows the margins of these major plates. **Tectonics** refers to the Earth's regional-scale structural and deformational features and evolution and to the processes that developed them. **Structural geology**, the subject of Chapter 3, deals with the local- and regional-scale features that would be important at the scale of engineering sites and urban areas.

Since the 1960s geophysicists have come to understand that continents and oceans are not permanently fixed in position but are moving extremely slowly (1–15 cm/yr) across the

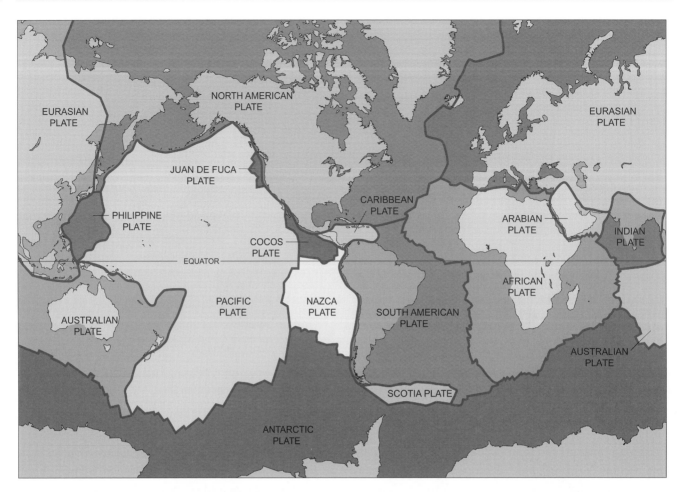

Figure 2.2 Margins of the major lithospheric plates, subduction zones, mid-ocean ridges and transform faults. (From Kious and Tilling (undated), *This Dynamic Earth*, US Geological Survey.)

surface of the Earth. What Wegener misunderstood was that it was really not the continents but the seafloor that was mobile and that caused continental separation by being continuously created at a **mid-ocean ridge** (MOR); the continents merely are like boxes moving on a conveyor belt. The young seafloor is generated by magma at a MOR due to convection currents in the Earth's mantle and consumed in old age by **subduction** at convergent plate boundaries beneath the lighter continents, giving rise to the term **seafloor spreading**.

The subducted plates are recycled into the Earth's mantle by a process called slab pull. The plate or slab descends into a trench of at least 10 km depth by virtue of its greater density than the continental rock that it under-rides. Figure 2.3 shows three cross-sections of the lithosphere, depicting (top) two different tectonic consequences produced by the recycling of seafloor and (bottom) continental materials in subduction zones. Areas of active subduction are distinguished by volcanic activity and earthquakes and, farther inland, rising mountain ranges. The Pacific Northwest of the USA, adjacent British Columbia and Alaska is a superb example of

the top schematic in Figure 2.3. There are about 170 active volcanoes in the USA and its Pacific island territories alone. Consequently the global threat of volcanic eruptions is real and the need for real-time monitoring should be evident to governments. Klein and Philpotts (2013) note that 9% of the world's population lives within 100 km of an active volcano, making eruptions a serious global threat.

Figure 2.4 shows volcanic Mount St. Helens after its 1980 eruption that killed 57 people, blanketed eastern Washington state with ash and caused more than $1 billion in damage. Mudflows following **volcanic eruptions**, known as **lahars**, from Mt. Rainier have been mapped down river valleys all the way to the Puget Lowland cities of Tacoma, Renton and Olympia in Washington state and thus pose a serious threat to life and infrastructure; these lahars are discussed in detail in Chapter 15. The monitoring of earthquakes in volcanic areas is important in that the seismicity associated with the movement of magma at depth provides a warning of impending eruptions. In fact, Mount St. Helens gave seven weeks of seismic and steam-venting warning before its catastrophic

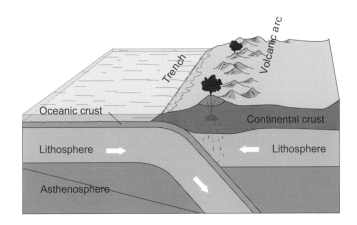

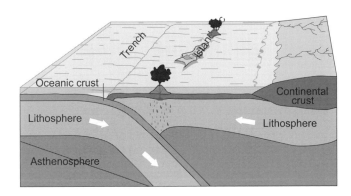

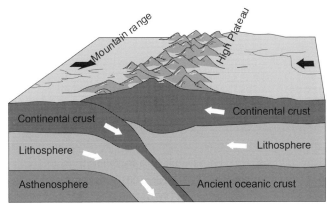

Figure 2.3 The subduction of lithospheric plates. (From Kious and Tilling (undated), *This Dynamic Earth*, US Geological Survey.)

Figure 2.4 Mount St. Helens, Washington, USA, photographed from the Pumice Plain on April 16, 1983, almost three years after its eruption. The Crater Glacier now wraps around the margins of the Lava Dome and extends northward to the lip of the crater floor. Here gas and ash are emitted by the volcano. The void in this crater was formed by a series of events in May 1980. First, a collapse was caused by a debris avalanche, and a blast produced a pyroclastic density flow that together removed 2.5 km^3 of volcanic debris. These were followed by a nine-hour eruption that ejected about 1 km^3 of tephra. (Photo courtesy of Patrick Pringle, Professor Emeritus of Earth Science, Centralia College, Centralia, Washington, USA.)

eruption in 1980. Figure 2.5 summarizes the various hazards associated with volcanic eruptions.

Mountain building is a result of subduction at convergent plate boundaries over geological time. Some mountain chains continue to grow, such as the Himalayas, which have risen over 10 km in the past million years due to the subduction of the Indian Plate beneath the Eurasian Plate. The Himalayas are formed from the leading edge of the Indian Plate, while farther north the Tibetan Plateau is part of the Eurasian Plate that has been forced upward by subduction. As one might expect, this convergent boundary is an area of considerable seismic activity; more than 70 000 people died in the 2008 Chengdu earthquake, which was a direct result of this convergence. Divergent boundaries – such as mid-ocean ridges and rift valleys – are where crust is created as plates separate from one another. A third boundary – a **transform fault** boundary (see Figure 2.6) – is of critical importance in geotechnical engineering in that it describes an area in which two plates slide past one another.

Transform boundaries were identified by J. Tuzo Wilson of the University of Toronto in 1965. These fracture zones are commonly found on the seafloor, but where they exist on land they are associated with destructive earthquakes and great loss of life and infrastructure. The San Andreas Fault in California and the Anatolian Fault in Turkey are the most prominent examples. Figure 2.7 shows how it is possible for the lateral offset of the two sides of the fault to be hundreds to thousands of kilometres long. At transform boundaries, the crust is neither subducted nor created. Rather, the two plates attempt to slip by each other, either because they are moving in opposite directions or because they are moving in the same direction but at different rates. Note the displacement of Los Angeles over the past 30 Ma in Figure 2.7. We shall return to this issue of seismicity and earthquakes in Chapter 14.

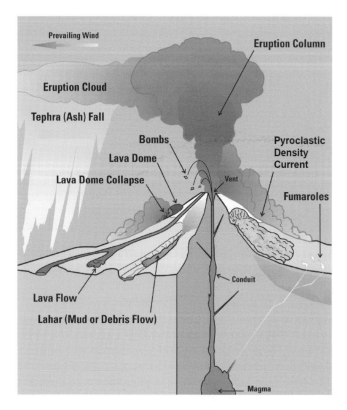

Figure 2.5 Proximal and distal volcanic eruption hazards. The proximal hazards – ash fallout and volcanic gases and projectiles – are more unpredictable than the distal ones, which include increased sediment loading and deposition in nearby rivers affected by fallout and lahars. The most destructive volcanic hazard is a pyroclastic density current, which is a hot volcanic flow containing a mixture of gases and fine and coarse particles, which can move at velocities of ~100 km/h, in contrast to slow-moving lava flows. (See National Academies (2017). Figure courtesy of the U.S. Geological Survey.)

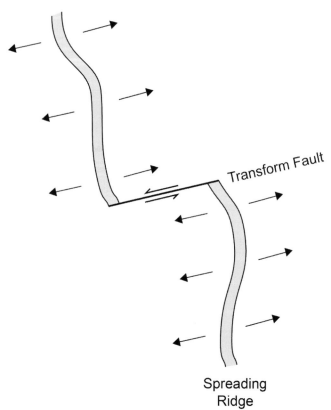

Figure 2.6 Transform fault. (From Mathez and Webster (2004), with permission.)

In addition to subduction, **isostasy** is another tectonic process involving vertical movement of the crust that is of importance to geotechnical engineers. The discovery that the Earth's crust was rebounding following deglaciation of northern Europe and North America caused the realization that the Earth's mantle was responding as a viscous fluid. Thus the raised shorelines along the Baltic Sea of Scandinavia and Russia and along the Hudson Bay in Canada indicate a rise in the land surface relative to sea level caused by this isostatic adjustment. Figure 2.8 shows the mechanism of post-glacial rebound by which the ice load depresses the crust into the mantle. With deglaciation, the mantle responds more slowly due to its viscosity ($\eta = 10^{21}$ Pa s) than to the elasticity of the lithosphere ($\mu = 10^{11}$ Pa). Consequently only after $\eta \div \mu$ or ~300 years is the viscous response of the mantle similar in effect to the elastic response of the lithosphere.

Isostatic equilibrium also affects mountain ranges and sedimentary basins. Rainfall, wind, ice and snowmelt result in the erosion of mountains and the development of deep valleys that drain the mountain slopes carrying heavy sediment loads. This loss of mass results in the steady rise of the mountain, exposing earlier layers of rock to subsequent erosion and so on. Differential erosion of the windward versus the leeward slopes of New Zealand's Southern Alps is caused by over 10 m of rain annually on the windward slopes. This causes the windward side to rise 5 mm/yr, while the leeward side of the Southern Alps rises less than 1 mm/yr, still a very substantial rate, decreasing to zero with distance from the mountains.

Similarly, sediments that are deposited in basins cause the depression of the basin as if it were ice, not sediments, that were loading the basin. This problem is of concern in heavily populated deltas such as those of the Mississippi (USA), Fraser (Canada), Pearl and Yellow (China) and Po (Italy) rivers. During glacial periods, the Mississippi downcut (eroded) an incised valley through its delta due to the lowered sea level caused by accumulation of water in glaciers as ice. The eroded sediment was transported into the Gulf of Mexico where it was deposited. During the interglacial warm periods, the

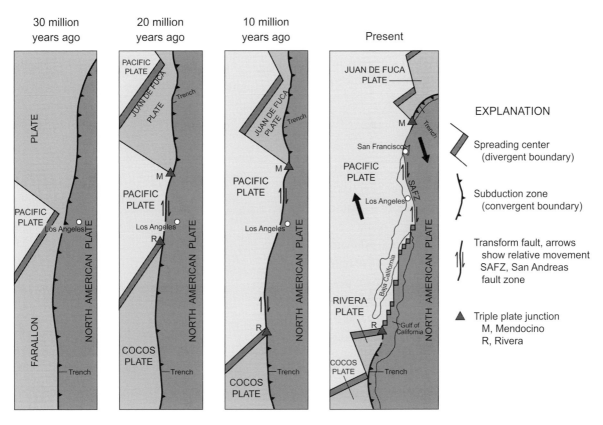

30 million years ago **20 million years ago** **10 million years ago** **Present**

EXPLANATION

Spreading center (divergent boundary)

Subduction zone (convergent boundary)

Transform fault, arrows show relative movement SAFZ, San Andreas fault zone

▲ Triple plate junction M, Mendocino R, Rivera

Figure 2.7 The evolution of the Californian coast by plate tectonics during the last 30 Ma (From Kious and Tilling (undated), *This Dynamic Earth*, US Geological Survey.)

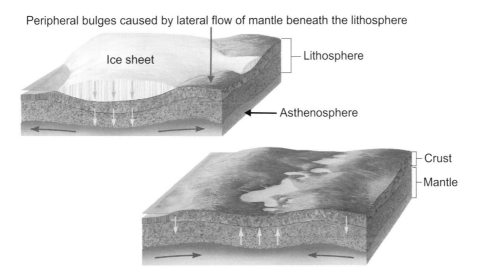

Peripheral bulges caused by lateral flow of mantle beneath the lithosphere

Ice sheet — Lithosphere

— Asthenosphere

— Crust

— Mantle

Figure 2.8 Isostatic rebound due to deglaciation. The weight of the ice depresses the mantle within the asthenosphere, causing it to flow laterally. Deglaciation causes unloading and isostatic rebound as the asthenosphere flows back. The rate of rebound is directly proportional to the density difference between mantle and meltwater and indirectly proportional to the viscosity of the mantle. Around the Baltic Sea in northern Europe and Hudson's Bay in Canada, the rate is roughly 1 m per hundred years. Note that the forebulge may still be subsiding in some areas and may not have reached lithostatic equilibrium. (Illustration by Gary Hincks, Post Glacial Rebound © Gary Hincks, with permission.)

raised sea level permitted the deposition of sediments within the river valley. The periods of incision and sedimentation resulted in 9 m of subsidence and rebound along the lower Mississippi Valley with lesser amounts of subsidence and rebound extending over 100 km along the coast.

2.2 Earth History

Four-and-one-half billion years of crustal evolution has created the lithosphere as described by the theory of plate tectonics. The first 700 million years has left little mark, e.g., the Acasta Gneiss in northern Canada. The earliest sedimentary rocks – those produced by the consolidation of sediments derived from the erosion of older rocks – have been discovered in Greenland and date to 3.8 billion years ago. About this time, microbial life is evident in sedimentary rocks from Western Australia; their habitat, which was probably marine, suggests that an atmosphere was beginning to form around the planet. What oxygen that was produced was used to oxidize dissolved iron in seawater and to create banded iron formations worldwide. Eventually, $\sim 2.4 \times 10^9$ years ago, an oxygen-rich atmosphere evolved and vast banded iron formations were deposited to form parts of the Shield terrains worldwide, e.g., the Mesabi Range in Minnesota, USA, and the Hammersley Range in Western Australia. The mining of such deposits is the source of most of the world's steel. The first 4 billion years are known as the Precambrian, and Precambrian Shield terrains form the **basement** rocks of the continents.

The global sedimentary record is continuous from around 540 million years before present (Ma) with the Cambrian period, a time of enormous evolution of life that contrasted with the almost complete absence of fossils and sediments in Precambrian time. Table 2.1 presents the geologic time scale from the Cambrian to the present, with abbreviations for ages before present given in "a" for annum following the SI system, so that one thousand years is "ka" and one million years is "Ma". The duration of time is, however, expressed in millions of years or Myr, e.g., the Cenozoic Era has lasted for 65.5 Myr and the Holocene epoch began 11.7 ka.

The Cambrian is the first in a series of sequentially younger periods of the Paleozoic Era: the Ordovician, Silurian, Devonian, Carboniferous and Permian periods. The Permian period ended approximately 250 Ma with the extinction of 90% of the Earth's species, and was followed by the Mesozoic Era, which contained the Triassic, Jurassic and Cretaceous periods. The final era began 65.5 Ma with the Cenozoic, which

Table 2.1 Divisions of geologic time (International Stratigraphic Chart, August 2009).

Era	Period	Epoch	Age (Ma)
Cenozoic (Cz)	Quaternary (Q)	Holocene	0.0117
		Pleistocene	2.6
	Neogene (N)	Pliocene	5.3
		Miocene	23.0
	Paleogene (Pε)	Oligocene	33.9
		Eocene	55.8
		Paleocene	65.5
Mesozoic (Mz)	Cretaceous (K)		146
	Jurassic (J)		200
	Triassic		251
Paleozoic (Pz)	Permian (P)		299
	Carboniferous (C)		359
	Devonian (D)		416
	Silurian (S)		444
	Ordovician (O)		488
	Cambrian		542
Precambrian			~4000

is divided into Tertiary and Quaternary periods. It should be noted that the International Commission on Stratigraphy no longer uses the term "Tertiary". Instead, it divides that time interval into the Paleogene period (i.e., Paleocene, Eocene and Oligocene epochs) and Neogene period (i.e., Miocene and Pliocene epochs). The US Geological Survey (2010) (USGS) retains the use of the term "Tertiary" and divides the Carboniferous into Mississippian (beginning 359 Ma) and Pennsylvanian (beginning 318 Ma) periods.

Consequently a purely imaginary borehole drilled initially into Quaternary sediments of recent origin may then exhibit an ever-deeper core indicating older Cenozoic formations, then Mesozoic and finally Paleozoic sedimentary rocks. At the bottom of the borehole will be the Precambrian basement. Sedimentary layers – i.e., **strata** – define a stratigraphic sequence of rocks. The Grand Canyon sedimentary sequence, shown in Figure 1.3, begins with the Permian Kaibab Limestone ($\sim$285 Ma) that forms the Canyon rim rock and ends with the Cambrian Tapeats Sandstone ($\sim$510 Ma) that rests on the Precambrian basement. Each formation within this sequence – e.g., the Kaibab Limestone, the Coconino Sandstone, the Bright Angel Shale – displays distinctive properties such as colour, grain size, mineralogy and fossil assemblage that separate each of them from adjacent formations. Each formation can be correlated with similar rocks over considerable distances that are found in the same ordered sequence and thus are considered to be the same sequence of formations.

These principles arise from certain general laws of **stratigraphy**:

- The **law of original continuity**: Strata deposited by water flow continue laterally in all directions until they thin or pinch out as a result of lack of sediment.
- The **law of original horizontality**: Water-laid sediments are deposited in strata that are horizontal and parallel to the Earth's surface.
- The **law of faunal succession**: Fossil organisms (fauna and flora) succeed one another in a definite and recognizable order, such that the relative age of rocks can be determined from their fossil content.
- The **law of superposition**: In any sequence of sedimentary strata that has not been overturned, the youngest stratum is at the top and the oldest is at the bottom. Each strata or **bed** is younger than the bed beneath it but older than the bed above it.

Geological periods are further divided into epochs and again into ages, usually on the basis of characteristic fossil assemblages (biostratigraphy) assisted by radiometric dating (chronostratigraphy). These shorter divisions of geologic time are seldom of importance to the geotechnical engineer because the rocks themselves have markedly different properties from adjacent rock strata that might affect their behaviour during or after construction. These shorter divisions can be seen in maps of all areas but they tend to be restricted to smaller geographic areas, although considerable effort has been undertaken in the last 30 years to bring them to a common stratigraphic standard. For example, Oxfordian sediments of the Upper Jurassic are identified in Switzerland as well as in the reference locality in southern England. Such shorter geologic time divisions may be only a rough guide to geotechnical properties because local weathering, glaciation or tectonics may have much altered their respective properties in Switzerland and England such that their common ancestry is no reliable guide to performance.

The **Quaternary period** – roughly the last two million years – has had the most profound effect on the geotechnical properties of the rocks, soils and sediments with which the engineer will work. Although the late Cenozoic ice ages began before the Quaternary began, probably in the late Eocene or 40 Ma when the Antarctic ice cap developed, it is the Quaternary glaciations that are best documented because each glaciation tends to erase the imprint of the previous one.

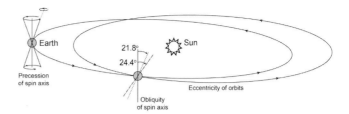

Figure 2.9 Milankovitch cycles caused by the Earth's orbit around the Sun. This theory, which is believed to explain at least 60% of the observed climatic fluctuation, postulates that ice ages are caused by the variations in three orbital features of the Earth with respect to the Sun and hence the flux of received solar radiation. The first of these is the precession of the equinoxes associated with a wobble in the Earth's axis of rotation with periods in the range 19 000–23 000 years. The second is the inclination of the Earth's axis with respect to the plane of planetary orbit with a periodicity of 41 000 years; this effect of obliquity is greatest at northern latitudes. The third is the long-term variation in the eccentricity of the Earth's orbit, i.e., its departure from a circular orbit, with a period of approximately 100 000 years. (After Lamb and Sington (1998).)

The cause of the glaciations is believed to be astronomical in origin and is associated with the Serbian astronomer Milutin Milankovitch, who identified the role of the Earth's orbit of the Sun that results in varying incident solar radiation at the Earth's surface (see Figure 2.9). According to Milankovitch's theory, the incident solar radiation varies by 20% in northern latitudes during the course of these cycles. Each glaciation typically lasted 90 kyr with short 10 kyr interglacial periods separating the various glaciations.

In summary, the age of rocks is only a rough guide to their geotechnical properties. It is reasonable to expect all Precambrian rocks to be crystalline and fractured. Similarly volcanic rocks formed from lava flows are **hard rocks** irrespective of their age. However, many volcanoes, especially those associated with subduction zones, erupt large volumes of ash, which can be extremely soft. The yellow tuff that emanated from Vesuvius (Naples, Italy), for example, was extensively tunnelled into by the Romans to build water reservoirs. This ash was used by the Romans to make the first "Portland cement" by adding it to lime mortar (Klein and Philpotts, 2016). Such "soft" volcanic rocks have therefore played important engineering roles.

Paleozoic formations are almost invariably well cemented – i.e., **lithified** into rocks – while Mesozoic formations may remain unconsolidated, i.e., not cemented or not indurated. Cenozoic formations are very often unconsolidated granular deposits of clay, silt, sand and gravel, except where calcium carbonate or silica cement has precipitated to form sandstone, limestone or other **soft rocks**. Note that care must be taken in the use of words such as **consolidation** because of its special meaning in soil mechanics with respect to the compressive testing of soils in an oedometer and time-dependent settlement of infrastructure founded on saturated clay soils. Generally speaking, Paleozoic sediments and those Mesozoic and Cenozoic sediments that were buried, consolidated and heated are much more likely to have become lithified, and thus giving them a degree of strength important to the geotechnical engineer.

2.3 Rock-Forming Minerals

The word **mineral** is defined as a naturally formed crystalline substance with a defined chemical composition found in the ground. Rocks are composed of an aggregate of different minerals, mostly compounds of silica, i.e., silicate rocks, and of carbonate minerals, for example, limestone. In the usual sense of the word, "mineral" means something extracted from the ground for its value, sometimes in metal content, such as gold, or its value as a construction material, e.g., silica sand or clay minerals. The geological use of this word includes all naturally formed crystalline substances that make up the composition of rocks.

Most rocks contain several minerals. Usually about 95% of a rock is composed of three or four minerals, with the remaining 5% containing many other identifiable minerals. Some rocks may be composed of just one mineral, such as quartz (SiO_2) in quartzite or calcite ($CaCO_3$) in limestone. Most igneous and metamorphic rocks contain a very small number of minerals, usually fewer than eight. This is a requirement of thermodynamic equilibrium, which is usually achieved in rocks formed at high temperature and pressure (Gibbs phase rule). Sedimentary rocks, formed at low temperature and pressure, do not achieve chemical equilibrium so they are not restricted by Gibbs phase rule and can contain more minerals. However, the weathering and transportation of sediments typically sorts the weathering products so that only a few more resistant minerals are found in any given sedimentary rock.

The principal rock-forming minerals are presented in Table 2.2 in terms of the mineral composition of the major rock types. The reader is urged to consult a handbook of colour photographs showing these minerals while reading this section (e.g., Bishop et al., 2001).

The major minerals are also the important ones from a geotechnical perspective. Although thousands of minerals are known, a knowledge of only the common rock-forming minerals is important for the geotechnical engineer. These

Table 2.2 Mineral composition (vol.%) of rocks in the Earth's crust (from Wenk and Bulakh, 2004).

Mineral	Sedimentary	Granitic–metamorphic	Continental basalt
Feldspars	17.3	52.2	45.7
Pyroxenes	4.8	3.4	23.8
Quartz	18.4	22.5	11.6
Olivene	0.5	0.4	7.6
Amphiboles	<0.1	9.8	4.7
Micas	<0.1	5.6	3.3
Clay minerals	32.8	0	0
Other silicates	0.1	3.7	1.2
Carbonates	19.2	1.5	0.5
Ore minerals	0.6	0.7	1.6
Phosphates	0.2	0.2	0
Hydrous Fe oxides	2.0	0	0
Evaporite minerals	1.0	0	0
Organic material	0.4	0	0
Volcanic glass	2.3	0	0
Volume fraction of rock types	10.4%	30.4%	41.8%

Note: Oceanic basalt is 17.4% of rock types, which is the balance of the last row of this table.

minerals should be studied from the point of view of their chemical and mineral stability because the strength of intact rock is dependent upon these properties. The overall strength of any structure is dependent upon the strength of its weakest member; this principle is also true for rocks. Some minerals, especially those containing much iron or magnesium silicate and most carbonates, weather more quickly when exposed to the atmosphere or to fresh groundwater than do most other minerals. The iron and magnesium silicates are formed at high temperature and are unstable at the conditions of the Earth's surface. Carbonate and sulfate minerals, such as calcite and gypsum, are soluble in rainwater and dissolve readily, especially if the rain is acidic, as found in many urban areas. Weathering reduces the strength of the rock even though other minerals may not have decayed.

The chemical behaviour of minerals in any rock that is used for **aggregate** in concrete may be very important because of alkali–aggregate reactions that damage the concrete by the formation of new minerals occupying a greater volume than the original mineral, thus causing expansion and fracturing of the concrete. The best known of these reactions is the alkali–silica reaction between alkali oxides such as Na_2O and K_2O with forms of silica such as opal and chalcedony.

The specific gravity (SG) of any rock is controlled by the weighted sum of the specific gravities of the individual minerals comprising the rock. The common rock-forming minerals have specific gravities in the range of 2.3 to about 3.0, e.g., gypsum ($CaSO_4 \cdot 2H_2O$: 2.35), orthoclase feldspar ($KAlSi_3O_8$: 2.62), quartz (SiO_2: 2.65), calcite ($CaCO_3$: 2.71), dolomite ($CaMg(CO_3)_2$: 2.87) and biotite ($K_3Si_3O_{10}(OH_2)$: 2.9+). Minerals containing heavy-metal ions such as iron have higher specific gravities: pyrite (FeS_2: 5.1) and goethite ($FeOOH$: 4.3).

Bear in mind that minerals have defined internal structures that arise from the combination of atoms into crystals. The way in which the atoms are arranged into crystals has some important effects on the mechanical strength of the minerals when forming rocks. When the mineral structure is in the form of a sheet, the mineral can be easily split and is said to have **cleavage**. Other minerals have internal structures that only allow them to break into irregular shapes. Thus the carbonate minerals are built from rhombohedral shaped crystals and the mineral will cleave into similarly shaped fragments.

2.3.1 Silicate Minerals

With the exception of carbonate minerals, the major rock-forming minerals are silicates, whose crystalline structure is described in terms of a lattice of silicon and oxygen atoms in the form of a tetrahedron. A tetrahedron is a polygon with four faces, four corners and six edges. Each silicon atom is associated with four oxygen atoms set in the corners at the end of each tetrahedral arm, i.e., SiO_4^{4-}. The hole created by the tetrahedron of oxygen ions is perfectly sized to accommodate the silicon atom. Should the mineral dissolve in water it forms $Si(OH)_4$(aq), which is a neutral species, where (aq) indicates an aqueous or dissolved species. Note that the net negative charge of 4− is balanced by a hydrolysis reaction that adds four hydrogen ions to the oxygens.

Silica tetrahedra can be joined together in several different configurations. In the case of feldspars, the most common minerals in the Earth's crust, the Al^{3+} occurs, like Si^{4+}, in tetrahedral coordination, whereas the K^+, Na^+ and Ca^{2+} occur in large irregular nine-fold coordination sites. The **coordination number** is the number of closest neighbours that surround a central atom or ion.

Figure 2.10 shows how tetrahedral and octahedral layers are formed into crystals of kaolinite and montmorillonite or smectite. Cations with similar ionic radii, like aluminum, can substitute for the silicon in a solid solution that is denoted

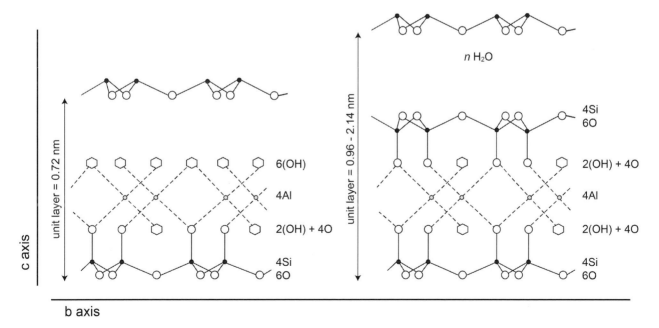

Figure 2.10 The molecular structures of (left) kaolinite ($Al_4Si_4O_{10}(OH)_8$) and (right) montmorillonite ($Al_4(Si_4O_{10})_2(OH)_4 \cdot nH_2O$), in which n water molecules exist in the interlayer spaces. Each crystal structure repeats its unit layers. (From Turekian, K., *Chemistry of the Earth*, PB, 1E, © 1972 Brooks/Cole.)

by parentheses, e.g., (Si,Al) or (Mg,Fe). The cations therefore act as staples that link the tetrahedra together and neutralize their negative charge. However, it is also necessary that the substitution that occurs maintains charge balance. In coupled substitution, pairs of ions substitute within the mineral structure to maintain charge balance. For example, a charge imbalance would result in the plagioclase feldspars if only Na^+ substituted for Ca^{2+}. However, if the substitution involves Na^+Si^{4+} replacing $Ca^{2+}Al^{3+}$, charge balance is preserved. So aluminum can substitute for silicon only if some coupled substitution maintains the charge balance. The tetrahedra can also satisfy the charge requirements by forming polymers, which are referred to as chain, sheet and framework silicates. We now consider the major rock-forming silicate minerals.

Olivine

$(Mg,Fe)_2SiO_4$: Magnesium and iron are interchangeable within the lattice. Usually occurs as isolated grains within igneous rocks. Light green, but darker with higher iron content. SG: 3.2–3.6. Easily decomposed at the Earth's surface.

Pyroxenes

General formula: $(X)_2Si_2O_6$, where X is Mg, Fe^{2+}, Mn, Li, Ti, Al, Ca or Na. SG: 3.1–3.5 depending on amount of Fe. Found in igneous rocks such as gabbro, while diopside $(CaMgSi_2O_6)$ is found in metamorphic rocks. Imparts a green colour to rocks and is easily weathered.

Amphiboles

Hornblende $(Ca_2(Mg,Fe^{2+})_4(Al,Fe^{3+})Si_7O_{22}(OH)_2)$ is the most common member of this group of chain silicates. SG: 3.0–3.5. Green, common in igneous and metamorphic rocks. Weathers easily. Crystals are rod-shaped and in metamorphic rocks are often aligned, as in hornblende schist.

Micas

Three principal groups of sheet silicates exist. Those rich in Fe^{2+} and Mg are dark micas such as biotite, while white mica is rich in Al. Biotite: $K(Mg,Fe^{2+})_3(Al,Fe^{3+})Si_3O_{10}(OH,F)_2$. SG: 2.7–3.3. Dark brown to black. Biotite weathers to vermiculite and is not often found in sedimentary rocks. Muscovite: $KAl_2(AlSi_3)O_{10}(OH,F)_2$. Colourless, six-sided, flat-shaped mineral. It reflects light well and is often seen sparkling in rocks like sandstones and siltstones in which it is found as a detrital mineral. Like biotite, muscovite is common in granite and mica schist. Chlorite: $(Mg,Fe,Mn,Al)_{4–6}(Si,Al)_4O_{10}(OH,O)_8$. SG: 2.6–3.2. Chlorite,

a green flaky mineral, is common in slate, schist and hydrothermal veins.

Feldspars

The most abundant of all minerals in the Earth's crust (43%). These are aluminum silicates with the general formula: $X(Al,Si)_4O_8$, where X is Na, Ca, K or Ba. Feldspar weathering produces clay minerals. There are three major types of feldspars: albite, $NaAlSi_3O_8$; orthoclase, $KAlSi_3O_8$, and anorthite, $CaAl_2Si_2O_8$. Albite and orthoclase can form mixtures known as **alkali feldspars**. Albite and anorthite form mixtures known as **plagioclase feldspars**. Feldspars are common in igneous rocks like granite of which they comprise about 75%. Usually white in colour but may be stained with other colours, e.g., labradorite, which is blue-green. SG: 2.6–2.8. Albite and anorthite mix to form plagioclase, which makes up about 50% of igneous rocks such as basalt. Feldspar crystals cleave easily along smooth planes and, as a result, are easily seen in rocks because of light reflection from these surfaces.

Quartz

SiO_2 is a hard and chemically resistant mineral that cannot be scratched with a knife blade, which is the common test for this mineral. SG: 2.65. Forms beautiful clusters of crystals in rock cavities and occurs in many different colours. A major constituent of igneous, metamorphic and sedimentary rocks; 12% of the Earth's crust and 18% of all sediments is quartz. During the crystallization of granite, the elements combine with silica until they are all located in crystal lattices; any silica remaining forms quartz. Rocks that are rich in quartz are known as siliceous. Quartz is often seen as white veins in rocks that contain it as a detrital mineral, for example, sandstones. Because it is resistant to weathering, it is often found as pebbles on beaches. Note that the white mineral forming in veins in carbonate rocks is usually calcite, but may be quartz from high-temperature hydrothermal intrusion. A rock that is composed almost entirely of quartz is called quartzite. Chert, also known as flint by archeologists, is a cryptocrystalline variety of quartz, commonly found as nodules in carbonate rocks, such as the chalk in the white cliffs of Dover, England, and on the north coast of France. Its fine crystal size makes it extremely durable.

Clay Minerals

These weathering products of feldspar are sheet silicates with a structure consisting of alternating tetrahedral (SiO_4) and octahedral (AlO_6) sheets linked by hydroxyl groups as in Figure 2.10. Important clay minerals include kaolinite ($Al_2Si_2O_5(OH)_4$), halloysite ($Al_2Si_2O_5(OH)_4 \cdot 2H_2O$) and smectite or montmorillonite ($Al_4(Si_4O_{10})_2(OH)_4 \cdot nH_2O$), in which n water molecules exist in the interlayer spaces. SG: 2.6–2.9. Clay minerals are of very great importance in geotechnical and geoenvironmental engineering as they possess swelling and shrinking properties that can damage buildings and infrastructure, have high adsorption capacity for dissolved contaminants in groundwater and have been associated with landslide failure surfaces in many regions. The structure and composition of clay minerals is discussed in Chapter 7 with reference to their role in weathering, soil heaving and landslides.

2.3.2 Carbonate Minerals

Calcite ($CaCO_3$) is the major mineral in limestones. It reacts readily with dilute hydrochloric acid (HCl) giving off bubbles of CO_2 gas, which is the standard test for calcite (SG: 2.71). **Dolomite** ($CaMg(CO_3)_2$) is formed during evaporation of seawater when brines with a high Mg/Ca ratio react with calcite (SG: 2.90). Dolomite reacts with warm HCl when ground to powder, but does not react strongly with cold HCl, providing a distinction from calcite. **Siderite** ($FeCO_3$) forms in a reducing (oxygen-free) environment and is sometimes found with clays and shales forming clay ironstones (SG: 3.89).

2.3.3 Evaporite Minerals

Gypsum ($CaSO_4 \cdot 2H_2O$) is a white evaporite mineral that is precipitated from evaporating seawater, followed by **anhydrite** ($CaSO_4$) and then **halite** (NaCl) (SG: 2.16) and other halide salts. Gypsum is common in clay formations as veins or thin beds. Gypsum (SG: 2.32) and anhydrite (SG: 2.9) are of economic importance in building construction as plasterboard. The presence of evaporite minerals in rocks indicates an absence of groundwater flow through that particular rock zone because their high aqueous solubility would cause them to dissolve into the groundwater. The expansive alteration of anhydrite to gypsum and its subsequent dissolution leading to rock failure is discussed later in Box 2.2.

2.3.4 Oxide Minerals

Most important among the oxides are the **ferric oxides**: hematite (Fe_2O_3), magnetite (Fe_3O_4) and goethite (FeOOH). Hematite and magnetite (SG: 5.2) are important ore minerals

in steel-making, and goethite (SG: 4.3) is important as an adsorbent of contaminants in groundwater. Lepidochrocite and limonite are forms of goethite. Various forms of hydrous ferric oxide (HFO) exist and are transformed in oxygenated environments into goethite. Magnetite may be identified because it is one of the few magnetic minerals; it is often present in igneous rocks as small black crystals and distinctive as a dense detrital mineral in sandstones. Hematite gives rocks a red or pinkish colour and is responsible for the red sandstones of the Colorado Plateau known as **red beds**. Other important oxides include manganite ($Mn^{2+}Mn^{4+}O_2(OH)_2$), gibbsite ($Al(OH)_3$), ilmenite ($FeTiO_3$), the ore of titanium, and uraninite (UO_2), also known as pitchblende, the ore of uranium.

2.3.5 Sulfide Minerals

Pyrite (FeS_2) is a brass-coloured mineral when found as a crystal and known as "fool's gold", but is black when it occurs dispersed in rocks as a particulate (SG: 5.1). It occurs as a precipitate in all types of rocks and those sediments with strongly reducing conditions, i.e., as a result of sulfate reduction. Modern pyrite forms initially as a monosulfide (FeS) that is transformed into pyrite over time. This mineral transformation involving crystalline restructuring occurs in anoxic low-temperature environments, e.g., wetlands and mudflats.

Many sulfide minerals are the principal ore minerals that are mined for metallurgical use (see section 2.4.4), e.g., chalcopyrite ($CuFeS_2$), galena (PbS), sphalerite (ZnS) and molybdenite (MoS_2). A particularly important sulfide is **arsenopyrite** (FeAsS), which is a source of arsenic in groundwater when oxidation causes mineral dissolution. Oxidation of sulfides also creates a strongly acidic water and releases heavy-metal contaminants that were previously trapped in the sulfide mineral lattice; this phenomenon is referred to as acid-mine (or acid-rock) drainage and plagues former mining districts with acidic streams and water contaminated with copper, iron, lead, zinc, etc., as discussed in Chapter 7.

2.3.6 Mineral Identification

If minerals can cause unstable soils through swelling, landslides or acid drainage from mines and tailing piles, then the complete identification of minerals is an important task that the geotechnical or geoenvironmental engineer may wish to have undertaken by a competent geoscientist and analytical laboratory. In his *Introduction to Rock Mechanics*, Goodman (1989) indicates that the geotechnical engineer needs to

BOX 2.2	**THE MINERALOGY OF ROCK FAILURE IN THE TAILRACE OF A HYDROELECTRIC POWER DAM**

The Churchill Falls Hydroelectric Project in Labrador, eastern Canada, was commissioned in 1971 and was at the time the largest hydroelectric power generator in the world, with an installed capacity of 5428 MW. This was more than twice the size of the 1936 Hoover Dam, near Las Vegas, Nevada, but much less than the far newer Three Gorges Project on the Yangtze River in China (18 200 MW). The electricity generated is transmitted through Québec to the northeastern USA.

Water descends 300 m from the huge reservoir behind the dam to the powerhouse chamber and then is discharged through two 1.7 km long tailrace tunnels to the Churchill River. The tailrace tunnels are 13.7 m wide, 18.3 m high and have a design flow of over 700 m^3/s. The tunnels are unlined for 80% of their length but are supported in the remainder by rock bolts and gunite, which is a mixture of Portland cement, sand and water, applied pneumatically. The western tailrace tunnel was inspected in 1999. While most of the strengthened sections of the tunnel had performed well over 28 years of service, one part of the tunnel – referred to as "altered rock" – was found to be partially collapsed; 5 m long rock bolts were missing and about 450 m^3 of rock-fall debris was detected under water. Figure 2.11 shows the altered rock on the west wall of the tunnel.

Figure 2.11 Altered rock of the west wall of the West Tailrace Tunnel, Churchill Falls, Labrador, Canada. (Photo courtesy of Boro Lukajic; used with permission of NALCOR Energy, Churchill Falls, Newfoundland and Labrador, Canada.)

It was determined from the mineralogy and the structure of the altered rock that erosion of the tunnel walls and arch would continue unless remediation was undertaken. The tunnel was out of service for 54 days, during which time it was partially dewatered and reinstallation of rock bolts and the application of shotcrete (sprayed concrete) were conducted from a barge.

The engineering company responsible for remediation retained a geologist to characterize the altered rock. The geologist examined thin sections of recovered intact and altered rock with a petrographic microscope. Figure 2.12 shows a photomicrograph of unaltered rock from the eastern wall of the tunnel. It is identified as a **hornfels**, which is a high-temperature facies associated with contact metamorphism. Figure 2.13 shows the unaltered part of the opposite western wall that was probably a granite before it was metamorphosed into a gneiss during the Precambrian era.

The altered rock is distinguished mineralogically by the presence of muscovite, clay minerals and anhydrite; anhydrite can be found in hydrothermal veins as well as in evaporite deposits. Some micaceous minerals have been altered to clays (see Figure 2.14). Anhydrite alteration to gypsum, which was subsequently dissolved by

BOX 2.2 | (CONT.)

Figure 2.12 Hornfels. Medium-grained quartz (light colour) with fine-grained interstitial biotite (red). Width of field, 4 mm; cross-polarized light. (Photo courtesy of Eva Schandl, Ph.D., Geoconsult, Toronto.) See colour plate section.

Figure 2.13 Granitic gneiss. Striated grains are plagioclase feldspar with polysynthetic twinning surrounded by quartz grains. Width of field, 4 mm; cross-polarized light. (Photo courtesy of Eva Schandl, Ph.D., Geoconsult, Toronto.)

Figure 2.14 Highly altered mica-rich gneiss. Coarse-grained crystalline mica (biotite and muscovite) is replaced by fine-grained clays, probably illite. Width of field, 4 mm; cross-polarized light. (Photo courtesy of Eva Schandl, Ph.D., Geoconsult, Toronto.)

BOX 2.2 | (CONT.)

Figure 2.15 Gneissic texture. Barely visible relict anhydrite aggregates (arrow) in large vug (black) are surrounded by slender prisms of amphibole and coarse-grained quartz grains. Width of field, 4 mm; cross-polarized light. (Photo courtesy of Eva Schandl, Ph.D., Geoconsult, Toronto.)

the seepage, explains the presence of voids known as **vugs**, as shown in Figure 2.15. The most weathered parts of the rock face are those containing large amounts of mica – biotite and/or muscovite – and other rocks in which the micas occur in relatively wide, parallel bands. The least affected rocks are the mica-poor granitic gneiss and the hornfels, in which the mica is dispersed within the rock matrix and the bands are composed of much less soluble minerals such as quartz, feldspar and mphibole. The sheet layering of the micas allows water penetration and progressive weathering to clays, whereby clays can be eroded, thus increasing the hydraulic conductivity of the weathered zone and further enhancing weathering.

The vugs created by the hydration of anhydrite to gypsum further weaken the rock. Geotechnical testing found that the dry bulk density of the altered rock was generally 5% less than that of the intact rock due to these chemical and physical weathering reactions. Ultimately, the volume change in the rock associated with anhydrite hydration, gypsum dissolution and the exposure of the clays led to the erosion of the rock wall by the tailrace discharge, which had a velocity of 4 m/s. This weakening would progress along the joint planes, which were spaced 100–400 mm apart in the altered rock zone as compared with 400–3000 mm in the intact rock. Consequently, progressive alteration would weaken the support of the weathered rock and it would collapse into the tunnel.

Sources: From Stevenson et al. (2004) and Schandl (2001).

---

be able to identify approximately 16 minerals and 40 rocks in hand specimens and to be familiar with their occurrence and properties. These minerals are identified in the preceding sections and the rocks are described in the following section.

Hand-specimen petrology involves the use of basic mineral identification procedures that rely on a determination of the **hardness** of a mineral, its cleavage, colour and lustre using a hand lens or a binocular microscope. Figure 2.16 presents Goodman's flowchart for the identification of the 16 minerals introduced earlier. The first division is on the basis

of hardness, i.e., whether the mineral can be scratched with one's fingernail or by a penknife. Mohs' hardness scale rates a fingernail as having a hardness of 2–2.5 and a penknife that of 5–5.5. **Cleavage** and **fracture** indicate how a mineral will break apart when struck with a hammer. Cleavage describes how a mineral will split along planes defined by the mineral's crystal structure, such as mica; fracture describes the irregular surfaces produced by some minerals upon breakage. Cleavage surfaces are both smooth and reflect incident light uniformly at one orientation.

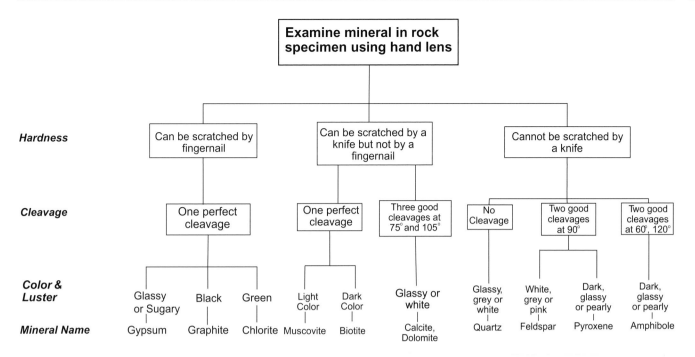

Figure 2.16 A simplified mineral identification flowchart showing the common rock-forming minerals. On the left-hand side of the flowchart are minerals that can be scratched by a fingernail, which has a Mohs' hardness of 2–2.5. On the right-hand side are those that cannot be scratched even with a knife, which has a Mohs' hardness of 5–5.5. (After Goodman (1989), with permission.)

Using Figure 2.16 we may discriminate between three common minerals that may appear similar to the untrained eye: quartz, feldspar and calcite. Quartz has no cleavage and its hardness of 7 does not allow it to be scratched with a knife. Feldspars have very well-defined cleavages that distinguish them from quartz. Calcite is readily scratched with a knife and can be confirmed with HCl acid. Differentiating between calcite and dolomite in the field requires caution and the application of the acid to a freshly exposed surface. Calcite reacts readily with HCl, dolomite less so. The dye Alizarin Red-S can also differentiate between the two using a hand lens: a fresh calcite surface will be stained pink whereas the dolomite will remain clear.

The petrographic microscope, such as that shown in Figure 2.17, permits the examination of mineral specimens and a more definitive identification of minerals, e.g., quartz from feldspar when both are glassy. For optical analysis, either mineral grains or a thin (i.e., 30 μm) section of core are mounted on a glass slide with the aid of resin. With the exception of opaque minerals, such as pyrite and magnetite, that must be viewed with reflected light, minerals prepared in this manner transmit light and can be viewed at various magnifications. Because minerals often have optical properties that vary with the direction at which light is transmitted through the specimen, the sample should be viewed in polarizing light with a microscope capable of rotating the sample 360° around the axis of the microscope. The resin has a refractive index of 1.538 and minerals with a higher refractive index appear as peaks on the slide while minerals with lower refractive index appear as pits. In this manner, improved identifications may be made. Thin-section photomicrographs of typical igneous, metamorphic and sedimentary rocks are presented in Figure 2.18.

Instrumental analysis of minerals includes methods of (1) elemental analysis to assist in the confirmation of mineral identifications, in particular electron microprobe analysis (EMPA) and X-ray fluorescence (XRF), (2) imaging analysis by scanning electron microprobe (SEM) and (3) mineral identification by X-ray diffraction (XRD).

Electron microprobes are ubiquitous in geochemistry laboratories because of their power to quantify zones of minerals in terms of the elemental oxides by virtue of their emitted X-ray spectra. The traditional method of elemental analysis in geology has been the use of an XRF instrument to provide ten or more measurements of elements as oxides, e.g., SiO_2, Fe_2O_3, Al_2O_3, MgO, CaO, Na_2O, K_2O, TiO_2, P_2O_5, MnO, etc. This is referred to as **whole-rock analysis** and is a standard and inexpensive basic measurement of the geochemical composition of a rock.

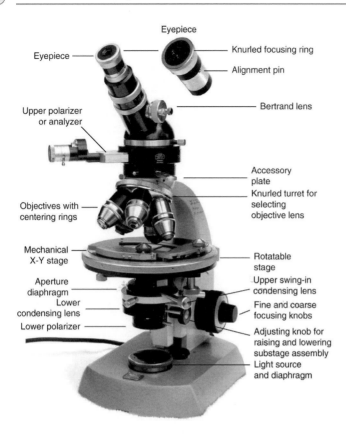

Figure 2.17 A petrographic microscope manufactured by Carl Zeiss. The instrument consists of objective lenses, a light source and a rotating stage on which the thin-section specimen is placed and viewed through the eyepiece. Because many minerals display different properties when light is transmitted in varying directions through the crystals, polarized light is used for mineral identification. The polarizer is positioned in the light path somewhere before the specimen together with an analyzer (a second polarizer), placed in the optical pathway between the objectives and the eyepiece. A camera may be located adjacent to the eyepiece to capture the image, such as is shown in Figure 2.18. (From Klein and Philpotts (2016), used with permission of Cambridge University Press.)

SEM machines now are fitted with X-ray spectrometers so that they have the capabilities of an electron microprobe. Therefore images of rocks and sediments can be interpreted with the aid of an elemental analysis. An SEM image of quartz and feldspar minerals from a sandstone petroleum reservoir is shown in Figure 2.19.

XRD analysis is a common method for detecting minerals in specimens and complements the use of the petrographic microscope. The rock sample is crushed into a powder and then subjected to X-ray analysis that produces an X-ray diffraction pattern in which each peak corresponds to a particular crystalline lattice plane of a particular mineral. The detector rotates at an angular velocity of 2θ and the plot of the peaks with respect to this angle is known as the diffractometer pattern or diffractogram. We will return to this topic in considering clay minerals in Chapter 7.

2.4 Rocks as Mineral Assemblages

Three distinct processes assemble **rock-forming minerals** in various proportions into rocks: (a) **igneous**, (b) **sedimentary** and (c) **metamorphic**. The reader is urged to consult a handbook of colour photographs of these rocks in reading this section (e.g., Bishop et al., 2001). Figure 2.18 illustrates these rocks in thin section.

Igneous rocks are formed either in the crust by the crystallization of melted silicate materials known as **magma** or on the Earth's surface after volcanic eruption. Flow of magma through pre-existing fractures forms **dikes** and **sills** (see Figure 1.2). Magma that solidifies underground produces an **intrusive rock**, such as granite. Extensive granitic bodies, like the Sierra Nevada in California or those of Devon and Cornwall in southwest England, are known as **batholiths**. Flows that appear on the Earth's surface produce **extrusive rocks**, such as basalt, andesite and rhyolite. Huge eruptions of basalt known as **large igneous provinces** have occurred throughout the world – e.g., the Columbia Plateau in the US northwest, the Deccan Traps in India and the Caribbean Plateau – and have been linked with the mass extinction of species through the release of volcanic aerosols (Saunders, 2005).

Please note that geologists have a particularly misleading tradition of characterizing igneous rocks as "acidic" for silica-rich rocks or "basic" for those poor in silica. Considerably better terms are **felsic** for light-coloured rocks and **mafic** for dark-coloured rocks, i.e., shorthand for rocks that contain feldspar or silica (felsic) or magnesium or ferric iron (mafic) minerals, respectively.

Weathering and erosion of igneous and metamorphic rocks create particulate matter that can be transported downslope and deposited as sediments. These are referred to as **clastic** or **detrital** sediments, e.g., sandstone or siltstone, whereas sediments that form by the precipitation of minerals from fresh water or seawater are referred to as **chemical sediments**. A third type of sedimentary rock is of biogenic or **biochemical** origin in that marine organisms build calcite or silica skeletons that are preserved in marine sedimentary deposits as limestones. Chemical sediments form during the evaporation of seawater in warm shallow seas, creating a series of evaporite minerals that precipitate as the volume of water decreases

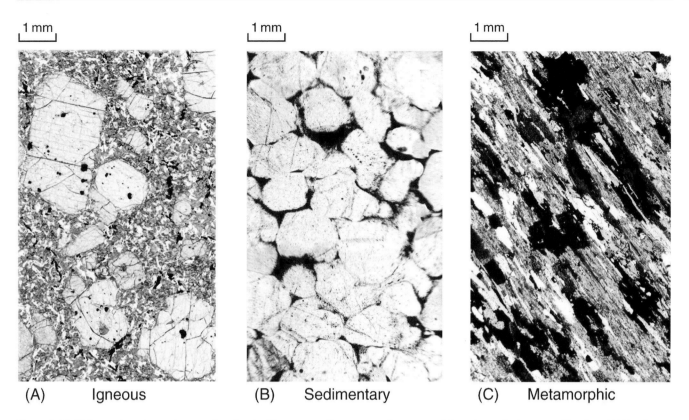

Figure 2.18 Thin-section photomicrographs of typical igneous, sedimentary and metamorphic rocks under low magnification. (A) Basalt from a Hawaiian lava lake showing phenocrysts of clear olivine set within a groundmass of brown pyroxene and opaque ilmenite ($FeTiO_3$); plane-polarized light. (B) Sandstone with sub-rounded quartz grains with dark clay particles and quartz cement filling the pore spaces; plane-polarized light. (C) Schist, a strongly foliated rock, caused by metamorphism of muddy sedimentary rocks that resulted in the bright blue and red mica crystals, which are preferentially oriented perpendicular to the maximum compressive stress; cross-polarized light. (From Klein and Philpotts (2013).) See colour plate section.

Figure 2.19 SEM image of sandstone showing quartz grains enclosed by pyramidal overgrowth cements filling pore space, and typical conchoidal fracture, with (bottom left) weathered potassium feldspar. (Magnus oilfield, North Sea, at 4 km burial; by courtesy of S. Haszeldine, University of Edinburgh; reproduced from Reed (2005) with permission of Cambridge University Press.)

relative to the dissolved solids. Carbonates generally form first, followed by gypsum and other sulfates, and finally halite and other halide salts; the sequence is dictated by mineral solubility. Often clays and clastic sediments are transported into areas of biogenic or chemical deposition and thus the resulting sediment is somewhat mixed. **Lithification** occurs by the cementation of the sediment grains, with calcite or silica being the usual cement.

The intrusion of a magma into rock causes profound petrological changes, e.g., limestone is converted to marble, sandstone to quartzite, etc. The effect of heat and pressure on the rock adjacent to a dike is known as **contact metamorphism**, as discussed in Box 2.2. The effects of subduction are known to cause a rising mountain chain inland of the subduction zone and to raise the temperature and pressure regionally beneath the mountain chain. This lower-temperature metamorphism is referred to as **regional metamorphism**, which produces rocks known as gneiss, schist and slate. Should the degree of metamorphism be sufficiently high, then quartzite and marble will also be produced by regional as well as contact metamorphism.

The three types described above – igneous, sedimentary and metamorphic – provide the most frequently used classification of rock types, e.g., that prepared by the Geological Society of London's Engineering Group Working Party (1977). However, it is possible to use a classification system that stresses the behavioural rather than the genetic attributes of rocks, i.e., crystalline texture, clastic texture, very fine-grained rocks and organic rocks (Goodman, 1989). In such a system, both granite and limestone are crystalline rocks, and chalk and slate are very fine-grained rocks. In the traditional genetic system, granite is an igneous rock, limestone and chalk are sedimentary rocks, and slate is a metamorphic rock, which is the system followed in the rest of this chapter.

2.4.1 Igneous Rocks

Granitic rocks

Granite is a coarse-grained igneous rock composed essentially of the minerals feldspar (60–80%), quartz (10–25%), mica (2–5%) and some minor constituents, including tourmaline. The feldspars are typically alkali feldspars, e.g., orthoclase, with some plagioclase. When there is more plagioclase than alkali feldspar, the rock is known as **granodiorite**; when alkali feldspar predominates and quartz is minor or absent, the rock is referred to as **syenite**. The mica may be either colourless but sparkling muscovite or dark brown to black biotite – see Figure 2.20. Tourmaline, a boron silicate, is usually blue-black and is found in clusters of needle-shaped crystals or in veins.

Granite is formed by slow crystallization from a magma under mountain ranges that are being uplifted by tectonic forces in the Earth's crust and upper mantle. If this magma erupts onto the surface, it may form a **rhyolite** lava flow or explode to form volcanic ash, which forms the rock known as **tuff**. The evolution of granite requires millions of years of heating of rock and slow crystallization when the heat supply ends. Thus large crystals have time to form and create a coarse-grained rock. During the later stages of granitic intrusions, small volumes of residual granitic liquid penetrate the earlier-formed mass as veins, cooling relatively quickly and forming **micro-granite**. This medium-grained relative of granite is made of crystals in the range 0.05–1 mm, but the rock may also contain feldspar crystals up to 1 cm in size (**phenocrysts**), and is then said to have a porphyritic texture. **Pegmatite** is a type of granite in which the quartz and feldspar crystals are very large, often 5 cm or greater.

Figure 2.20 Biotite granite exposed in a newly blasted road cut, Eldora, Colorado. Hammer for scale. (Photo by the author.)

Weathering of granite has very important implications for engineering projects. The superheated steam that is associated with magma converts feldspars to kaolinite. In humid tropical climates, granite is often completely decomposed, including the quartz, because its constituent minerals are leached by infiltrating rainwater, producing a reddish-brown soil rich in hydrous Fe, Mn and Al oxides known as **laterite**, which is further discussed in section 7.8.2. In temperate climates, weathering also occurs by infiltration of rain and snowmelt, causing erosion such that blocks of granite stand like towers on the landscape, known as tors in England or "inselbergs" in South Africa. Prominent examples are found throughout the Rocky Mountains of Colorado, such as the bald granitic knobs of Lumpy Ridge at Estes Park shown in Figure 2.21.

The important engineering factors related to granite are (1) the state of **kaolinization** of the rock mass and (2) the distribution, frequency and orientation of the joints or discontinuities. Granite within 5–10 m of the surface can be expected to be highly weathered, block-jointed and unsuitable for foundations for large structures. A completely weathered zone at

Figure 2.21 Cross-section through Lumpy Ridge, near Estes Park, Colorado. Note the domal structure of this batholith which covers 1500 km^2 in Rocky Mountain National Park. Weathering caused by deep penetration through vertical fractures, several of which are seen in the left centre of the photograph, results in the granite eroding in sheets. The sheet joints form in response to stress relief from erosion of overlying rocks and is common in massive granites.

the top of the rock mass might consist of residual boulders of still decomposing rock, mixed with sand-sized quartz and feldspar, and brown iron-stained clay. In humid temperate climates, a weathered zone above the **top-of-rock** can be a few metres thick. (Note that, in geotechnical usage in the UK, **rockhead** defines "a boundary between soil-like material and rock that is stronger and more resistant" (Hencher, 2012).) In tropical climates, the surficial lateritic soil can be so thick that building foundations might have to be designed as if it were a clay deposit if the unweathered bedrock is too deep to excavate. We return to this topic in Chapter 7.

Pockets of highly kaolinitized granite could be found at depth hidden below alluvium in river valleys. This was the experience discovered during the building of the Fernworthy Dam on Dartmoor in southwest England. Other dam sites in the area encountered kaolinitized granite beneath good-quality granite, presumably produced by deeply penetrating fractures such as those shown in Figure 2.21; we return to the issue of kaolinitized granite in Chapter 7. Under these conditions it is necessary to grout the weathered zone in an attempt to improve the foundation. It should be kept in mind that dam sites in river valleys might encounter fault zones or veins of weakened rock, as at the Fernworthy Dam site, because river valleys often form in the weakened rock of faults that may or may not be still seismically active (see Chapter 14).

Rhyolite is a fine-grained product of the granitic magma that has surfaced through volcanic eruption. The magma could emerge as a lava flow that forms rhyolite upon cooling and condensation or could explode because of the pressure drop in surfacing that produces pyroclastic rocks such as tuffs, which are commonly layered and contain rhyolite as well as glassy fragments, and **ignimbrites**, which are composed of pumice or vesicular glass that flow down the flanks of volcanoes. Pumice is highly vesicular or porous, very erodible and abrasive; it is useful as a lightweight aggregate material. Usually tuffs are low-density rocks, grey to yellow in colour but high density if the silica grains are welded together. Depending upon its degree of cementation, tuff may be easily eroded or resistant to erosion.

Diorite and Andesite

Diorite is the term for a dark-coloured, coarse-grained igneous rock composed chiefly of plagioclase feldspar, pyroxene and hornblende with minor quartz. **Andesite** is the fine-grained equivalent of diorite, frequently porphyritic in texture with phenocrysts of plagioclase, biotite mica or hornblende.

Obsidian and Pitchstone

Obsidian and **pitchstone** are very fine-grained, glassy igneous rocks that have solidified immediately after eruption. Both are dark-coloured or opaque glass and are very hard but brittle. Pitchstone is similar to metallurgical slag and contains numerous phenocrysts. Obsidian is a natural form of glass with the same shiny surface but black or dark grey.

Gabbro, Dolerite/Diabase and Basalt

These three igneous rocks are referred to as "basic" for historical not scientific reasons. **Dolerite** is the British term while **diabase** is the American term for this rock. They contain a higher content of Mg, Ca or Na in comparison with the silica-rich "acid" igneous rocks such as granite. They also contain significantly more Fe that with the Mg produces their green colour; thus ferromagnesian minerals such as olivine, pyroxene and hornblende comprise half the mineral content, with plagioclase and some quartz. The grain sizes vary from coarse-grained **gabbro** (with crystals greater than 0.5 mm), through medium-grained dolerite/diabase (0.05–0.5 mm) to fine-grained basalt (< 0.05 mm). The ferromagnesian minerals weather to hydrous ferric oxides and clays that line the joints formed by dikes and sills in the rock. Because of their good compressive strength characteristic,

Figure 2.22 Basalt, Grand Mesa, Colorado. Hammer for scale. Note blocky structure.

dolerite/diabase and basalt are used as crushed rock aggregate for concrete.

Gabbro and dolerite/diabase occur as sills and dikes, while basalt also occurs as lava flows. Some basalts contain tunnel-like discontinuities from gas flow during formation and there also can be vugs or pores that might or might not be filled with clay and be interconnected. Furthermore, basalts tend to fracture into blocks during cooling. Figure 2.22 shows the basalt that caps the Grand Mesa of western Colorado in which the blocky structure is evident. Therefore basalts are typically productive groundwater reservoirs, e.g., the Snake River Plain aquifer in Idaho. The best known examples of columnar basalt are the Giant's Causeway in Northern Ireland and the Devil's Postpile in California.

Basalt is strong and difficult to excavate, requiring blasting. However, pockets of highly weathered basalt or **breccia** – angular broken rock fragments bonded by a mineral cement – may be expected to be encountered during engineering operations. The foundation of the 180 m high Itaipu Dam in Brazil was constructed on basalt flows within which there were brecciated zones that presented the possibility of sliding failure. Tunnels were constructed in the basalt to link zones of good quality rock by the emplacement of concrete shear keys to improve the stiffness of the foundation.

Peridotite and Serpentinite

Peridotite is a rock consisting chiefly of olivine and other ferromagnesian minerals and which weather to serpentine, a highly undesirable secondary mineral that can be found in thick beds (**serpentinite**). Serpentinite should be regarded with caution because of its tendency to disintegrate into incompetent material during weathering and its low shear strength resulting in a low friction coefficient (see equation (1.8)).

2.4.2 Sedimentary Rocks

Previously we defined the general laws of stratigraphy using the Grand Canyon of the Colorado River in Arizona as a guide to sedimentary rocks. In introducing rock types, we defined three classes of sedimentary rocks: clastic or detrital, chemical and biochemical. We can use these classes to identify both types of sediment and the corresponding rock type as in Table 2.3. Lithification occurs by **diagenesis**, during which the deposited sediment undergoes compaction, cementation, dissolution and recrystallization as it is transformed from sediment to sedimentary rock (see Bridge and Demicco, 2008).

Because the seas and lakes in which these sediments were deposited had rivers draining into them carrying detrital or clastic sediments, there is often a mixture of clastic sediments with other sediments formed chemically or biochemically and acting as a cement binding the detrital or clastic grains into a solid. Therefore, there are many transitional types between the detrital and chemical and biochemical materials identified in Table 2.3.

Conglomerate and Breccia

Conglomerate and breccia are two types of coarse-grained sedimentary rocks consisting of pebbles (i.e., 4–64 mm sized particles), cobbles (64–256 mm) or boulders (>256 mm) cemented together by a fine-grained material – clay, silt, sand or a chemical cement such as calcite, silica or iron oxides. Figure 1.6 shows a typical river gravel that upon cementation will become a **conglomerate**. The source rock of this sediment underwent weathering in place, then erosion, followed by stream transportation causing rounding of the material and

Table 2.3 Characterization of sediments and sedimentary rocks (after Bishop et al., 2001).

Origin	Type of sediment	Type of sedimentary rock
Clastic or detrital	Gravel	Conglomerate, breccia, tillite
	Sand	Sandstone, arkose, orthoquartzite
	Silt or mud	Siltstone, mudstone, shale
Chemical	Calcareous	Calcareous mudstone, dolostone, travertine
	Siliceous	Flint and chert nodules
	Ferruginous	Ironstone
	Evaporite	Rock salt, gypsum rock
Biochemical or biogenic	Calcareous	Biochemical limestone
	Carbonaceous	Coal

finally deposition as a sediment. Sediments that have been rapidly deposited in thick beds without prolonged periods of abrasion remain angular and form **breccia**, which term is also used to describe angular clastic sediments that have not been transported by water, e.g., volcanic breccia, talus breccia, fault breccia or landslide breccia. If there is a long period of in-stream rolling and abrasion, the breccia will form a conglomerate following diagenesis.

Because natural processes are variable in efficiency, being dependent on time and local conditions, there is a continuous sequence from unconsolidated gravel to lithified conglomerate. This range of consolidation is important in geotechnical engineering because it controls the bearing strength of the rock and its hydraulic conductivity. Figure 1.6 shows how the pore spaces tend to become filled with sand, silt and mud over time, reducing the porosity and hydraulic conductivity of the deposit but allowing the in-fill materials to act as a cement binding the whole mass together to form a rock. Depending on the type of cementing material, the conglomerate may be disintegrated easily by weathering and subsequently eroded. If the conglomerate is well cemented, it could also be fractured, thus distinguishing itself from unconsolidated gravel and perhaps making it unsuitable as a source of aggregate.

Sandstones

Sandstone is an indurated sand deposit consisting of grains of quartz, with feldspar, mica and fragments of fine-grained rocks in the particle-size range 62 μm–2 mm. The cements are often quartz itself, perhaps calcite or hematite or other iron oxide. Sandstone forms in environments in which streams flow or wind or tides sort and deposit particles of this size range. Figure 1.4 shows a faulted sandstone along the beach

Figure 2.23 Dipping sandstone and siltstone layers of the Fountain Formation, Pennsylvanian–Permian age, photographed in the Amphitheatre, Red Rocks State Park, Denver, Colorado. The less resistant layers are siltstone, the more resistant are sandstone, which protrude. Swallow holes line the siltstones. These are typical "red beds", so-called from their hematite staining. (Photo by the author.)

near San Diego, California. Figure 2.23 shows the tilted sandstones and siltstones along the Front Range of the Colorado Rockies.

Sandstone forms as weathered detrital sediments accumulate in sedimentary basins and then undergo diagenesis, including compaction causing loss of porosity, cementation and various water–rock interactions leading to lithification. The detrital grains may have a small range of particle size, which is referred to as well **sorted** or poorly **graded**, or may vary to include sand grains through silt- (4–62 μm) to clay-sized (<4 μm) particles and be considered well graded or poorly sorted. The result is that sandstones are variable in geomechanical properties, porosity and colour, depending upon how they have been formed.

Sandstones in rock exposures, such as in Figure 2.23, often weather so that the bedding is etched out and gives the rock a distinct layered appearance. Thin layers of finer-grained material, perhaps clay or silt, will weather more readily and accentuate the bedding planes such that the sand-rich layers protrude. Any clayey material between the beds of sandstone will have an important control on the overall strength of the rock. The bedding planes may be regular in shape or may be variable, curving and intersecting each one; this structure is called **cross-bedding**. Cross-bedding is caused by alternation in direction of water or air currents during deposition of the sediment.

A variety of sandstone types exist that reflect their mineralogy. A pure quartz sandstone is an **orthoquartzite** whereas one with 25% or more of feldspar is known as an **arkose**. Sandstones that were deposited by turbidity currents offshore tend to have angular grains, are poorly sorted and are known as **greywacke** because of their colour. By convention, greywackes have 15% fine-grained matrix while those with <15% are referred to as **arenite**.

Siltstone, Mudstone and Shale

Siltstones are hard, coherent rocks that tend to be massive rather than laminated. Figure 1.5 is an example of a Precambrian siltstone. They have a gritty feel or taste to them and are usually defined as containing two-thirds of silt-sized particles. Depending upon the cementing mineral, they may weather easily and disintegrate into silt deposits; therefore they are not expected to be highly durable when used as a rockfill.

Finer-grained than siltstones are **argillaceous rocks**, which include shale and various mudstones. One means of discriminating amongst them is by whole-rock analysis to identify the amount of aluminum versus silicon; clay minerals are rich in aluminum while silts contain silica and are rich in silicon and poor in aluminum. A second discriminator is the **fissility** of the rock. Shales are fissile in that they break along surfaces that are associated with mica minerals that formed during diagenesis. Mudstones or claystones lack fissility.

Mudstones originate as fine-grained sediments in lacustrine (lake) or marine environments (see Box 2.3). The individual particles that comprise mudstone include clay and carbonate minerals, some quartz and pyrite and other sulfide minerals. Organic matter and pyrite tend to control the colour of the rock; the more organic matter, the more likely the rock is black. In outcrops, mudstone weathers easily and breaks up into small particles with the pyrite being oxidized,

producing an acidic water (section 7.8). Because mudstone is fine-grained, it is of low hydraulic conductivity and is of great importance in the development of groundwater flow patterns in sedimentary basins and for the disposal of toxic or radioactive wastes. Claystone is a non-fissile mudstone comprising at least two-thirds clay-sized particles.

Shale is a fine-grained sedimentary rock consisting of clay-sized particles, including clay minerals, mica, chlorite, quartz and carbonate and sulfide minerals, particularly pyrite. Shale forms from mud deposits in lakes, estuaries and seabeds where the absence of currents allows the clay particles to accumulate in thickness. Shales are fissile and are readily eroded. Figure 1.3 shows the presence of shale beds in the Grand Canyon in Arizona as slopes of weathered debris known as **talus**. Because they are more easily eroded than limestones or sandstones with which they are often formed, the alternation of shale with other rocks creates a landscape of alternating valleys and low hills.

The laminated structure and low hydraulic conductivity of shale are important geotechnical properties. The first means that it has **rippability** and can be machine excavated by means of a hydraulic back-hoe excavator or by a tractor with a ripper blade, without the use of explosives. The second affects groundwater conditions and surface-water runoff at engineering sites. Shale can provide a solid rock foundation but it usually has low shear strength and therefore must be regarded with caution. That said, many dams, tunnels and embankments encounter shales and have been engineered to address the weaknesses that the rock poses in construction.

Goodman (1993) divided shales into two engineering geological classes: (1) cemented shales are indurated and behave like rocks in that they have the strength of concrete; and (2) compaction shales are readily weathered and compressed and lack shear strength. A particularly important form of shale or mudstone weathering is known as **slaking** in which the rock undergoes wetting-induced disintegration and which is associated with slope stability and excavation difficulties (Shakoor and Erguler, 2009).

Calcareous Mudstone, Dolomite and Travertine

Marl is a form of calcareous mudstone with relatively small amounts of clay (<35%). It has been observed forming in lakes in the north-central of the USA. **Travertine** is a rock formed at springs, at geysers and in caves where calcite precipitates from warm water. **Dolomite** is a rock that is composed of at least 50% dolomite rather than calcite. The rock is occasionally referred to as dolostone. It appears that most

| BOX 2.3 | THE STICKY BUSINESS OF TUNNELLING BENEATH AUCKLAND, NEW ZEALAND |

Flysch sequences of rocks are thick accumulations of alternating marine sandstones and mudstones, one of which forms the regional bedrock beneath Auckland, New Zealand. The rapid growth of metropolitan Auckland means that tunnelling is occurring within the flysch rocks as sewage, road and rail links are built or modernized. This has led to the pronounced problem of *sticky spoil*, i.e., the adhesion of the waste rock (spoil) to tunnel boring machines (TBMs). This results in expensive periods of downtime for cleaning the face of the TBM.

The sedimentary formation responsible for this problem is the partly lithified East Coast Bays Formation (ECBF) of Miocene age. These are alternating sandstones and siltstones deposited by marine turbidity flows (**turbidites**) and inter-turbidite mudstones. A typical exposure is shown in Figure 2.24. The formation is described by Black et al. (2010) as "characteristic alternating, decimeter-bedded graded sandstones and laminated mudstones", in which the sandstones contain abundant fragments of previously lithified rock known as **lithic grains**. In the lower ECBF these are derived from the mudstones, although in the shallower sequence rocks volcanic grains are more common due to the volcanic eruptions during Auckland's Cenozoic past.

Figure 2.24 A cliff-section exposure of flysch sediments of the East Coast Bays Formation, Auckland, New Zealand, with a rock hammer for scale. The prominent beds are sandstone, whereas the recessed beds are mudstones. (Photo courtesy of Dr Philippa Black, University of Auckland; see Black (2010).)

From a geotechnical perspective the ECBF sedimentary rocks are considered to fall into the classification of **weak rocks**, that is, their uniaxial compressive strength is of the order of only 5 MPa. Such soft rocks make for easy excavation and, except where fractured and thus weakened, the ECBF rocks support most structures and form steeply cut slopes as shown in Figure 2.24.

Figure 2.25 presents a thin-section photomicrograph of the sandstone conducted during the petrographic examination of the cause of spoil adhesion. It shows a rock that is clast-supported, i.e., the large fragments form the skeletal backbone of the rock. These mudstone clasts comprise about 30% of the mass of the petrographic sample and are quite spherical, indicating that they survived transportation in a debris flow well enough to form the major component of both the turbidite sandstones and siltstones. Furthermore, the rock is moderately well sorted such that the grains are of similar size; this is equivalent to saying that it is poorly graded in geotechnical terms. The ratio of quartz to feldspars is around 2 and the total of these two minerals is roughly 25% of the total mass of the sample. The cementing minerals within the sandstones are both calcite and silica.

What, then, is the source of the *sticky spoil*? The thin section shown in Figure 2.25 contains both a muddy matrix comprising about 25% of the rock and the mudstone clasts which make up another 30%, both of which

BOX 2.3 | (CONT.)

contain clay minerals predominantly of the smectite class but also chlorite, illite and the zeolite clinoptilolite (2%). **Zeolites** are a class of alumino-silicate minerals associated with volcanic rocks; in the present case, the zeolite acts as a cement to strengthen the mudstone clasts during their transport as a debris flow. The zeolite is shown in Figure 2.26 set in a clay mineral matrix. Therefore the total clay mineral content of the rock is >50%, which is sufficient to produce a plastic, swelling material that will adhere to the face of the TBM and complicate the removal of the waste rock (i.e., *mucking*).

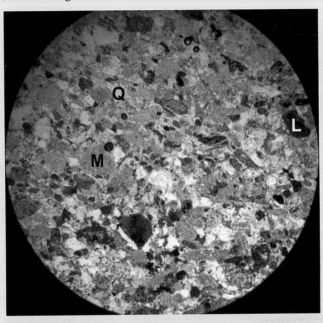

Figure 2.25 Photomicrograph of a thin-section sample of the ECBF sandstone collected approximately 40 m below ground surface. Width of photo, 4 mm; polarized light. The major components of this sample are mudstone fragments or "clasts" (M, comprising ∼30% of grains), lithic clasts derived from oceanic igneous rocks (L, ∼10%), quartz (Q) and feldspar (∼25%), a muddy matrix (∼25%) and other unidentified minerals (∼5%). (Photo courtesy of Dr Philippa Black, University of Auckland; see Black (2010).)

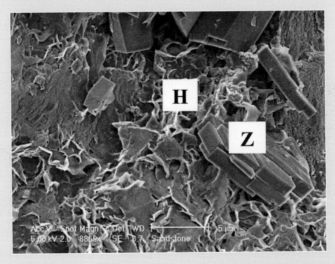

Figure 2.26 SEM photo of ECBF sandstone; scale shown on photo. Prominently displayed are zeolite (Z) crystals, identified as clinoptilolite, surrounded by a clay mineral matrix. Some of the clay has a honeycomb (H) structure, while the lower part of the photo shows a lamellar structure, which indicates that the area sample by the SEM is a mudstone clast within the sandstone. (Photo by H. Higham, courtesy of Dr Philippa Black, University of Auckland; from Black (2010).)

BOX 2.3 | (CONT.)

The problems posed by swelling, plastic clays are quite common. A similar problem was reported during tunnel construction in the Mercia Mudstone beneath Leicester, UK, by Atkinson et al. (2003) and has the same origin, i.e., highly plastic, swelling smectite clays. In Chapter 7 we shall return to the geotechnical issues raised by swelling clays and anhydrite in relation to clogging of TBMs.

Source: From Black et al. (2010).

dolomite rock has formed by the reaction of high-Mg-content brines with calcite in limestones, i.e., diagenetically. Dolomite has also been observed to occur as an evaporite mineral (see below) in coastal flats above the tide range where magnesium replaces calcium in calcite to form dolomite with the displaced calcium ion subsequently forming gypsum. While usually of high strength because it is formed from carbonate minerals, dolomite can dissolve to create caverns and therefore dolomite sites must be carefully characterized (see discussion of karst in section 13.3).

Chert

Chert is a form of precipitated silica, often referred to by archeologists as **flint**. Chert nodules are found in carbonate rocks such as the Cretaceous Chalk of England and France as shown in Figure 2.27. Chert is sometimes present as thin beds, but also as nodules in rocks of all ages. Chert is very hard and cannot be scratched with a steel blade. Rocks containing chert are notoriously difficult to drill through. But chert has little strength under shock forces, such as hammer blows, and breaks into small pieces with sharp edges when struck; hence it was the preferred material for making tools during the Stone Age and is used today to make pestle and mortars for grinding material.

Because of its hardness, chert-rich rocks resist weathering and erosion and form gravel and coarse sand deposits in alluvial valleys and along coasts, where they may be extracted for use as concrete aggregate. Although thin-bedded chert nodules can be easily extracted, thick beds of chert will need to be blasted for excavation. Being a hard and brittle rock, chert usually contains many joints, especially when the rock formation has been subject to folding. The Chalk of southern England, in which chert nodules are found in bedding planes of this soft limestone rock (Figure 2.27), can be easily broken up by excavators and the nodules can be readily extracted.

Ironstone

Ironstone is an iron-rich sedimentary rock found in France, England and from New York to Alabama. The principal iron

Figure 2.27 The English Chalk in Sussex showing bedding-plane fractures occupied by chert nodules and small plants. Note vertical stress-release fractures and rock-fall evidence of cliff failure. Hiker for scale. (Photo courtesy of E.H. Everett.)

oxide mineral is goethite, while siderite, hematite and pyrite are also present with cementation by calcite and dolomite.

Evaporite

Evaporite rocks are formed from the precipitated salts that were previously dissolved in seawater until evaporation caused their aqueous solubility limits to be exceeded. In theory, the evaporation of seawater should cause the sequential precipitation of calcite, dolomite, gypsum and anhydrite, halite and finally various other chloride and sulfate salts. Modern evaporites are forming today in shallow-water lagoons and on coastal salt flats known as sabkhas in the Persian Gulf. Because of their high aqueous solubility, the presence of evaporite rocks or minerals indicates an absence of fresh groundwater flow. They are of great importance as **cap rocks** or seals for petroleum reservoirs in that they are essentially impermeable.

Biochemical Limestone

Limestone is used in the manufacture of Portland cement and is typically hard and massive. Biochemical limestones form from the skeletons of marine organisms that secrete calcium carbonate; consequently, most limestone is white in colour when freshly broken. Limestone is composed mainly of calcite, but it contains minor amounts of clay minerals, such as illite or smectite, quartz, dolomite and pyrite. The texture, however, may comprise numerous fossil fragments that are mainly calcite. These organisms may only become apparent under a petrographic microscope; the organisms may be particularly important in dating the age of the limestone.

Limestones are typically identified by the well-known acid test, i.e., the application of dilute hydrochloric acid to the sample to produce the effervescence of CO_2. Carbon dioxide dissolved in rainwater also forms carbonic acid (H_2CO_3), which can also dissolve limestone and other carbonate rocks, producing solution channels, sinkholes and caves that are progressively enlarged. Thus limestone terrain always requires site characterization to determine if karst conditions are present (see section 13.3).

Most limestones are hard rocks but this does not apply to **chalk** and to **oolitic limestone**. Chalk is formed from the skeletal remains of plankton and algae and typically contains flint nodules and clay minerals. It is soft, friable and very porous. Like other limestones it forms clearly defined bedding planes – see Figure 2.27. When the chalk contains more than 13% clay, it is considered a marl, and is classified as a calcareous mudstone when the clay content exceeds 35%. Oolitic limestone is formed in warm seas when calcite precipitates on a silica sand nucleus and builds up a concentric series of spheres. The Jurassic limestones of the Cotswold Hills in England are formed from oolites of approximately 1–2 mm diameter.

Coal

Coal is a deposit of carbon found in layers referred to as "seams" that are interbedded with mudrocks and sandstones in formations of mainly Pennsylvanian age, i.e., Late Carboniferous in British terminology (300 Ma). The types of plants that form peat bogs, the source of coal, did not appear until the Devonian period, so coal is only found in younger rocks. Coal originally formed as a thick deposit of peat, a mass of plant remains accumulating under stagnant water and anoxic conditions, a necessary requirement for the preservation of the plant material; otherwise it would have been oxidized and thus destroyed. The coastal swamps of Virginia, North Carolina and Sumatra are modern equivalents.

The Pennsylvanian-aged coals were buried by deltaic sands as the seas rose and fell, most likely due to continental glaciations. This pattern of rising and falling sea level caused a cyclic depositional sequence of marine and nonmarine sediments known as a **cyclothem**, in which the coal stratum was underlain by nonmarine mudrocks and overlain by shallow marine sandstones and conglomerates (Klein and Philpotts, 2016). This sequence was repeated many times, with the coal gradually being formed by **coalification**, the dewatering, compaction and lithification of the sediments under increasing temperature and pressure.

Kesler and Simon (2015) indicate that a peat layer of approximately 100 m thickness was required to produce a 10 m thick coal seam. The Pittsburgh coal seam, which stretches from Pennsylvania southwestwards through West Virginia and Ohio, covers over $50\,000\,km^2$. Coal is ranked from lignite (73–78 wt.% carbon; specific energy $<20\,MJ/kg$), which is widely mined in Germany, through sub-bituminous and bituminous coal (90% carbon; 27–32 MJ/kg) to anthracite ($>90\%$ carbon; $>32\,MJ/kg$). Graphite is pure black carbon with a dull metallic lustre and a specific gravity of 2.2; it is the "lead" in pencils.

2.4.3 Metamorphic Rocks

Metamorphic rocks are formed from other types of rock by the action of heat ($>150^{\circ}C$) and pressure, both lithostatic and directed. Furthermore, metamorphism may involve high-temperature pore fluids. It should not be confused with lower-temperature processes such as weathering or diagenesis. The particle size of metamorphic rocks varies from fine-grained slates through fine- to medium-grained phyllite to medium- to coarse-grained schist, marble and gneiss. The increasing grain sizes reflect the increase in temperature of the various metamorphic grades, which are defined by distinctive mineral assemblages. There are two textures of metamorphic rock: foliated and massive.

The minerals in foliated rocks are aligned in certain directions rather than randomly. Sometimes they are segregated into groups of different kinds so that the rock has a striped appearance, e.g., gneiss, displaying segregated bands of quartz, light-coloured feldspar and dark mica. Alignment of minerals is a characteristic feature of **regional metamorphic rocks**. Such alignment, referred to as **foliation**, can affect the geomechanical properties of the rock mass and often occurs from the growth of micas in a preferred direction.

Second, there are massive varieties having randomly oriented but anisotropic texture, which is the result of the rock material having been affected by uniform pressure fields rather than shearing stresses. **Contact metamorphic rocks** are formed by heat from an intrusion, but there is no directed pressure and as a result the rock lacks any planar feature.

This produces hornfels (Figure 2.12), which typically is a very strong rock precisely because of the lack of alignment of mineral grains.

Slate, Phyllite and Schist

Slate, phyllite and schist may all have the same mineralogy; however, they will differ by the size of the crystals, which typically increase with higher-temperature metamorphism.

Slate is a hard rock, resistant to weathering, hence its use in roofing slates. They are very fine-grained and develop slaty cleavage. The orientation of the cleavage and joint systems are the most important geotechnical factors. Figure 2.28 shows the development of slaty cleavage in a road cut in the Canadian Rocky Mountains. Quite clearly, the stability of such a rock face is very much dependent on the angle that it makes with the cleavage and joint system. In surficial weathering, as with the road cut, slate will break into a large number of pieces, some very large and some of clay size, in response to the release of residual stresses no longer imposed on the exposed rock. The weathered zone, evident in the top left-hand corner of the photograph, usually has variable depth because of local factors and on many construction sites the depth to this **rockhead** will need to be determined by drilling or trenching during site investigation. Were such a rock encountered during tunnelling, it would be necessary to map the distribution of joints and cleavage planes so that groundwater flow could be estimated and precautions taken to prevent rock failure.

Phyllite typically contains chlorite, sericite and perhaps graphite or talc. It is green when chlorite is present and has a silvery sheen from the sericite. It is the result of the metamorphism of shale or mudstone at relatively low temperature, about 400°C, and in a highly anisotropic stress field. Smaller crystals form at these lower temperatures of regional metamorphism; the rocks are fine-grained and tend to split into flat pieces. Phyllite is a type of slate that has a texture characterized by the parallel orientation of mineral flakes and by oval-shaped crystals that stand out from the finer-grained matrix, i.e., porphyroblasts. While resistant to weathering, phyllite tends to split upon excavation due to the release of stresses imposed when the rock was originally subject to compression.

Schist is a product of regional metamorphism and consists of coarse silicate crystals, which can be chlorite, muscovite or its fine-grained equivalent sericite, biotite, garnet, hornblende and especially quartz. Flat and rod-shaped crystals are aligned so that the rock splits easily into flat or bar-shaped pieces caused by mineral alignment at temperatures of up to 700°C in a strongly anisotropic stress field. The resulting alignment

Figure 2.28 Slates exposed in a road cut in the Trans-Canada Highway near Lake Louise, Alberta. Face shown is approximately 2 m in height and is caused by breakage along slaty cleavage planes. (Photo by the author.)

causes planes of **schistosity** to form that are of lower strength. In contrast to slate, a rock formed under similar but lower-temperature conditions, the cleavage surfaces of schist are rougher because of the larger crystals that comprise the rock.

The geomechanical properties of schist are therefore controlled by the mineralogy and the stress-induced alignment. Higher-temperature minerals, such as hornblende, will tend to weather more rapidly than quartz and mica and thus be weaker in outcrop samples. Because of the strong forces that create the foliation, there can be shear planes between laminations, in which case the rock will be hazardous if encountered in slopes and tunnels.

Gneiss and Migmatite

Gneiss (pronounced "nice") has a distinctive banded appearance caused by the segregation of minerals at high temperatures. Figure 2.13 shows a granitic gneiss displaying the mineral composition of a metamorphosed granite, i.e., quartz, feldspar, mica and hornblende. These are not randomly distributed, as in granite, but are segregated into bands of one mineral or another, i.e., the rock displays foliation and its geomechanical properties are directional not isotropic. Gneiss that is rich in mica could rapidly slake and have easy cleavage. Figure 2.15 shows typical gneissic texture.

Migmatite is a mixed igneous and metamorphic rock. It is formed at high temperature when the rock begins to melt. The first material to melt is of granitic composition and this liquid segregates into sheets. The result is a layered rock consisting of light-coloured sheets of granite (of igneous origin) separating darker refractory material, the metamorphic part.

BOX 2.4 | **GEOTECHNICAL ENGINEER WANTED**
(JOB DESCRIPTION IN A CANADIAN NEWSPAPER)

Reporting to the Senior Geotechnical Engineer, the Geotechnical Engineer will be responsible for geotechnical and hydrogeological support to ensure safety and stability of existing and proposed open pits, underground excavations, water retention dikes and collection ponds. The successful candidate, who will have a B.Sc. in civil, geological or mining engineering and a minimum of three years experience, will:

- Inspect open pits, dikes, dams, underground excavations;
- Analyze relevant data to ascertain the stability of the terraced slopes of the open-pit mines;
- Initiate actions to stabilize slopes where necessary;
- Coordinate a geotechnical monitoring program;
- Promote a "safety culture" in mining operations;
- Provide geotechnical support to mine operation groups to ensure an understanding of geotechnical hazards and design requirements.

Mylonite and Pseudotachylite

Mylonite is the product of intense metamorphism due to recrystallization of rock within a fault zone under shear. The rock is fine-grained and will likely be foliated due to the shear stresses. We shall return to the topic of mylonites in the discussion of faulting (see Figure 3.12). Pseudotachylite is a black fine-grained or glassy rock formed by frictional fusion in a fault zone. It is associated with seismogenic zones (see Chapter 14).

2.4.4 Ore Deposits

Mineral deposits rich enough for commercial development for one or more elements are referred to as ore deposits. The extraction of ores occurs either by underground or open-pit mining. Mines are geotechnical creations of great complexity and sophistication; it is regrettable that they are not perceived to be "high tech". Deep rock mining in places such as South Africa and Australia had a profound influence on the development of rock mechanics as a discipline. The South African Au and Pt deposits and the Sudbury, Ontario, Ni–Cu deposits are associated with Precambrian rocks and operations at depths >1 km. The copper mines of Chile, Utah and Arizona are mostly open-pit mines. The Bingham Canyon mine in Utah produces ore that is today only 0.6% Cu and not the 2% when the mine began producing in 1906; consequently, it produces large volumes of processed tailings that must be carefully managed. The treatment of mine tailings and waste rock is the responsibility of geoenvironmental engineers. Open-pit mines, like Bingham Canyon and the Chilean copper mines, can be >3 km in diameter at the surface. The duties of geotechnical engineers in open-pit mines – particularly ensuring the stability of the terraced benches – are illustrated in Box 2.4, which presents a job description that appeared in a Canadian newspaper for work at a diamond mine in northern Canada.

A number of mineral deposits have already been discussed in this text, such as banded iron formations. These layered rocks are composed of centimetre-scale bands of magnetite and quartz known as taconite that precipitated from ocean waters during the Precambrian. The bulk of US iron ore is mined in Minnesota and Michigan from banded iron formations; similar Brazilian and Australian ores are shipped to Chinese and Japanese steel mills, although China is itself a major iron ore producer from its own banded iron formations. The Jurassic-age ironstones of Lincolnshire, England, and Lorraine, France, are relatively small compared with these Precambrian deposits and supply local mills.

Metallic elements in the magma, such as tin, copper, lead and zinc, do not fit in the crystalline lattices of the silicates that comprise granite but remain in the **hydrothermal brine** as it cools and precipitate out as sulfide minerals. Examples of legendary mineral deposits associated with granitic and volcanic rocks are the Colorado Mineral Belt, including Aspen, Telluride and Boulder, and the Cornish mining district in England.

The development of plate tectonics has allowed the development of a better understanding of the concentration of mineral deposits. Subduction offshore of California and Mexico produced volcanism onshore and the development of batholiths (i.e., **plutonism**) such as the Sierra Nevada further inland. The Californian coastal range is dotted with mercury deposits, while the Sierra Nevada foothills are host to the gold belt. Beyond the Sierras are the gold deposits of Nevada and the copper deposits of Utah and Arizona. The hydrothermal solutions that developed because of subduction apparently dissolved certain metals in the lithosphere and then precipitated them from solution when volcanism or plutonism caused these solutions to rise to levels of lower temperature and pressure. "Black smokers" – seafloor hydrothermal vents along the MOR – were responsible for Cu–Zn–Pb deposits such as those of Cyprus and Timmins, Ontario; these are known as volcanogenic massive sulfides.

Most ore deposits of industrial metals such as Pb, Zn and Cu are associated with hydrothermal solutions because these brines contain sufficient chloride ion to complex the metal ions and prevent the precipitation of metal sulfide ores during the migration of the brines. Eventually, geochemical conditions result in mineral deposition. For example, lead and zinc deposits in the Ozarks region of Missouri, known as Mississippi Valley type deposits, and the vast uranium deposits in the Athabasca Sandstone of Saskatchewan are both believed to have been deposited following long brine migration.

Finally, we should add mention of what are often referred to as industrial minerals or **aggregates**. These can be sand and gravel quarried from some fluvial or glaciofluvial deposit (section 10.4) or hard sedimentary or igneous rock that is subsequently crushed (section 5.2).

2.5 Summary

We have compared the Earth to an egg, with a dense yolk or core, a mantle or egg white, and a shell of continental and oceanic crust. Processes occurring within the crust are the subject of this book. Geothermal energy produced within the interior of the Earth drives convection within the mantle that in turn drives the mechanism of plate tectonics, which is responsible for the creation and motion of the continents. The continents are moved and reassembled over geologic time by the motion of the seafloor; therefore plate tectonics is synonymous with global crustal evolution.

The oldest rocks so far identified on the Earth's surface are found in the Shield terrains of Canada and Greenland and are about four billion years old, while the Earth itself appears to be of the order of 4.5 billion (4.5×10^9) years old. These Precambrian rocks form the backbone of the continents and contain vast mineral resources, particularly iron ore. The sedimentary record can be traced into the Precambrian; however, sediments began to display abundant fossils about 540 million years ago with the onset of the Cambrian period. The Paleozoic (542–251 Ma), Mesozoic (251–66 Ma) and Cenozoic (since 66 Ma) comprise the vast majority of the sedimentary record, the rocks and sediments of which contain most of the world's hydrocarbon and groundwater resources. The surface of the continental crust of the Earth, as it appears today, has been much altered during the last two million years, i.e., the Quaternary period consisting of the Pleistocene and Holocene epochs. Interpretation of stratigraphy is guided by the four general laws of stratigraphy, i.e., those of original continuity, original horizontality, faunal succession and superposition.

The most important of the rock-forming minerals are the silicates, carbonates and oxides. The geotechnical and geoenvironmental engineer needs to appreciate the nature of such minerals in terms of their density, solubility and potential to weather to clay minerals. Often 95% of a rock is composed of three or four minerals, while the remaining 5% might comprise 10–20 other identifiable minerals. Mineral identification is conducted using a petrographic microscope aided by geochemical, SEM and X-ray diffraction methods.

Rock-forming minerals are assembled by igneous, sedimentary and metamorphic processes into rocks of the same classes. Igneous rocks are formed either in the crust by the crystallization of silicate minerals in a magma or on the Earth's surface as volcanic rocks. The elevated temperatures and pressures associated with the intrusion of a magma into existing rocks cause their alteration and the creation of metamorphic rocks, such as those discussed at Churchill Falls in Labrador, eastern Canada. The physical and chemical weathering of existing rocks and the subsequent transport of the weathered rock materials by stream, wind and ice eventually leads to the deposition of sediments and, ultimately, the formation of sedimentary rocks.

In addition, we have illustrated the importance of the rock-forming minerals in geotechnical engineering by examining two case histories: first, the alteration of a mineral assemblage within Precambrian gneiss causing tunnel wall collapse and, second, the inhibiting effect of clay minerals on tunnel boring machines. In Chapter 3, we proceed to the nature of geological structures, such as faults and folds, and the way that geologists depict these structures in maps.

2.6 FURTHER READING

Bishop, A.C., Woolley, A.R. and Hamilton, W.R., 2001, *Cambridge Guide to Minerals, Rocks and Fossils*. Cambridge, UK: Cambridge University Press. — A useful companion.

Goodman, R.E., 1993, *Engineering Geology: Rock in Engineering Construction*. New York, NY: John Wiley and Sons. — A detailed account of the engineering geology of rock masses encountered in construction projects in the USA.

Jerram, D. and Petford, N., 2011, *Igneous Rocks in the Field*, 2nd edn. Chichester, UK: Wiley-Blackwell. — One of a series of pocket-book guides, in this case providing assistance to engineers dealing with construction within igneous rocks.

Klein, C. and Philpotts, A.R., 2016, *Earth Materials: Introduction to Mineralogy and Petrology*, 2nd edn. Cambridge, UK: Cambridge University Press. — This text constitutes the most useful summary of basic mineralogy and petrology presently in print. It is a valuable reference for all engineers working in geotechnical and geological engineering.

Mathez, E.A. and Webster, J.D., 2004, *The Earth Machine: The Science of a Dynamic Planet*. New York, NY: Columbia University Press. — A superb book on geology, as it is presently understood, written for the scientifically inquisitive individual by two members of the staff of the American Museum of Natural History in New York City. The permanent exhibition there – the Gottesman Hall of Planet Earth – presents the ideas documented in this book. Definitely worth a visit.

National Academies of Sciences, Engineering, and Medicine, 2017, *Volcanic Eruptions and their Repose, Unrest, Precursors, and Timing*. Washington, DC: National Academies Press. See https://doi.org/10.17226/24650. — An up-to-date review of the science of vulcanology and the issue of forecasting volcanic eruptions.

Ridley J., 2013, *Ore Deposits*. Cambridge, UK: Cambridge University Press. — A necessary guide to the geotechnical or geoenvironmental engineer working in the mining industry.

Tucker, M.E., 2011, *Sedimentary Rocks in the Field: A Practical Guide*, 4th edn. Chichester, UK: Wiley-Blackwell. — A very affordable pocket-book guide to the characterization of sedimentary rocks.

2.7 Questions

Question 2.1. In 2009, the Governor of Louisiana criticized the first State of the Union speech by President Obama for spending $140 million for volcano monitoring, which the Governor, a Rhodes Scholar, considered "wasteful spending". Are there indeed any benefits to such government expenditures when geologists cannot predict the long-term occurrence of such events?

Question 2.2. For the region in which you live, consider how plate tectonics has created the landmass and identify the principal features of its structural geology. Is the region tectonically active in terms of earthquakes and volcanoes? When was the last earthquake in your region and what was its magnitude? What continuing lateral and vertical processes are occurring that contribute to your region? What are their rates of movement in m/ka?

Question 2.3. It has been suggested that nuclear waste might be deposited in subduction zones that would carry the waste deep into the lithosphere. Would there be risks to such an approach that would appear to be relatively inexpensive compared with the mine-like deep geological repositories being constructed in Europe and North America?

Question 2.4. Steel and concrete are the essential building blocks of civil engineering. Pick a building, bridge or other example of local infrastructure near where you live and determine where the iron ore from which the steel was produced and the aggregate might have come from to build it.

Question 2.5. Cambridge University Press has printed and published books like this continuously since 1584. Which era, period and epoch contains this year? How many years in the past separates 1584 from year 2000?

Question 2.6. The early Cambrian period is one of the extraordinary evolutionary events in Earth history in that many hard-bodied animals appeared in the fossil record. If we compress 4.5×10^9 years – the age of the Earth – into just one year, when did this "Cambrian explosion" of evolutionary life begin in this compressed time scale, which ends on December 31 at midnight?

Question 2.7. Using this same compressed year, on which day of our geologically compressed year and when on that day did the Pleistocene glaciations begin?

Question 2.8. What causes a granitic pluton – part of a batholith similar to that shown in Figure 2.21 – to be dome-shaped? [*Hint*: Think about the physical properties of granitic magmas in the crust.]

3 Geological Structures and Maps

3.1 Introduction

On the east side of Las Vegas, Nevada, lies Lake Mead, which is skirted by the North Shore Road. Travelling this route, the engineer will cross faults, dry streambeds and then encounter the Tertiary-age Horse Spring Formation of the Muddy Mountains (13 Ma), which is shown in Figure 3.1 (Ball, 2005). How might we define the orientation of the tilted and folded Bitter Ridge limestone member, which forms the

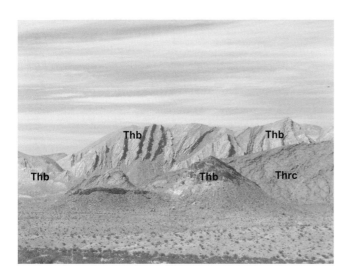

Figure 3.1 The Muddy Mountains along the North Shore Road, near Lake Mead National Recreation Area, Nevada. The Tertiary-age Bitter Ridge (Thb) limestone member of the Horse Spring Formation steeply dips along the skyline. Thrc represents the conglomerate lithofacies of the Rainbow Gardens Member of the Horse Spring Formation (Tertiary). (Formation identifications courtesy of Norma Biggar, University of Nevada Las Vegas.)

skyline along this part of North Shore Road, in a manner that is reproducible and can be represented on a map? Alternatively, how might we quantify the orientations of the fractures shown in Figure 2.21? In this chapter we shall explore how geological maps and geometric figures are employed to provide quantitative information about geological structures.

We begin this chapter by identifying basic structural measurements in a folded rock mass and then proceed to discuss fault types. We proceed to discuss the nature of fault zones, joints and other discontinuities that are important to the stability of engineered structures. Sedimentary structures are then considered, followed by an introduction to the various types of geological maps. Our goal is to teach the engineer to be able to read a geologic map and to make valid conclusions about local geologic structures.

Geologic structures may be described in terms of geometric elements, i.e. points, lines and planes. For example, faults may be conceptualized as planar structures, and, as such, may be described in terms of their orientation with respect to a compass direction and dip with respect to a horizontal plane. This method for describing geologic structures facilitates the use of graphical construction in descriptive geometry for solving basic structural geology problems such as the dip of a subsurface layer (such as a coal bed) or the apparent thickness of a dipping aquifer. This chapter ends with a discussion of stereographic projection to illustrate three-dimensional (3D) structures in two-dimensional (2D) space.

The structure of a sedimentary rock comprises units called **beds** or **strata**, as introduced in section 2.2, which are layers of a particular type of rock and, although the term should only be applied to sedimentary rocks, they are sometimes

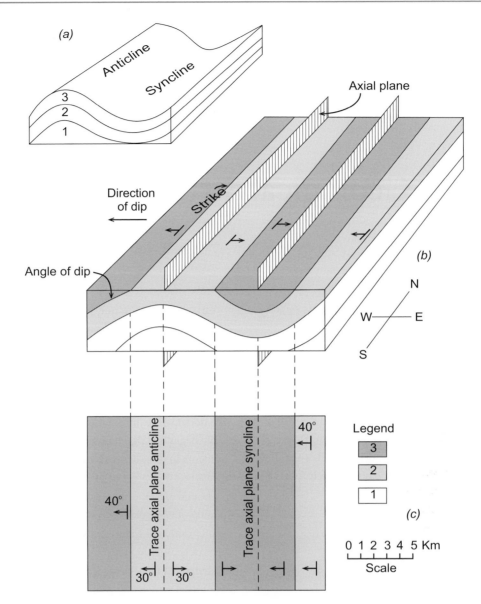

Figure 3.2 (a) A simple fold in sedimentary rock shown as a block diagram of an anticline and syncline. (b) The fold is eroded to a horizontal surface illustrating the outcropping formations in cross-section. Note the axial planes, dip and strike symbols and sequence of the outcropping strata. (c) The eroded folds are shown as they would be displayed on a geologic map. The legend shows the strata arranged in order of decreasing age upward, while dip angles are expressed in degrees. (From Dapples (1959).)

used for any rock type that has a layered structure. Sediments are typically deposited horizontally – see the law of original horizontality (section 2.2). A series of beds that have been similarly deposited is called a **formation** (Fm), which is defined as the smallest mappable unit, and may consist of a variety of rock types, e.g., sandstone, shale and carbonates. A **group** is an assemblage of formations, while a **member** is a formal unit identified within a formation. Figure 2.23 shows the Fountain Formation in Colorado and Figure 2.24 shows the East Coast Bays Formation in New Zealand. These surface exposures are known as **outcrops**. Note how the first letters of any group, formation or member are capitalized when they

refer to specific units that are formally recognized by national, state or provincial geological organizations.

Tectonic uplift of the Earth's crust, as well as crustal extension, crustal subsidence, shear and compression, may cause tilting and fold structures to develop in groups of formations, as in Figure 3.2. **Folds** are the typical result of compressive forces causing the strata to become deformed into corrugated structures. Folding intensity varies from shallow folds to very intense folding resulting in rock strain and fracture yielding discontinuities known as tectonic joints. Compression or extension in strata can result in the shearing of a stratum and its displacement, known as a **fault**.

3.2 Basic Structural Measurements

3.2.1 Strike and Dip in Folded Rocks

Figure 3.2(a) is a 3D block diagram of folded strata, which shows, according to the law of superposition (section 2.2), that the bed at the bottom, labelled 1, is the oldest and the bed on top, labelled 3, is the youngest. Beds 2 and 3 are exposed as outcrops. Figure 3.2(b) is a second block diagram depicting erosion of the folded strata to a horizontal plane. The axial planes of the folds are indicated, as are the dip and strike of the various beds.

The strike of each formation is measured first, then the dip is measured at right angles to the strike. The azimuth, or compass reading relative to due north, of the horizontal line on the dipping surfaces is called the **strike** of the bed. The dip of a bed is the angle, relative to the horizontal, at which the bed slopes downward. The **true dip** is measured as the maximum angle of inclination that the bed makes with the horizontal; any angle less than the maximum is an **apparent dip**. Thus dip and strike form a unique pair of related measurements,

although they may be reported in various formats because there are two possible directions for dip for a given strike.

One convention is known as the strike/dip/quadrant of dip so that the direction of the dipping bed is specified (Figure 3.3a). Thus a bed with a measurement of 015/40E has a strike 15° east of north (i.e., the azimuth measured as a 360° angle using three digits) and dips at 40° (two digits) to the east. Alternative conventions are the British (Figure 3.3c) and American (Figure 3.3b) right-hand rules and the dip direction/dip convention shown as Figure 3.3(d) (see Walker and Cohen, 2009; Lisle et al., 2011). The International Society of Rock Mechanics follows the American right-hand rule shown in Figure 3.3(b), in which the dip of the bed is to the right (ISRM, 1978).

Returning to Figure 3.2(b), strike and dip symbols are shown where beds 2 and 3 outcrop on the eroded surface and indicate the place of their measurement. If we project the observed field measured locations of the various beds onto a map surface – the plan view – we develop a **geological map**, as shown in Figure 3.2(c). The map view shows the geometry

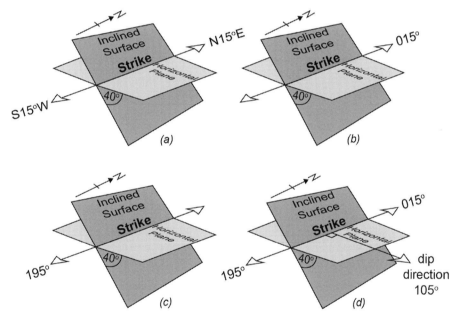

Figure 3.3 Strike, defined as the direction of a horizontal line on an inclined surface, is illustrated by the line marking the intersection of the light grey horizontal plane with the darker grey dipping surface. In this example, the strike direction is either 015° (N15°E) or 195° (S15°W) and the dip, measured in the direction perpendicular to the strike, is 40° to the east-southeast (in the direction 105°). Note that the orientation of an inclined surface, such as a bedding plane, fault or fracture, can be recorded in different ways as shown in (a) through (d). In (a), quadrant notation is used with the orientation recorded as strike/dip, i.e., N15°E/40°ESE. In (b) and (c), azimuth notation is used in which the strike is given in terms of the 360° that describe a complete circle. The American right-hand rule (b) is illustrated with the strike recorded such that the direction of dip is always to the right of the compass bearing. The strike is 015° because the surface is inclined to the right, or towards the east, at 40°, and the notation is 015/40 or 015°/40°. The British right-hand rule (c) uses the opposite direction, i.e., 195°, because the British rule is based on extending the thumb of the right hand in the direction of dip, and recording the direction of the extended fingers as the strike, yielding 195°/40°. The dip/dip direction notation is shown in (d) and is indicated by 40°/105°. (Figure used with permission of the estate of the late J. Wallach.)

BOX 3.1 | SOME RULES FOR INTERPRETING GEOLOGIC MAPS

- The law of original horizontality dictates that strata are initially deposited in a gently dipping configuration; lacustrine and deep marine sediments could be horizontal, but fluvial sediments would have an original gradient at the time of deposition.
- The law of superposition dictates that strata follow one another in chronological, but not necessarily continuous, order. Where strata are not overturned, the oldest beds in a sequence are on the bottom and the youngest beds are on the top.
- Outcrops of the same lithological sequence that are separated but aligned imply stratigraphic continuity.
- Strata occur in laterally continuous and parallel layers in a region.
- Sharp discontinuities in lithologic patterns imply the presence of faults, unconformities or intrusive contacts.
- Where erosion has exposed folds (see Figure 3.2), anticlines have their oldest beds in the centre of the anticline, while synclines have their youngest beds in the centre.
- Anticlines plunge towards the nose (i.e., closed end) of the structure, whereas synclines plunge towards the open end (see Figure 3.4).
- Fold orientation may be described by the trend and plunge of the trace of the axial surface and the strike and dip of the limbs.
- On geological maps showing streambeds, the lines of contact representing surfaces of contact (**contacts**) between horizontal beds, or beds that have a dip lower than the streambed gradient, V upstream, i.e., they form a V pointing upstream with respect to the topographic contours.
- Contacts of beds that have a dip greater than the streambed gradient V downstream.
- Contacts migrate downdip upon erosion, except for vertical beds.
- True dip angle of a planar structure can be seen in cross-section only if the cross-section is oriented perpendicular to the strike of the structure. For bedding dip, the cross-section would be oriented perpendicular to the strike of the beds; whereas, for a fault dip, the cross-section would be oriented perpendicular to the strike of the fault plane.

Sources: From Hozik et al. (2003) and van der Pluijm and Marshak (1997).

---

of the outcropping surfaces, the directions of strike and dip of the beds and the traces of the axial planes of the anticline and syncline. The map view indicates that bed 2 must be present beneath bed 3 on the western edge of the map. Likewise, considering the eastern edge of the map, the outcrop of bed 2 indicates that bed 3 has been eroded away. The orientation of the dip symbols identifies the presence of a syncline and an anticline. Box 3.1 lists a set of useful rules for interpreting geologic maps and cross-sections.

3.2.2 The Plunge of a Fold

Figure 3.4 (Dapples, 1959) illustrates how fold structures are mapped when the **crest** of an anticline or the **trough** of a syncline is not horizontal but is said to **plunge**. The plunging fold is inclined along its crest line as shown in Figure 3.4(a), which reveals the fold prior to erosion that resulted in the horizontal surface shown in Figure 3.4(b). The axial plane

shown in Figure 3.4(b) defines the lines of reference by which the fold structure of the anticline is characterized; the axial plane is a geometric construct, not a physical entity. Line *AB* defines the horizontal line from which the angle of plunge is measured downwards to the crest of the anticline. The direction of the horizontal line is the trend or bearing of the fold, i.e., *CB*, and the angle *BCD* is the plunge of the fold. The same measurements may be made along the axial trace of the trough of the syncline. Note that the strike of a fold is not the same as the strike of individual beds within the fold.

A geologic map of this fold structure in Colorado is shown as Figure 3.2(c). The various formations are identified in the map legend, with the first letter indicating geologic age and the second its formal name according to the Colorado Geological Survey. The oldest formation is the Triassic Chugwater Formation, which is overlain by the Jurassic Sundance and Morrison Formations and the Cretaceous Dakota Sandstone and Mancos Shale. Note that in the eastern part of

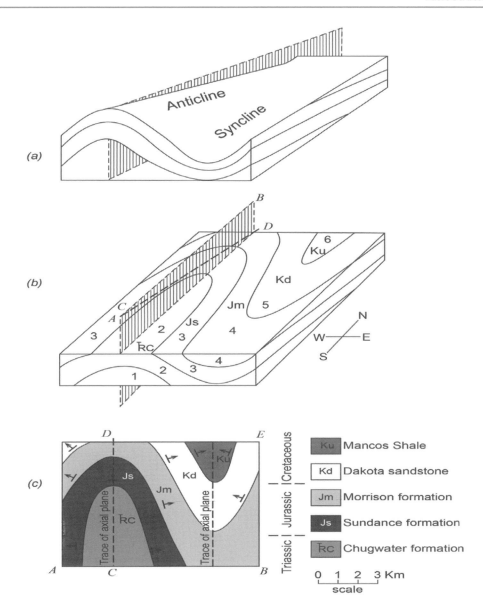

Figure 3.4 Similar to Figure 3.2, but with the anticline plunging to the north. The angle of plunge is that subtended by *BCD* as noted in (a). If the axial plane was not north–south, the angle of the axial plane with the north direction would be the rake or pitch. (b) Block diagram of (a) eroded to a horizontal surface exposing formations lower in the stratigraphic sequence. (c) Geologic map of (b) showing the axial traces for both anticline and syncline with formation symbols. (After Dapples (1959).)

the map the shale outcrop is surrounded by the Dakota Sandstone, while in the western part the Jurassic and Triassic formations crop out.

3.2.3 Cross-Sections

Cross-sections are vertical planes through the Earth's crust. An engineer drilling into this sequence of rocks needs such a geologic map to understand what he or she will encounter. An extensively drilled area, or one with much tunnelling, will likely have cross-sections already prepared by the responsible Geological Survey that will improve the predictability of the

formations that a drilled borehole will encounter. Figure 3.5 presents cross-sections along three profiles – *AB*, *DC* and *AE* – identified in Figure 3.4(c). The preparation of cross-sections requires knowledge of (i) the width of the outcrop along the profile, (ii) the dip and strike of each formation displayed, (iii) formation thicknesses and (iv) the topographic contours of the mapped area (Dapples, 1959).

3.2.4 Field Measurements

Measurements of strike are made with **compasses**, while dip and plunge are measured with a **clinometer**. The Brunton

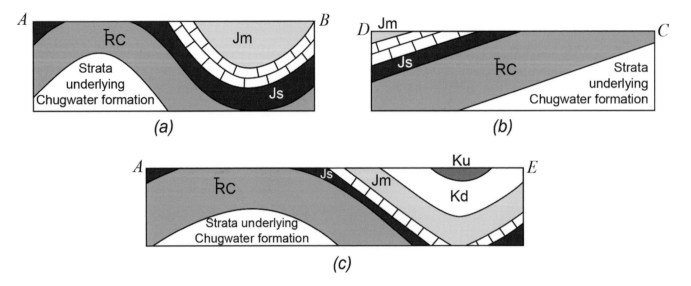

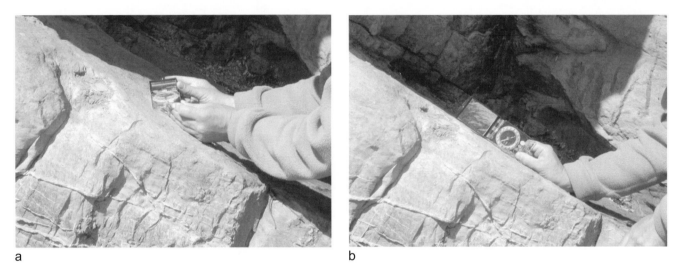

Figure 3.5 Cross-sections drawn along the lines indicated on Figure 3.4(c), i.e., (a) along *AB*, (b) along *DC* and (c) along *AE*. Compare with the block diagram in Figure 3.4(b). (After Dapples (1959).)

a

b

Figure 3.6 The use of a clinometer to measure the dip of a sloping plane relative to the strike, which is measured as the azimuth or compass bearing. The direction of the dip – at right angles to the strike – must be noted so that the direction of the slope plane is unambiguous. (From Lisle and Leyshon (2004); reproduced with the permission of Cambridge University Press.)

compass is one example, widely used in North America, of a geologist's compass that has a built-in clinometer. Digital clinometers are available to map location as well as strike and dip, and even cell phones now have applications that can be downloaded and used to measure dip and strike. Much less expensive mirror compasses are often used with a spherical bubble for levelling the compass. Figure 3.6 shows the use of a clinometer to measure strike and dip of a bed on an outcrop, i.e., a planar feature, using the built-in spirit level to align the instrument. The strike is the compass bearing obtained when the clinometer is lain along a horizontal line (Figure 3.6a),

while the dip angle (Figure 3.6b) is measured perpendicular to that horizontal line.

Planar features, such as joints, fractures or faults, form linear features or traces where they intersect the Earth's surface. These structures are measured by their plunge and pitch, as in Figure 3.7. In this case the clinometer is held upright along the linear joint trace (Figure 3.7c), with its edge aligned with the trace; now imagine a vertical plane (shown as the thick line in Figure 3.7a) passing through this trace. The angle that the linear trace is tilted in the vertical plane is the angle of plunge (30° in Figure 3.7a). The direction of

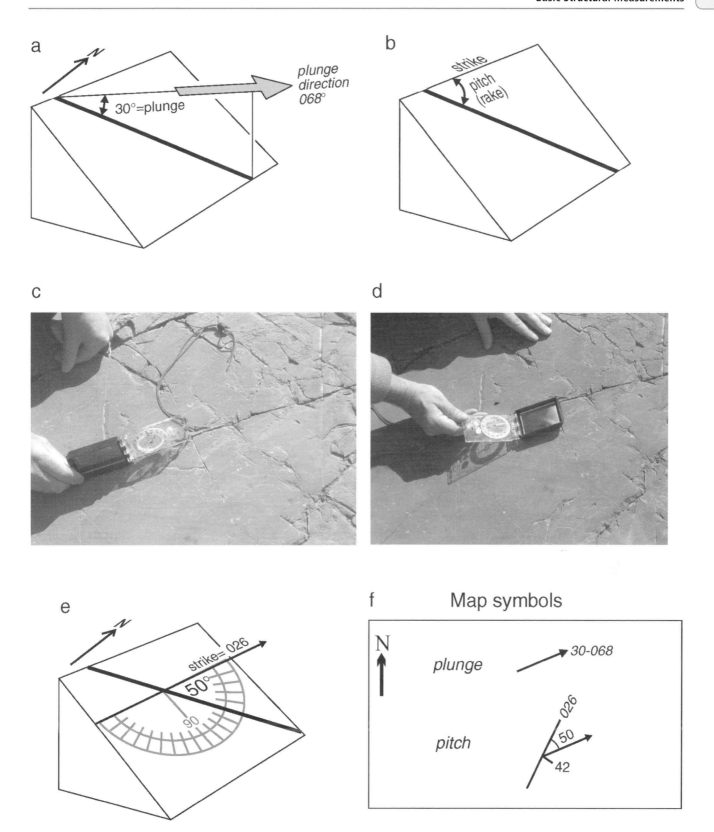

Figure 3.7 Orientation of linear structures, in this case the trace of a joint for which the plunge and pitch are measured. (From Lisle and Leyshon (2004); reproduced with the permission of Cambridge University Press.)

plunge is measured parallel to the strike of this imaginary vertical plane as shown in Figure 3.7(a). Figure 3.7(d) shows the clinometer with the compass being held horizontal and the mirror aligned along the edge of the trace; this is the plunge direction. The measured value is recorded as "angle of plunge–plunge direction", where the direction is the compass azimuth; in this case the measurement is 30–068. The **pitch** of this trace is the angle on the dipping plane that the plunge direction makes with the strike of the trace (Figure 3.7b and e). The map symbols record the results as azimuthal bearings (Figure 3.7f).

3.3 Faults and Shear Zones

Few geologic structures are as important to the engineer as **faults**, which are the result of the brittle fracture of rocks. Faults were introduced earlier, such as those shown in Figure 1.4, where the horizontal component of displacement or offset is readily visible across the faults by noticing the relative elevation of the marker beds (shown beneath the cap). Figures 2.6 and 2.7 illustrate other large-scale examples of faults in their plate tectonic setting. Adjacent and sub-parallel faults, such as those in Figure 1.4, define a **fault zone**.

To be classified as a fault, the walls of the fault must display evidence of displacement parallel to the fracture. Further evidence includes **offset beds**, cohesionless debris in the fault zone known as **breccia** and polished surfaces known as **slickensides** (Gudmundsson, 2011). If plastic (i.e., non-recoverable) deformation of the rock occurs prior to its failure, the deformed zone is known as a **shear zone** and the failure is said to be due to ductile processes, i.e., continuous deformation at the scale of observation. We shall return to this matter in section 4.3.2.

Figure 3.8(a) illustrates the principal fault types. The **wall** of a fault is the block of rock adjacent to the fault plane; the **hanging wall** is the rock above the fault plane, whereas the **footwall** is the lower block. In a **normal fault**, the hanging wall slips to a lower elevation relative to the footwall. **Strike-slip faults** slip along the direction of the fault strike. In a **reverse fault**, the hanging wall moves up the fault plane relative to the footwall. The **fault block** is that rock mass that slipped because of the fault; however, it may not be clear which block moved, only that there is a relative displacement. Figure 3.8(b) shows how displacements such as heave and throw are defined in faults.

The dependence of fault type on the principal stresses was first identified by the Scottish structural geologist E.M.

Anderson around 1905 and has proven to be of lasting significance (see Healy et al., 2012). Anderson (1951) first established that the ground surface should be considered a boundary condition of zero shear stress from the perspective of continuum mechanics. Therefore, in the triaxial system of stresses employed in rock mechanics and structural geology (see Figure 1.10 with the third stress, σ_y, being into the plane of the page), the vertical stress (σ_v) would be a principal stress perpendicular to the surface boundary condition, i.e., σ_z in Figure 1.10. The nature of a particular fault would then depend on how the vertical stress compared with the other principal stresses, i.e.,

1. if $\sigma_v = \sigma_1, \Rightarrow$ normal fault;
2. if $\sigma_v = \sigma_2, \Rightarrow$ strike-slip fault;
3. if $\sigma_v = \sigma_3, \Rightarrow$ thrust or reverse fault.

Anderson's theory of faulting is based on his observation that faults occur by shear failure during the brittle deformation of rocks, a consequence of which is that we may apply the Mohr–Coulomb strength criterion occurring under compressive stress (see Figure 1.13; and Turcotte and Schubert, 2002).

Returning to Nevada, the engineering visitor to the Hoover Dam is able to see some fine examples of normal faults along the walls of the canyon downstream of the Dam, as shown in Figure 3.9. Here we see the eastern (Arizona) wall of the canyon with extrusive volcanic tuffs penetrated by **mafic** dikes, so-called because of their composition of ferromagnesian minerals such as pyroxene, olivine, biotite and amphiboles. Figure 3.9 shows three Tertiary-age bedrock units described by Ball (2005). At the base of the photo is the Dam Conglomerate (T1), a well-stratified unit of clast-supported conglomerate and coarse-grained sandstone. Overlying T1 are two tuffs formed from ash-flow deposits, T2 and the Upper Dike, that are described as **lithic**, that is, composed of fragments of older rocks welded into a tuff. The dikes are composed of olivine basalt and basaltic andesite. The extrusive volcanic rocks dip to the northeast at 12–50° whereas the high-angle normal faults shown dip to the southwest.

The bridge shown in Figure 3.9 links Arizona and Nevada and was part of a major (US$240 million) engineering project to build the Hoover Dam Bypass during 2001–10. Site characterization involved extensive studies by engineering geologists, including 3D laser scanning for topographic mapping of the canyon walls, core drilling, optical televiewer borehole logging to acquire fracture data, downhole seismic surveys, laboratory rock-strength testing as well as various geologic

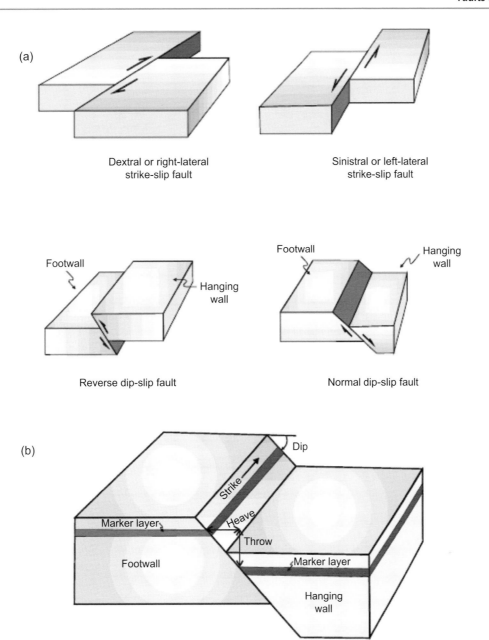

Figure 3.8 (a) Principal types of faults. (b) Displacements as measured for a normal dip-slip fault when a marker bed is present. (From Gudmundsson (2011); reproduced with the permission of Cambridge University Press.)

mapping methods. Figures 3.10 and 3.11 illustrate some of the structural features observed during the engineering geological mapping.

If faults are the result of brittle or ductile failure, then shear zones (Figure 3.11) represent plastic deformation and displacement. This plastic deformation occurs at great depths within the Earth's crust, i.e., below the brittle–ductile transition zone at 10–15 km, whereas brittle failure occurs above this transition zone. Fossen (2016) defines a shear zone as "a tabular zone in which strain is notably higher than in the surrounding rock" and reports that the strain in a shear

zone results in a displacement approximately the same size as its thickness, whereas in faults the displacement is typically 10–100 times the deformation zone thickness. Furthermore, Fossen (2016, p. 211) notes that, whereas in porous rocks faults tend to be relatively less permeable than the rock matrix, the fault zone is usually more permeable than the matrix in the non-porous rocks.

Seeking a deep geological repository for its nuclear waste, the Swiss examined the potential for disposal in granitic rocks by constructing an underground research laboratory at Grimsel, Switzerland (Nagra, 1992). The granite was

Figure 3.9 A mafic dike (arrows) intrudes into welded ash-flow tuff (T2) on the Arizona side of the Hoover Dam. Note close spacing of fractures and how various fault blocks have slipped from left to right towards the power house. The Upper Dike is also a tuff, whereas T1 is a conglomerate. (Geologic description by Ball (2005); photo by Jeff Keaton, Amec Foster Wheeler, Los Angeles, California, and reproduced with his permission.)

Figure 3.10 Exposed vertical fault surface showing inclined grooves (arrow) from displacement, Hoover Dam, Nevada, USA. (Photo by Jeff Keaton, Amec Foster Wheeler, Los Angeles, California, and reproduced with his permission.)

emplaced at depth about 10 Ma and then was progressively uplifted. While at depth, the rock was deformed at high temperatures and pressures, which produced ductile structures such as shear zones and the strongly deformed rock known as **mylonite**, shown in Figure 3.12. Following uplift and cooling of the rock (~1 Ma), the deformation characteristics were further complicated by brittle fracturing, yielding fault gouges and **breccias**, which are angular rock fragments embedded in a fine- to medium-grained matrix. Therefore, the brittle and ductile phenomena shown in Figure 3.12 reflect processes occurring over ten million years within the Earth's crust.

Figure 3.11 A 2 m high shear zone with clay core, and pen (arrowed) for scale, near Hoover Dam, Nevada, USA. (Photo by Jeff Keaton, Amec Foster Wheeler, Los Angeles, California, and reproduced with his permission.)

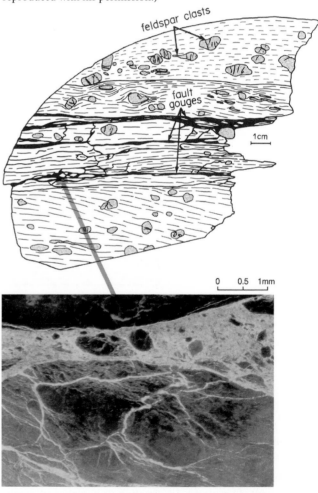

Figure 3.12 Mylonite, formed by plastic deformation in a shear zone in tight granite, Underground Research Laboratory, Grimsel, Switzerland. Note that the term "clasts" is normally restricted to rock fragments within sedimentary rocks rather than igneous or metamorphic rocks.(From Nagra (1992); reproduced with the permission of Nagra.)

BOX 3.2 | THE SIGNIFICANCE OF WEAKNESS ZONES IN ROCK TUNNELLING

The Norwegian engineering geologists Palmstrom and Berthelsen identified a number of zones of rock weakness that were evident in tunnelling. They defined a weakness zone as a layer, zone or vein in which the rock-mass properties are significantly poorer than in the surrounding materials. These zones of weakness require careful consideration during the design phase because they cause safety problems with tunnelling and are often the cause of costly delays. Examples include:

- faults and shear zones;
- layers of weak rock due to the presence of mica, soapstone, talc, gypsum or chlorite;
- altered rock zones, e.g., Figures 2.16 and 2.19 and Box 2.2; and
- dikes or veins, e.g., highly jointed dolerite or breccia.

The length of weakness zones in Norwegian tunnels excavated into Precambrian basement rocks varies from 1 to 20% and rock-supporting operations (e.g., shotcrete, structural timber or steel, etc.) can cause a 5–200% increase in the projected cost of excavation and mucking out. Rock support of weak zones typically amounts to 30–60% of these costs.

Figure 3.13 shows a map view of the pattern of weakness zones associated with a hydroelectric power project for which a tunnel is to be constructed from Lake B to Lake A. The weakness zones shown were identified by geological mapping and air-photo interpretation. The shortest tunnel ("Tunnel alt") would be 2.8 km long but would intersect 500–600 m of weak rock. The selected route is 300 m longer but passes through only 70–100 m of weak rock.

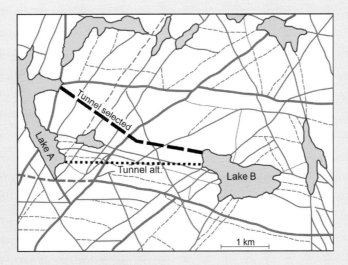

Figure 3.13 Pattern of weakness zones for a hydroelectric power project based on geological mapping and air-photo interpretation. (From Palmstrom and Berthelsen (1988).)

Figure 3.14 illustrates how the structural orientation of a weakness zone only 1 m thick influences the total length of weak rock encountered during tunnelling. The selected route provides a much less risky design by limiting exposure to weakness zones. Note that a 1 m thick weakness zone with a strike of 40° and a dip of 60° will cut a tunnel along ~22 m for an excavation of 6 m height and 10 m span. The structural orientation of the weakness zone influences construction safety and cost.

BOX 3.2 | (CONT.)

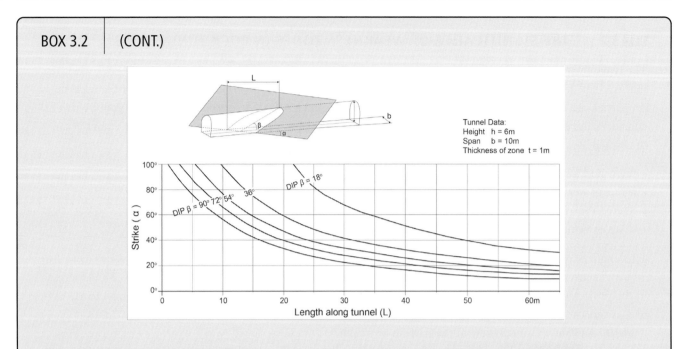

Figure 3.14 The structural orientation of a weakness zone to the alignment of a tunnel will profoundly influence how much weak rock a tunnel excavation will be exposed to. If we consider a weakness zone of 1 m thickness that is oriented vertically (dip = 90°) with a strike of 100°, then the zone of weak rock exposed will be no more than 1 m. But a zone with a strike of 40° and a dip of 54° will expose ∼25 m of weak rock. (From Palmstrom and Berthelsen (1988).)

Source: After Palmstrom and Berthelsen (1988).

3.4 Joints

A **joint** is a rock fracture with a small displacement or opening normal to the fracture plane but no visible displacement along the fracture plane (Gudmundsson, 2011). Figure 3.7 shows the mapping of a joint in the field. In geological terms, a joint is an **extension fracture**. Veins are joints filled with secondary (precipitated) minerals while fissures have large apertures (Fossen, 2016). In rock mechanics terms, a joint is considered a mechanical discontinuity with no visible shear displacement (ISRM, 1978). Figures 2.27 and 2.24 show vertical joints in chalk and flysch sediments that do not persist through neighbouring beds. Box 3.3 summarizes how field measurements of joints and other fractures are made. Section 15.5 on slope failure in rocks addresses the critical nature that joints play in landslides and section 4.3.3 considers joints in a geomechanical context.

In the Alberta foothills of the Western Canada Sedimentary Basin, the Paskapoo Formation is an important aquifer providing a groundwater supply to ∼64 000 water wells (Grasby et al., 2008). Figure 3.15 illustrates the interbedded nature of the sandstone–mudstone sequence of sediments that comprise the Paskapoo. The sandstone facies are relatively

Figure 3.15 The Paskapoo Formation in outcrop, Cochrane, Alberta, Canada. (Photo courtesy of Glen Stockmal, Geological Survey of Canada; reproduced with permission.)

resistant to erosion and are prominent whereas the mudstone–siltstone facies weather more readily and are recessed just as with similar facies in the Grand Canyon of Arizona (see Figure 1.3).

BOX 3.3 | FIELD MEASUREMENT OF FRACTURES

The principal parameters measured are as follows:

- *Attitude or orientation*: the strike and dip of the fracture.
- *Displacement*: the aperture when the fracture is open, or the thickness when the fracture is filled.
- *Trace length*: the length of the line of intersection of the fracture plane (**persistence**) with the rock surface where the scan line or profile is conducted.
- *Type of fracture*: identification of joint, mineral vein, dike, sill or fault.
- *Infill material*: identification of mineral infill, breccia, mylonite or fault gouge (often weaker than host rock).
- *Fracture frequency*: number of fractures measured along a scan line or profile per unit area or per unit volume of rock.
- *Fracture spacing*: the reciprocal of fracture frequency indicating distance between fractures.
- *Seepage*: rate of flow of water from individual fractures.

Sources: After ISRM (1978) and Gudmundsson (2011).

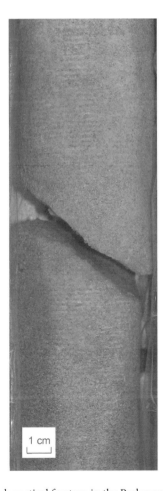

Figure 3.16 A sub-vertical fracture in the Paskapoo Formation. (From Grasby et al. (2008); Geological Survey of Canada; reproduced with the permission of NRC Research Press.)

Joints within the sandstone units (Figure 3.16) enhance the permeability of the Paskapoo. Their orientation is related to the regional tectonic stress regime and is aligned (Figure 3.17a) with the maximum principal horizontal stress (i.e., σ_{Hmax}, see Chapter 4) for southern Alberta rather than any alignment of the sandstone channel facies (Grasby et al., 2008). Rose diagrams, such as the radial histogram shown in Figure 3.17(a), are commonly used methods for displaying joint or fracture orientations. Figure 3.17(b) demonstrates that the thinner sandstone beds have a higher fracture density, i.e., closer spacing, which enhances permeability in these thinner beds.

The ratio of joint spacing (S) to layer thickness (T) in layered sedimentary rocks is of considerable interest to structural geologists, hydrogeologists and geotechnical engineers. Bai and Pollard (2000) report that many values indicate $S/T \sim 0.8$–1.2, which appears to reflect "well-developed" fracture sets. In the numerous cases where S/T is measured in the range 0.3–0.8, the normal stress acting perpendicular to the fractures changes from tensile to compressive, inhibiting the development of further fractures within the unfractured rock volume. Values of $S/T > 1.2$ indicate the likelihood of further fracture initiation and propagation until $S/T \to 0.8$–1.2.

3.5 Unconformities

It is not unusual to find a **hiatus** in sedimentary deposition and lithification indicating an interruption in an otherwise continuous stratigraphic record of rocks during which

(a)

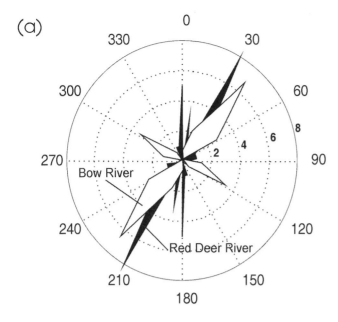

(b)

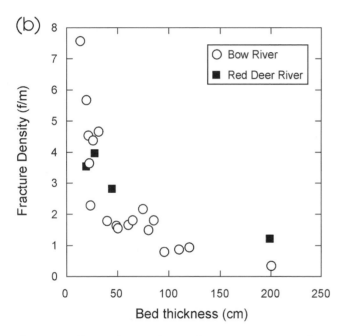

Figure 3.17 (a) Rose diagram showing fracture orientations in the Paskapoo Formation measured in the Red Deer and Bow River valleys. (b) Inverse relationship of fracture density to bed thickness in the same valleys. (From Grasby et al. (2008); Geological Survey of Canada; reproduced with the permission of NRC Research Press.)

erosion of sediments or no deposition has occurred. We define an **unconformity** in the stratigraphic record as a geological structure in which one set of beds lies on another set but there is a hiatus at their contact.

The major divisions of geological history shown in Table 2.1 are based on unconformities evident in some part of the world, the existence of which is used to indicate

Figure 3.18 The angular unconformity at Box Canyon Falls, near Ouray, Colorado, USA. Sub-horizontal Devonian sandstone overlies the vertical Precambrian quartzite.

the standard divisions of geological time. Such breaks are important geotechnically in that they may represent zones of weak rock strength or very different strengths on either side of the unconformity.

Van der Pluijm and Marshak (1997) define several types of unconformity. The most distinctive of these is the **angular unconformity**, in which beds below the plane of unconformity are truncated and have a different attitude (i.e., orientation) than beds above the unconformity. Figure 3.18 shows the angular unconformity in Box Canyon, Colorado. The lower Precambrian beds were metamorphosed, tilted and eroded before the depositional episode that emplaced the Devonian sands above the unconformity.

Angular unconformities are present in two of geologic history's most famous locations. At Siccar Point, Scotland, the angular unconformity observed by James Hutton in the 1790s was critical in his development of modern geological thinking. Vertical Silurian shales are capped by gently dipping Devonian sandstone, similar to that shown in Figure 3.18. In studying this unconformity, Hutton grasped the essence of the **uniformitarian principle** of geological processes – ancient processes are no different from ones we observe today. The roles of these processes would be determined by their rates, often very slow. Thus the vertical Silurian shales had to be first deposited, lithified, folded, elevated above sea level and partially eroded before the sediments that became the Devonian sandstone could be deposited on this eroded surface after it subsided below sea level.

At the bottom of the Grand Canyon of the Colorado River in Arizona (Figure 1.3), an angular unconformity separates the Precambrian Vishnu Schist from the overlying sedimentary Grand Canyon Series, also of Precambrian age, and another angular unconformity separates these Precambrian sedimentary rocks from the Cambrian Tapeats Sandstone.

Other types of unconformity identified by van der Pluijm and Marshak (1997) are the **disconformity**, in which beds above and below the unconformity are parallel, i.e., a hiatus with either erosion or no deposition or both, and the **nonconformity**, which represents an unconformity where sedimentary beds were deposited on basement rocks, such as igneous or metamorphic rocks. Figure 3.18 represents a nonconformity as well as an angular unconformity.

3.6 Introduction to Geological and Engineering Geology Maps

In preparing Figures 3.4(c) and 3.5, we were creating the elements of a geological map, i.e., the plan view and some cross-sectional images. Both figures also identify a **scale**, which is the relationship between distance on the map and distance on the ground. We refer to typical regional geological maps with scales of 1 : 1000 000 as small scale and more local maps having a large scale, such as 1 : 10 000, which is a scale much more suited to engineering work. Various scales allow the presentation of useful geological information important in engineering projects; however, engineering geology maps are typically published at larger scales.

In the USA, the US Geological Survey topographic map series divides the USA into quadrangles bounded by two lines of latitude and two of longitude; for example, the 7.5-minute maps have a scale of 1 : 24 000 (1 inch : 2000 ft). Engineering geology maps may be superimposed on geologic maps of this scale. In the UK, the British Geological Survey publishes a series of geological maps at scales of 1 : 625 000 and 1 : 50 000. It has recently created a set of national engineering geological maps at a scale of 1 : 1000 000 to show the surficial and bedrock geology of the UK (Dobbs et al., 2012). These scales are usually too small for many geotechnical purposes because they cannot show the necessary detail; thus roads and other critical features have to be exaggerated to display clearly their location. Site surveys employing boreholes and trenches are conducted to document the subsurface soil and rock conditions and the resulting "site plans" have scales of the order of 1 : 5000 or larger.

In studying a geologic map, the engineer should note the following:

- Title and serial number: the title is likely chosen from the name of the principal municipality on the map; the serial number relates to a national grid.
- Surficial or bedrock geology: either or both may be presented on the map.
- Stratigraphic column: the various rocks and surficial deposits will be identified with a key to the colour code used on the map.
- Lettering system: lettering will define the rock formations on an age basis and provide a means by which colour-blind people can use geologic maps.
- Strike and dip symbols define the attitude of sedimentary rocks.
- Faults: these will display the sense of movement along a fault i.e., up-, down- or along-strike.
- Unconformities: these are shown as geological boundaries meeting at an angle.
- Igneous intrusions: plutons, dikes, sills and volcanic rocks are distinguished in the legend of the map from sedimentary rocks.
- Cross-sections: as in Figure 3.5, cross-sectional views of sections, the ends of which are plotted on the map (see Figure 3.4b), are used to assist the reader to interpret the geologic structure portrayed on the map.
- Surficial deposits: whether called "surficial" or "alluvial" as in North America, "superficial" as in the UK or "drift", these deposits are usually distinguished in the legend of the map.

Figure 3.19 is based upon a representation prepared by the US Geological Survey (Bernknopf et al., 1993) of typical information shown on a engineering geology map. (Note that the geology in Figure 3.19 is illustrative rather than realistic.) It employs the **Unified Engineering Geology Mapping System** (UEGMS) of map acronyms (Keaton and DeGraff, 1996; Keaton and Rinne, 2002), as shown in Table 3.1, to identify soil and rock information. Maps like Figure 3.19 provide information on where one might site a sanitary landfill or other waste repository in such a way that it is removed from

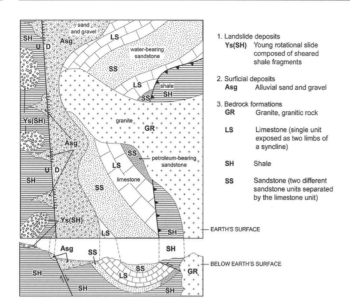

Figure 3.19 A schematic example of an engineering geology map using the Unified Engineering Geology Mapping System (Keaton and DeGraff, 1996; Keaton and Rinne, 2002). The landslide deposits are shown as colluvium and the acronyms are based on the Unified Landslide Classification System (see Keaton and DeGraff, 1996). (After Bernknopf et al. (1993); US Geological Survey.)

geologic hazards, such as earthquakes, landslides and flood-prone areas. Geological maps may also provide information on seismic risks by displaying the presence and outline of faults or on groundwater or energy resources, e.g., the presence of water-bearing and petroleum-bearing sandstones.

Engineering geology maps must be of a large scale to be useful. Price (2009) indicated that they will usually be between 1 : 5000 and 1 : 20 000 for detailed engineering location planning, while those for dam sites may be 1 : 1000 or 1 : 50 for rock slopes or tunnels. Invariably engineering geology maps will use the zoning concept – see discussion by Dearman (1991) – to identify zones in which the lithology or some engineering property is more-or-less homogeneous for the purpose of the map, e.g., occurrence of karstic features. Any engineering geology map must also exhibit clearly recognizable features through the use of a standardized set of mapping acronyms and symbols. Consequently, following the practice of the International Association of Engineering Geology (Matula, 1981), the following information might be recorded on an engineering geology map.

- **Geological data** such as mappable geologic units (using a protocol), geologic boundaries (with uncertainty indicated) and descriptions of (i) soil and rock units (using a protocol), (ii) exposures and outcrops, (iii) the state of rock weathering and alteration, and (iv) discontinuities.

Figure 3.20 Air photo of the re-aligned US Highway 93 or Hoover Dam Bypass near Boulder, Nevada. See Figure 3.21 for an engineering geology map of this area. (Photo prepared with Google Earth Pro.)

- **Hydrogeological data** such as (i) piezometric surfaces, (ii) groundwater flow patterns, (iii) hydraulic conductivities, and (iv) discharge areas, e.g., springs and seeps.
- **Geomorphological data** such as ground morphology, landslide scars and scarps, subsidence features, solifluction lobes and cambering.
- **Geological hazard potential** by which the following are noted: mass movements due to landslides, flooding of flood-plains and of dry alluvial fans, coastal erosion, seismicity and volcanic hazards, e.g., evidence of lahars.

Figure 3.20 is an air photograph of the completed new section of a highway in Nevada near the Hoover Dam. Figure 3.21 is the engineering geology map for this area developed during the planning stage of the re-aligned highway and illustrating the use of the UEGMS summarized in Table 3.1. Instead of a geological base map as in Figure 3.19, this figure uses the air photograph upon which are superimposed zones of similar lithology. Maps like this will be accompanied by a report that contains descriptions of outcrops and other geomorphic features, and the opinions, conclusions and recommendations of the engineering geologist.

Engineering geomorphology maps are typically developed during the reconnaissance of an area in which development is to occur, e.g., a highway or a pipeline is to be constructed or a geological hazard such as a landslide needs to be identified (e.g., Griffiths and Stokes, 2008; Griffiths et al., 2015). The compilations by Griffiths (2001, 2002) comprise several examples of these maps that originated from engineering

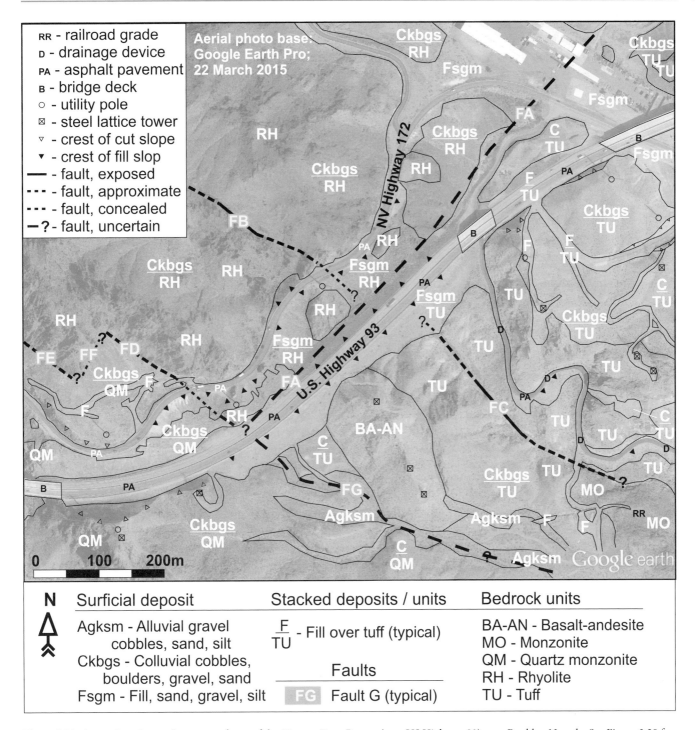

Figure 3.21 An engineering geology map of part of the Hoover Dam Bypass (new US Highway 93) near Boulder, Nevada. See Figure 3.20 for air photo. BA-AN is a basaltic and andesite lava; TU is a welded lithic ash-flow tuff; MO is a monzonite pluton (monzonites have equal amounts of alkali feldspar and plagioclase); QM is a quartz monzonite pluton; and RH is a rhyolitic lava with interbedded tuffaceous sedimentary rocks. See questions 3.6–3.9, which concern this map. (Prepared by Jeff Keaton, Amec Foster Wheeler, Los Angeles, California.)

geomorphology practice. As the name implies, their focus is on the ground surface, e.g., landside scars, outcrops and other morphological features.

Of particular practical importance is the use of geologic maps in land-use planning and regional economic development. The British Geological Survey (Smith and Ellison, 1999) prepared a series of thematic maps for such purposes. These included "opportunity maps" showing potential for development of mineral resources, groundwater or farm land, as well as "constraint maps" that identify

Table 3.1 Acronyms used in the Unified Engineering Geology Mapping System (UEGMS). Examples of this system include: (i) Agsm(f) ≡ alluvial material composed of gravel, sand and silt forming an alluvial fan; (ii) Crm-b(ta) ≡ colluvial deposits composed of rock rubble in a matrix of silt to boulders forming talus; (iii) Gm-b(t) ≡ glacial till composed of silt to boulders; (iv) SS-ST-SH ≡ interbedded sandstone–siltstone–shale. (Modified from Keaton and Rinne (2002).)

	Name (acronym)		
Surficial deposits			
Genetic classification	Alluvial (A) Fill (F) Marine (M)	Colluvial (C) Glacial (G) Residual (R)	Eolian (E) Lacustrine (L) Volcanic (V)
Lithologic classification	Clay (c) Gravel (g) Rock rubble (r) Peat (p)	Silt (m) Cobbles (k) Trash or debris (t) Organic matter (o)	Sand (s) Boulders (b) Erratic blocks (e) Diatomaceous earth (d)
Qualifier designations			
Alluvial deposits	Fan morphology (f) Pediment (p)	Floodplain (fp) Debris fan (df)	Terrace (te)
Colluvial deposits	Slope wash (sw)	Talus (t)	Creep deposits (cr)
Eolian deposits	Dune morphology (d)	Loess (l)	
Fill deposits	Uncompacted (u)	Engineered (e)	
Glacial deposits	Till (t) Esker (e)	Moraine (m) Kame (k)	Outwash (o) Ice contact (ic)
Lacustrine and marine deposits	Beach (b) Tide channel (tc)	Delta (d)	Marsh (ma)
Residual deposits	Saprolite (sa)	B horizon (bh)	Calcic horizon (ch)
Volcanic deposits	Air fall (af) Pyroclastic cone (pc)	Pyroclastic flow (pf) Lahar (la)	Surge (s)
Bedrock formations			
Sedimentary rocks	Sandstone (SS) Conglomerate (CG)	Siltstone (ST) Limestone (LS)	Claystone (CS) Shale (SH)
Igneous rocks	Granite (GR) Syenite (SY)	Andesite (AN) Rhyolite (RH)	Basalt (BA) Diorite (DI)
Metamorphic rocks	Quartzite (QT) Slate (SL)	Schist (SC) Marble (MA)	Gneiss (GN) Serpentine (SE)

potential problems in developing the mapped area, e.g., soft ground (compressible soils), faults and landslide scars, mine-waste dumps and other industrial wastes, and abandoned mine workings. Artificial-ground maps identify areas of "fill" or "made ground" that may indicate old waste disposal areas or backfilled ground that may hide an underlying aquifer.

In a similar vein, the California Geological Survey (Bedrossian et al., 2014) prepared surficial geology maps at a scale of 1 : 100 000 to allow the identification of alluvial-fan deposits on which recent suburban housing development is threatened by flash floods emanating from adjacent mountain ranges.

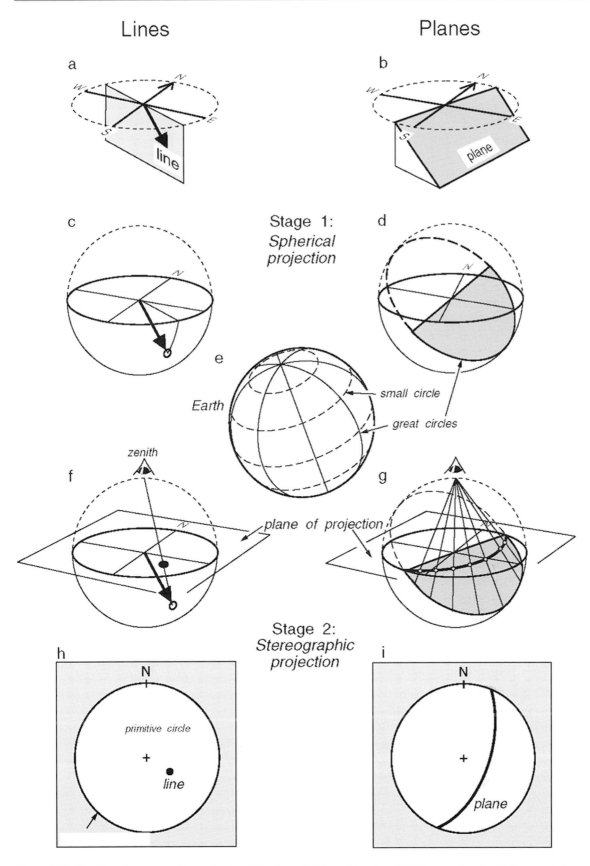

Figure 3.22 The idea of stereographic projection. (Courtesy of Lisle and Leyshon (2004); reproduced with the permission of Cambridge University Press.)

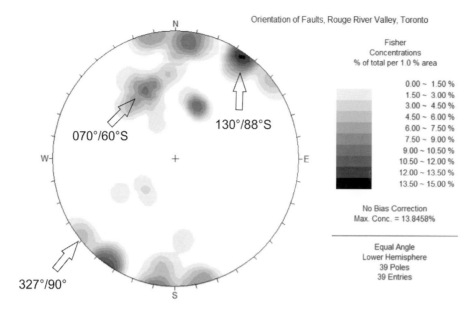

Figure 3.23 A stereogram illustrating the fault orientation data collected in the Rouge River Valley near Toronto, Ontario, and listed in Table 3.2. (From Wallach et al. (1998).)

3.7 Stereographic Projection

The reader is familiar with the representation of the Earth both as a sphere or globe and by two-dimensional maps showing the continents and the polar regions. There are many types of projection by which the spherical Earth may be represented in two dimensions, e.g., the US Geological Survey uses the Lambert equal-area projection for the US National Atlas. In Lambert projections, areas on a map are shown in true proportion to the same areas on Earth; however, directions are true only from the centre point. We also use Lambert equal-area projections to display features of interest in rock-mass characterization.

Planar and linear features may be represented as two-dimensional patterns by stereographic projection of their coordinates as in Figure 3.22 (Lisle and Leyshon, 2004). Beginning with a line on a plane passing through the centre of a sphere, we may project that line onto the surface of the lower hemisphere (Figure 3.22c) as a point. Similarly for a plane (Figure 3.22d), which is also projected onto the lower hemisphere as a great circle. This is known as **spherical projection**. Lines are reduced to points and planes to circles, thus allowing two-dimensional representation.

The second stage of the process, **stereographic projection**, is to project the point and the circle onto a horizontal plane of projection passing through the centre of the sphere. The point (Figure 3.22c) and the great circle (Figure 3.22d) on the lower hemisphere are then mapped onto the surface of the plane of projection from the perspective of the zenith

Table 3.2 Neotectonic faults mapped within the Rouge River Valley, Toronto, Ontario, and displayed as a stereogram in Figure 3.23. American right-hand rule employed. (From Wallach et al. (1998). © 2008 Canadian Science Publishing or its licensors. Reproduced with permission.)

Station R-1			
Normal faults	120°/47°S	108°/52°S	265°/88°N
	118°/50°S	065°/55°S	286°/83°N
	310°/87°N	335°/55°E	270°/65°N
	130°/83°S	076°/60°S	064°/72°S
	106°/86°S	130°/88°S	087°/65°S
	127°/47°S	073°/65°S	055°/56°S
	067°/34°S	078°/60°S	050°/60°S
	120°/47°S	075°/47°S	065°/63°S
	280°/80°S	100°/82°S	325°/89°S
	095°/70°S		
Reverse faults	310°/87°N	080°/88°S	102°/86°S
	095°/85°S	110°/85°S	
Vertical faults	084°/90°	330°/90°	
Station R-2			
Normal fault	132°/82°S		
Station R-3			
Normal faults	312°/40°N	330°/36°E	
Vertical fault	310°/90°		

of the sphere (Figure 3.22f,g). The plane of projection produces a great circle known as the primitive circle within which are plotted the line (Figure 3.22h) and the plane (Figure 3.22i) as **stereograms**.

Like the rose diagram in Figure 3.17, stereograms provide concise means of displaying orientation data, such as the surface expression of faults mapped by Wallach et al. (1998) near Toronto, Ontario, which are listed in Table 3.2 and displayed in Figure 3.23. The data represent mostly normal faults that originate in the bedrock and cause displacement along the bedrock–Quaternary sediment interface at the three locations shown in the table. These faults trend to the northwest.

3.8 Summary

The engineer engaged in infrastructure construction or geoenvironmental management needs to be able to read a geological or a geotechnical map and infer from it critical properties of the landscape and the subsurface geology. To achieve this goal, we have introduced the basic geological structures, e.g., folds, faults, shear zones, joints and unconformities, and the manner in which they are measured and presented, e.g., rose diagrams and stereograms. Structural measurements lead to the preparation of maps that represent the critical features of the terrain. The reader is reminded of Karl Terzaghi's warning (see Chapter 1) that the complexity of geological phenomena affects the design and longevity of infrastructure. We proceed next to discuss the geomechanical properties of rocks.

3.9 FURTHER READING

Dearman, W.R., 1991, *Engineering Geological Mapping*. Oxford, UK: Butterworth-Heinemann. — An exhaustive account of the preparation of engineering geological and geomorphological maps from a British perspective with many international examples.

Fossen, H., 2016, *Structural Geology*, 2nd edn. Cambridge, UK: Cambridge University Press. — A rigorous geomechanical introduction to the topic. Chapters 7 and 9 describe brittle deformation and faults concisely and with excellent illustrations.

Griffiths, J.S., 2002, *Mapping in Engineering Geology*. London, UK: The Geological Society. — A compilation of articles on engineering geology practice by British workers over the period 1972–98.

Gudmundsson, A., 2011, *Rock Fractures in Geological Processes*. Cambridge, UK: Cambridge University Press. — A splendid integration of field observations and theoretical concepts concerning the development of rock fractures and fluid transport within them. See Chapter 11 for a discussion of joints and Chapter 12 on the field analysis of faults.

Lisle, R.J. and Leyshon, P.R., 2004, *Stereographic Projection Techniques for Geologists and Civil Engineers*, 2nd edn. Cambridge, UK: Cambridge University Press. — A strongly recommended introduction to stereographic projection with application to geotechnical engineering at a modest price.

Lisle, R.J., Brabham, P. and Barnes, J., 2011, *Basic Geological Mapping*, 5th edn. Chichester, UK: Wiley-Blackwell. — The general principles of geological mapping, including the development of cross-sections, are described in this pocket book.

van der Pluijm, B.A. and Marshak, S., 2003, *Earth Structure: An Introduction to Structural Geology and Tectonics*, 2nd edn. New York, NY: W.W. Norton. — A well-illustrated guide to the subject written in a less demanding style than Fossen.

3.10 Questions

Question 3.1. Inspect Figure 3.1 of the Muddy Mountains in Nevada. In the centre of the photo are some steeply dipping beds that are partially obscured by the Thb label. Estimate the strike and dip of the bed beneath the label assuming you are looking north.

Question 3.2. Strike and dip conventions. The fault orientations listed in Table 3.2 employed the American right-hand rule, which is shown in Figure 3.3(b). (i) For a reported measurement of 310°/87°N, what would be the reported value in the strike/dip/quadrant convention? (ii) For 095°/86°S, report in British right-hand rule? (iii) For 270°/65°N, report in dip direction/dip convention?

Table 3.3 Relative ages of formations shown on Figure 3.21.

Relative age	Radiometric dates	Bedrock units	Lithology
Youngest	13.0 Ma	BA-AN	Basaltic and andesite lava
—	13.9 Ma	TU	Welded lithic ash-flow tuff
—	Not dated ~13.9 Ma	MO	Monzonite pluton, locally intrudes TU and may be its source
—	14.0 Ma	QM	Quartz monzonite pluton
Oldest	15.2–18.5 Ma	RH	Rhyolitic lava (with interbeds of tuffaceous sedimentary rocks)

Question 3.3. For the tunnel orientation shown in Figure 3.13 and the tunnel dimensions shown in Figure 3.14, how much weak rock will you have to tunnel through if the weakness zone has (i) a strike = 20° and dip = 80°, and (ii) a strike = 65° and dip = 70°.

Question 3.4. If a "well-developed" fracture system has an *S/T* ratio of 0.8–1.2 (see section 3.4), then what can be concluded from the fracture density plot shown in Figure 3.17(b)?

Question 3.5. Consider the engineering geological map in Figure 3.19. What are the types of faults that are shown in the cross-section? What is the sedimentary structure shown that has been intruded by the granite?

Preamble to questions 3.6 and 3.7. Consider the engineering geological map in Figure 3.21. Seven faults (labelled FA, FB, etc.) are shown in the engineering geology map of part of the Hoover Dam Bypass in an igneous rock terrain with bedrock units labelled by dominant rock type. For many engineering geology projects, the ages of the rock units are not substantially relevant; however, rock formation ages are relevant in some engineering geology projects, such as those that involve fault displacement. Faults in seismically active areas can be relevant if they pose a displacement hazard at the ground surface. Seismically active faults are present in the region around the US Highway 93 project and are sources of earthquake ground motion. Geoseismic hazard studies have been performed in the area of the Hoover Dam Bypass because of the importance of Hoover Dam. The faults that are considered to be active or potentially active are not in positions to pose surface fault displacement hazards in the area shown in the engineering geology map. The seven faults mapped in the area of Figure 3.21 are approximately located, concealed or uncertain, for the most part. Two locations along the faults have exposed fault planes with appearances similar to what is displayed in Figure 3.10. These exposures are on fault FB and fault FC at the locations of the fault designation symbols.

Question 3.6. Based on what you can interpret directly from the engineering geology map in Figure 3.21 and considering Figure 3.20, list what evidence you see or can infer that suggests the presence of these faults.

Question 3.7. Identify candidate locations on the engineering geology map in Figure 3.21 where examination of the rocks might produce further evidence that the mapped faults were actually faults or that they might extend along a projection farther than shown.

Preamble to questions 3.8 and 3.9. A published geologic map of the area (US Geological Survey Open-file Report 2013-1267A) indicates the radiometric ages of the bedrock units in the engineering geology map (Figure 3.21) as shown in Table 3.3. (Remember from section 2.2: Ma denotes million years as an age; an interval of 13.0 million years is denoted 13.0 Myr.)

Question 3.8. Given the age information in Table 3.3, what type of fault would FA likely be, if it is really a fault? On what evidence is your answer based?

Question 3.9. Given the age information in Table 3.3, are faults FD and FG likely to exist and be continuous? How might faults FE and FF be related to FD and FG, assuming that they really exist?

4

Rock Mechanics

4.1 Introduction

A number of elementary principles relevant to the mechanical properties of rocks and rock masses were introduced in Chapter 1, including (i) Amonton's law for sliding friction of a rock mass on an inclined plane, (ii) the Mohr–Coulomb strength criterion for defining a failure or yield envelope of a rock or soil under compression, (iii) Hooke's law for elastic behaviour, (iv) Terzaghi's principle of effective stress and (v) Darcy's law for flow through permeable media. In Chapter 2, plate tectonics was introduced to explain crustal evolution and the global forces that largely create continental stress regimes in rocks, the effects of which are evident in the structural features illustrated in Chapter 3.

In this chapter we will emphasize the significant differences between the *in situ* **rock mass** and intact **rock material** to distinguish fractured from unfractured rock conditions. It is the rock mass that is the subject of engineering works, but its characterization will depend on the testing of intact specimens of rock material from the fractured rock mass complemented by field tests of the rock mass.

The response of rocks to stress is the first subject of this chapter. We begin by considering the generation and measurement of stresses in hard and soft rocks. We then discuss strain and deformation of a rock sample and the strength of rock samples with particular reference to underground excavations and dam foundations. Finally we consider two more issues of great geotechnical concern: (i) groundwater flow in jointed rock masses and (ii) rock-slope failure.

4.2 Stress

Stresses are of critical importance in geotechnical engineering, mining and oil and gas production because they constrain what the engineer can and cannot accomplish safely and efficiently. For example, unloading through glacial and postglacial erosion of flat-lying sedimentary rocks in Appalachia has caused heave in valleys that has had an important effect on the geotechnical engineering of dam foundations (Ferguson and Hamel, 1981). Hydraulic fracture stimulation (**fracking**) by oil and gas companies is constrained by the stress regime such that drilling of horizontal wells takes place parallel to the axis of the minimum principal horizontal because hydraulic fractures develop perpendicular to this axis. While stress may be estimated at a specific location in a borehole or on a rock face, stress is neither homogeneous nor isotropic.

4.2.1 Origin of Stress Regimes

Stress regimes are the result of several tectonic processes, including (i) ridge push, (ii) slab pull, (iii) collisional resistance and (iv) basal drag (Fossen, 2016). A spreading ridge (see Figure 2.6) will push oceanic and continental crust laterally outwards (see Figure 2.3) and generate mountain ranges where other crust is encountered. As crust and lithosphere are subducted, regional stresses increase from slab pull. Collisional resistance occurs between subducting plates and the overlying crust and along transform faults (see Figure 2.6), while basal drag is associated with shear forces developed at the interface between the lithosphere and the asthenosphere (see Figure 2.1).

Such tectonic forces result in quite regular stress regimes across large areas of the Earth. The orientation of the maximum stress is often horizontal because of compression or erosional effects that lead to thrust fault regimes or strike-slip regimes, but in crustal spreading regions, e.g. rift zones, it will be vertical (normal fault regimes). The World Stress Map Project at the Helmholtz Centre near Berlin, Germany (www.world-stress-map.org/) is the global database on the present-day stress fields of the Earth's crust, with over 20 000 stress data. The reader can view regional stress patterns that have been developed from seismic, borehole, neotectonic, mining and other sources.

Secondary processes arise from (i) glacial-isostatic rebound that is still occurring in Canada, Russia and Scandinavia (see Figure 2.8), (ii) crustal loading from thick sedimentary sequences, e.g., the Gulf Coast of North America, and (iii) unloading from erosion. Franklin and Dusseault (1989) refer to **fossilized stresses** that were created by either glacial loading or uplift and/or erosion of sediments from a sedimentary basin. Numerous basins in North America – e.g., the Western Canadian, Michigan and Appalachian basins – were subject to erosion over geologic time, subsequent uplift (unloading) as well as loading from multiple glaciations. These basins still have complex stress regimes, often with very high horizontal stresses. Thus, while the isostatic rebound relieved the previous ice loading (9 MPa/km) effect, the

horizontal stresses would not be similarly relieved and thus appear elevated.

4.2.2 Stress Regimes and Indicators

Figure 4.1 illustrates the ways in which stress directions may be estimated and their magnitude approximated (Goodman, 1989). The first set of drawings (Figure 4.1a–c) represent relative displacement along fault planes, a subject introduced in Figure 3.8. In the case of faults, three regimes of principal stresses may be defined following Anderson's theory of faulting introduced in section 3.3.

- **Normal fault** (Figure 4.1a): These have a maximum principal stress (σ_1) in the vertically downward direction and a minimum principal stress (σ_3) pointing directly (normal) to the strike of the fault but at an angle to the plane of the fault, resulting in the down-thrown block slipping along the footwall; therefore, $\sigma_v = \sigma_1$.

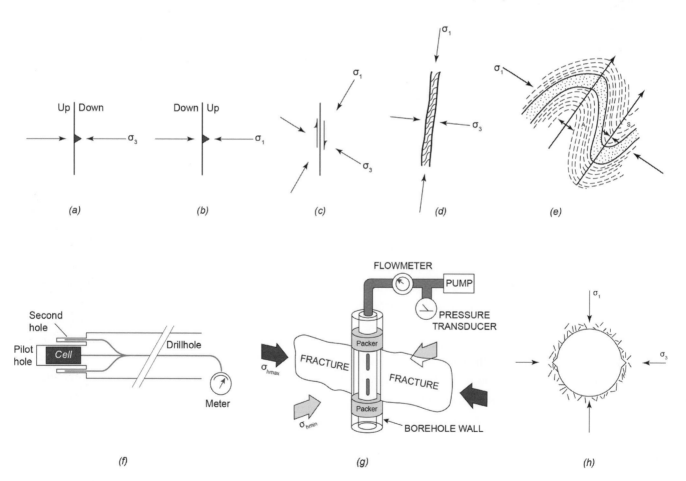

Figure 4.1   A collage of stress indicators and measurements: (a)–(c) indicators based on fault displacements, (d) an extension fracture, (e) an anticline and syncline pair in folded rock, (f) overcoring in a borehole, (g) hydraulic fracturing, and (h) borehole breakout. (Parts (a)–(e), (h) from Goodman (1989), (f) from Fossen (2016) and (g) from Bell (1996).)

- **Reverse fault** (Figure 4.1b): For these, the opposite must be true and the vertical stress must be the minimum principal stress, otherwise there can be no overthrusting, i.e., σ_3 is vertical, σ_1 is horizontal and $\sigma_v = \sigma_3$.
- **Strike-slip fault** (Figure 4.1c): For this case, the maximum principal stress is that imposed horizontally at an angle of $\sim$20–30° to the fault trace and the vertical stress has an intermediate value lying between the values of the two horizontal stresses, i.e., $\sigma_v = \sigma_2$.

Another indication of the horizontal stress regime is the presence of discontinuities (Figure 4.1d), which may provide an indication of the orientation of σ_3, while folds clearly indicate the direction of σ_1 (Figure 4.1e). Boreholes provide windows into stress regimes by direct measurements such as overcoring (Figure 4.1f) or hydraulic fracturing (Figure 4.1g), or by observation of breakouts (Figure 4.1h). These are discussed in detail by Doe et al. (2006), who point out the uncertainties in the various methods used; their summary of over- and understressed conditions in rocks is reproduced as Figure 4.2.

Overcoring involves first drilling a small-diameter borehole into which is inserted a deformation gauge (the cell in Figure 4.1f), which has 6–12 strain gauges that are affixed in different orientations so that the strain relief in all directions can be determined. The outer, larger-diameter hole is then drilled concentrically with the first. Thus an *in situ* cylinder

of rock is created that is now isolated from the rock mass itself and is consequently free of the stress associated with the rock mass. The deformation gauge can be rotated in the pilot hole and, if the rock is under compression, the change in diameter can be measured at different rotations, giving an indication of major and minor stresses in the plane that is perpendicular to the borehole. Elastic theory may then used to estimate the stresses involved (Goodman, 1989; Hudson and Harrison, 1997).

Hydraulic fracturing is widely used in geological and petroleum engineering to estimate σ_{Hmin}, the minimum horizontal stress, and, in certain cases, σ_{Hmax} (see Zoback, 2010) using Terzaghi's law of effective stress (section 1.3.4). The technique involves using a downhole fluid injection tool that is lowered on a drill string into the uncased borehole with inflatable packers above and below the tool; this is referred to as a "minifrac" test. A length of the borehole is selected for fracturing and it is sealed off by inflating the packers above and below the test section. By steadily raising the fluid pressure inside the packed-off zone, a fracture will be induced perpendicular to the minimum principal stress. The fluid injection pressure is then typically maintained for some time at the fracture propagation pressure. The fluid pressure is then "shut in" by closing the valve without releasing the pressure, which then decays slowly as the fracture closes. The fracture closure pressure is identified as the pressure drops, indicating σ_{Hmin}.

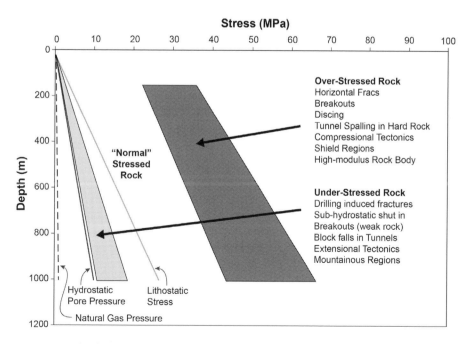

Figure 4.2 Pressure or stress versus depth and the possible geotechnical complications that arise from over- and understressed rocks. (Reproduced with the permission of Thomas W. Doe.)

Figure 4.3 Core disking in shale indicating high *in situ* stresses recorded in an oil and gas well, southwestern Ontario, Canada. The circled T indicates top of core at elevation 896.4 m below ground surface; the circled B indicates bottom of core at 901.3 m below ground surface. The dark grey shale is part of the Blue Mountain Formation that is shown to overlie the light grey fossiliferous, semi-nodular limestone of the Cobourg Formation with the contact at an elevation of 899.46 m. (From Armstrong and Carter (2010), © Queen's Printer for Ontario, 2010. Reproduced with permission.)

A related test is the leak-off test (Raaen et al., 2006), which is performed after a casing string has been cemented into place and the casing shoe then drilled through to about 3 m penetration into the sedimentary rocks below the shoe. Drilling is temporarily ceased, the blow-out preventers at ground surface are then closed around the drill pipe, and the drilling fluid pressure is increased as slowly as possible (80 L/min rate maximum) until a peak pressure is surpassed, indicating the formation has fractured. Pressure–time curve analysis is used to estimate the point at which fracture propagation occurred, which is approximately equal to σ_3, and generally taken to be σ_{Hmin}. This value is also used to limit the fluid pressure of the drilling mud that generally should not be exceeded during further drilling.

Borehole breakouts were originally observed by monitoring the diameter of a borehole completed to its total depth with a caliper tool to determine if the diameter is constant throughout (Bell, 1996). Sometimes a drill-pipe string may erode one wall of the borehole but usually not the opposite wall, which may be confused as a breakout. Breakouts, however, occur where opposite walls of a borehole show spalling or caving, indicating the orientation of σ_{Hmin}. More recently, borehole imaging tools (Tingay et al., 2008) have become available to measure breakouts and drilling-induced

fracturing of a borehole wall to provide stress orientation estimates.

The core recovered from a borehole may exhibit high *in situ* stress upon its recovery by splitting into small disk-shaped pieces, referred to as **core disking** (see Figure 4.3). Lim and Martin (2010) correlated core disking in granitic rocks with the ratio of maximum principal stress to the tensile strength (σ_t) and determined that disking was pronounced when $\sigma_1/\sigma_t > 6.5$, and the recovered core became progressively less intact with increases in this ratio.

It is to be expected that an engineer will become familiar with the geological features of a particular site when working as a project engineer. Site inspection might reveal surficial indications of the contemporary stress regime such as those shown in Figure 4.4. Glacial striations often mark bedrock outcrops in North America and Scandinavia, indicating the direction of ice advance, but have no direct relevance to stress patterns. However, arrays of neotectonic faults or fractures that offset the striations will have formed parallel to σ_{Hmax}.

Quarries are excellent sites to observe indicators of the maximum principal stress, such as anticlinal **pop-ups** on the floor of the quarry caused by unloading of the overlying rock or blast-induced fractures. Borehole traces may indicate post-drilling offset, i.e., a neotectonic effect, along some bedding

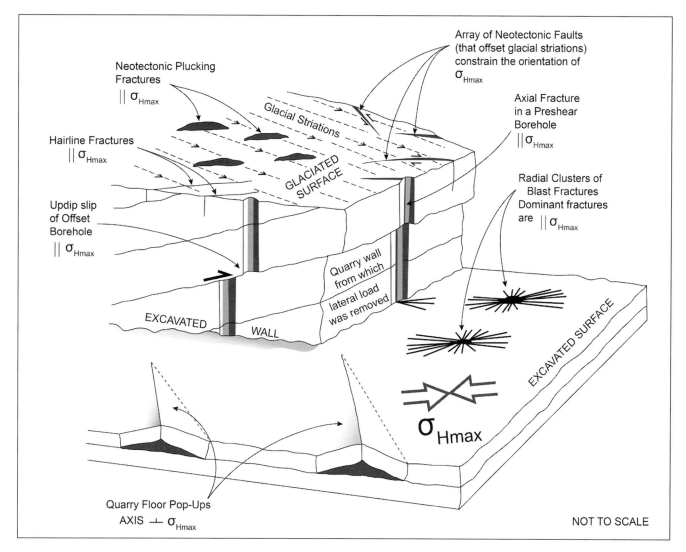

Figure 4.4 Schematic representation of surficial features diagnostic of contemporary stress directions. All diagnostic features are parallel (‖) to σ_{Hmax} except for the neotectonic fault arrays and the pop-up anticlines, which are perpendicular (⊥) to σ_{Hmax}. (After Bell (1996).)

plane that will occur parallel to σ_{Hmax}. However, offset in borehole drilling traces may also be caused by blast effects, such as those reported by Meinardus et al. (1993) in a road cut in West Texas, where neotectonic and stress relief hypotheses were also considered but rejected.

4.2.3 Variation of Stress with Depth in Hard Rocks

While we may estimate vertical stress quite simply as $\sigma_v = \rho g z$, where ρ is the average rock density of the column above the point of measurement, g is the acceleration of gravity and z is the depth of the point of measurement, specific information is needed on the principal stresses in matters of engineering design and implementation. Here we consider measured lithostatic stresses in mines and underground

excavations, such as in deep geological repositories for nuclear waste and in tunnels. We begin by assuming elastic rock responses to stress, for which Young's modulus and Poisson's ratio (Table 4.1) were introduced in section 1.3.3.

The lithostatic stress gradient is estimated on the basis on one's choice of average rock density. Thus van der Pluijm and Marshak (1997) choose $\rho = 2700 \, \mathrm{kg/m^3}$, which is the density of most common igneous rocks and yields a gradient of 26.5 kPa/m or 1.2 psi/ft. Thus at a depth of 1 km we may expect a lithostatic stress of ~26.5 MPa. However, Herget (1988) notes that in magmatic rocks, such as basalt and gabbro, the average density is close to $3000 \, \mathrm{kg/m^3}$, for which the gradient will be 29.4 kPa/m, i.e., 1.3 psi/ft.

The average ratio of horizontal to vertical stress (K) can be estimated for rocks or sediments that behave as elastic materials under one-dimensional loading (Goodman, 1989) by

Table 4.1 "Typical" values of Young's modulus of elasticity (*E*) and Poisson's ratio (*v*). (From Gudmundsson (2011).)

Rock type	E (GPa)	v
Limestone	20–70	0.22–0.34
Dolomite	30–80	0.10–0.30
Sandstone	20–50	0.10–0.30
Shale	1–20	0.10–0.25
Chalk	1–15	0.15–0.30
Gypsum	15–35	0.18–0.30
Conglomerate	30–40	0.22–0.23
Sand	0.05	0.45
Clay	0.05–0.1	0.4
Basalt	50–90	0.23–0.27
Diabase	55–100	0.24–0.28
Granite	20–50	0.15–0.24
Tuff	0.5–1.0	0.25–0.30
Gneiss	20–60	0.1–0.2
Marble	40–90	0.24–0.30
Schist	20–50	0.1–0.2
Quartzite	40–60	0.15–0.20
Water (ice)	9–10	0.33

$$K = \frac{\sigma_H}{\sigma_v} = \frac{v}{1-v}, \qquad (4.1)$$

where v is Poisson's ratio (equation (1.10)). However, the term $v/(1-v)$ does not apply to those rocks that have undergone cycles of erosion leading to a reduction in their original thickness Δz. Then the vertical load will have been reduced by $\rho g \Delta z$. Nor does it apply to situations of crustal extension or compression that may have strained the rocks. Consequently, horizontal stresses must be measured or estimated from geological features and history. Brown and Hoek (1978) report that observed values of K for deep rock mines worldwide fall in the range

$$\frac{100}{z} + 0.30 \leq K \leq \frac{1500}{z} + 0.50, \qquad (4.2)$$

where z is the depth in metres.

Sheorey (1994) developed a relatively simple equation for K based on the elastic properties of rocks, their densities and thermal expansion coefficients:

$$K = \frac{(\sigma_{Hmax} + \sigma_{Hmin})/2}{\sigma_v} = 0.25 + 7E\left(0.001 + \frac{1}{z}\right), \qquad (4.3)$$

where E is the average Young's modulus for the upper part of the Earth's crust measured in a horizontal direction. This equation has provided reasonable estimations of K in the Scandinavian Shield of Sweden and can thus yield preliminary estimates of horizontal stress (Martin et al., 2003).

Equations (4.2) and (4.3) allow approximation of the *magnitude* of horizontal stresses within broad limits. However, we are much better able to estimate their *direction* through observations such as those shown in Figures 4.1 and 4.4 (Goodman, 1989). Because civil engineering practice is preoccupied with the first several hundred metres depth, the average of the two horizontal stresses is generally larger than the vertical stress (Hudson, 1989).

4.2.4 Variation of Stress with Depth in Sedimentary Basins

According to Poston and Berg (1997) the average density of water-saturated sedimentary rocks in the US Gulf Coast basin is ~2300 kg/m³; consequently the vertical stress at 1 km depth would be only 22.5 MPa. Sedimentary rocks tend to be lighter because of variations in porosity; thus chalk has a mean density of 1850 kg/m³, sandstone of 2370 kg/m³ and limestone of 2430 kg/m³ (Mavko et al., 2009). However, the geomechanical properties in sedimentary basins for oil and gas production are more strongly influenced by the fluid pressures within these sediments than is the case with hard-rock mines.

The drilling engineer has to be aware of the downhole fluid pressure and must adjust the weight of the drilling mud accordingly to prevent a blowout, which implies that the fluid pressure causing formation fluids to invade a borehole must be repeatedly measured and continually controlled. Fluid pressures must also be known to design the cementing fluid pumped down the borehole and up the annulus between rock and casing to seal or "complete" the well. **Well completion** now not only means simply cementing the casing in place to prevent gas migration up the annulus, but also implies perforating the steel casing and (usually) hydraulic fracturing the well to open up fractures in low-permeability rock to allow better extraction of hydrocarbons.

Assuming that hydrostatic conditions exist, the fluid pressure at depth is given by reference to the hydrostatic gradient of 9.79 kPa/m (0.433 psi/ft) for fresh water. For saline water with a density of 1.07 g/mL, derived from seawater and geochemical reactions over geologic time, the fluid pressure gradient would be 10.5 kPa/m, while for petroleum (0.88 g/mL) it is 8.58 kPa/m and for natural gas (0.1 g/mL) it is 0.97 kPa/m. Any departure from the hydrostatic pore-pressure gradient is considered to be abnormal. Sedimentary rocks with pore pressures greater than hydrostatic are considered **overpressured**, while those less than hydrostatic are **underpressured**. In many sedimentary basins, there are lengthy intervals where the water is saturated with NaCl, generally taken to be the highest fluid density at a value of 1.2 g/mL.

Fluid pressure gradients for the Gulf Coast basin can vary from 10 to 16 kPa/m over a depth of 5000 m (Poston and Berg, 1997). However, gradients conceal the compartmentalization of high pressures in shale seals. Fluid pressures in the seals are considerably higher than in the adjacent sandstone reservoirs (Zoback, 2010), probably because the low permeability of the shales prevents their drainage when loaded by rapid rates of sedimentation, thus leading to overpressures.

In the Gulf of Mexico, sedimentation loading rates vary from 600 to 3000 m/Myr whereas in the North Sea of Europe the rates are 35–55 m/Myr. Swarbrick et al. (2002) consider that such rapid sediment burial without compensating drainage at depth, which is known as **disequilibrium compaction**, is the most potent generator of overpressures in sedimentary basins. However, there are other origins for high fluid pressures, including tectonic compression by which regional compressive forces transfer stress to the pore fluids.

Also fluid expansion may be caused by (i) clay transformation as smectite is converted to illite expelling the interlayer water and thus raising the pore pressure, (ii) hydrocarbon generation, in particular kerogen transformation and gas generation, and (iii) thermal expansion of water from the increase in geothermal temperature with depth.

Underpressured conditions are also common in sedimentary basins (Ingebritsen et al., 2006). Such basins are being exploited for hydrocarbons, e.g., the Appalachian basin that includes the Marcellus Shale, and the Western Canada Sedimentary basin. The development of nuclear-waste repositories (e.g., Clark et al., 2013) is also attractive where hydrocarbon exploration has not created abandoned boreholes into the underpressured formations. The origin of underpressured formations is uncertain; however, it is noteworthy that the principal underpressured zones are usually shales and mudrocks, just as was the case of overpressured formations. Obviously, the erosion of overlying sediments would reduce the vertical stress; and because the period over which fluid pressure equilibration occurs in shales is very long, underpressured conditions persist. Furthermore, ice loading during glaciation – and its subsequent removal during deglaciation – may play an important role in generating underpressured conditions.

If we define $K' = \sigma'_{Hmin}/\sigma'_v$ as the ratio of the minimum horizontal to the vertical effective stresses (K being the total stress ratio in (4.1)), then we may expect a stress distribution similar to that shown in Figure 4.5. Here we

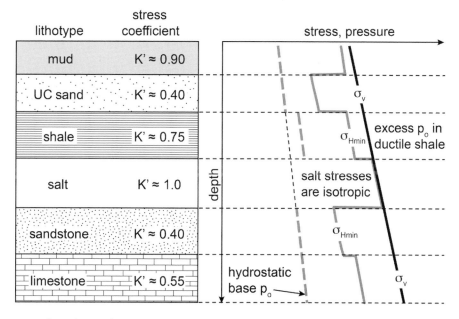

Figure 4.5 Stress coefficients $K' = \sigma'_{Hmin}/\sigma'_v$ and vertical stress distribution in an idealized sedimentary basin from which no erosion has occurred and the horizontal stress is less than the vertical stress, σ_v. Here σ_{Hmin} is the minimum horizontal stress and p_0 is the pore pressure. (From lecture notes on petroleum geomechanics, courtesy of M.B. Dusseault, University of Waterloo.)

assume no tectonic stresses causing significant compression nor other lateral stresses nor significant erosion. The sedimentary sequence includes stiff rocks at the base of the sequence, i.e., high Young's modulus (Table 4.1), with low-stiffness sediments near the ground surface. The ductile nature of the shale and its deformation during its compaction by overlying sedimentation causes σ_{Hmin} to approach σ_v, while the salt exhibits an isotropic stress regime because it was deposited as a viscous fluid (i.e., under hydrostatic conditions). While we might expect similar values of K' for sandstones and carbonates, diagenetic reactions in soluble carbonate and sulfate rocks (e.g., smectite–illite transformation, dolomitization) will affect local values. The value of K' reflects the degree to which σ_H approaches σ_v.

4.2.5 Erosional Unloading

If we allow erosion of the uppermost sediments of a sedimentary basin to occur over geologic time, then the present *in situ* stress regime will exhibit the fossilized stresses mentioned earlier. Let us assume a basin of horizontally deposited sedimentary rocks, such as is exposed in the Grand Canyon (see Figure 1.3), and further assume that the rocks behave in an elastic manner to erosion of the top beds, so that then erosion must remove a component of vertical stress, i.e., $\Delta\sigma_z$. If there is no lateral expansion normal to this vertical direction z – because of the confined conditions created by the continuity of the beds – then there is no elastic strain and $\epsilon_x = 0$ and $\epsilon_y = 0$. This set of assumptions yields the **Poisson effect** for the stress condition after erosion (see Fossen, 2016) that we write as

$$\frac{\Delta\sigma_H'}{\Delta\sigma_v'} = \frac{\nu}{1 - \nu}. \tag{4.4}$$

If the sandstone ($\nu = 0.25$) in Figure 4.5 is at a depth of 3 km, assume $\sigma_v = 70\,\text{MPa}$, $\sigma_H = 55\,\text{MPa}$ and $p_0 = 35\,\text{MPa}$, then $\sigma_v' = 35\,\text{MPa}$ and $\Delta\sigma_H' \approx 0.33 \cdot \Delta\sigma_v'$. If we allow erosion to remove 2.2 km of the basin, the sandstone is now overlain by 800 m of rock not 3 km and now $\sigma_v = 19.6\,\text{MPa}$ and $p_0 = 7.8\,\text{MPa}$ and $\sigma_v' = 11.8\,\text{MPa}$. Therefore $\Delta\sigma_v' = -23.2\,\text{MPa}$ and, correspondingly, $\Delta\sigma_H' = -7.7\,\text{MPa}$.

Therefore, in comparison with the original pre-erosional conditions, the effective horizontal stress is now $\sigma_H' = 12.3\,\text{MPa}$ (originally 20, minus 7.7) with $p_0 = 7.8\,\text{MPa}$ and $\sigma_H = 20.1\,\text{MPa}$. Consequently, now $\sigma_H > \sigma_v$. Thus our hypothetical example of erosion has caused the stress regime to change from one of normal faulting to one of reverse or thrust faulting according to Anderson's theory of

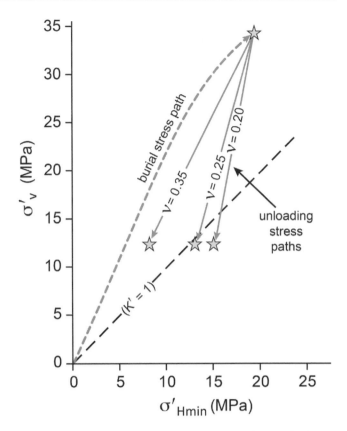

Figure 4.6 Stress path for burial and subsequent elastic unloading arising from erosion. (From lecture notes on petroleum geomechanics, courtesy of M.B. Dusseault, University of Waterloo.)

faulting, in that now $\sigma_H > \sigma_v$. To considerable depths in the Michigan and Williston Basins of North America, $\sigma_v = \sigma_3$ rather than $\sigma_v = \sigma_1$. The stress paths from the initial pre-erosion condition to the present stress regime are shown schematically in Figure 4.6 for $\nu = 0.20, 0.25$ and 0.35.

4.3 Strain and Deformation

4.3.1 The Load–Deformation Curve

We observed in section 1.3.3 that the **stiffness** of rocks as measured by Young's modulus reflects the difference between hard and soft rocks. Thus samples of chalk or claystone have values of Young's modulus less than 10 GPa, while fresh samples of granite, basalt and other hard rocks have values above 50 GPa. Common sedimentary rocks fall in the intermediate range (see Table 4.1).

In the elastic region, the **load–deformation** or stress–strain curve of a rock sample is linear, as shown in Figure 4.7. Somewhere along this linear curve will be located the maximum design load that might be considered for this particular rock; the slope is the Young's modulus. This load–deformation

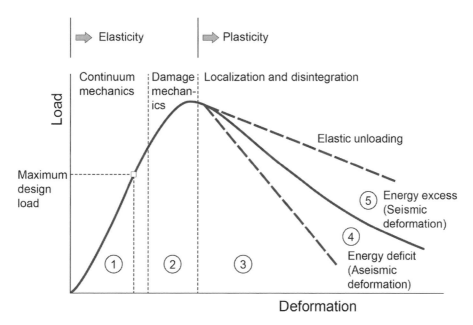

Figure 4.7 Load–deformation curve for rock loaded to destruction. (From Fairhurst (2003).)

or stress–strain curve was developed under compression in the laboratory. It should be noted that "many fresh, hard rocks are elastic when considered as laboratory specimens" (Goodman, 1989) but the presence of discontinuities or soft clay infill within the fractures causes specimens to behave with imperfect elasticity. Discontinuities or fractures distinguish an intact rock sample from the **rock mass**; the engineer will use information from the former to address issues of design, construction and operation of projects within rock masses, but the two should not be conflated.

In the context of foundation design, Goodman (1989) and Wyllie (1999) have argued that the potential settlement of rock foundations should involve estimation of the **deformation modulus** or E/q_u rather than simply E, where q_u is the unconfined compressive strength. This modulus is a measure of both elastic and inelastic behaviour. Therefore in the construction of dams across a valley or large high-rise buildings, the foundations of which may be underlain by rocks of differing stiffness, the deformation modulus would provide a better design guide to differential settling.

Deformation can be measured as **strain** as in (1.9), where $\epsilon_a = \Delta L/L$, in which ΔL is the deformation length; this expression is the common understanding of strain in the rock mechanics literature (e.g., Goodman, 1989; Hudson and Harrison, 1997). In structural geology, strain has a broader sense: "any change in shape, with or without change in volume, is referred to as strain, and it implies that particles within a rock have changed positions relative to each other" (Fossen, 2016).

The simple elastic phenomena described by Young's moduli are inadequate to describe the progressive failure of a rock sample under increasing confining stress. The deformation of the sample under stress is the result of the fracturing, rotation, distortion and translation of grains within the sample often resulting in some volumetric change, partly from closure of fractures and compression of any soft infilling materials. Figure 4.7 illustrates that the elastic region of deformation is followed by a non-elastic region, in which microcracks expand and the disintegration of part of the sample begins (region 3).

The sample tested in Figure 4.7 disintegrates through either seismic or aseismic deformation. The distinction is necessary because disintegration will occur locally depending on the **microstructure** of the rock sample, which creates a heterogeneous stress field at the grain scale of the sample. Thus part of the rock sample may disintegrate while an adjacent part will remain elastic and may continue to be described by continuum mechanics. It is the interaction between these two zones of deformation that determines whether the deformation is aseismic, in which case the elastic region can absorb the released energy from the disintegrating zone, i.e., region 4, or seismic, in which case the deformation of the sample is violent, i.e., region 5 (Fairhurst, 2003).

4.3.2 Brittle and Ductile Deformation

Removing compressive stress after such rock failure will not lead to the recovery of the original sample shape or volume.

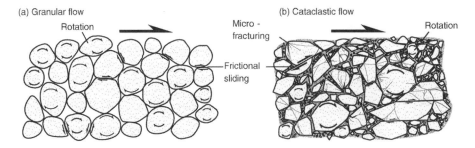

Figure 4.8 Two mechanisms of brittle-rock deformation. (a) Granular flow, which is common in the shallow deformation of porous rocks and sediments. (b) Cataclastic flow, which occurs in well-consolidated sedimentary and rocks of low porosity. (From Fossen (2016).)

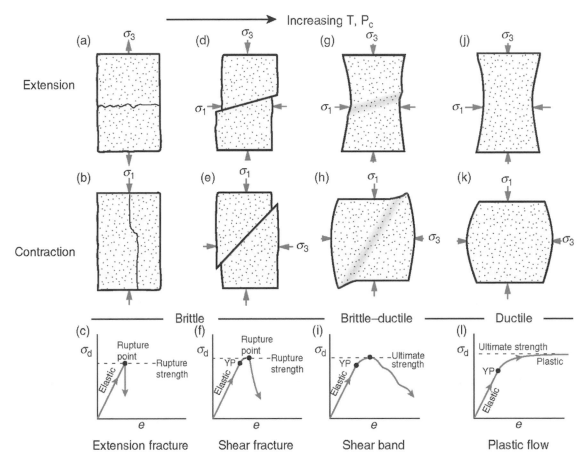

Figure 4.9 The development of the principal deformation structures under extension and compression: T, temperature; P, confining pressure; YP, yield point. (From Fossen (2016).)

This is the hallmark of **brittle deformation**, which includes a number of micro-mechanical processes causing the sample to yield and then fail, including (i) tensile failure, (ii) granular flow and (iii) cataclastic flow. **Tensile failure** develops first – even in a rock sample under compression loading – while shear mechanisms develop later. The overall strength and geomechanical properties of a rock sample are controlled by the geometry of the grains, i.e., the microstructure. It is the heterogeneities in the microstructure, irrespective of the apparent homogeneity of a rock sample to the human eye,

that cause local concentrations of tensile stress to develop, which in turn cause microcracking (Blair and Cook, 1998; Lan et al., 2010a). **Granular flow** occurs by frictional sliding and grain rotation in porous rocks and sediments, while **cataclastic flow** implies grain fracturing, crushing, rotation and frictional sliding. These are shown schematically in Figure 4.8.

Figure 4.9 summarizes the nature of experimental fractures developed during compression and extension loading of rock samples. Box 4.1 outlines the experimental methods

BOX 4.1 | GEOMECHANICAL TEST METHODS

Figure 4.10 illustrates the general methods of experimental testing in a rock mechanics laboratory that produce the deformation patterns of Figure 4.9. The methods are discussed in detail in standard rock mechanics texts and further discussion may be found in Chapter 5.

- The simplest geomechanical measurement is that shown in Figure 4.10(a), which is the basis of the **point-load strength** measurement. A portable testing machine (see Chapter 5) allows this test to be conducted in the field to provide a compressive-strength index of rock quality based on a standard 50 mm diameter core sample or irregular-shaped rock lumps (ASTM, 2005a). The results are reported in megapascals (MPa) given by $I_s = P/D_e^2$, where P is the load at failure and D_e is the equivalent core diameter. Measured values are in the range 0.25–8 MPa for limestone to 0.05–1 MPa for sandstone.

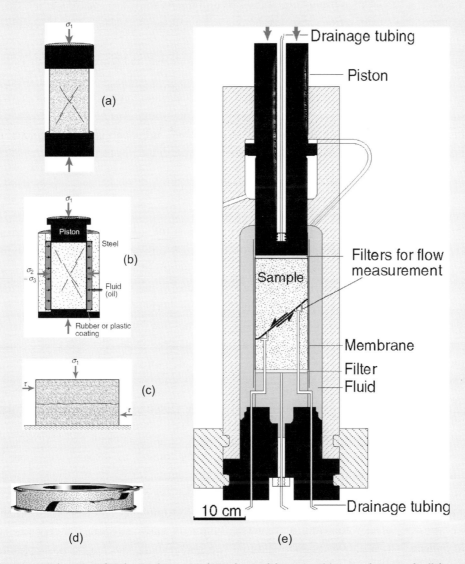

Figure 4.10 Experimental testing of rock samples in a rock mechanics laboratory. (a) A simple uniaxial cell for point-load strength tests. (b) A simple triaxial test rig allowing all three principal stresses to be specified. (c) A simple shear box that permits the resistance to shear (τ) to be measured for a particular vertical confining stress. (d) A ring-shear apparatus for sediment testing. (e) A more sophisticated triaxial test rig allowing fracture behaviour to be monitored while measuring the pore pressure in the fracture. Orientation of different fracture types with respect to the principal stresses are shown in each. (From Fossen (2016).)

BOX 4.1   (CONT.)

- In the testing laboratory, Figure 4.10(a) would apply to an **unconfined compression test** to measure the **unconfined compressive strength** (UCS) of a rock sample. This may also be referred to as the uniaxial compressive strength. Unlike the point-load sample, a core with a length-to-width ratio of $\sim$2 and flat smooth ends cut perpendicular to the core axis would be tested. The compressive strength is measured as the peak load divided by the initial cross-sectional area: UCS = P/A (ASTM, 2007). Values of UCS are provided by Goodman (1989) and Gudmundsson (2011). Hoek (1999) has defined a rock as "weak" when the *in situ* UCS is less than the *in situ* stress.

- The **triaxial compression test** shown in Figure 4.10(b) measures response to stress when $\sigma_1 > \sigma_2 = \sigma_3$. The axial stress σ_1 is applied to the rock sample via a ram and the confining pressure is applied as a fluid that is compressed against a rubber membrane to maintain σ_2 and σ_3. As shown in Figure 4.10(e), the pore pressure within a fracture can be monitored in some triaxial rigs. Axial stresses of >100 MPa and confining pressures of 50–100 MPa may be imposed (ASTM, 2007).

- **Shear testing** can be conducted using a direct shear box, as shown in Figure 4.10(c), in which the normal stress is maintained constant (ASTM, 2008). The objective of such a test is to determine the behaviour of the specimen, which may or may not contain a fracture, to an applied shear force. The shear strength is determined during the test at various applied normal stresses and at various shear displacements, which yields the shear stiffness. The apparent shear stress (τ) is defined as P_s/A, where P_s is the applied shear load and A is the cross-sectional area of the sample. Figure 4.10(d) illustrates a ring-shear device for measuring shear in clays and other sedimentary materials.

---

that reproduce such deformation patterns. The fractures that develop are characteristic of the stress regime imposed and the geomechanical properties of the rock sample. Initially, all samples show elastic behaviour; however, brittle failure and plastic flow (i.e., deformation without failure) occur with increasing compression and temperature. Extension fractures, which form perpendicular to σ_3, occur first, indicating tensile fracturing that grows from microcracks to fissures. Subsequently, shear fracturing occurs at 20°–30° to σ_1 and at a relatively low imposed stress, with axial cracks forming obliquely to σ_1. Shear bands are created at higher imposed stresses when friction between the crack walls is overcome. As the imposed stress is further increased, the shear bands become stronger than the adjacent rock, indicating ductile rupture. Fossen (2016) notes that shear bands are one member of a class of deformation bands generated by granular flow, cataclastic flow, the smearing of clay minerals, and mineral dissolution and cementation. While shear bands tend to decrease permeability, fractures usually increase permeability.

Thus, the mechanisms of rupture and yield are a consequence of the imposed stresses and can be segregated into several regions, as shown in Figure 4.9 – see discussions in Franklin and Dusseault (1989) and Fossen (2016). Brittle failure is most obviously illustrated by a fault zone (see Figure 3.10), where slip has occurred and one surface is displaced with respect to the other. **Ductile deformation** is permanent and unrecoverable, such as is seen in folded rocks. The shear zone shown in Figure 3.11 reflects ductile deformation, likely involving cataclastic flow. Goodman (1989) indicates that the brittle-to-ductile transition load varies from <20 MPa for shales, through 20–100 MPa for limestones to >100 MPa for sandstones and granite.

4.3.3 Discontinuities and Fractures

In the words of Hudson and Harrison (1997): "in the engineering context ... discontinuities can be the single most important factor governing the deformability, strength and permeability of the rock mass". In this engineering context, a **discontinuity** is any separation in the rock mass having zero tensile strength (or very low strength because of re-cementation) and having no implication as to its origin. The most prominent of discontinuities are **faults** and fault zones (e.g., Figure 1.4), which are parallel or sub-parallel sets of fractured rock (or cohesive soil) along which displacement of one side versus the other has clearly occurred; and **bedding-plane fractures** (e.g., Figure 2.27) are particularly common in rocks that have undergone stress relief.

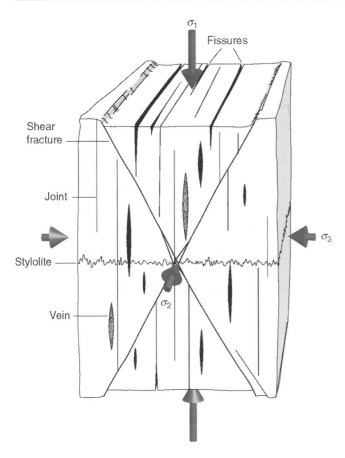

Figure 4.11 Orientation of different fracture types with respect to the principal stresses. (From Fossen (2016).)

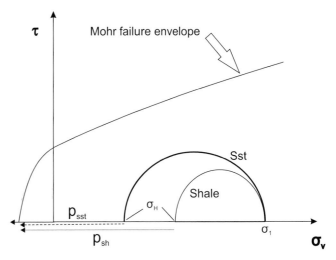

Figure 4.12 Vertical and shear stresses in a sandstone (Sst)–shale sequence resulting in tensile failure. Here p_{sst} and p_{sh} are the pore pressures required to cause tensile fracturing in the sandstone and shale, respectively. Note that tensile failure may also occur because of a decrease in σ_v from uplift or erosion of the overlying bedrock.

However, the origin of discontinuities is a critical field for structural geologists, who identify two broad types of fractures in the field: (i) **shear fractures**, such as faults, and (ii) **joints** and **fissures** (van der Pluijm and Marshak, 1997; Fossen, 2016). Figure 4.11 illustrates how these are perceived to occur with respect to the principal stresses. We have already noted above that shear fractures tend to form at an angle of 20–30° to σ_1 and that extension fractures – illustrated as joints and fissures in Figure 4.11 – form perpendicular to σ_3. Joints and fissures are extension fractures that can be up to hundreds of metres in length but with no visible lateral displacement. Fissures are defined by geologists as fluid-filled extensional fractures, while veins are extension fractures that are infilled by secondary precipitated minerals. Stylolites, which form by pressure solution of carbonate or quartz minerals in deep compacting sedimentary sequences, are serrated horizontal seams formed perpendicular to σ_1.

Vertical or sub-vertical extension fractures can be seen in the Paskapoo Formation, shown in Figure 3.15, and they appear more numerous than in the shale interbeds, which are partly covered with debris from weathering. The sandstone fractures probably result from tensile fracturing, which occurs commonly in rocks with high Young's modulus and differential stress (i.e., $\sigma_1 - \sigma_3$). We might expect that the Paskapoo sandstone beds are significantly stiffer than the mudstone or shale interbeds (see Table 4.1) and consequently the sandstone and shale beds will differ in their geomechanical responses to uplift or erosion or pore-pressure changes.

This is illustrated in Figure 4.12, in which the minimum horizontal stress for the shale is greater than that for the sandstone, although both rocks are assumed to have the same maximum principal stress because they are both subject to the same vertical load of overlying rock. If the rocks in Figure 3.15 were uplifted or erosion occurred above them, σ_v would be decreased and the minimum horizontal stress shown (σ_H) would then effectively move towards the failure envelope on the left of the figure. Tensile fracturing would occur when σ_H became sufficiently negative to reach the failure envelope. A relatively small pore pressure in the sandstone (p_{sst}) would cause the minimum horizontal stress of the sandstone to reach the failure envelope, whereas a larger pore pressure (p_{sh}) would be required to cause shale units to fracture by tensile fracturing.

4.4 Strength

Different rock-strength parameters are required for different geoengineering tasks. The stability of underground excavations such as tunnels is strongly influenced by the shear strength of the discontinuities within the rock mass being excavated (Martin et al., 2003). In the characterization of rock foundations, the shear strength would again be important in understanding the potential for sliding at the interface between any structure, such as a dam, and the rock foundation.

However, it is the **compressive strength** of the rock mass that would be important in the bearing capacity of the spread footings and bond strength of socketed piers and tension anchors (Wyllie, 1999). It is usually denoted by the symbols C_0, q_u or σ_c, in which case σ_{ci} refers to the intact laboratory specimen and σ_{cm} the estimate of rock-mass strength. Summaries of **unconfined compressive strength** or **uniaxial compressive strength** (UCS) values (i.e., σ_{ci}) vary considerably, particularly for sedimentary rocks because of their wide range of porosities. Thus limestone strengths vary over the range 4–250 MPa, dolomite 80–250 MPa and sandstone 6–170 MPa (Goodman, 1989; Gudmundsson, 2011). Igneous rock strengths typically exceed 100 MPa and can exceed 300 MPa, although tuff is typically very low (0.4–44). Point-load testing (Box 4.1) is a field test for UCS.

Figure 4.13 illustrates the range of UCS values for common rock types. Very strong rocks are those with strengths greater than 100 MPa, such as igneous and metamorphic rocks, and which are typically high in quartz content. These require a geological hammer to break or chip them. Weak rocks (see discussion in Chapter 5) are those with strengths less than 5 MPa and can be cut or peeled with a steel knife blade. Price (2009) draws attention to the rock fabric as an indicator of strength. Mineral orientation causes the strength to vary anisotropically. Rocks that have a well-defined grain (similar to that of wood) can be more easily split along the mineral grain than across it. This property of cleavage indicates that the oriented minerals are flat-shaped mica and chlorite crystals.

The average **tensile strength** of rocks is estimated to be ~3 MPa (Gudmundsson, 2011) although it has been measured by hydraulic testing to be as much as 9 MPa in a 9 km deep borehole in Germany. It is usually denoted by the symbols T_0 or τ. Because it is small compared to other strength measurements, it is often dismissed as being unimportant in most engineering applications; however, one exception might be where a stiff but thin load-bearing rock is underlain by a weaker rock and the load causes flexural failure of the load-bearing rock (Wyllie, 1999).

A basic measure of rock strength was provided earlier in section 1.3.2, in which the Mohr–Coulomb envelope was

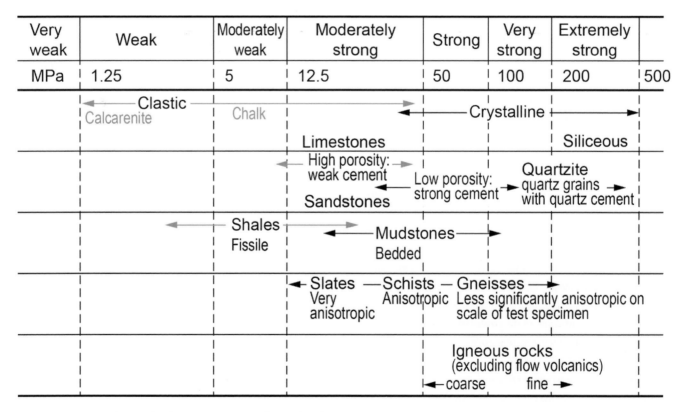

Figure 4.13 Range of unconfined compressive strength estimates for common rock types. (From Price (2009); reprinted by permission from Springer, © 2009.)

Table 4.2 Representative triaxial strength data with values for the peak shear-strength parameters based on laboratory tests on intact rock specimens. (After Franklin and Dusseault (1989).)

Rock	c (MPa)	ϕ (°)
Crystalline limestone	35	26
Porous limestone	30	26
Chalk	1	39
Sandstone	30	31
Quartzite	42	50
Slates and high-durability shales	30	39
Low-durability shales	0.2	26
Coarse-grained igneous rocks	56	45
Fine-grained igneous rocks	35	26
Schists	46	26

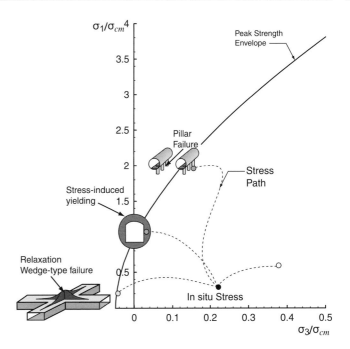

Figure 4.14 Stress paths that might occur following excavation of intact rock for a tunnel or other opening leading to different modes of instability. Loading leads to pillar failure as stresses are redistributed and to spalling of the tunnel walls. Wedge failures occur from unloading of the rock allowing blocks to fall. (From Martin et al. (2003).)

developed. Noting the development in Figure 1.12, we may rewrite (1.8) as

$$\tau = c + \sigma_n \tan \phi, \qquad (4.5)$$

where τ is the peak shear stress or **shear strength**, c is the cohesion intercept on the τ axis (τ_0 in Figure 1.12) representing the inherent shear strength, σ_n is the normal stress and ϕ is the angle of internal friction or shearing resistance, as shown in Figure 1.12. Note here that we are dealing with total stresses not effective stresses because in underground excavations dewatering would occur and the pore pressures reduced to zero. Obviously shear strength varies with the normal stress. Cohesion may be due to the tensile strength of the fracture surface or to the roughness of that surface, which we return to later in this chapter in considering the shear strength of discontinuities, or both. Franklin and Dusseault (1989) have provided a summary of representative triaxial strength data in terms of c and ϕ, which is reproduced as Table 4.2.

4.4.1 Design of Underground Excavations

The design of underground structures, such as a nuclear-waste repository or deep tunnels, must prevent the overstressing of the rock mass containing the underground excavation. This is accomplished by careful analysis of the *in situ* stress state with respect to the rock strength. Martin et al. (2003) noted that there would be two principal causes of instability of any hard-rock underground excavation: (i) **wedge failures** associated with gravity-driven processes associated with the geological structure of the underground

opening; and (ii) stress-induced failure or **yielding**. Their results are shown in Figure 4.14 where the Mohr–Coulomb envelope is presented in terms of the principal stresses rather than shear and normal stresses.

Wedge failures are gravity-driven collapses of the overhead rock mass representing a shear failure (see section 4.6). In underground excavations wedge failures occur from the roof ("hanging wall") or the sidewall of a tunnel. The strength of the wedge block is a function of its geometry and the confining stress, which may be partially relieved by nearby mining excavations. Figure 4.14 illustrates the *in situ* stress paths that might occur when excavating a tunnel in granite or similar such hard rock; note that the principal stresses are normalized by dividing them by the unconfined compressive strength of the rock mass (σ_{cm}). Wedge-type failures occur as the horizontal stress is reduced by excavation and σ_3/σ_{cm} goes to zero; the tensile strength shown in Figure 4.14 may in fact be an overestimate.

Stress-induced yielding occurs when the *in situ* stresses are altered by mining and the stress path then leads towards the peak-strength envelope shown in Figure 4.14. In weak and soft rocks, yielding will be reflected in plastic mechanisms of deformation of the walls by cataclastic flow and/or grain-sliding effects causing the convergence of the opposite walls. Hoek (1999) has noted that instabilities occur in tunnels of most shapes when the rock-mass strength is less than 10–20% of the maximum *in situ* stress. In hard and strong rocks under high stress, yielding can even be violent as stored strain energy is released (Martin et al., 2003). In mined excavations, the load is carried by pillars, which may fail from inclined shear fractures or internal splitting (Brady and Brown, 1993). The exposed tunnel wall and the pillars may further undergo stress-induced brittle failure by spalling as cracks behind the rock wall coalesce as the stresses are redistributed around the tunnel in the absence of the confining stress of the excavated rock (Ghazvinian et al., 2011).

4.4.2 Assessment of Dam Foundations

Geological conditions at dam sites pose geotechnical concerns related to sliding failures involving the shear strength of the

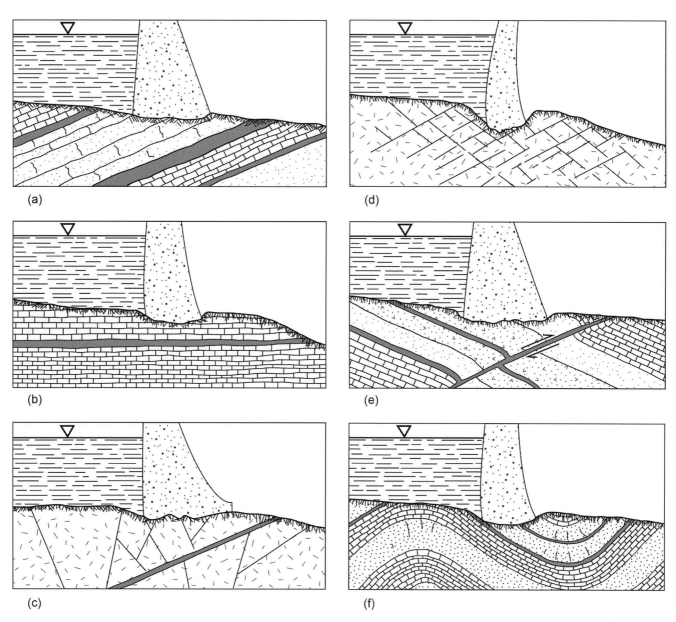

(a)

(b)

(c)

(d)

(e)

(f)

Figure 4.15 Geologic conditions promoting sliding failure of foundations of concrete dams, in which the dark grey unit indicates shale or clay gouge. (a) Brittle, fractured sandstones rest on a weak shale bed, which is dipping upstream. (b) Horizontally bedded limestone rests on a weak shale bed that extends downstream to a steep slope in the valley floor. (c) Fractured crystalline rock containing a fault, which contains clay infill of low shear strength. (d) Shear dislocations are likely in this scenario showing conjugate joint sets in crystalline rock. (e) Sedimentary rocks that dip downstream and which are intersected by a fault that dips upstream and contains infill of low shear strength. (f) Folded sedimentary rocks containing thin, weak beds of shale. (From Wahlstrom (1974); with the permission of Elsevier.)

rock foundation and its discontinuities (Wyllie, 1999). For example, Goodman and Ahlgren (2000) describe the problems caused by the mélange rock foundation of the Scott Dam in northern California. Mélange rock is common to northern California and consists of blocks of shale and basalt set within a fine-grained sheared matrix. In their words: "It was asserted that assigning an average shear strength value for such a material was guesswork and therefore a questionable basis for engineering calculations." Consequently, the safety of the dam had to be reassessed in the 1990s – 70 years after its construction – on the basis of detailed studies of the mélange foundation.

The impounded water in the reservoir will exert a horizontal stress on the dam itself that is restrained by the shear strength of the rock foundation and the weight of the dam. As the reservoir fills, the forces on the foundation incline from vertically downwards to include a vector in the downstream direction. Wyllie (1999) noted that sliding of gravity dams may occur either (i) by sliding of the base of the dam itself or (ii) in a planar fracture or fault in the foundation that daylights downstream of the dam. Consequently, foundation studies of dams must consider (i) the cyclic fluctuations in the reservoir behind the dam causing changing stress conditions and (ii) the likelihood that the large footprint that the dam makes on the underlying rock will mean rocks of various elasticity, strength and permeability will be affected, which may lead to differential movement. Rogers' (1992) reassessment of the failure of the St. Francis Dam in 1928, which killed 450 people in southern California, is a fine example of how a major dam can be founded on different rock types each of poor quality.

Wahlstrom (1974) noted that dam failures in flat, horizontally bedded foundations beneath earth and rockfill dams were more likely caused by seepage through the foundations than inadequate shear strength. Figure 4.15 shows six conditions that he identified for which sliding stability is of concern. In Figure 4.15(a), a sequence of sedimentary rocks includes a clay-rich shale, both of which are dipping upstream and daylighting downstream of the dam. In Figure 4.15(b), horizontally layered limestone beds contain a clay-rich shale interbed that daylights downstream of the dam. In Figure 4.15(c), fractured crystalline rock contains a fault zone with low-strength clay-rich infill material dipping upstream and daylighting downstream of the toe of the dam. In Figure 4.15(d), the fractured crystalline rock contains a conjugate joint pattern with an orientation susceptible to shear failure. In Figure 4.15(e), a sequence of sedimentary rocks dipping downstream contains a fault zone that daylights downstream of the toe of the dam. In Figure 4.15(f), a sequence of folded

sedimentary rocks contains clay-rich shale beds. In each case the fractures and the clay-rich shales reduce the overall stability of the dam and increase the likelihood of failure.

Deng et al. (2001) illustrate how geophysics was used to characterize the foundations of navigation structures (a shiplock and a shiplift) of the Three Gorges Project in China. These structures were to be separated by a rock barrier or island. Concern was raised about the stability of the adjacent vertical walls and the rock barrier after the excavation of the structures, in particular the stress-relieved and locally weakened zone referred to as the disturbed zone. The granitic foundation was tested by ultrasonic measurement of rock cores and by *in situ* testing, also with P-wave velocity measurements. Downhole sonic logging and cross-hole seismic testing indicated that the P-wave velocity of the disturbed zone was reduced by 34–38% and the borehole elastic modulus by 12–31%. These effects were the result of the development of extension fractures in the primary structural planes or perhaps sliding of local rock blocks along primary structural planes.

4.5 Flow in Fractured Rocks

Groundwater flow in discontinuities is a hazard in the development of underground excavations such as tunnels or nuclear-waste repositories. In tunnels, there must be drainage; in deep geological repositories, groundwater flow indicates an unsuitable rock for nuclear-waste disposal. In mines and quarries, groundwater has to be controlled, collected and discharged in such a way that it causes no further problems.

4.5.1 The Cubic Law

The hydraulic conductivity of a fracture in a rock mass is given by Witherspoon et al. (1980) as

$$K_f = (2b)^2 \rho g / 12 \mu, \tag{4.6}$$

where $2b$ is the apparent width of the fracture aperture, ρ is the fluid density, g is the acceleration of gravity and μ is the "flow" or dynamic viscosity. This expression assumes that the flow is laminar in a fracture that may be represented by a pair of parallel plates with smooth walls and an aperture of $2b$. This Darcy-like term is composed of two parts: (i) the capacity to transmit fluid, $(2b)^2/12$; and (ii) the fluid itself, ρ/μ.

The **cubic law** relates the flow rate (Q) per unit hydraulic head drop at steady state, $Q/\Delta h = C(2b)^3$, where C is a constant. We may rewrite the cubic law for the case of radial flow to a well in a fractured rock mass by

$$\frac{Q}{\Delta h} = \left(\frac{2\pi}{\ln(r_e/r_w)}\right)\left(\frac{\rho g}{12\mu}\right)(2b)^3, \qquad (4.7)$$

where r_e is the outer radius in the rock mass from which groundwater flow occurs and r_w is the wellbore radius. For the case of linear flow to a fracture, the cubic law becomes

$$\frac{Q}{\Delta h} = \left(\frac{W}{L}\right)\left(\frac{\rho g}{12\mu}\right)(2b)^3, \qquad (4.8)$$

where W is the fracture width and L its length (Witherspoon et al., 1980).

In the case of uniformly fractured rocks, the cubic law may be written in terms of the permeability (Ingebritsen et al., 2006) as $k_f = N(2b)^3/12$, where N is the number of fractures per unit distance. This leads to an expression of the hydraulic conductivity of such uniformly fractured rocks in which $K_f \approx \rho g (2b)^3/12\mu\Delta L$, where ΔL is the spacing between uniform fractures. Figure 4.16 illustrates this effect of joint spacing and aperture widths on the hydraulic conductivity of a zone isolated between two packers. For a rock mass with joints spaced evenly every metre, a joint aperture of 0.1 mm will yield a hydraulic conductivity of the packed-off zone of 10^{-6} m/s. This is typical of many fractured sedimentary rocks (see Table 11.1). If we consider the hydraulic conductivity of well-sorted sand and gravel as 10^{-3} to 10^{-4} m/s (Table 11.1), then an equivalent hydraulic conductivity will be measured in a rock mass with a joint spacing of 0.1 m (i.e., ten joints per metre) and joint apertures of 0.2–0.6 mm.

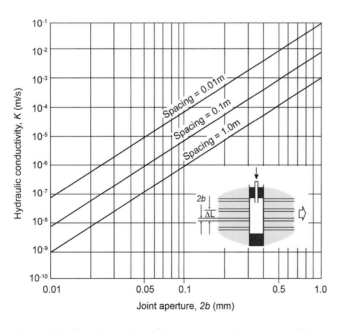

Figure 4.16 The relationship of joint spacing and aperture width on the measured hydraulic conductivity of a zone of rock isolated by straddle packers. (From Wyllie (2018); with permission.)

The joint spacing interval, i.e., whether there is single joint or ten or 100 per metre, affects the hydraulic conductivity of the isolated zone differently than does variability in the aperture width. Note that, as the joint spacing (ΔL) decreases, there is a proportional increase in the hydraulic conductivity because of the increase in the number of joints in the isolated section. However, as the aperture increases in a zone by an order of magnitude, the hydraulic conductivity of the interval increases by three orders of magnitude following the cubic law.

4.5.2 Groundwater Inflow into Tunnels

Deep excavated tunnels in fractured rock are likely to produce significant volumes of groundwater over time with varying inflow rates. The development of tunnels beneath the Swiss Alps in recent years (see Loew et al., 2010) resulted in the intensive measurement of inflow rates. Masset and Loew (2010) discuss the various forms of the cubic law used to evaluate the inflows into the Gotthard and other tunnels connecting Switzerland with Italy. The producing fractures in the crystalline rocks (e.g., granitic gneiss) were mostly brittle–ductile shear zones and brittle fracture zones associated with faults. Flow rates were measured over 100 m transects and varied from dripping conditions, i.e., 10^{-4} L/s, to 10^4 L/s in a log-normal distribution with a mean inflow rate of approximately 2 L/s per 100 m interval, with the highest rates below 1 km.

Just as the shear strength of a rock mass is dependent on the confining vertical stress, so too is the fractured-rock hydraulic conductivity (K_f). Raven and Gale (1985) subjected five samples of fractured granite to three uniaxial loading cycles up to 30 MPa. Each loading cycle caused the closing of the single fracture in the specimens from an initial aperture of 100–200 µm to a final estimate of $\sim$10 µm with a commensurate drop in flow rate as the fractures were irreversibly closed. Wright et al. (2002) used a special triaxial cell to test a specimen of a Tertiary limestone aquifer from South Australia. They measured a decline in hydraulic conductivity of 0.001–0.32 (m/day)/MPa from the closure of microfractures that developed during core recovery and relief of the *in situ* stresses ($\sim$1–1.5 MPa).

It is well understood that deep dewatering of sedimentary rocks and unconsolidated sediments leads to ground subsidence or settlement (see Chapter 13). Zangerl et al. (2003) show that tunnel dewatering during construction of the Gotthard A2 highway tunnel in crystalline rock caused a settlement of the ground surface several hundred metres above the tunnel. They concluded from numerical simulation that the measured settlement of ~10 cm in the granite above the tunnel was not caused by fracture deformation alone, but was a consequence of the combined influence of vertical and horizontal fracture deformation coupled with expansion of the intact rock matrix. Thus dewatering can cause hydromechanical effects leading to permanent changes even in crystalline rock.

4.6 Rock-Slope Failure

Finally, we return to the intersection of structural geology and rock mechanics and identify how discontinuities are associated with potential slope failure. For this work it is most helpful to use stereographic projection, which is discussed in Chapter 5. The potential failure of unstable rock slopes presents a **rock-fall hazard** along transportation corridors through mountainous terrain and in quarries and open-pit

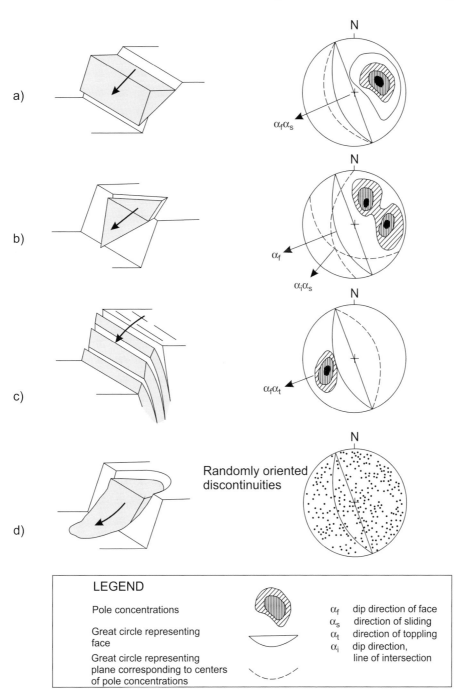

Figure 4.17 Principal types of block failures in slopes and their stereographic representation, where α is the dip direction. (From Wyllie (2018); with permission.)

mines. A full discussion of these slope failures may be found in Wyllie (2018).

Figure 4.17 illustrates the four principal modes of failure of unstable rock slopes and their representation by stereograms. Opposite each failure is the equivalent stereogram showing the dip/dip direction data for the four modes with contoured plots of the data points. Each stereogram plots the great circle representing the slope face, as well as the great circle for each plane corresponding to the centres of the pole concentrations. Angle α represents the dip direction measured as an azimuth, i.e., the direction of the horizontal trace of the line of dip as measured clockwise from the north.

The criteria for failure may be defined in terms of structural geology, and have been summarized by Hudson and Harrison (1997). In the simplest case (Figure 4.17a), **plane failure** occurs when a discontinuity has a strike that is parallel to the rock slope and a dip angle that is greater than the angle of friction of the discontinuity. With plane failure, the dip direction of the face (α_f) and the direction of the sliding surface (α_s) are the same.

Figure 2.23 shows an apparent planar failure in the sandstone outcrop at Red Rocks State Park in Colorado. The upper block of the outcrop at its downdip edge has failed along the bedding plane of the siltstone and slid away. Note

Figure 4.18 A wedge failure in a rock slope, Ouray Canyon, Colorado. Note that the dip of the slope exceeds the dip of the line of intersection of the two discontinuity planes thus making the wedge unstable. (Photo by author.)

that the dip of the sliding or slip plane along the siltstone is less than the dip of the exposed slope face. Hudson and Harrison (1997) identified the criteria for this failure mode as follows:

1. the dip of the slope must exceed the dip of the potential slip plane in order that the appropriate conditions for the formation of a discrete rock block exist;
2. the potential slip plane must daylight on the slope plane;
3. the dip of the potential slip plane must be such that the strength of the plane is reached; and
4. the dip direction of the sliding plane should lie within approximately ±20° of the dip direction of the slope.

Wedge failures occur (Figure 4.17b) when a second discontinuity plane is added to the rock face. A wedge failure in the rock face in Ouray Canyon, Colorado, is shown in Figure 4.18 with the two fracture planes clearly visible. The stereogram of the wedge failure mode has two pole concentrations representing the two discontinuities that form the wedge; the direction of sliding is determined by the intersection of the two discontinuities (α_i). Hudson and Harrison (1997) describe the criteria necessary for this failure mode as follows:

1. the dip of the slope must exceed the dip of the line of intersection of the two discontinuity planes associated with the potentially unstable wedge;

2. the line of intersection of the two discontinuity planes associated with the potentially unstable wedge must daylight on the slope plane; and

3. the dip of the line of intersection of the two discontinuity planes associated with the potentially unstable wedge must be such that the strengths of the two planes are reached.

Toppling failures (Figure 4.17c) occur where two sets of fracture planes dip into the slope of the rock mass. The first must dip steeply into the rock face, with the second set forming the base of the toppling block; the dip direction of the face and the direction of sliding are the same. Direct toppling failures occur "where individual columns are divided by widely spaced joints. The toe of the slope, with short columns, receives load from overturning, longer columns above. This thrusts the toe columns forward, permitting further toppling". Flexural toppling failures occur when "continuous columns break in flexure as they bend forward. Thinner layers transfer load to thicker ones. Sliding, undermining, or erosion of the toe lets the failure begin and it retrogresses backwards, with wide, deep tension cracks" (Goodman and Bray, 1977).

Circular failures were identified by Hoek and Bray (1981) as occurring in extremely weak rock as a result of numerous randomly oriented joint planes – see Figure 4.17(d). A circular failure will yield a randomly oriented set of discontinuities. The circular failure plane is a common assumption in soil mechanics, e.g., the limit equilibrium analysis of Figure 15.13.

4.7 Summary

Rock mechanics complements structural geology, which describes the observable product of geomechanical processes. Rock mechanics has its focus on quantification of the stresses imposed on rock masses and the resulting deformations given the measured rock properties. Rock mechanics constitutes the engineering science for design and construction on and within rock masses, e.g., foundations for buildings and dams or underground excavations for mines or tunnels.

We have considered the various indicators of stress as they are measured by structural geologists, e.g., fault and fold structures, and by geotechnical engineers, e.g., overcoring and hydraulic fracturing. Furthermore, we have noted that engineering problems occur in both overstressed and understressed rocks. The variation of stress with depth has been discussed for both hard-rock environments, such as deep mines, as well as sedimentary basins, in which the fluid pressure profile is critically important to engineering practices. We have discussed how erosional unloading of rocks has caused fossilized stresses to become locked in.

We began a discussion of strain and deformation by considering purely elastic responses. However, rocks under stress yield and deform by brittle and ductile processes, which can be investigated in the laboratory. Deformation creates discontinuities in the form of shear fractures, joints and fissures. These features are likely to be the most important elements in any geotechnical engineering study of a rock mass because they govern rock-mass strength, deformation and hydraulic conductivity.

The strength of rock specimens is typically assessed by triaxial testing and plotting the resulting data as a shear-strength envelope in a Mohr diagram, which we used to consider the stability of underground excavations. We introduced the cubic law to account for groundwater flow in fractured rocks and finally considered the structural conditions causing rock-slope failure. We will proceed in Chapter 5 to describe how rocks are characterized by geotechnical and geomechanical engineers, including the measurement of index properties and the concept of the geological strength index.

4.8 FURTHER READING

Brady, E.H.G. and Brown, E.T., 1993, *Rock Mechanics for Underground Mining*, 2nd edn. London, UK: Chapman and Hall. — An indispensable guide to rock mechanics for engineers developing underground excavations.

Chapman, D., Metje, N. and Stärk, A., 2010, *Introduction to Tunnel Construction*. London, UK: Spon, Taylor and Francis. — A comprehensive and well-illustrated account of the geotechnical engineering of tunnels in both hard-rock and soil conditions.

Franklin, J.A. and Dusseault, M.B., 1989, *Rock Engineering*. New York, NY: McGraw-Hill. — A text developed for final-year civil and geological engineering students at the University of Waterloo in Canada that is admirable in its scope and clarity.

Goodman, R.E., 1989, *Introduction to Rock Mechanics*, 2nd edn. New York, NY: John Wiley and Sons. — A standard reference on the topic with a civil engineering perspective by a legendary member of the University of California's rock mechanics faculty.

Hudson, J.A., 1989, *Rock Mechanics Principles in Engineering Practice*. London, UK: Butterworths. — A short and simplified introduction to the subject that is recommended to those wishing to gain a basic understanding of rock mechanics.

Hudson, J.A. and Harrison, J.P., 1997, *Engineering Rock Mechanics: An Introduction to the Principles*. Oxford, UK: Pergamon, Elsevier. — Based on the rock mechanics courses given at Imperial College, London, providing a comprehensive overview of the subject.

Turcotte, D.L. and Schubert, G., 2002, *Geodynamics*, 2nd edn. Cambridge, UK: Cambridge University Press. — A textbook for undergraduates learning the physics of the Earth with many solved examples and a clarity of discussion.

Wyllie, D.C., 1999, *Foundations on Rock*, 2nd edn. London, UK: Spon, Taylor and Francis. — Elegant and concise, written for graduate students in geological engineering and rock mechanics with excellent illustrations and worked examples.

4.9 Questions

Question 4.1. Develop a "rule of thumb" for the number of metres of rock per unit increase in MPa in the shallow crust of the Earth in which civil engineering structures are built. Assume that the average rock density is 2500 kg/m^3.

Question 4.2. According to Hoek (1999), Sheorey's equation (4.3) is a good representation of the *in situ* stress measurements for deep mines worldwide. If we consider a mine gallery at a depth of 1100 m with an estimated $K = 1.4$, calculate the value of Young's modulus from Sheorey's equation.

Question 4.3. Assume conditions in an inactive fault at a depth of 5 km are estimated as $\sigma_v = 112.5$ MPa with a pore pressure $p_0 = 52.5$ MPa. Compute σ_h' for values of K' given in Figure 4.5 for the fault in sandstone and one in shale.

Question 4.4. If the fault in question 4.3 were a strike-slip fault, in which direction would any induced hydraulic fractures develop?

Question 4.5. For the conditions shown in section 4.2.5, instead of a sandstone with $\nu = 0.25$, determine the post-erosion conditions for a shale with $\nu = 0.15$. Assume $\rho = 2500$ kg/m^3 and the fluid density is 1070 kg/m^3.

Question 4.6. For a joint aperture of 50 μm and a joint spacing of one fracture per 10 cm, what is the expected hydraulic conductivity in a 1 m isolated zone of borehole?

Question 4.7. If it were subsequently determined that the average fracture aperture was 80 μm and the joint spacing of one fracture per 10 cm was confirmed, how much larger is the expected hydraulic conductivity in a 1 m isolated zone of borehole?

Question 4.8. Assuming the dynamic viscosity of water at 20°C is 1 mPa s, compute the hydraulic conductivity of a fractured rock mass with a nominal aperture of 10 μm, i.e., under a unit head gradient.

5 Characterization of Rocks and Rock Masses

In the previous chapter, we identified the properties of intact rock and rock masses that control engineering design and construction – e.g., stress, deformation and strength. In this chapter, the discussion turns to the methods and measurements that are relevant to these properties and the particular features of rock masses that concern the geotechnical engineer. We define some of the tools of the engineering geologist and hydrogeologist. The reader is referred to section 5.9 at the end of this chapter for useful further reading.

Box 5.1 addresses the issue of uncertainties that must be considered during site characterization. Such uncertainties underlie all geotechnical and geoenvironmental investigations because the subsurface is invariably heterogeneous at the scale the engineer must work with. The degree of heterogeneity must be determined by site characterization.

We begin this chapter by describing the basic elements of site characterization where rock masses outcrop or are close to the ground surface; such characterization typically involves the adaptive phasing of tasks as shown in Table 5.1. Rock aggregate in construction and cement production is discussed and the quantification of potential aggregate source rocks is considered as an example of site characterization. Then we discuss the tools of site characterization for rocks and rock masses, including various drilling techniques, geophysics, coring, core logging and the characterization of rock aggregate sites. We proceed to discuss how the basic geomechanical properties of rocks are measured and reported. Then various rock-mass classification schemes are described.

We conclude the chapter with descriptions of hydraulic and geomechanical test methods used in boreholes, the concept of an engineering stratigraphic column to integrate data, and the use of terrestrial digital photogrammetry to present the data.

5.1 The Elements of Site Characterization

The acquisition of information concerning rocks and rock masses begins in the field with an engineering geologist undertaking a descriptive survey of an outcrop or exposed face in an open-pit mine or quarry. For example, the sequence of Figures 3.15 to 3.17 illustrate the mapping of structural features in the Paskapoo Formation of Alberta.

But, for many engineering applications, it will be insufficient to examine only surface exposures. Typically site investigation will involve the incremental steps outlined in Table 5.1, including detailed *in situ* and laboratory testing of the rock cores, which topics are considered in sections 5.3 and 5.5. In this section we consider the use of geophysical tools and the procedures involved in core acquisition and logging.

5.1.1 The Use of Geophysics

Reconnaissance mapping will yield engineering geology maps of the kind discussed in section 3.6. These will likely be complemented by geophysical surveys involving the methods discussed in Box 5.2.

Engineering geologists employ geophysical surveys for two main purposes: (i) to identify anomalous features prior to reconnaissance drilling, and (ii) to correlate strata between the boreholes subsequently drilled (Fookes et al., 2015). Mussett and Khan (2000) pose the following questions to those wishing to use geophysical tools:

1. *Does the problem have geophysical expression?* That is, is the feature that the engineer wishes to detect or investigate one that has a geophysical property that is amenable to

BOX 5.1	SITE CHARACTERIZATION IN THE PRESENCE OF UNCERTAINTIES

Although widely attributed recently to a US politician, the origin of the phrase "unknown unknowns" dates back many years. A famous geotechnical engineer, Dr. Elio D'Appolonia, the founder of D'Appolonia Consulting Engineers of Pittsburgh, Pennsylvania, wrote the following in 1979 in the *Proceedings of the British Columbia Royal Commission of Inquiry into Uranium Mining* (D'Appolonia, 1979):

> Site conditions always pose unknowns, or uncertainties, which may become known during construction or operation to the detriment of the facility and possibly lead to damage of the environment or endanger public health and safety. The risk posed by unknowns is somewhat dependent on the nature of the unknown relative to past experience. This has led me to classify unknowns into one of the following two types: (i) **known unknowns** (i.e., expected or foreseeable conditions), which can be reasonably anticipated but not quantified based on past experience as exemplified by case histories, and (ii) **unknown unknowns** (i.e., unexpected or unforeseeable conditions), which pose a potentially greater risk simply because they cannot be anticipated based on past experience or investigation. Known unknowns result from phenomena which are recognized, but poorly understood. On the other hand, unknown unknowns are phenomena which cannot be expected because there has been no prior experience or theoretical basis for expecting the phenomena.

Source: From D'Appolonia (1979) and Shields (2015).

Table 5.1 Phases of site investigation. (After Franklin and Dusseault (1989).)

Primary phase
Study topographic, air photo and geological maps
Investigate regional seismic and ground stress data
Walk the site to examine outcrops and exposures in tunnels and mines
Prepare 3D block diagrams of geology and preliminary report
Secondary phase
Measure discontinuities on outcrops and exposed faces of mines
Create test pits and trenches to collect samples
Conduct index testing in field or laboratory
Drill reconnaissance boreholes and acquire core
Conduct preliminary geophysical surveys
Classify rock mass
Refine 3D block diagrams of geology and revise report
Tertiary phase
Conduct further core drilling, perhaps inclined or horizontal
Collect jointing information from downhole TV cameras, straddle packers and cores
Conduct downhole geophysical logging to identify rock-mass quality and any faults
Conduct downhole hydraulic testing to measure hydraulic conductivity
Install piezometers and periodically monitor to define the groundwater flow system and head changes
Conduct *in situ* stress, strength and deformability measurements
Complete evaluation of soil, rock, groundwater and stress regimes
Submit final site characterization report

BOX 5.2	GEOPHYSICAL METHODS FOR ROCK-MASS INVESTIGATION

Gravity methods: Gravity methods are used to measure variations in the Earth's gravitational field that arise from differences in the density of rocks and sediments. Consequently, they have been used to identify the igneous bedrock surface beneath unconsolidated basin sediments in the US southwest by virtue of the substantial difference in their respective bulk densities, i.e., granite 2.5–2.8 Mg/m^3 versus alluvial fill ~2.0 Mg/m^3. Micro-gravity surveys are particularly important in identifying natural and excavated cavities in the shallow subsurface, particularly former coal-mine excavations that may cause subsidence.

Magnetic methods: Magnetometer surveys may be used to identify ferrous metal objects such as abandoned oil and gas wells or subsurface pipes and in mineral exploration and structural geology. Under favourable circumstances, it may be possible to discriminate between two types of rocks based on differences in their magnetic susceptibility and, in some cases, identify offsets in layering.

Electrical resistivity: The electrical resistivity, measured in ohm metres (Ω m), may be used to identify the depth to the rockhead where there is weathered material overlying parent rock, e.g., weathered granite (75 Ω m) versus fresh granite (3500 Ω m). In many cases, fractures create characteristically anomalous responses that can be interpolated between parallel lines of data.

Seismic refraction: Based on the seismic velocity within individual rock units, seismic refraction is used to investigate rock strength and rock **rippability** (for excavation purposes). It can also provide important information on structural features in rock masses, such as faults and weathered zones close to the ground surface.

Seismic reflection: Reflection surveys can achieve deep characterization of the subsurface based upon many years of application in the search for oil and gas. However, recent technological advances have made reflection surveys attractive in defining structural and stratigraphic features at depths less than 200 m (see Figure 10.10).

Ground-penetrating radar: GPR is widely used to investigate the rock masses to depths of 15–20 m for features such as faults and fracture zones, natural cavities and evidence of mine-induced subsidence and voids in low-electrical-conductivity materials.

Borehole geophysics: Wireline tools are used downhole to examine the lithology (gamma-ray logging to identify shale), porosity (neutron log), bulk density (density log), dip (dipmeter) and fracture zones (flowmeter and temperature profiling (Pehme et al., 2013)). Advanced orientated optical and acoustic imaging techniques can be used to identify fractures, measure their orientation and aperture to a few millimetres and in some cases assess stress effects on the boreholes.

Because several methods may be applicable to a particular problem, the advice of a professional geophysicist is necessary when selecting techniques.

Source: Modified from Reynolds (2011).

detection by one geophysical means or another at the scale of the investigation. For example, (i) a granitic pluton is detectable gravimetrically because the pluton itself yields an anomaly, whereas (ii) a fault in sedimentary rock may be detectable by seismic reflection, in which case the seismic reflection survey will yield an offset pattern indicating the presence of a fault.

2. *Is the variation lateral or vertical?* The granitic pluton is identifiable because of its lateral continuity resulting in a continuous value of *g*, the acceleration due to gravity.

Geophysical methods employing diffraction patterns or waveforms – seismic and GPR – require a vertical section of horizontal or sub-horizontal beds in order to identify the discontinuity that might be associated with a fault.

3. *Will the signal be detectable?* Assuming that the two previous conditions are met, the engineer should ask his/her geophysical consultant to confirm that the signal of the feature under investigation will be sufficiently strong to be detectable. That is, will the signal-to-noise ratio be satisfactory? In the case in which the engineer wishes to

ensure that a feature such as a karstic cavern of a certain size is absent, then the design of the geophysical survey must lead to the resolution of such a feature.

4. *Will the result be sufficiently unambiguous to be useful?* The engineer needs to be aware that geophysical techniques are limited in their precision and that certain subsurface features – other than those of concern to the engineer – may produce a similar outcome to the feature that the engineer seeks to identify.

5. *Is the survey practicable in terms of the time and money involved?* This issue requires an understanding of the uncertainties involved in the site geological or ground model and the risks associated with those uncertainties.

5.1.2 Drilling, Cores and Core Logging

The second and third phases of site characterization will be directed to acquiring basic properties of the rock mass, whether from outcrops, quarries or drilling. These rock samples will be tested for index properties by point-load, shear-box and other measurements. This phase of site investigation proceeds into detailed investigations involving rock drilling, usually by rotary or percussion drilling, perhaps coupled with further, more detailed geophysical surveys.

Rotary drilling is the basic drilling method for both hard- and soft-rock applications when cores are to be collected. Figure 5.1 shows a multi-purpose drilling rig that can undertake rotary drilling for geotechnical or mineral exploration purposes. A diamond drilling bit is attached to the end of the drill pipe that in turn is attached to the swivel (shown in Figure 5.1 with the coupling threads exposed). The bit, now at the end of a drill string, is then driven into the rock and the rotary action of the bit causes the rock at the bottom of the borehole to be milled and abraded. The circulation of a drilling fluid (clay slurries known as **drilling muds** or air or water) cools the bit and removes the cuttings. In soft or fractured hard rock, the drilling mud holds the borehole open. Cores collected from diamond drilling can vary in size; Table 5.2 lists standard sizes used in North America. For purposes of blasting, drainage and rock-bolt installation, **percussion drilling** is used, in which the bit is typically made of tungsten carbide; however, no core can be collected.

Core recovered from a borehole should be logged, photographed and characterized by index testing while still in the field in order to minimize changes that might occur due to stress relief, weathering and damage during transportation. Figure 4.3 shows core stored in a core box prior to logging

Table 5.2 Standard rock core sizes for diamond coring. (From Boart Longyear (2014).)

Size symbol	Borehole diameter (mm)	Core diameter (mm)	Borehole volume (L/100 m)
BQ	60	36.4	282
NQ	76	47.6	451
HQ	96	63.5	724
PQ	123	85.0	1180

Figure 5.1 A multi-purpose drilling rig designed for use in water well, mineral exploration and geotechnical investigations. It is capable of performing various drilling techniques, including conventional mud, auger, diamond wireline coring and standard penetration testing (SPT). Note that the mast can be tilted to allow angled drilling. (Photo courtesy Dando Drilling International Ltd.)

indicating extensive disking in the shales. This core will have been marked to indicate the top and bottom of the core and then photographed as in Figure 5.2 to show the true depths

Figure 5.2 Measurement of RQD index for the limestone core segment shown within the black outline. The core has been slabbed, which means that it has been cut down its length to reveal a flat surface. The core run outlined has a total length of 3 × 0.63 m or 1.89 m, and the RQD = 94%. The arrow points to a drilling-induced break in the core known as a "mechanical break", which is disregarded in measuring the length of core pieces. (From Armstrong and Carter (2010), © Queen's Printer for Ontario, 2010. Reproduced with permission.)

associated with the various rock segments. The core is then described as outlined in Box 5.3 and core sections identified, logged and removed for index testing, e.g., point-load testing (see section 5.3.1). Figure 5.3 illustrates a borehole log following principles recommended by engineering geologists in the UK, from which a core was recovered from the base of the glacial till to the bottom of the borehole.

Measurement of the core leads to the estimation of the rock quality designation (RQD) index, which was an early index of rock quality developed by Professor Deere at the University of Illinois (Deere and Deere, 1988), that has been incorporated into subsequent classification schemes. Figure 5.2 illustrates how the RQD index is calculated by counting core sections of at least 100 mm or 4 inches in length along the centreline of the core, while ignoring drilling-induced fractures:

$$RQD = \frac{\sum \text{length of core pieces} > 100 \text{ mm}}{\text{total core run length}} \times 100\%. \quad (5.1)$$

The core shown in Figure 5.2 has an RQD index of 90%, which is considered good rock quality; the same cannot be said for the shale at the top of the core box, which is shown in greater detail in Figure 4.3. Rock quality is designated by RQD index values as follows: 0–25%, very poor; 25–50%, poor; 50–75%, fair; 75–90%, good; and 90–100%, excellent.

The borehole log illustrated in Figure 5.3 is a simple example of how information might be presented. Black-bourn (2009), Norbury (2010) and Hencher (2012) discuss borehole and core logs in detail. The description of any borehole and the core from such a borehole may be specified by the jurisdiction in which the borehole is being drilled; consequently its format may be dictated. The end use of a borehole or core log will likely define what is included

in the log. Goodman (1993) points out that, in logs for foundations, as opposed to logs for mineral exploration, the critical information relates to the "character and location of the weak clay seams and zones of closely fractured rock or crushed rock".

Day et al. (2015) discuss the information required from core logging for use in numerical geomechanical modelling of complex rock masses. In particular, they note that, as both civil and mining underground excavations now occur in deep high-stress environments with engineering requirements related to complex excavation geometries and associated stress paths, the rock-mass behaviour depends to a significant degree on its failure mode under compressive stress. In such cases, fractures that are healed with secondary mineral fillings or mineral veins may be very influential in determining rock-mass behaviour. They discuss how core log information can be used to determine rock-mass strength parameters defined by the geological strength index that is explained later in this chapter and subsequently in numerical modelling of excavations created in such rock.

5.2 Characterization of an Aggregate Quarry

As an example of a typical characterization project conducted by an engineering geologist, we consider the investigation of an existing quarry near Ottawa in Canada to determine the volume of remaining high-quality mineral aggregate for (i) construction of foundations, highways and railroad tracks, (ii) building stone and (iii) hydraulic cement. In this context, aggregate refers to both crushed rock from quarries, and sand and gravel from pits, which topic is discussed in section 10.4.

BOX 5.3 | CORE LOGGING NOTATION

1. *Major rock type*: Identify the major rock type – limestone, granite, etc. – at the start of this description, followed by the other descriptors listed below.

2. *Colour*: In addition to the colour of the rock specimen, list other visual attributes, such as whether the rock is banded, mottled or stained.

3. *Grain size or texture*: Classify on the basis of visible grains, i.e., <0.06 mm, very fine-grained, in which individual grains cannot be seen with a hand lens; 0.06–0.25 mm, fine-grained, individual grains are barely visible under hand lens; 0.25–0.50 mm, medium-grained, grains clearly visible under hand lens and just visible to naked eye; 0.5–2.0 mm, coarse-grained, grains are clearly visible to the naked eye; and >2.0 mm, very coarse-grained, grains are of a measurable size. When the crystal diameter of an igneous or metamorphic rock is >10 mm, the specimen is referred to as **pegmatitic**.

4. *Rock hardness*: The degree to which a knife blade will cut the rock, e.g., soft, can be easily gouged or carved with a knife; and hard, can be scratched with a knife with difficulty and the scratch produces little powder and is often faintly visible. Figure 2.16 illustrates the role hardness plays in mineral identification.

5. *Solution and void features*: A rock may be described as solid (without voids), porous (voids <1.0 mm diameter), pitted (voids 1.0–6.0 mm diameter), vuggy (voids >6.0 mm diameter) and as a cavity, in which the voids are larger than the core. Rocks with numerous cavities are referred to as honeycombed. Small voids in volcanic rock formed by gas bubbles trapped during lithification are known as **vesicles**.

6. *Mineral identification*: **Mohs' scale of hardness** indicates a range of common minerals and their relative scratch hardness value, e.g., quartz (7), feldspars (6), calcite (3). A steel knife blade has a hardness of 5.5 and a fingernail of 2.5. See section 2.3.6 for a discussion of mineral identification.

7. *Bedding or foliation*: Bedding and foliation are described on the basis of thickness or spacing visible in the core, e.g., thinly bedded rock would indicate a thickness of 30–100 mm, while laminated or banded would indicate spacings of <10 mm.

8. *Fracture inclination*: Fractures are defined by their dip, e.g., gently (5–20°) or moderately (20–45°) or steeply (45–85°) dipping.

9. *Degree of fracturing/jointing*: The discontinuity spacing would be indicated by a standard index, e.g., RQD.

10. *Fracture roughness*: This measure would vary from **planar** (smooth surface) to **slickensided** (evidence of previous shear displacement along the discontinuity) to **stepped** (discontinuity exhibits angular steps).

11. *Fracture infill*: Goodman (1993) noted that the presence of weak clay seams may be "the most important part of the geologic record" in geotechnical engineering works. In hydrogeology, the presence of soluble minerals in fractures indicates the absence of groundwater flow, which may be important in site selection.

Sources: See US Bureau of Reclamation (1998), Blackbourn (2009) and Norbury (2010).

Mining of aggregate is strongly constrained by environmental and land-use regulations, and the volume of rock aggregate produced each year varies with economic conditions. Furthermore, the extractable rock may not be of the quality needed for the desired uses. Therefore, we will first consider what makes suitable aggregate.

Concrete requires mixing 60–75% aggregate with 15–20% water, 10–15% cement and 5–8% entrained air to produce concrete, of which about three-quarters of US production is employed in ready-mix concrete for foundations, highways, etc. Hydraulic (Portland) cement production requires limestone to be mixed with silica, alumina and iron oxide source rocks to produce a calcium aluminum silicate; fly ash or shale may provide the Si and Al oxides. The source-rock quality is critical. The presence of magnesium at concentrations greater than $\sim 5\%$ MgO_2 inhibits the formation of the calcium aluminum silicates; consequently dolomite cannot normally be substituted for limestone. Similarly, shales can provide calcite, silica and alumina, although an abundance of pyrite may cause problems with the production of SO_2 during the heating

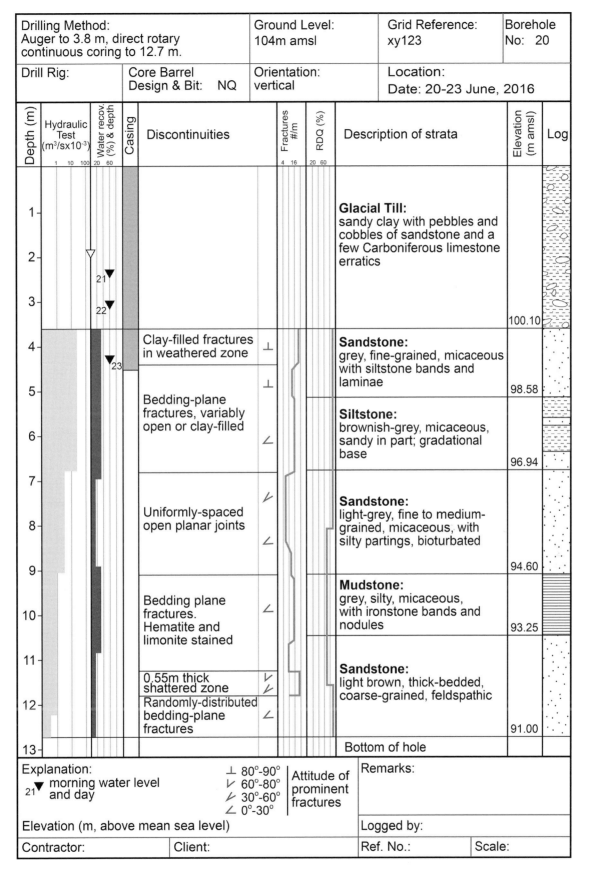

| Drilling Method: Auger to 3.8 m, direct rotary continuous coring to 12.7 m. | | Ground Level: 104m amsl | Grid Reference: xy123 | Borehole No: 20 |

| Drill Rig: | Core Barrel Design & Bit: NQ | Orientation: vertical | Location: Date: 20-23 June, 2016 |

Figure 5.3 Core log following the format proposed by the Geological Society (UK) Engineering Group. In addition, soil sample depths may be identified and a plot of core recovery may be added. (From Knill et al. (1970); with permission of The Geological Society of London.)

of the crushed rock in rotary kilns used to produce the cement (Wenk and Bulakh, 2004; Kesler and Simon, 2015).

Because transportation of aggregate incurs considerable cost in terms of fuel, wear and tear of highway pavements, greenhouse gas emissions and traffic congestion, there is a strong incentive to obtain aggregate from within 50 km (30 miles) of the site requiring the aggregate. Testa (2016) indicates that transporting aggregate 50 km in California increases its cost by US$4.50/ton over and above the nominal cost of the aggregate (~US$10/ton); typically, transportation costs are ~60% of the total cost of aggregate. (Most commonly the source of the aggregate is a quarry, i.e., a surface mine; however, underground mines do exist for the purpose of aggregate production.)

However, the *Los Angeles Times* reported that it was cheaper to import to Los Angeles a basalt aggregate by ship from a quarry on Vancouver Island in British Columbia 2330 km away than to truck aggregate from nearby southern Californian pits and quarries. The shipment costs, when coupled with trucking from Long Beach harbor to downtown construction sites, were US$16/ton, whereas trucking the aggregate 104 km from a quarry in Palmdale cost $22.75/ton. Furthermore, in such urban areas there is considerable resistance to quarrying by nearby neighbourhoods because quarries are associated with noise, pollution and traffic.

Here we are concerned with rock aggregate used to produce crushed stone and concrete. Any rock aggregate should exhibit sufficient strength (UCS > 100 MPa) to withstand crushing; that is, it must be durable. The principal technical concerns with choice of aggregate may be summarized as (i) its physical soundness and (ii) its hardness, strength and toughness. Physical soundness implies that, when employed in construction, the aggregate material must resist swelling and slaking from water uptake and weathering from freeze-dry and wetting–drying cycles. Weak rocks obviously make poor aggregate. Similarly, the aggregate must display good mechanical properties with respect to abrasion, loading and impact to avoid degradation (Langer and Knepper, 1995; Smith and Collis, 2001).

Furthermore, the presence or absence of some minerals is important in the use of either crushed stone or concrete. When mixed with bitumen, quartz in dolomite inhibits skidding on highway pavement; however, other forms of silica can react with alkalis in cement to damage concrete. Silica as opal, chalcedony, chert, volcanic glass, etc., may react with alkali ions, such as Na^+ from highway de-icing salt, causing an alkali silicate gel to form that produces local stresses in the pore structure of the concrete and a volume change followed by cracking of the concrete. Also the presence of sulfate minerals can inhibit the development of good cement concrete or asphalt–concrete mixes.

The application of de-icing salt can cause severe electrochemical damage to steel-reinforced concrete in bridges by corroding the steel rebar, leading to cracking of the surrounding concrete and its failure by spalling. The structure shows signs of failure in 5–10 years involving a loss of tensile strength of the steel reinforcing member and of the compressive strength of the concrete (Wenk and Bulakh, 2004). This problem has plagued many cities in North America and led to many expensive projects rebuilding highway bridges and overpasses.

The constraint of quarry proximity means that a variety of igneous, metamorphic and sedimentary rocks are used. Basalt, similar to that shown in Figure 2.22, which is referred to as trap rock, is a preferred source of aggregate because of its strength and resistance to abrasion. It is widely used to produce aggregate for asphalt production because the even distribution and random orientation of its feldspar and pyroxene crystals results in an absence of planes of weakness (Klein and Philpotts, 2013).

While basalt and dolerite (i.e., diabase) are widely distributed and quarried throughout much of the UK, this is not the case in the area around London where the greatest demand for aggregate occurs; consequently flint gravels in the River Thames partially substitute for the absence of bedrock sources (Smith and Collis, 2001). Many North American cities – such as Ottawa, Toronto, Montréal (e.g., Figure 5.9) and Chicago – rely on the presence of nearby Paleozoic dolomite bedrock for quarried aggregate, while coastal cities, such as Los Angeles, are increasingly reliant on marine shipping of hard rock.

Smith and Collis (2001) present the following outline of tasks necessary to identify potential aggregate sources:

- desktop study of air photos and geologic maps, followed by field sampling and lab testing;
- investigation of nearby aggregate sources to determine the suitability of that aggregate for the purposes required by your client;
- conduct full-scale site investigation by drilling and trenching, with more detailed laboratory analysis, and determine the need for site drainage;
- conduct aggregate production trials to determine extraction methods and production procedures; and
- establish a quality control program to monitor aggregate quality and variability.

Figure 5.4 Extraction issues in rock quarrying caused by structural features in Paleozoic carbonates near Ottawa, Ontario, Canada. (Reproduced with the permission of Rob Blair, R. Blair Geoscience Consulting Inc., Mississauga, Ontario, Canada.)

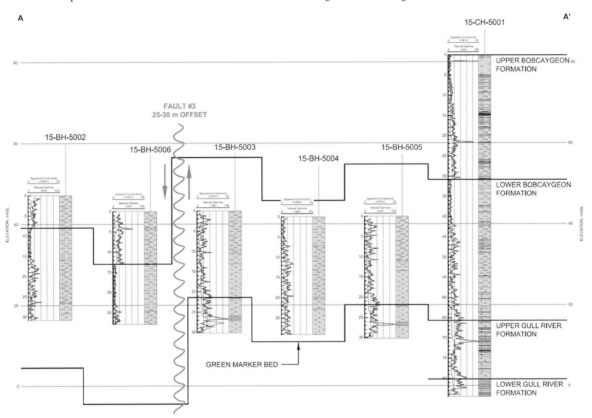

Figure 5.5 In order to estimate the volume of usable aggregate rock beneath the quarry floor shown in Figure 5.4, a program of drilling and downhole geophysical surveying was undertaken. The quarry floor is at the top of the borehole logs. The borehole profiles illustrate the gamma logging response in blue, the apparent conductivity log in red and the interpreted lithologic sequence of rocks on the right-hand side of the scale. (Reproduced with the permission of Rob Blair, R. Blair Geoscience Consulting Inc., Mississauga, Ontario, Canada.) See colour plate section.

Consider the case of Ottawa in eastern Canada, where mineral aggregate is extracted from nearby quarries that mine Paleozoic carbonate bedrock. Figure 5.4 shows a quarry face (∼30 m high) exposing the high-quality Bobcaygeon Fm limestone. However, a fault zone, associated with a block of rotated fractured ground outlined in yellow in the figure, presented difficulty in conventional quarry mining. The fault has 25–30 m of vertical offset as indicated by drilling. In order to define the aggregate resource beneath the floor of the existing quarry, a drilling, coring and borehole geophysical survey was undertaken.

The characterization program is illustrated in Figure 5.5, which shows a borehole log on the right-hand side (15-CH-5001) that was drilled from the original ground surface and used for stratigraphic control to define the Bobcaygeon Fm limestone and underlying Gull River Fm limestone. The basal Gull River Fm exhibits alkali reactivity problems associated with dolomite layers; therefore the base of the extractable rock is defined as the base of the Lower Bobcaygeon Fm. Five rotary percussion boreholes were drilled into the floor, followed by downhole gamma and conductivity logging. They indicated a 25–30 m fault offset as identified by the dolomite layer shown as the "Green Marker Bed", which indicates the top of the Gull River Fm. The recoverable Bobcaygeon Fm concrete-quality aggregate is identified between the two black stepped lines.

5.3 Field and Laboratory Measurements

Basic descriptions of rock materials – rock type, colour, grain size, hardness and fracturing – are made from an outcrop or by core logging as is discussed in Box 5.3 and illustrated in Figure 5.3. We shall discuss rock weathering in Chapters 6 and 7. Here we consider a number of index tests that provide insight into the properties of rock specimens, in particular strength parameters.

5.3.1 Index Tests

- **Hardness** is measured by the degree to which a steel hammer will rebound from a prepared rock surface. The Schmidt rebound hammer test, designed for concrete testing purposes, gives a rebound number which can be correlated with the unconfined compressive strength when the dry density is taken into account. Price (2009) indicates that the L-type hammer is best suited to rocks. The equipment is lightweight and can be easily set up in the field to allow real-time testing.
- **Point-load strength testing** was introduced in Box 4.1 as an index test for strength classification of rock specimens. Figure 5.6 shows a core specimen about to be tested to (tensile) failure in a field lab. This instrument can be readily set up to test in the field.
- **Ultrasonic pulse velocities** employ both primary (P) or compression and shear (S) waves to indicate the dynamic elastic constants of the rock, e.g., Young's modulus,

Figure 5.6 A point-load testing apparatus loaded with core between the two platens. The protective transparent plastic casing is removed to show the core specimen. The hydraulic jack and the digital pressure gauge are shown.

Poisson's ratio, the shear modulus and the bulk modulus (ASTM, 2005b). A specimen, which is at least twice as long as its diameter, is placed in a holder and an ultrasonic pulse is transmitted along the axis of the specimen and recorded at the opposite end. Sonic velocity tests are particularly useful in determining the state of microfracturing within the rock specimen (Goodman, 1989). Lee (1987) – cited by de Freitas (2009) – used P-wave testing, referred to as the compression-wave velocity, to test Korean granites with varying states of weathering from "fresh" to "highly weathered". Lee showed that the P-wave velocity decreased significantly with the degree of weathering as is shown in Figure 5.7. This was interpreted to indicate the increasing water saturation of the specimen, i.e., the fraction of the pore volume occupied by water, produced by weathering processes.

- **Slake durability testing** (ASTM, 2004) is conducted on weathered or clay-rich rocks, such as shales (see Goodman, 1989). These are generally considered **weak rocks**. Figure 5.8 shows the twin test drums that are rotated by the motor drive unit in the black box. Prior to testing, the specimens are weighed and oven-dried to determine their moisture content. The specimens are subjected to long cycles of oven drying and then short cycles of rotation in the test drums, which are partially submersed in water. The slake durability index (I_d) represents the percentage of dry mass of the sample retained on a 2 mm sieve following the wetting and drying cycles.

5.3.2 Strength

It is important to emphasize and distinguish between the strength of a single specimen of rock and that of the whole rock mass, which is largely determined by the frequency and orientation of its discontinuities. Thus laboratory tests are used to measure rock strength, in particular the unconfined or uniaxial compressive strength (UCS), which is the commonly quoted value of rock strength. Ranges of UCS for various rock types are shown in Figure 4.13.

Rock-mass strength must be deduced on the basis of both the principles of rock mechanics discussed in the previous chapter and the measured lab values, because the size of the specimen needed to be representative of *in situ* conditions would be at least 1 m in length. The measured lab value likely represents the maximum strength that might be expected (Wyllie, 1999). For other purposes, other measurements are important (see Goodman, 1989), e.g., (i) shear strength, the critical strength in the stability of slopes and in underground excavations, which is measured by triaxial testing (Figure 4.10b,e), (ii) tensile strength, which can be measured on core specimens by the Brazilian test method (ASTM, 2005c), and (iii) the point-load strength index (Box 4.1).

5.3.3 Porosity and Permeability

Porosity and permeability, which we define rigorously in section 11.1, refer to the presence of void space in the rock

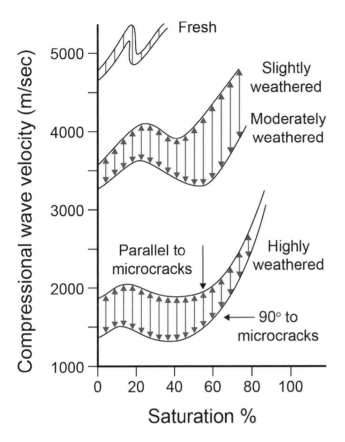

Figure 5.7 Variation in P-wave velocity as a function of water saturation for weathered Korean granites (Lee, 1987). (From de Freitas (2009), with permission of The Geological Society of London.)

Figure 5.8 Slake durability testing apparatus showing the two test drums containing the rock specimens that are rotated through the water baths.

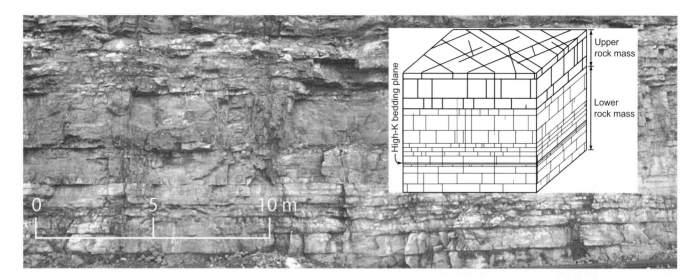

Figure 5.9 A 10 m scan line in a quarry, St. Eustache, Québec, Canada. The inset shows the conceptual model of the fracture network geometry with a block size having each side 5 m long and with high K_f fractures shown in bold. (From Lemieux et al. (2009); © 2008 Canadian Science Publishing or its licensors. Reproduced with permission.)

(**porosity**) and the ease of fluid flow through interconnected voids (**permeability**). The specimen for testing – a cylindrical core plug (usually about 38 mm × 50 mm) drilled out of the core sample – is first "cleaned and dried" so that any oil and residual water are removed from the core plug. Then the lab will measure the bulk density of the core plug using the calipered bulk volume, and then the permeability of the plug by a gas-flow technique. The total porosity is measured as the gas pore volume of the plug, usually by a technique based on Boyle's law of gas expansion.

These parameters are typically measured as standard **petrophysical tests** developed by the oil industry (Lucia, 1999; Tiab and Donaldson, 2004) and may also include measurement of residual fluid saturations, i.e., the fraction of oil, water or gas present per unit pore volume. In such cases where more than one fluid phase is present in the pores, it may be necessary to measure the capillary pressure versus fluid saturation distribution by high-pressure mercury intrusion to describe the pore-size and fluid distributions. This is important not only in petroleum reservoir engineering but also in understanding the potential for migration of gaseous or liquid radionuclides away from a deep geological repository or non-aqueous phase liquid migration through fractured rock.

5.3.4 Discontinuities

The importance of discontinuities in geotechnical engineering was emphasized in section 4.3.3. Norbury (2010) identifies many types of discontinuities: faults, fault zones, joints, fissures, bedding-plane partings, cleavage fractures, shear surfaces, etc. Box 3.3 summarizes the principal parameters that should be measured and their field description.

An index of the frequency of discontinuities in an exposed face of rock or outcrop is obtained by counting the number that intersect a **scan line** of suitable length to give a measure of the mean spacing between discontinuities. Figure 5.9 illustrates the measurement of the fracture pattern in a quarry using a 10 m scan line in a dolomite exposure north of Montréal, Québec, Canada.

Lemieux et al. (2009) divided the exposed quarry face into upper and lower rock masses and estimated a mean fracture aperture of 0.45 mm for the upper rock mass and 0.045 mm for the lower rock mass. Both of these measurements were considerably lower than the bedding-plane fracture (0.7 mm) shown in the inset to Figure 5.9. They then employed the **cubic law** (section 4.5.1) to generate K_f estimates of 0.17 m/s for the upper section and 0.0017 m/s for the lower; the bedding-plane fracture yielded $K_f = 0.4$ m/s.

The fracture porosity (n_f) is estimated by

$$n_f = \frac{\sum m_i b}{L}, \quad i = 1, \dots, j, \qquad (5.2)$$

where m_i is the number of fractures from the ith fracture set among a total of j fracture sets within a domain of length L and b is the fracture aperture. The mean fracture spacing yielded fracture porosity values of 0.3% (upper rock mass) and 0.03% (lower). However, if this rock face were submerged beneath the water table, the number of hydraulically active fractures

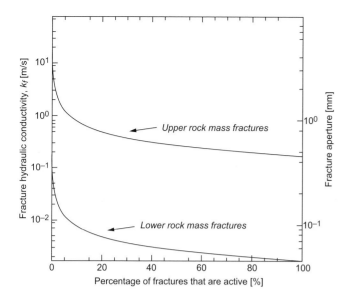

Figure 5.10 Single fracture hydraulic conductivity and aperture as a function of the percentage of hydraulically active fractures (vertical and horizontal), St. Eustache, Québec, Canada. (From Lemieux et al. (2009); © 2008 Canadian Science Publishing or its licensors. Reproduced with permission.)

would be unknown because any aperture may represent a fracture with no persistence, i.e., it pinches out or is filled by clay or other minerals. Figure 5.10 illustrates how the fracture hydraulic conductivity and the fracture aperture vary as a function of the percentage of hydraulically active fractures.

5.4 Rock-Mass Classification

Classification is the process by which the described rock characteristics are identified with respect to predefined classes. The original rock-mass classification schemes were based on concepts such as "stand-up time", which is a measure of the time-dependent stability of the tunnel or slope face, and the RQD system described in section 5.1.2.

Here we discuss three commonly used classification systems (see ASTM, 1995) and conclude with comments on weak and weathered rocks. It should be noted that three of the classification systems – RQD, RMR and Q – yield a minimum value of zero (0.01 for Q) and a maximum value of 100, which indicates high rock-mass strength, stiffness, block size, structural integrity and service life.

5.4.1 Rock-Mass Rating (RMR)

Bieniawski (1978) introduced the **rock-mass rating** (RMR) to characterize rock-mass deformability, i.e., the relationship

between the load applied to the rock mass and the resulting strain or deformation. His motivation in developing the RMR system was "for estimating the in-situ modulus of rock masses during the planning stage of a design investigation for a rock engineering project". Goodman (1989) and Wyllie (1999) provide good summaries of RMR, which Bieniawski (1978) referred to as a **geomechanics classification system**.

The RMR value is determined by allocating values to each of the following parameters:

- **A1**, strength of intact rock material based on point-load testing or UCS (maximum 15%);
- **A2**, drill core quality, e.g., RQD (20%);
- **A3**, joint and fracture spacing (30%), e.g., >3 m rates highly, <50 mm rates poorly;
- **A4**, joint characteristics, e.g., roughness, slickensided, etc. (25%);
- **A5**, presence of groundwater, with lower ratings for increasing pore pressures (10%); and
- **B**, orientation of joints (favourable or not) with respect to foundations, tunnelling and mining (up to −25%).

For example, parameter A2, drill core quality, allows a maximum value of 20%, indicating an RQD of 90–100%. Parameter B is used to reduce the RMR in cases of unfavourable orientation of joints. The various components are summed to obtain 100% by computing RMR = A1 + A2 + A3 + A4 + A5 + B.

Using results from three major engineering projects in South Africa, Bieniawski (1978) found a strong correlation between the RMR and the *in situ* modulus of deformation, E_M, measured in GPa: $E_M = 2 \times \text{RMR} - 100$.

5.4.2 The Q system

The Q or NGI system was developed at the Norwegian Geotechnical Institute (Barton et al., 1974) to describe the peak shear strength of tension fractures. It has been revised (Barton, 2002) and is frequently applied for tunnel construction and general underground excavation (Chapman et al., 2010). The index Q is calculated as a function of six variables:

$$Q = \frac{\text{RQD}}{J_n} \times \frac{J_r}{J_a} \times \frac{J_w}{\text{SRF}}, \tag{5.3}$$

where

- RQD is the rock quality designation index (0–100) discussed earlier in this chapter;

- J_n is the number of joint sets (0.5–20), which is a measure of the sizes of the blocks separated by joints;
- J_r is the joint roughness number, which varies from 0.5 for planar slickensided joints to 4 for discontinuous joints;
- J_a is the joint alteration number, which indicates whether the joint is healed or filled with impermeable material (0.75) or is infilled by crushed rock or clay (20);
- J_w is the joint water reduction number, which varies from 1 for <5 L/min to 0.05 for high inflow rates; and
- SRF is the stress reduction factor, which varies from 10 for loose rock containing numerous weak zones of clay or disintegrated rock to 0.5–2.0 for competent rock under high stress.

Barton et al. (1985) discuss joint geomechanical and hydraulic behaviour in the context of the Q system. Subsequently Barton (2002) showed that the Q index might be correlated with the P-wave velocity (v_P) by $v_P \approx 3.5 + \log Q$, where v_P is in units of km/s, and with the UCS, $Q_c = Q \times \sigma_c/100$.

Barton et al. (1974) developed the concept of an equivalent dimension (D_e) to identify the need for **tunnel support**, which refers to the need to reinforce an underground excavation with supporting materials such as concrete, shotcrete (concrete sprayed into a steel mesh) or rock bolts. This D_e is the ratio of the span of the excavated room to the excavation support ratio (ESR), which varies between 2–5 for temporary mine openings and <0.8 for underground excavations in use by people over the long term, e.g., railroad stations, underground nuclear-power plants and sports arenas. Thus plots of Q versus D_e identify where tunnel support might involve rock bolting (high Q and $D_e = 2$–50) or require reinforcement with concrete or shotcrete ($Q < 1$ and $D_e < 30$).

5.4.3 Geological Strength Index (GSI)

Hudson (1989) warned that the unique qualities of each rock formation inevitably means that "classification schemes are thus empirical and project specific". This led to the development of a new classification method – the **geological strength index** (GSI) – that incorporates geological features while based on a failure criterion for jointed rock masses (Hoek, 1999), such as shown in Figures 2.21 or 2.27. Embedded within this index is the Hoek–Brown failure criterion, which has been adopted in rock engineering to permit scaling intact rock measurements from the laboratory to the field scale.

The generalized Hoek–Brown failure criterion for a jointed rock mass is (Hoek and Brown, 1997)

$$\sigma_{e1} = \sigma_{e3} + \sigma_{ci}\left(m_b\frac{\sigma_{e3}}{\sigma_{ci}} + s\right)^a, \qquad (5.4)$$

where σ_{e1} and σ_{e3} are the maximum and minimum effective stresses at failure, m_b is the value of the Hoek–Brown constant m for the rock mass, s and a are constants dependent on the rock-mass characteristics and σ_{ci} is the UCS of the intact rock. The value of m_b is (Wyllie, 2018)

$$m_b = m_i \exp\left(\frac{\text{GSI} - 100}{28 - 14D}\right), \qquad (5.5)$$

where D is the disturbance factor indicating the extent of damage to the rock mass from blasting and stress relief during excavation. Table 5.4 in Wyllie (2018) provides photographs illustrating the suggested values of D from $D = 0$ for excellent quality rock to $D = 1.0$ for soft rock or rock that has suffered severe blast damage as in an open-pit mine. In equation (5.4), s and a rely on the GSI value such that the structure and surface properties of that rock mass are implicitly incorporated in the classification process:

$$s = \exp\left(\frac{\text{GSI} - 100}{9 - 3D}\right)$$

and

$$a = \frac{1}{2} + \frac{1}{6}(e^{-\text{GSI}/15} - e^{-20/3}).$$

Values of m_i, a constant identifying the intact rock properties, are presented in Figure 5.11, while values of GSI are presented in Figure 5.12.

The UCS of the rock mass (σ_c) is determined by equating $\sigma_{e3} = 0$, so that

$$\sigma_c = \sigma_{ci}s^a, \qquad (5.6)$$

while the tensile strength of the rock mass is obtained from (5.4) by equating $\sigma_{e1} = \sigma_{e3} = \sigma_t$:

$$\sigma_t = -\frac{s\sigma_{ci}}{m_b}. \qquad (5.7)$$

Therefore, to use the Hoek–Brown failure criterion in order to estimate the strength and deformability of a jointed rock mass, three parameters must be estimated:

1. the UCS value of the intact rock specimen (σ_{ci}), from which to estimate the UCS and tensile strength of the rock mass;
2. the value of the constant m_i for the same rock pieces; and
3. the value of GSI for the rock mass.

The GSI has been described as "a system for estimating the reduction in rock mass strength for different geological conditions as identified by field observations" (APEGBC, 2016). The GSI technical literature should be consulted to appreciate how it is applied. For example, Hoek (1999) summarizes

Rock type	Class	Group	Texture			
			Coarse	Medium	Fine	Very fine
IGNEOUS	Plutonic	Light	Granite 32 ± 3 Granodiorite 29 ± 3	Diorite 25 ± 5		
		Dark	Gabbro 27 ± 3 Norite 20 ± 5	Dolerite 16 ± 5		
	Hypabyssal		Porphyries 20 ± 5		Diabase 15 ± 5	Peridotite 25 ± 5
	Volcanic	Lava		Rhyolite 25 ± 5 Andesite 25 ± 5	Dacite 25 ± 3 Basalt 25 ± 5	
		Pyroclastic	Agglomerate 19 ± 3	Breccia 19 ± 5	Tuff 13 ± 5	

Rock type	Class	Group	Texture			
			Coarse	Medium	Fine	Very fine
METAMORPHIC	Non Foliated		Marble 9 ± 3	Hornfels 19 ± 4 Metasandstone 19 ± 3	Quartzites 20 ± 3	
	Slightly Foliated		Migmatite 29 ± 3	Amphibolites 26 ± 6	Gneiss 28 ± 5	
	Foliated			Schists 12 ± 3	Phyllites 7 ± 3	Slates 7 ± 4

Rock type	Class	Group	Texture			
			Coarse	Medium	Fine	Very fine
SEDIMENTARY	Clastic		Conglomerates 22 Breccias 20	Sandstones 17 ± 4	Siltstones 7 ± 2 Greywackes 18 ± 3	Claystones 4 ± 2 Shales 6 ± 2 Marls 7 ± 2
	Non-Clastic	Carbonates	Cristalline Limestone 12 ± 3	Sparitic Limestones 10 ± 2	Micritic Limestones 9 ± 2	Dolomites 9 ± 3
		Evaporites		Gypsum 8 ± 2	Anhydrite 12 ± 2	
		Organic				Chalk 7 ± 2

Figure 5.11 Values of the intact-rock variable m_i for various rock types. (From Hoek (1999); reproduced with permission of The Geological Society of London.)

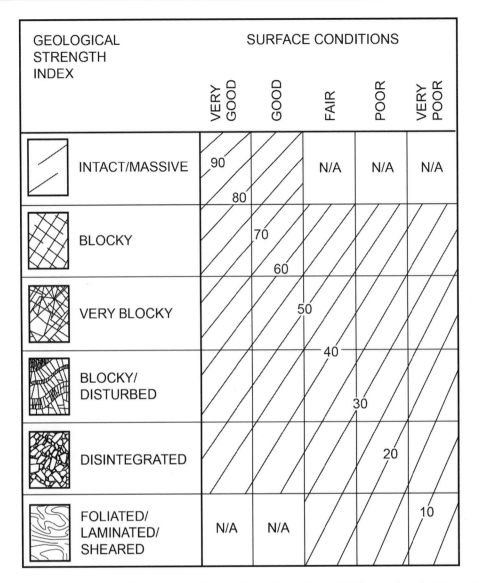

Figure 5.12 The geological strength index as a function of surface conditions. (From Hoek (1999); reproduced with permission of The Geological Society of London.)

its application in engineering geology, Martin et al. (1999) examine the extent of brittle fracture in tunnelling, Marinos et al. (2006) discuss GSI in the context of tunnelling through ophiolites (deep oceanic crust) in Greece, and Day et al. (2015) describe its use in geomechanical modelling of deep limestone excavations for nuclear-waste disposal.

5.4.4 Weak Rocks

Weak rocks form a special classification of rock masses that have attracted much attention by engineering geologists because of the hazards they pose (e.g., Santi, 2006). These include a variety of mudrocks, flysch deposits (Figure 2.21) and weathered rocks (Figure 6.9). Table 5.3 summarizes some of the engineering properties of weak rocks. Hathaway (1999),

who summarized weak-rock occurrences across the USA, provided the following definition of a weak rock:

> A consolidated earth material possessing an unusual degree of bedding or foliation separation, fissility, fracturing, weathering, and/or alteration products, and a significant content of clay materials, altogether having the appearance of a rock, yet behaving partially as a soil, and often exhibiting a potential to swell or slake, with the addition of water. Some weak rocks are also subject to time-dependent release of stored tectonically-induced stress. When weak for reasons other than weathering or alteration, weak rock is generally Cretaceous or younger in age.

Santi (2006) identified several types of weak rock:

- Rocks with high clay content: mudstones, flysch, shales and overconsolidated clays, etc. will exhibit low friction

Table 5.3 Engineering properties of weak rocks. (From Santi (2006).)

Test or property	Value or range for weak rock
Compressive strength	1–20 MPa
Standard penetration test (SPT)	15–90 blows/m
Rock quality designation (RQD)	<25–75%
Seismic wave velocity	<2100 m/s
Ratio of weathered matrix to unweathered blocks	>75%
Slake durability ($I_d(2)$)	<90%
Free swell	>3–4%
Natural moisture content	>1% for igneous and metamorphic rocks 5–15% for clay-rich rocks
Dearman weathering classification	Category 4 or greater
Rock-mass rating (RMR)	<36–60%
Q rating	<2

angles and may contain swelling clay minerals (smectites). Frequently these rocks are poorly cemented.

- Rocks with soluble minerals: chalk, anhydrite and gypsum-bearing rocks may undergo significant karst-like dissolution.
- Rocks with highly weathered materials: saprolite and weathered igneous and metamorphic rocks may have very little residual mineral cement from weathering reactions and therefore poor cohesion and low friction angles.
- Metamorphosed rocks: metamorphism can result in weak rocks exhibiting schistosity or other anisotropic properties and block-in-matrix rocks.

A number of test methods are available to characterize weak rocks. The slake durability tests shown in Figure 5.8 and discussed in section 5.3.1 is one example. Nickman et al. (2006) reviewed the parameters affecting durability and identified compressive strength, grain-size distribution and pore volume as being significant. They recommended an improved testing method and classification scheme based on three cycles of wetting and drying.

Lo et al. (1978) described a deformation gauge shown in Figure 5.13 that permitted semi-confined, time-dependent testing of shales during **free swell tests**. The sample is first submerged in water – or subjected to 100% humidity – and then the change in dimension of the two horizontal and one

Figure 5.13 University of Western Ontario deformation gauge used to measure the dimensions of two horizontal and a vertical directions in free swell tests as described in Lo et al. (1978). (From Lo et al. (1978). Reproduced with permission.)

vertical (axial) directions is measured for 100 days. When the calcite content in the shales tested exceeded ~15%, the shales appeared to be non-swelling due to the cementing action of the calcite.

5.5 Hydraulic and Geomechanical Testing in Boreholes

5.5.1 Hydraulic Testing

A deep borehole is a window into the subsurface, and numerous "downhole" tools have been developed to measure physical or chemical properties. In the petroleum industry, large companies specialize in borehole testing. Table 5.4 identifies the downhole tools used to characterize granitic rocks in Canada for their potential to host nuclear wastes. Many of

Table 5.4 Characterization methods used in deep boreholes. (Modified from Davison et al. (1994).)

Activity/survey	Information
Logging drill core	Geologic description of lithologic variations Location, orientation, geometric characteristics of fractures Nature of fracture infillings and alteration
Thermal logging of borehole fluid	Geothermal gradient Location of hydraulically active fractures
Flowmeter logging	Identification/quantification of hydraulically active fractures
Acoustic televiewer logging	Location and identified of fractures in borehole wall Inspection of stress-induced breakouts (Figure 4.1h)
Borehole TV camera survey	Location and orientation of fractures Character of fracture infillings Lithologic contacts and variations
Standard geophysical logs	Fracture locations Salinity of borehole fluid Lithologic variations and contacts
Single-hole radar survey	Location and orientation of large fractures (within 70 m)
Hydraulic fracturing	Magnitude of σ_3
Groundwater quality (GWQ)	Sample collection within a straddle-packer isolated interval
Hydraulic testing	Transmissivity and storativity of near field, isolated interval
Hydraulic head and GWQ monitoring	Installation of multi-level monitoring instrument

Table 5.5 Physical characteristics of core breaks used to determine permeability ranking in approximate order of importance. (From West et al. (2005).)

Less likely to be permeable	More likely to be permeable
Fresh crystalline appearance of break surfaces	Weathered rounded appearance
No calcite on break surface	Calcite on break surface
Complete closure of two surfaces possible	Surfaces cannot be close manually
Similar break not apparent at depth in other boreholes	Similar break apparent at same depth in other boreholes
Oriented perpendicular to borehole	Oriented with bedding
Core edges which match perfectly	Core edges which are rounded
No staining on surfaces	Staining on surfaces
Clay or mud present	No infilling
Previously classified as a machine break	Previously classified as a vertical or bedding plane fracture

the entries in Table 5.5 refer to the identification and testing of hydraulically active fractures, i.e., ones that produce groundwater and may interfere with geotechnical operations or conduct contamination.

Hydraulic testing of these discontinuities begins with an examination of the core from the borehole; thermal and flowmeter testing will add information. West et al. (2005) investigated the usefulness of core logging to identification of hydraulically active fractures and listed characteristics that indicate varying degrees of permeability. Their ranking of physical characteristics is shown in Table 5.5.

Anomalies in temperature profiles reveal seepage zones in fractured rock. Pehme et al. (2013) have shown how to identify seepage into a borehole (~100 m deep) from individual hydraulically active fractures by sealing the borehole with a removable liner. The liner prevents cross-flow within the open borehole and thus allows static hydraulic conditions to prevail while the temperature is increased by heating and temperature profiles are recorded as the heat dissipates from groundwater inflow. Using a sensitive temperature probe with an accuracy of 0.001°C, they demonstrated how hydraulically active fractures might be identified.

Borehole flowmeters have long been used to estimate groundwater flow in fractured rock. Typically a submersible pump is set just below the water surface in the borehole and the flowmeter is lowered to the base of the borehole. The pump is run to induce flow up the borehole and the flowmeter is steadily raised up the borehole and the flow rate is measured at each elevation, giving a plot of total flow rate, Q, versus elevation. The particular importance of this technique is that it allows the identification of high-permeability zones in the borehole that may be associated with coarse sands or bedding plane partings that are responsible for much of the flow into the borehole.

Wilson et al. (2001) evaluated three types of borehole flowmeter testing instruments to measure groundwater flow in carbonate bedrock and found considerable differences in the measured values. Their report indicated that vertical flow in the borehole may have caused the differences, which would require the use of a packer system to ensure that only horizontal flow is measured.

Hydraulic testing of fractured rock typically seeks to identify flow rates and fracture aperture widths within specified intervals of depth within a borehole, similar to that illustrated by the estimates of hydraulic conductivity versus fracture spacing in Figure 5.10. In the case of a borehole hydraulic test, a zone of the borehole is isolated by a pair of inflatable rubber packers that straddle a particular section of the borehole, as shown in Figure 5.14. The hydraulic parameter that is estimated is the **transmissivity** ($T(a, b)$) for the isolated section between a and b (West et al., 2005),

$$T(a, b) = \sin(\phi) \int_a^b K(z)\, dz, \qquad (5.8)$$

where $K(z)$ is the effective hydraulic conductivity of the fracture zone and ϕ is the dip angle of the borehole. The transmissivity is the volumetric flow rate of groundwater transmitted per unit width of the isolated zone when subjected to a unit hydraulic head gradient. We assume this flow to be essentially horizontal within the packed-off section.

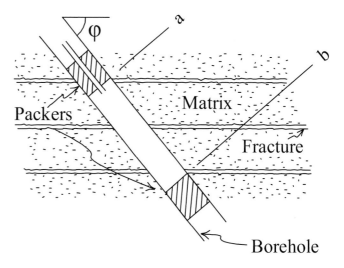

Figure 5.14 Schematic of downhole hydraulic testing system showing inflatable rubber packers isolating a zone of the borehole (a–b) that is penetrated by a fracture. (From West et al. (2005).)

The transmissivity is usually estimated by a constant-head injection test by which water is injected into the isolated zone of the borehole. The flow rate typically decreases monotonically to a steady-state rate of Q at which time a change in hydraulic head versus ambient conditions is recorded, i.e., Δh, which allows the transmissivity to be estimated by the Thiem equation (Todd and Mays, 2005),

$$T(a, b) = \frac{Q}{\Delta h} \frac{\ln(r_e/r_w)}{2\pi}, \qquad (5.9)$$

where r_e is the radius of influence of the hydraulic test and r_w is the radius of the well (West et al., 2005). Quinn et al. (2012) describe four types of hydraulic test methods for fractured rocks and the test equipment and procedures used to depths of $\sim$240 m. They note that under carefully controlled conditions the results of the four methods converge to very similar estimates of transmissivity.

5.5.2 Geomechanical Testing

In section 4.2.2, geomechanical testing of discrete intervals in boreholes and the measurement of *in situ* stress was discussed. The fluid pressure is monitored by a pressure transducer as fluid is pumped into a vertical borehole interval. Figure 4.1(g) shows the setup for such downhole testing, and the nature of the results of such testing is illustrated in Figure 5.15. In particular, this figure illustrates how the *in situ* tensile strength of a hydraulically fractured rock may be estimated (see Goodman, 1989).

With time, the **fracture pressure** (p_f) is reached and a hydraulic fracture of the bedrock occurs; this pressure is

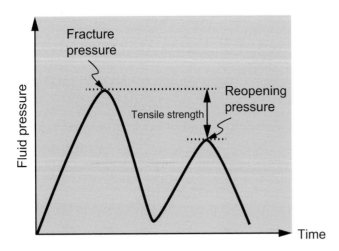

Figure 5.15 The downhole measurement of fracture pressure, reopening pressure and tensile strength in a discrete interval of a borehole isolated by straddle packers as in Figure 5.14. (From Gudmundsson (2011); reproduced with the permission of Cambridge University Press.)

referred to in petroleum engineering as the formation breakdown pressure (Zoback, 2010). The hydraulic fracture most likely will occur where some pre-existing joints occur, perhaps induced by stress relaxation following drilling operations. The newly created fracture allows fluid entry, thus causing a drop in the fluid pressure from p_f. The fracture will develop perpendicular to the minimum principal stress, σ_3, as discussed in section 4.2.2.

Further fluid injection – as shown in Figure 5.15 – causes the reopening of the fracture but at a lower imposed fluid pressure, known as the **reopening pressure**, p_r. The fracture is an extension fracture because it was created by reducing the compressive stress around the fracture zone to such an extent that only the tensile strength of the rock maintains the structure. The difference between the fracture and reopening pressures is the *in situ* **tensile strength** of the rock at the location of the hydraulic fracture, which is usually of the order of 3 MPa (Gudmundsson, 2011).

5.6 The Engineering Stratigraphic Column

Stratigraphic columns are used by geologists to capture sedimentary features in outcrops or cores of rock. This information is of marginal interest to the geotechnical engineer as such. However, Santi and Gregg (2002) show that it is possible to summarize certain details of the information discussed in this chapter through an engineering stratigraphic column that would include information on:

- rock-slope stability and anticipated modes of failure;
- groundwater flow or contaminant transport along bedding-plane features, such as is shown in Figure 5.9;
- rock quality or suitability as aggregate;
- rock strength and intensity of weathering; and
- thickness and continuity of surficial soils.

Figure 5.16 provides an example of such an engineering stratigraphic column from Missouri. The quartz sandstone unconformably overlies a dolomite in Figure 5.16(a). Important sedimentological information is imparted that identifies the texture of the sandstone and sedimentary features such as cross-bedding. The dolomite is identified in terms of textural and paleontological features, e.g., chert pebble conglomerate, stromatolites, etc. Such information is of little practical interest to a geotechnical engineer who must consider the issues listed above. Thus Figure 5.16(b) introduces geomechanical properties (RMR) and hydraulic conductivity (K) information, as well as symbols that identify the orientation and spacing of discontinuities, seepage, vegetation and susceptibility to rock fall. Such engineering stratigraphic columns therefore capture essential geotechnical information and aid the engineering characterization of rock masses.

5.7 Digital Photogrammetry and Monitoring

Digital photogrammetry and **terrestrial laser scanning** have rapidly become essential tools for geotechnical engineers requiring three-dimensional (3D) visualization of rock masses. **LiDAR** is the acronym for light detection and ranging, which is synonymous with terrestrial laser scanning. The usual objective of such work is to characterize the geometry of discontinuities in rock masses and identify zones that require rock engineering solutions, e.g., bolting, shotcrete, webbing, etc.

Haneberg (2008) presents a simple example of the 3D mapping of a rock face using a digital single-lens reflex camera coupled with commercial software to acquire 800 000–900 000 (x, y, z) data points from pairs of photographs similar to that shown in Figure 5.17. Thus the 2D photographic images are converted into a 3D surface model of the rock-mass exposures, in which the fracture orientations may be displayed and presented as stereographic projections, as with the rock slope shown in Figure 5.18. LiDAR scanners are increasingly used to characterize discontinuities in rock exposures, from which the geometry of the fracture system can be computed (Maerz et al., 2013).

(a)

(b)

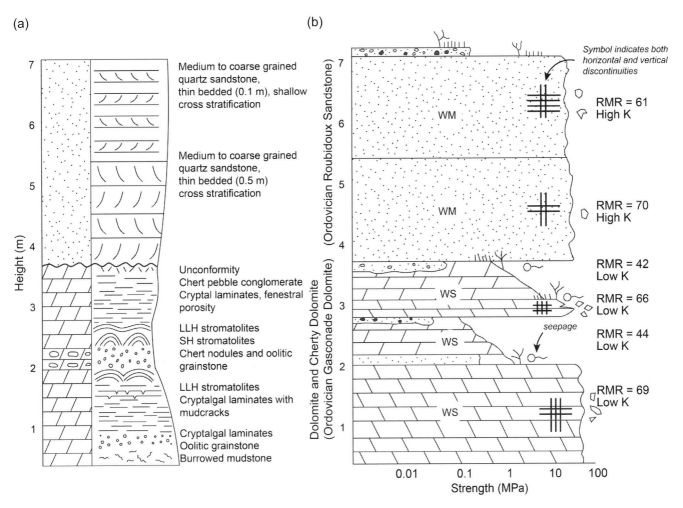

Figure 5.16 A stratigraphic column showing the contact of a sandstone with a dolomite in traditional geological format (a) and as an engineering stratigraphic column (b) emphasizing rock quality (RMR) and hydraulic conductivity, where high K indicates $K > 10^{-5}$ m/s and low K indicates $K < 10^{-7}$ m/s. (Modified from Santi and Gregg (2002); reproduced with the permission of the Association of Environmental and Engineering Geologists.)

Engineers and geomorphologists have become much better able to identify natural hazards along transportation corridors and earthquake zones using LiDAR. Kemeny and Turner (2008) describe rock-fall hazard assessment along highway routes with static LiDAR systems, while Lato et al. (2009) describe similar mapping of highway and railroad corridors by coupling mobile LiDAR systems with the **global positioning system** (GPS) or, to use the international term for GPS, the global navigation satellite system (GNSS).

Most recently, structure from motion (SfM) photogrammetric technology has proven to be a low-cost technology that combines the use of consumer-grade digital cameras with inexpensive software to create 3D topographic images of landslides (Stumpf et al., 2015) or river banks or alluvial fans (Micheletti et al., 2015). When employed with cameras

mounted on unmanned aerial vehicles or "drones", the potential for surveying or monitoring geomorphic features is greatly increased and is much less expensive than LiDAR.

GPS also allows accurate monitoring of critical infrastructure such as dams. Goodman (1993, p. 316) has illustrated wedge failures above the left abutment of the Libby Dam in Montana, USA, and the need for displacement monitoring. Rutledge and Meyerholtz (2005) describe how a differential GPS network was established on and adjacent to the Libby Dam to monitor both horizontal and vertical movements. The network revealed a normal annual displacement cycle of uplift and subsidence at the dam's crest indicating that the dam was deforming elastically. The further application of these tools is presented in Chapter 6 with reference to terrain analysis.

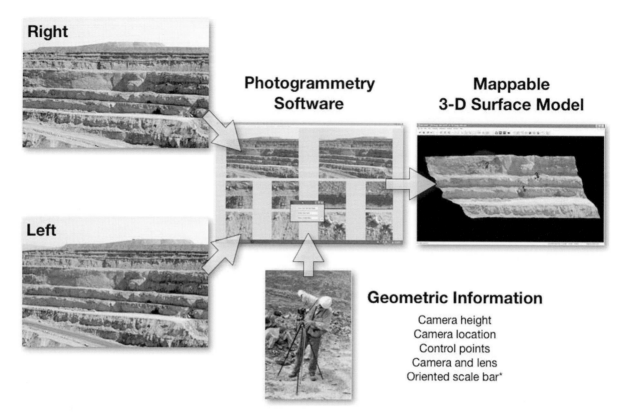

Figure 5.17 Schematic illustration of the digital photogrammetric process by which 2D photographs of the rock benches are used to generate a 3D surface model using appropriate photogrammetric software. (Photos reproduced with the permission of Bill Haneberg.)

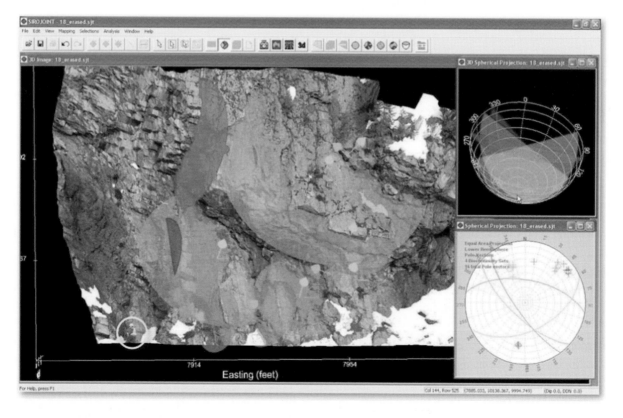

Figure 5.18 Screen capture showing the mapping of individual fractures or discontinuities and the associated stereographic plots. (Photo reproduced with the permission of Bill Maneberg.)

5.8 Summary

In this chapter on the engineering characterization of rocks, the difference between the laboratory testing of intact rock samples and the subsequent estimation of rock-mass characteristics is implicit throughout. Site characterization by geophysical surveys, test pitting, drilling and coring, and hydraulic testing are required to develop a preliminary geological model. As an example of a simple site characterization, we discussed the requirements for rock aggregate for use in highway and building construction and the production of concrete, and how suitable remaining rock aggregate was identified in an existing quarry.

Scan lines are established to measure patterns of discontinuities in exposed rock faces, although digital photogrammetric methods are increasingly used to capture this information. Basic intact rock properties are measured by index tests, but laboratory testing is required to estimate strength, porosity and permeability.

Hoek (1999) emphasizes that "the geological model is a dynamic tool that changes as more information is exposed". Classification schemes – RMR, Q and GSI – provide empirical means of characterizing rock masses for design purposes. The most sophisticated of these schemes – GSI – incorporates geological information while based on a failure criterion for jointed rock masses.

Hydraulic and geomechanical test methods were outlined that allow estimation of groundwater flow and strength parameters within fractured rock masses. Information of this kind can be summarized in engineering stratigraphic columns. The methods of digital photogrammetry and monitoring are increasingly used in rock engineering to increase the productivity of engineers.

In conclusion, Part I of this book has identified the heterogeneous nature of rock masses that are frequently dominated in a geotechnical and hydrogeological sense by discontinuities. These are the materials with which the engineer must work. We remind him or her to heed the advice recorded in Box 5.1: "Site conditions always pose unknowns, or uncertainties, which may become known during construction or operation to the detriment of the facility and possibly lead to damage of the environment or endanger public health and safety." These uncertainties may be unknown but foreseeable, or unknown and unexpected. Characterization projects should seek to minimize risks posed by the unknowns through the development of the robust geological model.

5.9 FURTHER READING

Blackbourn, G.A., 2009, *Cores and Core Logging for Geoscientists*. Caithness, UK: Whittles. — A guide to core logging with an emphasis on sedimentary rocks.

Franklin, J.A. and Dusseault, M.B., 1989, *Rock Engineering*. New York, NY: McGraw-Hill. — Chapter 6 presents the methods of site investigation, while chapter 14 provides a thorough description of methods of rock drilling and excavation for hard and soft rocks.

Goodman, R.E., 1993, *Engineering Geology: Rock in Engineering Construction*. New York, NY: John Wiley and Sons. — The standard American text on the subject, with valuable discussions of the engineering properties of different rock types.

Hencher, S.R., 2012, *Practical Engineering Geology*. London, UK: Spon, Taylor and Francis. — A most useful guide to engineering geology for the geotechnical engineer, with excellent material on weathered rocks.

Hoek, E., 2007, *Practical Rock Engineering*. Toronto, Canada: Rocscience Inc. — Hoek's definitive course notes on the subject, based on many case histories and theory developed by Hoek himself, now reproduced on the Rocscience website and provided free: https://www.rocscience.com/learning/hoek-s-corner/books.

ISRM, undated, *Geophysical characterization of fractures, Dam foundations affected by geological aspects* and *Challenges in rock mass strength determination for the design of underground excavations*. — Online one-hour lectures such as those listed may be viewed on the website of the International Society of Rock Mechanics and Rock Engineering: https://www.isrm.net/.

Norbury, D., 2010, *Soil and Rock Description in Engineering Practice*. Caithness, UK: Whittles. — A handbook on the engineering geological description of soils and rocks in the field with an emphasis on British and European standards.

Price, D.G., 2009, *Engineering Geology: Principles and Practice* (ed./comp. M.H. de Freitas). Berlin, Germany: Springer. — A text from the British school of engineering geologists that elegantly summarizes UK and European practice.

Reynolds, J.M., 2011, *An Introduction to Applied and Environmental Geophysics*, 2nd edn. Chichester, UK: John Wiley and Sons. — A comprehensive guide to the application of geophysical methods and their applications for the geotechnical or geoenvironmental engineer.

Smith, M.R. and Collis, L. (eds), 2001, *Aggregates: Sand, Gravel and Crushed Rock Aggregates for Construction Purposes*, 3rd edn. Engineering Geology Special Publication 17. London, UK: Geological Society. — A comprehensive account of the occurrences, field investigation, processing, extraction and description of aggregates and their use in concrete, mortar and asphalt products.

Turner, A.K., Kessler, H. and van der Meulen, M.J., 2019, *Applied Multidimensional Geological Modelling: Informing Sustainable Human Interactions with the Shallow Subsurface*. Chichester, UK: John Wiley and Sons.

Wyllie, D.C., 2018, *Rock Slope Engineering. Civil Applications*, 5th edn. Boca Raton, FL: CRC Press. — This fifth edition of Hoek and Bray's classic is strongly recommended for those working in rock mechanics and rock engineering.

5.10 Questions

Question 5.1. The RQD rating for the siltstone in Figure 1.5 is 80–90%, which is considered good. On this basis, how would you characterize the sandstone and mudstone beds in Figure 3.15 if they were present as (i) only sandstone and (ii) only mudstone? Assume you drilled a vertical borehole through the sequence of sandstones and mudstones. What RQD would you expect?

Preamble to question 5.2. The rock-mass deformation modulus, E_m, is a critical parameter where a foundation settlement is to be heavily loaded and the rock quality varies across a particular site, e.g., a high-rise building or a dam site (Wyllie, 1999). It is defined as "the ratio of stress to corresponding strain during loading of a rock mass including elastic and inelastic behaviour", whereas the elastic or Young's modulus is the stress-to-strain ratio for purely elastic behaviour. Wyllie (1999, table 3.2) indicates that E_m is measured *in situ* by plate-load tests and varies between 10 for sandstone with shale interbeds to <1 for poor-quality shales. If the Hoek–Brown parameters are known, it is possible to estimate E_m for depths less than 100 m by (Martin et al., 2003):

$$E_m = \left(1 - \frac{D}{2}\right)\sqrt{\frac{\sigma_{ci}}{100}} \times 10^{(\mathrm{GSI}-10)/40}.$$

Question 5.2. Estimate E_m for a rock mass that is considered to be strong with interlocking blocks and good surface quality. Assume that the disturbance factor, $D = 0.2$.

Preamble to question 5.3. A constant-head test is conducted in a fractured rock mass and analyzed by equation (5.9). The test section contains numerous small-aperture fractures rather than a single fracture. Thus the computed transmissivity is referred to as that of an **equivalent rock mass** (erm) value. From (5.9), we estimated the hydraulic conductivity of this erm by dividing the transmissivity by the length of the test section, i.e., $a - b$ in Figure 5.14. Let $a - b = L$, thus we may write the hydraulic conductivity of the erm as

$$K_{erm} = \frac{Q}{\Delta h L}\frac{\ln(r_e/r_w)}{2\pi}.$$

Question 5.3. In a test section of $L = 2$ m, the flow rate per unit length ($Q/\Delta h$) is measured at 10^{-6} m^2/s. The borehole radius is 10 cm and the area of test influence is 10 m. Compute K_{erm}.

Question 5.4. Would the bedrock units shown in Figure 3.22 produce good-quality aggregate?

Question 5.5. Would the bedrock units shown in Figure 3.22 produce good-quality Portland cement concrete?

Question 5.6. Compile a list of preliminary testing procedures for identifying weak rocks.

PART

II Soils and Sediments

The geomorphic processes and products of weathering and glaciation, the development of slopes and floodplains, and the transport and deposition of sediments are the subjects of Part II of this book.

First a note of caution. The word *soil* in geotechnical engineering means any unconsolidated material in the ground, and is not used by engineers in the same sense as it is used by soil scientists and geologists. In soil science, soil is a life-supporting material enabling the growth of agricultural produce and is classified as occurring in a profile characterized by different physical and chemical properties. The word *sediment* is used in these pages to indicate particulate matter that has been transported by water, ice or wind action and subsequently deposited or is still being transported. Thus the science of soil mechanics is also applied – quite appropriately – to the geotechnical behavior of sediments.

The point of departure for Part II is the Earth's energy budget that is the forcing function for the hydrologic cycle, and atmospheric and oceanic circulations. This leads us to a quantification of the hydrologic cycle, including the volumetric flux of water through the cycle, the distribution of global average annual precipitation, and the origin of arid and humid regions of the Earth. Regrettably, space does not permit us to consider atmospheric and oceanic transport of heat and moisture that is so important in a period of climate change. But climate and the hydrologic cycle are directly involved in the global weathering process that produces characteristic soil profiles, slopes and floodplains, which are the subjects of Chapter 6.

Generally speaking, it is the natural tendency of rock near the surface to break into smaller particles with greater surface area per unit mass or volume, i.e., greater specific surface. Where remaining in place and undergoing biogeochemical weathering, soil and new minerals are formed. The interaction of the particles with rain and snowmelt produces natural waters that reflect the properties of the minerals with which the water comes into contact. This leads us to Chapter 7 on environmental geochemistry and mineralogy, which includes a short review of aqueous environmental chemistry that is then employed to consider the critical matter of acid-rock drainage. Therefore Chapter 7 considers matters of great concern to geoenvironmental engineers; it concludes with a discussion of the geotechnical importance of clay minerals and soils and rocks rich in clay minerals.

Glacial erosion and the deposition of glacial sediments is the subject of Chapter 8. These sediments, the effects of glaciation on the pre-existing bedrock and the after-effects of deglaciation have shaped much of the landscape of North America, northern Europe and mountainous areas throughout the world. This chapter concludes with a discussion of frozen soils – permafrost – that challenge those building pipelines across frozen terrain to bring oil and gas to markets further south. The remoteness of untapped hydrocarbon reservoirs north of the Arctic Circle means that the behaviour of permafrost will remain of importance to the geotechnical engineer throughout this century.

While fluvial landforms are introduced in Chapter 6, the nature of fluvial sediments is specifically considered in Chapter 9. The erosion and transport of fluvial sediments is critical to the geotechnical design of bridge foundations that will remain resistant to scour and bridge collapse, as is shown in Figure 9.1. Sedimentary deposits, including stream deposits (alluvium), alluvial fans and deltaic deposits, are then described. We discuss the sediment yield of a major river – the Fraser of British Columbia in Canada – and consider how the bedload and suspended loads change with distance from its source to its outlet near Vancouver, and how the delta of the Fraser has built itself into the sea. The process of sedimentary diagenesis yielding sedimentary rocks completes the discussion of fluvial sediments and considers the loss of porosity and hydraulic conductivity in sediments by consolidation and cementation.

Part II ends with a review of the characterization of soils and surficial sediments in Chapter 10. As with rock-mass characterization, we begin by reviewing the workflow involved in site characterization, including core acquisition and logging, drilling methods, and cone penetrometer and geophysical testing. We then turn to the textural analysis of soils and sediments, which begins with grain-size analysis, and allows inferences to be drawn as to the processes that resulted in sedimentary deposition. There are both geotechnical and geological styles of description, and both need to be appreciated. Then the geotechnical properties of a soil or sediment sample – principally its hydraulic conductivity, shear strength and consolidation potential – by laboratory testing are described. Finally, we conclude with a discussion of the characterization of aggregate sources – sand and gravel – that are used in construction.

CHAPTER

6   Terrain Evolution and Analysis

We begin this part of the book by considering the nature of those climatic, hydrologic and geomorphic processes that shape terrain and conclude with a discussion of terrain analysis methods.

The driving forces behind terrain evolution are tectonics, which cause uplift of land, the climate, the hydrologic cycle and gravity. We have discussed tectonics all too briefly in Chapter 2 and will return to it in Chapter 14, where we consider tectonics in the context of seismicity and earthquakes.

Climate and the hydrologic cycle are the natural subjects of our first section. We consider the processes acting upon rocks that create the weathered materials that form soils and sediments and those that produce streamflow and sediment transport. Therefore, an understanding of the Earth's energy balance and the hydrologic cycle are essential to understand soils and sediments, and the effects of climate change (see Box 6.1).

Weathering is the subject of the second section, in particular physical weathering processes; we will defer discussion of chemical weathering until Chapter 7. The role of weathering in slope stability follows, which is a prelude to further consideration of slope stability in Chapter 15. Then in section 6.4 we consider the nature of alluvial channels and their floodplains, where so much infrastructure is created and must be maintained by geotechnical and geoenvironmental engineers. We conclude with section 6.5 on terrain analysis, i.e., the interpretation of landscapes using remotely sensed images by which we are able to assess potential threats to life and infrastructure from landslides, earthquakes, floods and other geohazards.

6.1 Climate and the Hydrologic Cycle

The incoming solar radiation at the top of the Earth's atmosphere is 341 W/m^2 (Trenberth et al., 2009), which drives the hydrologic cycle as well as the global atmospheric circulation. This radiative flux vastly exceeds that from the convection and conduction associated with the Earth's geothermal heat flux, which is of the order of 87 mW/m^2 (Turcotte and Schubert, 2002), which is referred to in Box 2.1. We shall begin with an energy budget for the Earth and then turn to discussions of the hydrologic cycle and streamflow generation.

6.1.1 Energy Budget of Solar Radiation

Figure 6.1 shows how this radiative flux is proportionally employed in heating the Earth's surface. Of this total incoming flux, 23% is absorbed by atmospheric aerosols, ozone, water vapour and CO_2. Some 30% of the total incoming solar radiation is reflected back to space and thus defines the albedo of the Earth. Less than 50% (i.e., 161 W/m^2) is absorbed at the Earth's surface. This division of the incoming short-wave solar radiation is shown on the left side of Figure 6.1.

The right side of Figure 6.1 shows how the absorbed surface radiation is transferred back to the atmosphere or directly back into space via the atmospheric window. Thus the majority of solar energy absorbed is transmitted to the atmosphere by infrared surface radiation, latent heat and sensible heat transfer. The sensible heat transfer is in the form of thermals, i.e., rising air currents sought by glider pilots, due to variable surface heating, while the latent heat takes the form of evapotranspiration. Figure 6.1 shows that the incoming

BOX 6.1 | EFFECT OF CLIMATE CHANGE ON THE HYDROLOGIC CYCLE

It is predicted by the scientific community that the Earth is warming, perhaps solely due to greenhouse gases, perhaps partly by solar variability. How will this warming affect the hydrologic cycle in ways that affect engineers? We can point to two effects that obviously matter: (1) melting ice caps will raise sea levels and future storm surges will be more likely to damage coastal infrastructure and (2) changing patterns of atmospheric precipitation – rain and snow – may affect water supplies, cause increased soil erosion and landslides, and result in lengthy droughts. We shall pursue the first topic in Chapter 16, but address the second here.

With climate warming we may expect to see pronounced changes in the character of precipitation, i.e., the intensity, frequency and duration of rainfall and snowfall events, because the amount of water vapour that air can retain increases by 6–7%/°C according to the Clausius–Clapeyron equation (Trenberth, 2005). Thus increased evaporation, due to increased temperature, will cause moisture convergence into storms and also increase the intensity of precipitation. In turn, the frequency and duration of precipitation events are likely to decrease, thus promoting drought or exacerbating an existing drought. It is not surprising therefore that drought has noticeably increased since the 1970s in areas such as Africa, Australia, the Mediterranean and southwestern USA, while precipitation has increased at higher latitudes (Trenberth et al., 2003, 2007).

In this context, considerable concern exists among water resources managers about extended severe droughts that may severely curtail water supplies in arid areas. A prime example is that of the US Southwestern states that extract water from the Colorado river basin and rivers draining the Sierra Nevada mountains separating California and Nevada. Tree-ring cores from the upper Colorado and Sacramento river basins were used to extend streamflow records to AD 800, i.e., over 1200 years of paleoclimatic and instrumental record. Woodhouse et al. (2010) discovered that a drought lasting from AD 900 through AD 1300 in southwestern North America greatly exceeded subsequent droughts, such as the "Dust Bowl" of the 1930s. However, the driest decade of this medieval drought was not as warm as the recent drought in the US Southwest.

But this medieval drought, which Woodhouse et al. (2010) argue requires water management agencies to consider in their plans, was much more persistent and far more widespread than the recent drought, which began in the Colorado basin in 1999. The Hoover Dam on the Colorado River created Lake Mead that began to fill in 1937 when Lake Mead's elevation was below 1100 ft (above mean sea level) but was over 1200 ft in 1999. However, by January 2015 the level of Lake Mead was below 1090 ft and it was only 1081 ft in late 2017. The paleoclimatic reconstruction of the medieval drought indicates that the recent ten-year drought may be far from over. You can follow the ups and downs of Lake Mead by viewing http://www.usbr.gov/lc/region/g4000/hourly/hourly.html.

solar radiation is balanced by reflected solar radiation and outgoing long-wave radiation with an imbalance of 0.9 W/m^2 (Trenberth et al., 2009).

This energy budget shows that half of the solar radiation absorbed by the Earth is used in latent heat transfer from the Earth's surface to the atmosphere in the form of evapotranspiration of water. The transfer of surface radiation, sensible and latent heat implicit in Figure 6.1 drives atmospheric circulation. In the mid-latitudes, i.e., 20–60° north and south of the equator, energy transport is mainly through the atmosphere rather than the oceans (Fasullo and Trenberth 2008). For example, Britain and northwestern Europe is warmed predominantly by atmospheric transport of heat associated with the mid-latitude westerly air masses rather than oceanic

heat transport by the Gulf Stream, as is popularly thought (Seager, 2006). However, in the tropical latitudes the surface flux of energy is mainly through the transport of heat in the oceans.

6.1.2 The Hydrologic Cycle

Only 3% of the Earth's water resources are fresh water. The vast majority is seawater or other saline water that is unfit to drink or use in agriculture or industry without expensive treatment. A hydrologic budget for the Earth is shown in Table 6.1. Note that approximately 70% of fresh water is stored in glaciers and permafrost that are currently being reduced by global warming and, in the case of ice caps, these volumes are added to the seawater total. Table 6.1 makes clear that the volumes of fresh water in the atmosphere and rivers are minute and that groundwater and, to a lesser extent,

lakes and runoff are the most accessible sources of fresh water available in most parts of the Earth. (Note that soil moisture is considered unavailable for extraction except by plants.)

Trenberth et al. (2007) have estimated the fluxes between hydrologic compartments by reference to their energy budget discussed above. Their estimate of the evaporation from oceans is $413\,000\,km^3/yr$ of which $373\,000\,km^3/yr$ is returned to the oceans by precipitation; an additional $73\,000\,km^3/yr$ evapotranspires from land surfaces. The annual average precipitation – rainfall and snow – falling on the land surface of the Earth is $113\,000\,km^3/yr$, of which 65% is evapotranspired, with the remainder running off as streamflow, i.e., a runoff ratio of 35% involving $40\,000\,km^3/yr$. This streamflow discharge to the oceans balances the *net* water vapour transport from the oceans to land.

Therefore, $113\,000\,km^3/yr$ of water vapour is precipitated on land surfaces comprising $1.48 \times 10^{14}\,m^2$, yielding an areal average precipitation of 760 mm. Regions that receive

Table 6.1 Global water distribution. (Data from Trenberth et al. (2007).)

Water source	Volume (km^3)	Percentage of fresh water	Percentage of total water
Oceans	1 335 040 000	—	97.0
Ice caps, glaciers and permanent snow	26 350 000	62.8	1.9
Groundwater	15 300 000	36.4	1.1
Rivers and lakes	178 000	0.42	0.013
Soil moisture	122 000	0.29	0.009
Permafrost	22 000	0.05	0.002
Atmosphere	12 700	0.03	0.0009

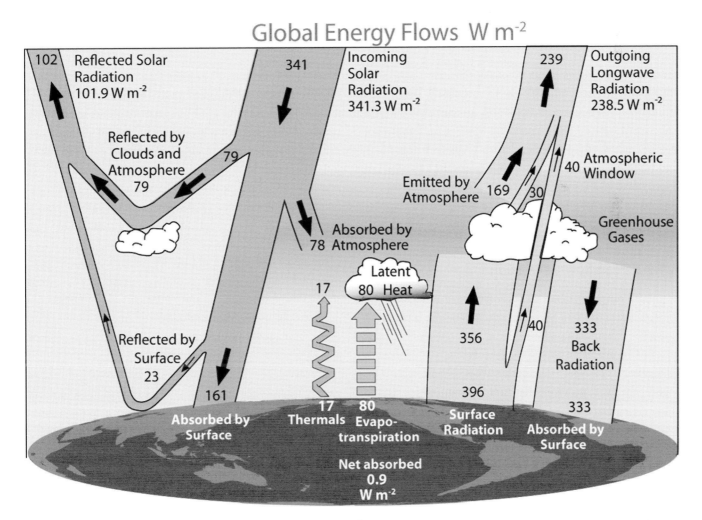

Figure 6.1 The global annual mean energy budget of the Earth in W/m² for the period March 2000 to May 2004. The broad arrows indicate relative importance of energy flows. (From Trenberth et al. (2009); © American Meteorological Society. Used with permission.)

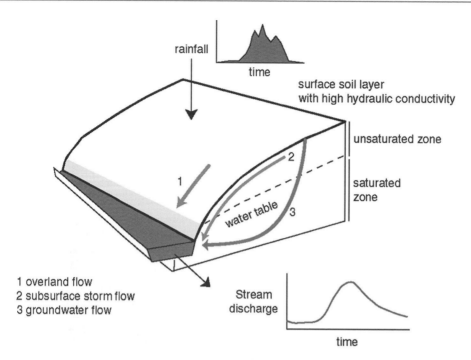

Figure 6.2 Hydrologic mechanisms (1, 2, 3) by which streamflow is generated by rainfall on a permeable hillside in a small tributary watershed. (After Freeze (1974); reproduced with permission of the American Geophysical Union. From Anderson and Anderson (2010), courtesy of Cambridge University Press.)

more than this average are those equatorial areas, such as the Amazon Basin and southeast Asia, and the west coasts of mid-latitude continents, e.g., Britain, Ireland, British Columbia and the US Pacific Northwest. All these areas experience the migration of low-pressure storm cells bringing much atmospheric moisture. By contrast, descending air masses at approximately 30° north and south of the equator lead to regions of significantly below-average precipitation, e.g., the US Southwest, Saharan Africa and Australia.

6.1.3 Streamflow Generation

Moving from the global scale to that of a watershed, we consider the processes by which precipitation is transferred downslope from high ground to a river channel. Runoff generation processes are shown in Figure 6.2. Later, in Chapter 11, we shall define the cross-section shown in Figure 6.2 as a **groundwater flow system**, which is the fundamental concept of hydrogeological systems (Freeze, 1974; Freeze and Cherry, 1979). We note that the rainfall associated with the storm shown in the rainfall hydrograph either infiltrates as **groundwater recharge** or runs off as **overland flow**.

The concept of overland flow was conceived by the American hydrologist Robert Horton as occurring when the rate of rainfall exceeds the infiltration capacity of the soil. Overland flow – pathway 1 in Figure 6.2 – is now believed to be a rare event restricted to small areas of a watershed and, where it does occur, involves only 10–30% of the rainfall events (Freeze, 1974). While it is believed to be limited in importance in humid regions, Hortonian overland flow may play a more important role in arid-zone watersheds where soils can be relatively impermeable due to the presence of a calcium carbonate pavement (caliche). But when the water table rises to the ground surface in floodplain areas in humid regions, then saturated overland flow occurs in limited contributing areas of the watershed. Dunne and Black's (1970) work in Vermont indicated that this process is of considerable importance in streamflow generation.

Rainwater that infiltrates into the soil may either move laterally as **subsurface storm flow** or vertically to the water table and then laterally as **groundwater discharge** to the stream (pathway 3). The relative amounts of each are dependent on the soil's hydrologic properties, the location of the area within the hillslope and the rates of rainfall. Freeze (1974) believed that, in most humid regions, storm flow in streams in forested areas is the result of subsurface storm flow that develops in source areas of variable size close to the stream channels themselves. This feature is shown as pathway 2 in Figure 6.2. Groundwater discharge – pathway 3 in Figure 6.2 – provides the **baseflow** of streams, thus sustaining streamflow between periods of storm runoff.

6.2 Weathering of Rock

Rocks may be exposed at the surface of the Earth by intrusion or uplift or by the erosion of overlying rocks. Irrespective of the cause, the newly exposed rock will therefore be in an initial state of disequilibrium with the surficial stresses and geochemical environment. The release of the *in situ* stresses upon exposure will promote fracturing, while geochemical alteration will promote the development of clay minerals and other weathered mineral products. Weathering creates pore space, allows water to penetrate rock and thus acts to promote plant growth and ecosystem development. Weathered rock is also of great importance in geotechnical engineering in terms of mineral extraction, slope stability and foundation engineering. In this section the nature of physical weathering processes is introduced and discussed; we shall return to the subject and address geochemical weathering in the next chapter.

6.2.1 The Weathered Profile

Figure 6.3 presents an ideal weathering profile that might be expected to develop over time on a granitic rock subject to prolonged physical and chemical weathering (Migon, 2010). The depth of such weathering profiles may be 70–90 m, and not just in tropical, high-rainfall areas. Ollier (2010) points out that oxide and sulfide zones in mineral deposits may be several hundred metres deep and these are also the results of geochemical weathering aided by deep fracturing.

An obvious feature of Figure 6.3 is the presence of **corestones** throughout the weathered profile; these unweathered or partially weathered blocks of parent rock tend to be from 20–30 cm in width up to 5 m in diameter. It is important to notice that the profile begins with unweathered rock at the base and then progresses upwards through increasingly weathered materials to saprolite and the mobile regolith.

With Figure 6.3 in mind, it is helpful to define some terms (modified after Wald et al., 2013).

- **Weathered bedrock**: Bedrock that is altered from its original state by chemical weathering processes (oxidation and hydrolysis in particular).
- **Saprock**: Weathered rock that retains the original rock fabric, but crumbles by hand to individual grains often referred to as **grus**. Mineral grains are not significantly geochemically weathered; consequently clay mineral content is low. Has sufficient porosity to hold significant plant-available water and to cause the rock to lose mechanical strength.
- **Saprolite**: Highly weathered rock, while retaining the original rock fabric, more than 20% of weatherable minerals

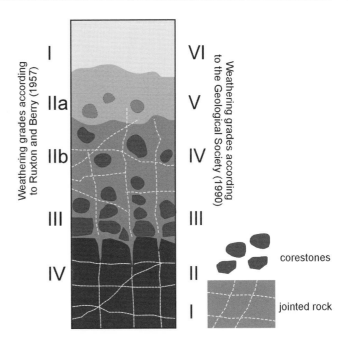

Figure 6.3 An idealized weathering profile developed on granitic rock. (From Migon (2010); with permission of The Geological Society of London.)

(biotite, feldspars, olivine, hornblende, etc.) have altered to clay minerals. Easily crumbles in the hand and becomes plastic when wet.

- **Soil**: A natural medium of granular solids comprising mineral grains, organic materials, water and gases that occurs on the land surface and has horizons distinguishable from the parent material. Has the ability to support rooted plants.
- **Regolith**: Unconsolidated earth material above hard bedrock including soil, saprock, saprolite, loess, colluvium, alluvium, etc. A mobile regolith, which is soil-like, organic-rich and detached from the weathered rock from which it developed, undergoes hillslope creep due to the gravitational forces exerted on it.
- **Lithic contact**: The boundary between soil and underlying bedrock that is sufficiently lithified to inhibit hand excavation. Similar in meaning to top-of-rock or rockhead in engineering geological terminology.

Saprock is shown in Figure 6.4 and may be contrasted with saprolite, shown in Figure 6.5. Neither saprock nor saprolite retains significant mechanical strength. Both can be readily augured or excavated with a shovel, while retaining the relict fabric of the parent rock. Anand and Paine (2002) identify saprock in Western Australian regolith as directly overlying bedrock and underlying saprolite; they refer to saprock as slightly weathered rock of low porosity with <20% of the weatherable minerals having been altered. The studies of

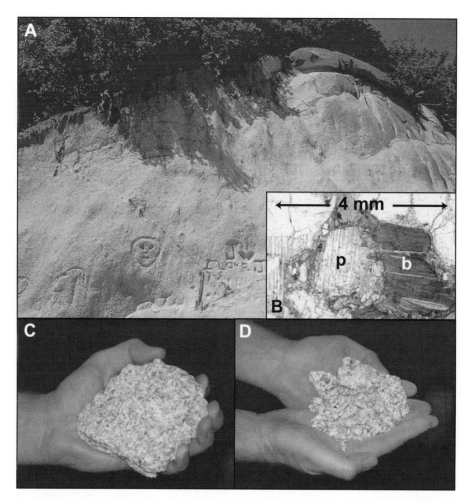

Figure 6.4 Saprock, Sierra Nevada mountains, California. (A) Friable saprock outcrop that is easily vandalized with graffiti; 1.15 m tile spade for scale. (B) Thin-section photomicrograph illustrating pore volume (blue), plagioclase (p) and partially weathered biotite (b). (C) While sample maintains integrity when held, it is easily crushed with bare hands (D). (From Graham et al. (2010); with permission of The Geological Society of America.)

saprock and saprolite in California and North Carolina by Graham et al. (2010) indicate that saprock may often be mistaken for saprolite.

Figure 6.6 illustrates the development of **macropores**, large soil voids, in temperate-region soils, in this case in southern Ontario, Canada (Cey and Rudolph, 2009). Macropores are typically created by the action of worms, small mammals, etc., or by root penetration that produce pathways for rapid infiltration of rain and snowmelt into the subsurface in soils of otherwise lower hydraulic conductivity.

Table 6.2 presents the two sets of weathering grade schemes developed over the past 60 years and in widespread use today. Ruxton and Berry (1957) developed their five-class scheme based upon their observations in Hong Kong. However, this scheme has been replaced there because it has been found to be inadequate in the critical aspect of core logging. Hencher (2012, appendix C) presents a number of alternative

standards tested for the highly weathered igneous rocks of Hong Kong, and the interested reader is referred to his discussion and that of Norbury (2010) for European schemes. Also presented is the scheme of the Geological Society of London's Engineering Group Working Party (1990) and that of the International Society for Rock Mechanics (1978), which reverses the numbering sequence. US practice tends to follow this latter scheme (e.g., Goodman, 1993).

6.2.2 The Effects of Climate

The weathering profile that is observed on a rock slope in humid temperate or tropical climates will have developed over long periods of geological time. For example, Ollier (2010) indicates that the deep weathering profiles of Brazil and Australia date back 10–70 Ma, i.e., throughout the Cenozoic Era. Such weathering alters both the strength and hydraulic

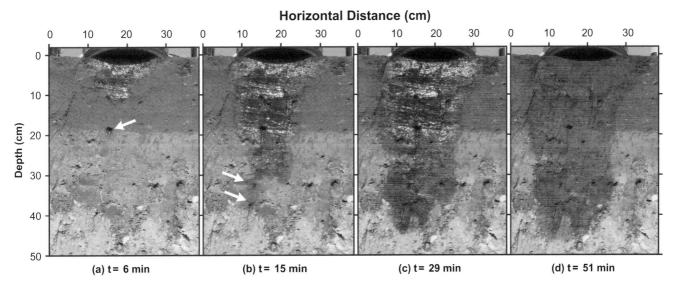

Figure 6.5 Saprolite from the North Carolina Piedmont, USA. (A) This outcrop preserves the structural features of the parent rock, yet the saprolite is soft and easily excavated by the bulldozer shown in the lower right and which provides scale. (B) Thin-section photomicrograph shows thorough alteration of the weatherable primary minerals to clay minerals. (C) The extensive weathering and generation of clay minerals causes the saprolite to yield as a plastic material upon wetting. ((A) Photo credit: G. Simpson. (B) Photo credit: M. Vepraskas. From Graham et al. (2010); with permission of The Geological Society of America.)

Figure 6.6 Flow through a worm-burrow macropore illustrated by dye migration and time-lapse photography. Infiltration occurs (a, b) from a surface source with a −0.4 cm pressure head over a period of 29 minutes (c). Arrows indicate early breakthrough of the dye solution in worm burrows. Dye redistribution is photographed (d) at 51 minutes after infiltration began. (From Cey and Rudolph (2009); reproduced with the permission of John Wiley and Sons Ltd.)

Table 6.2 Weathering grades for weathered basement rocks, i.e., igneous and metamorphic rocks, based on the classification schemes by Ruxton and Berry (1957), International Society for Rock Mechanics (1978) and Geological Society of London's Engineering Group Working Party (1990). (Modified from Migon (2010).)

Scheme of Ruxton and Berry (1957)			Scheme of GSL (1990) and ISRM (1978)		
Description	Formal name	Zone no.	Zone no.	Formal name	Description
Structureless sand–clay mixture, ≤30% clay, quartz and kaolin dominant	Residual debris	I	VI	Residual soil	Rock completely converted to soil; no evidence of original structure and fabric; increase in volume and porosity
Rounded corestones are <10% of area with <5% clay	Residual debris with corestones	IIa	V	Completely weathered	All rock decomposed and disintegrated but original structure and fabric intact
Original rock structure largely preserved, while corestones occupy 10–50% of area		IIb	IV	Highly weathered	>50% of rock decomposed and disintegrated; fresh rock occasionally present as corestones
Dominance of rectangular corestones (50–90%) surrounded by residual matrix	Corestones with residual debris	III	III	Moderately weathered	<50% of rock decomposed and disintegrated; rock discoloured and partly weakened; corestones may be common.
>90% is intact rock, minor amounts of debris along fractures, iron staining may be present	Partially weathered rock	IV	II	Slightly weathered	Discolouration is the main sign of change, mainly along discontinuities, some weakening of fabric
No visible sign of rock weathering	Bedrock		IB	Faintly weathered	Slight discolouration along major discontinuities
			IA	Fresh	No visible sign of weathering

conductivity of the rock-slope materials, thereby increasing their susceptibility to slope failure and landsliding.

While deep weathering may be a special feature of tropical climates, it is not restricted to them. Ollier (2010) points to the granites of Sweden and southwest England for evidence of such weathering in the form of weathered joint sets and the formation of the "china clay" deposits of Devon and Cornwall, respectively. In such humid climates, rainfall accelerates the rate of weathering at the ground surface and in open joints by continually removing the product of weathering, thus forcing geochemical reactions to continue operating in one direction. Rainwater is slightly acidic (pH 5.6) because it dissolves atmospheric carbon dioxide: $CO_2 + H_2O = H_2CO_3$. In addition, plant respiration adds CO_2 to the soil atmosphere and humic acids are also formed, causing the weathering of soil minerals to produce plant nutrients, phosphorus, calcium, potassium, etc., all of which are necessary for plant growth. Water acts as the solvent in these biogeochemical processes in which minerals are weathered – i.e., converted to simpler minerals such as clays and oxides and to dissolved ions – and pore space is increased due to the mass loss of

ions. We consider geochemical weathering in detail in the next chapter.

In the cold climates of high latitudes and in mountainous areas everywhere, these processes are much retarded because of lower temperatures and lower water fluxes through the soil profile. In these environments, the cyclic process of freeze and thaw of water in rock joints accelerates the breakdown of rocks through the expansion of water upon freezing (~11% per unit volume) that penetrates deeper into joints as the rock progressively degrades. The jointing shown in Figure 6.7 is likely a consequence of such freeze–thaw cycles as well as thermal expansion and contraction of the rock by the daily progress of solar radiative heating about each particular rock face.

While thermal stresses cause diurnal expansion and contraction of rock, depending on the angle of the Sun, cold temperatures require the action of freezing water in cracks within rocks to cause strain. Such **frost cracking** is due to the migration of liquid water towards a freezing front at temperatures in the range of −3°C to −10°C. Once in this frost window, the water molecules freeze and expand in the

Figure 6.7 Jointing in Precambrian granites, Colorado, USA.

Figure 6.8 Convex topography and frost action, Rocky Mountain National Park, Colorado, USA.

shelter of microcracks. Anderson and Anderson (2010) note that most rock strain of this kind occurs due to the annual frost cycle that provides sufficient time for water migration as thin films.

Extremely cold climates often produce a patterned landscape of shattered rock with creep and solifluction leading to slopes that are convex in shape with rounded ridge crests, such as that shown in Figure 6.8. **Solifluction** was defined by Verhoogen et al. (1970) as "the slow downslope sludging over a frozen substratum of the surface soil of arctic and alpine regions, rendered liquid during the annual thaw by its high water content and disaggregated clay particles". This mobile regolith may be expected to undergo slope movements of centimetres to metres per year.

In arid climates, the rock type has a profound influence on weathering, as was shown in Figure 1.3, resulting in bedrock outcrops of weathering-resistant rocks. Vegetation is sparse and slopes tend to be linear and marked by complex gullies. Desert pavements develop due to the erosion and transport of fine-grained soils by wind action known as **eolian deflation**. Such pavements are predominantly composed of pebble-sized gravels, while only ∼20% of the Earth's deserts are covered with sand. Arid soils are characterized by a surface of **calcrete** or "caliche" that yields a white to reddish-brown colour indicating the precipitation of calcium carbonate and other soluble minerals (Walker, 2000).

6.2.3 Mechanical Properties of Weathered Rocks

The overall effect of weathering of rock is to fragment it and thus increase the total surface area and porosity of the original mass of rock, which aids in further weathering. This fragmentation is coupled with changes in rock strength, volume and moisture content as shown in Table 6.3 for measurements of rock properties from the weathering of granites from southwest England. Note that, with increasing elevation and weathering grade, the uniaxial compressive strength decreases steadily from 262 MPa for fresh, unweathered rock to 5 MPa for the residual soil. The results demonstrate that a cubic metre of fresh rock has a moisture content of 1% of a similar volume of the most weathered material derived from it and a substantially higher saturated bulk density.

Thuro and Scholz (2004) undertook an analysis of granites of the Königshainer Berge in southeastern Germany following tunnelling in 1995–97. This 3.3 km autobahn tunnel was driven through mica granites that had been subjected to hydrothermal alteration (Hecht et al., 1999) after emplacement as plutons at 300 Ma. Hydrothermal fluids dissolved much quartz from the rock, thus increasing its porosity, which was subsequently filled with illite and chlorite clays. Much later, ∼65 Ma, the granites were exposed to weathering that promoted mechanical disintegration and geochemical decomposition. The uniaxial compressive strength measured by point-load testing varied from 120–250 MPa for grade II slightly weathered rock to 0.5–25 MPa for grade IV highly weathered rock. The porosity was 1–2.5% for grade II and 5–13% for grade IV rock using the GSL/ISRM weathering scheme.

Table 6.3 Properties of a weathered granite from southwest England. (Modified from Smith and Collis (2001).)

Description of rock sample	Weathering grade	Saturated uniaxial compressive strength (MPa)	Saturated bulk density (kg/m^3)	Saturation moisture content (%)
Weakly cemented soil (coreable)	V	5	2240	10.0
Core of block	IV	26	2440	4.13
Core of block	III	46	2550	1.97
Completely stained block	II	105	2560	1.52
Whole sample, 90% stained	II	163	2580	1.09
Stained rim of block	II	232	2620	0.35
Fresh	I	262	2610	0.11

6.3 Weathering and Slope Movement

6.3.1 The Failure of Slopes

Landslides threaten life and property and challenge the geotechnical engineer's professional responsibility to protect public welfare. We will begin this section by discussing how weathering affects slope movements as a preparation for a lengthier study of landslides in Chapter 15.

Geotechnical engineers (e.g., Coduto, 1999: Briaud, 2013) regard weathered soil materials as subject to the Mohr–Coulomb failure criterion as was previously discussed for rocks (i.e., section 1.3.2). Using effective stresses, the shear strength (s) of a residual soil was expressed in (1.13) in terms of the effective normal stress on the plane of failure (i.e., σ_e):

$$s = c' + \sigma_e \tan \phi', \qquad (6.1)$$

where s is in units of kPa (sometimes written as τ_f, the shear stress at failure), c' is the effective cohesion and $\tan \phi'$ is the effective friction angle. Box 6.2 presents measurements of these parameters observed due to slope failure in weathered rocks.

Deere and Patton (1971) presented an analysis of the effect of weathering on the shear strength of a relict joint in an inclined weathered rock zone, shown in Figure 6.9. Figure 6.9(a) presents a weathered profile of rock that uses the weathering scheme of Ruxton and Berry (1957), i.e., the residual soil is grade I, while Figure 6.9(b) shows the shear strength envelope for three samples collected from the joint. Sample III is the least weathered and its shear strength exhibits the highest angle of internal friction (ϕ). Samples II and I were collected from increasingly weathered parts of the profile and

(a)

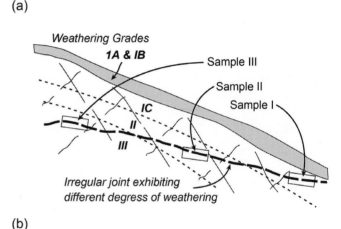

(b)

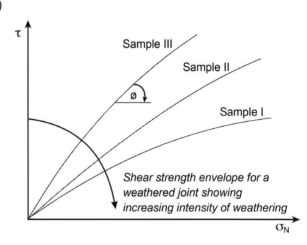

Figure 6.9 Effect of weathering on the shear strength of a irregular joint. (a) Note that the weathering grades employ the scheme of Ruxton and Berry (1957). (b) Mohr–Coulomb failure envelope for the three samples identified in (a). (After Deere and Patton (1971); with permission from ASCE.)

BOX 6.2   SHEAR STRENGTH OF RESIDUAL SOILS AND WEATHERED ROCKS

Relict structures were noted by Deere and Patton (1971) as being responsible for slope failures in residual soils or saprolite that developed on granites in North Carolina. They noted that joints formed planes of weakness and, where the joints dipped towards the plane of excavation, slope failure was often observed. Vertical joints often were observed to contain clay that had been transported from the overlying residual soil and blocked drainage through the fractures, thus causing an increase in head and a greater potential for slope failure.

Also manganese and iron oxide coatings of fractures are evident on many joint planes in the weathered granites of the Piedmont of the southeastern USA and of Hong Kong. The $Fe(\text{III})$ and $Mn(\text{IV})$ ions may be produced by (1) the oxidation of sulfide minerals (e.g., FeS_2, MnS) and their subsequent precipitation as amorphous or semicrystalline oxide coatings or (2) the dissolution of primary oxides (e.g., Fe_3O_4, MnO_2) and their subsequent precipitation as oxide coatings. The shear strengths of such secondary minerals are much less than that of the saprolite and often are the locus of slope failure.

The list below presents results of the Mohr–Coulomb strength parameters for a variety of weathered rocks and residual soils gleaned from Deere and Patton (1971) review of reports. The shear strength parameters are given for both undrained conditions (c = cohesion; ϕ = friction angle) and drained conditions (c' = effective cohesion, ϕ' = effective friction angle). See section 10.3.2 for details.

- Decomposed micaceous gneiss measured by direct shear tests: $c = 30$ kPa, $\phi = 37°$.
- Partly weathered and strongly fractured mica schists and phyllites: $c = 70$ kPa, $\phi = 35°$, Caracas, Venezuela.
- Residual soil derived from phyllite: $c = 0$ kPa, $\phi = 24°$ when perpendicular to schistosity; and $c = 0$ kPa, $\phi = 18°$ when parallel to schistosity. This orientation was determined to be a critical variable in the failure of residual soils derived from schists, Caracas, Venezuela.
- Decomposed Hong Kong granitic soil: $c = 0$ kPa, when saturated; however the angle of internal friction varies greatly, from $\phi = 25$–$34°$ in fine-grained soils to $\phi = 36$–$38°$ in coarse-grained soils.
- Mercia Mudstone Group (UK): highly weathered with only 2% carbonate, $c' \leq 10$ kPa, $\phi' = 25$–$32°$; unweathered with 20% carbonate, $c' \leq 30$ kPa, $\phi' > 40°$.

Source: From Deere and Patton (1971).

have lower angles of internal friction. Thus the saprolite in grade IC rock has the lowest friction angle and therefore is most prone to slope failure at any particular normal stress.

Consequently, weathering has two principal effects on the failure envelopes: (1) a decrease in the shear strength (s, τ_f) at high normal stresses (σ_N); and (2) a decrease in the normal stress at which the initial mode of failure changes from sliding to shearing. Therefore, attention must be paid to **relict structural features**, such as the joint in Figure 6.9, that persist in the rock slope and are inherited from the unweathered bedrock, e.g., faults, foliation, bedding planes and joints. Box 6.2 presents further details on this subject, which in turn leads us to a fuller consideration of how weathering affects slope stability.

Figure 6.10 shows the morphology of an idealized landslide, which we will identify in Chapter 15 as a complex earth slide–earth flow. The term **earth** is used to identify that slope material is at least 80% composed of sand-, silt- and clay-sized particles. Slope movements of materials larger than earth-sized particles, i.e., larger than 2 mm, are known as **debris** or rock slides. Beneath the crown of the slope is the **main scarp**, which is distinguished by its steep surface from which displacement has occurred. Above it, on the crown, may be tensile fractures referred to as "crown cracks".

Returning to Figure 6.10, further down the slide we encounter the displaced material that reaches down to the **toe** of the slide, which is usually a curved margin of displaced material. The distance from the toe to the main scarp comprises a zone of depletion in the upper reaches of the slide and a zone of accumulation at the foot of the slide. The **surface of rupture** forms the lower boundary of displaced material below the original ground surface; the idealization of this surface in soil mechanics is known as the **slip surface** or **failure surface**. Note the presence throughout

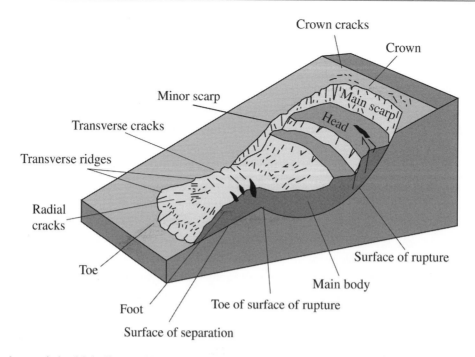

Figure 6.10 Nomenclature of a landslide illustrated by an earth flow. (From Highland (2004); courtesy US Geological Survey.)

the displaced material of transverse cracks or minor scarps that allow infiltration of rain or snowmelt and thus may aid in the reactivation of the slide. The engineer should be aware that scarps are also created by active faulting (see Box 6.3).

Weathered granitic rocks are the scene of frequent slope failures of the kind shown in Figure 6.12. Migon (2010) has pointed out that the early stages of weathering of such rocks (zone IV of Ruxton and Berry (1957) scheme in Table 6.2) involve the opening of large fractures due to tensile stresses and geochemical alteration along such fractures. These processes result in rock slides and rock falls. Debris slides occur in zone III or moderately weathered materials on relatively shallow slopes. In completely weathered rock (zones I and IIa), where few corestones remain and there are significant amounts of clay, rotational slides and earth flows are common because of the absence of relict structures such as fracture planes.

We shall now consider how such parameters vary in (a) the weathered clays and mudrocks of the UK, (b) the weathered igneous rocks of Hong Kong and (c) the rocks that form the flanks of active volcanoes.

6.3.2 Clays and Shales

Over the seasons, in stiff clays and shales, the pore pressure within a slope will vary in a cyclic pattern – high in the rainy season or during snowmelt and low in the dry season. This pore-pressure cycling is hypothesized to cause a weakening of the slope material (fatigue) due to both mechanical and chemical processes (Picarelli et al., 2004; Picarelli and Di Maio, 2010), such as vertical fissures, cracks, formation of thin shear bands, i.e., weathering. In this context, rainfall and snowmelt are the triggers for slope failure; however, the predisposing factor is the general weakening of the clay or shale by cyclic pore-pressure changes.

An example of this deterioration is in the shear strength of clays that was measured for the London Clay by Chandler and Apted (1988). London Clays were formed during the Eocene epoch and have been exposed due to considerable erosion ($\geq 150\,m$) such that the overconsolidation ratio is 20–40. Weathering, in the form of fracturing to 5–10 m depth and oxidation to a yellowish brown colour, has weakened the clay such that the cohesion of samples from partially and fully weathered zones ($c' = 8\text{–}18\,kPa$) is much less than was measured on unweathered samples ($c' = 28\,kPa$).

This comparison assumes no change in the friction angle between these two sets of results for London Clay, which is usually taken to be $\phi' = 20°$. Chandler and Apted (1988) concluded that the geotechnical expression of the weathering of the London Clay (and the Mercia mudrocks of the English Midlands) is as a reduction in cohesion and thus weathering contributes to slope failure in clays, mudrocks and shales. The weathering of the London Clay and its geotechnical properties were discussed in more detail by Chandler (2000), where he indicates the importance of topography and prior water-table elevations on the depth of weathering.

BOX 6.3   SCARPS: FROM LANDSLIDES OR ACTIVE FAULTING?

A controversy raged in northern California for six years through to 1983 concerning the safety of the first commercial US nuclear reactor licensed for the production of medical diagnostic isotopes. The controversy was sparked by a conclusion of the US Geological Survey (USGS) in 1978 that the nuclear reactor was sited on or near an active fault system. Meehan's (1984) description of this controversy is worth the reading of anyone who is interested in the applied earth sciences and in the nature of geoscientific evidence in an adversarial hearing.

Intensive investigations were undertaken by General Electric, the reactor operator, and various geological agencies. Three slip surfaces were identified, one less than 100 m from the reactor. The USGS thought that these slip surfaces were indicators of active faults. General Electric's consultants disagreed and thought them to indicate failure surfaces of landslides, not of faults. The US Nuclear Regulatory Commission's decision of 1983 resulted in the shutdown of the nuclear reactor. Hart et al. (2012) explain why slope failure was more likely than recent faulting at this reactor site and others in the western USA, including that shown in Figure 6.11.

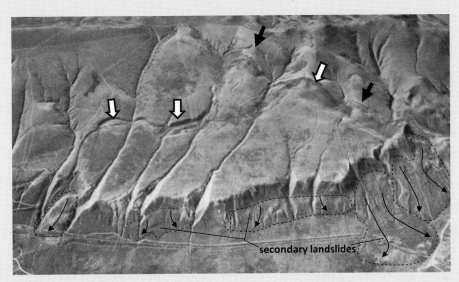

Figure 6.11 The Smyrna Bench, part of the Saddle Mountains, Washington state, exhibits some prominent scarps that were originally identified in 1970 as landslide scarps then reclassified in 1996 as fault scarps before being most recently identified as indicating a translational landslide (Hart et al., 2012). The bold black arrows indicate the evidence that suggested an active fault trace to geologists in 1996. The bold white arrows indicate the trace of the head-scarp of the landslide. (Google Earth image; reproduced with the permission of the Association of Environmental & Engineering Geologists.)

Consequently, it is necessary to issue a warning about the similarity of scarps from landslides and those that originate with earthquakes, such as those shown in Figure 14.1. It has been observed in areas with active faults that geomorphic patterns produced by seismic faults are very similar to those produced by landslides. Indeed, large landslides and faults can occupy the same location and landslides may be triggered by faults (see Chapter 15).

Hart et al. (2012) argue that, in such cases, it is critical that the engineers and geoscientists involved in the review of the site location consider four lessons learned:

1. What is the geologic context? Does the area or the formation have a history of slope failure? Does the suspected fault have a length, orientation and age consistent with the local tectonic setting?
2. It is important to develop and test multiple working hypotheses that are consistent with the field information and the geological context of the site.
3. Large-scale landslides produce geomorphic and geological features that are similar to those produced by tectonic events.
4. It is critical that perspective be maintained on the ultimate use or significance of the geological interpretation.

Source: From Hart et al. (2012).

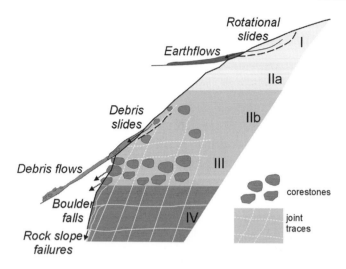

Figure 6.12 Weathering grade and mass movement relationships in basement rocks. (From Migon (2010); with permission of The Geological Society of London.)

6.3.3 Igneous Rocks

The residual soils of Hong Kong are subject to intense tropical storms leading to continual problems with slope failures in the region. Hong Kong suffered several large landslides in the 1990s in areas in which kaolin-rich zones were present in the failed slopes. This prompted an assessment of the problem by Campbell and Parry (2002). The kaolin develops from the alteration of feldspars and biotite, which are both common in the igneous rocks of Hong Kong (see Chapter 7). Kaolin and manganese oxides are commonly observed in both volcanic and granitic saprolite in the hillsides of Hong Kong.

Figure 6.13 shows veins of kaolin beneath the failure surface in a trench dug into the Fei Ngo Sham Road landslide of June 1998, which involved 2500 m³ of material with the surface of rupture being 5 m below grade. Campbell and Parry (2002) noted that the abundance of kaolin in the slope probably reduced the mass strength of the saprolite. The weathered rock in this area is tuff with interbedded siltstone, most likely underlain by granite. As shown in Figure 6.13, the kaolin veins dip into the slope of the saprolite and are of two kinds. The first is a three-dimensional lattice-like network in which the veins are closely spaced and referred to in Figure 6.13 as a **stockwork**. The second is kaolin-infilled **relict joints**. It is noticeable that the veins dip into the slope at an angle of ~43°.

Similar failures in other kaolin-rich slopes have occurred across Hong Kong. Campbell and Parry (2002) indicated that most kaolin veins are millimetres to hundreds of millimetres thick and may be continuous for up to 60 m. A summary of the peak shear strength measurements by undrained triaxial testing for the saprolite indicate $c' = 5\text{--}8$ kPa, while $\phi' = 33\text{--}38°$.

Table 6.4 Properties of rocks from the summit of Mount Rainier. (From Zimbelman et al. (2003).)

Alteration intensity	Friction angle range (°)	Unconfined compressive strength (MPa)	Bulk density (kg/m³)
0–1	38–46	85–207	2.5
2	30–40	40–120	2.4
3	22–28	28–56	2.2
4	16–23	8–21	1.1–1.7

Jiao et al. (2005) remarked that the fine-grained texture of the saprolite and its vertical distribution on hillsides in Hong Kong allows it to act as an aquitard for underlying confined groundwater; thus the presence of a kaolin-rich saprolite allows hydraulic heads to build up deeper in the profile, enhancing the effects of pore-pressure changes. Figure 10.14 illustrates the texture of poorly sorted residual soil that develops on granite. Hencher and Lee (2010) have written an interesting account of the hydrogeological processes that account for the landslides of Hong Kong.

6.3.4 Volcanic Rocks

A general principle of geochemical weathering (see Chapter 7) is that it is most notable in materials that were formed furthest from equilibrium with the temperature and pressure of the Earth's surface. Quite obviously, rocks forming the flanks of volcanic summits qualify as being in a state of disequilibrium.

Zimbelman et al. (2003, 2004) have studied the mechanical properties of volcanic rocks, in particular the altered rocks, faults and fractures within the summit regions that possess the highest potential energy for catastrophic volcanic collapses. Their data for rock strengths, friction angles and densities are presented as a function of the alteration intensity in Table 6.4. The alteration intensity reflects the degree of hydrothermal alteration of these volcanic rocks towards clay minerals. This topic of failure of volcanic slopes will be revisited in Chapter 15 when volcanic triggering of landslides is considered.

6.4 Alluvial Channels

We begin this section by defining two new parameters that are critical to understanding streamflow, sediment transport and bank erosion: bed shear stress and stream power. We

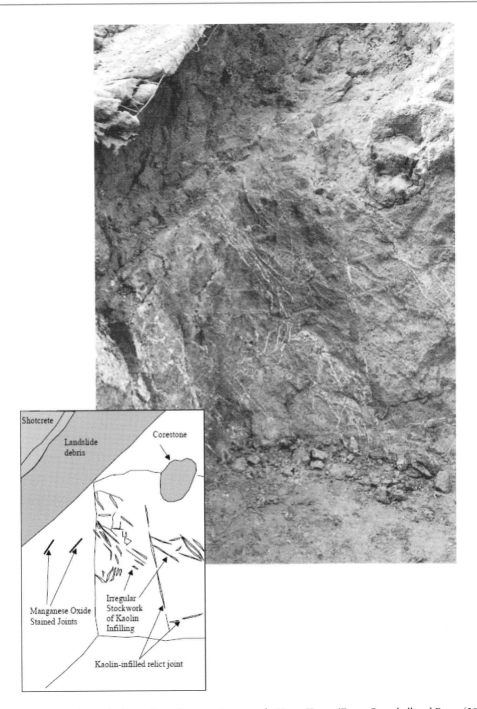

Figure 6.13 Pattern of kaolin veins beneath the surface of rupture in a trench, Hong Kong. (From Campbell and Parry (2002).)

shall confine our discussion to alluvial channels, that is those channels that flow through sediments that were previously deposited by the same channel or its predecessors and which (sediments) it is capable of remobilizing. Our discussion then turns to channel properties, then channel morphology, then channel switching or avulsion and finally to paleoflood hydrology. We shall return in Chapter 9 for a fuller discussion of fluvial processes and sedimentation.

6.4.1 Stream Power

The **bed shear stress** is that stress required to entrain a sediment particle of maximum diameter, D, by a particular flow. The drag force that entrains the sediment particle may be simplified as follows (Robert, 2003):

$$F_d = \rho g \, vol \, S_f, \tag{6.2}$$

where ρ is the particle density, g is the acceleration due to gravity, and *vol* is the volume of water entraining the particle along a river of friction slope S_f. The bed shear stress is the force per unit area that the volume of water of depth z exerts on the bed, i.e.,

$$F_d/A = \tau_0 = \rho g z S_f. \tag{6.3}$$

The interested reader is referred to Julien (2002, pp. 25–29), who develops this equation from the equations of motion.

A commonly used parameter by geomorphologists is **stream power**, which represents the rate of transporting or eroding sediment (Robert, 2003). The stream power per unit bed area is

$$\omega = \tau_0 V, \tag{6.4}$$

where V is the average stream velocity. The units of stream power are watts, those of the rate of doing work, or force $\times$ distance per unit time = watts (1 W = 1 J/s). Questions 6.4 and 6.5 address the development of this equation and a similar equation for stream power per unit length of the channel.

6.4.2 Hydraulic Geometry of Channels

The general properties of a drainage basin and the stream channels contained within it are represented in Figure 6.14, in which the parameters are shown varying from upland to floodplain. In the upland valley there is a net erosion of soil material that is transported down-valley and deposited in the floodplain or carried out to sea. The bed material grain size and channel gradient decline coherently down-valley, while streamflow and stored sediment increase. Stream power – roughly streamflow $\times$ channel gradient – typically reaches a maximum in the upland valley.

Most of the drainage basin comprises upland areas that provide runoff and sediment to the lower basin. In the upland areas, the adjacent hillslopes are coupled to the streams by virtue of the streams' capacity to immediately remove sediment due to the high channel gradient and stream power. In the floodplain of the basin, sediment deposition occurs and the volume of stored sediment increases rapidly. The hillslopes of the floodplains are isolated or uncoupled from the stream and may deposit their sediment as alluvial fans on the borders of valleys. The stream channels themselves may erode the stored sediment in the channel banks and the streambed and then transport it for subsequent deposition in marine deltas or further offshore (see Chapter 9).

Church (2002) suggests that stream channels function as "sediment sorting machines" in that sediment size varies systematically along the length of the drainage basin. Table 6.5

(a)

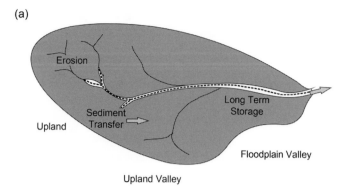

(b)

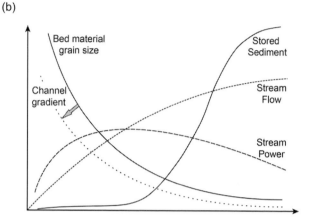

Figure 6.14 Variation in channel properties within a drainage basin. (From Church (2002).)

illustrates size classes commonly used to categorize stream sediments; a more comprehensive classification can be found in Figure 10.12. In the upland areas where there is a large supply of sediment from adjacent hillslopes, all but the largest particle sizes are transported downstream. In the floodplain, however, the **competence** of the stream to entrain and transport sediment of larger sizes decreases and suspended and bed sediments become smaller.

The streamflow of a river varies with the hydraulic characteristics of any particular cross-section of the stream. Leopold and Maddock (1953) of the US Geological Survey referred to these relationships as the **hydraulic geometry** of the stream channel. As streamflow discharge (Q) increases downstream, the channel characteristics vary according to the power functions:

$$w = aQ^b,$$
$$z = cQ^f,$$
$$V = kQ^m,$$
$$S = gQ^z,$$

with

$$w \times d \times V = Q = aQ^b \times cQ^f \times kQ^m,$$
$$b + f + m = 1,$$
$$a \times c \times k = 1.0,$$

Table 6.5 Grain-size scales and conversion table. (After Church (2003).)

US sieve	Size		Size class
mesh no.	(mm)	(μm)	(Wentworth scale)
	4096		Very large boulder
	1024		Medium boulders
	256		Large cobble
	64		Very coarse gravel
	32		Coarse gravel
	16		Medium gravel
	8		Fine gravel
5	4		Granules
10	2	2000	Very coarse sand
18	1	1000	Coarse sand
35	0.5	500	Medium sand
60	0.25	250	Fine sand
120	0.125	125	Very fine sand
230	0.0625	62.5	Coarse silt
	0.0312	31.2	Medium silt
		15.6	Fine silt
		7.8	Very fine silt
		3.9	Clay

where w = width, z = depth, V = velocity, Q = discharge, b, f and m are exponents, and a, c and k are coefficients. Robert (2003) suggested that characteristic values of the exponents are $b = 0.50$, $f = 0.40$ and $m = 0.1$, implying that channel widths increase at a rate greater than that of channel depths with increasing distance downgradient.

6.4.3 Channel Morphology

Previously we defined stream competence as the ability of a stream to entrain and transport sediment of a particular size. We shall pursue this issue in greater detail in Chapter 9; however, it should be apparent that the competence of a stream is closely related to the bed shear stress it exerts and the associated stream power. The morphology of channel bedforms is governed not only by competence but also by sediment supply and is described using terms such as straight, meandering and braided; channels that are actively depositing sediment are considered **aggrading** while those with a net sediment loss, due to erosion, are termed **degrading**. Figure 6.15 illustrates the various domains of river channel morphology that may be observed as a function of channel gradient and stream discharge.

Similarly, Figure 6.16 shows the morphological characteristics of alluvial river channels and their controlling variables. Fluvial sediment transport is strongly linked to the morphology of the stream channels, in particular their hydraulic gradient and their sediment supply. Church (1992) uses the terms **bed material** to describe the sediment that forms the bed and lower banks of an alluvial channel and largely governs the morphology of that channel and **wash material** as that sediment that, having been entrained, is transported in

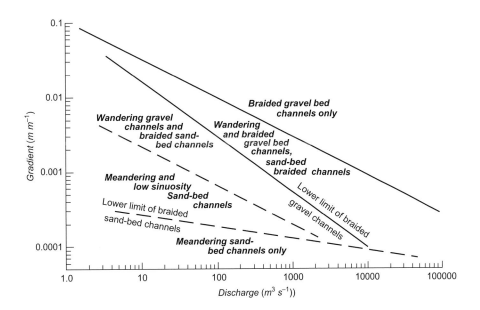

Figure 6.15 Gradient versus discharge relationship. (From Church (2002).)

Decreasing Channel Stability ⇨

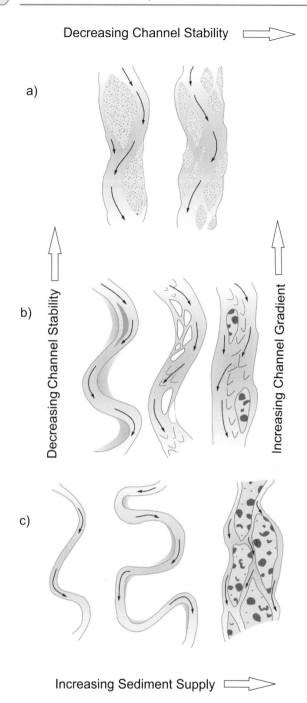

Increasing Sediment Supply ⇨

Figure 6.16 Alluvial river channel forms and their principal governing factors: (a) bed-material-supply-dominated channels; (b) wandering and braided channels; (c) wash-material-dominated channels. Large stable sediment bars show development of vegetation patterns. (Modified from Church (1992); reproduced with the permission of John Wiley and Sons.)

suspension. The patterns in Figure 6.16(a) are of channel forms that are dominated by bed material (high channel gradients), whereas the patterns in Figure 6.16(c) are of forms characterized by wash material (low channel gradients). Meandering and braided channels lie somewhere in between

these two end members. As sediment supply increases, channels become less stable and develop meandering and braided patterns; in-stream sediment storage in the form of bars becomes increasingly evident.

6.4.4 Avulsion

One feature of aggrading stream channels and their floodplains that is of considerable engineering and social importance is that of channel diversion, known as **avulsion**. Slingerland and Smith (2004) define avulsion as "the natural process by which flow diverts out of an established river channel into a new permanent course on the adjacent floodplain". They recount how avulsions of the Yellow River in China have occurred seven times in its recorded 2000-year history, each time with great loss of life and crop destruction. Leeder (1999) notes that the frequency of a river's avulsion may be as little as ten years or as much as 6000 years "with no clear underlying control". In fact, as long as the river is aggrading, avulsions may be quite frequent as is the case with braided glacial outwash streams and alluvial fans. The trigger for an avulsion event is most likely to be a flood but the event may also be triggered by earthquakes, ice jams, log jams, beaver dams or migration of bars that temporarily block the throat of a branched channel (Slingerland and Smith, 2004).

Figure 6.17 shows the avulsion of the Saskatchewan River in western Canada in the 1870s that occurred over a period of about ten years (Smith et al., 1998); this gradual avulsion contrasts with the sudden and catastrophic avulsions of the Yellow River. The floodplain here is known as the Cumberland Marshes, which is an area of anastomosing channels (see Figure 6.16), shallow lakes and wetlands. As with many avulsions, the old channel of the river was lined with natural levees, the crests of which were elevated above the adjacent floodplain, probably by 3–4 m as would occur with the continuing aggradation of the river over several thousand years.

The chronology of fluvial events that occurred prior to, during and following the avulsion are illustrated by the five panels of Figure 6.17; the age dates were determined by radiocarbon dating of peat samples from channel cutbanks and auger boreholes. In panel (a), the Ancestral South Angling (ASA) Channel is shown approximately 2000 years before present (BP) at which time it was no longer the main channel of the Saskatchewan River, which was off the map and to the south. One thousand years later (see panel (b)), the river follows a new course that is now known as the Old Channel.

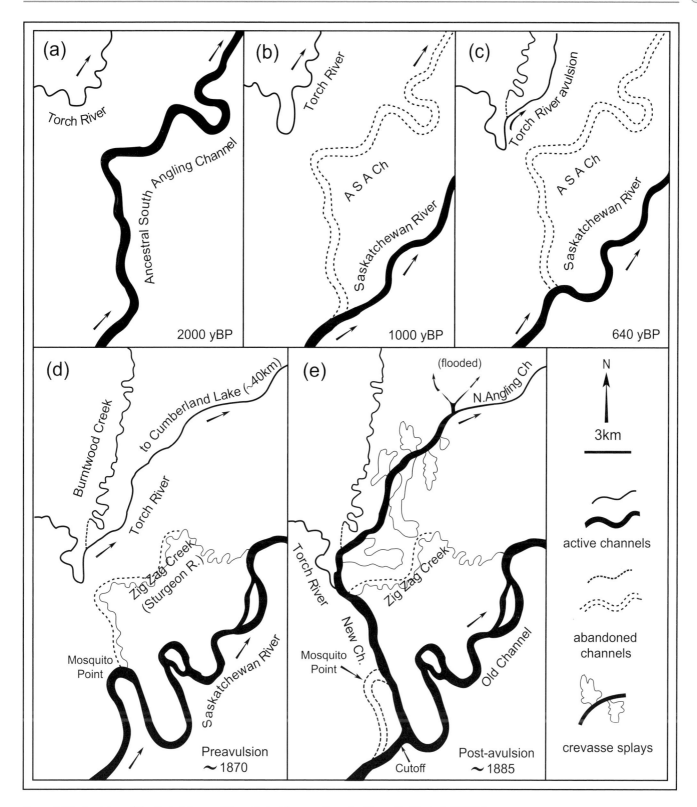

Figure 6.17 Avulsion of the Saskatchewan River, Saskatchewan, Canada. (From Smith et al. (1998).)

By 640 years BP, the small Torch River to the north had formed a new channel aligned with and only 1 km from the ASA channel (see panel (c)). Shortly before the avulsion occurred, the meander at Mosquito Point (see panel (d)) migrated northwards, and the old ASA channel reconnected with the Saskatchewan River as well as with the Torch River. The avulsion then employed these existing conduits to deepen and widen the path of the redirected Saskatchewan River to the north (panel (e)) as the Torch channel captured the river. Today only 5% of the annual flow of the river passes down the Old Channel shown in Figure 6.17(e).

6.4.5 Paleoflood Hydrology

Avulsion is thought to be most frequently triggered by floods. Anderson and Anderson (2010) point out that the geomorphic effects of large floods can dominate the appearance of landscape for centuries or even tens of thousands of years. Consequently the prediction of the peak discharge and frequency of large floods is of paramount importance to civil engineers.

However, the estimation of the peak discharge of a flood with a return period of once every hundred years – the 100-year flood or one with an **annual exceedance probability** (AEP) of 10^{-2}/yr – is usually beset with problems of inadequate hydrologic records that can provide a statistically reliable sample size from which to estimate this flood; i.e., each sample would be a 100-yr hydrograph, and few rivers have more than a few hundred years of record, especially in North and South America. But by combining geomorphic evidence of the highest floodwater elevation (stage) in the

floodplain with the principles of open-channel hydraulics, it has been possible to provide estimates of peak discharge in drainage basins; this discipline is known as paleoflood hydrology.

Paleoflood hydrology, as it has developed since the 1970s (see House et al., 2002), identified **slackwater flood deposits** as the primary indicator of a historically high stage of a river, i.e., a **paleostage indicator**. Such deposits are believed to have formed in eddies at the mouths of tributaries of rivers that caused sedimentation of sands and silts in topographically elevated locales relative to the present position of such rivers. If undisturbed by even larger floods, there is a reasonable likelihood that pedogenic processes will create a soil horizon within the slackwater flood deposit that can be age dated using radiocarbon. Thus an approximate age of the paleoflood can be estimated as well as the paleostage, i.e., the maximum flood stage.

Figure 6.18 presents the various types of geomorphic and stratigraphic evidence for paleofloods, in particular the identification of a stable terrace whose soils can be radiocarbon dated and thus provide an upper age limit to an elevation that has not been eroded by floodwaters (Levish, 2002; O'Connell et al., 2002). Using hydraulic models of bedrock channels

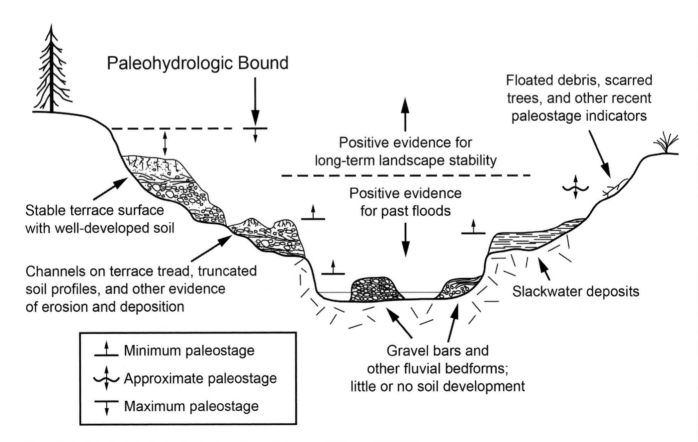

Figure 6.18 Paleohydrologic flood indicators. (From Ostenaa and O'Connell (2005).)

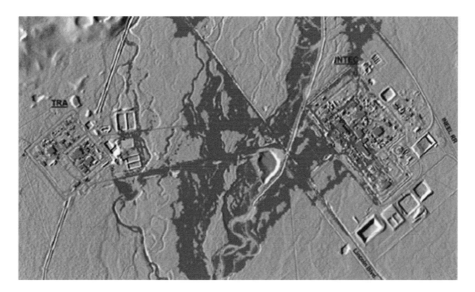

Figure 6.19 Potential inundation of the Idaho National Laboratory by the Big Lost River determined by paleoflood hydrologic analysis. (From Ostenaa and O'Connell (2005).)

and estimates of bed roughness described by Manning's coefficient, the paleoflood hydrologist employs the same tools as used by engineering hydrologists to estimate peak flood discharges, e.g., the step-backwater method.

Despite early skepticism from the hydrologic engineering community that is commonplace with new and somewhat exotic techniques introduced into well-established fields, paleoflood hydrology gained acceptance because it successfully addressed a particularly critical issue for the US Bureau of Reclamation (USBR) in the 1990s. The USBR needed to understand the peak discharge of low-probability floods at its dams in the western USA (Levish, 2002); this discharge from a probable maximum rainfall was identified as the **probable maximum flood** (PMF). Many of the USBR dams were unable to accommodate the estimated PMF for their watershed and thus were identified as in need of costly retrofitting to avoid dam failure. However, it was determined (Harris and Feldman, 2002) that "recent paleoflood evidence in the western United States indicates that the largest floods occurring in the past 10 000 years are significantly smaller than PMF estimates". (The importance of the period of 10 000 years is that it excludes glacial meltwater episodes that are no longer relevant because of glacial retreat from the watersheds involved; glacial meltwater floods were of a much greater discharge, see Chapter 8.)

Evidence of the utility of paleoflood hydrology was reported by Ostenaa and O'Connell (2005) for the Big Lost River in Idaho, USA, to determine its potential for inundation of research facilities, including nuclear reactors, at the Idaho National Laboratory (INL). Using only stream-gauge data

from the Big Lost River and the surrounding region, the US Geological Survey estimated the discharge of a flood with a 100-year return period (AEP = 10^{-2}/yr) as 106 m^3/s. However, Ostenaa and O'Connell (2005) undertook geomorphic mapping, which was supported by trenching and radiocarbon dating of organic samples at several localities, and two-dimensional hydraulic modelling based on accurate topographic maps of the drainage basin. They based their results on probability density functions for bed shear stress and stream power for the channel reach through the INL site in order to allow the erosion of surficial loess and stream-bank deposits. Their approach yielded a mean peak discharge for the 100-year flood of 87 m^3/s.

Figure 6.19 is a schematic of their predicted inundation of the INL by the Big Lost River and depicts a flow rate near 106 m^3/s with allowance for the effects of culverts and infiltration. The Test Reactor Area (TRA) and the Idaho Nuclear Technology and Engineering Center (INTEC) are shown not to be inundated by the predicted flood.

6.5 Terrain Analysis

Terrain analysis is the process of interpreting stereographic images of landscapes to undertake site assessment or to determine the potential for damage to human life and infrastructure arising from natural and man-made geohazards. Traditionally hazard-mapping tasks, such as mapping landslides and active fault zones and their proximity to highways, pipelines and other infrastructure, have been conducted by air-photo analysis and surficial geology maps. For project

Figure 6.20 Digital elevation model of the Zymoetz landslide, near Terrace, British Columbia, Canada. The landslide was initiated by a rock avalanche of one million cubic metres of bedrock in the headwaters of a tributary of the Zymoetz River. The long-runout (4 km) debris flow ruptured a gas pipeline (dashed line in valley) and created a fan deposit. See Boultbee et al. (2006) for an analysis of the landslide. (LiDAR image from Geertsema and Clague (2011); with permission.)

work, the scales involved in analyzing air photos (1 : 30 000) or maps (1 : 250 000) may be insufficient for site assessment or project design. Consequently, with the introduction of high-resolution digital photogrammetry, there is increasing use of softcopy mapping methods for these purposes. Now investigations can use structure from motion (SfM) techniques aided by unmanned aerial vehicles (see section 5.7) or softcopy mapping in conjunction with **LiDAR** (light detection and ranging) or optical satellite imagery to develop a **digital elevation model** (DEM). The many technical aspects of this topic are reviewed by Griffiths (2017).

The DEM can be viewed from vertically above as a digital raster image or as a three-dimensional (3D) surface viewed from some perspective angle, a distinct advantage it holds over topographic maps. Geertsema and Clague (2011) discuss the development of DEMs for the transportation corridors through the mountain ranges of British Columbia in western Canada that accommodate many existing and planned natural-gas pipelines connecting gas fields in the western Canada sedimentary basin to tidewater ports on the Pacific coast of British Columbia. Figure 6.20 is a DEM of the 2002 Zymoetz rock slide–debris flow, near Terrace, British Columbia, described in detail by Boultbee et al. (2006). The landslide travelled 4.3 km down the tributary valley shown and ruptured the gas pipeline (indicated by the dashed line in Figure 6.20) near the Zymoetz River. Typically, only a 1 km wide pipeline corridor is studied for slope stability, i.e., 500 m on either side of the proposed pipeline.

6.5.1 Air-Photo Analysis

Terrain analysis was first practiced during World War II for investigation of military targets. Post-war applications included studies of fracture lineaments to identify high-yield wells in carbonate aquifers (Lattman and Parizek, 1964) or to assess landslides and dam-site foundations (Mollard, 1988). The technique involves pairs of air photos that overlap by showing ~60% of the same area on each air photo. The overlap produces a stereoscopic or 3D perspective of the common area when viewed through a stereoscope as shown in Figure 6.21. This figure shows a stereo pair being viewed for a desktop aggregate study using an inexpensive stereoscope, shown in the inset. The accompanying photo shows its ready use in the field. Stereoscopes come in various sizes and cost from $25–80 for a pocket instrument to about US$2000 for a professional mirror stereoscope.

An exemplary pair of stereo images is shown in Figure 6.22. This stereo pair of air photos is of the Peterborough drumlin field, approximately 100 km northeast of Toronto, Ontario, Canada. These glacial landforms, which may rise from 1 m to 10 m above the adjacent land, show the direction of glacial movement by the crescent-shaped scours (s) and long narrow flutes (f) identified on the stereo pair (Gilbert, 1994). We shall return to discuss the origin of these landforms in Chapter 8.

The application of air-photo analysis to identifying potential landslide crossings of transportation corridors remains a major role of the technique. This is particularly the case

Figure 6.21 Air-photo interpretation for an aggregate study. (Courtesy J.D. Mollard, 2010.)

when repeat air photos are studied of a particular landslide-prone area over the course of tens of years. When non-experts undertake aerial photograph interpretation, Hart et al. (2009) have recommended subdividing the landscape into a number of geomorphic features, e.g., plateaux, incised river channels, river terraces, etc., in order to increase the likelihood of identifying landslides. They stress the importance of field verification of the results. Analysis of vertically oriented stereographic images allows the identification of many important geomorphological features for landslide characterization as shown in Table 6.6.

6.5.2 LiDAR

The most significant development for mapping of terrain features of recent years is the use of airborne or satellite-borne LiDAR yielding high-resolution DEMs based on laser scanning, which generally means pixel dimensions less than 2 m (Crosby et al., 2011). This utilizes a pulsed laser ranging

system mounted in either an aircraft or a satellite with a global positioning system (GPS) to record precise positioning information. The orientation of the aircraft or satellite is monitored by an inertial measurement unit, and the integration of data from this unit, from the GPS and from the laser ranging system, yields millions to billions of 3D coordinates of the ground surface, vegetation and buildings in a particular surveyed area. The LiDAR measurements are produced as shaded relief images that can be interpreted just like air photographs.

One of the earliest adoptions of LiDAR was in the Puget Sound region of Washington state (USA), where local and federal governments collaborated to understand geohazards such as unmapped, active fault zones and landslide-prone areas (Haugerud et al., 2003) following the discovery of a Holocene fault scarp on Bainbridge Island in Puget Sound. This scarp was oriented in the opposite orientation to other scarps of the Seattle fault and led to the identification of two earthquakes approximately 1100 years ago. Furthermore,

Table 6.6 Information from interpreting vertical stereographic images. (Modified from Hart et al. (2009).)

Topic	Information available by interpretive analysis
Topography	Areas of mountains, hilly or hummocky terrain; scarps and cliffs; floodplains; coastline
Land cover	Forest; woodland; grassland; bare ground; highway infrastructure; quarries
Drainage features	Watersheds; lakes, reservoirs, canals; stream networks
Terrain features	Volcanic: lahar tracks, lava fields, crater Landslides: rock falls, debris flows, lateral spreads Eolian: dunes, desert pavement Coastal: cliff face, offshore bar, beach Glacial: eskers, till plains, drumlins, kames Periglacial: patterned ground, pingos, outwash plains Fluvial: meanders, oxbow lakes, alluvial fans Karst: pavement, dolines, caves
Structural features	Fracture lineaments; faults; folds

Figure 6.22 Air-photo stereo pair showing Peterborough drumlin field. (From Sharpe (1994). Reproduced with the permission of Natural Resources Canada, photo A24314-30.)

because LiDAR readily penetrates vegetative cover, the Puget Sound LiDAR Consortium identified twice as many landslides with LiDAR as had been identified by conventional air-photo analysis and also eliminated other suspected landslides from the inventory kept by local government. Figures 6.23 and 6.24 illustrate the use of LiDAR imaging in developing maps that are improvements on those that can be obtained by traditional photogrammetric methods.

6.5.3 Softcopy Mapping

Air photos represent hardcopies; however, geomorphologists increasingly analyze terrain using **softcopy photogrammetry** and specialized 3D glasses so that measurement of terrain features and image inspection is on a computer monitor rather than hardcopy. Such analysis complements LiDAR in that LiDAR provides limited information on soil materials and drainage conditions in upland areas. The principal advantage of softcopy mapping is its ability to enhance the image of an area (provided modern high-resolution digital images are available) by zooming to scales at a higher resolution than the original image (1 : 24 000 or 1 : 40 000), e.g., to a scale of 1 : 2000 or better (O'Leary and Isidoro, 2016). By so doing, terrain features of importance in project design or site assessment may be inspected at high resolution. This improved understanding can lead to better appreciation of geohazards and an improved appreciation of where monitoring instrumentation and boreholes should be sited.

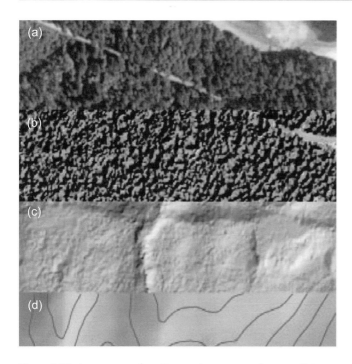

Figure 6.23 A sequence of contiguous image maps from north to south of part of the Puget Lowlands, Washington state, USA, illustrating a 600 m wide area. (a) Digital orthophoto (1 m pixels) showing shoreline of Puget Sound. (b) LiDAR first-return surface (2 m postings) showing continuation of regional road crossing NW to SE through forest. (c) LiDAR bare-earth model after post-processing to segregate ground returns (2 m postings) showing a fault scarp along the east–west direction. (d) US Geological Survey 10 m digital elevation model with 10 ft 1 : 24 000 scale contour intervals, in which the stream location disagrees with the LiDAR image of (c). (From Haugerud et al. (2003), courtesy US Geological Survey; courtesy of The Geological Society of America.)

6.5.4 Satellite Imagery

According to Petley (2012), optical satellite images, while relatively expensive, remain important in developing landslide susceptibility and hazard maps and landslide inventories. The spatial resolution of these images is now equal to conventional air photographs; however, they do not typically provide stereoscopic imagery (the SPOT system being an exception). A further constraint is the need for clear-sky conditions. Table 6.7 presents a summary of the capabilities of some satellite-based remote sensing systems, including Landsat (USA), SPOT (France) and IKONOS (USA). Omitted from this table is Google Earth, which has archived old images as well as new ones, so that landform changes can be detected. Radar-satellite images based on synthetic aperture radar have been used to study ground motion due to earthquakes but not landslides for reasons discussed by Petley (2012).

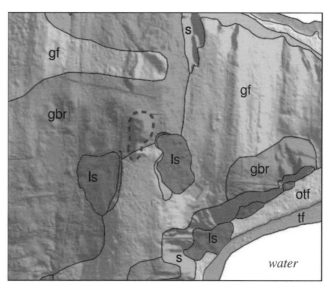

Figure 6.24 A geomorphic map of part of Bainbridge Island, Washington state, USA, based on high-resolution (1–2 m) LiDAR topographic imaging. Symbols: s, scarp; ls, landslides; tf, tidal flat; otf, old tidal flat; gf, fluted glacial surface; gbr, glaciated bedrock surface. Red lines indicate bedding traces; blue dashed line is a relict shoreline from a late-glacial lake. View is 1.5 km wide. (From Haugerud et al. (2003), courtesy US Geological Survey; courtesy of The Geological Society of America.) See colour plate section.

Table 6.7 Satellite imagery resolution and equivalent map scale. (Modified from Hart et al. (2009).)

Satellite and Sensor	Period of service	Resolution (m)	Equivalent map scale for interpretation
Landsat 1	1972–78	80	
MSS			1 : 250 000
Landsat 7	1999–		
ETM+		30	1 : 100 00
Panchromatic		15	
Thermal IR		60	1 : 50 000
SPOT 2, 4	1986		
Multispectral		20	1 : 50 000 to
Panchromatic		10	1 : 25 000
IKONOS	1999		
Multispectral		4	1 : 10 000
Panchromatic		1	1 : 5000

Notes: MSS, Multispectral Scanner; ETM, Enhanced Thematic Mapper; IR, infrared.

6.6 Summary

We have begun this second part of the book with an all-too-brief introduction to the study of terrain evolution and analysis. The reader is encouraged to consult section 6.7 for useful further reading. The energy budget arising from solar radiation is identified as the driving force of terrain evolution after global and local tectonics have played their part. Energy is redistributed on the Earth's surface propelling the hydrologic cycle that in turn promotes rock weathering, soil formation, and sediment erosion and transport, and leads to streamflow generation.

We have identified the nature of the weathered profile of rocks and their geotechnical properties in terms of the Mohr–Coulomb failure parameters – cohesion and the angle of internal friction – and thus have identified the origin of landslides, to which we return in Chapter 15. Measured values of cohesion and the friction angle are reported for clays, shales, and igneous and volcanic rocks.

The bed shear stress and stream power exerted by flowing water in a channel and the hydraulic geometry of a channel were defined. Channel morphology was defined in terms of sediment supply, channel stability and channel gradient. Bed material in the form of gravel was identified as the principal textural component of high-gradient channels, whereas wash material in the form of silts and fine sands dominate low-gradient channels. When the sediment supply was large, channels have changed course into a new or reworked channel – a process known as avulsion. Frequently a flood was the trigger of such change. Slackwater flood deposits and other paleostage indicators have provided information on low-probability flood events known as paleofloods; such extreme events can guide flood control management in floodplains but only where paleostage indicators are identified.

While air-photo analysis continues to provide useful information on landforms and potential threats to life and infrastructure, softcopy mapping provides enhanced information to the engineer from high-resolution aerial photographs. Furthermore, airborne LiDAR has become a powerful new tool in terrain analysis to inspect landforms by virtue of its ability to create digital elevation models that complement topographic maps and air photos. LiDAR has proven particularly important in identifying landslides and active faults. Finally optical and radar satellite images continue to play important roles in identifying geohazards, in particular in seismically active areas.

6.7 FURTHER READING

Anderson, R.S. and Anderson, S.P., 2010, *Geomorphology: The Mechanics and Chemistry of Landscapes*. Cambridge, UK: Cambridge University Press. — An excellent introduction to geomorphology that stresses the physical and chemical basis for geomorphic processes. Particularly suitable as a reference volume for geotechnical engineers.

Bull, W.B., 1991, *Geomorphic Responses to Climatic Change*. New York, NY: Oxford University Press. — An analysis of how climatic change, lithology and uplift interact with fluvial processes in various desert and humid regions of the Earth.

Burbank, D.W. and Anderson, R.S., 2011, *Tectonic Geomorphology*. Chichester, UK: Wiley-Blackwell. — An impressive summary of how field methods are used to deduce erosion rates, assess landform evolution and date earthquakes.

Calcaterra, D. and Parise, M. (eds), 2010, *Weathering as a Predisposing Factor to Slope Movements*. Engineering Geology Special Publications No. 23. London, UK: Geological Society. — An international review of how weathering has predisposed slopes to ultimate failure. This informative monograph discusses weathering in various rock types and the geotechnical and hydrogeological processes resulting in slope failure.

Cruden, D.M. and Thomson, S., 1987, *Exercises in Terrain Analysis*. Edmonton, Canada: Pica Pica Press (University of Alberta Press). — An introduction to air-photo analysis by an engineering geologist and a civil engineer with experience in identifying geohazards to transportation corridors in western and northern Canada. Includes numerous stereo pairs of river valleys, permafrost, glaciated terrain and landslides.

Easterbrook, D.J. and Kovanen, D.J., 1999, *Interpretation of Landforms from Topographic Maps and Air Photographs: Laboratory Manual*. Upper Saddle River, NJ: Prentice-Hall. — A companion lab manual for Easterbrook's *Surface Processes and Landforms*. It couples air-photo analysis with inspection of the topographic maps for the same areas.

Fookes, P.G., Pettifer, G. and Waltham, T., 2015, *Geomodels in Engineering Geology: An Introduction*. Caithness, UK: Whittles. — A useful introduction to the development of three-dimensional block diagrams applied to civil engineering practice complemented by many photographs showing various geomorphological features.

House, P.K., Webb, R.H., Baker, V.R. and Levish, D.R. (eds), 2002, *Ancient Floods, Modern Hazards: Principles and Applications of Paleoflood Hydrology*. Washington, DC: American Geophysical Union. — This volume on paleoflood hydrology demonstrates how historical flood records might be augmented by geomorphic measurements.

Rees, W.G., 2013, *Physical Principles of Remote Sensing*, 3rd edn. Cambridge, UK: Cambridge University Press. — A review of the scientific basis of aerial photography, electro-optical systems, laser profilers and microwave systems in environmental sciences.

Walker, M.J. (ed.), 2012, *Hot Deserts: Engineering, Geology and Geomorphology*. Engineering Group Working Party Report. London, UK: Geological Society. — An account of the geomorphological characteristics of desert landforms throughout the world as they relate to engineering practice.

6.8 Questions

Question 6.1. Estimate the evapotranspiration (ET) and runoff ratio for a watershed near where you live. (Make sure that there are no significant water imports or exports from the watershed so that ET can safely be assumed the principal loss from the watershed.)

Question 6.2. A stream draining a granitic catchment of $30 \, km^2$ has had an average sediment erosion rate of $2.0 \times 10^6 \, kg/yr$ over 30 years. Identify the assumptions that you must make in calculating the erosion rate on an annual basis. Give the answer in $\mu m/yr$.

Question 6.3. How does weathering of fresh rock affect the rock's Mohr–Coulomb parameters c' and ϕ' in (6.1)?

Question 6.4. Derive an equation for stream power per unit length of a channel (Ω) in terms of the friction slope S_f and discharge Q, where the drag force is given by (6.2), the channel length is L and time is t.

Question 6.5. Now continue this development to derive ω in (6.4). Allow $vol = wLz$, where w is channel width, L is channel length and z is channel depth.

Question 6.6. It has been observed that the average velocity of a stream channel tends to increase with drainage basin area. Under what conditions can we infer this from the hydraulic geometry relationships for stream channels?

7 Environmental Geochemistry and Mineralogy

7.1 Introduction

Recharge waters from rain or snowmelt to soil have only a few milligrams per litre of dissolved solids – such as sodium and chloride from the sea and from dust particles – and dissolved gases (indicated by the symbol (g)) – such as oxygen, $O_2(g)$, and carbon dioxide, $CO_2(g)$. However, infiltration of soil water yields groundwater that evolves in terms of total dissolved solids (TDS) and acquires a distinct geochemical signature. In this section we consider the fundamental physical-chemical processes controlling natural water quality as it evolves from rain or snowmelt initially through the process of chemical weathering in the unsaturated zone. We will follow this evolution in Chapter 12 as recharge evolutes in the saturated zone of a groundwater flow system and finally discharges as streamflow. In Chapter 12 we also consider how these same physical-chemical processes affect contaminants in contact with groundwater and cause a degradation in groundwater quality. Box 7.1 provides a reminder of how chemical concentrations that are used in this book are presented and converted.

The interaction of rainwater or snowmelt with soil and rock minerals describes the process of chemical weathering. We may think of it as a gigantic acid–base titration on the global scale in which rock-forming minerals are the bases and the infiltrating rain or snowmelt are the acids:

$$\text{rocks} + \text{rain and snowmelt} \Rightarrow \text{weathered rocks} + \text{natural waters.} \quad (7.1)$$

The products of weathering in (7.1), i.e., disaggregated rock and soil plus groundwater or river water, are the results of chemical reactions that are **irreversible reactions** in the sense that, at standard temperatures and pressures (STP), nature does not take the products and reproduce the reagents. This does, however, occur through subduction and the recycling of plates, but that is obviously not a process occurring at STP! We consider these irreversible reactions first, and then the **equilibrium reactions** ($A+B \rightleftharpoons C+D$) such as solubility and sorption that limit ion concentrations in groundwaters, under the heading of the **law of mass action**, which is discussed in section 7.3.

7.2 Chemical Weathering

Why do some compounds dissolve in water readily and not others? What controls the concentration of these compounds that we find in natural waters? Why do some minerals, such as the K-feldspar in Figure 2.19, weather while the neighbouring quartz is unaltered? The answers to these questions start with an examination of the water molecule.

The water molecule is a good solvent for ions that are contained within minerals because of its polar structure (Figure 7.1a). A large negatively charged oxygen atom projects two bonds with hydrogen atoms that are separated by an angle of 105°. The hydrogen atoms are positively charged, and thus the whole molecule exhibits electroneutrality. The two positively charged hydrogen atoms thus neutralize the double negative charge on the oxygen. If the groundwater was not electrically neutral, it would be like a battery! Therefore all chemical analyses that you receive from a laboratory should be electrically balanced – usually within 5–10% – with the same amounts of positively and negatively charged ions as measured in meq/L.

BOX 7.1 | CONCENTRATION UNITS

We may express the concentration of solutes in groundwater is several ways.

- Weight of solute per litre (L) of solution, e.g., mg/L or μg/L.
- Molarity (m): number of moles of solute per litre of solution (often this will appear as millimoles per litre or mmol/L).
- Molality (M): number of moles of solute per kilogram of solution (necessary for high TDS solutions and modelling).
- Mole fraction: for a solute B in a solution of A, the mole fraction, X_B = (moles of B)/(moles of A + moles of B).

Gram formula weights (GFW) are obtained by consulting the Periodic Table of elements that may be found in any introductory chemical textbook. The GFW for calcium (Ca) is 40.08, which means that the mass of one mole of Ca is 40.08 g. For the sulfate ion (SO_4^{2-}), we take the GFW for sulfur and add four times the GFW for oxygen (O) such that 1 mol SO_4^{2-} = 32.06 g from sulfur, plus 4 × 15.9994 g from oxygen = 96.06 g total. One millimole is obviously one-thousandth of a mole. Because we are discussing concentrations in mg not g, it follows that

$$mmol/L = \frac{mg/L}{gram\ formula\ weight}. \tag{7.2}$$

A sample of groundwater from a shallow aquifer containing 24 mg/L of calcium has 0.60 mmol Ca^{2+}. A solution of 1 mol/L when multiplied by Avogadro's number (6.022×10^{23} molecules per mole) expresses the number of ions or molecules in the litre volume. But it does not express the constraint of **electroneutrality**. Therefore we need to acknowledge that the electrical balance of the samples requires us to consider the charge of each ion. Thus we define the number of equivalents (eq) per litre that relates concentration to charge:

$$mmol/L = \frac{meq/L}{charge\ of\ ion}. \tag{7.3}$$

Therefore our sample with 0.6 mmol/L of calcium has 1.2 meq/L of Ca^{2+} and will require balancing by 1.2 meq/L of anions. Finally, we relate molarity to molality:

$$mmol/L = molality \cdot density \cdot \left(\frac{weight\ solution\ -\ weight\ solutes}{weight\ solution} \right) \cdot 1000. \tag{7.4}$$

Source: From Appelo and Postma (2005).

Each water molecule is therefore an electric dipole, and a network of water molecules is formed due to the electrostatic attraction of one molecule for another. The association of these water molecules is by weak hydrogen bonds that can be easily broken or distorted by ions, such as Na^+ and Cl^-. The dissolution of halite – the NaCl mineral – merely releases ions to groundwater; it does not create them. Sodium chloride easily dissolves in groundwater because the attraction of the Na^+ and Cl^- ions to the dipoles formed by water molecules is much greater than the inter-ionic attraction within the halite crystal. Many other simple minerals also dissolve like halite, and we call this **congruent dissolution** (Figure 7.1b). However, this is not true with rock-forming silicate minerals such as the feldspars. Plagioclase dissolves incongruently,

allowing some Ca^{2+} or Na^+ to dissolve, but the silicate structure is preserved and slowly weathers to a clay mineral. Furthermore, most organic molecules do not readily dissolve in water because they are rather non-polar. Remember the old adage: "like dissolves in like".

A typical soil zone in the recharge area of a groundwater flow system might contain the following minerals – some in trace amounts – that are sources for the major-ion solutes listed in parentheses:

- alumino-silicate minerals, in particular the feldspars (H_4SiO_4, HCO_3^-, Ca^{2+}, Na^+ and K^+);
- carbonate minerals, in particular calcite and dolomite (Ca^{2+}, Mg^{2+} and HCO_3^-);
- sulfate minerals, in particular gypsum (Ca^{2+}, SO_4^{2-});

a)   Polarity of H-O bonds

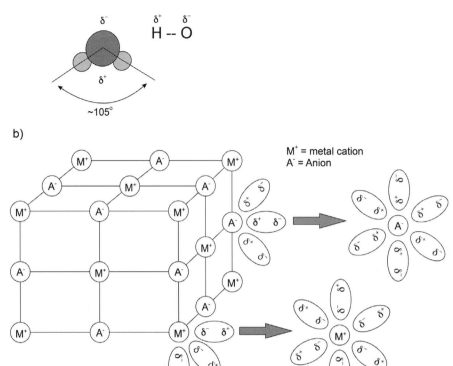

b)

M$^+$ = metal cation
A$^-$ = Anion

Figure 7.1 Dissolution from an ionic crystal into water. (From Moeller and O'Connor (1971); reproduced with the permission of The McGraw-Hill Companies, Inc.)

- sulfide minerals, perhaps pyrite (Fe^{2+}, SO_4^{2-}, H^+);
- oxides, in particular silica, ferric oxide and manganese oxide; and
- organic matter, including humic and fulvic acids that dissolve sparingly and are measured as dissolved organic carbon (DOC).

One of the most important weathering reactions that releases ions to soil waters (and eventually groundwater and streams) is the dissolution of calcite, which produces the bicarbonate anion that is a powerful neutralizing agent:

$$CaCO_3 + H_2CO_3 = Ca^{2+} + 2HCO_3^-. \qquad (7.5)$$

We shall discuss this further in section 7.5, but note that the weathering agent here is carbonic acid that is formed by the dissolution of carbon dioxide in the infiltrating water ($CO_2 + H_2O = H_2CO_3$). Secondly, the weathering reactions for alumino-silicate minerals have the general form

$$\text{cation–Al-silicate} + H_2CO_3 + H_2O$$
$$\Rightarrow \text{cations} + HCO_3^- + H_4SiO_4 + \text{Al-silicate.} \qquad (7.6)$$

The dissolved products are cations, bicarbonate and silicic acid (H_4SiO_4), which, like water, is only slightly ionizable. The solid mineral phase remaining is an alumino-silicate mineral, typically a clay mineral. We will consider three examples (Stumm and Morgan, 1981). We will take plagioclase and then biotite and produce the clay mineral kaolinite from each. Then we will take kaolinite and produce gibbsite, the oxide of aluminum and the source of bauxite for aluminum production.

4 millimoles of plagioclase $\Rightarrow$ 3 millimoles of kaolinite:

$$4Na_{0.5}Ca_{0.5}Al_{1.5}Si_{2.5}O_8 + 6H_2CO_3 + 11H_2O$$
$$\Rightarrow 2Na^+ + 2Ca^{2+} + 6HCO_3^- + 4H_4SiO_4 + 3Al_2Si_2O_5(OH)_4 \qquad (7.7)$$

1 millimole of biotite $\Rightarrow \frac{1}{2}$ millimole of kaolinite:

$$KMg_3AlSi_3O_{10}(OH)_2 + 7H_2CO_3 + \frac{1}{2}H_2O$$
$$\Rightarrow K^+ + 3Mg^{2+} + 7HCO_3^- + 2H_4SiO_4 + \frac{1}{2}Al_2Si_2O_5(OH)_4 \qquad (7.8)$$

1 millimole of kaolinite $\Rightarrow$ 1 millimole of gibbsite:

$$Al_2Si_2O_5(OH)_4 + 5H_2O \Rightarrow 2H_4SiO_4 + Al_2O_3 \cdot 3H_2O \qquad (7.9)$$

Table 7.1 Synthesis of soil water in millimoles. (After Stumm and Morgan (1981).)

Reaction	H_4SiO_4	Na^+	Ca^{2+}	Mg^{2+}	K^+	HCO_3^-	SO_4^-	Cl^-
Calcite dissolution	—	—	0.10	—	—	0.20	—	—
Plagioclase weathering	0.04	0.02	0.02	—	—	0.06	—	—
Biotite weathering	0.02	—	—	0.03	0.01	0.07	—	—
Halite dissolution	—	0.01	—	—	—	—	—	0.01
Pyrite oxidation	—	—	—	—	—	—	0.02	—
Composition (millimoles)	0.06	0.03	0.12	0.03	0.01	0.33	0.02	0.01
Composition (mg/L)	6	0.7	4.8	0.75	0.40	20	2	0.4

Finally, we will consider a reduction–oxidation or **redox** reaction, in which electrons are transferred from a source to a donor. In this case our source will be pyrite, which is present in many mineral deposits and whose oxidation leads to acid mine waters. The mineral product here is ferric hydroxide, an important sorbent for trace metals and contaminants dissolved in groundwater.

$$3\tfrac{1}{2}H_2O + FeS_2 + 3\tfrac{3}{4}O_2 = Fe(OH)_3 + 4H^+ + 2SO_4^- \quad (7.10)$$

Using the above stoichiometric relations – those defining the relation of elemental weights of reactants and products – we may synthesize a natural water from the reactions involving 10^{-1} millimoles of calcite, 4×10^{-2} millimoles of plagioclase, 10^{-2} millimoles of biotite, 10^{-2} millimoles of pyrite and 10^{-2} millimoles of halite particulate matter in the rainfall. Table 7.1 shows the synthesis of this natural water that is typical of soils in temperate northern latitudes.

The yield of solutes from a watershed is controlled by the minerals in the soil profile (see Box 7.2) and the climate, in particular the amount of precipitation and temperature (Carey et al., 2005). Precipitation, whether in the form of rain or snowmelt, is essential to promote weathering reactions that require water to dissolve, oxidize and transfer solutes, while increasing temperature promotes the rates of reaction. Thus gibbsite (equation (7.9)) is a product of weathering in tropical climates where all cations are leached from the soil by the abundant rainfall and the warm temperatures promote reaction rates. An additional factor in weathering is the rate of uplift of a landmass by tectonic processes that creates new landforms, which can be weathered both physically and chemically to produce fresh mineral surfaces capable of solute and particulate production.

Table 7.2 presents solute weathering yields from watersheds in the USA, Puerto Rico and Taiwan that experience very different climates. The three US watersheds are in the Western

Cascades of Oregon, the southern Appalachians of North Carolina and the White Mountains of New Hampshire. The Oregon watershed is underlain by andesite of Cenozoic age with Douglas fir, hemlock and western red cedar forests. The watershed in the southern Appalachians is characterized by mixed hardwood forests of oak and maple with bedrock of Paleozoic metamorphic rocks. Paleozoic quartz mica schists underlie the White Mountains watershed in New Hampshire that is vegetated by sugar maple, beech and birch. The Puerto Rican watershed is a dense tropical rainforest situated on Cretaceous andesite with interbedded marine mudstones. Lastly, the Taiwanese watershed is a subtropical mixed evergreen hardwood forest underlain by Cenozoic sedimentary rocks that are being uplifted at a rate of 6 mm/yr. Uplift in the Puerto Rican (0.04 mm/yr) and Oregon (~1 mm/yr) watersheds is much less, while it is essentially absent in the watersheds of New Hampshire and North Carolina.

It is not surprising that landslides are a major source of mass wasting or physical weathering in the watersheds in Taiwan, Oregon and Puerto Rico, where uplift is measurable. These three watersheds receive high rainfall amounts from onshore storm systems. Topographic elevation is not the determining factor because parts of the White Mountains watershed in New Hampshire are the highest of the five – it ranges from 488 to 792 m above sea level (asl), while the Taiwanese watershed is at ~700 m asl. Temperature obviously plays an important role as well as precipitation. The Taiwanese watershed is the most productive because uplift exposes new mineral surfaces continuously that are exposed to high rainfall amounts in subtropical conditions. The lowest solute production yields are associated with the southern Appalachian watershed of North Carolina and the White Mountains watershed of New Hampshire; both are situated in old Paleozoic rocks with no significant uplift from tectonic forces. While the denudation rate of the Taiwanese watershed

BOX 7.2 | SOIL HORIZONS AND SOIL COLOUR

Soil scientists identify five master soil horizons as occurring in a well-developed profile.

- *O horizon*: Uppermost zone, dominated by plant litter and **humus** (decomposed organic matter) that display various stages of decomposition.
- *A horizon*: (Ah in the Canadian soil classification system) Dark topsoil layer that is rich in humus and mineral grains exhibiting pronounced weathering and fine granular structure from cultivation and pasturing.
- *E horizon*: (Ae in the Canadian soil classification) Light-coloured zone exhibiting removal of soluble minerals and ions (**eluviation**) leaving quartz as the residue; found mainly in forest soils.
- *B horizon*: Zone that has accumulated soluble minerals and ions from the E horizon, rich in clay, humus and Al, Fe and Mn oxides.
- *C horizon*: Lowermost zone, which exhibits weathered bedrock material or sediments with evidence of the parent geologic structure and fabric, including saprock or saprolite.

Soil colours are a result of the material coating the soil and are described using the Munsell colour chart system. Black or brown soils indicate humus or ones rich in Mn oxides. Brown and yellowish brown soils are typical of ferric oxides, which have ratings of the **hue** or chromatic composition of the soil: R = red, Y = yellow, such that hue ranges from 2.5 to 10, e.g., 10R, 2.5YR, 5YR, 10YR, etc. White soils may indicate sodium salts or carbonate or sulfate minerals. Light grey soils are likely rich in quartz grains (see Figure 2.19), which are only soluble as H_4SIO_4.

Source: From Schaetzl and Thompson (2015).

Table 7.2 Solute weathering yields from five watersheds in moles per hectare per year. Mean annual temperature, *T*; mean annual precipitation, Ppt. (After Carey et al. (2004).)

Watershed	T (°C)	Ppt (mm)	H_4SiO_4	Cations	HCO_3^-
Taiwan	18	4290	7750	7320	5160
Puerto Rico, USA	23	3500	5316	8461	7707
Oregon, USA	8.5	2300	3250	4598	—
North Carolina, USA	13	1780	1357	922	747
New Hampshire, USA	14	1400	627	1049	126

is similar to its uplift rate, i.e., 6 mm/yr, the denudation rates of the US watersheds are much less. The Oregon watershed has the largest of the three and it is only 0.01–0.2 mm/yr.

7.3 Ionic Strength and Activity

Equation (7.5), the dissolution of calcite, is an example of a mineral–water chemical equilibrium reaction. The general equation for such reactions is

$$aA + bB = cC + dD \tag{7.11}$$

for which the forward rate of reaction is $k_f[A]^a[B]^b$ and the reverse reaction rate is $k_r[C]^c[D]^d$, where k_f and k_r are rate constants, and the square brackets denote concentrations. At equilibrium, the forward reaction rate is equal to the backward reaction rate, and the equilibrium constant, K_{eq}, is given by

$$K_{eq} = \frac{k_f}{k_r} = \frac{[C]^c[D]^d}{[A]^a[B]^b}, \tag{7.12}$$

which defines the **law of mass action**.

In this case A and B are reactants and C and D are products, which may be minerals or dissolved ions, i.e., **solutes**, in which the solvent is water. Physical chemists consider such solutions as being aqueous electrolytes, in which the dissolved ions are capable of conducting charge. Natural waters are composed of both dissolved ions (e.g., Ca^{2+}, Cl^-) and neutral species such as dissolved organic carbon (DOC), silica (H_4SiO_4) and gases, all maintaining electroneutrality within the water sample.

Because of strong inter-ionic forces, the effective concentrations of ions are less than the analytical concentration. A new variable known as the **activity** of an ion is defined on the basis of its chemical potential (Langmuir, 1997), which is analogous to the mechanical potential (gh) and the electrical potential (V). The chemical potential of each component will be equal in all phases of solution at equilibrium. Pure solids and liquids have activities of unity. We refer to the activity of an ionic or neutral species "i" as a_i and the relationship between activity and the analytical concentration is

$$a_i = \gamma_i \cdot m_i, \tag{7.13}$$

where γ_i is the activity coefficient of ionic species i and m_i is its molality.

Therefore, if you wish to convert water quality data from a lab to the effective concentrations – from which you can calculate mineral solubilities – you must first calculate the activity coefficients of all the species considered important in your water quality sample. Software does this for you using one of several relationships between the activity coefficient and the **ionic strength** of the solution:

$$I = \tfrac{1}{2}\sum(m_i z_i^2), \tag{7.14}$$

where m_i in the analytical concentration of component i expressed in moles and z_i is its electrical charge. For the soil water in Table 7.1, we can calculate the following ionic strength:

$$\begin{aligned} I = \tfrac{1}{2}\sum &(0.00006 \cdot 0)(0.00003 \cdot 1)(0.00012 \cdot 4)(0.00003 \cdot 4) \\ &\times (0.00001 \cdot 1)(0.00033 \cdot 1)(0.00002 \cdot 4)(0.00001 \cdot 1) \end{aligned}$$

$$= 0.0011 \text{ m}. \tag{7.15}$$

We may define two relationships between the ionic strength and the activity coefficients for various common ionic solutes. At low ionic strength (i.e., $I < 0.001$ m), we can use the Debye–Hückel equation,

$$\log \gamma_i = -Az_i\sqrt{I}, \tag{7.16}$$

where A is a temperature-dependent constant; at 25°C, $A = 0.5092$. At intermediate ionic strengths, such as that of seawater ($I = 0.7$ mol/kg), it is necessary to use a relationship that reflects the more complicated ion interactions that occur, such as the Davies equation,

$$\log \gamma_i = -Az_i^2\left(\frac{\sqrt{I}}{1+\sqrt{I}} - 0.3I\right), \tag{7.17}$$

which is valid for ionic strengths between 0.1 and 0.7 mol/kg.

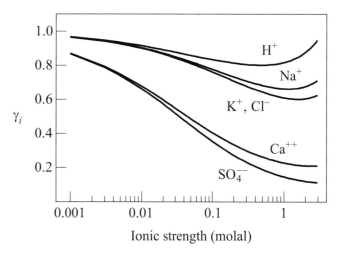

Figure 7.2 Activity coefficients of common ions as a function of ionic strength. (From Bethke (2008); reproduced with permission of Cambridge University Press.)

Langmuir (1997) notes that the ionic strength can be estimated from both the total dissolved-solids (TDS) concentration in mg/L and the specific electrical conductance (SEC) for waters (µS/cm), in which the predominant cation–anion pair is calcium and bicarbonate, using the following identities:

$$I = 2.8 \times 10^{-5} \times \text{TDS}, \tag{7.18}$$

$$I = 1.9 \times 10^{-5} \times \text{SEC}. \tag{7.19}$$

Figure 7.2 shows how activity coefficients vary with ionic strength. You can now calculate the activity coefficients of the calcium and bicarbonate ions of the soil water shown in Table 7.1 as question 7.2 at the end of this chapter.

7.4 Solubility of Minerals

Solubility is the concentration of a solute (Ca^{2+}, Cl^-, etc.) in an aqueous solution containing a second phase, such as a solid or a gas, in equilibrium with the solution. At equilibrium, the solution is called a saturated solution. Supersaturated solutions may occur if there are kinetic controls on the rate of new mineral precipitation or if nuclei, about which crystallites might form, are absent. Undersaturated solutions are those in which the mineral or gas phase has not yet reached equilibrium by dissolving into the aqueous phase and further dissolution may occur. The solubility product is an equilibrium constant that relates concentration, i.e., solubility, to activity. For sodium chloride dissolution,

$$NaCl = Na^+ + Cl^-, \tag{7.20}$$

BOX 7.3	COMPUTING THE SOLUBILITY PRODUCT FROM MEASURED SOLUBILITY

The experimentally determined solubility of the mineral gibbsite $Al(OH)_3$ at $25°C$ is 3×10^{-7} g/L as the aluminum ion. Compute K_{s0}.

Start with the equation:

$$Al(OH)_3 = Al^{3+} + 3OH^- \tag{7.21}$$

The solubility $S = [Al^{3+}]$ or $3S = [OH^-]$ because there are three moles of hydroxide for each mole of aluminum in solution. The formula weight of gibbsite is 77.98 g/mol. Therefore the solubility (S) may be written as 3.85×10^{-9} mol/L and the ionic strength of this solution is

$$I = \tfrac{1}{2} \sum (S(3)^2 + 3S(-1)^2) = 2.3 \times 10^{-8}. \tag{7.22}$$

Therefore we can see that the activities of the aluminum and hydroxide ions are unity because of the low ionic strength. The solubility product of gibbsite at zero ionic strength is given by

$$K_{s0} = (a_{Al^{3+}})(a_{OH^-})^3 = (S)(3S)^3 = 27S^4 = 5.89 \times 10^{-33}. \tag{7.23}$$

We write this solubility product as the negative logarithm, $pK_{s0} = 32.2$ (the symbol "p" in front of K_{s0} indicates $-\log_{10}$ of K_{s0}). This low solubility product is typical of oxide minerals. Amorphous iron hydroxide has a solubility product of $pK_{s0} = 38.0$ at $25°C$, while the more crystalline form, goethite, is less soluble ($pK_{s0} \sim 41$).

Sulfates and carbonates are relatively soluble minerals. For example, the solubility product for gypsum ($CaSO_4 \cdot 2H_2O$) is $pK_{s0} = 4.8$, while that for calcite ($CaCO_3$) is $pK_{s0} = 8.4$. Such solubility products can only be written for minerals that dissolve congruently. Therefore it is not possible to write similar values for silicates, such as plagioclase ($(Na,Ca)(Al,Si)AlSi_2O_8$) or the clay mineral kaolinite ($Al_4Si_4O_{10}(OH)_8$). Their solubility is low and strongly dependent on the pH of the water.

---

we may write the solubility product, K_{s0} or K_{eq}, as

$$K_{s0} = K_{eq} = \frac{a_{Na}a_{Cl}}{a_{NaCl}}, \tag{7.24}$$

where K_{s0} indicates that the solubility product is for an ionic strength of zero. Because the activities of pure solids are assumed to be unity, $a_{NaCl} = 1.0$ and (7.24) simplifies to the product of the activities of the two solutes. We conclude from this discussion that the solubility of any mineral is dependent on the activities of the solutes that form the mineral and, because activities are a function of ionic strength, the solubility product of any mineral phase is also a function of ionic strength. In practice, solubilities are computed through entering water quality data into software that computes the ionic strength of the sample, the relevant activities and then the solubility of a particular mineral for those conditions. Box 7.3 illustrates how a solubility product might be computed from experimental data.

Therefore we may write the general expression for the solubility product for any solid M_xN_y that dissolves congruently with the equilibrium reaction $M_xN_y = xM + yN$ as

$$K_{s0} = a_M^x \cdot a_N^y, \tag{7.25}$$

where the term on the right-hand side of the equation is known as the **ion activity product** (IAP). If ions M and N exist in a groundwater and their ion activity product is greater than the K_{s0}, the groundwater is considered saturated with respect to the solid M_xN_y. In order to determine if saturation of a particular solid is obtained in any sample of groundwater or streamflow, we can compute the **saturation index**:

$$SI = \log_{10}\left(\frac{IAP}{K_{s0}}\right). \tag{7.26}$$

Therefore if $SI = 0.0$, the water sample is considered saturated (i.e., in equilibrium) with the solid with solubility product K_{s0}. If $SI > 0.0$, the sample is said to be supersaturated, while if $SI < 0.0$, the sample is said to be undersaturated with that same solid, in which case a net dissolution of this solid is expected to occur. Slow reaction kinetics may hinder the precipitation of a solid that is supersaturated in an aqueous sample.

7.5 Alkalinity and Carbonate Mineral Dissolution

Water ionizes slightly, i.e., $H_2O = H^+ + OH^-$, for which we can write

$$K_w = a_{H^+} \cdot a_{OH^-}. \qquad (7.27)$$

Note that the denominator (a_{H_2O}) of this reaction is unity because liquid water is a pure substance. At 25°C the ion activity product of this reaction is equal to 1.00×10^{-14} or $pK_w = 14.00$ (the symbol "p" in front of an activity indicates $-\log_{10}$ of that activity). When the acidity of water is neutral, $a_{H^+} = a_{OH^-}$, and it follows from (7.27) that $a_{H^+}^2 = 10^{-14}$. Therefore, $a_{H^+} = \sqrt{10^{-14}} = 10^{-7}$. The negative logarithm of the activity of the hydrogen ion is written as $-\log_{10} a_{H^+} = $ pH = 7. At 10°C the IAP of this reaction is $pK_w = 14.53$, and neutral pH = 7.27. But, it turns out, neither of these values reflect the pH of uncontaminated rainwater.

The gaseous (g) dissolution of atmospheric CO_2 in rain and snow gives them an original acidity:

$$CO_2(g) + H_2O = H_2CO_3 = H^+ + HCO_3^- \qquad (7.28)$$

Thus atmospheric carbon dioxide hydrolyzes rainwater, splitting it into hydrogen and bicarbonate ions. This process occurs whenever carbon dioxide dissolves into water and we will revisit it in Chapter 12. Suffice it to say, the dissolved carbon dioxide forms a weak acid (H_2CO_3) for which the distribution diagram for carbonate species is shown in Figure 7.3. The equivalence points are where $m(H_2CO_3) = m(HCO_3^-)$ and where $m(HCO_3^-) = m(CO_3^{2-})$, where m represents the moles of each ion. These are shown in Figure 7.3 as the first and second dissociation constants (pK_1 and pK_2) of carbonic acid.

Using these expressions for the dissociation of carbonic acid H_2CO_3 into HCO_3^- and then bicarbonate's dissociation into a proton, H^+, and the carbonate anion, CO_3^{2-}, we begin to solve for the pH of rainwater. But we need also to employ the equations for the dissociation of water (see equation (7.27)) and the electroneutrality equation, which says that the sum of charges on cations must be balanced by the charges on the anions, i.e., $m(H^+) = m(HCO_3^-) + 2m(CO_3^{2-}) + m(OH^-)$. After much arithmetic and trial-and-error substitution, we obtain the equilibrium pH of rainwater as pH = 5.7. Acid rain, with pH < 5.7, refers to the dissolution of sulfur dioxide from fossil-fuel combustion and its oxidation and hydration in the atmosphere to sulfuric acid.

We may now define alkalinity as the capacity of a volume of water to accept acids (i.e., protons) due to the presence of

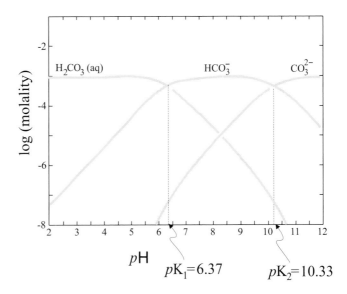

Figure 7.3 Distribution of dissolved CO_2 species at 10^{-3} M. (From Zhu and Anderson (2002); with the permission of Cambridge University Press.)

dissolved bases within that volume (Langmuir, 1997). The most significant base is usually the bicarbonate ion; less important bases at near-neutral pH values include the carbonate and hydroxyl ions. The total alkalinity is given by

$$alk = HCO_3^- + 2CO_3^{2-} + OH^- - H^+ \qquad (7.29)$$

The standard method of alkalinity measurement reports values in mg/L or meq/L of calcium carbonate. To convert mg/L as $CaCO_3$ to a bicarbonate concentration, multiply by 1.22; for meq/L as $CaCO_3$, the conversion factor is meq/L $\times$ 50.04. You can compute these conversion factors by solving the questions at the end of this chapter.

What then neutralizes the natural or anthropogenic acidity of rainfall or snowmelt? Equation (7.6) indicated that carbonic acid might be converted to bicarbonate by the chemical weathering of silicate minerals. Thus acid neutralization is accomplished by the weathering of plagioclase and biotite, as shown in Table 7.1. This table indicates that calcite dissolution (see (7.5)) is also a source of acid neutralization – i.e., bicarbonate and/or carbonate production – and is usually considered the principal source of neutralization in soils containing carbonate minerals. In fact, Langmuir (1997) indicates that as little as 1% calcite or dolomite in silicate soils is sufficient for these carbonate minerals to dominate the soil-water chemistry.

The critical variable in calcite (or dolomite) dissolution is the partial pressure of dissolved carbon dioxide (P_{CO_2}), which we estimate by measuring the alkalinity and pH of the water sample, i.e.,

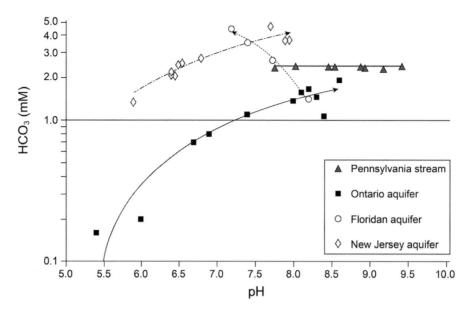

Figure 7.4 Evolution of natural waters in three groundwater flow systems and a stream, where arrows indicate flow directions. Initial partial pressures (P_{CO_2}) of recharge in the New Jersey and Ontario aquifers are $\sim 10^{-1}$ atm, while the value for the Floridan aquifer is $\sim 10^{-3}$ atm. Sources of data: Ontario aquifer, from Jackson and Patterson (1982), Lower Sands Aquifer; Pennsylvania stream and New Jersey aquifer, from Langmuir (1997); Floridan aquifer, from Plummer and Sprinkle (2001), path V.

$$CaCO_3 + CO_2(g) + H_2O = CaCO_3 + H^+ + HCO_3^-$$
$$= Ca^{2+} + 2HCO_3^- \qquad (7.30)$$

This equation is equivalent to (7.5) in that $H_2CO_3 = CO_2(g) + H_2O$ as in (7.30). Therefore the dissolution of carbonate minerals along a flow line in the subsurface is a function of P_{CO_2}. As calcite is dissolved by carbonic acid (H_2CO_3), H^+ and P_{CO_2} must be depleted according to this equation and the amounts of dissolved Ca^{2+} and HCO_3^- must increase, as must the pH. According to Appelo and Postma (2005), the initial P_{CO_2} in the upper metres of an unsaturated soil zone varies somewhere between the atmospheric concentration of 0.03%, i.e., $P_{CO_2} = 10^{-3.5}$, and 3%, or $P_{CO_2} = 10^{-1.5}$, with the value increasing as a function of temperature and mean annual evapotranspiration associated with biological activity in the soil.

Figure 7.4 shows the relationship between the master variables pH and P_{CO_2} with the measured bicarbonate concentrations along flow paths in groundwaters from New Jersey and Florida in the USA, Ontario in Canada and a stream in Pennsylvania, USA. While the groundwater flow system is a closed system with respect to increases in P_{CO_2}, the stream is an open system by virtue of its contact with the fixed supply of atmospheric carbon dioxide. The patterns exhibited by the Ontario and New Jersey groundwaters show a typical closed system response with an increase in bicarbonate and pH along the flow paths reflecting differences in carbonate content in the aquifers. The Floridan aquifer, however, responds with an increase in bicarbonate and a decrease in pH due to the dissolution of dolomite and anhydrite and the precipitation of calcite (Plummer and Sprinkle, 2001).

7.6 Redox Processes

Perhaps the reader remembers watching a chemistry instructor dropping metallic sodium (Na^0) into a beaker of water, then standing back to watch the explosive reaction as the sodium dissolves and is oxidized in the water: $Na^0 \Rightarrow Na^+ + e^-$. **Oxidation** involves the removal of an electron from an atom or group of atoms, in this case metallic sodium, while **reduction** is the addition of electrons to an atom or group of atoms. By convention we write ox $+ ne^- =$ red, where "ox" refers to the oxidant, "red" to the reductant, n to the number of moles of electrons (e^-) transferred to the oxidant from the reductant and the equilibrium constant is $\log K^*$. Therefore the appropriate way to write the metallic sodium **redox** reaction is $Na^+ + e^- = Na^0$.

The most commonly used parameter for redox processes is provided by the Nernst equation in the form of the redox potential, Eh or E_H:

$$E_H = E_H^0 + \frac{2.3RT}{nF} \cdot \log \frac{ox}{red}, \qquad (7.31)$$

where E_H is the redox potential (volts), E_H^0 is the standard redox potential of the reaction ox $+ ne =$ red at 25°C and unit activity, R is the gas constant (8.314 J/K mol), T is the absolute temperature, n is as above and F is the Faraday constant indicating the electric charge of 1 mol of electrons. The fraction in the third term reduces to 0.059 at 25°C and one atmosphere pressure.

Standard redox potentials of common geochemical reactions can be looked up in references such as Stumm and Morgan (1981) or Langmuir (1997). These will yield values of E_H^0 varying from -2.7 V for $Na^+ + e^- = Na^0$ to $+1.8$ V for $Co^{3+} + e^- = Co^{2+}$. Think of this relation by referring to (7.31): the sodium metal dissolves and is fully oxidized; therefore the ratio $Na^0/Na^+ \ll 1$ and E_H^0 must be very small, like -2.7 V. The standard potential of 0.0 V is that of the standard hydrogen electrode (SHE): $2H^+ + 2e^- = H_{gas}$. Thus the "H" in redox potential, E_H, refers to this SHE potential of zero volts; sometimes it is written Eh.

In the field or the lab it is hazardous to employ hydrogen gas in measurements of the redox potential of a water sample. Instead, we use a platinum (Pt) electrode with an internal reference electrode of known potential versus the standard hydrogen electrode. The measured potential relative to this reference electrode is usually referred to as the ORP or oxidation–reduction potential. However, it can be readily converted to an E_H value by adding the potential between the reference electrode and the SHE (refer to chapter A6 of the US Geological Survey's *National Field Manual for the Collection of Water-Quality Data*, http://water.usgs.gov/owq/FieldManual/). This measured Pt electrode potential, E_H, in natural or contaminated waters is a mixed potential that is not strictly interpretable by chemical thermodynamics (see Langmuir, 1997, ch. 11). However, waters with dissolved oxygen typically have high E_H values, i.e., $E_H > 300$ mV, while waters that have undergone reduction through the presence of hydrocarbons have values $E_H < 0$ mV.

Just as there are open and closed systems of dissolved carbon dioxide that control pH, so there are also open and closed redox systems controlling E_H. In redox geochemistry, an open system is one that is open to a continuous supply of dissolved oxygen (DO) or other oxidant that prevents the river, lake or groundwater becoming a closed system. In open systems, reduced compounds (e.g., hydrocarbons, dissolved iron or hydrogen sulfide) become oxidized in a predictable sequence. When reductants, such as petroleum hydrocarbons and other forms of dissolved organic carbon (DOC), become the dominant or excess compounds in such waters, then a closed system sequence of reduction reactions occurs that is also predictable by chemical thermodynamics and is also

catalyzed by microbial processes. By excess, we mean the soluble reductants that exist after all DO is reduced or the soluble oxidant that is present after all DOC or other reductants are oxidized. A fast-flowing river into which there is discharge of acid-rock drainage waters is an example of an open redox system provided that DO is always in excess. A groundwater contaminated by a large release of gasoline that consumes all oxidants is an example of a closed redox system.

The open and closed redox sequences are shown in Figure 7.5. The upper half of this figure shows the closed system sequence of reduction reactions that begins with the reduction of oxygen as DO at a relatively high E_H, followed by denitrification (nitrate → ammonium), manganese oxide to dissolved manganese (Mn^{2+}), ferric oxide to dissolved ferrous iron (Fe^{2+}), sulfate reduction and methane fermentation in that order of occurrence.

The lower half of the figure indicates the open system sequence of oxidation reactions beginning at a relatively low E_H with the oxidation of DOC by excess DO or any other of the oxidation reactions shown, followed by the oxidation of dissolved sulfide, then ferrous iron, ammonium and finally manganese. Any reduction reaction shown in the lower half can be oxidized, i.e., is thermodynamically allowable, by an oxidation that has its origin at a higher E_H, e.g., biotite containing ferrous iron oxidation by DO to produce vermiculite.

Note that all oxidation reactions produce protons and therefore lower the pH, while all reduction reactions consume protons and raise the pH. For example,

$$Fe(OH)_3(s) + 3H^+ + e^- = Fe^{2+} + 3H_2O \qquad (7.32)$$

$$FeS_2(s) + \tfrac{7}{2}O_2 + H_2O = Fe^{2+} + 2SO_4^{2-} + 2H^+ \qquad (7.33)$$

where (s) indicates a solid phase. Equation (7.32) shows the reduction of ferric hydroxide, a common coating causing sand to be yellow or red, and the consumption of three moles of protons in the oxidative dissolution of the hydroxide. Equation (7.33) illustrates the critical acid-rock drainage reaction: the oxidation of the mineral pyrite produces two moles of protons for each mole of pyrite oxidized by DO. Equations (7.10) and (7.33) present the same reaction, written slightly differently so that in (7.10) pyrite is replaced by a secondary mineral; local conditions will dictate which reaction is more likely.

Figure 7.6 shows the cross-sectional profile of a 1 km long fluvial sand aquifer that illustrates a closed system redox sequence. Groundwater recharge occurs on the high ground on the left side of the figure (distance 0 m), flows left to right through the upper and middle sands as well as through a

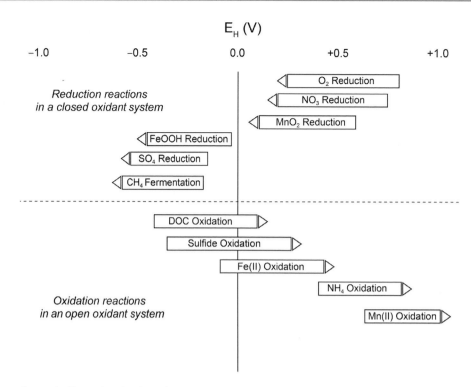

Figure 7.5 Sequences of microbially mediated redox reactions at pH 7 in natural and contaminated waters. (After Stumm and Morgan (1981); with the permission of John Wiley and Sons.)

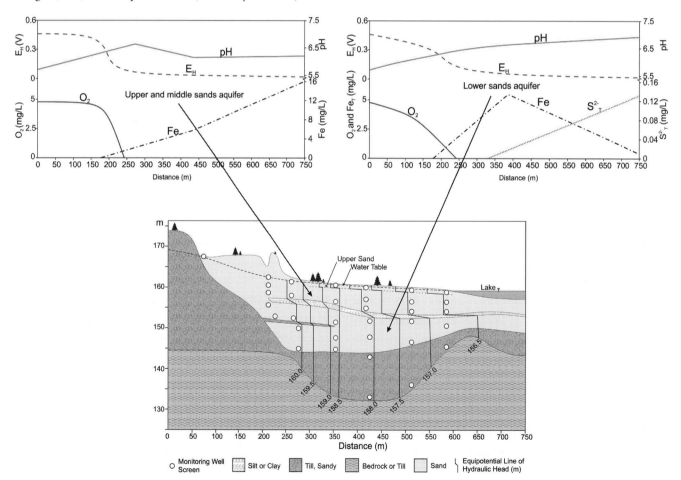

Figure 7.6 Redox sequences in a short (1 km) groundwater flow system, Chalk River, Ontario, Canada. (After Jackson and Patterson (1982); with the permission of John Wiley and Sons.)

Table 7.3 Geochemical classification of sedimentary environments by Berner (1981) based on molar concentrations of dissolved oxygen and sulfide.

Type	Environment	Characteristic phases
1	Oxic ($O_2 \geq 10^{-6}$ m)	Hematite, goethite, MnO_2; no organic matter
2	Anoxic ($O_2 < 10^{-6}$ m)	
2A	Sulfidic ($H_2S \geq 10^{-6}$ m)	Pyrite, marcasite, rhodochrosite, alabandite; organic matter present
2B	Non-sulfidic ($H_2S < 10^{-6}$ m)	
	○ Post-oxic	Glauconite and other Fe^{2+}/Fe^{3+} silicates (also siderite, rhodochrosite, vivianite); no sulfide minerals; minor organic matter
	○ Methanic	Siderite, rhodochrosite, vivianite; earlier formed sulfide minerals; organic matter

deeper sand aquifer, both of which discharge to the lake on the right. The groundwater quality and hydraulic head data are from 0.6 m screened intervals in numerous monitoring wells (well screens shown). Dissolved oxygen (DO) disappears in the first 250 m downgradient; nitrate is also reduced in this zone of the aquifer. Thereafter ferrous iron appears in both upper/middle and lower sand aquifers. However, in the lower sand aquifer only, the increase in ferrous iron is halted by the presence of sulfate-reducing bacteria (Jackson and Patterson, 1982); these bacteria catalyze the reduction of sulfate to aqueous sulfide ~500 m downgradient of the recharge area. Thus, the ferrous iron has been removed from deep groundwater and precipitated as iron sulfide. The E_H exceeds 300 mV or 0.3 V while DO is present in the groundwater and then decreases to ~100 mV when sulfide is present.

On a historical note, this redox sequence was first observed in the early 1970s by W.M. Edmunds of the British Geological Survey in the Lincolnshire Limestone (Edmunds and Walton, 1983) and later in the Sherwood Sandstone (Smedley and Edmunds, 2002). Simultaneously the US Geological Survey (Thorstenson et al., 1979; Chapelle and McMahon, 1991) observed the sequence in large US aquifers. The closed system redox sequence no doubt occurs elsewhere in the world; however, in the US and British aquifers much larger distances of ~20 km or more were required for the redox sequences to develop rather than the short distance in Figure 7.6.

Similarly, Berner (1981) had also conducted early studies of redox processes although in anoxic marine sediments and had noted the same sequence. He constructed a geochemical redox classification scheme that was based on the concentrations of dissolved oxygen and sulfide with which he associated various characteristic minerals, as shown in Table 7.3. Figure 7.6 illustrates how groundwater flow systems can encompass both

oxic and sulfidic environments within a few hundred metres of one another, although this separation may be only a few vertical metres in marine sediments.

7.7 Sorption and Sorbents

Sorption is a broad term describing the removal of dissolved ions from aqueous systems onto a mineral or organic **sorbent**. The **sorbate** is the dissolved ion or organic molecule that is sorbed; it might be a metal ion released by mining operations or an organic contaminant dissolved by groundwater from a gasoline release beneath a fuel terminal. Sorption is important in that it stores and thus limits the concentration of many ions and molecules in aqueous systems. Furthermore, sorption can create ore deposits by trapping dissolved ions and concentrating them into valuable minerals, e.g. the role of organic matter in creating the reducing conditions that trap uranium in roll-front deposits. In fact, Ingebritsen et al. (2006) state explicitly: "Most economically significant ore deposits exist because of the advective transport of solutes and heat by flowing groundwater." Sorption may allow sorbed contaminants to be transported significant distances in rivers as the suspended-sediment load or may greatly retard the transport of dissolved contaminants in groundwater.

We may simplify the process of sorption to a discussion of (a) how the aqueous geochemistry affects sorption and (b) how the various sorbents affect sorption. Then we shall consider several quantitative models of sorption that are used to represent the process using software tools. The interested reader is referred to Langmuir (1997) for a rigorous account of the subject.

Aqueous geochemical conditions determine the extent of sorption of metal ions to oxide-mineral surfaces such as $Fe(OH)_3$, FeOOH or MnO_2 or oxide minerals with

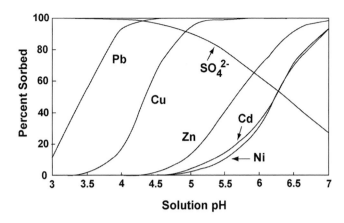

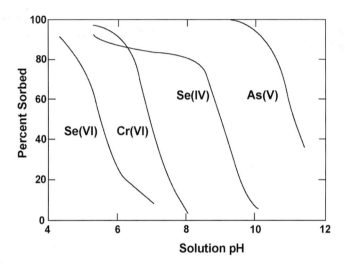

Figure 7.7 Adsorption edges for cations (above) and anions (below) over the pH range of natural and contaminated waters. (From Smith (2007); courtesy of the Geological Society of America.)

these general formulas (see Smith, 2007). The upper part of Figure 7.7 shows the pronounced pH dependence of various cations and sulfate on sorption by hydrous ferric oxide (FeOOH). Sorption of cations (Pb^{2+}, Cu^{2+}, etc.) increases with pH because the proton (H^+) is a competitor with the metal cation for sorption sites on the sorbent, while sulfate sorption decreases with pH because of the increase in the competing hydroxide ion at high pH.

Similarly, the lower part of Figure 7.7 shows there is strong pH dependence on sorption by hydrous ferric oxide of the **oxyanions** of selenium, chromium and arsenic, so called because they exist in dissolved form as stable, negatively charged anionic species such as $HSeO_3^-$ or $HAsO_4^{2-}$. The critical pH ranges over which sorption of these cations and oxyanions occur are known as their **adsorption edges**. For Pb and Cu the critical range is pH 3–5, while for alkaline-earth cations such as Ca, Mg and Sr it is pH 6.5–9.

The sorbents of greatest interest are those with high specific surface areas (measured in m^2/g) because large areas provide numerous sorption sites (see Langmuir, 1997). Interlayer clay minerals, such as montmorillonite, have specific surface areas of 600–800 m^2/g, while the simple clay mineral kaolinite has only 10–38 m^2/g. In each case, there may be only a single sorption site per unit square nanometre – mostly negatively charged – but the large surface area of clay minerals means that they are efficient sorbents. Oxide minerals, such as hydrous ferric oxide ($Fe(OH)_3 \cdot nH_2O$), have surface areas of 250–600 m^2/g with site densities of as much as 20/nm^2. Sandy soils, however, have low specific surface areas because clay minerals and oxide-mineral sorbents are present in limited amounts and quartz has a specific surface area of only ~0.1 m^2/g.

The aquifer sands shown in Figure 7.6 have specific surface areas of only ~1 m^2/g. The upland area in the top left corner of Figure 7.6 was used in the 1950s for disposal of radioactive liquids, including the hazardous radionuclide ^{90}Sr and the radioactive isotope of water, tritium. In the years following these disposals, it was observed from soil cores and monitoring wells that ^{90}Sr was migrating at only ~3% of the speed of tritium. Because the groundwaters were undersaturated with calcium carbonate, it was most unlikely that the ^{90}Sr would have been retarded by precipitation as a carbonate mineral, thus sorption appeared responsible for this retardation reaction (Jackson and Inch, 1983).

Soil cores collected beneath the uplands area (distance 200 m; Figure 7.6), where the ^{90}Sr was disposed, had a specific surface area of 1.2 m^2/g, while cores from the middle-sand aquifer at a distance of 350 m had specific surface areas of only 0.4 m^2/g. The difference in specific surface areas arose from the intense weathering of the biotite to vermiculite beneath the uplands area where DO was present in the groundwater. As Figure 7.5 shows, oxygen reduction can cause the oxidation of minerals containing Fe^{2+}, such as biotite mica. In this weathered zone, the vermiculite was present at 2–4% whereas it was present at 0.5% only 100 m away in the unweathered zone.

Figure 7.8 presents the relationship of ^{90}Sr between aqueous and sorbed states, the plot of which is known as an adsorption isotherm for reasons that have little to do with aqueous sorption but refer back to gas adsorption experiments where temperature effects are significant. Two linear isotherms are shown, with each indicated by a distribution coefficient (K_D), which is the slope of the linear isotherm. The weathered soil exhibited an isotherm with distribution coefficient $K_D = 9.8$ mL/g, while the isotherm slope for the unweathered zone was only $K_D = 5.9$ mL/g. The competing cations were slightly higher in the weathered zone (Ca $\sim$ 1.2 meq/L; pH $\sim$ 5.9) than in the unweathered zone (Ca $\sim$ 0.9 meq/L; pH $\sim$ 6.3); however, the abundance of vermiculite with its high specific surface area was sufficient to cause a higher degree of sorption in the weathered zone.

A general partitioning model was developed to assess how the various sorbents contributed to the sorption of a metal ion like ^{90}Sr when the distribution coefficients for individual minerals are known. Thus, consider the sorption of a dissolved metal ion, Me, by a solid S, $Me_T + S_T = MeS$, which yields the conditional equilibrium constant (in L/g)

$$K_{AD} = \frac{[MeS]}{[Me_T][S_T]},\qquad(7.34)$$

where [MeS] is the concentration of the sorbed metal ion in moles per litre, $[Me_T]$ is the total concentration of dissolved metal ion in moles per litre, and $[S_T]$ is the total concentration of the sorbent in grams per litre. The K_{AD} is a conditional equilibrium constant because it is conditional on the specific aqueous system in terms of temperature, sorbents and water chemistry.

If the conditional equilibrium constants and concentration of each of the k sorbents are known, i.e., $K_{AD}(k)$ and $S_T(k)$, then we may write (Jackson and Inch, 1983)

$$\frac{[MeS_{ST}]}{[Me_{ST}]} = \frac{\sum_k K_{AD}(k)S_T(k)}{1 + \sum_k K_{AD}(k)S_T(k)},\qquad(7.35)$$

where $[MeS_{ST}]$ is the total metal sorbed to all the various sorbents, $[Me_{ST}]$ is the total metal – both dissolved and sorbed – and $K_{AD}(k)$ and $S_T(k)$ are the conditional equilibrium constants and concentrations of the kth sorbent, respectively. Note that $[Me_{ST}] = [Me_T] + [MeS_{ST}]$. While the terms become somewhat mind-numbing, perhaps the reader will have noticed that the fraction of the metal sorbed to the kth sorbent must be

$$\frac{K_{AD}(k) \cdot S_T(k)}{\sum_k K_{AD}(k)S_T(k)}.\qquad(7.36)$$

An assumption of this partitioning model is that the sorption of any metal ion by these sorbents is described by a linear adsorption isotherm. For the case of ^{90}Sr discussed above, we have already established a linear sorption isotherm in Figure 7.8, thus $K_{AD} = K_D$. We may proceed to use this partition model to estimate the fractional distribution

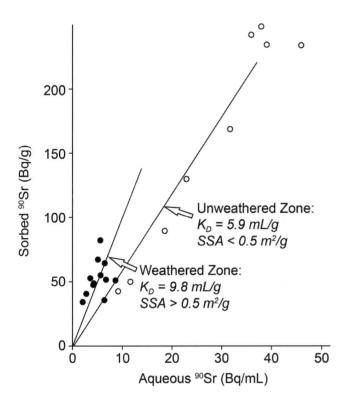

Figure 7.8 Adsorption isotherms for ^{90}Sr for Chalk River aquifer. Sorbed and dissolved concentrations in becquerels. (From Jackson and Inch (1989); with permission from Elsevier.)

Table 7.4 Partition model of ^{90}Sr in a contaminated sand aquifer, for which bulk density = 1.7 g/cm^3 and average porosity = 0.4. (From Jackson and Inch (1983).)

Sorbent	Abundance (%)	$10^3 K_{AD}$ (L/g)	S_T (g/L)	$K_{AD}S_T$	$K_{AD}S_T/\Sigma$
Quartz	45	0.4	1900	0.8	0.05
Feldspar	38	4.7	1600	7.5	0.45
Biotite	4	3.7	200	0.7	0.04
Vermiculite	4	37	200	7.4	0.44
Muscovite	3	2.6	100	0.3	0.02
	94			$\Sigma = 16.7$	1.00

of the radionuclide by mineral sorbents. Table 7.4 shows the fractional sorption of ^{90}Sr to the five identified mineral sorbents within the sand aquifer of Figure 7.6. While quartz comprises 45% of the aquifer minerals, it accounts for only 5% of the sorption of ^{90}Sr, which is consistent with our earlier discussion of its low specific surface area. Feldspar, biotite and muscovite all sorbed amounts of ^{90}Sr proportional to their concentration in the aquifer. By contrast, the 4% of vermiculite accounted for 44% of the sorbed radionuclide. The disproportionate effect of vermiculite shown in Table 7.4 is due to its high specific surface area, which indicates a very high density of sorption sites, many of which will be similar to hydrous ferric oxide because of the oxidation of the ferrous iron in the parent biotite mineral. The material balance on the model accounts for 94% of the ^{90}Sr.

7.8 Acid-Rock Drainage

7.8.1 Metal-Sulfide Mineral Oxidation

Acid-rock drainage (ARD), often referred to as **acid-mine drainage** (AMD), poses a particularly acute problem of aquatic contamination in mining regions. Sulfide minerals exposed either naturally or through mining or construction operations cause acidification of stream waters and the subsequent transport of suspended particulates with sorbed metal contaminants causing aquatic pollution and the destruction of fish habitat. In the USA there are between 100 000 and 500 000 abandoned or inactive mine sites, mainly in the western states, over 50 of which are identified for Superfund restoration under US environmental law. In British Columbia, Canada, over 60 major mine sites are believed to pose a significant ARD risk to surface waters (Price, 2003).

Open-pit mines or tailings, which are piles of waste or milled rock, expose these sulfide minerals, in particular pyrite, to atmospheric precipitation containing DO. We may write the chemical reactions associated with the release of dissolved iron by pyrite oxidation as follows (Seal and Hammarstrom, 2003):

$$FeS_2 + \tfrac{7}{2}O_2 + H_2O \rightarrow Fe^{2+} + 2SO_4^{2-} + 2H^+ \quad (7.37)$$

$$FeS_2 + 14Fe^{3+} + 8H_2O \rightarrow 15Fe^{2+} + 2SO_4^{2-} + 16H^+ \quad (7.38)$$

Whether reaction (7.37) or (7.38) will dominate the oxidation of pyrite depends on the pH of the aquatic system. Near-neutral conditions will favour reaction (7.37), while reaction (7.38) will likely dominate at low pH because of microbial

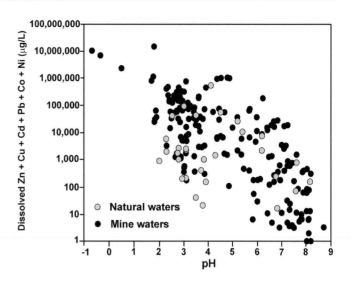

Figure 7.9 Ficklin diagram for ARD waters. (From Smith (2007); courtesy of the Geological Society of America.)

catalysis by *Acidithiobacillus ferrooxidans* that causes oxidation reactions to occur at rates orders of magnitude faster than reaction (7.37). Similar reactions may be written for the release of copper and acidity from chalcopyrite (CuFeS$_2$), zinc from sphalerite (ZnS), lead from galena (PbS), arsenic from arsenopyrite (FeAsS) and mercury from cinnabar (HgS). The net result of metal-sulfide mineral oxidation is shown in Figure 7.9, which illustrates how the dissolved load of metal ions released from sulfide minerals varies with pH in natural and mine-influenced waters draining various mineral deposits.

7.8.2 Mineral Deposits

The metal-sulfide minerals are mined from **hydrothermal** or **base-metal** ore deposits that were formed by high-temperature fluids in three geologic environments (Ridley, 2013): (I) areas of magmatic activity, in particular at plate boundaries; (II) regions undergoing magmatic and tectonic activity; and (III) sedimentary basins invaded by metal-transporting fluids that have migrated over distances of hundreds of kilometres.

The porphyry copper deposits of type I are located near modern plate boundaries, as in Arizona and Chile. The type II deposits include many legendary gold lode deposits, such as those of Kalgoorlie, Western Australia, and the gold-mining camps of the western Sierra foothills in California, which were formed in association with the intrusion of the Sierra Nevada batholith. The recently developed Carlin-type gold deposits in Nevada are also type II. In contrast to types I and II in which fluid temperatures typically exceed

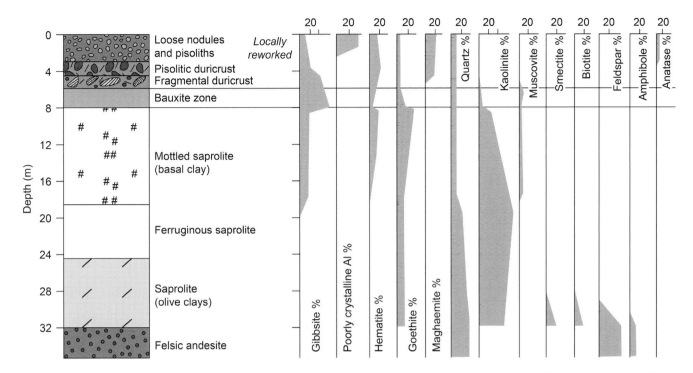

Figure 7.10 Lateritic soil profile showing the mineralogy of a bauxite profile in Western Australia that developed from the weathering of andesite. (From Ridley (2013), after Anand (1998), with permission of CRCLEME and Cambridge University Press.)

400°C, type III deposits were formed at lower temperatures (60–250°C) and the hydrothermal fluids were often transported as high-salinity waters in sedimentary basins and thus are referred to as **strata-bound** deposits. These include the lead–zinc deposits found in carbonate rocks of the US Midcontinent Basin that are known as Mississippi Valley-type (MVT) deposits, hydrothermal fluids that "exhale" from sedimentary rocks into bottom waters of marine troughs (SEDEX deposits) and the red-bed strata-bound copper deposits of Germany and Poland known as "Kupferschiefer", i.e., copper-shale deposits.

As opposed to these hydrothermal sulfide-mineral deposits, **supergene** deposits are formed at the surface under oxidizing conditions. Referring back to Chapter 6 on geomorphology, the weathering of rocks is promoted by the flow of rainwater and snowmelt through the soil horizon and the underlying regolith as shown in Figure 6.2. Geologists often refer to water that is an active part of the hydrologic cycle as "meteoric water", which is a misleading term now embedded in the literature. It is this process of groundwater recharge and percolation that drives the development of supergene ores.

Ridley (2013) has identified several types of supergene deposits of which bauxite deposits are perhaps the most extreme examples due to deep weathering (section 6.2) in a seasonally humid environment. The bauxite deposits of Australia are residual hydrous aluminum oxide deposits

(gibbsite, boehmite) developed from the intense weathering of Al-rich source rocks, such as granites, through the formation of **lateritic weathering profiles**. Figure 7.10 illustrates these mineralogical profiles in a former bauxite mine in Western Australia containing gravel-sized concretions known as pisolites that overlie a ferruginous zone of ferric oxide and silica concretions known as a duricrust, which in turn overlies a thin bauxite zone, clay layer, ferruginous saprolite and, finally, unaltered andesite bedrock (Anand, 1998). Therefore the bauxite ore is close to the top of the lateritic profile and usually of several metres thickness covering an area of tens of square kilometres. With depth to the ferruginous upper saprolite, the mineralogical profile shows an increase in kaolinite and a decrease in gibbsite.

The schematic of a typical open-pit mine shown in Figure 7.11 illustrates the principal features of such mining operations and the volumes of rock requiring removal, whether it is enriched ore, sub-economic ore or waste rock. Once the overburden is removed, any supergene ore can be mined. The ore body itself comprises reserves and resources. **Reserves** are legally defined in many nations and are extractable on the basis of current economic, engineering and regulatory conditions. **Resources** are ore bodies that have the potential for future extraction.

Large volumes of waste rock are identified in Figure 7.11 that must be mined to allow extraction of the ore body. This

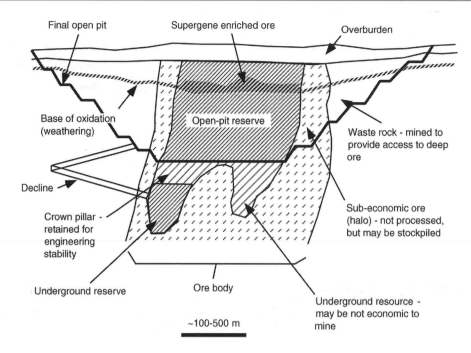

Figure 7.11 Schematic cross-section through an open-pit copper mine identifying the geological, economic and engineering definitions of ore. (From Ridley (2013); with the permission of Cambridge University Press.)

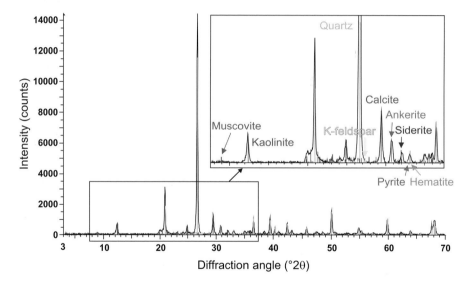

Figure 7.12 X-ray diffractogram of waste-rock minerals. (From Raudsepp and Pani (2003); with the permission of the Mineralogical Association of Canada.)

waste rock and tailings from milling operations to concentrate the ore must be stored near the mine or mill. Such waste-rock and tailings piles frequently contain significant amounts of sulfide minerals as well as **gangue minerals**, which are those minerals in rock that surrounds the ore but which have no economic value. Quantitative analysis of the mineral fraction of waste rock and mill tailings is undertaken by X-ray diffraction analysis of powdered materials from the piles (Raudsepp and Pani, 2003). Figure 7.12 presents an X-ray powder pattern of waste rock indicating the presence

of calcite, quartz, kaolinite, ankerite ($CaFe(CO_3)_2$), siderite, hematite, K-feldspar, muscovite and pyrite.

7.8.3 Reactions in Mine-Waste Tailings

In sulfide-containing piles of waste rock and tailings, the initial oxidation reaction usually involves pyrite (Jambor, 2003), although pyrrhotite (Fe_7S_8) may be the principal sulfide mineral present and is also rapidly oxidized. Pyrrhotite oxidation leads to its replacement by native sulfur (S) and marcasite (also FeS$_2$) and subsequently to the formation of goethite (FeOOH) and various ferrous sulfate minerals such as jarosite (KFe$_3$(SO$_4$)$_2$(OH)$_6$). pH buffering by various minerals occurs in the predictable sequence shown in Table 7.5 following contact with DO from rainwater or snowmelt.

Figure 7.13 demonstrates the progressive acidification of pore water in tailings containing carbonate minerals, which provide most of the neutralization capacity (Blowes et al., 2003):

$$CaCO_3 + H^+ \rightleftharpoons Ca^{2+} + HCO_3^- \qquad (7.39)$$

Dissolved iron from sulfide oxidation becomes available to form secondary siderite:

$$Fe^{2+} + HCO_3^- \rightleftharpoons FeCO_3 + H^+ \qquad (7.40)$$

Any primary siderite, as well as the secondary siderite, will then be dissolved:

Table 7.5 The pH buffering ranges of various minerals observed in acid-neutralization sequences. (From Blowes et al. (2003).)

Mineral	pH range
Calcite	6.5 to 7.5
Siderite	4.8 to 6.3
Al(OH)$_3$	4.0 to 4.3
Fe(OH)$_3$	2.5 to <3.5

$$FeCO_3 + H^+ \rightleftharpoons Fe^{2+} + HCO_3^- \qquad (7.41)$$

Finally, hydrous oxides of Al^{3+} (or Fe^{3+}) are formed and subsequently dissolved should acidification continue:

$$Al^{3+} + 3(OH)^- \rightleftharpoons Al(OH)_3 \qquad (7.42)$$

Similarly goethite may form from jarosite. As oxygen enters the tailings or waste rock pile and is dissolved by the percolating pore water, mobile ions (As, Zn, Ni, SO$_4^-$) will be transported into the deeper tailings. As the acid neutralization capacity of the tailings is progressively depleted with depth, an acidic pore water, rich in dissolved metals, seeps into the underlying sediments or bedrock and ultimately discharges to a stream (see Figure 7.14), causing the aquatic contamination shown in Figure 7.9.

7.9 Clay Minerals

Clay minerals, which were introduced in section 2.3.1, form either through alteration of rocks by weathering or by precipitation from hot seawater (Wenk and Bulakh, 2004). They comprise five groups with similar repeating structures of aluminum octahedra and silicon tetrahedra, which constitute a unit cell, as shown in Figure 2.10. By ionic or **isomorphous substitution**, in which a silicon ion (4+) may be replaced by an aluminum ion (3+), clay minerals develop a net negative charge and thus a positive cation exchange capacity. This is possible because of the similar ionic radii of Al (0.053 nm)

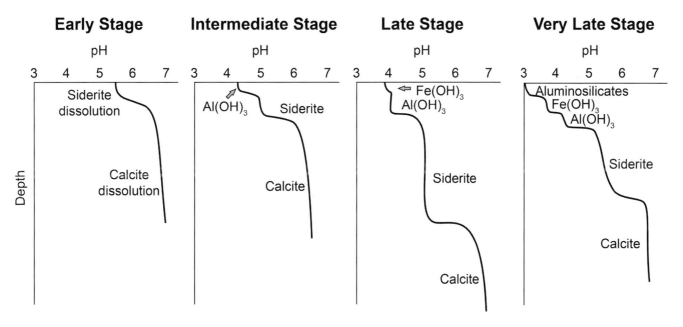

Figure 7.13 Schematic showing the progressive change in pore-water pH over time in waste-rock tailings. (From Blowes et al. (2003); with the permission of the Mineralogical Association of Canada.)

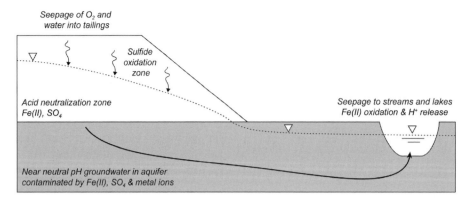

Figure 7.14 Schematic showing the transport of dissolved components along the groundwater flow path from the vadose zone of a tailings deposit to surface-water discharge. (From Blowes et al. (2003); with the permission of the Mineralogical Association of Canada.)

and Si (0.040 nm). Other pairs of ions may also be responsible for ionic charge development, usually within the aluminum octahedral layer. Figure 7.15 illustrates the structure, composition, surface areas (SA), basal spacing of their unit cells and cation exchange capacities of the principal clay mineral groups.

The simplest of these groups is that of **kaolinite**, which is shown in Figure 2.10 in atomic form, and in which the unit cell is 0.7 nm thick. In many references, the unit cells are reported in angstroms (1 Å = 0.1 nm); thus kaolinite may be referred to as a "7 Å mineral". Halloysite has the same structural formula as kaolinite but also contains attached water molecules: $Al_2Si_2O_5(OH)_4 \cdot 4H_2O$. These minerals have a 1 : 1 layer structure, i.e. single tetrahedral and single octahedral layers.

Clay minerals with a 2 : 1 layer structure, i.e. two tetrahedral layers and a single octahedral layer, are much more geochemically reactive. The **smectite** group and **vermiculite** have large specific surface areas, which will contain many charged sites, and consequently high cation exchange capacities. The most common of the smectite group is montmorillonite (see Figure 2.10). Smectite minerals have a unit cell basal spacing that varies between 9.6 and 21.4 Å, but is nominally 14.2 Å. The interlayer spaces are occupied by exchangeable ions. Geotechnical engineers make frequent use of a particular montmorillonite-rich clay material known as bentonite for sealing piezometers and annular spaces around shafts to prevent water flow.

Mica minerals are layer silicates that are often associated with clay minerals. These minerals, such as muscovite, illite and biotite, have unit cells of 10 Å. Muscovite is a white or grey-coloured mineral with a typical chemical formula of $KAl_2Si_3AlO_{10}(OH)_3$, whereas biotite is usually brown

or black due to the presence of Mg^{2+} and Fe^{2+}. Chlorite has a green appearance associated with an interlayer unit of $Mg(OH)_2$ and a unit cell of 14 Å. In the slow process of mineral alteration known as diagenesis, illite is formed in sandstones from kaolinite by the addition of K^+ and silica and the loss of alumina and water, while chlorite is formed from smectite by adding Fe^{2+} and Mg^{2+} and losing silica, water, Na^+ and Ca^{2+}. In petroleum geology, illite and chlorite are considered to be indicators of deep burial (>1 km) and thus indicative of conditions under which oil and natural gas may form.

7.9.1 Heave in Colorado Claystones

Heave can occur without a primary role being played by pyrite oxidation as mentioned in Box 7.4. Near Denver, Colorado, heaving bedrock is associated with (i) the hydration and subsequent swelling of bentonitic beds within the claystone exposed at surface and (ii) hydration swelling of bedrock blocks due to infiltration events (Noe et al., 2007). These authors note that present-day heaving is associated with the uppermost 20 m of weathered bedrock that was heavily fractured by past rebound that allowed deep infiltration and oxidation in the past.

In the steeply dipping bedrock shown in Figure 7.16, it is possible that water may infiltrate to considerable depth via the fractures, causing the saturation of adjacent claystones and thus some heaving in the variably weathered and unweathered bedrock zones portrayed. However, rebound and gypsum crystallization from pyrite oxidation were not identified as important to present-day heaving, although swelling of rocks by the hydration of anhydrite to gypsum is a well-known concern in tunnelling projects (Rauh et al., 2006).

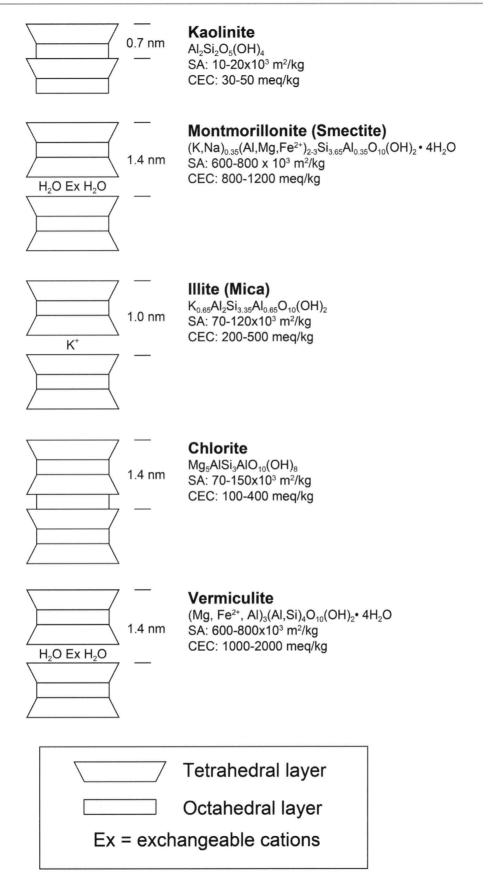

Kaolinite
$Al_2Si_2O_5(OH)_4$
SA: $10\text{-}20 \times 10^3$ m²/kg
CEC: 30-50 meq/kg
0.7 nm

Montmorillonite (Smectite)
$(K,Na)_{0.35}(Al,Mg,Fe^{2+})_{2\text{-}3}Si_{3.65}Al_{0.35}O_{10}(OH)_2 \cdot 4H_2O$
SA: $600\text{-}800 \times 10^3$ m²/kg
CEC: 800-1200 meq/kg
1.4 nm
H_2O Ex H_2O

Illite (Mica)
$K_{0.65}Al_2Si_{3.35}Al_{0.65}O_{10}(OH)_2$
SA: $70\text{-}120 \times 10^3$ m²/kg
CEC: 200-500 meq/kg
1.0 nm
K^+

Chlorite
$Mg_5AlSi_3AlO_{10}(OH)_8$
SA: $70\text{-}150 \times 10^3$ m²/kg
CEC: 100-400 meq/kg
1.4 nm

Vermiculite
$(Mg, Fe^{2+}, Al)_3(Al,Si)_4O_{10}(OH)_2 \cdot 4H_2O$
SA: $600\text{-}800 \times 10^3$ m²/kg
CEC: 1000-2000 meq/kg
1.4 nm
H_2O Ex H_2O

Tetrahedral layer

Octahedral layer

Ex = exchangeable cations

Figure 7.15 Structure and properties of clay minerals. All data from Wenk and Bulakh (2004) except cation exchange capacity from Appelo and Postma (2005).

BOX 7.4 | THE EFFECTS OF PYRITE OXIDATION ON ENGINEERED STRUCTURES

Pyrite oxidation poses problems of ARD or AMD for mining engineers. However, civil and geological engineers face other problems created by geochemical weathering of pyrite. Chigira and Oyama (1999) list the following.

Heave in black shales. Black shales are known to heave upon oxidation, causing structural damage to buildings built above such shales. The dark colour of these shales is due to their high organic carbon content. Structural damage caused by this process has occurred in cities as far apart as Oslo in Norway, Cleveland and Pittsburgh in the USA, and Ottawa and Québec in Canada. The actual mechanism of heave in black shale beneath building foundations was perhaps first confirmed by Quigley et al. (1973). They identified the growth of gypsum ($CaSO_4 \cdot 2H_2O$) crystals as the primary cause of heave, with the calcium being produced by calcite dissolution following microbially catalyzed pyrite oxidation; these crystals grew in bedding plane fractures within the shale bedrock. Bérubé et al. (1986) discussed a somewhat different occurrence at a college near Québec City, in which both gypsum and fibroferrite ($FeSO_4 \cdot 5H_2O$) crystallization were thought to be responsible for the 10–20 cm of heave that occurred over ten years. These authors discuss possible construction and remedial approaches for heave-induced building damage. Matheson and Quigley (2016) have discussed the numerous cases of structural damage caused by pyrite oxidation in Dublin, Ireland, mudrocks and backfill aggregate composed of the mudrock, and have outlined an instructive sequence of mineralogical, petrologic and chemical testing procedures to identify the presence of the pyrite and the gypsum, which seldom amounted to more than 2% of mineral grains in the fractures of the aggregate.

Embankment failure. Evidence of extremely rapid weathering of mudrocks emerged in a study of the partial collapse of a dam embankment in Derbyshire, England. The failure was caused by pyrite oxidation of shale fill on the downstream face of the dam that occurred rapidly after emplacement. Pye and Miller (1990) reported that 4–10% volume loss of the fill occurred without heave. Jarosite and gypsum were the principal secondary minerals formed.

Weathering of unlined tunnel walls. The ventilation of unlined tunnels and rock excavations allows oxygen diffusion into the drying rock. Oyama and Chigira (1999) describe the formation of gypsum and an Al-sulfate on the surfaces of such rock walls. The oxidation fronts had penetrated 2–65 cm into the mudstone over a period of 45–85 years. These authors measured diffusion coefficients for oxygen, which were directly proportional to the hydraulic conductivity of the weathered rock. In such oxidized zones of sedimentary rock, sandstones may be somewhat strengthened by the formation of ferric oxide cements while oxidized mudrocks are significantly weaker than fresh rock zones.

Corrosion. Chigira and Oyama (1999) indicate that corrosion of iron and steel pipes and sheet pile will likely occur in groundwater acidified by sulfide-mineral oxidation. Thus iron water and sewer pipes emplaced in pyrite-bearing soils or rock may corrode given that backfill will allow oxygenated conditions to develop. Similarly, concrete footings may become damaged by sulfate attack.

Noe et al. (2007) recommend that, in situations such as in the Denver area, it is important for geotechnical engineers and engineering geologists to carefully note in logging their cores both the depth of weathered bedrock (i.e., depth of weathered claystone and of fractures) and the penetration depth of oxidized fractures into the variably weathered bedrock. The source of the infiltrating water may be rain,

snowmelt, irrigation water or the leakage of water from water-supply or sewer pipes that have failed from corrosion or heave.

7.9.2 The Hong Kong Landslides

The steep hillsides of coastal China and Japan are associated with landslides due to high rainfall falling on weathered

(a) Regional Framework Characteristics

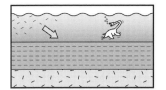

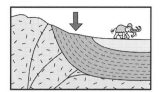

Deposition and Burial:
Clay-Rich Sediments

Uplift, Faulting, Folding:
Steeply Dipping Beds

Erosion and Unloading:
Overconsolidated

(b) Site-Specific Framework Characteristics

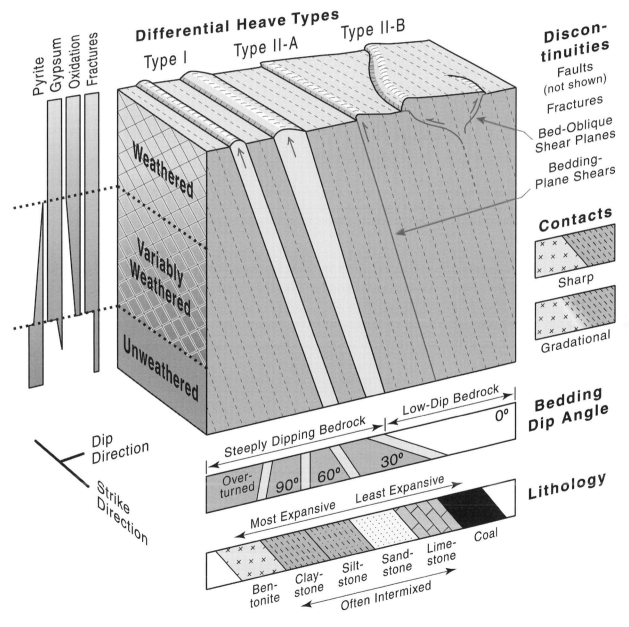

Figure 7.16 Conceptual 3D model of site-scale characteristics of steeply dipping claystone showing interpreted relationships between heave features and the underlying bedrock framework. (From Noe et al. (2007); with the permission of the Association of Environmental and Engineering Geologists.)

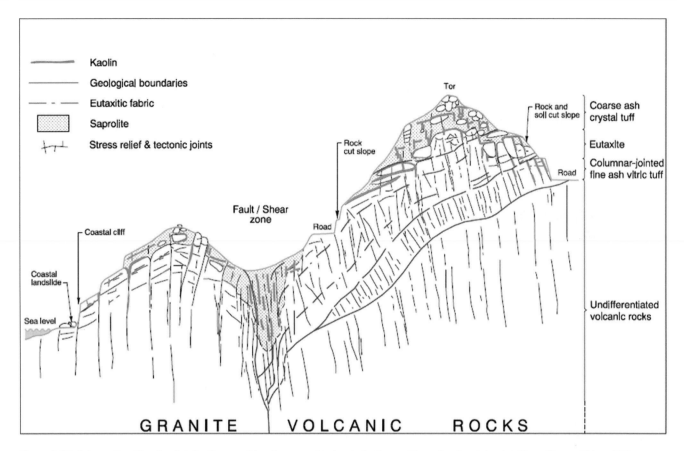

Figure 7.17 Schematic of kaolin distribution resulting from weathering in both granitic and volcanic rocks, Hong Kong, China. (After Campbell and Parry (2002).)

bedrock. On the other side of the Pacific Ocean, similar problems are associated with coastal hillsides in Oregon, Washington and British Columbia. Rock type alone does not explain landslides, but in Japan certain rocks are associated with landslides because of their weathering: granitic bedrock, volcanic rock known as ignimbrite, mudstones and sandstones. The weathering reactions break down the minerals in the unweathered rock into secondary ones and the fragmented and unconsolidated soil that develops (the **regolith**) lacks the shear strength of the parent minerals. The regolith is often rich in clay minerals and Fe and Mn oxides. Sidle and Ochiai (2006) write that it is the interaction of the weathering of bedrock with the infiltration of rainwater and the subsequent rise in pore pressure – and decrease in effective stress – that most accurately describes the susceptibility of these rocks to slip.

The Geotechnical Engineering Office of Hong Kong (2007) is famous for its expertise in the subject of landslides and engineering geology in general. In 1995, Typhoon Helen triggered a series of 70 landslides, two of which were large and deep-seated and caused several fatalities and much property damage. The bedrock in Hong Kong is a mix of granitic and

volcanic rocks that have weathered into a **saprolite** containing clays and hydrous Fe–Mn oxides. It was discovered that the failure planes along which the landslides slipped downslope were rich in kaolinite, either as a low-permeability layer parallel to the bedding planes or as kaolinite-filled joints in volcanic rocks (Kirk et al., 1997). Using electron microscopes, it was determined that much of the kaolinite had been converted by the curling of the kaolinite "books" into tubes of halloysite, a fibrous, hollow clay mineral with a layer of weakly bound water. Halloysite has also been implicated in rupture surfaces associated with landslides in the Oregon coastal ranges and in Kyushu, Japan. Figure 7.17 is a schematic illustration of the presence of kaolinite clays in weathered granitic and volcanic rocks in Hong Kong, China.

Campbell and Parry (2002) undertook a geotechnical analysis of the clay-rich zones in the weathered rocks associated with the 1995 landslides. These included infill zones, in which the clays were deposited in pre-existing discontinuities, and veins, in which the clays were less frequently deposited in discontinuities that may have a modern origin, e.g., stress relief. The kaolinite was found to be commonly associated with black deposits of manganese oxides in relict joints that

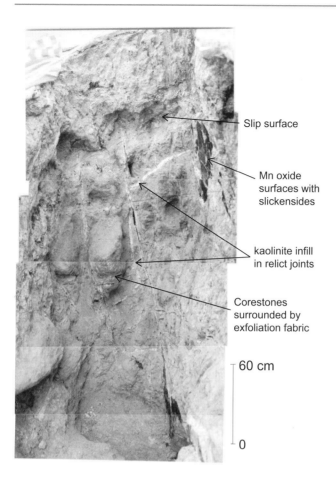

Slip surface

Mn oxide surfaces with slickensides

kaolinite infill in relict joints

Corestones surrounded by exfoliation fabric

60 cm

0

Figure 7.18 Kaolinite (white) and manganese oxide (black) infillings of relict joints in a trench excavated following the 1998 Sai Sha Road Landslide, Hong Kong, China. (From Campbell and Parry (2002).) See colour plate section.

clays, especially montmorillonite, talc, chlorite, serpentine, micas, graphite and molybdenite.

7.9.3 Clogging of Tunnel Boring Machines in "Sticky" Clay Soils

In Box 2.3 we discussed how swelling of clay minerals contained within the mudstone rocks beneath Auckland, New Zealand, hinders the operation of tunnel boring machines (TBMs). Much research has been conducted by German engineering geologists on this subject. Thewes and Burger (2005) reported that the advance rate of TBMs in clay formations can be just two-thirds that in granular formations and thus can cause cost overruns and contractual disputes. The clogging occurs at the cutting wheel and the suction inlet area that are not effectively flushed. Spagnoli et al. (2011) examined how different electrolytes (NaCl and $CaCl_2$) affected the clogging potential, while Sass and Burbaum (2009) discussed the development of a method for assessing the adhesion of clays to TBMs.

However, it is not just clay soils and mudstone that create clogging problems for TBMs. Rauh et al. (2006) point out that anhydrite swelling causes similar problems during TBM operation, as gypsum is produced upon wetting. Clogging in this case is related to the crystallinity of the anhydrite particles, which they indicate is a function of the overburden thickness. Older rocks, such as the Permo-Triassic Haselgebirge, contain massive anhydrite formed under 6 km of overburden, thus allowing time, pressure and temperature effects to create large anhydrite crystals. Such crystals have relatively low swelling capacity compared with the Gypsum Keuper Formation of the Upper Triassic that had a rock cover of only ~1 km and contained fine-grained anhydrite interlayered with black claystone.

also indicate shear surfaces (see Figure 7.18). They determined that the platy morphology of kaolinite caused it to have both lower peak and residual shear strengths than those of the halloysite. Goodman (1989) has identified the following minerals as having a low coefficient of friction (see equation (1.3)) and therefore may present a problem of slope stability:

7.10 Summary

The focus of this chapter is the varied and important effects of chemical weathering, which directly causes (i) soil formation by weathering bedrock, (ii) groundwater to develop chemical properties by reacting with minerals and gases, (iii) the formation of ore deposits by leaching and concentrating minerals, and (iv) the development of landslide-prone terrain, as in Hong Kong. These products of chemical weathering are intimately coupled with mechanical weathering, which was discussed in the previous chapter.

While inorganic water quality data are typically examined today using commercial software, the engineer should understand the basis of the reactions considered by the software. Fundamentally these involve understanding the ionic strength and ion activities of an aqueous sample and their effect on mineral solubilities. Dissolution of oxygen and carbon dioxide gases control the basic properties of natural and contaminated waters through their ability to cause mineral dissolution and precipitation.

Various minerals form as acid–base and redox properties change, some of which are powerful sorbents of contaminants and some of which are weathering products that concentrate minerals. Mining operations expose sulfide ores to oxidation, resulting in extremely acidic waters that cannot support aquatic life.

Clay minerals are the result of extensive weathering of primary rocks. Most usefully, they provide the ore for the production of ceramic materials. They are essential to geotechnical engineers in that they permit structures to be sealed against water circulation and thus isolated for their intended purpose, e.g., piezometer seals or nuclear-waste repositories. At the same time, they may cause hazards by swelling and thus heave of foundations. Furthermore, their weak shear strength causes problems with slope stability.

7.11 FURTHER READING

Appelo, C.A.J. and Postma, D., 2005, *Geochemistry, Groundwater and Pollution*, 2nd edn. Leiden, The Netherlands: Balkema. — A detailed introduction to groundwater chemistry that uses the PHREEQC code, which may be downloaded without cost from the USGS's website.

Clark, I.D., 2015, *Groundwater Geochemistry and Isotopes*. Boca Raton, FL: CRC Press. — The best introduction to groundwater quality available for the beginning engineer or geoscientist.

King, T.V.V. (ed.), 1995, *Environmental Considerations of Active and Abandoned Mine Lands*. USGS Bulletin 2220. Denver, CO: US Geological Survey. — An introduction to the issue of acid-rock drainage with reference to the USGS's work at the Summitville Mine in Colorado.

Langmuir, D., 1997, *Aqueous Environmental Geochemistry*. Upper Saddle River, NJ: Prentice-Hall. — More rigorous than Appelo and Postma, with particular focus on the physical chemistry of natural waters, including the computation of activities, redox processes and sorption reactions.

Ridley, J., 2013, *Ore Deposit Geology*. Cambridge, UK: Cambridge University Press. — The various types and origins of ore deposits are described in a style that is readily understood by engineering students lacking formal geological training.

Schaetzl, R.J. and Thompson, M.L., 2015, *Soils: Genesis and Geomorphology*, 2nd edn. New York, NY: Cambridge University Press. — A well-illustrated guide to all aspects of soil science.

Zhu, C. and Anderson, G., 2002, *Environmental Applications of Geochemical Modeling*. Cambridge, UK: Cambridge University Press. — Geochemical modeling of environmental problems are presented and discussed; suitable as a guide for using PHREEQC or MINTEQ or similar software.

7.12 Questions

Question 7.1. The alkalinity of a water sample is reported as 120 mg/L as $CaCO_3$. What is its concentration in terms of the bicarbonate ion?

Question 7.2. What are the activity coefficients of calcium and bicarbonate of the soil water shown in Table 7.1?

Question 7.3. Cores of rock are recovered from a series of shallow boreholes in which the groundwater is known to be relatively dilute, i.e., TDS < 500 mg/L. The following minerals are identified within the cores: (1) gypsum in a core section containing no fractures; (2) reddish goethite crystals present in fractures; (3) pyrite in a deeper fracture; and (4) kaolinite in shallow fractures. Indicate what the presence of these minerals indicates about the geochemical environment in which these minerals were identified.

Question 7.4. According to Figure 7.5, why is it likely that reduction reactions in which organic carbon molecules are oxidized result in higher TDS concentrations in groundwaters?

Question 7.5. According to Figure 7.5, why is it that oxidation reactions result in lower pH?

Question 7.6. The equilibrium redox relationship between hydrous ferric oxide and the ferrous ion may be written as

$$Fe(OH)_3 + 3H^+ + e^- = Fe^{2+} + 3H_2O, \qquad E_H^0 = 1.06\,V. \tag{7.43}$$

Assume your water sample has a measured pH of 6.5 and contains 0.56 mg/L of dissolved iron. What is the estimated E_H?

CHAPTER

8 Glacial Sediments and Permafrost

In 2009, the *New York Times* reported that a Mr. DeBoer had opened a nine-hole golf course at the mouth of Glacier Bay near Juneau, Alaska, in 1998 on land that was regularly inundated by tides when his family settled in the area in the 1940s. Glacial retreat of ~10 m/yr and isostatic rebound of ~7 cm/yr has provided the golfers of Juneau with a recreational benefit from global warming!

In the same year, Moore et al. (2009) noted that the mountain glaciers of western North America have retreated since the "Little Ice Age" of the nineteenth century and now present a variety of **geomorphic hazards**, including (i) the potential failure of outburst floods from moraine-dammed lakes, (ii) landslides on oversteep slopes and (iii) debris flows on moraines. Similarly, *Der Spiegel* reported in 2013 that, in Switzerland, the melting of the glaciers is creating numerous lakes and also the threat of outburst floods and landslides in populated areas. The melting and retreat of the world's glaciers from China to Alaska, particularly from 2000 to 2015, is unprecedented (Goudie and Viles, 2016).

Glaciers are composed of perennial snow and ice masses that are continuously, if slowly, migrating by gravity. Over time, snow falling on the glacier is compressed and recrystallized into ice. The ice mass undergoes **ablation**, which is loss of mass due to melting and runoff, evaporation, sublimation and calving, i.e., collapse at its leading edge.

In this chapter we consider first the history and physical nature of glaciation during the Quaternary, then the nature of glacial sediments, and close by considering permafrost, periglacial environments and their geotechnical importance.

8.1 Glaciation during the Quaternary

In section 2.2 the Quaternary period of the Cenozoic Era was identified as comprising the Holocene epoch, beginning 11.7 ka, and the Pleistocene epoch, that began the Quaternary at 2.58 Ma (Gibbard et al., 2009). Thus the reader lives in an interglacial epoch that is nearly 12 000 years old. Furthermore, we noted that the origin of ice ages has been associated with Milankovitch cycles caused by the Earth's orbit around the Sun (Figure 2.9) and that continental glaciation had resulted in isostatic rebound during deglaciation (Figure 2.8).

The most recent glaciation of eastern North America – the **Wisconsin glacial stage** – began approximately 90 ka with the radial expansion of the Laurentide Ice Sheet from northern Québec towards the Great Lakes Basin but did not cross the Great Lakes until about 23 ka. The **Last Glacial Maximum** (LGM) occurred in the Great Lakes Basin at 18 ka, as shown in Figure 8.1. At this time, the volume of the Laurentide Ice Sheet was ~ 33×10^7 m^3 or enough to lower the mean sea level by 120 m. It stretched from the Rocky Mountains to the Atlantic coast and north to the Arctic.

Contemporaneous continental glaciations occurred in northern Europe (Anderson et al., 2007), where it was known as the Weichsel, and western North America, with ice domes in Scandinavia and the Canadian Cordilleran mountains, respectively. Elsewhere, regional glaciations occurred in the European Alps, the Andes, Antarctica and mountain ranges in southern California, New Zealand and Australia. Mean temperatures in the mid-northern latitudes were 5–15°C cooler than at present.

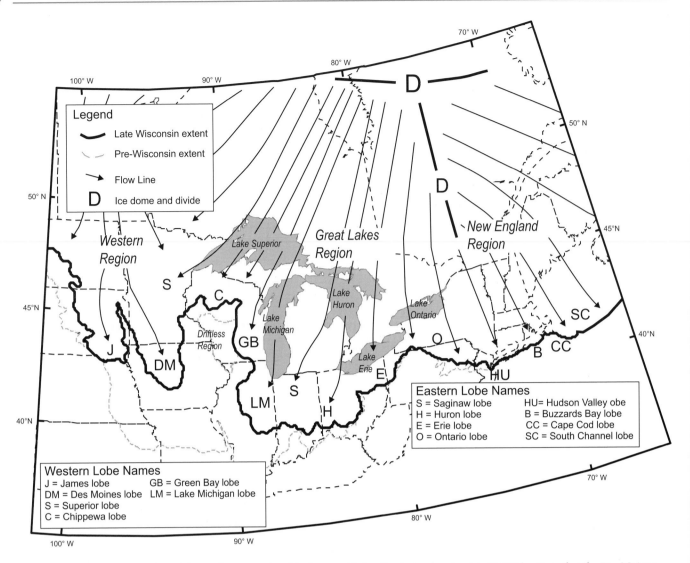

Figure 8.1 The maximum extent of the southern margin of the Laurentide Ice Sheet at approximately 21 000 ka. Note that the Des Moines lobe developed much later (14 000 ka), while the New England lobe was earlier (23 000 ka). Ice flow directions are shown relative to the ice dome in the Laurentian highlands in eastern Canada. (From Colgan et al. (2003).)

These cooler temperatures were estimated using the $^{18}O/^{16}O$ environmental isotope ratio (see Box 8.1 and Figure 8.2) of deep cores of glacial ice and seabed foraminifera microfossils ("**forams**"). The $^{18}O/^{16}O$ in the calcite of forams indicates the progress of glacial and interglacial events in the Quaternary period. During glacial events, large volumes of seawater are evaporated and fall as snow in high latitudes, resulting in growing snow packs and ice sheets. During evaporation, the lighter oxygen molecule (^{16}O) is preferentially volatilized, leaving an enriched $^{18}O/^{16}O$ signature in the seawater, which is incorporated in the calcite of the forams. During **interglacial** events, significant warming occurs and there is ice retreat and re-vegetation of the tundra by forest. Furthermore, there is a net flux of ^{16}O to the oceans, such that the $^{18}O/^{16}O$ ratio decreases and is recorded again in the calcite of the forams. An **interstadial** event is a period of somewhat greater warmth during a glacial cycle. Box 8.1 describes how stable environmental isotopes are used to measure climatic conditions.

Figure 8.3 shows the isotopic stratigraphy of $^{18}O/^{16}O$ in the Santa Barbara channel off the Californian coast. Note that the difference between the glacial maxima (stages 2 and 4) and the interglacial stages (1 and 5) are of the order of $1-2\,^0/_{00}$ (per mil). Anderson et al. (2007) note that a change of $0.1\,^0/_{00}$ is believed to indicate a 10 m change in sea level. By using spectral (Fourier) analysis, it has been shown that these time series are consistent with the Milankovitch cycles illustrated in Figure 2.8 and concluded that the sequence of Quaternary ice ages are a consequence of the Earth's solar orbit.

BOX 8.1 | STABLE ENVIRONMENTAL ISOTOPES AS CLIMATIC INDICATORS

Stable isotopes are identified by their atomic mass, or the sum of protons and neutrons, in atomic mass units (amu). Thus the nuclide with 8 protons and 8 neutrons identifies one with 16 amu, which is ^{16}O. However, about 0.2% of all oxygen atoms have 10 neutrons, and the ratio of $^{18}O/^{16}O$ is indicative of the temperature at which the water molecule containing this oxygen atom underwent isotopic partitioning in the hydrologic cycle. This partitioning is known as **fractionation**.

Fractionation causes small changes in the isotopic ratios within a water mass. Consequently the isotopic ratio of a winter rain may have a slight but significant difference from that of a summer rain. Isotopic differences are measured in parts per thousand (per mil, $^0/_{00}$) and the results reported as

$$\delta^{18}O_{sample} = \left[\frac{(^{18}O/^{16}O)_{sample}}{(^{18}O/^{16}O)_{reference}} - 1 \right] \times 1000\,^0/_{00}\ VSMOW, \qquad (8.1)$$

where VSMOW refers to Vienna Standard Mean Ocean Water, the reference standard maintained by the International Atomic Energy Agency (IAEA) in Vienna, Austria. Positive values of $\delta^{18}O$ indicate that a sample has an excess of ^{18}O compared with the reference and is referred to as "enriched", while negative values indicate a depletion relative to the reference, e.g., a sample with $\delta^{18}O = -10\,^0/_{00}$ has 1% less ^{18}O than VSMOW. Similar expressions are used for deuterium (D), the heavy isotope of hydrogen in the water molecule, thus $\delta D \equiv \delta^2H$, and also $\delta^{34}S$, $\delta^{13}C$, etc., each with its own reference standard.

Cold regions exhibit isotopically depleted waters while warm regions are noted for isotopically enriched samples. Seawater evaporation produces a depleted water vapour with $\delta^{18}O_v \sim -10\,^0/_{00}$ VSMOW. Figure 8.2 shows this relationship in the form of the predicted isotopic ratios of $^{18}O/^{16}O$ in the water vapour and the rainfall during orographic precipitation. The vapour mass is more depleted than the rainfall derived from the vapour because condensation during **rainout** enriches the precipitation in ^{18}O and D.

The relationship of these two isotopic ratios is given by the **global meteoric water line** for global atmospheric precipitation otherwise known, confusingly, as **meteoric water** (Clark and Fritz, 1997): $\delta D = 8.17\delta^{18}O + 11.27\,^0/_{00}$ VSMOW. Local meteoric water lines (LMWLs) can be drawn to demonstrate how climatic effects create specific relationships. Thus modern groundwater in San Diego, California, is described by $\delta D = 8.0\delta^{18}O + 2$, while precipitation at Resolute Bay in northern Canada is given by $\delta D = 7.8\delta^{18}O + 5.3$. The interested reader may consult the Global Network of Isotopes in Precipitation (http://www-naweb.iaea.org/napc/ih/IHS_resources_gnip.html) for other LMWLs.

Remenda et al. (1994) showed that ancient groundwater trapped in glaciolacustrine clays in North Dakota and adjacent Manitoba could be used to establish the late Pleistocene climate of the area. While modern precipitation in the area averages -13 to $-14\,^0/_{00}$, the pore water from the clays averaged $-25\,^0/_{00}$, which corresponds to an air temperature of $-16°C$, compared with an average air temperature today of $0°C$. Ferguson and Jasechko (2015) have shown that groundwater ^{18}O values in North America vary from -12.5 to $-25.3\,^0/_{00}$ and become less negative towards the southern extremity of glaciation. They further note that groundwater values are typically higher (less negative) that the average $\delta^{18}O$ value for the Laurentide Ice Sheet.

Interglacial 5e, which occurred 120 ka, resulted in the marine inundation of coastal southern California (Sengebush et al., 2015) that has been associated with sea-level rise ($\sim$6 m) onto a landmass undergoing tectonically induced uplift ($\sim$17 m). Karrow et al. (2000) have correlated the stratigraphy of glacial sediments in the Great Lakes area of North America with these same marine oxygen isotope stages back to 150 ka. The interglacial that preceded the Wisconsin glaciation (marine isotope stage 5 in Figure 8.3) is known as the Sangamon in North America, the Ipswichian in the UK and the Eemian in Northern Europe. Anderson et al. (2007) provide a full account of global patterns and cycles of climatic change throughout the Pleistocene and Holocene.

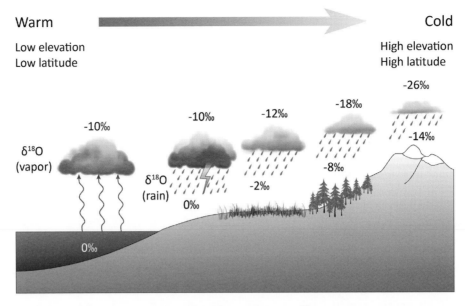

Figure 8.2 Isotopic evolution of "meteoric water" during orographic precipitation as ^{18}O. Water vapour evaporating from the ocean is shown to be depleted relative to ocean water ($-10\,^0/_{00}$) and becomes increasingly depleted with elevation (i.e., lower temperature and pressure). Condensation during rainout causes isotopic enrichment of ^{18}O; thus atmospheric vapour (-10 to $-26\,^0/_{00}$) yields enriched precipitation (0 to $-14\,^0/_{00}$). (Courtesy I.D. Clark, 2015; reproduced with the permission of the Taylor and Francis Group LLC Books.)

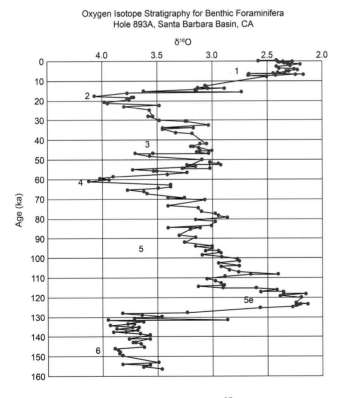

Figure 8.3 Vertical profile of benthic foram $\delta^{18}O$ data against age for cores recovered from borehole 893A, Santa Barbara Basin, California. Interstadial 5e at 120–125 ka is associated with sea-level rise and the subsequent marine inundation of coastal valleys in San Diego. (From Kennett (1995).)

8.2 Ice Flow and Glaciotectonics

The gravity-induced movement of glacial ice downslope comprises three modes: (i) the plastic deformation of the ice itself; (ii) the sliding of the ice over the glacier bed material; and (iii) the deformation of the bed. The second and third modes comprise the **basal slip** of the glacier (Cuffey and Paterson, 2010). The glacial beds are termed hard, i.e., bedrock, or soft, i.e., glacial sediment. The sedimentary materials most commonly in contact with the ice are known as **subglacial till**, which is discussed in the section 8.4. Figure 8.4 illustrates the various modes of glacial deposition.

The geotechnical engineer working in glaciated terrain is often concerned with the properties of the glacial sediments. The flow of glacial ice results in **glaciotectonic deformation**. This deformation is of two types (Lee and Phillips, 2013). In front of the glacier, proglacial deformation produces structures that are characterized by shear and compressional forces. This leads to a forebulge ahead of the ice front that may contain extruded subglacial materials known as a push moraine or thrust blocks of bedrock or till (Van der Wateren, 2005). Beneath the glacier, subglacial deformation produces strong shearing and extensional features (Hooke, 2005). Figure 8.5 illustrates the various geomechanical processes occurring at the glacier bed–till interface.

For basal slip to occur – and thus for the glacier to advance – the subglacial till must be fully water-saturated and

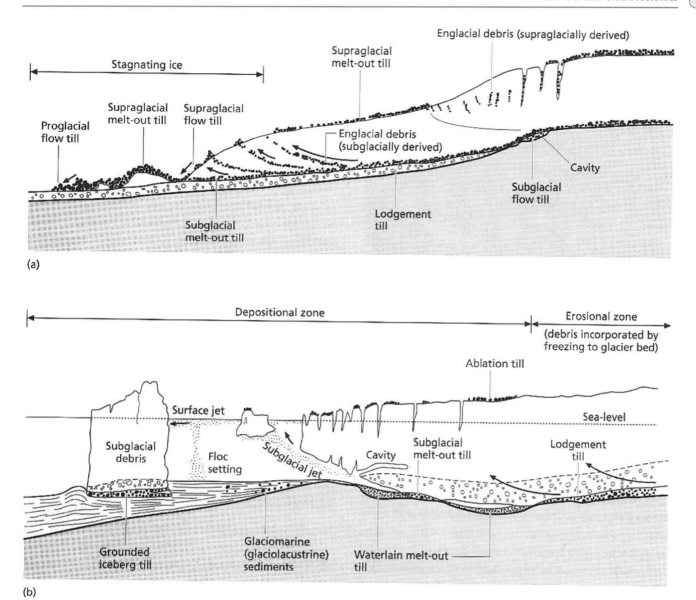

Englacial debris (supraglacially derived)

Supraglacial
melt-out till

Stagnating ice

Supraglacial   Supraglacial
melt-out till   flow till

Proglacial
flow till

Englacial debris
(subglacially derived)

Cavity

Subglacial
flow till

Lodgement
till

Subglacial
melt-out till

(a)

Depositional zone

Erosional zone

(debris incorporated by
freezing to glacier bed)

Ablation till

Surface jet

Sea-level

Subglacial
debris

Floc
setting

Subglacial jet

Cavity

Subglacial
melt-out till

Lodgement
till

Grounded
iceberg till

Glaciomarine
(glaciolacustrine)
sediments

Waterlain melt-out
till

(b)

Figure 8.4 Cross-sectional profiles of glaciers illustrating processes by which till and glaciomarine or glaciolacustrine sediments are deposited. Panel (a) shows the case of a terrestrial glacier, while (b) shows that of a grounded coastal glacier. Outwash of sediment occurs from subglacial channels at the base of glaciers or adjacent to them producing kame and esker deposits. (From Leeder (1999); reproduced with the permission of John Wiley and Sons.)

maintain an elevated pore pressure similar to the total normal stress imposed by the overlying ice (density $\sim 900\,\text{kg/m}^3$), i.e., this condition requires that subglacial temperatures should have reached the melting point temperature and water is present at the glacier–bed contact to permit slip. Rates of basal slip vary substantially, from 30 m/a for the Storglaciären in northern Sweden to 1000–2000 m/a for the Columbia glacier in coastal southern Alaska. However, glacier surges of 10–15 m/day have been documented by Kamb et al. (1985) that they associate with increases in basal water pressures caused by a transition from the normal drainage tunnel configuration to a linked cavity system that restricts flow. The

critical variable is referred to as the **effective pressure** (p_e) at the base of the glacier rather than the effective stress, such that $s_r = c_0 + p_e \tan \phi_u$, where s_r is the ultimate or residual shear strength of the till, c_0 is the cohesion intercept on a Mohr–Coulomb failure plot and ϕ_u is the ultimate friction angle (Iverson et al., 1998a).

For the glacier, decreases in the subglacial effective pressure cause increased rates of sliding by separation of the basal ice from the bed and by generating a down-glacier force on ice in contact with the bed. Within the subglacial till, increases in effective pressure cause it to deform with dilation of the pore volume. Moore and Iverson (2002) showed that, under constant shear stress imposed by a ring-shear device (Figure 4.10d) applied to an overconsolidated till, pore-volume dilation (Figure 4.8) caused weakening within the shear zone of the specimen and subsequent shear displacement. But drainage, which is a function of the permeability of the specimen, caused a decrease in the pore pressure and subsequent strengthening of the till. This cycle of weakening followed by strengthening under constant shear – each time causing slip – concluded with the failure of the till specimen after ten episodes of slip once the till had dilated to its steady, critical-state porosity.

In his summary of field and laboratory investigations of glaciers with deformable beds (i.e., subglacial till as in Figure 8.5), Iverson (2010) reported that basal slip rates were primarily a function of the effective pressure (p_e). The hydrogeomechanical process, in which pore dilation and drainage, and strengthening of the till occurs, is neither drained nor undrained in the geotechnical sense but transitory because the till texture limits the rate of drainage from till to adjacent subglacial channels at the ice–till interface. The extent to which bed deformation is continuous

Table 8.1 Residual strength (s_r) in kPa of subglacial tills. (Modified from Cuffey and Paterson (2010).)

Glacial till	s_r, where $\sigma_e = 50$ kPa	ϕ (deg)	Apparent c_0 (kPa)
Storgläcieren, Sweden	24	26	5
Black Rapids, Alaska	43	40	1.3
Breidamerkurjökull, Iceland	35	32	3.7

beneath the glacier likely determines its overall basal slip rate (Hooke, 2005).

Till samples behave as Mohr–Coulomb materials, as shown in Table 8.1. The residual strength for an effective normal stress on the slip plane of 50 kPa shown in Table 8.1 is that required before basal slip occurs. The cohesion value is considered an apparent cohesion in that it is the shear strength at zero normal stress for Mohr–Coulomb behaviour and reflects the frictional strength of the till sample rather than bonding between grains. Larger values of s_r indicate higher frictional strengths.

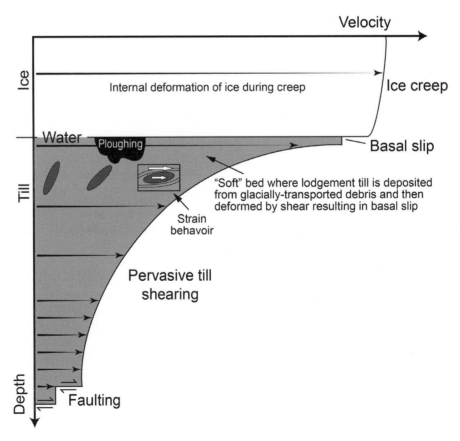

Figure 8.5 Glaciotectonic deformation of subglacial till, known as the deforming bed model. If the bed was exposed bedrock rather than till, the cavity produced by ploughing would be referred to as being caused by "quarrying". The strain fabric is that reported for basal till in Poland by Larsen and Piotrowski (2003). (After Lee and Phillips (2013).)

8.3 Glacial Erosion and Landforms

Glacial erosion of bedrock involves subglacial **abrasion** leaving a smoothed surface except for striations left by rocks embedded in the base of the glacier and cavities created by **quarrying** of bedrock (see Figure 8.5). Quarrying involves the dislodgement of blocks of the bedrock by sliding ice, which is facilitated by pre-existing fractures in the overridden rock mass. The quarried blocks then act as abrasive agents to erode the bedrock. Soluble rocks, such as limestones, may also undergo dissolution adding to the erosion of the subglacial bedrock. The various eroded materials and deposits are illustrated in Figure 8.4 for the cases of terrestrial and coastal glaciers.

Hooke (2005) notes that the rate of glacial erosion is likely to be high where basal melting is low and this meltwater refreezes to the till or bedrock. The rate of abrasion of crystalline bedrock may only be 1–2 mm/a, although it may be considerably higher in meltwater streams (up to 30 mm/a). Glacial erosion of Swedish crystalline bedrock

during a glacial cycle of ~100 000 years has been estimated at ~1 m (Påsse, 2004).

Erosion rates of sedimentary bedrock, particularly where fractured, were likely higher because of the effects of quarrying. Numerical simulations (Hildes et al., 2004) of regional erosion by the Laurentide Ice Sheet over a full glacial cycle of 120 000 years put total erosion of the area shown in Figure 8.1 at 0.4–0.6 m for a ground surface covered by a regolith to 1.6 m for exposed bedrock. Whole regions have had their sedimentary cover much reduced depending on the strength of the underlying bedrock. Shield areas of Canada have only a few metres of glacial overburden, whereas areas underlain by sedimentary rocks may have over 100 m.

Glacial erosion and the subsequent production of sediment entrained in meltwater produce differing landforms. These are referred to as landsystems by Colgan et al. (2003). Figure 8.6 illustrates a landform for areas of low topographic relief with prominent till plains, such as is found across the US Midwest from the Dakotas to the Great Lakes. The defining feature of this landsystem is that of subglacial sediment

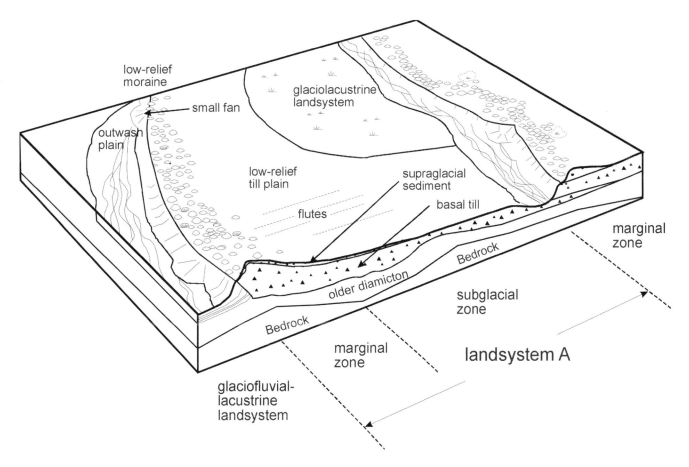

Figure 8.6 Glacial landforms associated with low-relief till plains and low-relief end moraines. The marginal zone shows a glacial till overlying earlier glacial sediment (diamicton). The subglacial zone is a till plain with flutes, grooves and ridges, showing the direction of ice movement. (From Colgan et al. (2003).)

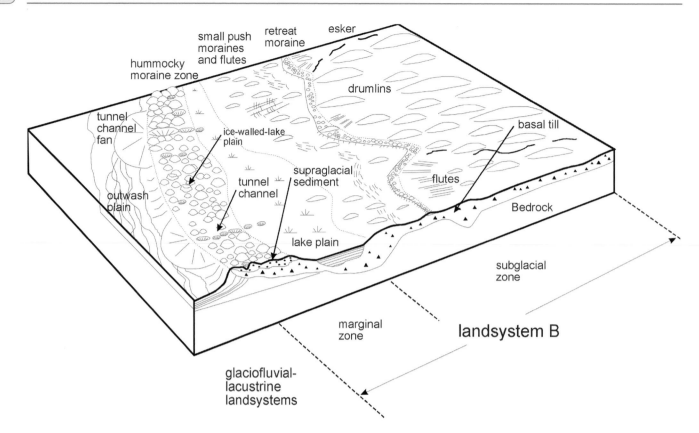

Figure 8.7 Drumlins and high-relief hummock end moraines. The ice marginal zone shows end moraine with hummocky topography deposited as supraglacial sediment. The subglacial zone is characterized by drumlins and flutes oriented in a direction parallel to the movement of the ice sheet. Eskers were deposited during glacial retreat. (From Colgan et al. (2003).)

transport and the subsequent deposition of a basal (lodgement) till unit. Their landsystem A is divided into zones that reflect depositional proximity to the ice margin and subglacial till deposition with no bedrock outcrops, which is typical of sedimentary basins where most rocks are erodible.

Areas of high topographic relief are shown in Figure 8.7. Such terrain, which may be found from New England to the interior plains of the Dakotas and Saskatchewan, includes high-relief hummocky terrain and drumlins (Figure 6.18), which are subglacially smoothed elongate hills of till or, occasionally, sand and gravel. Much of the terrain is subglacial in origin, while the hummocky terrain may indicate an ice margin zone.

Figures 8.6 and 8.7 provide a guide to terrain analysis where glaciated landforms are present. While the examples are drawn from the southern Laurentide Ice Sheet for areas in the USA from the Dakotas to New England, the features may be recognized in glaciated landscape typical of sedimentary basins. Figure 8.8 shows a lateral moraine in Rocky Mountain National Park, Colorado.

Colgan et al. (2003) also identify two other landsystems that are characterized by (i) bedrock-dominated glacial

Figure 8.8 Lateral moraine, Rocky Mountain National Park, Colorado. The photograph is taken from the end moraine looking up the Fall River Valley, which displays a classic glacially eroded U shape. The foreground is composed of glaciolacustrine sediment deposited from a lake dammed by the end moraine.

landscapes, such as is found in New England, New York and Pennsylvania, and (ii) low-relief terrain with aligned hummocks and ice-thrust masses, as found in the US

BOX 8.2	GLACIAL ERRATICS, CATASTROPHIC PALEOFLOODS AND THE ORIGIN OF STONEHENGE

Glacial **erratics** are ice-rafted rocks deposited distant from their original rock type. Large erratics form supraglacial deposits sometimes the size of houses, more often the size of an automobile or a cow. Such ice-rafted rocks are found along the Ice Age Floods National Geologic Trail in the US Pacific Northwest.

The origin of these Ice Age Floods is believed to be glaciers in Montana that produced exceptionally large breakout floods, including those known as **jökulhlaups**, an Icelandic word for glacial flood. Such rocks now sit ~250 m above the Columbia River, which indicates the huge volume of meltwater that drained down the Columbia. These catastrophic glacial breakout floods carved deep channels into the basalt bedrock, creating the Channeled Scabland of eastern Washington. O'Connor and Baker (1992) estimate that the maximum paleoflood discharge near the ice-dammed Glacial Lake Missoula exceeded $(17 \pm 3) \times 10^6 \, m^3/s$ over the course of a few days and was $(10 \pm 2.5) \times 10^6 \, m^3/s$ downstream at the head of the Columbia River gorge. (Note that the maximum discharge of the Amazon is ~200 000 m^3/s.) Remarkably, these paleofloods appear to have been repeated episodically on the basis of rhythmites found in valleys that were back-flooded during these cataclysmic floods (Bjornstad et al., 2002).

By virtue of their foreign lithology, glacial erratics also identify the migration path of paleo-ice streams during the Pleistocene and before. The glaciers leave **dispersal trains** of these detrital rocks that can be fingerprinted geochemically as well as lithologically; Ross et al. (2009) use isotopic data and airborne gamma-ray spectrometry together with a 350 km long dispersal train to map the path of Pleistocene glaciers across Saskatchewan, Canada, following the LGM.

Further west, along the foothills of the Alberta Rocky Mountains, is the Foothills Erratics Train of quartzite boulders that can be traced 580 km from the Jasper National Park to the Montana border of the USA. Although the erratics were transported in an easterly direction out of the Cordilleran Ice Sheet in the Canadian Rocky Mountains into the foothills, the dispersal train follows a southeasterly path to the Montana border because the Laurentide glacier with its dome in eastern Canada prevented the mountain glacier from transporting the debris eastwards. Therefore the convergence of the Cordilleran Ice Sheet and the Laurentide glacier caused the erratics to be transported to the southeast.

Stonehenge in southern England is the iconic prehistoric ruin probably built by neolithic Britons about 4500 years ago. The famous π-shaped outer circle of rocks is formed from 30–40 ton silicified sandstone rocks found only 30 km away, while the bluestones, which comprise a circle within the sandstones, weigh only four tons but originate in the Preseli Hills in Wales, ~250 km away.

As recently as 2017, the Wikipedia entry for Stonehenge indicated that the bluestones – so named because when wet they exhibit a blue-grey colour – were transported from Wales by prehistoric people as megaliths on a track of logs on which the large stones were rolled. Another megalith theory according to Wikipedia involves the use of a sleigh carrying each 40-ton rock. This mode of transport was successfully demonstrated near Stonehenge in 1995 when a team of more than 100 workers pushed and pulled the rock the 30 km from the area where the sandstone is found.

Always keen to spoil a good myth, geoscientists (John and Jackson, 2009) believe that glacial transport of these four-ton bluestones is a more likely cause of their transport to the area following the erosion of the bluestones in Wales. They point out that the pattern of ice flow from the Irish and Welsh ice caps would have caused the development of a dispersal train of erratics along the converging path of these two glaciers leading to their deposition near Stonehenge.

Midwest. These thrust blocks, evident also in southern Saskatchewan, Hamburg, Germany, and the coastline of East Anglia, UK, are known to pose slope stability problems created by ice-thrusted sediments with reduced shear strength (Sauer, 1978).

8.4 Glacial Sediments

We may identify six glacial sedimentary depositional systems (Eyles and Eyles, 1992; Leeder, 1999; Bridge and Demicco, 2008), most of which appear in Figures 8.4, 8.6 and 8.7:

- *Subglacial sediments*: Lodgement till is typically dense, over-consolidated sediment that may exhibit an internal fabric aligned in the direction of ice movement.
- *Supraglacial sediments*: A stagnating ice sheet will likely deposit melt-out till from entrained sediment within the glacier. Melt-out till is loose sediment and exhibits no internal fabric. Debris may be deposited at the glacier front as flow tills.
- *Glaciofluvial sediments*: Subglacial streams will develop channels or tunnel valleys beneath the ice sheet and create eskers, linear ridges of coarse materials, and glaciofluvial deposits, which are sediments deposited in contact with ice, typically composed of braided channels of sand and gravel. Outwash sand plains (sandar) may also develop.
- *Glaciolacustrine sediments*: In front of ice sheets, proglacial lakes developed from ponded meltwater. Seasonal patterns of sedimentation deposit light-coloured sandy or silty summer facies overlain by darker clays in the winter. These couplets are known as **varves**, a form of rhythmite inferred to consist of annual deposits and used to measure ages of Pleistocene deposits.
- *Glaciomarine sediments*: Meltwater released at the ice margin produces coarse-grained subaqueous deposits in fan-shaped ridges on the seabed. Subaqueous stratified mud containing dropstones from melting ice rafts may overlie earlier lodgement tills (Rust, 1977).
- *Eolian sediments*: Wind-blown silt, loess, forms thick eolian deposits that are typically yellow through oxidation. The Palouse loess of Washington state (Figure 1.7) developed from wind blowing across unvegetated outwash plains. Large eolian sand landforms form the Nebraska Sandhills that developed following the LGM (Bridge and Demicco, 2008).

For the geotechnical or geoenvironmental engineer, some glacial sediments are of particular importance, i.e., till, marine clays and outwash deposits. Till is often the foundation material at construction sites or landfills; marine clays are very susceptible to slope failure; while outwash is of great importance as a groundwater supply for industry and municipalities and for gravel aggregate used in construction. The old term **drift** is used to indicate glacially transported Pleistocene rock and sediment.

8.4.1 Till

Evans et al. (2006) have identified several classes of subglacial till including: (i) glacitectonite, which is rarely observed lithified or semi-lithified sediment retaining some fabric of the parent material while indicating shear-deformation structures, e.g., push moraines; (ii) subglacial traction till, which is deposited at the base of a glacier, e.g., lodgement and "deformation" tills; and (iii) melt-out till, e.g., end moraines. These authors note that observations of tills beneath contemporary glaciers indicate that they are likely to be hybrids of these classes; however, many tills show evidence of passive sedimentation and no evidence of disturbance of underlying laminated beds.

Till was once referred to as boulder clay, particularly in British literature, on account of its wide textural variation, i.e., poor sorting of grain size. This is evident in Figure 8.9, where mud, sand, gravel and clasts of rock are all evident in the exposure. This till contains both carbonate rock clasts from local bedrock as well as more distant boulders from the Canadian Shield. An unsorted mixture of such sediments is described as a **diamicton** indicating a glacial origin but of an uncertain depositional process.

Figure 8.9 A 2–3 m exposure of till, Peterborough, Ontario. Note the wide range of sediment sizes, the middle zone of clay that remains damp and the angular embedded clasts.

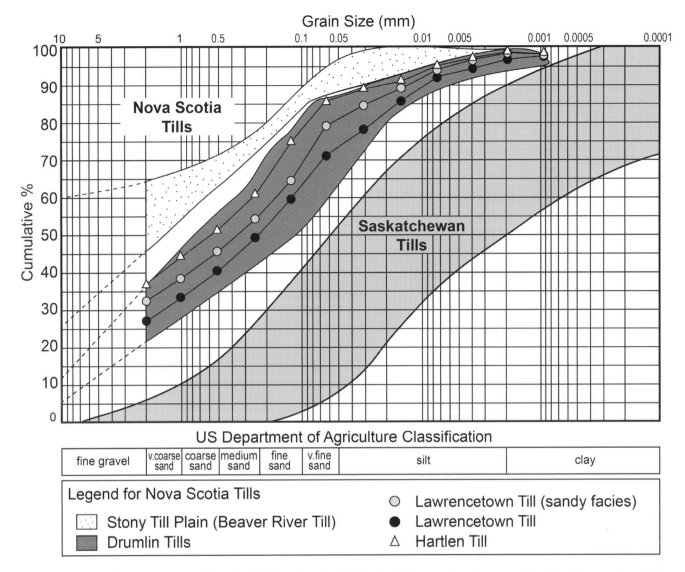

Figure 8.10 Textural variations in Canadian tills. The Nova Scotian tills developed from erosion of metamorphosed sandstone and granitic rocks and display a sandy texture. (From Lewis et al. (1998).) The Saskatchewan tills are finer-grained, reflecting the erosion and entrainment of the local bedrock composed of shale and sandstone. (From Scott (1971).)

Figure 8.10 illustrates the variation in texture of two Canadian tills. The Nova Scotian tills have developed from glacial erosion and deposition in areas of metamorphic and granitic rocks and have acquired a generally sandy texture. By contrast the Saskatchewan tills were generated by erosion and deposition of shales and sandstones and are thus generally finer-grained.

The Nova Scotian tills display significant textural differences (Lewis et al., 1998). The Hartlen and Lawrencetown Tills are finer-grained than the Beaver River Till, which is reflected in their undrained shear strengths: $s_u = 200\text{–}300\,$kPa (Hartlen Till) and $s_u = 100\text{–}250\,$kPa in the drumlin facies of the Lawrencetown Till. Measurements of the sandy Beaver River Till and sandy facies of the Lawrencetown Till yield

zero shear strength. The Hartlen and Lawrencetown Tills are overconsolidated from glacial loading and their low hydraulic conductivity ($< 10^{-8}\,$m/s) has proven them suitable for use as low-permeability liners for fuel tank farms and landfills.

The Saskatchewan tills have been exhaustively studied in the context of glacial loading and their geotechnical characteristics (Sauer et al., 1993; Christiansen and Sauer, 1998) such that they can often be differentiated by carbonate mineral content, texture and Atterberg limits. The thickness of the glaciers in southern Saskatchewan was estimated at 1150–2000 m depending on proximity to the ice front. Near Saskatoon, the total stress at ground surface was estimated at 14.4 MPa; however, the elevated subglacial water pressures appear to have limited the effective preconsolidation stress

Table 8.2 Weathering of lodgement tills with the weathering zone defined by the GSL/ISRM scheme of Table 6.2, where c' is the effective cohesion, ϕ' the effective friction angle and ϕ'_r the residual strength friction angle for drained (38 mm) samples. (Modified from Eyles and Sladen (1981).)

Weathering stage	Zone	Description	Maximum depth of zone (m)	Strength parameters
Highly weathered	IV	Strongly oxidized, leached of carbonates, boulders rotted	3	$c' = 0\text{–}25\,\text{kPa}$ $\phi' = 27\text{–}35°$ $\phi'_r = 15\text{–}32°$
Moderately weathered	III	Oxidized, increased clay content, carbonates still present	8	Same as Zone IV
Slightly weathered	II	Selective oxidation along fracture surfaces	10	
Unweathered till	I	No oxidation, no carbonate leaching		$c' = 0\text{–}15\,\text{kPa}$ $\phi' = 32\text{–}37°$ $\phi'_r = 30\text{–}32°$

to $1800 \pm 200\,\text{kPa}$. The shear strength parameters of the overconsolidated sandy tills are $c' = 20\text{–}30\,\text{kPa}$ and $\phi' = 30°$. By geotechnical arguments, Sauer et al. (1993) estimated that the hydraulic head within the ice mass over southern Saskatchewan was at an elevation of ~70% of the ice thickness.

A critical geotechnical feature of glacial tills is their **weathering profile**, which develops from oxygen-rich groundwaters infiltrating sand seams and fracture networks. The depth of weathering is typically of the order of 5–10 m, with fracture spacing decreasing with depth. Fractures are highlighted by the oxidation of the fracture surfaces and the presence of root holes. Their hydraulic continuity is revealed by fluctuating water levels in piezometers and tritium in groundwater.

Klint et al. (2013) reported a fracture spacing of >1 m in Danish tills, with fractures becoming sparse below 8 m. Similarly, in Canada, fracture networks in clay tills in Ontario were measured to a depth of 2.5–4 m, with some major fractures extending beyond 5.6 m (Ruland et al., 1991), while a Saskatchewan clay till is highly oxidized and fractured to a depth of 4.6 m, with some fractures extending to a depth of ~8 m (Keller et al., 1989).

Eyles and Sladen (1981) also noted that tills from Northumberland, England, exhibited fracturing to 8 m depth. These tills were developed by glacial erosion of coal and shale beds within the Carboniferous "Coal Measures". They reported that lodgement tills were highly weathered down to 3 m depth and then moderately weathered to 8 m. In the first 2 m, the

carbonate content of the till and its undrained shear strength were much reduced compared with values obtained from 2–5 m. The dissolution of the carbonate is related by Eyles and Sladen (1981) to the oxidation of pyrite and marcasite in the tills derived from the Coal Measures. The till was interlayered with sand and gravel lenses or overlaid laminated clays, which were extensively slickensided indicating shearing, both of which posed slope stability problems. Table 8.2 summarizes the measured geotechnical properties for drained samples using triaxial (c', ϕ') and shear-box or ring-shear (ϕ'_r) testing.

8.4.2 Glaciomarine Clays

In eastern Canada and Scandinavia glaciomarine clays that have undergone post-glacial isostatic rebound (Figure 2.8) pose significant slope stability problems, as is shown in Figure 8.11. The clays in eastern Canada – known as the Champlain Sea clays – were deposited at the ice front when seawater invaded the St. Lawrence and Ottawa valleys during the late Pleistocene, about 12 ka. Till from the glacial erosion of the Shield was the source material. The clay-sized fraction (<2 μm) of these glaciomarine clays is 40–90% of all particles; their mineral content includes plagioclase, quartz, microcline, hornblende, amphibole and some carbonate minerals from nearby sedimentary rocks. The principal clay mineral present is illite, a non-expanding clay mineral.

The rate of sedimentation of these clays was rapid, probably ~50 mm/a; thus bonding between clay particles is limited

Table 8.3 Physical properties of some Champlain Sea clays. The parameters are: s_u, undrained shear strength; s_{ur}, remoulded undrained shear strength; S_t, sensitivity of the sample; Sal, salinity of the pore water, i.e., total dissolved-solids concentration; w, water content; w_L, liquid limit; w_P, plastic limit; I_P, index of plasticity; and $<2\ \mu$m, clay-sized fraction. (From Locat et al. (1984).)

Site	s_u (kPa)	s_{ur} (kPa)	S_t	Sal (g/L)	w (%)	w_L (%)	w_P (%)	I_P (%)	$<2\ \mu$m (%)
Grande Baleine	32	0.1	282	0.7	58	36	24	12	74
Olga	23	1.0	19	0.3	83	73	26	47	94
St. Alban	80	2.2	37	2.0	40	36	21	15	43
Chicoutimi	245	0.5	532	—	33	28	18	10	55
Outardes	109	0.6	181	—	64	34	22	12	50

Figure 8.11 Earth flow in glaciomarine clays, Notre-Dame-de-la-Salette, Québec, Canada. (Reproduced with the permission of Natural Resources Canada 2011, courtesy of the Geological Survey of Canada, Photo 2011-062 by Greg Brooks.)

during flocculation and the clays would have been deposited at above their liquid limit (Scott, 2003). Therefore rapid shear forces, coupled with the absence of cohesion and pore pressures equal to the overburden load, can cause an otherwise solid clay sample to behave as a viscous fluid (see Mitchell and Soga, 2005).

Deltaic sands (10–20 m) overlie the Champlain Sea clays in the Ottawa Valley in eastern Canada and it is believed that the sands were transported into the area by outburst floods from glacial Lake Agassiz in Manitoba and caused preconsolidation of the clays (Quigley et al., 1983; Graham and Teller, 1984). Approximately 100 years later when the seas had retreated, further outburst floods from Lake Agassiz eroded through the clays leaving elevated clay terraces. This induced freshening of the clay pore waters while increasing the **sensitivity** of the clays, which is defined by the ratio of undisturbed to remoulded shear strength, i.e., $S_t = s_u/s_{ur}$ (Scott, 2003).

The Atterberg limit tests are routinely used by geotechnical engineers to evaluate the plasticity of clay samples. Table 8.3 provides a summary of such index and textural properties of Champlain Sea clays from Québec (Locat et al., 1984). **Quick clays** are those with the propensity to quickly lose their shear strength when remoulded and have sensitivities $S_t > 16$ (McCarthy, 1998) and remoulded shear strengths of $s_{ur} < 1$ kPa.

8.4.3 Glaciofluvial Sediments

Glaciofluvial landforms, such as the tunnel valleys and eskers shown in Figure 8.7, reflect mechanisms by which subglacial drainage of meltwater occurs. Beyond the ice front, outwash deposits of sand and gravel yield valuable construction aggregate, such as the terrace gravels of southern Ontario (Brennand and Shaw, 1994; see Figure 10.19) that were deposited where three lobes of the Laurentide Ice Sheet converged. All these glaciofluvial landforms are important for hydrogeological reasons as well, in that their sands and gravels form productive aquifers sometimes protected by overlying fine-grained tills or glaciomarine or glaciolacustrine clays.

Figure 8.12 shows glaciofluvial sand and gravel sets within the Oak Ridges Moraine, southern Ontario, Canada. The thickness, scale and sedimentary structure of these glaciofluvial sediments are typical of glacial flood deposits (jökulhlaups) in stratified moraines across southern Ontario. Their formation as channel, subaqueous fan sequences provides a predictive sediment-facies model that can be directly related to hydrofacies characteristics (Sharpe et al., 2013), and is thus important to groundwater assessment and management.

Figure 8.13 is a cross-sectional view of an esker outcrop that provides aggregate where exposed above the water table and a

Figure 8.12 Glaciofluvial sand and gravel sets within the Oak Ridges Moraine, Stouffville, Ontario. An active aggregate operation reveals a partial view of large-scale structures (∼20 m thick, 100 m wide and 500 m long). The sub-horizontal sand and gravel sets (grey), including occasional low-energy fine beds (brown set on pit face above the truck), are truncated at the top right of the sequence and overlain by yellow oxidized fine sand. Such large-scale structures relate to very high-energy sedimentation in deep water, in contrast to metre-scale structures in pro-glacial streams. (Reproduced with the permission of Natural Resources Canada, courtesy of the Geological Survey of Canada, Photo GSC-2017-045.)

Figure 8.13 Eskers are shoestring-shaped ridges of glaciofluvial sediments. This esker outcrop 30 km east of Ottawa, Ontario, Canada, is photographed in an aggregate quarry. It forms the Vars–Winchester esker aquifer further south where it is submerged beneath the water table and provides a municipal groundwater supply. The central ridge is a high-permeability cobble gravel deposited in a subglacial meltwater stream and is no more than 150 m wide. Seismic profiles indicate the esker is approximately 1 km wide with gently sloping flanks covered by sand and pebble gravel fans deposited in standing water at the ice front. It is overlain by laminated fossiliferous mud and sand deposited by the Champlain Sea. Geologic interpretation and photo by Don Cummings, DCGeo Consulting. (Reproduced with the permission of Natural Resources Canada.)

groundwater supply where submerged. Champlain Sea muds form a drape over the esker. The esker shown has a central core of high-permeability sand and gravel and rests on an eroded bedrock or till surface and is overlain by proglacial sand and mud deposited on the floor of the Champlain Sea. The esker core is believed to have been deposited by outburst floods with estimated discharges of several thousand cubic metres per second that occurred during glacial retreat (Cummings et al., 2011).

Buried valleys infilled with coarse sediments constitute major groundwater supply aquifers in the US Midwest and the Canadian Prairies. One example is the Teays–Mahomet bedrock valley system (Melhorn and Kempton, 1991), which has its origin in West Virginia and Kentucky and can be identified westwards through Ohio, Indiana and Illinois. Goldthwait (1991) traces the 2 km wide Teays valley beneath glacial sediment from an elevation of 225 m above mean sea level (amsl) in Charleston, West Virginia, to 107 m amsl in west-central Indiana where it joined the wider Mahomet drainage system of Illinois. Along its course, groundwater is produced to supply major municipalities and industry in Cincinnati and Dayton, Ohio, Lafayette, Indiana, and Champaign-Urbana, Illinois.

The buried-valley channels themselves may have been excavated prior to the Quaternary and subsequently filled during Quaternary time, i.e., it is unclear whether these valley drainage systems are of pre-glacial fluvial or glacial origin. Montgomery (2002), who has compared fluvial versus glacial erosion in the Olympic Peninsula of Washington state, reported that glacially eroded valleys in Alpine areas draining >50 km² have two to four times as much cross-sectional

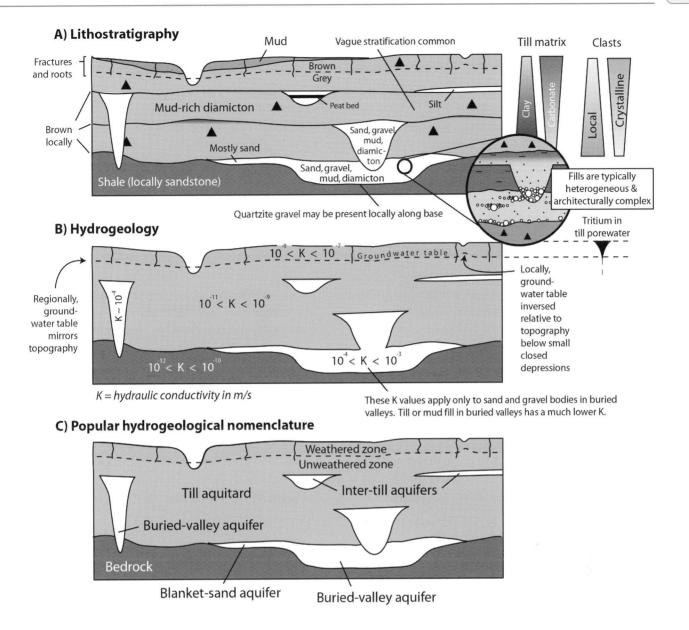

A) Lithostratigraphy

Fractures and roots

Mud

Vague stratification common

Brown
Grey

Brown locally

Mud-rich diamicton

Peat bed

Silt

Mostly sand

Sand, gravel, mud, diamicton

Sand, gravel, mud, diamicton

Shale (locally sandstone)

Quartzite gravel may be present locally along base

Till matrix

Clay

Carbonate

Clasts

Local

Crystalline

Fills are typically heterogeneous & architecturally complex

B) Hydrogeology

Tritium in till porewater

$10^{-9} < K < 10^{-7}$ Groundwater table

Regionally, groundwater table mirrors topography

$K \sim 10^{-4}$

$10^{-11} < K < 10^{-9}$

$10^{-12} < K < 10^{-10}$

$10^{-4} < K < 10^{-3}$

Locally, groundwater table inversed relative to topography below small closed depressions

K = hydraulic conductivity in m/s

These K values apply only to sand and gravel bodies in buried valleys. Till or mud fill in buried valleys has a much lower K.

C) Popular hydrogeological nomenclature

Weathered zone
Unweathered zone

Till aquitard

Inter-till aquifers

Buried-valley aquifer

Bedrock

Blanket-sand aquifer

Buried-valley aquifer

D) Popular genetic interpretations

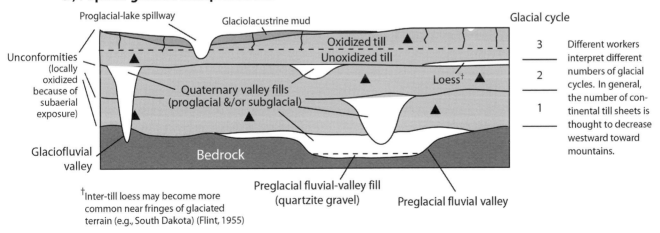

Proglacial-lake spillway

Glaciolacustrine mud

Glacial cycle

Oxidized till
Unoxidized till

Loess[†]

Unconformities (locally oxidized because of subaerial exposure)

Quaternary valley fills (proglacial &/or subglacial)

Glaciofluvial valley

Bedrock

Preglacial fluvial-valley fill (quartzite gravel)

Preglacial fluvial valley

3

2

1

Different workers interpret different numbers of glacial cycles. In general, the number of continental till sheets is thought to decrease westward toward mountains.

[†]Inter-till loess may become more common near fringes of glaciated terrain (e.g., South Dakota) (Flint, 1955)

Figure 8.14 Geology and hydrostratigraphy of Quaternary sediments overlying the sedimentary bedrock of the Western Canadian Sedimentary Basin, most often of Cretaceous sandstones and shales. (A) The stratigraphic relationships between bedrock valley and Quaternary sediments; (B) hydraulic conductivities in m/s; (C) hydrogeological terminology for the various units; and (D) interpretation of the glacial origin of the various sedimentary units. (From Cummings et al. (2012).)

area and 500 m greater topographic relief than comparable fluvial valleys.

Piotrowski (1997) concluded from his studies of German tunnel valleys that fine-grained glacial sediments such as till would not permit more than 25% of basal glacial meltwater to drain; consequently most drainage would occur episodically as outburst floods through tunnel valleys eroded into till or bedrock and subsequently filled with coarse-grained sediment. Therefore glaciofluvial processes may have exploited significant pre-glacial bedrock valleys and deepened and widened them. Work by the Geological Survey of Canada (Sharpe et al., 2004) in southern Ontario indicates that late-glacial outburst floods from the Laurentide Ice Sheet created a regional erosional surface and that the deep channelized meltwater flow in the Oak Ridges Moraine and Eastern Ontario landscape is responsible for the Finger Lakes troughs of upstate New York.

On the Canadian Prairies (see Figure 8.14), Cummings et al. (2012) report that most of the Prairie buried valleys were filled with heterogeneous sediment over multiple glaciations. Some of these valleys were pre-glacial and incised into the Cretaceous bedrock, others were glaciofluvial buried valleys that incised through tills with sediments deposited parallel to the ice front (see Figure 8.6) or as tunnel valleys that developed subglacially. These buried-valley aquifers have a unique hydrogeological response to pumping (van der Kamp and Maathuis, 2012).

8.5 Permafrost

In section 6.2.2, we discussed the effects of cold climate weathering on rock causing disintegration by frost action and solifluction of soils and sediments. We now return to those processes affecting soil, rock and infrastructure in perennially frozen ground or **permafrost**, defined as soil or rock that has remained frozen ($\leq 0°$) for at least two years.

The geomorphological features of such landscapes are referred to as **periglacial** in that they identify non-glacial freeze–thaw processes that may be identified in Alpine areas, such as the Rocky Mountains of Colorado (Figure 6.5), as well as now-temperate regions that were previously in close proximity to glaciated areas. Permafrost that occurs in regions beneath which >90% of the land surface is permafrost is known as continuous permafrost. Regions with lesser amounts of permafrost are known as regions of discontinuous permafrost (50–90%), or sporadic permafrost (50–90%) or isolated permafrost (<10%) (Anderson and Anderson, 2010).

Thawing of coastal-plain permafrost in northern Alaska, caused by the global climatic warming, has been well documented since the 1980s. The hummocky topography caused by melting ground ice produces a subsidence process known as **thermokarst**. Subsidence is initiated by the phase changes from ice to water that results in a loss of volume of a representative elementary volume (REV) of the frozen ground. This is associated with the compaction and settling of soil grains that are no longer supported by ice and a compaction of the soil matrix from the drainage of water and loss of pore pressure (Rowland and Coon, 2016).

Jorgenson et al. (2006) determined that the development of thermokarst terrain might affect 10–30% of the Arctic lowlands of Alaska, where the North Slope oil and gas infrastructure crosses the coastal plain, and thus alter Arctic ecosystems in a profound manner. For North America in general, Smith et al. (2010) have observed that the rates of ground warming, and therefore permafrost degradation, have been more pronounced north of the tree line. In the more southerly discontinuous (i.e., warmer) permafrost zone, latent heat effects have somewhat attenuated permafrost degradation particularly where vegetation is present; however, they warn that warmer permafrost is thawing at temperatures below zero.

We now consider the thermal regime of permafrost, then periglacial environments and conclude by reviewing the geotechnical issues associated with permafrost and its degradation by warming.

8.5.1 Thermal Regime

Ground temperature profiles, such as that shown in Figure 8.15, are the fundamental measurements of the geothermal regime controlling permafrost growth and decay. The shallow regolith or rock zone that experiences seasonal freezing and thawing is known as the **active layer**; the active-layer thickness is typically about 1 m deep although it can be thinner in the high Arctic. Because mineral–water interactions and the presence of solutes in the pores of the soil can depress the freezing point of water, the temperature of water freezing is shown in Figure 8.15 as the ice-nucleation temperature, which may be slightly below 0°C. The ground-surface temperature (i.e., at 0.1 m) of an undisturbed active layer may vary by 10–20°C during the summer warming season.

The depth of the active layer depends on the sinusoid in surface temperature, and the thermal and hydrologic properties of the regolith or the soil, which is likely to be a mixture of mineral and organic matter common to periglacial

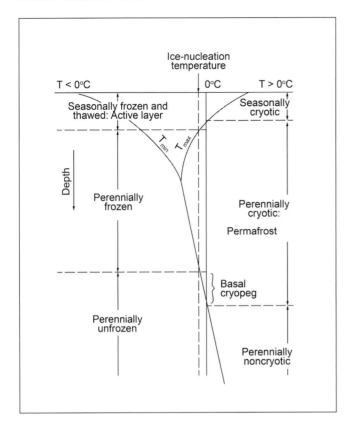

Figure 8.15 Schematic representation of the temperature profile with depth in a permafrost area, showing the terms used to indicate the seasonal or perennial temperatures relative to 0°C and terms indicating seasonal or perennial presence or absence of ice. (After van Everdingen (1987).)

environments. The subsurface temperatures will vary sinusoidally with depth over the course of the summer and a time lag will occur between maximum air temperature and depth of warming, usually 20–30 days after the maximum surface temperature. Turcotte and Schubert (2002) indicate that the time lag for a temperature change to propagate to a depth z is approximately z^2/κ, where κ is the thermal diffusivity of the regolith (~ 1 mm^2/s).

To estimate the maximum depth of the active layer, Anderson and Anderson (2010) define a depth scale $z_\star = \sqrt{\kappa P^\star/\pi}$, where $P^\star$ is the period of oscillation of the sinusoid in surface temperature, i.e., one year. The annual temperature amplitude, T_{amp}, will decline exponentially with depth and at $z_\star$ will be about one-third that at the surface (i.e., 1/e). They estimate the active-layer thickness by

$$z_{active} = -z_\star \ln\left[\frac{-\overline{T}_s}{T_{amp}}\right], \qquad (8.2)$$

where $\overline{T}_s$ is the mean annual temperature at ground level. The omission of the latent heat of ice (334 kJ/kg) from (8.2) means

that the equation overestimates the true depth of active-layer thawing.

The top of the perennially frozen soil is defined as the permafrost surface, at which elevation the surface temperature of the permafrost is measured. Where the surface temperature is colder than $-3°$C, the permafrost is considered unlikely to thaw. However, warmer ground-surface temperatures may induce thawing at permafrost surface temperatures in the range -2 to $0°$C. The surface temperature is measured at the base of a shallow casing in which a thermistor is installed 0.3–0.4 m into the permafrost surface and linked to a waterproof data logger. Vertical temperature profiles through the full thickness of the permafrost into deeper unfrozen material are measured using deep boreholes drilled in the spring while the ground surface remains frozen (Osterkamp and Jorgenson, 2009).

The permafrost zone itself may be as much as 600 m deep in the high Arctic at 75°N, decreasing to 15 m at 55°N (Lacelle et al., 2008; French, 1996). This depth reflects the thermal balance of the mean ground-surface temperature and the geothermal gradient of 25°C/km. Seismic methods are often used to investigate permafrost; LeBlanc et al. (2004) have shown how high-resolution vertical seismic profiling can be coupled with seismic cone penetrometer testing to image the cryostratigraphy of 20 m thick permafrost at 56°N and determine the dynamic response of the ground to loading by earthquakes or vibrating machinery.

8.5.2 Periglacial Environments

The geothermal regime of permafrost dictates the nature of hydrologic systems. Figure 8.16 illustrates the terms used to define aquifers in permafrost areas with respect to the thermal regime of Figure 8.15. Unfrozen zones of arctic landmasses exist as **taliks** that may be permanent or seasonal.

Ice may be present in the soil or regolith in three distinguishable forms (Anderson and Anderson, 2010): (i) **pore ice**, present in the pores of soils; (ii) **segregation ice**, which forms as the thin horizontal ice lenses in soils that are associated with frost heaving; and (iii) **massive ground ice**, which forms as metre-scale ice zones within shallow regolith and soils. Pollard and French (1980) estimated the volume fraction of pore and segregation ice in soils from the Mackenzie Valley delta in northern Canada as exceeding 60% at the permafrost surface (1 m depth) but decreasing to 30–60% at 2–10 m below ground surface.

Frost heave is of particular importance in the freezing of soil water. Not only does the water expand upon freezing,

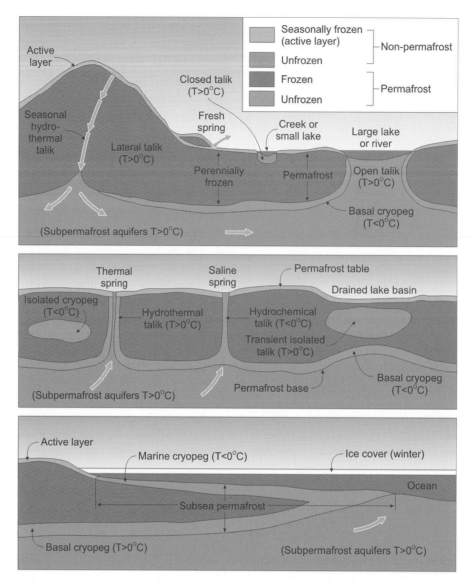

Figure 8.16 Schematic of aquifers in permafrost areas. Supra-permafrost aquifers exist in the active layer and in closed river and lake taliks. Sub-permafrost aquifers occur in open taliks and below the basal cryopeg. In the high Arctic, springs are often associated with salt diapirs and produce saline discharge. (After van Everdingen (1987); reproduced with permission of A. Rivera (ed.), *Canada's Groundwater Resources*, p. 609. Markham, ON: Fitzhenry & Whiteside Limited.)

but it also creates a suction potential causing water migration to the horizontal ice lens that is forming, thus promoting segregation ice in silts and fine sand. Konrad (1994) describes this phenomenon by a parameter known as the segregation potential (SP), where SP = ν/grad T, where ν is the rate of water intake into an ice lens and grad T is the temperature gradient. With the formation of the ice lens, the frozen soil above the lens cannot transmit water, consequently the ice lens is augmented by water migrating vertically from unfrozen soil below.

In perennially frozen soils, the hydraulic conductivity (K) would be extremely low because of the pore-ice content and low interconnected liquid water content. Burt and

Williams (1976) used a simple permeameter to measure the decrease in K from 0° down to −0.4°C. The hydraulic conductivity of fine sand without ice lenses decreased by five orders of magnitude to $<10^{-11}$ m/s as the temperature decreased to −0.2°C. The hydraulic conductivity of silts without ice lenses decreased by two orders of magnitude to $\sim 10^{-8}$ m/s at −0.2°C, while dense lenses of ice caused additional order-of-magnitude decreases in K to $\sim 10^{-10}$ m/s at −0.2°C.

Van Everdingen (1987), one of the few hydrogeologists ever to work in periglacial environments, observed that recharge to groundwater flow systems would be least affected where fractured or karstic bedrock outcropped, whereas frozen soils

Figure 8.17 Polygonal ice-wedge pattern, northern Yukon coastal plain, Ivvavik National Park, Yukon, Canada. (Photo courtesy of I.D. Clark, University of Ottawa.)

would inhibit groundwater recharge. While discharge from open taliks was perennial, discharge from closed taliks would likely end during wintertime. He reported that all groundwater discharge from springs with flow rates >5 L/s or water temperatures >5°C or dissolved-solids concentrations >1 g/L is most likely evidence of discharge from sub-permafrost aquifers through open taliks.

The landscape of much of the coastal plain of arctic North America is patterned as shown in Figure 8.17. The networks of ice-wedge polygons are believed to be due to horizontal tensile stresses related to thermal contraction that overcome the tensile strengths of the frozen ground causing ice-wedge cracking (Anderson and Anderson, 2010). In such areas the ground-ice content is typically ~10–20%. The polygons shown in Figure 8.17 have low centres of marsh-like materials and ridges that are dry, peat-rich and underlain by ice wedges. High-centred polygons also exist with moats of water underlain by ice wedges (French, 1996).

Weathering reactions in polar environments appear to transition from predominantly mechanical weathering at the ground surface, which is caused by frost action, wetting and drying cycles and thermal dilation of rock, to an increase in geochemical weathering with depth at the base of the active layer of permafrost. In the high Canadian Arctic (75°N) where the active-layer thickness is 0.3–0.9 m, Lacelle et al. (2008) report carbonate mineral dissolution and reprecipitation with total dissolved-solids concentrations reaching 400 mg/L.

8.5.3 Geotechnical Issues

Global warming and resource development cause profound geomorphological change and attendant geotechnical problems in the Arctic: thermokarst subsidence in areas of resource exploration and infrastructure development, coastal erosion and landslides.

Thermokarst subsidence was identified by Mackay (1970) as distinct from "thermal erosion" in causing permafrost degradation. Mackay stated that it was not simple melting of exposed ice-rich soil that was causing permafrost degradation, such as the scars left by seismic-exploration lines and off-highway road tracks. Rather, heat conduction into underlying ice-rich soils causes melting and subsidence by increasing the depth of thawing and releasing the melted ice. Consequently, melting of massive ground ice, e.g., ~14% volume fraction of the first 10 m of permafrost in the Mackenzie delta of Canada (Pollard and French, 1980), can yield subsidence of 1.4 m.

Coastal erosion in permafrost regions occurs where thawing massive ground ice is exposed along coasts subject to strong onshore wave action, particularly during storms. Overduin et al. (2014) describe how such erosion has occurred across the Arctic, from the North Slope of Alaska to the Laptev Sea in Siberian Russia. They indicate that local rates of erosion have doubled since year 2000 and reach 20 m/a locally, i.e., a retreat in the shoreline of 20 m. Along a 60 km stretch of Alaska's Beaufort Sea, coastal erosion increased between 2002 and 2007 from 8.5 to 14 m/a, making coastal communities and oil wells vulnerable (Jones et al., 2009). Failure of coastal blocks has been investigated by Hoque and Pollard (2009), including consideration of the intersection of ground-ice wedges with the failure plane.

Landslides in permafrost are commonplace and result from the seasonal melting of the soil raising water saturations and subsequently pore pressures leading to slope failure. The Zymoetz landslide (Figure 6.20) is an example, which began as a rock slide in permafrost terrain. A particular feature of **thaw flows** in permafrost is their retrogressive behaviour, i.e., the scarp of the landslide progressively migrates backwards from the initial slope failure until stabilized by vegetation that insulates the ice-rich scarp (Highland and Bobrowsky, 2008).

Figure 8.18 illustrates a retrogressive thaw flow. In fine-grained permafrost soils in the Mackenzie Valley of northern Canada, the rate of head-scarp retrogression may be correlated with the height of the scarp wall and the overall slope angle of retreat (Wang et al., 2009). The thaw flow process is continuous if ice-rich soil in the head-scarp is continually exposed to thawing, and slope retreat may be as high as 10 m/a; however the rate of retrogression is limited when the height is ~6–8 m.

The safety of transportation infrastructure and corridors in permafrost terrain is the most obvious source of concern

Figure 8.18 An active retrogressive flow slide, near Inuvik, Northwest Territories, Canada, is shown on the right of the photo with a test plot on the left from which the organic cover had been removed to investigate landslide triggering in fine-grained permafrost soils. The test plot is 20 m × 20 m and the removal of the organic cover allowed deeper penetration of the thaw depth, illustrating the importance of the vegetative cover in slope stability. (From Wang and Saad (2007); reproduced with the permission of the Association of Environmental & Engineering Geologists.)

Figure 8.19 Differential settlement caused by permafrost thawing at a culvert crossing, Alaska Highway, Yukon, Canada. (From "A Bumpy Road: Highways and Thawing Permafrost", 2012 McBride Lecture by Paul Murchison, Yukon Government, Canada.)

for geotechnical engineers. Darrow et al. (2012) describe a quite typical problem of slope stability along a transportation corridor but with the added complexity of frozen soil beneath the highway. They measured creep of 2.5 cm/a within ice-rich clayey soils (−0.5°C) beneath the Richardson Highway connecting Anchorage and the port of Valdez, Alaska, which created a potential failure surface made even more complex by deep groundwater drainage.

Further north in Alaska, the Dalton Highway is threatened by **frozen debris lobes**, which are masses of soil, rock, vegetation and ice moving down slopes affected by permafrost. Simpson et al. (2016) measured one such lobe moving with a surface speed of 1.2 cm/day and estimated cohesion and friction angles for the lobe. Inclinometer measurements indicated a shear zone at 20 m depth, approximately 6 m above bedrock,

with displacement of the lobe limited to the materials above this shear zone. Further instability in these frozen debris lobes will occur following further thawing of ice and frozen soil that will increase pore-water pressures and decrease the lobe's cohesion.

Highway pavement in permafrost terrain is susceptible to thaw settlement, soil consolidation and pavement degradation by potholes, cracks, etc. (see Figure 8.19). Experimental facilities have been established in discontinuous permafrost along the Alaska Highway in the Yukon and in Manitoba to study the thermal and hydrogeological environment affecting highway stability. The test site along the Alaskan Highway was designed to investigate the relationship of heat loss from groundwater flowing beneath the highway. De Grandpré et al. (2012) concluded that heat transfer from the groundwater promoted permafrost degradation and pavement subsidence. In Manitoba, Batenipour et al. (2014) describe the failure of a highway embankment caused by thawing of the permafrost resulting in pore-pressure changes leading to compressive and shear deformation of the embankment material.

8.6 Summary

The evidence for global climate warming in the twenty-first century is incontrovertible. Warming is particularly evident in the retreat of continental ice masses and coastal glaciers. We live in an interglacial period that has lasted 10 kyr and now will experience the hazards of this warming, including the threat of unstable melting slopes and the sudden drainage of lakes in alpine areas and the destruction of permafrost in arctic environments.

In this chapter we have reviewed the onset, migration and retreat of continental ice sheets during the Pleistocene epoch that has left sedimentary deposits across parts of northern Europe, North America, China, New Zealand and South America.

We have used ratios of the stable environmental isotopes of oxygen and hydrogen ($\delta^{18}O$ and δD) to identify past climates in pore waters trapped in glaciolacustrine clays and in the shells of forams.

Furthermore, we have discussed how the motion of glaciers in the form of basal slip is connected to the subglacial deformation of glacial till and the production of glacial sediments through abrasion, quarrying and ploughing. The motion of glaciers causes deep erosion of valley floors, while the melting of glacial ice and its drainage has produced distinct landscapes and subsurface features. These include sand and gravel outwash deposits used as construction aggregate and glaciofluvial aquifers used for groundwater supply. Till and glaciolacustrine and glaciomarine clays are often fine-grained materials and present significant problems associated with their use as foundations; we have summarized the shear-strength properties for tills and a glaciomarine clay.

We concluded with a discussion of the thermal regime of permafrost, the nature of periglacial environments in which warming is complicating resource development, and the geotechnical issues associated with this warming: coastal erosion, landslides along transportation corridors, frost heave and thermokarst subsidence.

8.7 FURTHER READING

Anderson, D.E., Goudie, A.S. and Parker, A.G., 2007, *Global Environments through the Quaternary: Exploring Environmental Change.* Oxford, UK; Oxford University Press. — Now in its second edition, this text provides much useful background information on all matters related to the Quaternary, including glaciation and related climatic issues.

Anderson, R.S. and Anderson, S.P., 2010, *Geomorphology: The Mechanics and Chemistry of Landscapes.* Cambridge, UK: Cambridge University Press. — This volume contains technically detailed chapters on both glacial and periglacial processes.

Cuffey, K.M. and Paterson, W.S.B., 2010, *The Physics of Glaciers*, 4th edn. New York, NY: Elsevier. — A comprehensive account of the physics of glaciers and ice sheets with discussions of Quaternary climate change and related geophysical forcing functions and sea levels.

Evans, D.J.A. (ed.), 2005, *Glacial Landsystems.* London, UK: Hodder Arnold. — A collection of 17 chapters describing the various glacial landforms with excellent illustrations.

French, H.M., 1996, *The Periglacial Environment*, 2nd edn. Harlow, UK: Longman. — While not technically explicit, this introduction to geomorphology and landscape evolution in polar latitudes is a most useful reference for the geotechnical engineer facing construction issues in permafrost regions. It is now available in electronic format as a third edition.

Hooke, R.LeB., 2005, *Principles of Glacier Mechanics*, 2nd edn. Cambridge, UK: Cambridge University Press. — A treatise on glacier mechanics that will appeal to geotechnical engineers with an interest in the properties of subglacial till.

Michel, F., van Everdingen, R.O., Woo, M.-K. and Dyke, L., 2014, Permafrost groundwater region. In *Canada's Groundwater Resources* (ed. A. Rivera), pp. 596–636. Markham, ON: Fitzhenry & Whiteside. — A recent review of hydrogeological processes in permafrost regions.

8.8 Questions

Question 8.1. How is glacial till differentiated from glacial lake clays?

Question 8.2. How are interglacial events identified in sediments, bedrock and groundwater?

Question 8.3. Estimate the residual strength, s_r, of a till sample from the Storglaciären, which was measured with the following parameters: $c_0 = 0\,\text{kPa}$, $p_e = 110\,\text{kPa}$ and $\tan\phi = 25.5°$.

Question 8.4. Compute the remoulded shear strength (s_{ur}, kPa) for the samples in Table 8.3 using the relationship (Scott, 2003):

$$s_{ur} = \frac{1}{(I_L - 0.21)^2},$$

where $I_L = (w - w_P)/I_P$.

Question 8.5. Estimate the "depth scale" of thermal warming of the regolith ($z_\star$) in permafrost and use this value to estimate the depth of the active layer (z_{active}) when the mean annual surface temperature is $-8°C$ and its annual amplitude is $20°C$.

CHAPTER

9 Fluvial Processes and Sediments

Previously, we introduced the concept of stream power in Chapter 6, which is integral to the erosion, transportation and deposition of sediment, and then discussed channel morphology and avulsion, the hydraulic geometry of alluvial channels and paleoflood hydrology. We now return to the matter of fluvial systems to consider the fluvial processes at work in sediment transport, the nature of fluvial sediments and their diagenesis or conversion into sedimentary rocks. This list of topics is the stuff of textbooks, not simple chapters; the reader is referred to the suite of references at the end of this chapter for further reading.

Of all fluvial processes, an engineer is most likely aware of the hazards posed by floods, which are considered the most costly natural hazard in the USA. However, the hydrology of floods, beyond that of paleoflood hydrology discussed earlier, is outside our scope of reference. Our context is with fluvial issues that have a particular focus on those most likely to involve geotechnical or geoenvironmental engineers, such as bridge scour, particularly during floods.

An example of the failure of a bridge during high flows is shown in Figure 9.1. In 2017, a one-hundred-year-old bridge across the Savitri River in India collapsed, most probably due to bridge scour, killing 20 people; while near Dublin, Ireland, in 2009, a railway bridge failed just after a train had passed over it. This problem is also acute in the USA, where, between 1989 and 2000, one-half of the 500 bridge collapses were attributed to scour-affected foundations. Whether one tries to instrument the bridge itself for possible failure, or the streambed around its piers, is a matter of current research (*The Economist*, October 15, 2016). A less dramatic but more commonplace issue for the engineer is the high rate of soil erosion at construction sites (1600–40 000 tonnes/km^2 yr;

Goudie and Viles, 2016). Such rates are typically orders of magnitude larger than rates from grassland, agriculture or urban areas and require mitigation practices on site.

However, we are also concerned with those fluvial processes implicit in more subtle and emerging issues such as dam removal, channel reconfiguration, the protection of aquatic habitats, fluvial transport of contaminated sediments and problems associated with sedimentation behind dams and in navigable waterways. Therefore we begin by considering sediment transport in open channels before turning to the nature of sedimentary deposits.

9.1 Sediment Erosion and Transport

9.1.1 Entrainment of Sediment

To understand erosion in alluvial channels, such as with bridge scour, we need to further develop the relationship between the bed shear stress and sediment displacement begun in section 6.4.1. We start by defining the mean flow velocity (V) at equilibrium in an open channel (Chanson, 2004) as

$$V = \sqrt{\frac{8g}{f}} \sqrt{\frac{D_H}{4} \sin \theta}, \qquad (9.1)$$

where g is the acceleration of gravity, f is the Darcy friction factor, θ is the channel slope, and D_H is the hydraulic diameter of the channel, which is approximated by $D_H = 4(A/L_P)$ in which A is the cross-sectional area of the channel and L_P is the length of the wetted perimeter.

Figure 9.1 Bank erosion along the concave or "cut" bank of a meander on the Rivière-à-Mars, Québec, Canada. Bridges crossing rivers at the apex of such bends are vulnerable to concave-bank erosion. A flood in the Saguenay region in July 1996 destroyed this railroad bridge by the erosion of about 75 m of concave bank. The erosion removed the left abutment and caused the left span to collapse into the channel. (Reproduced with the permission of Natural Resources Canada, 2011, courtesy of the Geological Survey of Canada, GSC photo 1998-015D by Greg Brooks.)

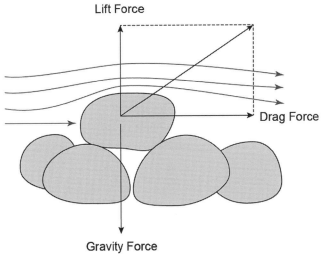

Figure 9.2 Gravity, drag and lift force components acting on a sand grain or pebble lying on a streambed showing resultant force entraining grain and incident force on grain. (After Middleton and Wilcock (1994); with permission of Cambridge University Press.)

Further, we define the Reynolds number (Re), which is the ratio of inertial to viscous forces acting on a fluid volume, as $Re = \rho(VD_H/\mu)$, where μ is the dynamic viscosity of the fluid and ρ the fluid density. Flow is considered turbulent when $Re > 2000$ and laminar below this value. The Darcy friction factor is written as a function of the Reynolds number (Roberson and Crowe, 1993).

Now consider the forces on sediment resting on the bed of a stream channel, as in Figure 9.2, which we assume represents a turbulent boundary layer caused by the irregularity of the streambed surface. Aside from gravity, we can identify two force components affecting the state of rest of the pebble shown in the figure. The first is the drag force F_D, which was introduced in (6.2) and is due both to resistance associated with sedimentary bedforms, such as dunes and bars within the channel, and to drag exerted by the individual grains of sand and gravel. We may write the drag force in the form (Middleton and Wilcock, 1994; Bridge, 2003)

$$F_D = C_D \frac{\rho V^2}{2} A_s, \qquad (9.2)$$

where C_D is a drag coefficient ($\sim$0.5) and A_s the cross-sectional area of the solid, i.e., the pebble. Likewise, we may write the lift force as

$$F_L = C_L \frac{\rho V^2}{2} A_s, \qquad (9.3)$$

where C_L is a lift coefficient. Note that both drag and lift forces are a function of the square of the stream velocity. The product of the two forces – and the restraining forces of gravity and those due to vegetation and cementation – are responsible for sediment entrainment into the flowing water, which is referred to as the **suspended load**, and for mobilizing bed sediment as the **bed-load**. Clearly bed material may become intermittently suspended during transport, in which case it is referred to as the **suspended bed-material load**. Fine-grained suspended sediment from bank erosion and overland flow that remains continuously suspended even at low flow velocities is referred to as the **wash load** (Bridge, 2003).

With these principles in mind, we may now consider the flow and boundary conditions necessary for sediment transport to be initiated on the basis of experimental evidence. Recalling equation (6.3), we may write the expression for the average bed shear stress on the wetted perimeter as (Chanson, 2004)

$$\tau_0 = \frac{f}{8}\rho V^2. \tag{9.4}$$

At a critical shear stress $(\tau_0)_c$, the bed-load begins to be displaced and suspension of sediment occurs. This shear-stress condition is defined (Chanson, 2004) by the **Shields parameter** (τ_*),

$$\tau_* = \frac{\tau_0}{\rho(s-1)gd_s} = \frac{V_*^2}{(s-1)gd_s}, \tag{9.5}$$

where s is the ratio of particle (solid) density to fluid density (i.e., ρ_s/ρ), d_s is the characteristic diameter or particle size that is displaced and V_* is the critical shear velocity defined by $V_* = \sqrt{\tau_0/\rho}$. Table 9.1 summarizes values of the critical

Table 9.1 Values of the critical shear stress $(\tau_0)_c$, critical shear velocity V_* and settling velocity w_0 for values of the characteristic sediment particle diameter d_s. Note the direct dependence of the parameter values on particle diameter. The equation used for settling velocity here is similar to but differs from that of Chanson (2004). (After Julien (2002).)

d_s (mm)	$(\tau_0)_c$ (N/m^2)	V_* (m/s)	w_0 (mm/s)
>512 (medium boulder)	447	0.67	2715
>64 (small cobble)	53	0.23	959
>8 (medium gravel)	5.7	0.074	338
>0.25 (medium sand)	0.194	0.0139	31
>0.016 (medium silt)	0.065	0.008	0.167

shear stress $(\tau_0)_c$ and critical shear velocity V_* for various characteristic sediment sizes d_s.

To describe the onset of sediment suspension and creation of the wash load, we must first identify the terminal fall velocity or **settling velocity** (w_0) of a suspended particle, which is given by Chanson (2004) as

$$w_0 = -\sqrt{\frac{4gd_s}{3C_d}(s-1)}, \tag{9.6}$$

where C_d is a drag coefficient and the other parameters are as previously defined. Values of w_0 are included in Table 9.1. Fluid turbulence normal to the streambed, roughly approximated by V_*, will promote suspension of sediment particles, which is experimentally observed when $V_*/w_0 > 0.2$–2.

We began this section with the goal of illustrating the desired relationship between bed shear stress and sediment displacement. This objective is achieved by correlating the Shields parameter with a dimensionless sediment particle number, $d_* = d_s\sqrt[3]{(s-1)g/v^2}$, where v is the kinematic viscosity (i.e., μ/ϱ). Figure 9.3 illustrates the relationship we have sought and shows the threshold above which bed-load and suspended sediments are entrained. The trough in the threshold line (solid line) is equivalent to $V_*d_s/v \simeq 10$, which parameter is known as the particle Reynolds number and is frequently used as the abscissa in portraying the relationship of critical shear stress with sediment motion (Chanson, 2004) rather than the dimensionless particle number of Figure 9.3.

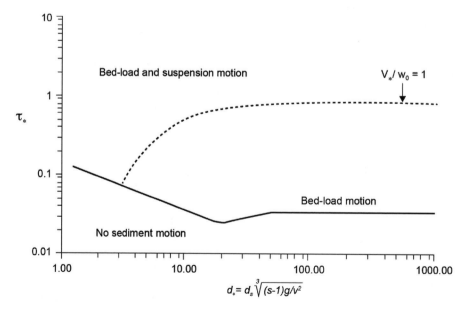

Figure 9.3 The Shields parameter, τ_*, as a function of the dimensionless particle diameter, $d_* = d_s\sqrt[3]{(s-1)g/v^2}$. (From Chanson (2004); with permission of Elsevier.)

Table 9.2 Fluvial sedimentary regimes associated with increasing values of the Shields parameter (τ_*); w/d indicates the width-to-depth ratio in the channel. (After Church (2006).)

τ_*	Sediment type	Sediment transport regime	Channel morphology
≤ 0.15	Sandy gravel to cobble–gravel	Bed-load-dominated	Single-thread to braided channel, $w/d > 40$
0.15–1.0	Sand to fine gravel	Mixed load, high proportion moving in suspension	Mainly single-thread $w/d < 40$
>1.0	Sandy channel beds, Fine sand to silt banks	Suspension-dominated	Single-thread, meanders, point bars, $w/d < 20$

As indicated in Chapter 6, Church (2006) refers to the ability of a particular streamflow to mobilize sediment of a particular grain size as **competence**, which is quantified by the Shields parameter. He reports that the critical value of the Shields parameter for sediment entrainment in a streamflow is $\tau_* \approx 0.045$ for typical heterogeneous mixtures of streambed sediment. Table 9.2 summarizes Church's guide to streambed sediments, channel morphology and the related Shields parameter. Bed-load-dominated sediment-transport regimes exhibit Shields parameters ≤ 0.15 and are dominated by cobble and gravel beds, with channels subject to avulsion and frequent shifting and with deep scouring possible at sharp bends. Suspended-sediment-dominated regimes ($\tau_* > 1.0$) have sandy streambeds with fine sand to silt banks and single-thread channels that are highly sinuous.

It is important to understand that, while peak discharge may progressively increase downstream within a watershed, the competence of a stream may not because of a reduction in stream power or change in channel morphology. The stream channel may widen or the stream power decrease (see Figure 6.14), thereby reducing the stream's competence, which in turn results in sediment accumulation within the channel, i.e., "storage". In a carefully monitored drainage basin in the desert southwest of the USA, Graf (1983) describes a situation in which the stream power even increases downstream but the competence is reduced because of a major increase in the channel's width/depth ratio. The pebbles and cobbles are not entrained and the sediment load is dominated by finer sediment. Graf concluded that "the prudent watershed analyst is one who employs a judicious mixture of engineering and geomorphic approaches" to his or her work.

The various alluvial channel morphologies associated by Church with streamflow regimes are shown in Figure 6.15. Bed-load-dominated streams are similar to that shown in Figure 9.4; the photograph illustrates one that is a cobble–gravel-dominated, single-thread channel and subject to major

Figure 9.4 Cobble–gravel bed in a steep channel in the Canadian Rocky Mountains, Banff, Alberta, Canada.

flood events. Suspension-dominated channels are single-threaded with meanders and point bars like that shown in Figure 9.5.

9.1.2 Estimating Erosion Potential

Erosion-induced failures of infrastructure elicit prominent news coverage. The failure of flood-protection levees in New Orleans during Hurricane Katrina in 2005 is one such

Figure 9.5 Meanders and point bars in the sandy channel of the Lower Wabash River, Illinois–Indiana state boundary, near Grayville, Illinois, USA. Streamflow is from upper right to left. Scale 1 : 30 000. The light-coloured, newly deposited sand and gravel beaches are lateral accretion surfaces.

example, which is discussed in Chapter 16. Failure of levees usually occurs following overtopping of the levee by flood waves and the subsequent erosional failure on the landward side of the levee that regresses through the body of the levee. A more common example of erosional failure of infrastructure is **bridge scour**, such as shown in Figure 9.1. The failure of bridge foundations and piers by fluvial erosion is responsible for about 60% of all highway bridge failures in the USA.

The collapse of a bridge near Amsterdam, New York State, USA, carrying the New York State Thruway (Interstate 90) across Schoharie Creek in 1987 caused ten deaths following the erosion of its foundation and a pier footing. This particular bridge had already successfully survived a flood in 1955 with a peak discharge of ∼2000 m³/s but failed 32 years later when the peak discharge was only 1800 m³/s. The bed material of the creek in glacial till was armoured by a protective layer of sandstone cobbles and boulders from the nearby Catskill Mountains; however, this armour layer could be eroded at discharges above ∼600 m³/s. A forensic assessment of the failure of the bridge discovered a 4.3 m deep scour hole within the till below the footing of one of the bridge piers and concluded that the failure was a result of the cumulative effects of numerous floods. It became apparent that erosion of such geologic foundation materials was progressive and

quite unlike the rapid erosion of sandy river beds; therefore a different approach would be required (Keaton et al., 2012).

Following the Schoharie Creek catastrophe, State departments of transportation in the USA and elsewhere investigated those bridges spanning rivers and determined that 25 000 of the 500 000 total were at risk. These were considered **scour critical**, i.e., if subjected to the design flood, the calculated scour depth would sufficiently erode the foundation so that the design load would not be supported. Preventative measures undertaken following the Schoharie Creek disaster resulted in a substantial reduction in scour failures between 1991 and 2005 (Briaud, 2006). Here we consider the theoretical basis by which the erosion potential of soil and rock streambed foundations is estimated.

Briaud (2008) defines **erodibility** as the relationship between erosion rate $\dot{Z}$ in m/s and the shear stress τ. A simple dimensionless parameter is defined for use on a site-specific basis:

$$\frac{\dot{Z}}{V} = \alpha \left(\frac{\tau - (\tau_0)_c}{\rho V^2} \right)^m, \tag{9.7}$$

where $\tau - (\tau_0)_c$ is the net shear stress in the horizontal direction causing erosion, and α and m are parameters that characterize the soil specimen investigated using a scour measuring instrument known as an erosion function apparatus (Briaud et al., 2001).

On the basis of these measurements, Briaud (2008) developed graphical relationships between the mean grain size displaced by erosion with the critical shear stress and velocity. Figure 9.6 illustrates the relationship for critical shear stress and mean grain size. (Note that shear stress, not velocity, is chosen to be displayed because there is a weak dependence between the velocity profile and shear stress, particularly at the sediment–water interface.) While the data in Table 9.1 indicate a direct dependence of the critical shear stress and the critical velocity on particle diameter, Figure 9.6 shows that this is true only for sand and gravel grain sizes. Briaud et al. (2001) observed that interparticle forces in fine-grained sediments increase the scour resistance; consequently the general direct dependence of mean particle diameter (d_{50} in Figure 9.6) with critical shear stress fails. Nevertheless, these same authors confirm that the shear stress applied at the water–sediment boundary is the principal parameter causing erosion and that fine-sand beds are the most susceptible to erosion.

Figure 9.7 identifies categories of erodible sediments based on shear stress. This figure includes information concerning the erodibility of rock masses, where erosion is due to either slaking of clay-rich rocks (see section 5.3.1) or plucking and

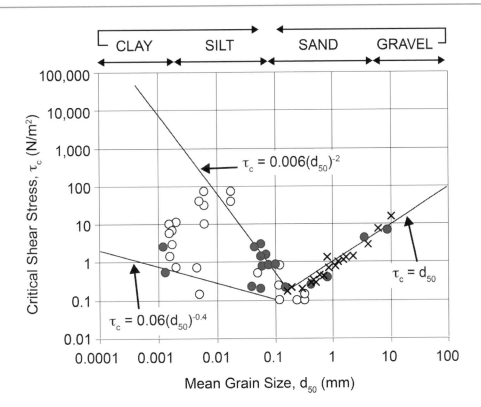

Figure 9.6 Critical shear stress as a function of mean grain size (d_{50}). (The experimental data used in the figure are from Briaud (2008), which article includes a similar figure for critical velocity versus mean grain size. Reproduced with permission of ASCE.)

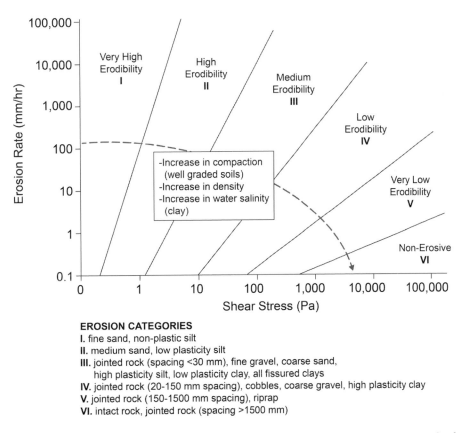

EROSION CATEGORIES
I. fine sand, non-plastic silt
II. medium sand, low plasticity silt
III. jointed rock (spacing <30 mm), fine gravel, coarse sand, high plasticity silt, low plasticity clay, all fissured clays
IV. jointed rock (20-150 mm spacing), cobbles, coarse gravel, high plasticity clay
V. jointed rock (150-1500 mm spacing), riprap
VI. intact rock, jointed rock (spacing >1500 mm)

Figure 9.7 Erosion categories for soils and rocks based on shear stress. (From Briaud (2008), which contains a similar figure for erosion rate versus velocity. Reproduced with permission of ASCE.)

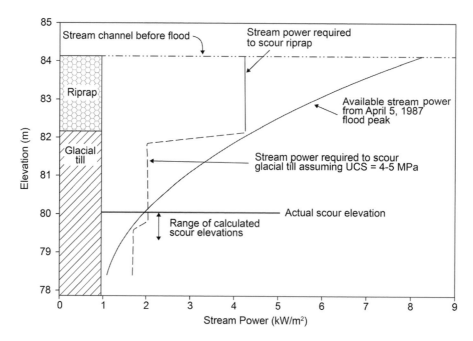

Figure 9.8 Relationship between available stream power at peak discharge in Schoharie Creek, New York State, and the erosion resistance of riprap and the glacial-till streambed foundation materials during the 1987 flood. (Modified from Annandale (2000).)

quarrying causing removal of whole blocks of rock during high-turbulence flows. **Hydraulic plucking and quarrying** require that the excess shear stress at the interface of the rock mass and stream be sufficient to exceed the frictional and gravitational forces working to keep the blocks in place. In this latter case, the orientation of the joints with respect to the turbulent flow direction is particularly important.

Only durable rocks with sufficient strength are considered suitable as bedrock foundations for bridges, e.g., sandstones and unweathered limestones among sedimentary rocks, and most igneous and metamorphic rocks. The special case of scouring of sedimentary rocks has been addressed by estimating the cumulative hydraulic loading causing scour such as that associated with the Schoharie Creek disaster. Because stream power ($\omega = \tau V$, (6.4)) integrates the hydraulic effects of flow velocity, flow depth and channel slope and can be accumulated over time, it is the principal variable associated with the progressive erosion of bedrock streambeds (Keaton et al., 2012). Figure 9.8 illustrates the application of stream power as a predictive tool in accounting for the erosion of both riprap and glacial till during the 1987 flood peak at the Schoharie Creek bridge.

Recently, an innovative approach has been developed to evaluate the erodibility of degradable sedimentary rocks, such as claystones, siltstones, sandstones, limestones and jointed bedrock. This technique involves the modification of the standard slake durability test (see section 5.3.1) to allow estimation

of a **geotechnical scour number** ($m/(W/m^2)$) in units of scour depth (m) per unit stream power (W/m^2). Based on these measurements, Keaton (2013) developed correlations between unit stream power and scour depth for any bridge site with known hydraulic parameters, as shown in Figure 9.9. The most durable rocks tested were some limestones, sandstones and blocky siltstones, while the least durable were thinbedded siltstones and claystones.

9.1.3 Sedimentation in Reservoirs

The building of dams to impound water for irrigation, power and/or flood control can be traced back to antiquity. The Egyptians built dams upstream of Cairo 5000 years ago, while Chinese engineers erected a 30 m high dam 1200 years ago that is still in use to divert irrigation waters. Unfortunately, this Chinese dam has suffered the fate all dams are subject to – its reservoir has filled with sediment (Collier et al., 2000).

The age of major dam building in the USA lasted from the 1930s New Deal era until the 1960s. In Canada two major hydroelectric power dams are under construction in British Columbia and Labrador, but their power generation and reservoir capacity are dwarfed by the Three Gorges Dam on China's Yangtze River that will eventually produce 22 GW of power and create a reservoir 600 km long.

After three years of monitoring sediment transport above and below the Three Gorges Dam, Yang et al. (2007) reported

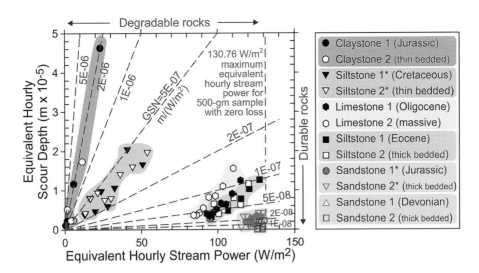

Figure 9.9 The geotechnical scour number in units of m/(W/m²) as a function of equivalent hourly stream power and scour depth. (After Keaton (2013).)

that 151×10^6 tons/a had been trapped by the reservoir or roughly two-thirds of that transported to the dam. The Yangtze River downstream of the dam had been transformed from a depositional system to an erosional one because of increased competence for erosion of most grain sizes. However, the deposition of sediment behind the dam caused a decrease of 85×10^6 tons/a in sedimentation in the estuary of the Yangtze, which threatens the sustainability of the rich fishery of the East China Sea.

The negative downstream consequences of dam building on rivers are now well appreciated and increasingly quantified in terms of declines in fish populations and habitat, while mean annual flood costs increase in the USA because of building in floodplains (Graf, 2001). Here we consider sediment production and the trapping of sediment in reservoirs behind dams within watersheds where agriculture and old mining practices are the principal sources.

- Grenada Lake in Mississippi has lost approximately 3% of its flood-storage capacity over a 50-year period, 1954–2003 (Bennett et al., 2005). The lake receives sediment from two tributaries of which one is forested and relatively stable in terms of sediment production (1127 km²), while the other is in agricultural production (corn, cotton, soybeans) and subject to high erosive losses of soil by overland flow into the tributary (1530 km²). Careful monitoring using acoustic surveying of the reservoir sediments and radiometric profiles (^{137}Cs) of cores has yielded estimates of annual point sedimentation rates in Grenada Lake that vary between 10 and 30 mm/a. This sedimentation is only 16% of the sediment eroded from the unstable channel; 75% of the eroded

sediment is stored upstream of the lake in the unstable tributary. While the reservoir itself has been little affected over 50 years, this stored sediment constitutes ten million tons that is available for remobilization into the lake.

- Englebright Lake in California perhaps constitutes a worst-case example for reservoir sedimentation. It has lost approximately 25% of its flood-storage capacity over a 61-year period (1940–2001) because of it location in the Yuba River watershed, which was the setting for intensive hydraulic mining of sediment to recover gold during the nineteenth century (Snyder et al., 2004). Consequently over 500×10^6 m³ of previously mobilized alluvial sediment is available for erosion into this reservoir, which drains a watershed of 2870 km². The average annual rate of sediment influx into the reservoir following dam completion in 1940 until bathymetric surveying in 2004 was 359×10^6 m³/a or a load of $\sim 400 \times 10^3$ tons/a. Two-thirds of this sedimentary infill was sand and gravel, the balance was silt and clay.

The removal of dams has accelerated in recent years as old dams are found to be unsound for structural or geotechnical reasons or are removed to improve aquatic habitat downstream. In such cases the releases of sediment from reservoirs cause problems. One such release of ~ 7000 m³ of alluvium from a reservoir in the Colorado Rocky Mountains could be traced for 12 km downstream and killed at least 4000 fish. The released sediment was deposited mainly within pools that filled with silt, sand and gravel. These pools developed in bedrock and are typically 52 m × 19 m × 3.5 m deep and spaced down the river at ~ 12 times the mean channel width. Wohl and Cenderelli (2000) determined that the sediment was subsequently and sequentially flushed from upstream to downstream pools. After one year of flushing ~70–80% of the released sediment had been transported downstream.

9.2 Fluvial Sedimentary Regimes

The nature of fluvial sediments, whether alluvium, alluvial fans or deltaic deposits, is a subject of fundamental importance to geotechnical and geoenvironmental engineers. Irrespective of flood-control measures upstream, cities and towns develop their floodplains for commercial, residential and industrial purposes. Residential building on alluvial fans exposes the residents to the fatal hazards of flash flooding. Deltas are subject to subsidence from overproduction of groundwater or hydrocarbons or from sedimentary compaction.

Fluvial sediments are distinguished by their heterogeneity and permeability and have been studied intensely by petroleum geologists as reservoir rocks, e.g., the Alaskan North Slope field, Brent in the North Sea and Daqing in China. They provide good water-supply aquifers but, if contaminated, can produce long plumes because of the presence of high-permeability sand and gravel zones. Here

we consider the general features of fluvial sedimentation, in particular their heterogeneous structure. We begin with a discussion of fluvial sedimentary structures associated with alluvial deposition that was introduced in section 6.4 and illustrated in Figure 6.12, then turn to the special case of alluvial fans, and end with consideration of deltaic deposits.

9.2.1 Alluvial Deposits

Sedimentologists (Bjørlykke, 1989; Miall, 1992; Leeder, 1999) characterize alluvial sediments with various scale-dependent terms:

- **Facies** describes a sedimentary feature on the basis of texture (e.g., "a sandy facies") or sedimentary origin ("deltaic facies") or lithology ("carbonate facies").
- **Architectural elements** are groups of sedimentary facies that describe distinctive shapes and internal geometries of fluvial deposits.
- **Bars and bedforms** are large-scale sedimentary features that fill an alluvial channel and are recognizable at a distance. For example, the Wabash River (Figure 9.5) contains numerous point bars, while the gravel-bed reach of the Fraser River (Box 9.1) contains braided channels with

---

BOX 9.1 | SEDIMENT TRANSPORT IN A LARGE RIVER

The Fraser River (Figure 9.10) rises in the Interior Plateaux and the Columbia and Rocky Mountains of British Columbia, Canada. The headwaters drain thick glaciated terrain into which the Fraser and its tributaries have incised. The three gauging stations shown in Figure 9.10(a), for which hydraulic and sedimentological data are summarized in Table 9.3, illustrate the nature of the hydraulic profile of the lower reaches of a major river.

The Fraser River emerges from a canyon north of Hope as a cobble–gravel bed channel confined on one bank by rock. The channel from 150 km to Sumas Mountain at 95 km, which contains the Agassiz gauging station described in Table 9.3, is characterized as a wandering, braided, gravel-bed reach with mid-channel vegetated islands (slope $= 4.8 \times 10^{-4}$; $w/d = 77$; $d_{50} = 25$ mm). Gravel bars are exposed at low water that are augmented by gravel waves transported down river and which, when incorporated into the gravel bars, cause flow diversion and erosion of adjacent banks.

At Sumas Mountain the river changes form to that of a single-thread, sand-bed channel of reduced slope (5.5×10^{-5}; $w/d = 43$; $d_{50} = 0.38$ mm). Measurements at the gauging station at Mission indicate that the gravel bed-load is deposited upstream. While tidal effects are measured at Mission, salt water does not migrate further than the Port Mann gauging station.

There are no significant tributaries of the Fraser River between the Agassiz and Hope gauging stations; consequently the mean annual and peak flood flows are similar. Between the Agassiz and Mission gauging stations, the Harrison and Chilliwack Rivers and smaller tributaries increase mean flow volumes modestly, about 18%, and suspended sediment loads 1.5%. For a 20-year period (1966–86) suspended-sediment sampling at the Agassiz and Mission stations showed $\sim 17 \times 10^3$ tons/yr transported downstream at each station with another 3 million tons/yr of suspended bed-material load (i.e., >0.177 mm) measured at the Mission station.

BOX 9.1   (CONT.)

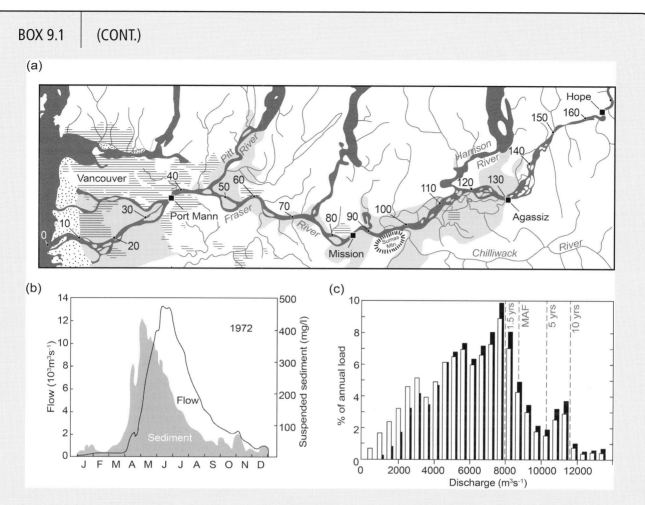

Figure 9.10 (a) Lower Fraser River valley, British Columbia, Canada, showing the principal gauging stations and kilometric distance from the Fraser River delta. (b) Seasonal pattern of suspended sediment transport, 1966–82. (c) Proportional distribution of daily suspended sediment load, 1966–82. (From McLean et al. (1999).)

Table 9.3 Hydraulic and sedimentological data for the Fraser River (see Box 9.1). The velocity is that associated with the mean annual flood. The slope is an average water-surface slope at the gauging station; w/d indicates width-to-depth ratio; d_{50} is the cumulative percentage finer value for submerged bed material.

Gauging station	Bed material	Mean annual flood (m^3/s)	Velocity (m/s)	w/d	Slope	d_{50} (mm)
Hope	Cobble–gravel	8766	3.2	265	6.0×10^{-4}	30
Agassiz	Gravel bed	8760	2.6	77	4.8×10^{-4}	25
Mission	Sand bed	9790	1.5	43	5.5×10^{-5}	0.38

The seasonal pattern of suspended-sediment transport, shown in Figure 9.10(b), is dominated by the spring freshet of snowmelt in April and May that entrains sediment and transports it past the Agassiz gauging station before flow peaks in June. The percentage of the annual load transported as a function of streamflow discharge is shown in Figure 9.10(c). The largest fraction of suspended sediment at Agassiz – and the other gauging stations on the Fraser River – is transported at flows ranging from 7500 to 8000 m^3, which is less than the mean annual flood. The bed-load percentage profile (not shown) versus discharge is similar. All of the suspended-sediment load at Agassiz (and Hope) gauging stations is considered wash load that is carried downstream even at low flows because sediment supply is sufficiently low to allow the available flow to suspend it.

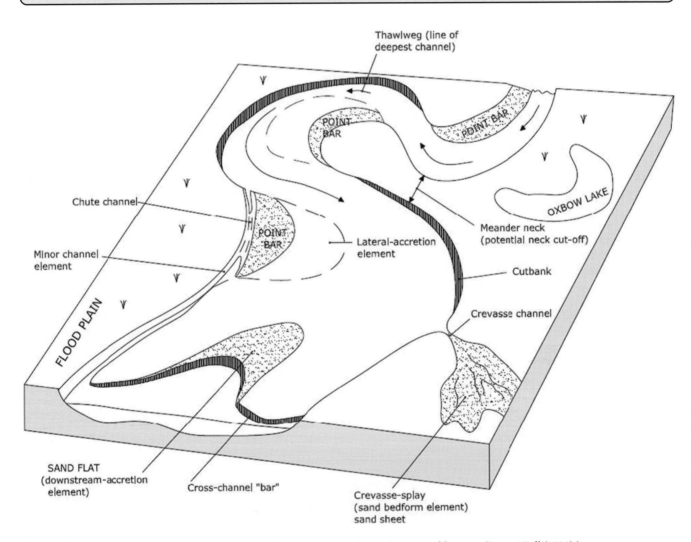

Figure 9.11 General features of floodplains and fluvial deposits. See text for explanation of features. (From Miall (1992).)

mid-channel islands that restrict navigation and divert flow (see Figure 6.16). In the sand-bed reach of the Fraser, bedforms include dunes that are associated with high flow velocities.

Figure 9.11 illustrates the general features of floodplains and associated fluvial deposits in a meander-belt system similar to that of the Mississippi River and many others. In the channel, **point bars** have formed on the interior of meander bends and **cut banks** on the exterior of the bend where active erosion occurs. Channels migrate laterally and become abandoned through avulsion (see section 6.4.4); meanders eventually become isolated from the thalweg of the channel and form oxbow lakes. Sand flats form downstream as the channel widens and the flow velocity decreases. Overbank flooding creates coarse-grained crevasse splay deposits and fine-grained levees.

In these fluvial systems, facies may be identified that provide a description of some particular feature of importance to the geologist. Those facies important to an engineer might be ones identifying hydraulic conductivity for application to dewatering or contaminant migration or to the presence of clays for slope stability purposes. Facies are typically identified by the principal grain size present (upper case) with lower-case modifiers attached to provide context. Some examples are: Gp = gravel facies with *planar* cross-beds; Sr = sand facies with *ripple* cross-lamination; and Fm = fine-grained facies exhibiting a *massive* overbank structure with desiccation cracks. The modifiers indicate important properties that differentiate various sediments, e.g., Gms is matrix-supported gravel, indicating that the pores between gravel clasts are filled with finer materials rather than the gravel being clast-supported, indicating an open framework (see Figure 1.6).

Miall (1992) has divided the channels and bars of rivers into a number of architectural elements, which are composed of assemblages of individual facies as shown in Figure 9.12. Architectural elements are of a scale of importance to the engineer or hydrogeologist in that they describe the dominant textures and structures present.

- **Channels**: Fl (laminated, fine-grained), St (sand with trough cross-beds, dunes); Sr (sand with ripples); Sh (sand, horizontal laminations); Gm (massive bedded gravels).
- **Lateral accretion**: Fl and Sl (sand, low-angle cross-bedding).
- **Sediment gravity flow**: Gms (massive, matrix-supported gravel); Sg (gravity-segregated sands).
- **Gravel bar and bedform**: Gp (stratified gravel, planar cross-beds).
- **Downstream accretion**: DA contains various sandy facies, e.g., St, Sr, etc.
- **Sand bedform**: channel fills or crevasse splay deposits containing St, Sr, Sl, etc.
- **Laminated sand**: sandy sheet deposits including Sh and Sl.
- **Overbank fines**: Fl and Fm.

Engineers and hydrogeologists use facies and architectural elements to develop flow models for groundwater extraction and petroleum reservoir management. In such modelling projects it is essential to capture both the overall structure as well as the heterogeneity of the sedimentary deposits and the degree of interconnection of permeable zones that dictates flow to water-supply or oil and gas production wells. The primary parameter controlling groundwater flow, contaminant transport, petroleum recovery or gas migration is the permeability/hydraulic conductivity structure of a sedimentary formation. Koltermann and Gorelick (1996) pointed out that permeability might vary over 13 orders of magnitude while porosity varies only over two orders of magnitude, and thus the spatial variation of permeability or hydraulic conductivity dictates fluid flow and ultimate fluid recovery.

The practice began in petroleum reservoir modelling and has been adopted by hydrogeologists. For example, Figure 9.13 illustrates the heterogeneity of a section of a Mississippi River meander belt showing the upper 15 m of the sedimentary sequence (see Jordan and Pryor, 1992; Bridge and Demicco, 2008). The open abandoned channel shown in the left-hand panel was filled episodically by sands from the river when still occasionally connected, but later by muds as it became isolated as an oxbow lake. The channel bed is analogous to a small petroleum reservoir known as an oil pool but could equally be an aquifer used for groundwater supply. The dimensions of this particular meander belt are approximately 8 km × 8 km × 30 m deep. The core on the right-hand panel indicates high hydraulic conductivity (i.e., $>10^{-3}$ m/s) associated with the point-bar deposit. The low hydraulic conductivity of the shallow interbedded muds would form a reservoir seal in an oilfield or aquitard in a groundwater flow system.

The permeable channel deposits illustrated by Figure 9.13 were formed by **lateral accretion** of the river. The sediments were deposited sequentially as the channel migrated laterally because of **cutting and filling** by the river, i.e., eroding and depositing. The upper part of this figure shows the 8 km cross-section through the meander belt formed by the continual accretion of point-bar deposits (see Figure 9.5). These deposits are the principal agents of lateral accretion shown in Figure 9.13. They are created by helical flow, as in Figure 9.14, in a meander bend where the outer bank is eroded (the cut or erosional bank), the inner bank is augmented (the point bar) and fresh sediment is deposited in a fining upward sequence on the bar (see Leeder, 1999). The annual rate of channel migration of the Helm meander bend shown in the lower left of Figure 9.5 was 10 m/a between 1903 and 1959, during which time the channel migrated southwards (down the page) to its present position (Jackson, 1976).

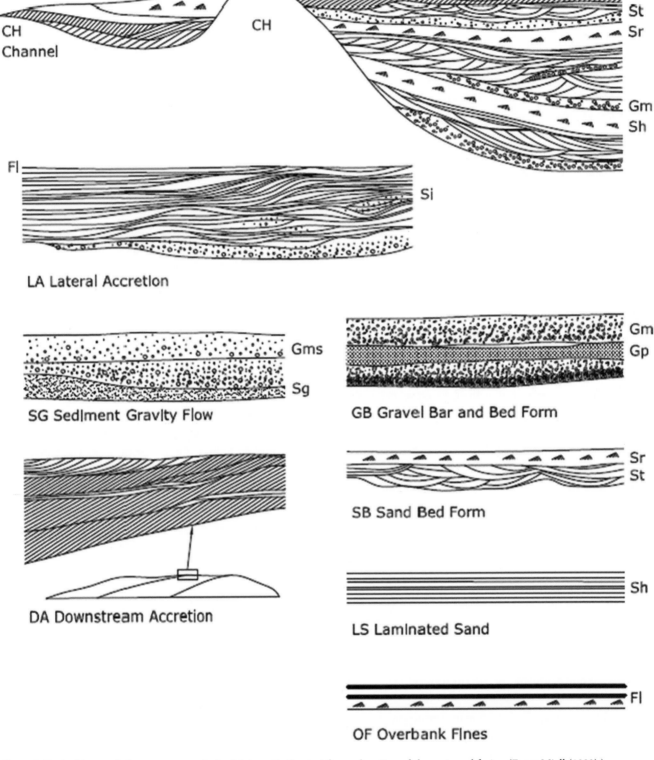

Figure 9.12 Architectural elements present in fluvial deposits. See text for explanation of elements and facies. (From Miall (1992).)

Heinz and Aigner (2003) investigated sedimentary facies within glaciofluvial gravels in the Rhine River valley to understand how the heterogeneity of these deposits might affect the hydraulic conductivity (K) of typical gravel aquifers. The

Rhine gravels were divided into five **lithofacies** based upon textural properties, and which define the hydraulic conductivity range. As is shown in Figure 9.15, this range of hydraulic conductivity is over seven orders of magnitude for the

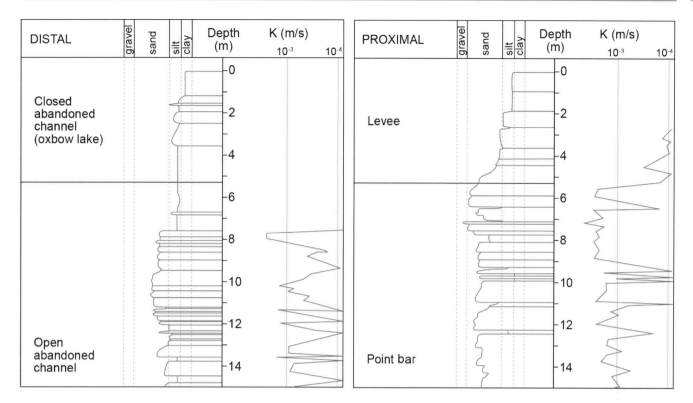

Figure 9.13 The fluvial sedimentary structure, texture and heterogeneity in hydraulic conductivity of two sections from a Mississippi meander belt. The left-hand (distal) stratigraphic sequence is from a core collected approximately 10 km from the present course of the Mississippi River, whereas the right-hand (proximal) core was collected from nearby the present channel.

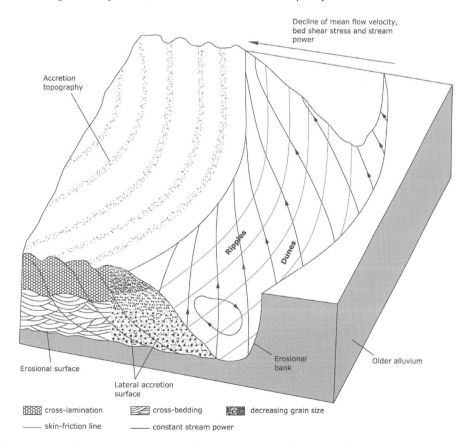

Figure 9.14 Helical flow in a meander bend resulting in lateral accretion of a point bar as the older alluvium is eroded on the cut bank. (From Leeder (1999).)

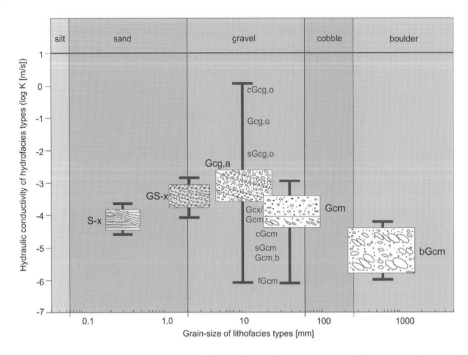

Figure 9.15 Sedimentary geologists define lithofacies as a mappable and laterally extensive part of a stratigraphic unit. This figure shows five such lithofacies in glaciofluvial sands and gravels in southwestern Germany, i.e., S-x, GS-x, Gcg,a, Gcm and bGcm. These lithofacies define the textural properties of the deposit but need to be further subdivided into 12 hydrofacies to identify meaningful hydrogeological facies with similar hydraulic conductivities. (From Heinz and Aigner (2003).)

lithofacies Gcg,a, which represents clast-supported gravels ($K = 10^0$ to 10^{-7} m/s). The sediments were then subdivided into 12 **hydrofacies** with more limited ranges of hydraulic conductivity and grain size (see Box 11.1 for a discussion of estimating K from grain-size distributions). Those gravel samples that were of uniform grain size, i.e., those that displayed the best **sorting**, have the shortest error bars on Figure 9.15. Similarly, Anderson et al. (1999) examined glaciofluvial gravels in Wisconsin, where the mean hydraulic conductivity was 10^{-5} m/s, despite the gravels being very poorly sorted.

The hydrofacies code in Figure 9.15 is defined by the symbols $i_1 I_1 i_2 i_3 i_4$, where i_1 is the minor component of the unit (b = boulder, c = cobbles, s = sand, f = fines (silt + clay)), I_1 is the major component (S = sand, G = gravel), i_2 is texture (e.g., c = clast-supported, m = matrix-supported), i_3 is stratification (e.g., x = stratified, m = massive (no bedding), g = graded), and i_4 is additional information (e.g., i = imbrication, a = alternation, such as o = open framework, b = bimodal, h = horizontally stratified, p = planar stratified, t = trough cross-stratified). Thus stratified sands (S-x) and stratified sands and gravels (GS-x) are much more tightly defined in terms of their ranges in hydraulic conductivity due to being relatively well sorted. An example of a poorly sorted

fluvial deposit is shown in Figure 1.6, although even here some open-framework gravel zones exist.

9.2.2 Alluvial Fans

By virtue of their elevated location above populated valleys, alluvial fans are much sought after areas for residential building in the US southwest, New Zealand, India, France, Spain, Italy, Japan and elsewhere. In San Diego, California, the football stadium and the regional gasoline-distribution tank farm both occupy an alluvial fan that was previously the site of San Diego's groundwater supply (Sengebush et al., 2015).

Comprehensive reviews of the structure and fluvial processes associated with alluvial fans may be found in Bull (1977) and Blair and McPherson (2009). The general features of alluvial fans may be summarized as follows (Bull, 1977).

- Fans occur along mountain fronts, usually uplifted by a fault, which may remain active.
- Fan sediments are deposited at steep slopes along the base of these mountains as streams become unconstrained from narrow mountain valleys.
- Fan deposition is a result of reduced competence due to the decrease in confinement of the stream exiting the mountain

(a)

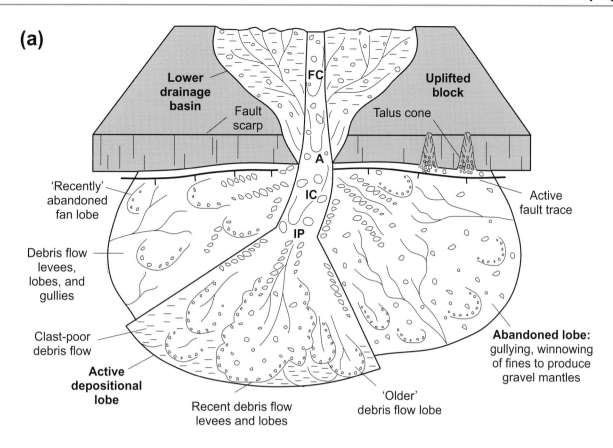

(b)

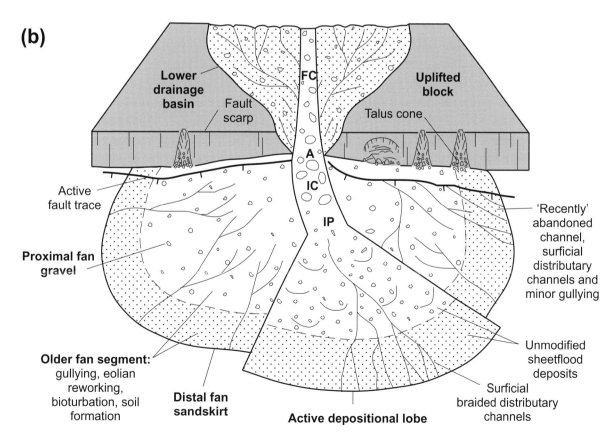

Figure 9.16 Schematic illustrations of the two types of alluvial-fan depositional features: (a) type 1 fans developed by debris flows; (b) type 2 fans developed by sheet floods. Abbreviations: A = fan apex; FC = drainage-basin feeder channel; IC = incised channel on fan; IP = fan intersection point. (From Blair and McPherson (1994); reproduced with the permission of SEPM, Society for Sedimentary Geology.)

source zone or reduced sediment load being supplied to the stream by the hill slopes of the source area.

- The morphology of alluvial fans is typically semi-conical, with the apex of the fan located where the stream exits the mountain front.
- Fans are unlikely to be in steady state. Most are growing (i.e., aggrading), some being eroded where the sediment supply is depleted.
- The proportion of an alluvial fan that is flooded by discharges of low return period varies with the size of the source area and the fan area. For fans with areas of >100 km^2, flooding in arid regions may inundate areas of <5% of the fan, while small fans may have 20% of their areas flooded.

Two types of alluvial fans have been identified by Blair and McPherson (1994) and are illustrated in Figure 9.16: type 1 fans are produced mainly by debris flows, while type 2 fans are produced dominantly by sheetfloods. All alluvial fans are composite deposits formed by multiple depositional and erosion events with high radial slopes compared with rivers, i.e., 2–12° compared with <0.5° in aggrading gravel-bed or sand-bed rivers and river deltas. The active depositional lobes on the fan surface aggrade mainly by debris flows (type 1) or sheetfloods (type 2). Fan texture is typically dominated by poorly sorted mixtures of sand, pebbles, cobbles and boulders, whereas debris-flow deposits contain mud. Type 2 fans exhibit some fluvial characteristics such as stratified layers of sandy gravel and tend to be more permeable due to the absence of mud.

Flash flooding and debris flows on alluvial fans claim lives and inflict property damage regularly. Field mapping of one particular debris flow in California by Blair and Cook (1998) illustrates the hazard caused by alluvial-fan flooding. An August 1984 thunderstorm destabilized colluvial slope materials resulting in the mobilization of ~50 000 m^3 of sediment. The debris flow produced two distinct pathways or tracts of sediment, with bouldery levees of 1–3 m height formed in the upper proximal reaches of the fan and 20–100 cm sediment lobes in the distal reaches. Scars of earlier debris flows indicated that this particular alluvial fan was progressively created by such events. Consequently, the California Geological Survey (Bedrossian et al., 2014) have mapped alluvial-fan deposits in California as an aid to local governments to identify areas threatened by flash floods emanating from adjacent mountain ranges.

A team of geomorphologists and engineers assembled by the US National Academy of Sciences (1996) wrote the following when asked to define the criteria necessary to establish 100-year recurrence interval alluvial-fan flooding for purposes of flood insurance:

> Alluvial fan flooding is a type of flood hazard that occurs only on alluvial fans. It is characterized by flow path uncertainty so great that this uncertainty cannot be set aside in realistic assessments of flood risk or in the reliable mitigation of the hazard. An alluvial fan flooding hazard is indicated by three related criteria: (a) flow path uncertainty below the hydrographic apex, (b) abrupt deposition and ensuing erosion of sediment as a stream or debris flow loses its competence to carry material eroded from a steeper, upstream source area, and (c) an environment where the combination of sediment availability, slope, and topography creates an ultrahazardous condition for which elevation on fill will not reliably mitigate the risk.

9.2.3 Deltaic Deposits

Deltas are sites of large cities, ecologically important wetlands and vast hydrocarbon reservoirs, e.g., the Ganges, Nile, Niger, Mekong, Mississippi and Po deltas. A particular feature of modern deltas is their subsidence, which is associated with compaction, faulting and tilting of the newly deposited sediments. With the steady rise of the sea level (~3 mm/a; Pugh and Woodworth, 2014), the land formed by deltas is threatened with inundation by storm surges, e.g., Hurricane Katrina's inundation of New Orleans in 2005.

Deltas form where alluvial sediment is discharged into a basin occupied by the sea or a lake. In the case of a marine delta front, the fresh water discharges as a buoyant plume onto the denser seawater, rapidly decelerates and sheds its sediment load. Sand bars form proximal to the point where the channel discharges to the sea and finer-grained pro-delta in distal areas. Here we consider the delta developed where the Fraser River (see Box 9.1 and Figure 9.10a) discharges to the sea in coastal British Columbia to the south of Vancouver, Canada.

Bedrock is typically at a depth of 500 m beneath the Fraser and is overlain by the sequence of Pleistocene and Holocene sediments shown in Figure 9.17. The principal texture in the section is silt and sand that occur as massive and stratified facies. The Holocene sediments coarsen upwards, reflecting the deltaic environment in which the beds deposited first, the **bottom-set beds**, are deposited offshore followed by the sandy **fore-set beds** that are deposited at an angle on the slope of the delta closer to the shoreline and then covered by the **top-set beds**. The Pleistocene deposits are separated from the Holocene by 7 m of sheared clayey silt, which is underlain by 120 m of interbedded massive to laminated clayey silt with

FD 87-1

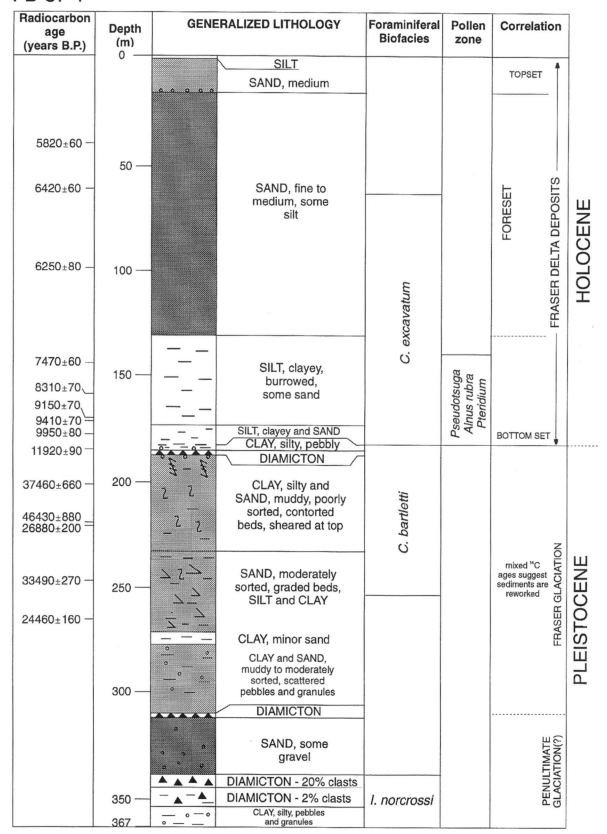

Figure 9.17 Stratigraphy, sedimentology and radiometric chronology of borehole FD 87-1 from the Fraser delta. (From Clague et al. (1998), courtesy of the Geological Survey of Canada, reproduced with the permission of Natural Resources Canada.)

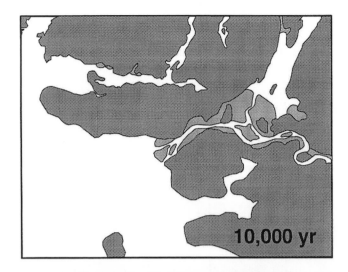

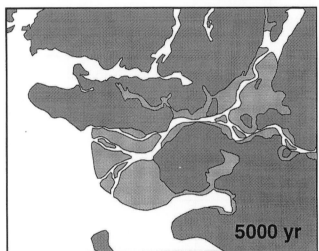

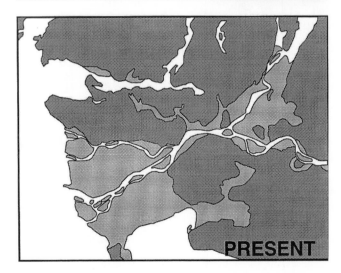

Figure 9.18 Holocene evolution of the Fraser delta. The dark shading is the pre-Holocene landmass, while the light shading indicates the Holocene floodplain, fans and peat bogs. (From Clague et al. (1998), courtesy of the Geological Survey of Canada, reproduced with the permission of Natural Resources Canada.)

sand and pebble layers. Seismic reflection and borehole investigations (Clague et al., 1991) have allowed the stratigraphy of the post-glacial sediments to be defined in detail.

Most of the radiocarbon (^{14}C) dates occur sequentially with depth, as would be expected; however, this is not always the case. Bioturbation can overturn sediments, **reworking** them, after they have been deposited and yield sample date sequences that appear inexplicable. However, the reliability of ^{14}C dating becomes poorer beyond 30 ka (Clark and Fritz, 1997) and this may also explain the non-sequential ages shown in Figure 9.17. It is probably more important to note the general increase in age with depth and the good discrimination throughout the Holocene.

The general growth of the Fraser delta is illustrated by Figure 9.18. The fore-set beds shown in Figure 9.17 were deposited after 10 ka, which is shown in the upper figure of Clague et al. (1998). By 5 ka the delta front had extended westwards to the modern delta shoreline. The present delta is highly urbanized, undergoing subsidence and threatened by rising sea levels. Neilson-Welch and Smith (2001) have described the intrusion of seawater into the deltaic sediments extending up to 500 m inland caused by the density difference of seawater to fresh water (see Chapter 12). While natural consolidation of the Holocene sediments is responsible for subsidence rates of 1–2 mm/a, anthropogenically induced rates are much greater. Mazzotti et al. (2009) estimated these rates at 3–8 mm/a, caused by large buildings erected on the compressible delta soils, including the airport, ferry and seaport terminals. However, Samsonov et al. (2014) analyzed synthetic aperture radar data from 1995–2012 and determined a maximum vertical deformation rate of ~20 mm/a caused, they hypothesized, by groundwater extraction (see Chapter 13) and construction on the soft deltaic sediments. Furthermore, they concluded that the subsidence was increasing rapidly, therefore posing significant future problems for infrastructure protection.

9.3 Diagenesis of Clastic Sediments

The mechanical and geochemical reactions that occur to sediment following deposition are **diagenetic reactions**. They are responsible for the primary permeability and porosity of sedimentary rocks and reflect their lithification. Thus sand becomes sandstone and clay becomes shale or mudrock. Subsequent alteration of these rocks, such as that due to faulting, are secondary processes. When post-depositional changes to rock fabric and physical properties occur at the

BOX 9.2   SUBSIDENCE AND LAND LOSS IN THE MISSISSIPPI RIVER DELTA

The Mississippi depositional basin, referred to as the Mississippi embayment, formed during the Cretaceous, about 100 Ma. By Plio-Pleistocene time, roughly 5 Ma, glacially induced incision of the Mississippi Valley caused the ancestral Missouri and Ohio Rivers to drain to the Gulf of Mexico via the ancestral Mississippi to form the present watershed. The basin stretches for 400 km along the Mississippi and Louisiana coasts. Fluvial or marine sediments are >500 m thick near New Orleans, but increase seaward to >4000 m at the margin of the continental shelf. The delta was created by cyclic growth of deltaic headlands, channel avulsion and deposition in different parts of the sedimentary basin, yielding the characteristic bird-foot shape of the delta headland (Blum and Roberts, 2012).

The delta of the Mississippi River has been closely monitored to determine the rate of wetland loss to seawater inundation following subsidence. This loss of wildlife habitat in the delta is complemented by the additional loss of protection from storms afforded by barrier islands along the coast. The American Society of Civil Engineers (Izzo, 2004) estimated the loss of land at 65–90 km^2/a, particularly in an area south of New Orleans and west of the Mississippi River outfall. The ASCE anticipated that by 2050 an additional 2500 km^2 (1000 square miles) will be lost over and above the 3900 km^2 (1500 square miles) lost in the previous 50 years.

Three processes appear responsible for this subsidence in the delta.

- *Hydrocarbon production from pre-Holocene sediments*: Deltas often host vast hydrocarbon reservoirs that formed prior to the Holocene. Morton and Bernier (2010) have shown that there is a strong correlation between the loss of wetlands due to their submergence with the rates of production of oil, gas and formation water from the oil and gas fields of the southern Mississippi delta. This is consistent with the findings of Zoback (2010), who describes how oil, gas and water production in one such field caused compaction and then faulting.
- *Compaction of Holocene strata*: Faulting and tilting of deltaic sediments is widely recognized and documented worldwide (Törnqvist et al., 2006). As long as sediment loading continues, there will be dewatering and compaction of the sediment. While subsidence along the Texas Gulf Coast and many sections of the Mississippi delta is minimal, these authors found that the area of the bird-foot of the delta is more rapidly subsiding.
- *Reduction in sediment supply*: Blum and Roberts (2012) argue that dam building along the Mississippi–Missouri river system has reduced sediment supply by 50% since 1950 and therefore this sediment was unavailable to replace the land lost to subsidence and sea-level rise. Before the age of dam building in the 1930s, the suspended load carried by the Mississippi River was ~400–500 MT/a (million tonnes per annum). However, dam building and the diversion of water into the Atchafalaya River greatly reduced this sediment transport. These two rivers now carry only 204 MT/a; consequently the delta is atrophying from a lack of sediment that is reflected in the seawater inundations and subsidence.

ground surface, they are usually referred to as weathering rather than diagenesis.

Bridge and Demicco (2008) provide a comprehensive account of diagenesis:

1. **Mechanical compaction**: Diagenesis begins in shallow parts of a sedimentary basin with the physical process of compaction during sediment burial causing the expulsion of pore water. Compaction operates on the initial grain packing of the recently deposited sediments that might yield an initial porosity of 48% for cubic packing, i.e., sediment grains are stacked such that their centres are orthogonal to their nearest neighbours as would happen with four spheres that fill a cube, or 26% for rhombohedral packing, which is that of a stack of billiard balls. The compression behaviour of sands begins at low stress (<5 MPa) with porosity decreasing due to the frictional slip and rotation of grains. Compaction is most pronounced in the stress interval of 5–25 MPa, in which

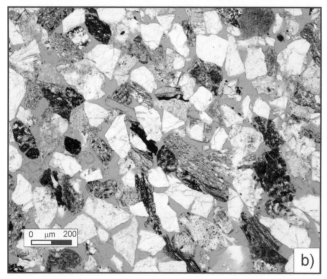

Figure 9.19 Photomicrographs of thin sections from cored sections of sandstone channels within the Paskapoo Formation, Alberta. (a) Photo from a massive, fine-grained, non-porous sample consisting of quartz, detrital carbonate clasts and chert. The grain-supported fabric has low porosity due to calcite cementation. (b) Photo shows a fine- to medium-grained sandstone, but lack of cementation and clay-mineral alteration has allowed the maintenance of an open pore network (grey spaces between grains). (From Grasby et al. (2008); Geological Survey of Canada.)

there is significant grain crushing causing substantial loss of porosity, with the effect being greater for coarser sands. According to Chuhan et al. (2003), the effects of grain slippage, rearrangement and crushing have mostly reached completion by 25 MPa.

2. **Cementation**: This stage is characterized by grain overgrowth by cementation, usually calcium carbonate, iron oxide or silica cements that precipitate from pore waters flowing through the sediment and create mineral bonds between grains. When cements form in pore throats, permeability is much reduced.

3. **Dissolution**: Geothermal heating and sediment loading cause elevated temperatures and pressures that result in mineral dissolution, e.g., feldspar dissolution in sandstones leading to the formation of kaolinite and quartz, which may not involve any change in porosity. Dissolution processes are intimately associated with mineral replacement processes (see next item).

4. **Geochemical diagenesis**: Mineral replacement reactions include those of carbonate rocks in which calcite may be replaced by dolomite or chert. Wood may become petrified through replacement by quartz. In clay mineral diagenesis, early-stage clay diagenesis creates kaolinite and smectite, while illite and chlorite are late-stage clays that indicate burial temperatures in excess of 100°C; consequently the presence of illite (and not smectite) in sedimentary rocks is associated with petroleum generation.

Figure 9.19 illustrates diagenetic changes in the sandstone channels of the Paskapoo Formation in Alberta, Canada, which was featured in Chapter 3. In the upper photomicrograph, porosity is destroyed, but it is maintained in the lower photograph. Therefore the process of diagenesis causes sedimentary rocks to become lithified, otherwise referred to as indurated or consolidated, the result of which is a loss of pore space and reduction in (primary) permeability as well as porosity. Similar processes of cementation, compaction and dissolution occur in carbonate rocks, for which Lucia (1999) should be consulted.

9.4 Summary

The processes and results of sediment entrainment have been reviewed. Bridge scour and other examples of sediment entrainment are considered in the context of the Shields parameter (equation (9.5)), which we have expressed as a function of a critical shear stress acting on the streambed boundary layer. We have related this critical shear stress to an erosion rate nomogram (Figure 9.7) that indicates the degree of erodibility of sediments and rocks in a streambed. Class I erosion category is that of very high erodibility in the cases of fine sand and non-plastic silts. Class V erosion category indicates that of very low erodible rock and riprap. Furthermore, we have recalled the concept of stream power introduced in Chapter 6 to explain the progressive effects leading to bridge scour and illustrated these effects by introducing the geotechnical scour number.

Sediment transport in watersheds leads to deposition and infilling of reservoirs behind dams and to reduced navigation of waterways. We have considered loss of reservoir capacity that varies from 3% behind a Mississippi dam to 25% in a northern Californian watershed much affected by mining. Also, we have reviewed sediment transport in the Fraser River over a 20-year period by identifying various sediment transport categories: the wash, suspended, bed and suspended bed-material loads.

Three sedimentary regimes have been considered in some detail: (i) alluvial deposits, which form floodplains through lateral accretion, (ii) alluvial fans, which are subject to devastating floods and debris flows, and (iii) deltas, which are subject to continuing subsidence. The sedimentologist uses facies and architectural elements to define various features of these sedimentary deposits, in particular features of distinct permeability important to the development of fluid flow models. We conclude with a discussion of mechanical and geochemical changes to sediments, referred to as diagenesis, that produce sedimentary rocks.

9.5 FURTHER READING

Anderson, R.S. and Anderson, S.P., 2010, *Geomorphology: The Mechanics and Chemistry of Landscapes*. Cambridge, UK: Cambridge University Press. — This text contains good descriptions of many of the processes discussed in this chapter, including stream power as the principal variable controlling streambed erosion, bedrock channels and modes of sediment transport.

Bridge, J.S., 2003, *Rivers and Floodplains: Forms, Processes, and Sedimentary Record*. Oxford, UK: Blackwell Science. — A rigorous treatment of the physical processes involved in sediment transport, sedimentary structures and floodplain development. Chapter 3 provides a useful summary of the basic theory of sediment transport.

Bridge, J.S. and Demicco, R.V., 2008, *Earth Surface Processes, Landforms and Sediment Deposits*. Cambridge, UK: Cambridge University Press. — This textbook combines an introduction to fluvial processes and geomorphology with traditional material on sedimentology, including a good discussion of diagenesis.

Chanson, H., 2004, *Environmental Hydraulics of Open Channel Flows*. Burlington MA: Elsevier Butterworth-Heinemann. — A text for those interested in open-channel hydraulics in rivers and estuaries coupled with material on turbulent mixing and dispersion, i.e., "environmental hydraulics".

Julien, P., 2002, *River Mechanics*. Cambridge, UK: Cambridge University Press. — A modern text on fluvial mechanics and river engineering suitable as a standard reference for geotechnical engineers interested in riverbank stability, flood-control measures and estuarine processes.

Leeder, M., 1999, *Sedimentology and Sedimentary Basins: From Turbulence to Tectonics*. Oxford, UK: Blackwell Science. — A broad overview of sedimentary processes with a bias to interpretation of sedimentary structures using fluid dynamics.

Robert, A., 2003, *River Processes: An Introduction to Fluvial Dynamics*. London, UK: Arnold. — A concise and informative account of sediment erosion and transport in a variety of channel types, bedforms and aquatic habitats.

9.6 Questions

Question 9.1. Assume that the critical Shields parameter is $\tau_* \approx 0.045$ and you wish to calculate the diameter of a quartz particle that is entrained at a critical shear velocity of $V_* = 0.05$ m/s.

Question 9.2. Why should the available stream power profile shown in Figure 9.8 have a maximum value at the stream surface and a minimum at the sediment–water interface?

Question 9.3. A stream-power analysis of the Schoharie Creek, New York, bridge scour disaster (Figure 9.8) indicated that the 4.3 m deep scour hole, which was associated with the failure of a bridge pier, was the result of a daily stream power of 1790 (ft-lb/s)/ft^2 × 24 hours, where the dimensions are of work (ft-lb) per unit time (s) per unit area (ft^2) of the streambed for one day. This daily stream power was accumulated over a period of 33 years (1954–87). If 1 ft-lb/s = 1.35 W, compute the scour number for this scour hole in metres of scour depth per unit of daily stream power in kW h/m^2.

Question 9.4. As an engineer, your task is to predict the design scour depth at a bridge site when the design life of the bridge is 20 years. Based on slake durability tests, the geotechnical scour number for the bedrock at the bridge site is determined to be 5×10^{-7} m/(W/m^2), the average annual cumulative stream power is estimated as 100 kW/m^2 and the design service life of the bridge is estimated as 20 years. What is the predicted scour depth after 20 years?

Question 9.5. Explain why helical flow (Figure 9.14) in a meander is associated with the development of floodplains.

Question 9.6. Figure 9.15 correlates 12 hydrofacies with their measured hydraulic conductivities. Explain why hydrofacies cGcg,o is so much more permeable than hydrofacies bGcm, even though bGcm contains boulders.

CHAPTER

10 Characterization of Soils and Sediments

Geoscientists and engineers will view field samples of soils and sediments differently as a result of their educational training and professional requirements. Only collaboration in practice and elaboration in textbooks bridge the differences in perspective. Unlike civil engineers, their geoscientific partners in site characterization – engineering geologists, hydrogeologists and geological engineers – acquire and report data that reflect their broad education in the geosciences and their practice in the field. The purpose of this chapter is to build this bridge between the two professional perspectives.

Geotechnical and geoenvironmental engineering students are likely to have completed an undergraduate course in soil mechanics or geotechnical engineering before encountering this book. Introductory geotechnical textbooks cover a wide variety of experimental techniques and general principles. Consequently, the elementary geotechnical laboratory topics of plasticity and moisture content, index tests, bulk and particle density, and compaction tests need not be considered explicitly in this chapter. Some geotechnical tests were introduced earlier involving testing with a triaxial cell (Chapter 4), a ring-shear device (Chapter 8) and testing of sediment erodibility (Chapter 9). Because the geological and geotechnical hazards discussed in this book typically involve hydrogeological, shear strength and consolidation processes, we will discuss permeameter, triaxial and oedometer testing briefly in this chapter for those unaware of the nature of the measurements.

As in our discussion on the characterization of rocks (Chapter 5), we begin this chapter by considering the elements of site characterization when the site of interest is composed of soils and sediments that may be of fluvial, glacial, lacustrine or other origin. The failure to properly conduct a site characterization is oftentimes the source of engineering failure and the lesson is then relearned by a new generation of engineers.

A recent example is the 2014 failure of the mine tailings dam at the Mount Polley Mine in British Columbia, Canada (Figure 10.1). The failure occurred through the breach of a 40 m high perimeter embankment dam emptying a large (4 km^2) tailings pond. The Mount Polley Review Panel (2015) that studied the failure concluded that the root cause of the breach was the undrained failure of a glaciolacustrine unit, which was part of the dam's foundation, under the imposed load of the perimeter embankment. The Panel further concluded that the design did not take into account the complexity of the subglacial and pre-glacial geological environment of the perimeter embankment foundation and that omissions associated with site characterization were similar to creating a loaded gun.

Therefore, we focus on the following topics to link geotechnical and geoscientific perspectives: (i) borehole drilling and coring methods in soils and the nature of core logs; (ii) the use of cone penetrometer testing; (iii) the use of surface and borehole geophysical tools; (iv) the sedimentological analysis of cores; (v) laboratory testing of soils and sediments critical to the analysis of geological hazards; and (vi) the characterization of sand and gravel deposits to determine their potential use as building aggregate. While this discussion cannot be exhaustive, we recommend the interested reader consult the section on further reading at the end of this chapter.

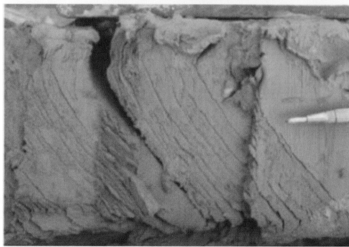

Figure 10.1 An oblique view of the breach of the Mount Polley embankment dam, British Columbia, August 2014, which was built to contain mine tailings. The breach is approximately 40 m high from the crest of the dam to the whaleback (W) formed by erosion-resistant foundation till that was upthrust by the shear failure of the glaciolacustrine unit (GLU). As such, it acts as a control section for the outflow from the breach. Eroded till blocks displaced by the flow are shown as J. Immediately beneath the till is the upper GLU (shown below) that failed by shear. Behind the dam, gully erosion of the tailings occurs by flow sliding. (Photos from the Mount Polley Independent Expert Investigation and Review Panel (2015), reproduced by courtesy of the Government of British Columbia.)

10.1 Elements of Site Characterization

The general principles of site characterization have been listed by Rogers (2006), who considered site exploration methods in use by engineering geologists. Adopting Rogers' principles and adding a number of additional features, it is the practice of engineering geologists that site reconnaissance should be devoted to developing a preliminary model of the geological structure and geomorphology of any site that would require identifying the following features:

- geological structure, including soil development;
- presence of colluvium, alluvium, glacial sediments, landslide debris and lacustrine sediments;
- the depth and nature of weathering;

- the stratigraphy of the underlying bedrock;
- evidence of joints, folds, faults, karst and disconformities;
- hydrogeological features, such as depth of water table, evidence of perched groundwater, overpressured zones and high-permeability units; and
- determination of appropriate soil and bedrock parameters, e.g., strength, compressibility and hydrologic variables.

This preliminary model is referred to by different names, such as **geomodel** (Fookes et al., 2015) or **ground model** (Hencher, 2012) in the UK or **site characterization model** in North America (e.g., APEGBC, 2016).

Box 10.1 outlines the professional practice guidelines that the Mount Polley Panel Report recommended be developed

BOX 10.1 | SITE CHARACTERIZATION MODELS

The premise of any **site characterization model** (SCM) is that it is based on adequate interpretations of geological, geomorphological and hydrogeological conditions, in which the adequacy is judged with respect to the engineering circumstances and the project requirements (Hencher, 2012).

Site characterization will begin with a review of published reports of the locality in which the site is to be developed, i.e., terrain analysis, as discussed in section 6.5. These may be extensive in the case of urban areas or negligible in the case of remote areas. Regardless, topographic and geological maps should be consulted and air photo, LiDAR and satellite images obtained. Because we have already considered characterization of rock masses in Chapter 5, let us assume that the properties of the site in question are dominated by its surficial geology, i.e., bedrock is at such depth as to make it a minor concern except for its potential seismicity. Unmanned aerial vehicles or drones may be used to develop three-dimensional images of the site. While expensive, LiDAR may provide the most useful information if vegetation is obscuring air-photo interpretation.

In that British Columbia was glaciated in the Quaternary, the glacial geology is particularly important and thus features prominently in the APEGBC guidelines. Drilling methods that produce undisturbed samples and continuous core are therefore important so that sedimentary features indicating potential shear zones can be identified in lacustrine, glaciolacustrine, glaciomarine and fine-grained alluvial overbank sediments. The same principle of identifying soils and sediments that are prone to lateral shearing would be critical in characterizing unglaciated areas.

An important product of this initial site characterization is a set of terrain maps that indicate topographic features, including hill slopes, gullies and landslides. These might represent terrain stability, terrain hazard and risk, sediment erosion potential and seismic risk. In alpine and northern regions, the presence of permafrost should be indicated. Where karst is present, special attention should be played to identifying this hazard and its surface expressions.

Geophysical surveys will be important in identifying anomalous surficial features and the spatial continuity of foundation materials, such as the glaciolacustrine unit responsible for shear failure at the Mount Polley dam site. Seismic methods may be used to map the bedrock contact and indicate the density of the surficial materials. Electrical and electromagnetic surveys can indicate clay units or soil salinity. Ground-penetrating radar may be useful in identifying permafrost or discontinuities in layered sediments. Magnetic and gravity methods may be used to define bedrock contacts and cavities.

The inadequate site characterization identified at the Mount Polley dam site led to the following guidance from APEGBC regarding intrusive geotechnical site investigations:

- The depth of the geotechnical investigation should be similar to the expected final height of the tailings embankment dam or deeper should conditions indicate.
- The spacing between boreholes should take into account the variability and continuity of the geological units that will comprise the foundation of the infrastructure to be constructed and could be a few metres apart for complex conditions to 100 m for continuous layers.
- The number of boreholes should be sufficient to draw reliable cross-sectional maps of the various soils and bedrock present in the foundation.
- Undisturbed sampling of cohesive soils should be undertaken for laboratory testing.
- The potential for artesian hydraulic heads in the foundation soils should be clearly identified.

Source: After APEGBC (2016).

to ensure adequate site characterization for future tailings dams, although the recommendations are sufficiently general to be appropriate to addressing the features identified above by Rogers (2006) for building, industrial plant, transportation or other infrastructure sites.

In the remainder of this section on site characterization, we consider the various methods available to engineers and geoscientists to acquire data to generate a site characterization model, including borehole drilling techniques and the use of cone penetrometer tests and surface and borehole geophysics. This discussion is then followed by discussions of borehole drilling methods, core acquisition and the application of cone penetrometer testing and geophysical surveying to site characterization.

10.1.1 Drilling Methods

A wide variety of drilling methods are available and their choice depends on the purpose of the borehole within a particular project and the nature of the geological materials that are expected to be encountered. Here we consider five methods that are commonly used.

Cable-tool drilling is a percussion method that drives a drill bit and casing into soils or poorly consolidated rock using a chisel-type bit. The bit is hammered into the soil exposed at the base of the casing, which is advanced with the bit (see Figure 10.2a). The drive force on the bit is a donut hammer, which is raised and dropped repeatedly, thus pushing the bit deeper into the soil, and which pulverizes any consolidated rock present. The displaced soil or pulverized rock – the **drill cuttings** – is removed by bailing, which means that the drill bit must be withdrawn.

If a split-spoon sampler is driven into the soil to collect a core sample, the number of hammer blows on the drill rod to project the sampler 150 mm or 6 inches into the soil or rock is recorded and is reported as the **standard penetration test** (SPT) N value. The method has the advantages of being able to drill to great depths and minimizes the possibility of cross-contamination into deeper units through bailing, but it is slow (5–8 m/day) and destroys any poorly consolidated rock it encounters. However, as the sample barrel approaches bedrock, it may fail to identify the bedrock contact because of cobbles from the colluvium or weathered bedrock above the contact plugging the barrel and thus misrepresenting the bedrock contact or **rockhead**. Similarly, the sample barrel will encounter increasing resistance as it is pushed

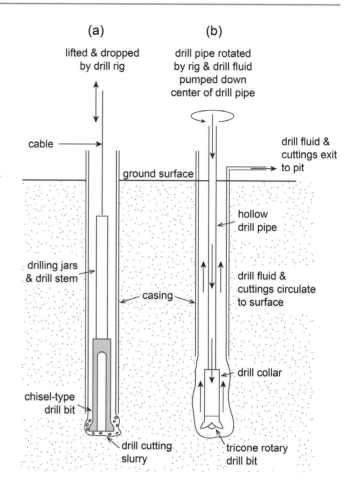

Figure 10.2 Schematic illustrations of (a) cable-tool drilling and (b) direct mud–rotary drilling operations. In direct rotary drilling, the drilling fluid is pumped down the drill stem and recovered from its return flow up the borehole annulus. Reverse rotary drilling has fluid circulation in the opposite direction. (From CCME (1984); with permission.)

through soft sediments towards a stiff layer and the blow counts will again increase rapidly. Rogers (2006) discusses the corrections that should be employed to properly present SPT data.

Another percussion method especially suited for coarse deposits, e.g., cobbles and boulders in alluvium or in glacial till, is the **Becker drill**. A diesel-driven hammer pushes the steel drill stem into the ground, with the number of blows recorded for each 300 mm or one foot of penetration. When Becker rigs are configured with inner and outer tubes, air may be pumped down the annulus of the tubes and recovered from the inner tube at ground surface and the drill cuttings recovered. The engineer or geologist on site records the texture of the cuttings, the diesel hammer's bounce chamber pressure and the hammer-blow rate.

Rotary drilling was introduced in the USA in the 1820s for water-supply well drilling and was used with a diamond bit in 1860 (Anderson, 1984). In the twentieth century it became widely used in oilfield drilling and further development thereafter was rapid. Figure 5.1 illustrates the general features of a rotary-drilling rig; Figure 10.2(b) shows the general principles of operation. A rotary drill bit, oftentimes the tricone bit with three coupled cones, and drill stem are driven into the soil or rock due to the rotation of the drill pipe by the drilling rig. In direct mud rotary drilling, drill fluid is pumped down the centre of the drill pipe, through the rotating tricone drill bit and entrains the drill cuttings, which are carried back up the annulus of the drill pipe; reverse rotary drilling circulates the drilling fluid in the opposite direction. The cuttings are circulated to the ground surface where they are retained on a screen placed in line and the drilling fluid is retained in a tank or pit.

The cuttings are examined to determine their texture, and perhaps mineralogy, and then logged. However, their exact depth of origin is often poorly known and therefore may not allow satisfactory logging of the stratigraphy. The drilling fluid is usually mud, but it can be air, which prevents drill fluid contamination of formations although the flour produced by the drill bit can coat fractures and the borehole wall. When mud rotary drilling is employed, the cuttings are entrained in a clay-rich slurry and it is difficult to identify the amount of fine-grained sediment present.

Sonic drilling is a recently developed technology that accelerates the rate of rotary drilling significantly and addresses some of these problems. Boreholes are drilled and cased by rotating and vibrating the drill rods as shown in Figure 10.3 and casing at sonic resonance frequencies causing the drill bit to vibrate up and down and rotate. Sonic drilling has a wide variety of uses and can drill soils and hard rock more rapidly than traditional mud–rotary methods. However, the heat produced appears to turn soft Floridan limestone into "oatmeal" and may cause some volatilization and loss of contaminants contained within the plastic-wrapped core. Sonic drilling has proven particularly useful for deep drilling and coring of alluvial basins and subsequent installation of monitoring wells.

Augers are another very old technology whose origin is obscure but perhaps developed from the Archimedes screw used for centuries to lift water. The rotation of spiral auger flights raises soils that have been displaced by a drill bit with protruding teeth that erodes the soil at the base of the auger flights. Solid stem augering is used for the installation of simple, shallow monitoring wells or piezometers.

Figure 10.3 A sonic drilling rig showing the drill head and casing. The drill head is slightly larger than a standard rotary drill head and imparts the rotary motion. It also contains an oscillator, which causes a high-frequency force to be superimposed on the drill string. As the drill bit physically vibrates up and down, it pushes the drill string down and rotates it. (Courtesy of Sonic Drilling Ltd., Surrey, British Columbia, Canada.)

Deeper boreholes (30–50 m) require a hollow-stem auger, which has hollow drilling rods on which the auger flights are welded (see Figure 10.4). The drill rig rotates the centre rod and augers, with the bottom of the centre rod plugged to prevent sand from entering and plugging the drill string. At the prescribed depth, the centre rods and plug can be pulled, a core sampler driven through the bottom hole now exposed and advanced into undisturbed soil beyond the bottom of the augers (see Figure 10.5). A monitoring well, such as shown in Figure 11.6, can then be set inside the auger flights, the annulus sealed and the casing removed around it.

Figure 10.4 An all-terrain hollow-stem auger is shown with a stack of 1.5 m long auger sections. The augers are fitted with a cutter head on the lead auger and the sections are connected together to form a string that lengthens as the borehole deepens. The centre rod of the auger is shown resting on the ground. It prevents soil from entering the hollow augers and can be raised to allow a soil sampler to be lowered and driven into the exposed soil at the bottom of the borehole. (Courtesy of Aardvark Drilling Inc., Guelph, Ontario, Canada.)

10.1.2 Core Acquisition

Drill cuttings, those fragments of the soil or sedimentary rock displaced by drilling, may not be of sufficient quality for many site investigations. Consequently, it is frequently deemed necessary to collect cores of soil from boreholes. Table 10.1 illustrates that the choice of a soil-core sampling device varies with depth and the cohesion of the soil, while rock samples are collected in core barrels as discussed in section 5.1.2. The various types of standard core barrels are discussed and photographed in Mayne et al. (2001).

Perhaps the simplest soil-core sampling device is the split-barrel, heavy-wall or "split-spoon" sampler that is threaded to a drill rod at one end and has a hardened drive shoe at the other (Figure 10.5a). It is hammered into the soil at the bottom of the borehole. The core barrel is made of two half-sections that may be separated after sample retrieval and the core examined within its inner liner. The ball valve in the split-spoon sampler cannot be relied upon to retain cohesionless soils, such as sands from below the water table, and cobbles and boulders may prevent a sample being collected, but its

Table 10.1 Use of soil-core samplers.

Material		Split spoon	Shelby tube	Piston sampler
Cohesionless soils	<20 m depth	5	3	1
	>20 m depth	4	2	1
Cohesive soils	<20 m depth	5	1	3
	>20 m depth	4	1	2

Notes: Adapted from *Subsurface Assessment Handbook for Contaminated Sites* (Waterloo Institute for Groundwater Research, 1994). Rating illustrates most appropriate (1) to least appropriate (5). Rating based upon completeness of recovery of undisturbed soils.

strong barrel is otherwise well suited to coarse soils. A Shelby tube is a thin-walled core sampler (Figure 10.5b) with a bevelled leading edge that is connected to the drilling rods by a threaded section. It too may resist sampling gravel and larger particles but is well suited to soft soils.

Mayne et al. (2001) report that the vast majority of geotechnical soil-core samples in the USA are collected with split-spoon samplers, which provide a disturbed soil sample of sand, silt and clay materials. As Bishop (1948) observed in the first issue of the new geotechnical journal *Géotechnique*, undisturbed samples are distinguished by three criteria: (1) the soil structure is preserved, (2) there is no change in water content or void ratio, and (3) there is no change in the constituents and chemical composition of the soil. When the target sample is a fine-grained soil, Shelby tubes may be used to recover essentially undisturbed samples. However, this leaves the problem of sampling cohesionless sands and other coarse-grained soils, in which case loss of sample from the core barrel and the disturbance of the sample are significant.

In recent years, there has been much interest by hydrogeologists in obtaining undisturbed cores of cohesionless sands using piston samplers. These are thin-walled tube samplers with a piston located at the base of the tube that is driven upwards by the soil as the core barrel is pushed downwards. Cores of cohesionless soils can be recovered by closing a valve at the top of the piston corer and securing the wireline to the piston to create a partial vacuum as the core barrel is retrieved. Zapico et al. (1987) describe one (Figure 10.5b) that has its origin in one invented by Bishop (1948). The US Geological Survey (Murphy and Herkelrath, 1996) has developed a sample-freezing drive shoe that can be used with the Zapico et al. sampler. McElwee et al. (1991) describe a more complex piston sampler for water-saturated cohesionless sand and gravel that avoids the need for drilling mud that is a feature of the Zapico et al. sampler.

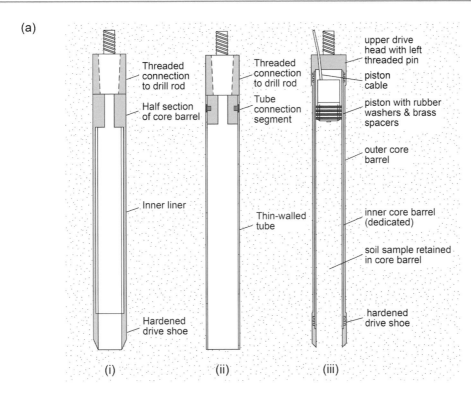

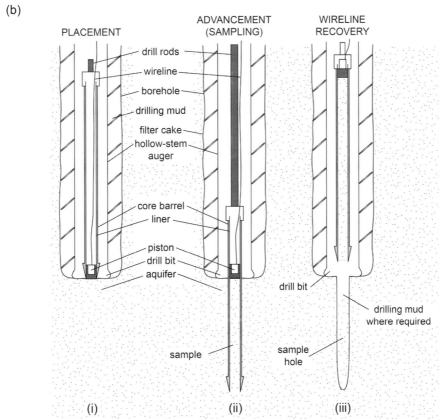

Figure 10.5 Core acquisition at bottom of augers. (a) Samplers: (i) a split-spoon core sampler, (ii) a Shelby tube sampler, and (iii) a piston core-barrel sampler. (From CCME (1984); with permission.) (b) The collection of a piston core sample showing advancement through the bottom of the auger flights into the aquifer below and recovery with the aid of a wireline. (After Zapico et al. (1987).)

10.1.3 Core Logs

As mentioned above, professional perspectives and academic education result in engineers and geoscientists observing and recording different information in core logs of soils and sediments. In section 5.1, the general principles of core logging were introduced and illustrated. Figure 10.6 shows core logs from archived water-well, geotechnical and hydrogeological records compared with that of a sedimentary geologist logging the same formation. This figure was developed by detailed sedimentological mapping of the Oak Ridges Moraine north of Toronto, Ontario, to investigate its hydrogeological potential (Sharpe et al., 2004).

Typically, the logging of boreholes by water-well drillers leaves much to be desired. The water-well driller quite naturally provides a simple untrained description of the soil recovered during drilling operations. A geotechnical survey is typically constrained to shallow depths and quite often provides only the barest of details about soil texture. Hydrogeologists, however, have geological training and will tend to see more detail in the same sediments recovered by coring. Nevertheless, sedimentary geologists are trained to identify much greater detail and will likely isolate features that illuminate fluvial depositional processes, e.g., cross-laminations

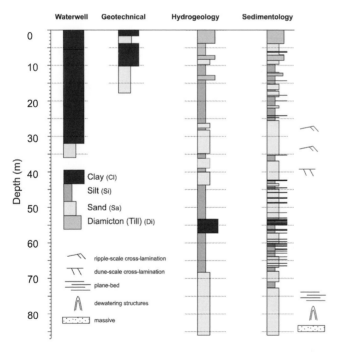

Figure 10.6 A comparison of archived core logs from water-well, geotechnical and hydrogeological studies of the same sedimentary formation with that provided by sedimentological logging. Note the contrast in reported unit thicknesses, sediment classification and sedimentary structures. (From Logan et al. (2006).)

and fine-scale properties of hydrogeological importance but not noted by the hydrogeologist. A good photographic record and preservation of soil samples will allow the later recovery of some of this information.

What Figure 10.6 makes clear is the poor quality of geoscientific information that is provided by water-well logs, which often afford the engineer an insight into regional or local geology when no other information is available. In Figure 10.6, the actual thickness of shallow clay is not 30 m but rather a few centimetres and the fine-grained soil is silt with sand interbeds not clay. Figure 10.6 suggests that an engineer working in a new area may find that earlier hydrogeological reports provide better resolution of individual units within the shallow subsurface than earlier geotechnical or water-well records and are thus more informative.

It is often the case that the project engineer is not present when samples of soil and rock are collected in the field and logged by a geologist. In such cases the engineer must rely on the core log to provide the necessary details of the soil or rock sampled, and the engineer must be capable of understanding the log and its geotechnical or hydrogeological implications. This places a significant responsibility on the geologist to identify critical features in the core and transmit them to the log.

Figure 10.7 shows a simple log of a core of coarse alluvium and clasts of poorly indurated sandstone with grain size identified using the Unified Soil Classification System (USCS). The USCS (ASTM, 2017) is discussed in all introductory geotechnical or soil mechanics textbooks in North America (e.g., Briaud, 2013) and by Norbury (2010). It divides soils and sediments into two general classes:

- *Coarse-grained soils and sediments* in which >50% of the particles are visible to the naked eye. In practice, this means that more than half the sample grains would be retained on a 0.075 mm sieve, i.e., the #200 US Standard Size Mesh. This fraction is further subdivided into gravel and sand on the basis of whether >50% of the particles are retained (i.e., gravel) or pass through (i.e., sand) a 4.75 mm sieve (#4 US Standard Size Mesh). Further subdivisions are made on the basis of the amount of fine-grained particles, such that gravel or sand samples with <12% fines are considered "clean" gravel or sand. Group symbols include: GW = clean, well-graded gravel; GP = clean, poorly graded gravel; GM = silty gravel; GC = clayey gravel; SW = clean, well-graded sand; SP = clean, poorly graded sand; SM = silty sand; and SC = clayey sand.

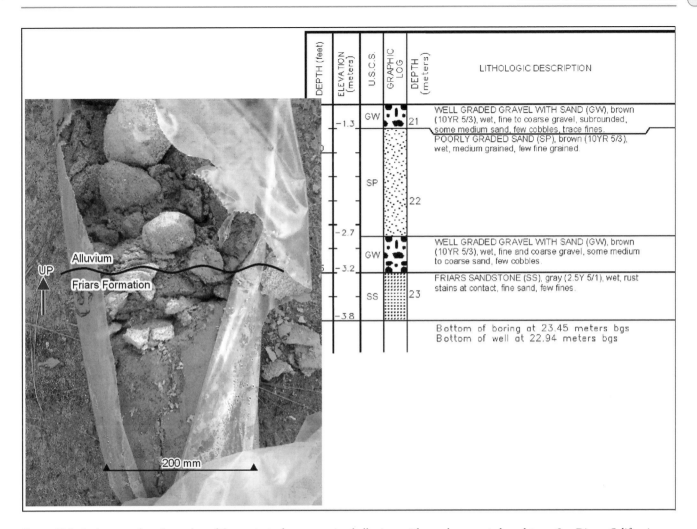

Figure 10.7 A photograph and core log of the contact of coarse-grained alluvium with poorly cemented sandstone, San Diego, California. The plastic core covering indicates recovery by sonic drilling. The textural properties of the various units are characterized using the Unified Soil Classification System (USCS). Thus SP is poorly graded sand and GW is well-graded gravel. The lithologic description characterizes the sample by texture and colour. Above the contact are the cobbles of the basal gravel unit, while below the contact are sandstone clasts. (Photograph courtesy of R.M. Sengebush, INTERA Inc.)

- *Fine-grained soils and sediments* in which <50% of the particles are visible to the naked eye. In this case, more than half of the particles pass through the 0.075 mm or #200 US Standard Size Mesh sieve. These samples are further subdivided on the basis of the Atterberg liquid limit test. Thus if $w_L < 50$ the sample is considered "lean" as in: ML = silt and rock flour; CL = clay; and OL = organic silt. When $w_L \geq 50$, the sample is considered "fat" as in: MH = silt with high plasticity; CH = clay with high plasticity; and OH = organic clay. Organic soils have the symbol Pt indicating peat.

10.1.4 Cone Penetration Testing

Cone penetration testing, known as CPT, was developed in the Netherlands in the 1930s and has become an indispensable tool for site characterization (Lunne et al., 1997; Rogers, 2006; Robertson, 2009). A cone-shaped tool, mounted on the end of a series of drilling rods, is driven into the ground and the response to its penetration is measured as both point and side frictional resistance. Truck-mounted CPT units are frequently used to characterize sand, silt, clay and other soft soils and sediments and mine tailings, although coarser sediments can be sampled with special rigs. In addition to its role in defining the soil stratigraphy, it is most useful in providing pore-pressure data through the piezocone test and information leading to estimation of geotechnical parameters, thus providing results that may be used in geotechnical design.

The tool is shown in Figure 10.8, where the principal components are identified. The total force acting on the cone is Q_c and the **cone resistance** is denoted by q_c. Similarly the total force acting on the sleeve is Q_s and the sleeve friction is

Table 10.2 Correlation of standard penetration test *N* and cone penetration test q_c values for granular soils of frictional resistance ϕ. (From Price (2009).)

Relative density (%)	*N*	q_c (MPa)	ϕ (°)
Very loose (0–15%)	<4	<2	<30
Loose (15–35%)	4–10	2–4	30–35
Medium dense (35–65%)	10–30	4–12	35–40
Dense (65–80%)	30–50	12–20	40–45
Very dense (80–100%)	>50	>20	>45

Note: Relative density is defined by Price (2009) as "a measure of the state of compaction which the soil has reached" and is defined in terms of void ratio as RD = $(e_{max} - e)/(e_{max} - e_{min})$ for a particular soil and is often related to an *N* value. Briaud (2013) refers to it as the density index.

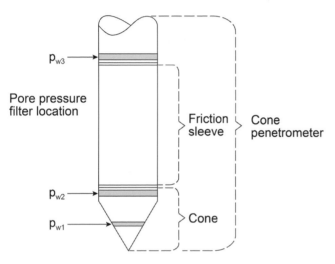

Figure 10.8 Terms used in cone penetrometer testing describing locations of pore-pressure measurement, i.e., p_{w1} is the pore pressure measured on the cone, p_{w2} is that behind the cone and p_{w3} is that measured behind the friction sleeve. (From Lunne et al. (1997).)

denoted by f_s. The CPTu test measures the pore pressure in as many as three locations: p_{w1} and p_{w2} refer to the pore pressures within and behind the cone tip, respectively, whereas p_{w3} is measured behind the friction sleeve. (Note that in the CPT literature these are referred to as $u_{1,2,3}$; however, we continue to use p_w as the term for pore pressure in this book.) The correlation between SPT *N* and CPT q_c measurements is shown in Table 10.2, where ϕ is the angle of frictional resistance.

The tool is driven into the ground either from a small portable unit that may be used in an area of limited access (height < 5 m) or more commonly from a truck or tracked

vehicle. The rate of cone penetration is typically of the order of 20 mm/s such that detailed profiling can proceed at a rate of 100–150 m/day. Figure 10.9 presents the typical results of a CPTu test showing profiles of corrected cone resistance (q_t), friction ratio (R_f), pore pressure ($u = p_w$), a soil behaviour type (SBT) index and the soil stratigraphy.

The **soil behaviour type index** is defined by the CPT measurements (Robertson, 2010):

$$I_{SBT} = \left[\left(3.47 - \log \frac{q_c}{p_a} \right)^2 + (\log R_f + 1.22)^2 \right]^{0.5}, \quad (10.1)$$

where q_c is the CPT cone resistance (or corrected cone resistance, q_t), p_a is the atmospheric pressure ($\sim$100 kPa), R_f is the friction ratio ($= (f_s/q_c) \times 100\%$) and f_s is the sleeve friction.

While a CPT rig does not provide a soil sample, it is common practice for operators of such rigs to employ a variety of complementary sampling tools to allow soil and water samples to be collected using the CPT installation equipment. In addition to pore-pressure measurements using CPTu tools, special CPT tools are available for areas subject to earthquakes by adding accelerometers to the tool (SCPTu) to measure compressional (P) and shear (S) waves (see Chapter 14). Also it is possible to map hydrocarbon contamination using an ultraviolet optical screening tool (UVOST/CPT) to define zones of non-aqueous phase liquid contamination (see Chapter 12).

10.1.5 Geophysical Surveys

Earlier in section 5.1.1 we noted that geophysical surveys are used by engineering geologists for two principal purposes: (i) to identify anomalous features prior to reconnaissance drilling, and (ii) to correlate strata between the boreholes subsequently drilled (Fookes et al., 2015). Geophysical surveys offer useful guidance in site characterization if the survey has been carefully planned but the results must be treated with care. Rapid developments in practices, software and sensors in geophysical surveying over the past 10–20 years indicate that only up-to-date references should be consulted for further information, e.g., Reynolds (2011) and Milsom and Eriksen (2011).

We may divide geophysical surveys into two kinds: surface and borehole geophysics. Surface geophysics deploys sensors or systems on the ground surface to map the subsurface with respect to a particular geophysical parameter, e.g., electrical conductivity (resistivity) or seismic waves. The surface geophysical techniques that have emerged as the most

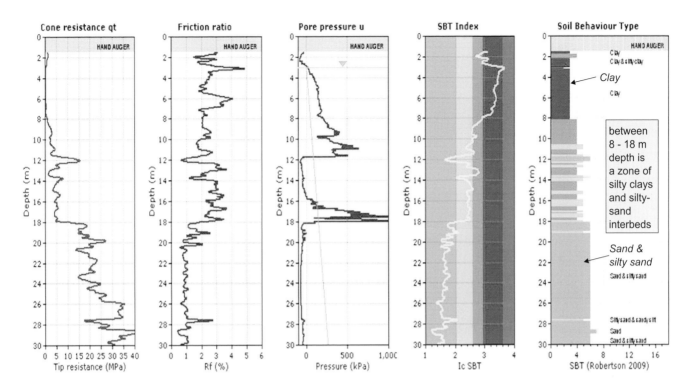

Figure 10.9 A CPTu profile of 30 metres of soil showing the variation in cone resistance, friction ratio and pore pressure with depth. The soil behaviour type index is used to define the textural class of soil. (From Robertson (2010).)

useful for modelling surficial geology are electrical resistivity tomography, ground-penetrating radar and seismic surveying. In borehole geophysics, downhole sensors are lowered by wireline to measure physical properties of the soils and sediments penetrated by the borehole. A selection matrix for choosing various geophysical methods to investigate soils and sedimentary basins is presented in Table 10.3.

Electrical resistance imaging (ERI) is used to profile the upper 20–40 m of the subsurface using parallel arrays of small metal electrodes (typically 48 to 96 at a time), which are inserted into the ground over typical distances of 100–500 m to make electrical contact and transmit current. The positioning of the spatial array of electrodes governs the area of investigation and the electrode spacing (e.g., 2–3 m) governs the resolution and depth of the survey. Alternatively, insulated wire cable "dipoles") are towed over the terrain and current transmitted, and the received voltage signal is measured by other received dipoles. Zones of low resistivity (high conductivity) might indicate clay-rich soils, the water table or saturated sediments, while relatively resistant zones might indicate coarse sediments. The bedrock surface may appear as the lower boundary in the cross-sectional images that are developed. The features identified by ERI can subsequently be compared with geological controls, such as are provided by a borehole log or an adjacent outcrop, to constrain the reported results.

Ground-penetrating radar (GPR) tools propagate electromagnetic (EM) waves into the shallow subsurface and record the reflected EM waves from partitioning by the shallow materials encountered as well as their change in characteristics. The portable or wheeled GPR instrument transmits radio waves into the subsurface and receives a return signal that may be displayed in real time to the operator in the field. The method is limited to a penetration depth of about 50 m when low radar frequencies (25–100 MHz) are used; however, the penetration depth is usually considerably less in cases of a shallow water table, i.e., 5–15 m. In electrically conductive materials, such as clays or where the pore water is conductive, the penetration of GPR is typically less than a metre and thus is of limited usefulness. If high-resolution shallow profiling is required, high-frequency (200–500 MHz) reflection surveys are conducted, yielding impressive resolution of the first 3–10 m of the subsurface. Reflections are most commonly caused by changes in water content associated with various petrophysical properties of soils and sediments; consequently the surfaces of clay beds, the bedrock contact and the water table are often identifiable. GPR has thus become the tool of choice for characterizing shallow sands and gravels

Table 10.3 A selection matrix for choice of geophysical techniques when considering targets in soils and sedimentary basins. (From Lucius et al. (2007).)

Geophysical method	Contact boundary	Top of bedrock	Water table	Fault detection	Rippability
Seismic refraction	S	W	S	S	W
Seismic reflection	W	S	NR	S	S
Direct-current resistivity	S	S	S	S	NR
Ground-penetrating radar	S	S	S	S	NR
Electromagnetic	S	S	NR	NR	NR

Notes: Matrix adapted from that originally developed by the US Army Corps of Engineers. W, the method works *well* in most materials and natural configurations; S, the method works only under *special circumstances* of favourable materials or configurations. NR, the method is *not recommended* for this task.

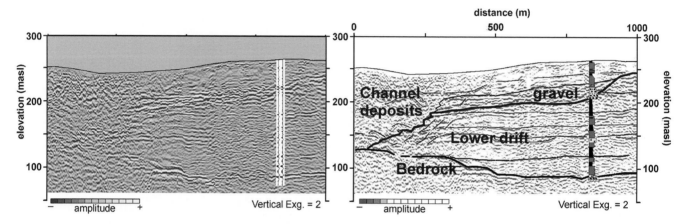

Figure 10.10 High-resolution seismic reflection profile of a section of the Oak Ridges Moraine, southern Ontario, Canada. Inflections in the seismic velocity log (left panel) identify important stratigraphic contacts that are interpreted in the right panel, where the bedrock profile is identified on the basis of the velocity log and the borehole core log shown in the right panel (see Figure 10.11 for details of this log). Above the bedrock, other sedimentary contacts are identified by the parallel reflectors and the information from the core log. (From Sharpe et al. (2004); © 2008 Canadian Science Publishing or its licensors. Reproduced with permission.)

(e.g., Aspirion and Aigner, 1999; Bristow and Jol, 2003; Baker and Jol, 2007). GPR surveys, however, are highly susceptible to cultural interference, but with shielded antennas they can be operated almost anywhere, including within buildings.

When deeper profiles are required, **seismic** surveys are conducted based on the elastic responses of sediments and rocks at depth to a stimulating force, such as a sledgehammer blow to a plate on the ground surface or a discrete explosive charge set in a shallow hole. The seismic waves, consisting of both compressional (P) and shear (S) waves (see Chapter 14), are detected in linear arrays of geophone detectors emplaced at ground surface. **Seismic refraction** surveys of soils and sediments yield cross-sectional images of seismic velocities by which the bedrock contact may be easily identified. Thus, in the 1960s, a seismic refraction survey was used to define the paleochannel and bedrock topography of the ancestral

Great Miami River valley between Dayton, Ohio, southwards to the Ohio River. The thickness and extent of the sands and gravels in the Quaternary buried-valley aquifer was used to develop a new groundwater supply wellfield for Cincinnati, Ohio (Watkins and Spieker, 1971); however, within the alluvium the definition of the coarser permeable units was poor. Although with several limitations of its own, the refraction technique is insensitive to electrical conductivity and therefore a complement to GPR.

Modern **seismic reflection** surveys can provide much greater textural detail than seismic refraction. Figure 10.10 shows a high-resolution seismic log and its interpretation of part of the Oak Ridges Moraine north of Toronto, Ontario. The bedrock is shown as a deep reflector while parallel reflectors indicate bedding planes within the lower sediment or drift unit, an interlayered sand and silt unit that functions as a confined aquifer with a hydraulic conductivity range of

2×10^{-3} to 1×10^{-7} m/s (Sharpe et al., 2003, 2004). The channel sediments are coarse glaciofluvial deposits known as the Oak Ridges Moraine that can be traced tens of kilometres and form a water-table aquifer ($K = 7 \times 10^{-3}$ to 3×10^{-6} m/s).

Borehole geophysics has been applied to sedimentary basins since 1911 when Conrad Schlumberger, Professor of Physics at Ecole des Mines in Paris, France, developed an electrical resistivity well logging tool (Anderson, 1984). Over time, these methods were vastly expanded by the petroleum industry and adopted by hydrogeologists to evaluate groundwater supplies. By the 1960s, the US Geological Survey had its own logging truck that was used to examine the response of a suite of borehole logging tools (caliper, gamma, spontaneous potential, resistivity, neutron and gamma–gamma logs) in mud-filled uncased boreholes in Saskatchewan, Canada, over depths of 200 m. The resulting profiles (Dyck et al., 1972) provide a useful reference guide to the nature of the geophysical responses of sands, silts, clays and tills to these logging tools.

Figure 10.11 shows the downhole geophysical logs of the borehole shown in Figure 10.10. The Oak Ridges Moraine and the underlying Scarborough (S) and Thorncliffe (T) Formations – the lower drift of Figure 10.10 – are well defined by the logs of conductivity (i.e., the inverse of electrical resistivity) and natural gamma, which identifies clay-rich zones, while the magnetic susceptibility log picks out the coarser units (Russell et al., 2007).

The use and analysis of low-frequency surface wave techniques, such as spectral analysis of surface waves (SASW) and **multi-channel analysis of surface waves** (MASW), are emerging as methods for assessing both P- and S-wave velocities of the near surface, i.e., a few metres to tens of metres. Although these techniques rely heavily on post-processing in the frequency spectrum and are evolving, they are increasingly being used by geotechnical engineers for characterization of the shallow subsurface, the water-table zone and in earthquake risk assessment. Park et al. (2007) provide a basic overview of MASW.

10.2 Textural Analysis of Soils and Sediments

Soil and sediment grains are broadly classified as detrital, clastic or chemical. **Detrital** or **clastic** mineral grains are derived from the weathering of continental rocks, i.e., they are **terrigenous sediments**, and are often rounded by sediment gravity flows or fluvial transport. The term "clastic" is derived

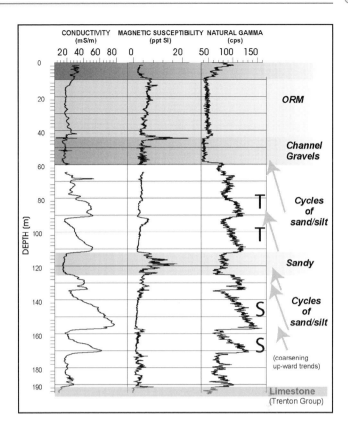

Figure 10.11 The borehole geophysical log for the borehole shown in Figure 10.10, which provided the basis for the interpretation of the seismic velocity profile. The units identified are the Scarborough (S) and Thorncliffe (T) Formations and the Oak Ridges Moraine (ORM). (From Russell et al. (2007); reproduced with the permission of Natural Resources Canada.)

from **clasts**, which are fragments of sediment or sedimentary rock such as are shown in Figures 1.6 and 10.7. Grains that exhibit interlocking crystal growth fabric acquired during diagenesis are referred to as **chemical** grains, e.g., carbonate minerals.

Texture refers to the grain size, roundness and shape of grains that comprise a soil or sediment sample, in particular a detrital sample. Texture and grain composition are related to the composition of the source rock or **provenance** and the physical and chemical weathering in the source area and subsequently during transport, deposition and diagenesis. Chemical weathering begins at grain edges and works inwards by diffusional processes, ultimately producing clay minerals in layer silicates; thus the outer weathered surfaces are susceptible to abrasion during transport and consequently grains become more rounded and less angular. Abrasion and breakage are purely physical processes occurring during gravity flows and sediment transport that create new grain surfaces, rounded or angular, respectively. The coarse-grained gravel shown above the alluvium–sandstone contact in Figure 10.7 varies from rounded to angular in shape. Helpful discussions of these topics may be found in Pettijohn et al. (1987) and Bridge and Demicco (2008).

In this section we discuss how geoscientists characterize the textural properties of a soil or sediment sample in terms of the statistics of particle size. These parameters are then used to obtain more insight into the textural properties than is captured by the Unified Soil Classification System (USCS) shown in Figure 10.7 and may allow identification of the depositional environment of the sample and certain physical properties of engineering importance.

10.2.1 Udden–Wentworth Scale

Norbury (2010) states that the most widely used particle classification system internationally is the USCS. However, the USCS provides only limited information on the engineering characteristics of the soil, which are perhaps better defined by the EN ISO 14688-2 classification (BSI, 2018). This critique is also obviously true with respect to the textural methods used by geoscientists, which, however, can be usefully interpreted by the engineer if he or she appreciates the information disclosed by sedimentological grade scales.

Grain-size classification in sedimentary geology pre-dates the introduction of the USCS in 1948. Johan Udden investigated the size distribution of eolian-sand particles in 1898 and established grade scales that varied by a factor of 2. Subsequently C.K. Wentworth defined a set of grade scales still in common use as the **Udden–Wentworth scale**. This system of grade scales is illustrated in Figure 10.12(a) with the size fraction >2 mm defined as gravel according to Blair and McPherson (1999).

Particle sizes are given in millimetres except for boulders, which are measured directly in the field in metres. Sieving can be used to separate size fractions in the range of pebbles (4 mm) to coarse silt (0.044 mm). Sedimentation or settling tubes may be used to separate sand fractions, and the pipette method allows discrimination of silts and clays. Geoscientists may also prepare thin sections of sands and silts, as in Figure 2.25, and study their size distribution by binocular microscope. Scanning electron microscopes allow examination of grain textures from sands to clays, as in Figure 2.19.

Particle sizes are frequently characterized by geoscientists using the phi scale:

$$\phi = -\log_2 d, \qquad (10.2)$$

where d is the particle length in millimetres. The phi scale yields a dimensionless number that is positive for particles of 1 mm and less and negative for particles larger than 1 mm, i.e., very coarse sands and gravels.

The phi scale is used to determine the mean, median, sorting and other moments. The mean is defined as (Folk, 1980; Pettijohn et al., 1987)

$$Me_\phi = \frac{\phi_{16} + \phi_{50} + \phi_{84}}{3}, \qquad (10.3)$$

where the median grain size is ϕ_{50}. Using cumulative percentages, ϕ_{16} and ϕ_{84} indicate the respective grain sizes for which 16% and 84% of the sample is coarser than these sizes. Thus 16% of this sample would be caught on a sieve with grain size ϕ_{16} and 84% would pass through it. As Folk (1980) noted, cumulative particle-size frequency curves represent grain-size distributions better than histograms because the frequency curve is independent of the sieve interval.

10.2.2 Inferences from Textural Analysis

Statistical measures such as equation (10.2) are used to characterize particle-size distributions and occasionally to infer environments of sedimentary deposition and hydrogeological properties. Figure 10.13 illustrates how field observation of cores (as shown in Figure 10.7) and textural analysis allow the geoscientist to identify two very different depositional environments in the same Californian basin. In this case, groundwater contamination migrated rapidly through the deep paleochannel sands bypassing the floodplain silts that act as aquitards to groundwater flow. The range of mean grain sizes of the paleochannel sands is from 0.2 to −1.3 ϕ, while that of the floodplain silts is 6.9 to 7.8 ϕ.

The grain-size heterogeneity in a soil or sediment sample is indicated in the USCS by terms such as "well graded" or "poorly graded". To geoscientists, a well-sorted sediment has grains of a uniform size, i.e., it would be termed "poorly or uniformly graded" by a geotechnical engineer. Thus the use of grading and sorting reflect the opposite connotations – grading is useful geotechnically in implying frictional strength, while good sorting is important in fluid flow applications in that the pore matrices of soils and sediments are relatively free of fine-grained particles (see Figure 1.6). The **sorting parameter** (s_I) used by geoscientists is Folk's (1980) inclusive graphic standard deviation,

$$s_I = (\phi_{84} - \phi_{16})/4 + (\phi_{95} - \phi_5)/6.6, \qquad (10.4)$$

(a)

PARTICLE LENGTH (d₁)			GRADE	CLASS	FRACTION	
m	mm	ɸ			Unlithified	Lithified
4.1	4096	-12	very coarse	Boulder		
2.0	2048	-11	coarse	Boulder		
1.0	1024	-10	medium	Boulder		
0.5	512	-9	fine	Boulder		
0.25	256	-8	coarse	Cobble	Gravel	Conglomerate
	128	-7	coarse	Cobble	Gravel	Conglomerate
	64	-6	fine	Cobble	Gravel	Conglomerate
	32	-5	very coarse	Pebble		
	16	-4	coarse	Pebble		
	8	-3	medium	Pebble		
	4	-2	fine	Pebble		
	2	-1		Granule		
	1	0	very coarse	Sand	Sand	Sandstone
	0.50	1	coarse	Sand	Sand	Sandstone
	0.25	2	medium	Sand	Sand	Sandstone
	0.125	3	fine	Sand	Sand	Sandstone
	0.063	4	very fine	Sand	Sand	Sandstone
	0.031	5	coarse	Silt		
	0.015	6	medium	Silt		
	0.008	7	fine	Silt		
	0.004	8	very fine	Silt		
	0.002	9		Clay	Mud	Mudstone or shale
	0.001	10		Clay	Mud	Mudstone or shale
	0.0005	11		Clay		
	0.0002	12		Clay		
	0.0001	13		?		

(b)

S₁ (ɸ)	SORTING TERMS
0.0	
	well sorted
0.5	
	moderately sorted
1.0	
1.5	poorly sorted
2.0	
2.5	very poorly sorted
3.0	
3.5	extremely poorly sorted
4.0	
4.5	unsorted or weakly sorted
∞	

Figure 10.12 (a) The Udden–Wentworth scale for grain-size and textural classifications as used by sedimentary geologists and extended by Blair and McPherson (1999). (b) Folk's (1980) index to sediment sorting terms. (From Blair and McPherson (1999); reproduced with the permission of SEPM, Society for Sedimentary Geology.)

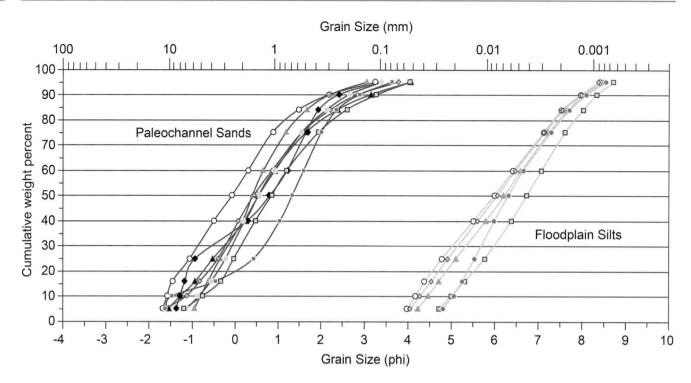

Figure 10.13 Field logs and textural analysis of cored soil and sediment samples allow geologists to identify the depositional environment of the sample, in this case fluvial (paleo)channel sands and floodplain silts at depths of 100–200 m beneath Llagas basin of central California.

which includes 90% of the distribution of grain sizes. Figure 10.12(b) summarizes Folk's (1980) classification scale for ranges of the sorting parameter. Sorting parameter values for the samples shown in Figure 10.13 average 2.0 ϕ for the sands and 1.8 ϕ for the silts, i.e., both sands and silts are poorly sorted. The sorting parameter in SI units is written as $s_I = (d_{84} - d_{16})/4 + (d_{95} - d_5)/6.6$.

Instead of the sorting parameter, a uniformity coefficient is used in soil mechanics, $C_u = d_{60}/d_{10}$, where the independent variables refer to 60% or 10% of all particles being finer than these diameters. Muir Wood (2009) indicates that residual soil developed from the weathering of granitic rocks might exhibit $C_u = 380$, whereas a well-sorted gravel from a river might approach a uniformity coefficient of 1.5. Figure 10.14 illustrates the preferred geotechnical presentation of soil texture using particle diameters and a cumulative percentage finer scale, whereas Figure 10.13 presents the preferred approach in sedimentary geology with the use of the ϕ scale and the cumulative percentage coarser than the ϕ value for that particular sieve size.

The effect of good sorting on hydraulic conductivity is illustrated by samples collected during the remediation of soils beneath a former oil refinery and tested to determine their ability to transmit injected remedial fluids or to recover

gasoline that had escaped from storage tanks over the years. Well-sorted sands, $s_I = 0.19$, exhibited better hydraulic conductivity, $K = 1.70d^{1.4}$, than did moderately sorted sands, $s_I = 0.76$, $K = 0.70d^{1.3}$. The sorting of these fluvial sands determines the hydraulic conductivity for any particular grain-size distribution.

In some cases the engineer might have only a grain-size analysis from which to determine the hydraulic conductivity of a soil sample. It is possible to estimate hydraulic conductivity for sand and gravels from such textural data, but estimates for silts and clays, which may be extensively penetrated by fractures or macropores, should be considered cautiously. Box 10.2 summarizes the results of studies on the (saturated) hydraulic conductivity by Chapuis (2008).

In summary, statistical measures are used by geoscientists for two principal purposes. The first of these is to identify the textural characteristics of a depositional environment – paleochannel sands and floodplain silts (Figure 10.13) – that may be applied generally to a particular study area. The second is to determine how non-uniformity in grain size affects a certain physical property, e.g., hydraulic conductivity in the above example, for a particular soil or sediment sample. These may have implications in calculations concerning soil or sediment compressibility, strength or permeability.

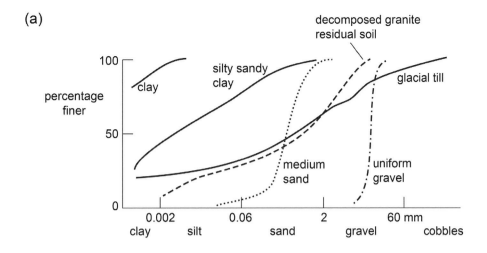

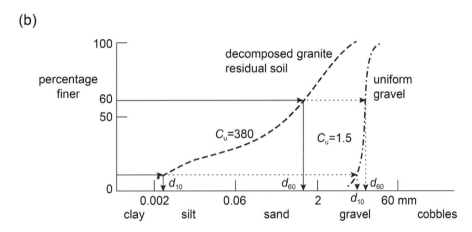

Figure 10.14 Geotechnical representation of textural properties of soil: (a) typical distributions of various soils; (b) grain-size distributions for a non-uniform residual soil and a uniform river gravel. (From Muir Wood (2009).)

BOX 10.2 | **ESTIMATING THE HYDRAULIC CONDUCTIVITY OF SOILS BY GRAIN-SIZE CURVES**

Chapuis (2008) has investigated the robustness of various prediction equations for the saturated hydraulic conductivity of soils, both cohesionless and cohesive. The reader is urged to consult the journal articles listed below before developing a spreadsheet program for personal use. The steps below summarize those of Chapuis (2008).

Step 1: Inspection of the soil sample. There is no good reason to submit a sample for grain-size or textural analysis if its quality has been compromised during collection. Make sure that the sample is from a single stratigraphic unit and that drilling fluids or muds have not contaminated it. Assuming that the sample meets these criteria, proceed to analyze for particle and dry bulk density, grain-size distribution and, for the Kozeny–Carman estimates, the Atterberg limit tests for plasticity. For cohesionless soils, there are four choices for estimating hydraulic conductivity: the Hazen method (see step 3), the NAVFAC method (step 4), the Chapuis method (step 5) and the Kozeny–Carman method (step 6). For plastic, unfractured clayey soils, only the Kozeny–Carman method can be used provided a complete grain-size analysis, including silt and clay fractions, has been conducted.

BOX 10.2 | (CONT.)

Step 2: Specific surface area (SSA). The specific surface of a soil is its surface area per unit weight. For non-plastic soils, SSA can be estimated by

$$SSA = \frac{6}{\rho_s} \sum \frac{P_D - P_d}{d}, \tag{10.5}$$

where ρ_s is the density of the solids (kg/m^3), and $P_D - P_d$ is the percentage in dry mass between consecutive sieves of nominal sizes D, the larger sieve size, and d, the smaller size. The smallest sieve size, d_{min}, is assumed to be 0.005 mm. If there is no estimate of the smaller grain-size percentages, i.e., 1–80 μm, then SSA can be significantly underestimated. Chapuis and Aubertin (2003) have also developed an estimator of SSA for plastic soils.

Step 3: Hazen's equation. Hazen's equation was developed in the 1890s for the purposes of selecting sand grain fractions for water filtration. It assumes that the sand is: (1) "loose", i.e., it should be close to a maximum void ratio or $e_{max} \sim 1$; (2) uniform, i.e., $U < 5$; and (3) has $0.1 \leq d_{10} \leq 3$ mm. The Hazen equation, for hydraulic conductivity in cm/s, is

$$K = C_H \cdot (d_{10})^2, \tag{10.6}$$

where the effective grain size, d_{10}, is expressed in mm and C_H is the Hazen coefficient. At a temperature of 20°C, $C_H = 1.50$; and at 10°C, $C_H = 1.16$.

Step 4: NAVFAC method. This approach appeared in 1974 in a US Navy design manual and is discussed by Chapuis (2004) in terms of its computation and its requirements. It provides a relationship between $\log K$ and $\log d_{10}$ that is a function of the void ratio for values of $0.3 < e < 0.7$.

Step 5: Chapuis' equation. For soils without plasticity, Chapuis (2004) developed the following expression for soils with effective diameter $0.13 < d_{10} < 1.98$ mm and void ratio $0.4 < e < 1.5$:

$$K = 2.4622 \left[\frac{(d_{10})^2 \cdot e^3}{(1+e)} \right]^{0.7825}. \tag{10.7}$$

Step 6: Kozeny–Carman equation. Chapuis and Aubertin (2003) found that the Kozeny–Carman equation could be used for soils in the range $1 \times 10^{-11} < K < 1 \times 10^{-1}$ m/s if and only if the full grain-size distribution is known, including a hydrometer analysis for clay particles, so that the SSA (10.5) can be specified, in which case the hydraulic conductivity in m/s is given by

$$\log K = 0.5 + \log \left[\frac{e^3}{G_s^2 \cdot SSA^2 \cdot (1+e)} \right], \tag{10.8}$$

where SSA is in m^2/kg and G_s is the specific gravity of the soil grains. This equation yields a value of K in m/s that is correct to 1/3 to 3 times the permeameter-measured value.

Source: After Chapuis (2008)

10.3 Laboratory Testing of Soils and Sediments

Earlier in this chapter, the difficulties in obtaining undisturbed soil samples were mentioned; consequently any laboratory measurement of critical parameters could suffer from sampling and handling disturbance modifying measurement of the parameter of interest. It is the responsibility of the geotechnical team charged with drilling and coring to bear this in mind and it is of special importance when evaluating geologic hazards or foundations in heterogeneous soils.

Porosity and bulk and grain densities are typically measured early in laboratory testing and allow the engineer to estimate the void ratio of a soil sample. The relationship

Figure 2.12 Hornfels. Medium-grained quartz (light colour) with fine-grained interstitial biotite (red). Width of field, 4 mm; cross-polarized light. (Photo courtesy of Eva Schandl, Ph.D., Geoconsult, Toronto.)

Figure 2.13 Granitic gneiss. Striated grains are plagioclase feldspar with polysynthetic twinning surrounded by quartz grains. Width of field, 4 mm; cross-polarized light. (Photo courtesy of Eva Schandl, Ph.D., Geoconsult, Toronto.)

Figure 2.14 Highly altered mica-rich gneiss. Coarse-grained crystalline mica (biotite and muscovite) is replaced by fine-grained clays, probably illite. Width of field, 4 mm; cross-polarized light. (Photo courtesy of Eva Schandl, Ph.D., Geoconsult, Toronto.)

Figure 2.15 Gneissic texture. Barely visible relict anhydrite aggregates (arrow) in large vug (black) are surrounded by slender prisms of amphibole and coarse-grained quartz grains. Width of field, 4 mm; cross-polarized light. (Photo courtesy of Eva Schandl, Ph.D., Geoconsult, Toronto.)

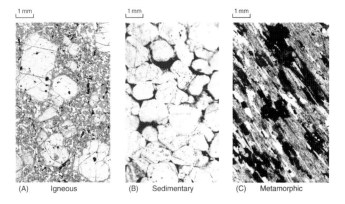

Figure 2.18 Thin-section photomicrographs of typical igneous, sedimentary and metamorphic rocks under low magnification. (A) Basalt from a Hawaiian lava lake showing phenocrysts of clear olivine set within a groundmass of brown pyroxene and opaque ilmenite ($FeTiO_3$); plane-polarized light. (B) Sandstone with sub-rounded quartz grains with dark clay particles and quartz cement filling the pore spaces; plane-polarized light. (C) Schist, a strongly foliated rock, caused by metamorphism of muddy sedimentary rocks that resulted in the bright blue and red mica crystals, which are preferentially oriented perpendicular to the maximum compressive stress; cross-polarized light. (From Klein and Philpotts (2013).)

Figure 3.1 The Muddy Mountains along the North Shore Road, near Lake Mead National Recreation Area, Nevada. The Tertiary-age Bitter Ridge (Thb) limestone member of the Horse Spring Formation steeply dips along the skyline. Thrc represents the conglomerate lithofacies of the Rainbow Gardens Member of the Horse Spring Formation (Tertiary). (Formation identifications courtesy of Norma Biggar, University of Nevada Las Vegas.)

Figure 3.9 A mafic dike (arrows) intrudes into welded ash-flow tuff (T2) on the Arizona side of the Hoover Dam. Note close spacing of fractures and how various fault blocks have slipped from left to right towards the power house. The Upper Dike is also a tuff, whereas T1 is a conglomerate. (Geologic description by Ball (2005); photo by Jeff Keaton, Amec Foster Wheeler, Los Angeles, California, and reproduced with his permission.)

Figure 3.10 Exposed vertical fault surface showing inclined grooves (arrow) from displacement, Hoover Dam, Nevada, USA. (Photo by Jeff Keaton, Amec Foster Wheeler, Los Angeles, California, and reproduced with his permission.)

Figure 3.11 A 2 m high shear zone with clay core, and pen (arrowed) for scale, near Hoover Dam, Nevada, USA. (Photo by Jeff Keaton, Amec Foster Wheeler, Los Angeles, California, and reproduced with his permission.)

Figure 3.15 The Paskapoo Formation in outcrop, Cochrane, Alberta, Canada. (Photo courtesy of Glen Stockmal, Geological Survey of Canada; reproduced with permission.)

Figure 3.20 Air photo of the re-aligned US Highway 93 or Hoover Dam Bypass near Boulder, Nevada. See Figure 3.21 for an engineering geology map of this area. (Photo prepared with Google Earth Pro.)

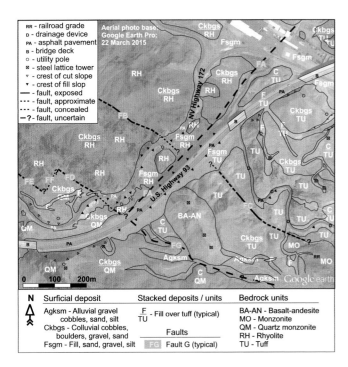

Figure 3.21 An engineering geology map of part of the Hoover Dam Bypass (new US Highway 93) near Boulder, Nevada. See Figure 3.20 for air photo. BA-AN is a basaltic and andesite lava; TU is a welded lithic ash-flow tuff; MO is a monzonite pluton (monzonites have equal amounts of alkali feldspar and plagioclase); QM is a quartz monzonite pluton; and RH is a rhyolitic lava with interbedded tuffaceous sedimentary rocks. See questions 3.6–3.9, which concern this map. (Prepared by Jeff Keaton, Amec Foster Wheeler, Los Angeles, California.)

Figure 4.3 Core disking in shale indicating high *in situ* stresses recorded in an oil and gas well, southwestern Ontario, Canada. The circled T indicates top of core at elevation 896.4 m below ground surface; the circled B indicates bottom of core at 901.3 m below ground surface. The dark grey shale is part of the Blue Mountain Formation that is shown to overlie the light grey fossiliferous, semi-nodular limestone of the Cobourg Formation with the contact at an elevation of 899.46 m. (From Armstrong and Carter (2010), © Queen's Printer for Ontario, 2010. Reproduced with permission.)

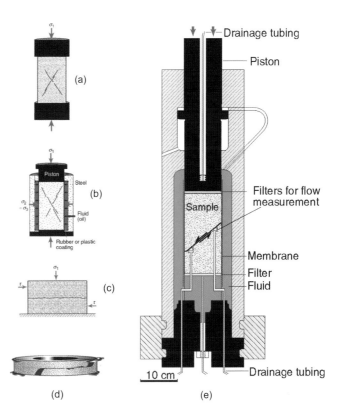

Figure 4.10 Experimental testing of rock samples in a rock mechanics laboratory. (a) A simple uniaxial cell for point-load strength tests. (b) A simple triaxial test rig allowing all three principal stresses to be specified. (c) A simple shear box that permits the resistance to shear (τ) to be measured for a particular vertical confining stress. (d) A ring-shear apparatus for sediment testing. (e) A more sophisticated triaxial test rig allowing fracture behaviour to be monitored while measuring the pore pressure in the fracture. Orientation of different fracture types with respect to the principal stresses are shown in each. (From Fossen (2016).)

Figure 5.4 Extraction issues in rock quarrying caused by structural features in Paleozoic carbonates near Ottawa, Ontario, Canada. (Reproduced with the permission of Rob Blair, R. Blair Geoscience Consulting Inc., Mississauga, Ontario, Canada.)

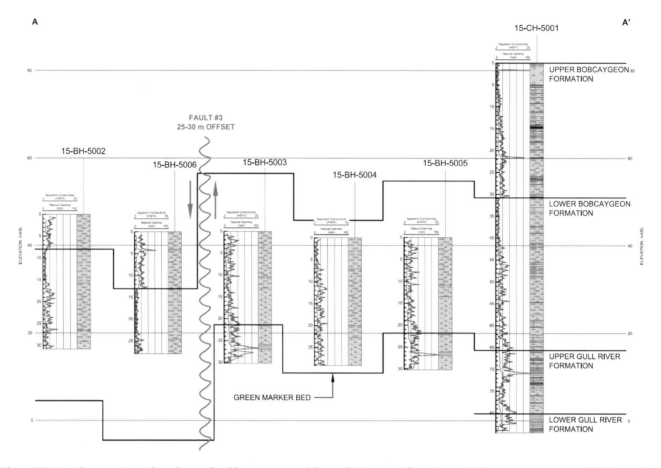

Figure 5.5 In order to estimate the volume of usable aggregate rock beneath the quarry floor shown in Figure 5.4, a program of drilling and downhole geophysical surveying was undertaken. The quarry floor is at the top of the borehole logs. The borehole profiles illustrate the gamma logging response in blue, the apparent conductivity log in red and the interpreted lithologic sequence of rocks on the right-hand side of the scale. (Reproduced with the permission of Rob Blair, R. Blair Geoscience Consulting Inc., Mississauga, Ontario, Canada.)

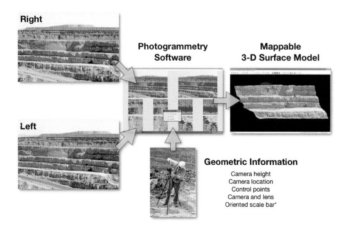

Figure 5.17 Schematic illustration of the digital photogrammetric process by which 2D photographs of the rock benches are used to generate a 3D surface model using appropriate photogrammetric software. (Photos reproduced with the permission of Bill Haneberg.)

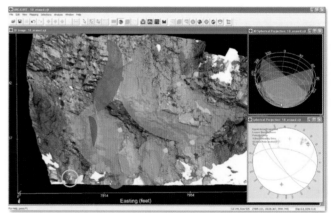

Figure 5.18 Screen capture showing the mapping of individual fractures or discontinuities and the associated stereographic plots. (Photo reproduced with the permission of Bill Haneberg.)

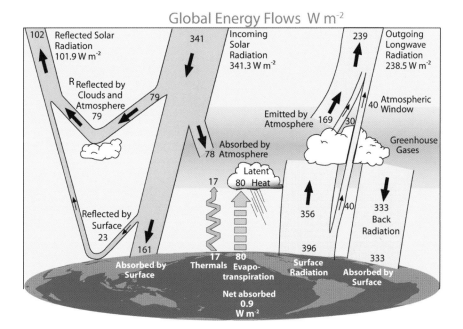

Global Energy Flows W m⁻²

Figure 6.1 The global annual mean energy budget of the Earth in W/m² for the period March 2000 to May 2004. The broad arrows indicate relative importance of energy flows. (From Trenberth et al. (2009); © American Meteorological Society. Used with permission.)

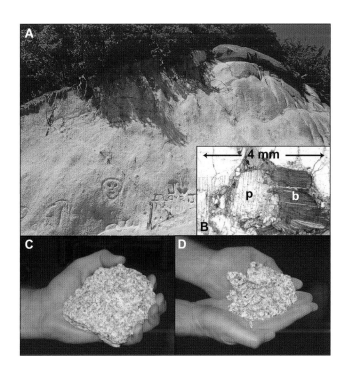

Figure 6.4 Saprock, Sierra Nevada mountains, California. (A) Friable saprock outcrop that is easily vandalized with graffiti; 1.15 m tile spade for scale. (B) Thin-section photomicrograph illustrating pore volume (blue), plagioclase (p) and partially weathered biotite (b). (C) While sample maintains integrity when held, it is easily crushed with bare hands (D). (From Graham et al. (2010); with permission of The Geological Society of America.)

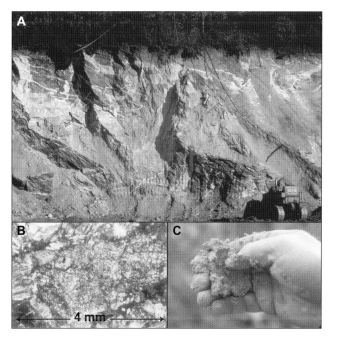

Figure 6.5 Saprolite from the North Carolina Piedmont, USA. (A) This outcrop preserves the structural features of the parent rock, yet the saprolite is soft and easily excavated by the bulldozer shown in the lower right and which provides scale. (B) Thin-section photomicrograph shows thorough alteration of the weatherable primary minerals to clay minerals. (C) The extensive weathering and generation of clay minerals causes the saprolite to yield as a plastic material upon wetting. ((A) Photo credit: G. Simpson. (B) Photo credit: M. Vepraskas. From Graham et al. (2010); with permission of The Geological Society of America.)

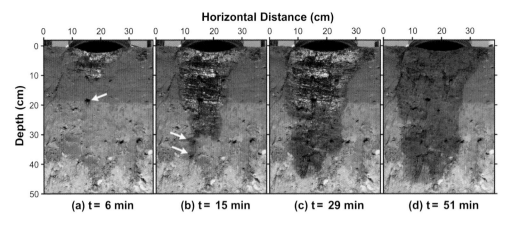

Figure 6.6 Flow through a worm-burrow macropore illustrated by dye migration and time-lapse photography. Infiltration occurs (a, b) from a surface source with a −0.4 cm pressure head over a period of 29 minutes (c). Arrows indicate early breakthrough of the dye solution in worm burrows. Dye redistribution is photographed (d) at 51 minutes after infiltration began. (From Cey and Rudolph (2009); reproduced with the permission of John Wiley and Sons Ltd.)

Figure 6.13 Pattern of kaolin veins beneath the surface of rupture in a trench, Hong Kong. (From Campbell and Parry (2002).)

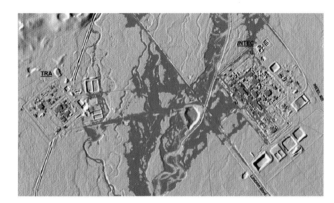

Figure 6.19 Potential inundation of the Idaho National Laboratory by the Big Lost River determined by paleoflood hydrologic analysis. (From Ostenaa and O'Connell (2005).)

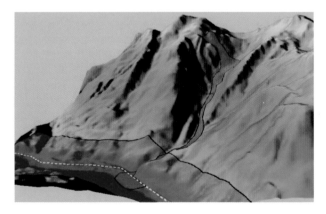

Figure 6.20 Digital elevation model of the Zymoetz landslide, near Terrace, British Columbia, Canada. The landslide was initiated by a rock avalanche of one million cubic metres of bedrock in the headwaters of a tributary of the Zymoetz River. The long-runout (4 km) debris flow ruptured a gas pipeline (dashed line in valley) and created a fan deposit. See Boultbee et al. (2006) for an analysis of the landslide. (LiDAR image from Geertsema and Clague (2011); with permission.)

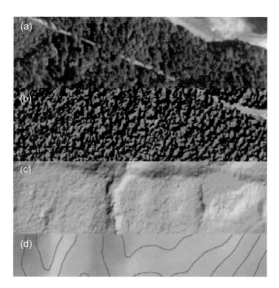

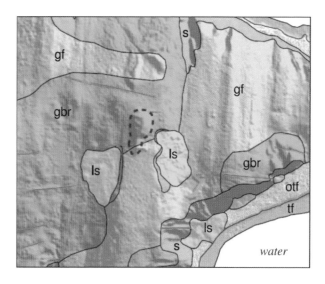

Figure 6.23 A sequence of contiguous image maps from north to south of part of the Puget Lowlands, Washington state, USA, illustrating a 600 m wide area. (a) Digital orthophoto (1 m pixels) showing shoreline of Puget Sound. (b) LiDAR first-return surface (2 m postings) showing continuation of regional road crossing NW to SE through forest. (c) LiDAR bare-earth model after post-processing to segregate ground returns (2 m postings) showing a fault scarp along the east–west direction. (d) US Geological Survey 10 m digital elevation model with 10 ft 1 : 24 000 scale contour intervals, in which the stream location disagrees with the LiDAR image of (c). (From Haugerud et al. (2003), courtesy US Geological Survey; courtesy of The Geological Society of America.)

Figure 6.24 A geomorphic map of part of Bainbridge Island, Washington state, USA, based on high-resolution (1–2 m) LiDAR topographic imaging. Symbols: s, scarp; ls, landslides; tf, tidal flat; otf, old tidal flat; gf, fluted glacial surface; gbr, glaciated bedrock surface. Red lines indicate bedding traces; blue dashed line is a relict shoreline from a late-glacial lake. View is 1.5 km wide. (From Haugerud et al. (2003), courtesy US Geological Survey; courtesy of The Geological Society of America.)

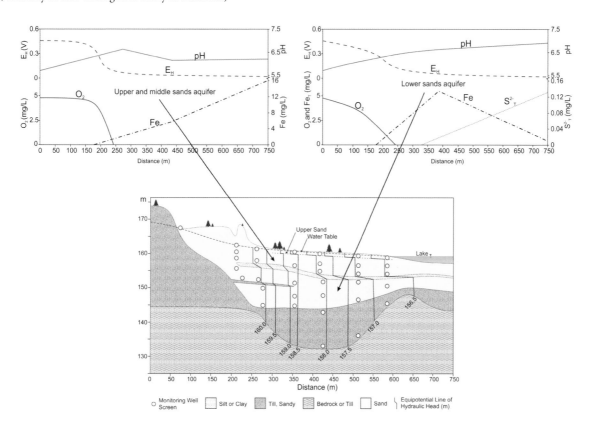

Figure 7.6 Redox sequences in a short (1 km) groundwater flow system, Chalk River, Ontario, Canada. (After Jackson and Patterson (1982); with the permission of John Wiley and Sons.)

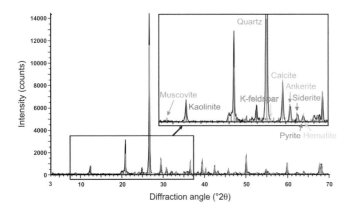

Figure 7.12 X-ray diffractogram of waste-rock minerals. (From Raudsepp and Pani (2003); with the permission of the Mineralogical Association of Canada.)

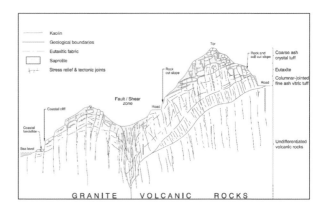

Figure 7.17 Schematic of kaolin distribution resulting from weathering in both granitic and volcanic rocks, Hong Kong, China. (After Campbell and Parry (2002).)

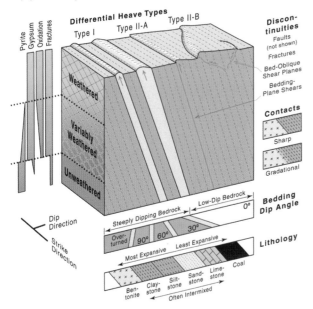

Figure 7.16 Conceptual 3D model of site-scale characteristics of steeply dipping claystone showing interpreted relationships between heave features and the underlying bedrock framework. (From Noe et al. (2007); with the permission of the Association of Environmental and Engineering Geologists.)

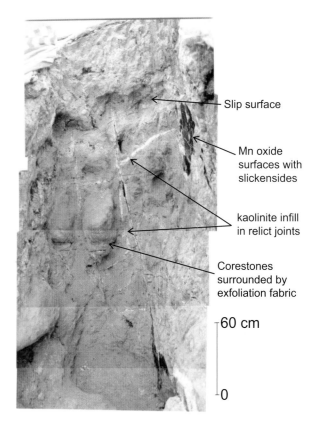

Figure 7.18 Kaolinite (white) and manganese oxide (black) infillings of relict joints in a trench excavated following the 1998 Sai Sha Road Landslide, Hong Kong, China. (From Campbell and Parry (2002).)

Figure 8.8 Lateral moraine, Rocky Mountain National Park, Colorado. The photograph is taken from the end moraine looking up the Fall River Valley, which displays a classic glacially eroded U shape. The foreground is composed of glaciolacustrine sediment deposited from a lake dammed by the end moraine.

Figure 8.11 Earth flow in glaciomarine clays, Notre-Dame-de-la-Salette, Québec, Canada. (Reproduced with the permission of Natural Resources Canada 2011, courtesy of the Geological Survey of Canada, Photo 2011-062 by Greg Brooks.)

Figure 8.12 Glaciofluvial sand and gravel sets within the Oak Ridges Moraine, Stouffville, Ontario. An active aggregate operation reveals a partial view of large-scale structures (~20 m thick, 100 m wide and 500 m long). The sub-horizontal sand and gravel sets (grey), including occasional low-energy fine beds (brown set on pit face above the truck), are truncated at the top right of the sequence and overlain by yellow oxidized fine sand. Such large-scale structures relate to very high-energy sedimentation in deep water, in contrast to metre-scale structures in pro-glacial streams. (Reproduced with the permission of Natural Resources Canada, courtesy of the Geological Survey of Canada, Photo GSC-2017-045.)

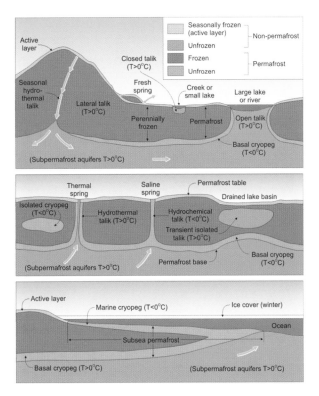

Figure 8.16 Schematic of aquifers in permafrost areas. Supra-permafrost aquifers exist in the active layer and in closed river and lake taliks. Sub-permafrost aquifers occur in open taliks and below the basal cryopeg. In the high Arctic, springs are often associated with salt diapirs and produce saline discharge. (After van Everdingen (1987); reproduced with permission of A. Rivera (ed.), *Canada's Groundwater Resources*, p. 609. Markham, ON: Fitzhenry & Whiteside Limited.)

Figure 8.17 Polygonal ice-wedge pattern, northern Yukon coastal plain, Ivvavik National Park, Yukon, Canada. (Photo courtesy of I.D. Clark, University of Ottawa.)

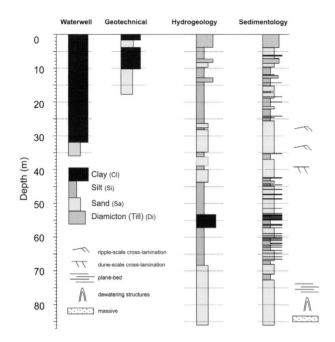

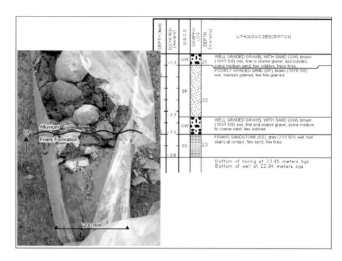

Figure 10.6 A comparison of archived core logs from water-well, geotechnical and hydrogeological studies of the same sedimentary formation with that provided by sedimentological logging. Note the contrast in reported unit thicknesses, sediment classification and sedimentary structures. (From Logan et al. (2006).)

Figure 10.7 A photograph and core log of the contact of coarse-grained alluvium with poorly cemented sandstone, San Diego, California. The plastic core covering indicates recovery by sonic drilling. The textural properties of the various units are characterized using the Unified Soil Classification System (USCS). Thus SP is poorly graded sand and GW is well-graded gravel. The lithologic description characterizes the sample by texture and colour. Above the contact are the cobbles of the basal gravel unit, while below the contact are sandstone clasts. (Photograph courtesy of R.M. Sengebush, INTERA Inc.)

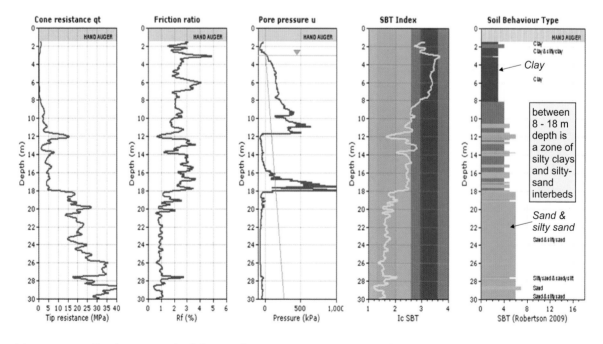

Figure 10.9 A CPTu profile of 30 metres of soil showing the variation in cone resistance, friction ratio and pore pressure with depth. The soil behaviour type index is used to define the textural class of soil. (From Robertson (2010).)

Figure 10.19 Glaciofluvial deposits, such as this kame near Heidelberg, Ontario, Canada, supply local urban areas with aggregate and are located with the help of surficial geology maps. This pit produces 3000 tons/day of sorted materials extracted by front-end loaders from the various excavations. The conical piles of sorted products are shown following washing and distribution by conveyor belt. Imported asphalt cement is used to produce asphalt at the plant in the centre of the photo, adjacent to the dark piles of asphalt, and ready-mix concrete is produced at the plant at the entrance to the pit. Trucking costs limit the distance that aggregate and construction products can be sold profitably; consequently this pit, and others like it, provide aggregate only locally (within 30 km) by truck. Therefore, regional land-use planning requires that such pits be permitted proximal to urban areas, such as Kitchener/Waterloo shown on the skyline, for infrastructure development if long-distance transport (train, barge, ship) is to be avoided and aggregate costs are to be minimized. (Photo courtesy of Jim Karageorgos, Steed and Evans, St. Jacobs, Ontario, Canada.)

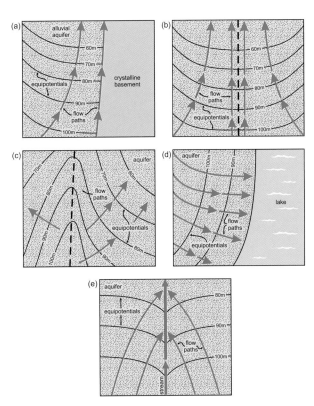

Figure 11.10 The water table is the regional piezometric surface at atmospheric pressure. This figure shows various hydraulic-head boundaries identifying the associated equipotential lines and groundwater flow lines. (a) The effect of a no-flow boundary created by the presence of crystalline basement rock of very low hydraulic conductivity adjacent to an alluvial aquifer. (b) The effect of a symmetry boundary on flow lines. (c) The effect of a linear trend of high hydraulic heads on the flow pattern. Because the water table typically follows topographic contours, it is likely that this ridge reflects such topography. (d) Flow pattern towards a lake or other body of (approximately) constant head. (e) A constant-head boundary represented by a stream into which the groundwater is discharging. (Reproduced from Rivera, 2014, Groundwater basics. In A. Rivera (ed.), *Canada's Groundwater Resources*, p. 45. Markham, ON: Fitzhenry & Whiteside Limited).

Figure 11.18 Glaciofluvial sand and gravel, Tacoma Harbor, Washington state. While this deposit is unsaturated, it illustrates the architecture of very permeable, semi-confined sand and gravel aquifers that are productive water-supply aquifers and vulnerable to contamination. The light-coloured Quaternary interglacial sediments at the top are remnant ash beds from 15–70 ka. Beneath them is the strongly oxidized glacial outwash gravel, Qpog or Quaternary pre-Olympia glacial, 70–780 ka. It overlies an oxidized silt that is partially eroded by the gravel channel. Geologic interpretation by Kathy Goetz Troost, University of Washington.

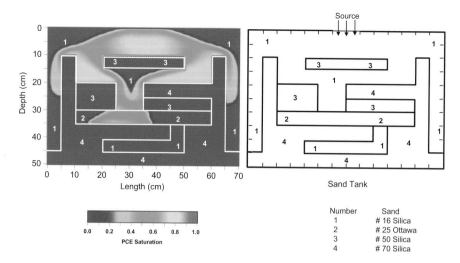

Figure 12.10 The Kueper sand-box experiment: simulation at 310 s after PCE release by M. Jin, formerly, INTERA, Austin, Texas. (Reproduced with permission from Elsevier.)

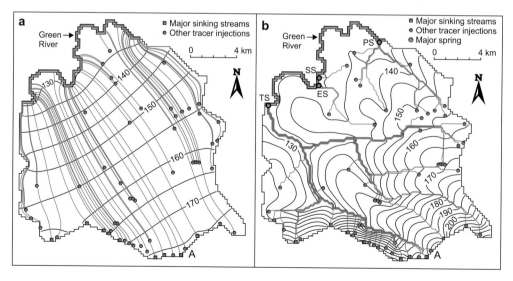

Figure 13.10 Karst terrain. (a) Ponds occupying sinkholes in the Leadville Limestone, San Juan Mountains, Colorado, USA. The limestone is Mississippian period but was not karstified until the Pennsylvanian. (b) Clint-and-grike feature that has developed on the exposed Amabel dolomite, Manitoulin Island, Lake Huron, Ontario, Canada. Rock hammer for scale is shown bridging the fissure, i.e., the "grike".

Figure 13.13 Simulated patterns of hydraulic head distribution in the karst aquifer at Mammoth Cave, Kentucky, USA, based on measured heads (metres above sea level) and spring-flow discharges and 58 tracer tests. (a) The distribution of head in a homogeneous porous-medium aquifer with $K = 1.1 \times 10^{-3}$ m/s. (b) A channel network within the karst aquifer that reproduces the principal features of the head distribution as interpreted by the field evidence. The hydraulic conductivity distribution in this simulation varies from 4×10^{-5} to 7 m/s, which was the simulated K value for the final 6.6 km of channel to Turnhole Spring, marked TS. (Figure from Worthington (2009); reprinted by permission of Springer from *Hydrogeology Journal* © 2009.)

Figure 13.15 Vertical and horizontal solution features, Haile Quarry, Florida. (a) A cover-collapse sinkhole. (b) The channels that drain the sinkholes and create the high hydraulic conductivities measured in karst aquifers. The photo scale indicates that the central aperture channel is approximately 10 cm wide.

Figure 14.12 Trenches in Coachella Valley, California. (Courtesy of Roy Shlemon.)

Figure 14.1 Fault scarps, Little Cottonwood Canyon, near Salt Lake City, Utah. The three fault scarps shown (arrowed) are the result of normal faulting during the late Pleistocene and Holocene. The individual scarps are 10 m or more high, with an estimated slip rate of 0.9 mm/yr for 16 000 years (Hanson and Schwartz, 1982). (Photo courtesy of the Utah Geological Survey.)

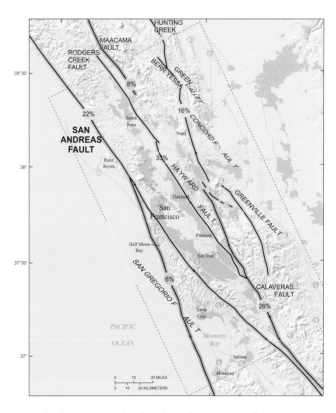

Figure 14.19 An example of a seismic hazard map of the San Francisco Bay area of California showing known active faults with a 72% probability of one or more $M6.7$ or greater earthquakes from 2014 to 2043. Yellow traces indicate 32 minor faults. (Courtesy of the US Geological Survey.)

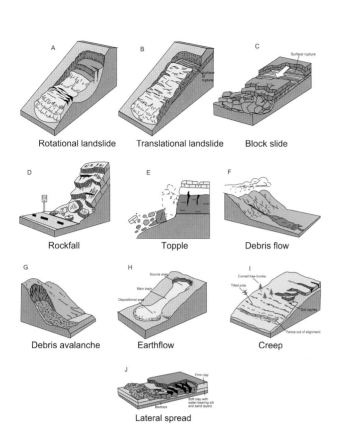

Figure 15.1 Major types of landslides described in this chapter. (After Highland (2004); courtesy of the US Geological Survey.)

Figure 15.5 The Oso landslide, Washington state, USA, March 22, 2014. (a) View of the Oso landslide and geologic exposures in the scarps. Timber harvesting had occurred in years previous to the landslide above the scarp and may have permitted increased groundwater recharge to the glaciofluvial outwash. Active seepage was observed at the outwash/till contact by the GEER team two months after the event. (b) Liquefied lacustrine sediments on the flank of the landslide. Note the rafted block set in the flow material. (Courtesy GEER, Geotechnical Extreme Events Reconnaissance, Jeff Keaton and Joseph Wartman, Team Leaders; Keaton et al. (2014).)

Figure 15.4 Toe erosion along the Athabasca River, Alberta, Canada.

Figure 15.8 Earth flow at Notre-Dame-de-la-Salette, Québec, triggered by an M_w5.0 earthquake on June 23, 2010 with an epicentre located about 14 km away. (Photo courtesy of Charles O'Dale, used with permission.)

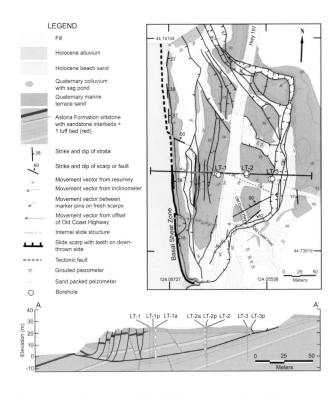

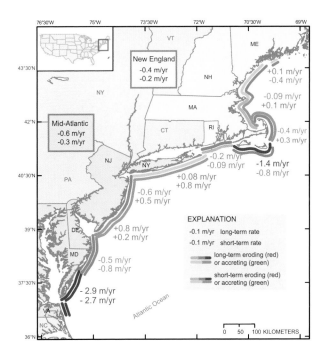

Figure 16.1 Erosion (−ve) and accretion (+ve) rates of shoreline change along the US northeast coast. Long-term rates were calculated over a period of approximately 150 years prior to year 2000, while short-term rates were calculated over a period of 1960–70 through year 2000. The values shown indicate regionally averaged net rates of shoreline change with about 60% of the shoreline indicating net erosion during the recent period. (From Hapke et al. (2013), courtesy of the US Geological Survey.)

Figure 15.11 Johnson Creek landslide, Oregon. (From Priest et al. (2011); with the permission of the Association of Environmental and Engineering Geologists.)

Figure 15.19 The Vaiont Dam and landslide, photographed in 2001. (Photograph courtesy of Paolo Semenza.)

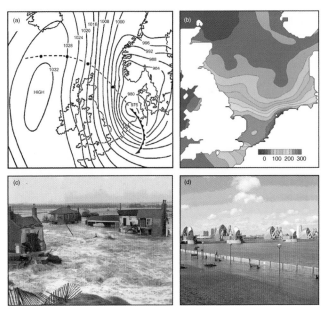

Figure 16.12 A 1953 storm surge in the North Sea inundated parts of Britain, Belgium and the Netherlands, which suffered the greatest loss of life. (a) The track of the storm centre from January 30 (00:00 h) to February 1 (00:00 h), with the black dots indicating the centre in 12 h intervals. (b) The maximum computed surge throughout the affected area in centimetres. (c) Flooding on the Norfolk coast of England. (d) The Thames Barrier that was completed in 1984 to protect central London from a similar storm surge. (From Pugh and Woodworth (2014), with permission of Cambridge University Press.)

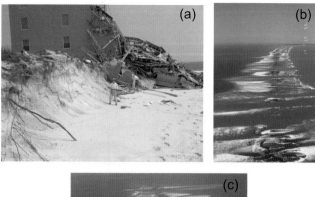

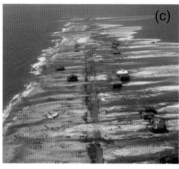

Figure 16.14 Examples of dune response along the Gulf of Mexico coastline described by the USGS storm-scaling model: (a) collision response, Orange Beach, Alabama; (b) overwash response, Dauphin Island, Alabama; and (c) inundation response, Dauphin Island, Alabama. (Photos courtesy of the US Geological Survey and US Army Corps of Engineers.)

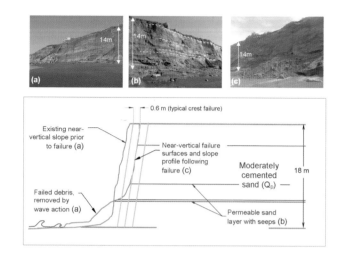

Figure 16.20 A geomorphic model of the failure of a sandstone bluff caused by groundwater flow through permeable layers within the sandstone of an otherwise moderately cemented sandstone, i.e., UCS ~ 340 kPa. Wave action removes the failed debris while loss of tensile strength causes slab failure and retreat of the bluff. (From Collins and Sitar (2008), courtesy of the US Geological Survey.)

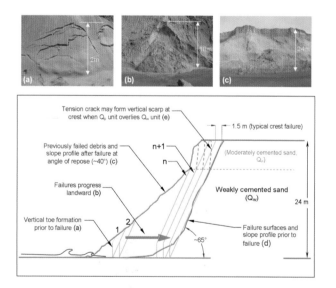

Figure 16.19 A geomorphic model of bluff failure caused by wave action on weakly cemented sandstones, i.e., UCS = 5–30 kPa. (a) Wave action on previously failed debris causes failure of small slabs less than 2 m high. (b) Shear failures increase landward with height as shown in the main panel, i.e., 1, 2, ..., n, but without the retreat of the crest. (c) Wave action on previously failed debris with an angle of repose of 40°. The erosion cycle begins again once the intact slope fails and a new debris deposit is created. (From Collins and Sitar (2008), courtesy of the US Geological Survey.)

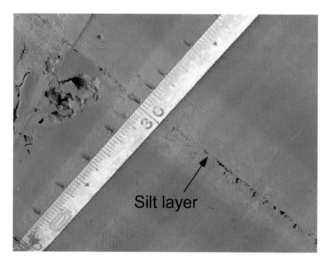

Figure 16.24 A thin silt weak layer, approximately 1 mm thick, in marine clay that failed in testing after remoulding with ~10 cm displacement. The red arrows indicate the failure-induced softening on either side of the silt layer. A Mohr–Coulomb failure analysis (equation (1.13)) of the silt would be quite different from that of the adjacent clay. This silt layer would be of lower effective cohesion and thus lower effective shear strength than the clay. (From Locat et al. (2014). Reprinted by permission of Springer, © 2014.)

BOX 10.3 | ESTIMATING POROSITY, VOID RATIO AND SAMPLE BULK DENSITY

The mass of a dry porous medium is restricted to the matrix of solid particles (Dullien, 1979):

$$m = \rho_s V_s = \rho_b V_b, \tag{10.9}$$

where m is the mass of the sample, ρ_s is the density of the solid particles in the sample, ρ_b is the bulk density of the sample, V_s is the volume of solid particles and V_b is the bulk volume. It follows that the **total porosity** (n) of the sample is therefore

$$n = 1 - (V_s/V_b) = 1 - (\rho_b/\rho_s). \tag{10.10}$$

The bulk density is estimated by oven drying and comparing the sample mass before and after, while the particle density is the oven-dried mass of solids per unit volume measured by water displacement.

If we consider the sample as partially or fully saturated with water, then the expression for the **bulk density** for a sample becomes (Muir Wood, 2009)

$$\rho_b = \frac{\text{mass of solid} + \text{mass of water}}{\text{total sample volume}} = \frac{m_s + m_w}{V_s + V_v}, \tag{10.11}$$

where m_s and m_w are the masses of solid particles and water, and V_v is the void volume. The void ratio is by definition

$$e = V_v/V_s. \tag{10.12}$$

The mass of water in the sample is

$$m_w = S_i e \rho_w V_s, \tag{10.13}$$

where S_i is the degree of water saturation. The mass of solid particles is

$$m_s = \rho_s V_s = G_s \rho_w V_s, \tag{10.14}$$

where G_s is the specific gravity of the soil, i.e., $G_s = \rho_s/\rho_w$.

Combining equations (10.12), (10.13) and (10.14), we obtain

$$\rho_b = \frac{S_i e + G_s}{1 + e} \cdot \rho_w. \tag{10.15}$$

In a fully saturated soil sample $S_i = 1.0$, in which case ρ_b is often referred to as the wet bulk density, i.e.,

$$\rho_{sat} = \frac{e + G_s}{1 + e} \cdot \rho_w, \tag{10.16}$$

and the dry bulk density is

$$\rho_{dry} = \frac{G_s}{1 + e} \cdot \rho_w. \tag{10.17}$$

Here $\rho_{dry} = \rho_b$ in equation (10.9).

---

between these variables is discussed in Box 10.3. The choice – void ratio or porosity – reflects the context of the study. The geotechnical engineer is often interested in soil strength, compressibility and settlement, in which the void ratio is often an independent variable. The sedimentary geologist may be working with hydrogeologists or geoenvironmental engineers who are interested in the measurement of the pore volume, and thus porosity is the variable of interest.

Ideally, laboratory measurements can support inferences made by field measurements and are useful in identifying small-scale phenomena observed in field measurements. Here we briefly consider laboratory testing involving three specific techniques that are of interest in geotechnical and geoenvironmental engineering: (1) permeameter testing to measure hydraulic conductivity; (2) triaxial testing to measure shear strength; and (3) oedometer testing to measure soil consolidation. Our inclusion of these tests here is for the purposes of completeness given their importance in the assessment of geotechnical and geoenvironmental hazards (see the section on further reading at the end of this chapter for additional information).

10.3.1 Hydraulic Conductivity

Hydraulic conductivity of soils and sediments can be determined in the laboratory with either constant-head or falling-head permeameters, which are discussed in standard texts on geotechnical engineering (e.g., Coduto et al., 2011; Briaud, 2013) and groundwater hydrology (Todd and Mays, 2005). The falling-head test, used normally with relatively low-K soils, involves the measurement of the falling head in a vertical tube, i.e., a piezometer, connected to a soil column of known dimensions through which flow occurs. The hydraulic conductivity is estimated by Darcy's law during a time interval over which the head drop is measured. For clays, the appropriate measurement is that obtained by a consolidation test using an oedometer, which is discussed below.

The constant-head test is preferred for cohesionless soils with a minimum hydraulic conductivity of 1×10^{-6} m/s. This test dates back to the original experiments by Henri Darcy (1856) in Dijon, France, around 1856 and involves a soil sample contained within a soil column of known dimensions

with water flowing through the column (see Figure 1.15). Hydraulic conductivity (K) is measured using "Darcy's law", which we have discussed in section 1.3.5. Equation (1.16) indicates that $K = Q/(A \cdot dh/dL)$.

Figure 10.15 illustrates a constant-head permeameter designed for upward flow through the soil column. Care must be taken to ensure that the soil column when tested is fully saturated with water with no significant air entrainment, as this will considerably lower the measured hydraulic conductivity through the relative permeability effect (see Chapter 12). Upward flow is a precaution to allow any residual entrapped air to drain from the column prior to testing. The hydraulic gradient through the column shown in Figure 10.15 can be measured across two equidistant zones (i.e., ab or bc), which should be the same. Any gradient difference across these zones would indicate poor column packing procedures.

Laboratory permeameter measurements must test appropriate soil zones, and therefore cored soil materials must be carefully chosen. The soil sample is most likely to be from a vertical core; therefore, if the soil column is repacked, there is the possibility of the loss of fine-grained silt or clay that may be important in vertical groundwater flow. If it is the horizontal component of groundwater flow (and therefore hydraulic conductivity) that is sought, it is important to carefully select specific core sections for testing that represent high-K zones.

For example, in field and lab tests of soils from a site in Kansas, Butler (2005) determined that the permeameter

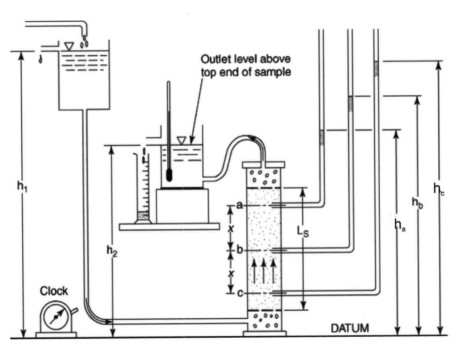

Figure 10.15 Measurement of hydraulic conductivity using a constant-head permeameter with upward flow through the soil column to displace any air bubbles trapped during packing of the column. (From Head and Epps (2011), courtesy of Whittles Publishing.)

analyses of the original and repacked cores were 16% and 39% of the hydraulic conductivity determined from a pumping test that stressed this cored zone. Furthermore, the hydraulic conductivity for the repacked core was a factor of 2.4 greater than that of the original core, indicating that the repacking caused a loss of thin layers of low-K soil material.

10.3.2 Shear Strength

The experimental testing of rock specimens was discussed in Chapter 4 and the general nature of unconfined, shear and triaxial testing of these specimens is illustrated in Figure 4.10. Soils and rocks typically fail diagonally by shear of the material rather than by compression, even when the specimen is unconfined with no lateral support, as is shown in Figure 4.10(a). With soils and sediments we are concerned with the resistance to shear failure, i.e., the slippage of one soil or sediment unit relative to an adjacent unit along a plane of failure resulting in instability detrimental to an engineering project or existing infrastructure.

The simplest measurements of shear in soils are conducted with a shear box (see Figure 4.10c). Shear-box experiments date back to the 1840s and the investigation of slope stability by the French engineer Alexandre Collin. Modern shear-box testing may be used to measure the shear strength of cohesionless soils, such as sands and gravels, and of cohesive soils, provided that the rate of displacement during the test permits drainage. Ring-shear testing (Figure 4.10d) was discussed in the context of glaciotectonics (section 8.2) and is used to measure the residual shear strength of cohesive materials.

Shear strength may also be measured in a triaxial cell, as shown in Figure 4.10(e), which is the standard method for developing the Mohr–Coulomb failure envelope. Triaxial tests are preferred to direct shear tests of clays because triaxial testing does not require the choice of a specific shear plane to be tested and are preferred to uniaxial compression tests because typically σ_3 cannot be assumed to be zero. Geotechnical reference texts, such as Head and Epps (2011, 2014) and Briaud (2013), describe the various approaches to measuring the shear strength parameters in equation (1.13), $s = c' + \sigma' \tan\phi'$, where s is the shear strength of the sample in kPa, c' is the effective cohesion, σ' is the effective stress normal to the plane of failure and ϕ' is the effective friction angle.

Triaxial testing of soils may be conducted under drained or undrained conditions, the choice being dictated by the context of the investigation. For those situations in which it may be assumed that an imposed load will not cause rapid pore-water

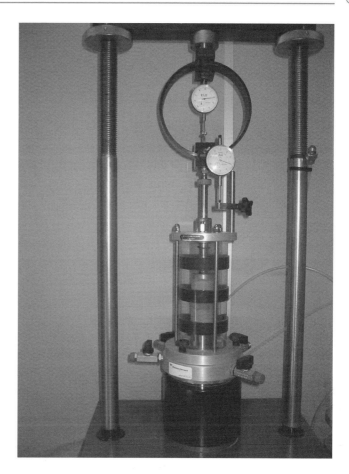

Figure 10.16 Soil specimen shown contained in a triaxial cell prepared for undrained testing within a load frame. The specimen is enclosed in a waterproof membrane and then placed in the cell that is flooded and maintained at σ_3. The upper dial gauge mounted in the load ring measures the normal force imposed on the specimen, i.e., the stress. The lower dial gauge measures the strain or displacement. For drained conditions, the pore pressure of the specimen would be measured by a pressure transducer connected to a tube exiting the base of the cell and seated into the base of the specimen. (From Head and Epps (2011), courtesy of Whittles Publishing.)

drainage – saturated silts or clays – it is felt appropriate to test a soil specimen under undrained conditions (i.e., total stresses) yielding the undrained shear stress.

Slope stability over longer periods, such as construction of embankments, earth dams and the like, more typically dictates the use of triaxial tests that simulate pore-water drainage and the resulting failure envelope. In this case, a minimum of three identical cylindrical soil-core specimens of diameter 38–110 mm and length twice the diameter are chosen. Each is tested independently in the triaxial cell (Figure 10.16) under increasing normal stress, with the lateral stress increased proportionally such that the deviator stress ($\sigma_1 - \sigma_3$) increases with each test. When the deviator stress is sufficient, the

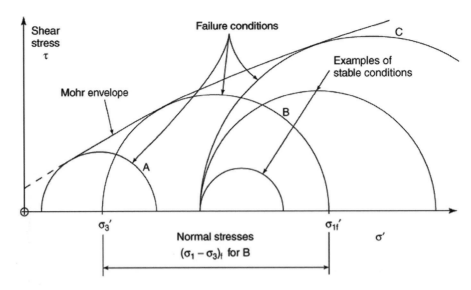

Figure 10.17 Mohr circles and the associated failure envelope based on tests A, B and C, for which shear failure of each specimen occurred. The ordinate is the shear stress, τ, and the abscissa is the effective normal stress, σ'. Examples of stable conditions, each of which has relatively small deviator stresses, are also shown. The effective principal major and minor axial stresses are shown on the abscissa, as is the deviator stress at failure $(\sigma_1 - \sigma_3)_f$ for test B. (From Head and Epps (2011), courtesy of Whittles Publishing.)

specimen will fail and the measurements of σ_1, σ_3 and the pore pressure p_w are plotted to allow c', σ' and ϕ' to be measured from the slope of the Mohr failure envelope (Figure 10.17).

10.3.3 Consolidation of Soils and Sediments

Some of the initial experiments in soil mechanics by Karl Terzaghi, its founding father, addressed the consolidation of soils, in particular clays, in order to understand the phenomenon of settlement of buildings and other infrastructure. Terzaghi's instrument also allowed him to measure the hydraulic conductivity of the soil sample in his laboratory overlooking the Bosphorus Strait in Turkey. Terzaghi's results appeared in the *Engineering News Record* in the early 1920s and prompted similar investigations in the USA and the UK. Terzaghi's life and work has been described by Professor Richard Goodman of the University of California at Berkeley (Goodman, 1998).

Modern investigations into soil consolidation employ an instrument known as an oedometer, an example of which is shown in Figure 10.18. The purpose of the oedometer consolidation test is to examine the compressibility of a soil sample and its rate of compression. The soil sample may be an overconsolidated clay from the shallow subsurface or a deep clay unit comprising an aquitard in a sedimentary basin.

In the oedometer consolidation test, the soil specimen is loaded into the oedometer and one-dimensional consolidation is measured as a function of the applied load. Typically,

four to eight different vertically imposed loads are each tested incrementally for at least 24 hours. The specimen, usually 75 mm diameter and 20 mm thick, is laterally constrained by a confinement ring and allowed to drain through porous plates positioned above and below the specimen. The loading cap transmits the applied load, P, to the soil sample through the porous stone shown. The specimen is allowed to drain, but not to become undrained. The water bath ensures that the soil sample remains fully saturated throughout the experiment and the dial gauge represents a linear variable differential transformer that records the change in sample dimension.

A number of important assumptions are involved in the analysis of the data. The specimen being tested is assumed to be homogeneous and fully saturated with pore water. Pore-water drainage is assumed to obey Darcy's law and is in response to the imposed load with vertical flow through the porous stone plates. The rate of consolidation is assumed to be a function of the hydraulic conductivity of the specimen, and the changes in effective stress cause a corresponding change in the void ratio of the specimen. Therefore the specimen volume changes with the transfer of total stress from pore water to effective stress, i.e., that component supported by the soil skeleton.

The tests result in consolidation curves showing various relationships, such as percent soil compression versus log time, and void ratio versus effective vertical stress (see Head and Epps, 2011; Briaud, 2013). The principal result sought is the coefficient of consolidation (c_v), which may be written

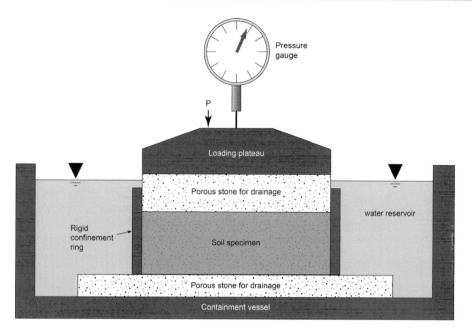

Figure 10.18 Oedometer for measuring one-dimensional consolidation of soils. (After Muir Wood (2009).)

(Muir Wood, 2009) as $c_v = KE_0/\rho g$, where K is the hydraulic conductivity and E_0 the one-dimensional stiffness of the clay sample. The units of c_v are m^2/yr. Consolidation of soils and sediments is dominated by compression of clay formations, while coarser zones play a passive role (see Mitchell and Soga, 2005). We will return to this topic in Chapter 13 when we discuss settlement induced by groundwater extraction in sedimentary basins containing significant thicknesses of clay strata.

10.4 Characterization of Aggregate Sources

The mining of sand and gravel deposits for use as construction aggregates in ready-mix concrete, pre-formed concrete products, highway foundations, etc., is a geotechnical task that involves the engineer in land-use planning issues that may lead to public controversy.

Because of high transportation costs for aggregate, sand and gravel pits (Figure 10.19) and rock quarries are usually located within 50 km of their urban market. Competition for land use is significant and access to good aggregate resources may be controlled by urban planning or by groundwater- and ecosystem-protection regulations that prevent them from being exploited. Alluvial deposits are obviously to be found in river valleys and their development can be damaged by river avulsion (section 6.4.4) drowning the pit. In-stream mining by dredging or dragline extraction is quite common but poses significant problems with respect to the sustainability of fish

habitat through increased streambed erosion and suspended sediment transport (Lingley and Jazdzewski, 1994; Norman et al., 1998).

Once a potential sand and gravel deposit has been located, whether by satellite or airborne imaging in the case of remote areas or by surficial geological maps in the case of developed areas, then trenching, test drilling and geophysical surveys follow. The coarseness of the aggregate requires that the drilling method be able to penetrate thick gravel zones and collect samples from them. Hollow-stem augers or cable-tool or other percussion drilling rigs may be employed, together with hardened sample core barrels (Smith and Collis, 2001). More recently, sonic drilling has provided a means to retrieve cores (7.6–25 mm in diameter) of cobbles, boulders and cohesionless sands, which may be flowing because of the high hydraulic heads associated with groundwater discharge areas in valley bottoms.

The presence of shallow groundwater may limit the suitability of a site. Sutphin et al. (2002) develop methods for estimating gravel resources in glaciofluvial deposits in a New England valley, where extraction is constrained by a variety of regulations referred to as sterilization of the resource, e.g., the prohibition of mining below the water table of an esker deposit such as that shown in Figure 8.13 or proximity to streams and wetlands or encroachment of urban areas. They estimated that only 75% of the sand and gravel in the esker deposits is exploitable, i.e., 25% is sterilized.

Lucius et al. (2007) provide a helpful review of surface geophysical tools for the characterization of sand and gravel

Figure 10.19 Glaciofluvial deposits, such as this kame near Heidelberg, Ontario, Canada, supply local urban areas with aggregate and are located with the help of surficial geology maps. This pit produces 3000 tons/day of sorted materials extracted by front-end loaders from the various excavations. The conical piles of sorted products are shown following washing and distribution by conveyor belt. Imported asphalt cement is used to produce asphalt at the plant in the centre of the photo, adjacent to the dark piles of asphalt, and ready-mix concrete is produced at the plant at the entrance to the pit. Trucking costs limit the distance that aggregate and construction products can be sold profitably; consequently this pit, and others like it, provide aggregate only locally (within 30 km) by truck. Therefore, regional land-use planning requires that such pits be permitted proximal to urban areas, such as Kitchener/Waterloo shown on the skyline, for infrastructure development if long-distance transport (train, barge, ship) is to be avoided and aggregate costs are to be minimized. (Photo courtesy of Jim Karageorgos, Steed and Evans, St. Jacobs, Ontario, Canada.)

aggregate deposits in North America, with the benefits and limitations of each method clearly explained and an estimate of data collection rates:

- *Seismic*: Lateral boundaries and depth to bedrock are readily identified; however, data processing will be time-consuming and contribute to increased project costs (0.2–0.6 line km/day).
- *Resistivity*: Depending on the equipment employed, data collection is straightforward and initial results may be interpreted in the field (1–2 line km/day).

- *Ground-penetrating radar (GPR)*: May be limited by depth of penetration (see section 10.1.5) but investigation times are brief (1–5 line km/day).
- *Electromagnetic (EM)*: While time-domain EM is weak at shallow depths (<10 m), site coverage is good and data are quickly processed; frequency-domain EM allows rapid site characterization with good resolution of bedrock or clay boundaries (1–5 line km/day).

The advice of a professional geophysicist should be obtained to allow cost-effective project management.

10.5 Summary

Earlier (Box 5.1), the advice of Elio D'Appolonia was used to introduce the importance of site characterization: "Site conditions always pose unknowns, or uncertainties, which may become known during construction or operation to the detriment of the facility and possibly lead to damage of the environment or endanger public health and safety." Chapter 10 has sought to explain how geotechnical and geoscientific approaches to site characterization are undertaken to reduce these uncertainties. Geotechnical failure within the foundation soils at Mount Polley (Figure 10.1) resulted in the development of guidelines for the preparation of dam-site characterization models. We have sought to illustrate how geotechnical and geoscientific approaches to site characterization are complementary. Site characterization begins with a review of the available technical literature, some of it specific to the region where the site is located, some of it unrelated to the project being undertaken but specific to the geologic materials on site. Topographic and geologic maps are consulted and remote sensing images acquired. The regional geology must first be reviewed, in particular the surficial geology. The depth and nature of weathering, bedrock stratigraphy and information concerning soil strength, compressibility and permeability are acquired.

We have discussed the use of drilling methods to develop the site characterization model through the acquisition of cores. This information is supplemented by the use of cone penetrometer and geophysical surveys. Core samples allow the engineer and geoscientist to undertake textural analysis of the cored material and laboratory testing of soils and sediments for shear strength, compressibility and permeability. The final product of site characterization – a site conceptual model or ground model – becomes the basis for project design and is updated as further information is acquired during construction and operations.

10.6 FURTHER READING

Fookes, P., Pettifer, G. and Waltham, T., 2015, *Geomodels in Engineering Geology: An Introduction*. Caithness, UK: Whittles. — Provides numerous site conceptual models developed by British engineering geologists. A particular strength of this book is its many coloured photographs of geologic conditions throughout the world.

Head, K.H. and Epps, R.J., 2006–14, *Manual of Soil Laboratory Testing*, vols 1–3. Caithness, UK: Whittles. — This is a three-volume set that clearly sets out the methods of soil laboratory tests used in North America and Europe. It is well illustrated and the theory of each method is discussed in detail. Volume 1 (Head, 2006) discusses soil classification and compaction tests. Volume 2 (Head and Epps, 2011) discusses hydraulic conductivity, shear strength and compressibility tests. Volume 3 (Head and Epps, 2014) discusses effective stress testing.

Hencher, S.R., 2012, *Practical Engineering Geology*. London, UK: Spon, Taylor and Francis. — An introduction to engineering geology for geotechnical engineers with excellent coverage of weathered rocks.

Mayne, P.W., Christopher, B.R. and DeJong, J., 2001, *Manual of Subsurface Investigations*. National Highway Institute FHWA NHI-01-031. Washington, DC: Federal Highway Administration. — Many of the tools discussed in this chapter are reviewed in this illustrated report and may be downloaded from the FHWA website.

Norbury, D., 2010, *Soil and Rock Description in Engineering Practice*. Caithness, UK: Whittles. — An indispensable guidebook prepared for geoscientists entering the field of engineering geology and required to log cores of soil or rock.

Price, D.G., 2009, *Engineering Geology: Principles and Practice* (ed./comp. M.H. de Freitas). Berlin, Germany: Springer. — Contains detailed and well-illustrated text on the preparation of site investigation methods written for those with geological training.

Reynolds, J.M., 2011, *An Introduction to Applied and Environmental Geophysics*, 2nd edn. Chichester, UK: John Wiley and Sons. — A comprehensive guide to geophysical surveys of the shallow subsurface particularly suitable for geotechnical and geoenvironmental engineers.

10.7 Questions

Question 10.1. Discuss how cores recovered during drilling operations might become disturbed.

Question 10.2. In Figure 10.9, the soil behaviour type profile indicates evolving sedimentary deposition. How would you describe them?

Question 10.3. Why might a geophysicist prefer to use seismic refraction rather than seismic reflection to examine a potential sand and gravel deposit for aggregate exploitation?

Question 10.4. Figure 10.11 illustrates the textural changes in a borehole in glacial sediments. How would grain size vary over the distance of one of the cycles of sand/silt shown for the S and T formations? Use ϕ units to identify the coarsening-upward sequence.

Question 10.5. What would be your drilling method of choice if you were required to obtain a continuous core of alluvial-fan materials?

Table 10.4 Estimation of specific surface area (SSA) of sample A (see Box 11.1).

Grain size (mm)	Passing (%)	$X = (P_D - P_d)/100$	SSA	$X \cdot$ SSA
0.0228	95	—	—	—
0.0184	90	0.05	?	?
0.0142	84	?	159.8	?
0.0096	75	0.09	?	?
0.0057	60	?	?	?
0.0042	50	?	?	?
0.0031	40	?	?	?
0.0019	25	?	?	?
0.0013	16	?	?	?
0.0010	10	?	?	?
0.0007	5	?	?	168

Question 10.6. Compute the specific surface areas of soil sample A in Table 10.4 by completing the table where "?" exist. Assume the density is $\rho_s = 2650\,\text{kg/m}^3$. Notice how the contribution of the different grain-size fractions increases with the fineness of the grain size. Sum each grain-size fraction to obtain the total SSA of the soil sample.

PART

III | Groundwater

Rocks, soils and groundwater form the trilogy of interconnected subjects of this book. The science describing groundwater occurrence and flow and the properties of permeable rocks and soils is known as hydrogeology to geoscientists or groundwater hydrology to engineers. This book presents a more complete discussion than is normally accorded hydrogeology in books on the applied earth sciences written for civil engineers. However, as Professor Burland has made clear (see the Preface), failure to properly assess groundwater conditions has led all too often to geotechnical failure. Throughout this book, we consider the hydrogeological conditions that geotechnical and geoenvironmental engineers will encounter in practice: dewatering of large sites, development of groundwater supplies, remediation of contamination, land subsidence, liquefaction and slope stability.

Chapter 11 presents the principles of flow through porous media and the main equations of flow, aquifer compressibility and storage. Hydraulic testing of aquifer and aquitard materials is outlined and groundwater flow systems are examined to allow the engineer to understand how pore pressures vary with time and location in such systems. Having introduced Darcy's law in section 1.3.5, we may define an **aquifer** as a soil or rock formation that is sufficiently permeable to allow transmission of significant quantities of groundwater through it under natural hydraulic gradients, which are typically <0.01, indicating relatively little resistance to flow. An **aquitard** is a formation that presents resistance to flow and requires a high hydraulic gradient for groundwater to flow through it. Aquifers provide water supplies; aquitards control flow patterns and often protect aquifers from contamination.

Chapter 12 describes briefly the evolution of groundwater quality and of groundwater contamination. Over the past 20 years, geoenvironmental engineers have been busy attempting to prevent the widespread contamination of groundwater that occurred in previous generations due to ignorance about the hydrogeological effects of waste disposal and fuel storage. We identify two kinds of contaminants – immiscible liquids such as gasoline and various solvents that fail to dissolve and become a separate non-aqueous phase in the subsurface – and other contaminants that are readily miscible in groundwater, e.g., road salt and nitrate from fertilizers. We shall consider how contaminated sites are characterized using drilling and geophysical tools. In addition, we shall consider certain special cases of groundwater contamination, e.g., from seawater intrusion and landfill leachate. Finally, we consider how public water-supply wellfields may be protected once the groundwater flow system is properly mapped.

CHAPTER

11 Hydrogeology

We now return to hydrogeological issues that have been considered in our discussion of Darcy's law (section 1.3.5), hydraulic testing of rock in boreholes (section 5.5), the estimation of hydraulic conductivity from textural analysis (Box 10.2) and permeameter testing of soils (section 10.3.1). Here we consider those basic parameters of importance in flow through porous media: porosity, velocity, permeability and hydraulic conductivity. On the basis of this analysis, we then develop the groundwater flow equation and define the principal parameters of importance: transmissivity and storativity. We place groundwater flow in the context of gravitationally driven flow systems involving recharge and discharge areas. Finally, we discuss the characteristics of aquifers and aquitards.

11.1 Flow through Porous Media

Porosity and permeability are the fundamental properties of porous media. A porous medium can be both porous and permeable, but high porosity does not necessarily indicate high permeability. A clay may be very porous but is usually of low permeability unless weathering has fractured it. This introduces the concept of effective porosity, i.e., that porosity controlling fluid flow in porous media. We then consider the relationship between porosity and hydraulic conductivity, allowing us to compute groundwater velocity. We then define the factors that affect hydraulic conductivity in the field and the hydraulic potential, which governs groundwater flow.

11.1.1 Porosity

In Chapter 1, a wide range of porosities was noted for recently deposited sediments and sedimentary rocks. Table 11.1 presents the total or physical porosity and hydraulic conductivity of a number of sedimentary rocks and sediments. Note the extremes in these values: some Dutch marine sediments have porosities of only ~10% while the lacustrine clays of Mexico City have porosities of over 70%. Sedimentary rocks generally have a lower average porosity than recently deposited sediments; e.g., compare the sands and sandstones in Table 11.1. Well-sorted sands have porosities of 37–51%, while those of poorly sorted gravels (see Figure 1.6) are much lower. The shape of flocculated clay platelets might initially cause porosities to be >50% but these are later reduced by compaction. The Mexico City lacustrine clay shown in Table 11.1 is undergoing compaction by groundwater pumping from beneath and infrastructure loading from above, and has not yet approached the porosity of typical lithified clays, i.e., shales and mudstones.

11.1.2 Velocity, Permeability and Hydraulic Conductivity

Darcy's law, introduced as equation (1.16), $q = Q/A = -K \, dh/dL$, has several significant implications.

First, even though the specific discharge (q) term in Darcy's law has units of length divided by time, it is not a velocity but a discharge per unit cross-sectional area, i.e., $m^3 \div (s \times m^2)$. The velocity is actually much greater because the groundwater must flow by an indirect path dictated by the interconnected

Table 11.1 Total porosity (n) and hydraulic conductivity (K) of sedimentary rocks and sediments.

Material	n (%)	K (m/s)
Sandstone, UK	15–35	1×10^{-4} to 6×10^{-6}
Sandstone, Canada	16	1×10^{-3} to 1×10^{-7}
Chalk, UK	25–45	2×10^{-9} to 6×10^{-13}
Limestone, UK	10–25	$\leq 6 \times 10^{-9}$
Limestone, USA	21–46	2×10^{-4}
Dolomite, USA	11–22	2×10^{-7} to 2×10^{-8}
Dolomite, Canada	4–11	2×10^{-3} to 2×10^{-10}
Shale, USA	30	1×10^{-13} to 1×10^{-14}
Shale, Switzerland	14–20	$< 1 \times 10^{-12}$
Gravel and sand, Germany	27	$(2-5) \times 10^{-3}$
Glaciofluvial sand, Germany	36	5×10^{-3} to 1×10^{-4}
Glaciofluvial sand and gravel, Canada	30–35	2×10^{-3} to 6×10^{-5}
Fluvial sand, Canada	38	$(1-5) \times 10^{-5}$
Fluvial sand, Ohio, USA	37	9×10^{-5}
Fluvial sand, Illinois, USA	45	8×10^{-5}
Beach and dune sand, US Gulf Coast	45–51	$(6-80) \times 10^{-5}$
Clay till, Canada	32	1×10^{-10}
Sandy silt till, Canada	15	8×10^{-10} to 3×10^{-11}
Lacustrine clay, Mexico	70–90	10^{-8} to 10^{-11}
Silts and clays, US Gulf Coast	~25	10^{-7} to 10^{-11}
Marine silts and clays, The Netherlands	8–34	2×10^{-5} to 2×10^{-8}
Clay, UK	5	2×10^{-8} to 5×10^{-12}

Notes and sources:

UK sedimentary rocks: Permo-Triassic Sandstones, Cretaceous Chalk and Jurassic Limestone (Lawrence et al., 1990).

Canadian sandstone: Prince Edward Island (Francis, 1989).

US limestone: tracer test results in Eocene Floridan aquifer (Robinson, 1995); see section 11.1.2 for the effective porosity of this formation.

Canadian dolomite: Silurian Lockport Fm, Niagara Region, Ontario (Novakowsi and Lapcevic, 2004).

US dolomite: Permian Culebra dolomite, WIPP site, New Mexico (Holt, 1997).

US shale: Pierre Shale, South Dakota (Neuzil, 1993).

Swiss shale: Opalinus Clay, n (Pearson, 1999) and K (Gautschi, 2001).

German sands and gravels: stratified glaciofluvial deposits, Rhine Valley (Heinz et al., 2003).

Canadian glaciofluvial sand and gravel: Gloucester, Ontario (Jackson et al., 1985).

Canadian fluvial sand: Chalk River, Ottawa Valley, Ontario (Parsons, 1960).

US sands: Whitewater River bar sand, Ohio, and Wabash River bar sand, Illinois (Pryor, 1973).

Clay till: Saskatchewan (Keller et al., 1989).

Sandy silt till: southern Ontario (Gerber and Howard, 2000).

Lacustrine clay: Mexico City (Vargas and Ortega-Guerrero, 2004).

US Gulf Coast silts and clay: fluvial, deltaic and marine origin, apparently fractured (Hanor, 1993).

Dutch marine silts and clays: (Bierkens, 1996).

UK clay: Oxford Clay, n is numerical model estimate (Sen and Abbott, 1991).

porosity or **effective porosity** (n_e) of the porous medium. Thus, $v = q/n_e$ defines the **average interstitial velocity** or **average linear velocity**. Because velocity is a vector, this relationship implies that the direction of groundwater flow is understood; strictly speaking, this equation identifies groundwater speed. Therefore, the lower the porosity, the higher the velocity for a fixed flow rate, Q. When the principal permeable pathways in a soil or rock are due to discontinuities that represent a small percentage of total porosity, the result is rapid flow through such features. Caution, therefore, must be used to choose an appropriate porosity value.

The entry for the Floridan limestone aquifer in Table 11.1 underlines this very important point, i.e., the total or physical porosity of the rock is not necessarily the effective porosity controlling groundwater flow and contaminant transport. A tracer test in the Floridan limestone aquifer at the Old Tampa wellfield involved tracer injection into one well and extraction from a second well. Although the matrix porosity of the limestone was 21%, the effective porosity of only 1%, which was associated with solution channels in the limestone, determined the first arrival of tracer at the extraction well (Robinson, 1995). The difference in magnitude is critical in contaminant transport because the first arrival of contamination is inversely related to the effective porosity, rather than the total porosity. The velocity within the rock matrix may be close to zero. Permeable formations that contain two very different porosity systems are referred to as **dual-porosity** aquifers.

Second, because $p_1 = \gamma(h_1 - z_1) > p_2 = \gamma(h_2 - z_2)$, we notice that the pore pressure in the column shown in Figure 1.15 does not control the fluid flow direction, thus $(p_2 - p_1)/\Delta L \neq (h_2 - h_1)/\Delta h$. Rather, it is the gradient in hydraulic head that provides the basis for determining the direction of fluid flow because the head incorporates pressure, density and elevation terms:

$$h = \frac{p}{\gamma} + z. \tag{11.1}$$

Third, Darcy's law is only valid for laminar flow. Bear (1972) presents a version of the Reynolds number for flow in porous media, i.e., $Re = qd/v$, where d is some grain-size dimension, such as d_{10}, which is the grain diameter that exceeds the size of 10% of the grains in the porous medium by weight, q is as previously defined and v is the kinematic viscosity of the fluid. Bear notes that Darcy's law is valid as long as Re, as defined above, does not exceed 1–10.

Finally, the hydraulic conductivity term in Darcy's law is not the permeability term that we are seeking because it is a function of both the porous medium *and* the fluid that is being transported through it. We seek a term that is only dependent upon the properties of the porous medium and expresses the ability of that medium to conduct a Newtonian fluid.

Hubbert (1940) pointed out that this term is known from experiment to be proportional to the square of the diameter (d) of the grain size of the particles in the soil column and to the specific weight (ρg) of the fluid and inversely proportional to the dynamic viscosity (μ, i.e., $v = \mu/\rho$) of the fluid moving through the column:

$$\frac{q}{dh/dl} = -Nd^2 \frac{\rho g}{\mu}. \tag{11.2}$$

It is convenient to combine N, another coefficient of proportionality, and d^2 to form a single material property known as the specific permeability or intrinsic permeability (k) with the dimensions of length squared, which is the term we have been seeking to describe the fundamental transport property of the porous medium. It is almost always referred to simply as **permeability**. Traditionally, in petroleum engineering, permeability is expressed in darcies (D); 1 D is defined as the property of a porous medium in which 1 mL of water of 1 cP viscosity flows in 1 s through an area of 1 cm^2 due to a pressure gradient of 1 atm/cm. In SI units, it is expressed in square micrometres, i.e., 1 μm^2 = 10^{-12} m^2. One darcy (1 D) is equal to 0.987 μm^2. A one darcy reservoir rock would constitute a very promising oilfield; however, the permeabilities of most reservoir rocks are less and are therefore expressed in milli-darcies (mD or 10^{-3} D).

Therefore, we can now define the **hydraulic conductivity** (K) as

$$K = \frac{k\rho g}{\mu} = \frac{k\rho}{v}. \tag{11.3}$$

The US Geological Survey adopted the following definition (Lohman et al., 1972): "the hydraulic conductivity of the [porous] medium is the volume of water at the existing kinematic viscosity (v) that will move in unit time under a unit hydraulic gradient through a unit area measured at right angles to the direction of flow".

Because geotechnical engineers usually encounter only water present in soils and rocks, not other liquids, "permeability" is often used in geotechnical engineering to mean hydraulic conductivity. It is recommended that the use of "permeability" be restricted to the intrinsic property of a soil or rock to transmit any fluid, and if groundwater is the relevant fluid then "hydraulic conductivity" be used and quantified in SI units.

For all practical purposes, the conversion between hydraulic conductivity and permeability is given by $1\,\mathrm{D} \sim 10^{-5}\,\mathrm{m/s} \sim 2.7\,\mathrm{ft/day}$ in US units. The range of hydraulic conductivities for various rocks and sediments is presented in Table 11.1. We may say that permeable geological materials are considered to be those with $k \geq 1\,\mathrm{D}$ or $K \geq 10^{-5}\,\mathrm{m/s}$.

11.1.3 Factors Affecting Hydraulic Conductivity

The measured hydraulic conductivity, whether measured *in situ* in the field or by laboratory permeameter, is subject to a number of influences. Here we consider only the **saturated hydraulic conductivity** of the soil or sediment, i.e., full water saturation is present throughout the total porosity. In sandy sediments and soils, the hydraulic conductivity is strongly affected by its sorting with respect to texture, as shown in Figure 9.15. Clayey sediments, while generally of low hydraulic conductivity, can contain discontinuities caused by weathering or **bioturbation**, e.g., the burrowing activities of worms, crayfish or mammals (see Figure 6.6). Such worm burrows and decayed root traces in loam soils permit vertical flow velocities exceeding 40 m/day (Cey and Rudolph, 2009). Sedimentary rocks may be fractured – particularly along bedding planes as in Figure 5.9 – or contain solution cavities in the case of carbonate and sulfate formations as in Figure 13.15.

In the case of clays and clay-rich sediments, the effective porosity may be that associated with fractures or root holes (**macropores**) that are again a small fraction of the total porosity. In southwestern Ontario, Canada, the clay tills are vertically fractured to a depth of 5–10 m with hydraulic conductivities in the range $>10^{-9}$ to 10^{10} m/s (Ruland et al., 1991). These same clay tills exhibit stress-dependent closure that is consistent with rough fractures. Below 12 m depth, i.e., below where the total stresses are $\sim$240 kPa and the effective stress is 120 kPa, the fractures are closed and the bulk hydraulic conductivity is similar to that of the clay till matrix (Sims et al., 1996). At a hazardous waste landfill in Louisiana, the fractured clays were determined by water balance to have an effective vertical hydraulic conductivity of 10^{-7} m/s. Only one laboratory estimate was close to this value; the remainder were in the range of unfractured clays, i.e., 10^{-8} to 10^{-11} m/s (Hanor, 1993).

Very few sediments or rocks are sufficiently homogeneous to be described as isotropic in permeability and most exhibit scale dependence. Depositional processes cause preferred anisotropic permeability patterns that influence the flow of fluids in the horizontal direction and are important even at small scale. Results from many permeability studies of sedimentary rocks indicated ratios of $k_h : k_v$ are typically from 1.5 : 1.0 up to 3 : 1 (Oelkers, 1996). Anisotropy in hydraulic conductivity measured in 11 pumping tests in sands and gravels determined that seven were in the range of $K_h : K_v$ between 2 : 1 and 10 : 1, with four in the range 10 : 1 to 50 : 1 (Bair and Lahm, 1996).

In some cases, vertical permeability or hydraulic conductivity values can exceed horizontal values because of **secondary permeability** features, i.e., those occurring after deposition of the formation. Such features would include **macropores** formed by weathering or bioturbation, vertical fractures in clays, tills and sedimentary rocks. Figure 11.1 shows vertical fracturing in a till that permits a relatively rapid (for these tight sediments) infiltration rate of $\sim$3 m/yr (Grisak and Cherry, 1975; Grisak et al., 1976) (see section 8.4.1).

11.1.4 Fluid Potential

Darcy's experiment inspired Hubbert (1940) to determine the fundamental variable dictating the direction of fluid flow in porous media. We have shown that neither elevation nor water pressure by themselves explain the direction of fluid flow in Figure 1.15 and therefore in the field. Rather, it is the gradient in the hydraulic head, which may be measured at every point in a flow system, such as Darcy's sand column, and which is expressed in equation (1.16). Hubbert showed

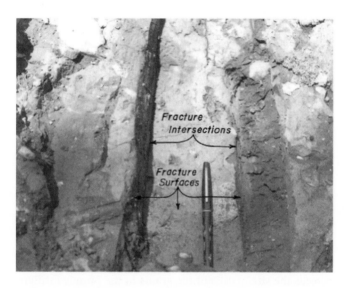

Figure 11.1 Heterogeneities and discontinuities dominate infiltration and other hydrogeological processes. Fractured till, Manitoba, Canada. The fractures shown allow rapid infiltration of rain and snowmelt. Pen for scale. (From Grisak et al. (1976); this content has been reproduced with the permission of the Royal Society of Canada.)

that the hydraulic head is an expression of the total potential energy of the fluid per unit mass (Φ) in units of m^2/s^2:

$$\Phi = gz + \frac{p}{\rho} + \frac{v^2}{2}, \qquad (11.4)$$

where Φ is defined as the **fluid potential**, p is the gauge pressure at a particular point in the flow system of elevation z, and v is the velocity at which the fluid is moving at that point. Hubbert (1940) noted that this is a generalization of Bernoulli's equation of 1738 relating the elevation, pressure and speed of a frictionless fluid along a flow line. Because groundwater and similar fluids experience friction in flow from a high elevation to a lower one in the flow system, the potential energy of the fluid must also decrease in the direction of fluid flow, as must the total hydraulic head (h_T) due to both elevation change and an amount representing the head loss due to friction.

The total head (h_T) of the flow system at any particular point is obtained by converting from units of energy per unit mass to energy per unit weight, i.e., dividing through by g. This yields h_T in units of length (m):

$$\frac{\Phi}{g} = z + \frac{p}{\rho g} + \frac{v^2}{2g} = h_T. \qquad (11.5)$$

The velocity term is inconsequential in groundwater flow systems, with the possible exception of karst rocks or open fractures with high flow rates. Therefore, the total head is effectively the sum of the elevation and pressure terms that we introduced as (11.1). The fluid potential, Φ, at any point in the flow system is simply the total head times the acceleration due to gravity:

$$\Phi = h_T \cdot g = gz + \frac{p}{\rho}. \qquad (11.6)$$

11.2 Compressibility, Storage and the Flow Equation

The steady-state flow of groundwater is represented by the classical flow-net analysis of seepage; the extension of this steady-state problem to regional groundwater flow is the subject of section 11.4. However, a general equation of flow through porous media must include non-steady-state or transient effects if we are to consider groundwater pumping or recharge. These are important in groundwater-resources evaluation as well as in understanding land subsidence (see Chapter 13).

11.2.1 Compressibility

In saturated porous media, the solid matrix of the soil or rock is compressed by the extraction of groundwater. Consider a volume of water (V_w) that is compressed by a change in pressure (dp) causing a change in volume ($-dV_w$). The compressibility of water (β) is

$$\beta = \frac{1}{E_w} = \frac{-dV_w/V_w}{dp} = 4.4 \times 10^{-10} \ (Pa^{-1}), \qquad (11.7)$$

where E_w is the bulk modulus of water. Note that the unit of compressibility is inverse pressure or m^2/N. A saturated porous medium of sand may be compressed through the compression of (a) the water, (b) the sand grains and (c) the porous medium through grain slippage and the rearrangement of the sand grains into a lower-porosity packing geometry.

The compressibility of the sand grains may reasonably be assumed to be even less than that of the water at the relatively small stresses we are considering (i.e., <2 MPa). Therefore, it is the compressibility of the granular skeleton of the porous medium that is important and it is the effective stress (σ_e) experienced by the skeleton that changes, i.e., from the change dh, and $d\sigma_e = -dp = -\rho \cdot g \cdot dh$.

We may now define the vertical compressibility of the porous medium as

$$\alpha = \frac{-dV_{pm}/V_{pm}}{d\sigma_e}, \qquad (11.8)$$

where V_{pm} is the volume of the porous medium and $-dV_{pm}$ is the change in this volume from the redistribution of the overburden load. Table 11.2 presents results from laboratory tests of soil and rock consolidation measured using an oedometer (see Figure 10.18 and section 10.3.3). Results may be presented in terms of effective stress versus strain or effective stress versus void ratio. Because we are interested in the change in hydraulic properties, we shall consider only the latter, thus

$$\alpha = \frac{-dL/L}{d\sigma_e} = \frac{-d\sigma_e/(1+e)}{d\sigma_e}, \qquad (11.9)$$

where e is the void ratio (equation (1.15)), L is the original thickness of the soil or rock sample placed in the oedometer and $-dL$ is the amount of consolidation due to the applied load on the test cell.

Therefore, at any location within our soil or rock, pumping will reduce the pore pressure and redistribute the overburden load partially to the skeleton of the porous medium. The soil or rock will compress and lose a fraction of its total porosity. Simultaneously, the groundwater will expand slightly with

BOX 11.1 | **MEASURING HYDRAULIC HEADS**

Equation (11.1) presents the basic definition of hydraulic head in cases in which the velocity term can be ignored, i.e., $h = p/\gamma + z$. The accurate and precise measurement of head is important in geotechnical and geoenvironmental engineering so that flow rates and directions and potential uplift on infrastructure may be reliably computed. The principal errors involved have been identified by Post and von Asmuth (2013) as follows:

- *Survey benchmark*: Head measurements must be referenced to a surveyed benchmark, usually identified on the rim of the casing of the monitoring well, so that all measurements are made to this reference location and elevation.
- *Errors in the measurement instrument*: Electrical probes are frequently used with calibrated water-level tapes that have in many cases replaced the earlier use of wetted chalk tapes. However, increasingly, pressure transducers are being used. Each method may produce measurement errors – stretch in the tape or instrument drift in pressure transducers. Consequently, a comparison of measurements is recommended.
- *Errors in correcting pressures to water levels*: The pressure head (p/γ) is dependent on the density of the groundwater in the monitoring well (see Box 11.2), which may vary with depth and not be that of fresh water.
- *Time lag effects*: A pore-pressure change in the groundwater adjacent to a monitoring-well screen will always require some time to equilibrate, particularly if the porous medium is of low permeability. Equilibration or response times are of the order of minutes for permeable media to months for tight formations.
- *Defects in the monitoring well*: When the piezometer or monitoring-well screen intersects formations with varying hydraulic conductivities, the measured water level will be a function of all formations intersected and may mask important information. If the bentonite well seal or the casing or casing couplings have cracked, the measured water level may reflect heads elsewhere along the well casing.

Table 11.2 Range in values for compressibility and specific storage. (From Domenico (1972).)

Material	α (Pa^{-1})	s_s (m^{-1})
Plastic clay	3×10^{-7} to 2×10^{-6}	2×10^{-4} to 2×10^{-3}
Stiff clay	1×10^{-7} to 3×10^{-7}	1×10^{-4} to 2×10^{-4}
Medium-hard clay	7×10^{-8} to 1×10^{-7}	9×10^{-5} to 1×10^{-4}
Loose sand	5×10^{-8} to 1×10^{-7}	5×10^{-5} to 9×10^{-5}
Dense sand	1×10^{-8} to 2×10^{-8}	1×10^{-5} to 2×10^{-5}
Dense sandy gravel	5×10^{-9} to 1×10^{-8}	5×10^{-6} to 9×10^{-6}
Jointed rock	3×10^{-10} to 7×10^{-9}	3×10^{-7} to 6×10^{-6}
Solid rock	$< 3 \times 10^{-10}$	$< 3 \times 10^{-7}$
Water (β)	4.4×10^{-10}	—

the reduction in pore pressure. This expansion of water and reduction in porosity releases water from storage. Conversely, if water is recharged to the soil or rock, the pore pressure must rise and the effective stress must be reduced. This discussion assumes that the water and the porous medium of soil or rock behave elastically; however, we shall see in the case of unconsolidated sediments in Chapter 13 that the reduction in porosity may be inelastic and thus irreversible, leading to land subsidence.

11.2.2 Storage and Transmissivity

Our goal in this section remains to define an equation for transient groundwater flow, i.e., that which is varying over time due to pumping or seasonal cycles of recharge and discharge. Consequently, we must define two additional parameters: storativity and transmissivity. Idealizations explaining storage and transmissivity are shown in Figure 11.2 for confined and unconfined aquifers. Assuming fully saturated conditions, we may define the **specific storage** (s_s) of an aquifer as the volume of water that a unit volume of aquifer releases from storage under a unit decline in hydraulic head. (In the case of artificial recharge, it is the volume of water a unit volume of aquifer adds to storage under a unit rise in hydraulic head.) We

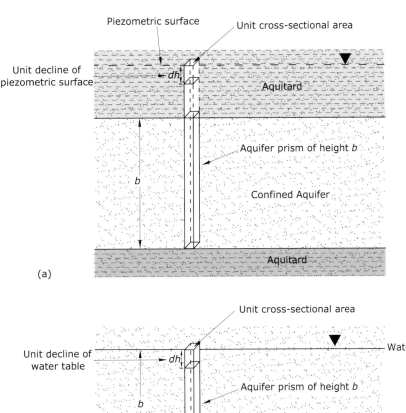

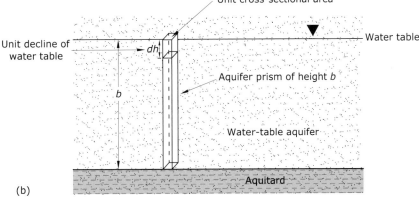

Figure 11.2 Definition schematic for transmissivity and storativity under (a) confined and (b) water-table conditions. (From Ferris et al. (1962); figure courtesy of the US Geological Survey.)

shall estimate the expressions that govern the production of water during pumping from, firstly, the expansion of the groundwater and, secondly, the compression of the aquifer.

The volume of groundwater produced by the expansion of water is given by (11.7): $dV_w = -\beta \cdot V_w \cdot dp$. Because a unit volume of porous medium means $V_{pm} = 1$, then the volume of groundwater in that unit volume must be $V_w = n \cdot V_{pm}$ and $dp = \rho \cdot g \cdot dh$, so that for a unit decline in head ($dh = -1$) we obtain

$$dV_w = \beta n \rho g. \tag{11.10}$$

The volume of groundwater produced by the compression of the porous medium is obtained from (11.9). Recall that $d\sigma_e = -\rho \cdot g \cdot dh$; therefore a unit decline in head means $dh = -1$ and

$$dV_w = \alpha \rho g. \tag{11.11}$$

The specific storage is the sum of these two terms, i.e.,

$$s_s = \rho g(\alpha + n\beta). \tag{11.12}$$

We have defined the specific storage to be the volume of groundwater that a unit volume of aquifer releases from storage because of expansion of the water and compression of the porous medium under a unit decline in average hydraulic head within the unit volume. It follows that the aquifer is acting elastically and that the specific storage is measured in units of inverse length, which a dimensional analysis of (11.12) will confirm. Table 11.2 lists values of s_s for soil and rock types.

Returning to Figure 11.2(a) we may define the **transmissivity** as the product of the hydraulic conductivity and the aquifer thickness (b),

$$T = K \cdot b, \tag{11.13}$$

Table 11.3 Specific yield of various soil materials. (From Johnson (1967).)

Material	Range (%)	Average (%)
Clay	0–5	2
Silt	3–19	8
Sandy clay	3–12	7
Fine sand	10–28	21
Medium sand	15–32	26
Coarse sand	20–35	27
Gravelly sand	20–35	25
Fine gravel	21–35	25
Medium gravel	13–26	23
Coarse gravel	12–26	22

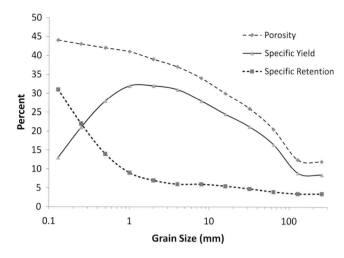

Figure 11.3 Relationship between porosity, specific yield and specific retention of sediments from South Coastal Basin, California. (After Johnson (1967); figure courtesy of the US Geological Survey.)

and the **storativity** of the aquifer as

$$S = s_s \cdot b. \qquad (11.14)$$

These two terms, T and S, are of particular importance in the hydraulic testing of boreholes and wells, which are typically treated as a two-dimensional system with a uniform aquifer thickness. In three-dimensional flow systems, they lack meaning and the equations are formulated in terms of K and s_s (and porosity in the case of contaminant transport). The transmissivity obviously has units of $(m/s) \cdot m$ (i.e., m^2/s), while storativity is dimensionless (i.e., $m^{-1} \cdot m$). Freeze and Cherry (1979) note that a productive aquifer will exhibit $T > 0.015\,m^2/s$, which, in traditional US engineering units, is equivalent to 100 000 US gallons per day per foot of drawdown over a one-foot width of the aquifer. An inspection of Table 11.2 indicates that values of storativity vary over the range 0.005 for clays to 0.00001 for a few metres of jointed or fractured rock, depending upon the compressibility of the material.

For unconfined aquifers (see Figure 11.2b) the transmissivity is poorly defined because pumping will change the saturated aquifer thickness. When such aquifers are dewatered to allow foundation work or to develop a groundwater supply, the relevant storage term is

$$S = s_s b + S_y, \qquad (11.15)$$

where S_y is known as the **specific yield** (see Figure 11.3). While specific storage may be of the order of $10^{-5}\,m^{-1}$ for a water-table aquifer, the specific yield may be 10–30%; thus water-table drainage dominates the storage term in

unconfined systems. Table 11.3 presents measurements of the specific yield of soil materials that were mainly measured in Californian groundwater basins by the US Geological Survey (Johnson, 1967). The difference between the specific yield of an unconfined aquifer and its saturated porosity is related to the capillary forces that might retain groundwater.

Because capillarity is an inverse function of pore-throat radius, the **specific retention**, S_r, of porous media increases with a decrease in particle size (see Figure 11.3). As dewatering proceeds, the water table drops and pore spaces drain but the rate of drainage is further delayed by the reduced hydraulic conductivity as flow occurs partly in the newly unsaturated zone. When drainage ceases, the specific yield and the specific retention are defined as

$$S_y + S_r = n. \qquad (11.16)$$

Note that S_y and S_r are water contents expressed as a fraction of the total soil or rock volume, whereas S and s_s are storage terms. In Chapter 12, the symbol, S_w, is used to express fluid saturations, i.e., fractions of the pore volume occupied by specific fluids, e.g., water or non-aqueous liquids, in which case the literature often uses S_r to indicate the residual saturation of the fluid in the pore.

11.2.3 The Groundwater Flow Equation

The derivation of the groundwater flow equation follows conservation of mass principles based upon the material balance for a control volume. Full developments of this equation can be read in the texts listed in the further reading at the end of this chapter. The general form of the transient groundwater flow equation is

$$\frac{\partial}{\partial x}\left(K_x\frac{\partial h}{\partial x}\right) + \frac{\partial}{\partial y}\left(K_y\frac{\partial h}{\partial y}\right) + \frac{\partial}{\partial z}\left(K_z\frac{\partial h}{\partial z}\right) = s_s\frac{\partial h}{\partial t}, \quad (11.17)$$

where K_x, K_y and K_z are the hydraulic conductivity values in the three dimensions of the control volume and $h(x, y, z)$ is the hydraulic head at any point in the porous medium represented by the control volume. Remember that the hydraulic conductivity is a function of the properties of both the porous medium and the water, e.g., $K_x = (k_x\rho_w g)/\mu$; therefore we assume that ρ_w does not vary throughout the control volume so that we can simplify the form of the hydraulic conductivity tensor to just these three components. Equation (11.17) is known as a diffusion equation in applied mathematics, hence T/S (or K/s_s) is defined as the **hydraulic diffusivity**.

Similar practical needs led to the employment of this equation in groundwater flow models, such as the USGS' MODFLOW simulator (Harbaugh, 2005). Equation (11.17) can be rewritten for a two-dimensional aquifer when we assume that T and S do not vary significantly with depth or, at least, can be reasonably approximated by depth-averaged values. Therefore, for the horizontal, x, and vertical, y, dimensions, we may rewrite the groundwater flow equation as

$$\frac{\partial^2 h}{\partial x^2} + \frac{\partial^2 h}{\partial y^2} = \frac{S}{T}\frac{\partial h}{\partial t}. \quad (11.18)$$

This equation is the basis for groundwater flow simulations and for hydraulic testing of wells in confined aquifers. To account for pumping or recharge, we add an additional term, $W(x, y, z, t)$, to the equation that represents a source (W negative) or sink (W positive) of water to the aquifer to complete the material balance:

$$\frac{\partial^2 h}{\partial x^2} + \frac{\partial^2 h}{\partial y^2} - W = \frac{S}{T}\frac{\partial h}{\partial t}. \quad (11.19)$$

11.3 Wells and Hydraulic Testing

Field hydraulic testing is the most reliable method for obtaining large-scale estimates of hydraulic parameters such as K, T and S. Hydraulic testing requires wells either for producing groundwater in aquifer tests or for monitoring water-level drawdown in adjacent monitoring wells that is the result of groundwater extraction. In geotechnical and geoenvironmental engineering, wells are needed for dewatering excavations, mines and tunnels during construction or operation, groundwater supply or contaminant plume capture, moni-

toring hydraulic heads beneath dams or plumes of dissolved contaminants, and injection of treated wastewater.

11.3.1 Wells

The basic elements of any such well are shown in Figure 11.4, which is a small-diameter (50 mm) monitoring well or piezometer designed to sample dissolved contaminants. Groundwater flows in through the well screen, the size of perforations of which are chosen to filter out most of the particles that can enter the well from the adjacent formation. Submersible pumps can lift the groundwater to the surface through a tubing string and an inflatable packer could be used to seal off the well screen to exclude standing water in the casing. The well screen, in this case, is built of stainless steel that is inert with respect to most contaminants and corrosion-resistant. Production wells, for high-rate extraction of groundwater, necessarily are of larger diameter with longer well screens, as in Figure 11.5.

While deep dewatering operations employ drilled wells, shallow dewatering of sandy soils is typically conducted using self-jetting well points installed by hydraulically driving a well point into the sand with a downward-directed jet of water that

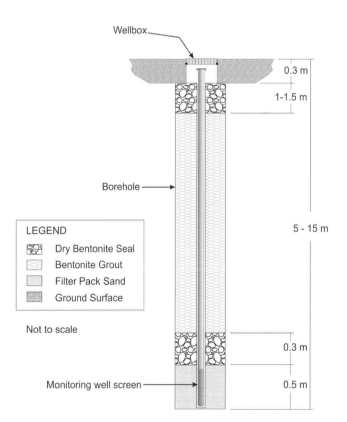

Figure 11.4 Schematic showing details of the construction of a monitoring well or piezometer.

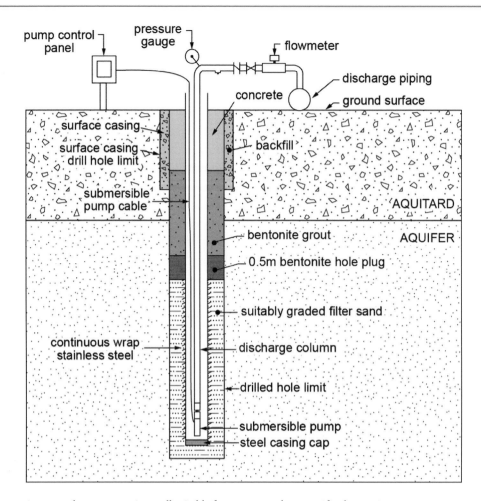

Figure 11.5 High-capacity groundwater extraction well suitable for water-supply or aquifer dewatering.

is contained within the well point. A typical site dewatering operation might contain tens to hundreds of low-yielding well points connected to header pipes that in turn are connected to a vacuum-assisted pump that can extract groundwater up to 7.5 m below ground surface at 100 L/s.

11.3.2 Hydraulic Testing

Slug tests are a widely used and inexpensive method of testing the local hydraulic conductivity adjacent to a well screen. They are "the everyday workhorses of site investigations associated with groundwater contamination … and ground engineering" (Black, 2010). Spreadsheet analysis software is available from the Kansas Geological Survey (see Butler et al., 2003) and the US Geological Survey (Halford and Kuniansky, 2002) that allows the method to be applied in both confined and water-table aquifers. Black (2010) identifies means by which the quality of slug tests can be optimized.

The general features of the slug test involve the near-instantaneous change in head of a well by the injection of a slug of known volume (either a solid or a gas) and the subsequent recovery of the initial head. Figure 11.6 presents the elements of the pneumatic slug-test method for formations of high hydraulic conductivity (Butler et al., 2003). The slug is injected as a nitrogen gas pulse with the initial water-level change estimated using the air-pressure transducer at the wellhead. The downhole pressure transducer records the response of the well–aquifer system to the pulse, which is stored by the data logger. The response and elapsed time data are analyzed in the spreadsheet (Butler et al., 2003) such that the head change is normalized to the total head change.

Slug tests and the testing done on fractured rocks described in section 5.5 provide only local estimates of hydraulic conductivity around any well screens. Consequently, if larger-scale estimates are needed for long-term groundwater capture, it is necessary to undertake a "pumping" or "aquifer test". Generally two hydraulic responses are observed, i.e., those of a **confined aquifer** and those of an unconfined or **water-table aquifer** as shown in Figure 11.7. The confining unit is typically a zone of silts and/or clays or mudrocks that

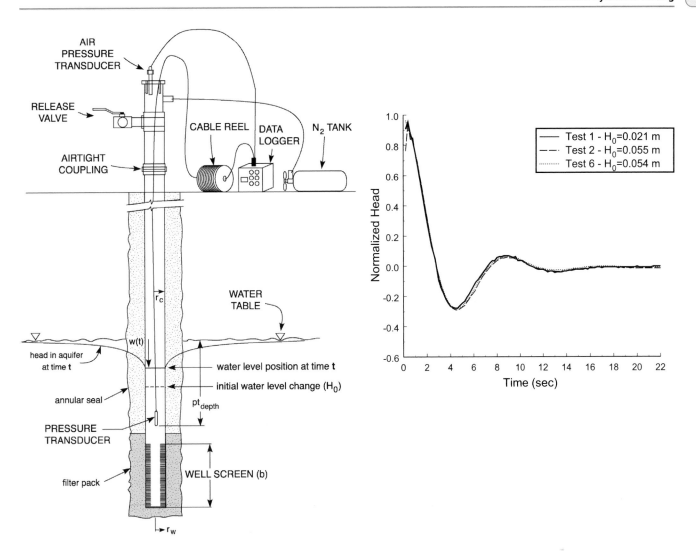

Figure 11.6 Slug test instrumentation (left) employing a pressure transducer and data logger installed in a monitoring well that partially penetrates a highly permeable aquifer. The hydraulic response (right) results in water-level oscillation from inertial effects. This response is normalized and analyzed by spreadsheet software. (From Butler et al. (2003); reproduced with the permission of John Wiley and Sons.)

constitute an aquitard and which offer some protection of the aquifer below to contamination from the ground surface. A water-table aquifer is a common feature of Quaternary sedimentary deposits, particularly those found in floodplains. A confined aquifer might be a buried-valley aquifer (Figure 8.14) or a fractured sandstone or carbonate formation. As pumping of an extraction well progresses, a cone of depression of hydraulic head spreads from the well, known as a **drawdown cone**, with non-steady radial flow occurring to the pumping well because of continuing releases from storage from overlying aquitards and drainage of water-table aquifers.

Aquifer testing involves the careful construction and development of extraction and monitoring wells, the creation of a hydrogeological model of the aquifer–aquitard system, the execution of the hydraulic test and finally the analysis of the hydraulic-head drawdown data based upon mathematical

models of the system, now conducted with the aid of software packages. Much of the history of hydrogeology in the twentieth century is associated with the evolution of aquifer-testing techniques, beginning in 1906 with the Thiem equation describing the steady radial flow of groundwater to an extraction well completed in a confined aquifer (equation (5.9)). In the 1930s, C.V. Theis of the US Geological Survey developed a mathematical solution to the typical case of the transient (i.e., non-steady-state) response of the piezometric surface to pumping based on a heat-flow analogy.

The Theis method assumes that the extraction well fully penetrates a confined aquifer, as is shown in Figure 11.7. We further assume that the aquifer is homogeneous, of constant thickness and infinite in areal extent, requirements that can be overcome in numerical flow models but not in our idealized world of exact analytical solutions. If we pump the well at

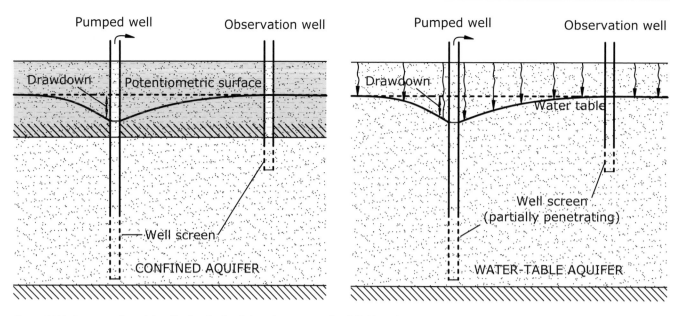

Figure 11.7 Conceptual models of hydraulic-head drawdown in confined (left) and water-table (right) aquifers. (Modified from Barlow and Moench (1999); figure courtesy of the US Geological Survey.)

a constant rate of Q, a nearby observation well will detect a drawdown in the head that extends outwards from the extraction well over time. Because the Theis equation is for non-steady, radial flow in a confined aquifer, we can proceed to analyze the cone of drawdown as it extends outwards. The rate of drawdown decreases with distance from the pumped well as an ever larger zone of the confined aquifer is captured by the drawdown cone. The polar-coordinate form of the transient flow equation (11.18) for radial flow in a confined aquifer is

$$\frac{\partial^2 h}{\partial r^2} + \frac{1}{r}\frac{\partial h}{\partial r} = \frac{S}{T}\frac{\partial h}{\partial t}. \tag{11.20}$$

The Theis solution yields the transmissivity as a function of flow rate (Q), storativity as a function of T, and the drawdown (dh) at an observation well a radial distance (r) from the pumped well as

$$dh = \frac{Q}{4\pi T}\int_u^\infty \frac{e^{-u}\,du}{u} = \frac{Q}{4\pi T}W(u), \tag{11.21}$$

where $W(u)$ is known as the well function of u, where $u = r^2 S/(4Tt)$. Hence, T and S are defined as

$$T = \frac{Q}{4\pi(dh)}W(u) \tag{11.22}$$

and

$$S = \frac{4Tu}{r^2/t}. \tag{11.23}$$

Theis found that T and S could be determined by superimposing a plot of drawdown versus elapsed time and finding a match point that would yield $W(u)$ and $1/u$ on the "Theis curve" and T and S on the drawdown curve. This curve-matching process is now replaced by software computations.

The drawdown curves shown in Figure 11.8 were prepared for an observation well sited 3.2 m from a fully penetrating extraction well in hypothetical confined and water-table aquifers. The drawdown from pumping a water-table aquifer is less than in a similar but confined aquifer. This reduced drawdown is due to the delayed drainage from an expanded capillary fringe created by pumping-induced drawdown (Bevan et al., 2005). However, the early-time elastic response of the water-table aquifer is identical to that of the confined aquifer and can be used to estimate the hydraulic properties as if it were a confined aquifer. The later departure from the Theis curve is from the delayed drainage that stabilizes the drawdown before the water-table response once again begins a Theis pattern after 3000 s. If the aquifer test is run for a sufficiently long time, the water-table aquifer response will eventually resume the Theis curve but offset as shown in Figure 11.8. Figure 11.9 illustrates the operation of an aquifer test using an extraction well such as that shown in Figure 11.5.

11.4 Groundwater Flow Systems

Groundwater flow nets are steady-state representations of (11.18), i.e., those obtained when $\partial h/\partial t = 0$. The simplest of all flow nets is the water table, which is the surface of saturated soil or rock where the pore pressure is atmospheric

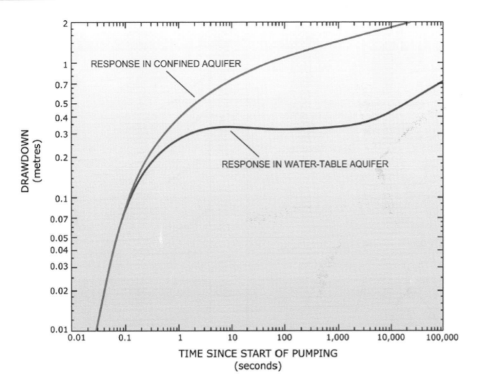

Figure 11.8 Drawdown curves prepared by Barlow and Moench (1999) for an observation well that is sited 3.2 m from a fully penetrating extraction well in hypothetical confined and water-table aquifers. These illustrate the initial confined-aquifer response of the water-table curve and the subsequent effect of delayed yield in flattening the response prior to further drawdown. (Figure courtesy of the US Geological Survey.)

Figure 11.9 Running an aquifer test to determine the hydraulic conductivity of a buried-valley aquifer, San Diego, California. The shut-off valve to a submersible pump, which has been lowered downhole, appears above the protective casing of the middle of three wells. The extracted groundwater is pumped to storage in the Baker tank in the background. The nearest casing protects a monitoring well with the cable to an electric water-level probe shown. This monitoring well is screened at the same depth as the pumping well. (Courtesy Dave Schug, AECOM, San Diego, California.)

(i.e., zero gauge pressure) and the hydraulic head is therefore equal to the elevation head. Examples of regional water-table flow patterns are shown in Figure 11.10. A local perched water table may exist above a regional water table if a confining layer prevents drainage; however, such perched water tables are not shown in this figure. The water table is usually a subdued replica of the land surface topography.

While well hydraulics identifies average properties of aquifers and aquitards, the numerical simulation of groundwater flow patterns focuses on understanding how the hydrogeological properties and architectural arrangement of various soil and rock types dictate hydraulic head patterns throughout the flow system. Although regional groundwater flow systems are most often studied as steady-state phenomena, we will also consider transient effects, because we wish to consider how groundwater flow patterns change with recharge, which is of direct concern in the stability of slopes.

11.4.1 Idealized Flow Patterns in Groundwater Basins

In the 1960s, Tóth, Freeze and Meyboom began using cross-sectional flow nets on a regional scale to identify flow patterns

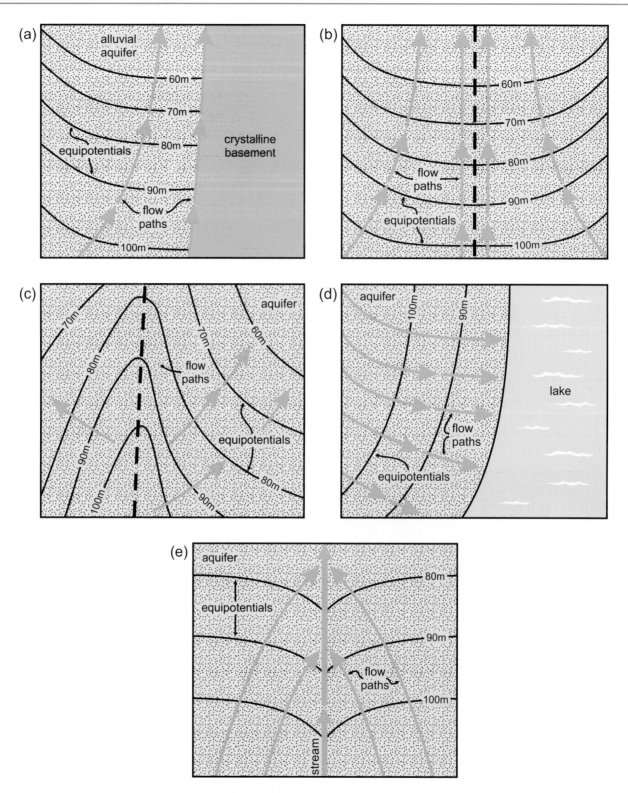

Figure 11.10 The water table is the regional piezometric surface at atmospheric pressure. This figure shows various hydraulic-head boundaries identifying the associated equipotential lines and groundwater flow lines. (a) The effect of a no-flow boundary created by the presence of crystalline basement rock of very low hydraulic conductivity adjacent to an alluvial aquifer. (b) The effect of a symmetry boundary on flow lines. (c) The effect of a linear trend of high hydraulic heads on the flow pattern. Because the water table typically follows topographic contours, it is likely that this ridge reflects such topography. (d) Flow pattern towards a lake or other body of (approximately) constant head. (e) A constant-head boundary represented by a stream into which the groundwater is discharging. (Reproduced from Rivera, 2014, Groundwater basics. In A. Rivera (ed.), *Canada's Groundwater Resources*, p. 45. Markham, ON: Fitzhenry & Whiteside Limited).

within groundwater flow systems on the Canadian prairies and understand how topography and hydraulic conductivity contrasts affect patterns of recharge and discharge (Freeze and Cherry, 1979). Tóth extended the work of Hubbert (1940) by recognizing that, for steady-state conditions, (11.18) is the Laplace equation, for which solutions were well established and could be adapted for flow in hypothetical groundwater basins. Such patterns of fluid flow and fluid dynamic behaviour in deep basins are important for mining, hydrocarbon extraction, and waste disposal.

Figure 11.11(a) illustrates the general patterns of hydraulic head variation in Tóth's unit basin of homogeneous, isotropic hydraulic conductivity with impermeable sides and base. The water table slopes upward to the right from the point of discharge at the extreme left of the ground surface, whose slope is c'. Piezometers (Figure 11.4) display the measured head from which the pore pressure is obtained. Therefore, the pattern of hydraulic head in this idealized basin shows a decrease with depth in the recharge area (i.e., piezometer nest 2), equal heads with depth in the midline area (piezometer nest 1) and an increase in head with depth in the discharge area (piezometer nest 3).

Figure 11.11(b) shows the pressure–depth $p(d)$ variation in which the static pressure head equals the depth below the water table. A $p(d)$ diagram for the midline area has the normal hydrostatic pressure–depth slope for fresh water (i.e., 9.8 kPa/m). In the recharge area, because of downward flow, the $p(d)$ slope must be slightly less than hydrostatic, e.g., slope 2 versus slope 1 in Figure 11.11(b), while the $p(d)$ slope in the discharge area exceeds the hydrostatic slope, e.g., slope 3 versus slope 1. The recharge and discharge areas can be identified on the basis of their $p(d)$ slopes – shown in Figure 11.11(b) as $\tan \alpha$, i.e., the vertical pressure gradient.

Tóth (2009) defined the dynamic pressure increment, Δp, as the pore-pressure difference between measured and hydrostatic gradients. Figure 11.11(b) illustrates that this must be zero for the piezometer nest 2 and be positive in the discharge area and negative in the recharge area. Tóth refers to the "supernormal" and "subnormal" pressures as defining the discharge and recharge areas, respectively (see Figure 11.11c), while the oil industry refers to them as "overpressured" and "underpressured" formations (see section 4.2.4).

Important examples of subnormal-pressured basins are the Denver Basin of northeastern Colorado and Nebraska (Belitz and Bredehoeft, 1988) and the Palo Duro Basin in west Texas (Senger and Fogg, 1987), whereas the Michigan Basin displays a supernormal pressure profile in central Michigan (Bahr et al., 1994) and southern Ontario (Raven et al., 1992), with

fluid densities of the order of 1150 kg/m^3, which is considered to be brackish water (i.e., total dissolved solids, TDS = 1000–10 000 mg/L). Box 11.2 explains how hydraulic heads are expressed when the groundwater in a borehole varies in density with depth, which is typical of large sedimentary basins.

11.4.2 Shallow Flow Systems

Most frequently, geotechnical and geoenvironmental engineers work in the environment of shallow groundwater flow systems that contain both saturated and unsaturated zones. Infiltration of rainfall and snowmelt occurs through the unsaturated or **vadose** zone; the transient effects of recharge are often associated with slope failures.

In the unsaturated zone, the pressure head of equation (11.1) is negative, i.e., it displays a soil-water tension or suction denoted by ψ. If $\psi > 0$, the soil is saturated and the pore pressure is $p = \rho g \psi$. Therefore, we will differentiate between heads above atmospheric pressure as h and those below by ψ, while remembering that this is really a continuum of pressures as shown in Figure 11.12. This figure shows the functional relationships between ψ and the hydraulic conductivity, K, and fractional moisture content, θ, for a hypothetical unsaturated soil of porosity n. The fractional water saturation of the pore volume is therefore $S_w = \theta/n$.

We refer to the differing paths of wetting and drying as **soil-water hysteresis**, which has its origins in the geometry of pores, entrapped air and other pore-scale effects. Note that the hydraulic conductivity is a function of the moisture content; consequently the term **saturated hydraulic conductivity**, K_0 in Figure 11.12, refers to the state of saturation when $\psi \geq 0$ and $\theta = n$. The surface $\psi = 0$, where the gauge pressure is atmospheric, is the surface of the water table. Because of capillarity, pores immediately above the water table are also saturated but under tension or negative pressure; this is the **capillary fringe**.

Figure 11.13 shows the vertical cross-section of a shallow groundwater flow system of 3 km in length and 110 m in depth with a steady-state flow pattern. The subsurface boundary ABCDEF is known as a no-flow boundary because all streamlines that develop within the flow system are contained within the cross-section and none flows across this boundary. The equipotential lines are contours of hydraulic head. The upper boundary, AHGF is a specified head boundary along which ψ is fixed from -1500 cm ($\theta = 0.02$) at point H to 0 ($\theta = 0.30$) at points A and G, where the groundwater

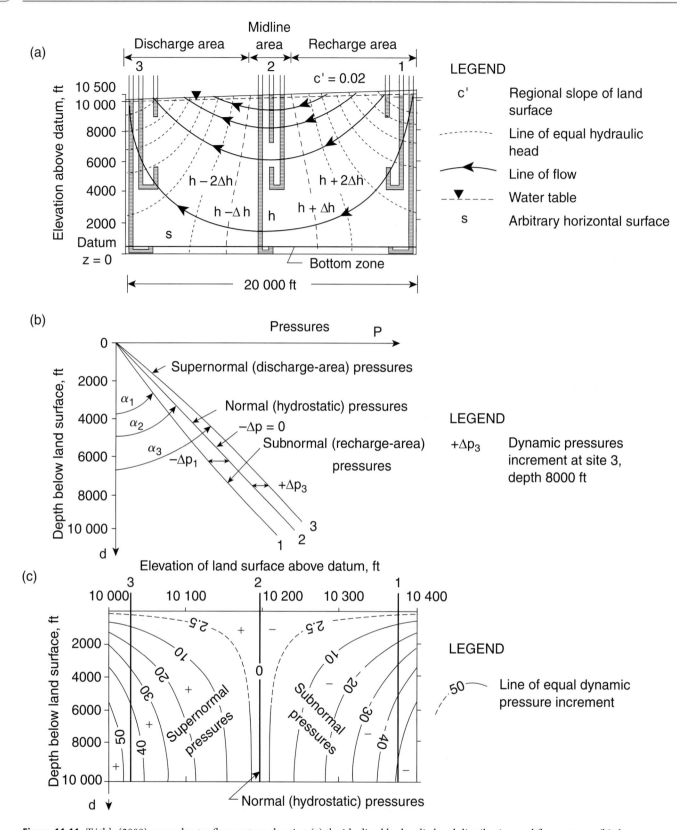

Figure 11.11 Tóth's (2009) groundwater flow system showing (a) the idealized hydraulic head distribution and flow pattern, (b) the pore pressure profile, $p(d)$, and (c) the distribution of head with depth for three pressure profiles, where + signs indicate upward flow in discharge areas and − signs indicate downward flow, i.e., recharge areas. (Reproduced with the permission of Cambridge University Press.)

BOX 11.2 | EXPRESSING HYDRAULIC HEADS IN FORMATIONS WITH VARIABLE TDS

During long residence times in deep basins, groundwater acquires dissolved mineral solids. These dissolved solids (**solutes**) are typically measured as total dissolved solids (TDS) and contribute to the increasing density (or specific weight) of the groundwater, e.g., the very old waters of the Michigan Basin that have fluid densities much greater than that of fresh water, which has a fluid density of $1000 \, kg/m^3$. We define waters on the basis of TDS as follows (Freeze and Cherry, 1979, p. 84):

- fresh water, $<1000 \, mg/L$;
- brackish water, $1000–10\,000 \, mg/L$;
- saline water, $10\,000–100\,000 \, mg/L$ (note that the TDS of seawater is $\sim35\,000 \, mg/L$); and
- brine, $>100\,000 \, mg/L$.

The simplest way of presenting hydraulic heads in formations with elevated TDS is to calculate the head from the *in situ* measured formation pressure and treat the pressure as if it were due to fresh water. This is known as the equivalent fresh-water head (h_f),

$$h_f = z + \frac{p}{\rho_f \cdot g}, \tag{11.24}$$

where z is the elevation of the measuring point, such as a piezometer screen, p is the formation pressure corrected for atmospheric pressure (i.e., gauge pressure), ρ_f is the density of fresh water and g is the acceleration due to gravity (note that $\rho_f \cdot g$ is the unit weight of water). Such measurement typically requires the use of downhole packer systems to isolate a test section of the borehole and to allow the pressure to be recorded in the isolated section. If the rock is of low permeability, this may require a very considerable period of time. Low-permeability rocks may also make it very difficult to collect a sample of the groundwater in order to measure TDS and fluid density.

However, this method assumes that the formation pressures measured are attributable to a single fluid density. With deep boreholes into sedimentary basins or along sea coasts, where seawater intrusion is possible, this is quite unlikely. Thus the fresh-water head needs to be corrected by considering the variation in density with depth. This yields the environmental-water head (h_e),

$$h_e = h_f - \left(\frac{\rho_f - \rho_{avg}}{\rho_f} \right) (z - z_r), \tag{11.25}$$

where ρ_{avg} is the average density of water between points z and z_r, in which z is as previously defined and z_r is a reference point above which the groundwater is fresh and below which it is of variable density. We define ρ_{avg} by

$$\rho_{avg} = \frac{1}{(z_r - z)} \int_z^{z_r} \rho \, dz. \tag{11.26}$$

The integral may be evaluated by simple numerical integration based upon a measured or even hypothetical fluid-density profile.

Sources: De Weist (1965, pp. 304–315) presents the theory, while Jorgensen et al. (1982) and Raven et al. (1990) present applications.

flow system discharges and streams are considered to flow perpendicularly to the figure. The discharge area identified at point G is known as a local discharge area, while that at A is the regional discharge area. Note that, below the upland recharge areas, the vertical hydraulic head is downwards while in the regional discharge area it is upwards.

Freeze (1972) assumed that the depth of water-table fluctuations would be small relative to the saturated thickness of the flow system and that the configuration of the water table would be stable over time. Water tables do fluctuate significantly in recharge areas because heads are always downward and droughts can cause a significant decrease in recharge.

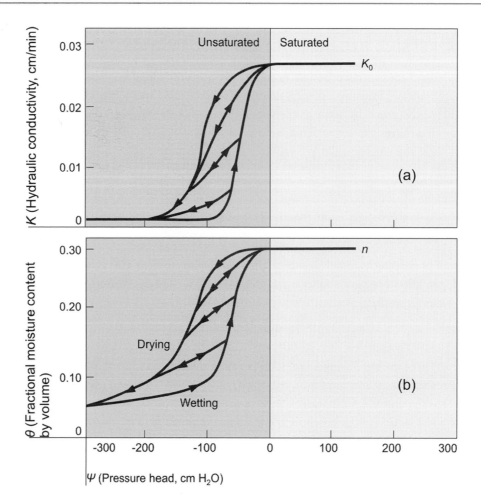

Figure 11.12 Functional relationships between pressure head (ψ) and (a) hydraulic conductivity and (b) fractional moisture (water) content (θ) for a hypothetical unsaturated soil. (After Freeze (1972); courtesy of International Business Machines Corporation, © 1972 International Business Machines Corporation.)

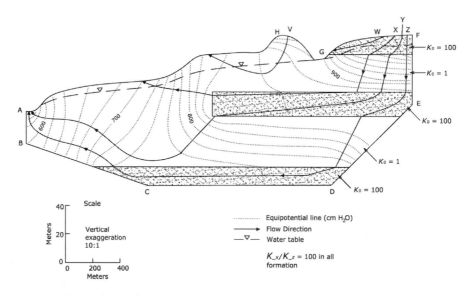

Figure 11.13 Steady-state regional groundwater flow system at a 10:1 vertical exaggeration with the orthogonality relation between equipotential lines and flow lines preserved. Note that very different flow paths are traced from closely neighbouring recharge locations. (After Freeze (1972); courtesy of International Business Machines Corporation, © 1972 International Business Machines Corporation.)

The fluctuation in any discharge area is much less because **baseflow** discharge to streams and evapotranspiration limit the water-table rise in the valley bottom and upward flow maintains the water table.

The **hydrostratigraphy** of the flow system in Figure 11.13 comprises a high-permeability facies with $K_0 = 100$ units within a flow system composed of a less permeable facies with $K_0 = 1$ unit. Both facies have wetting and drying soil moisture curves as shown in Figure 11.12 and horizontal-to-vertical anisotropy of $K_x : K_z = 100 : 1$. By plotting the flow system at a $10 : 1$ vertical exaggeration, Freeze maintained the orthogonal relationship between equipotential lines and streamlines.

The streamlines in the flow system are strongly affected by the location and geometry of the high-permeability units. These aquifers channel flow towards the discharge area, while the relatively impermeable zones act as aquitards, creating confined aquifers and large vertical hydraulic gradients. The results of Freeze's simulations – shown in Table 11.4 – illustrate how travel times vary within the flow system depending upon the location of recharge identified in Figure 11.13.

Table 11.4 Travel times, entry and discharge rates from simulation for various locations in the recharge area as shown in Figure 11.13. (From Freeze (1972).)

Parameter	V	W	X	Y	Z
Saturated hydraulic conductivity at surface (cm/day)					
K_x	50	5000	5000	5000	5000
K_z	0.5	50	50	50	50
Depth to water table (m)	20	12	12	12	12
Length of flow path (m)	540	360	1900	2800	2900
Travel time (yr)	330	1.02	123	660	226
Entry rate in recharge area (cm/day)	0.06	58	30	10	2.0
Discharge rate at exit point (cm/day)	0.75	78	1.5	3.5	5.0

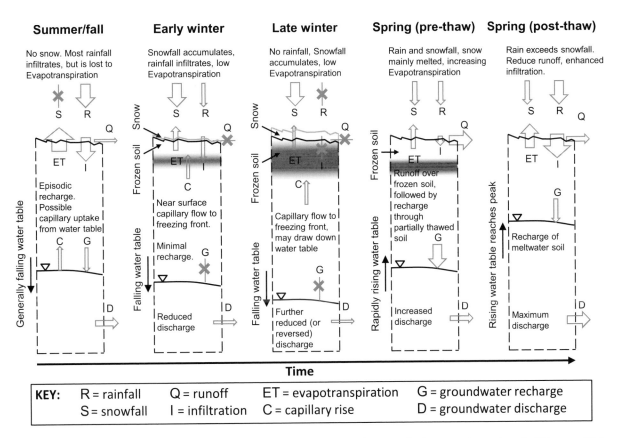

KEY: R = rainfall   Q = runoff   ET = evapotranspiration   G = groundwater recharge
S = snowfall   I = infiltration   C = capillary rise   D = groundwater discharge

Figure 11.14 Conceptual model of groundwater recharge occurring in a semi-arid, seasonally frozen region such as the US upper Midwest and the Canadian Prairies, with macropore-rich soils. The width of the arrows indicated the relative magnitude of the various fluxes. (From Ireson et al. (2013); reprinted by permission from Springer, copyright 2013.)

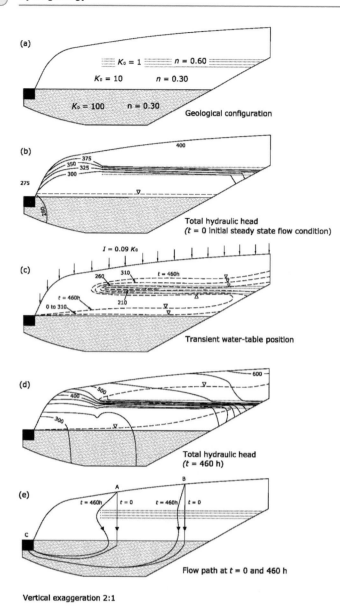

(a)

$K_0 = 1$ $n = 0.60$

$K_0 = 10$ $n = 0.30$

$K_0 = 100$ $n = 0.30$

Geological configuration

(b)

400

375
350
325
300
275

280

Total hydraulic head
(t = 0 initial steady state flow condition)

(c)

$I = 0.09\,K_0$

260   310   t = 460h

t = 460h
0 to 310   210

Transient water-table position

(d)

600

500
400

300

Total hydraulic head
(t = 460 h)

(e)

A   B

t = 460h   t = 0   t = 460h   t = 0

C

Flow path at t = 0 and 460 h

Vertical exaggeration 2:1

Figure 11.15 The transient development of a perched regional groundwater flow system showing the time-dependent shift in location of specific flow lines following a prolonged recharge event. (From Freeze (1972); courtesy of International Business Machines Corporation, © 1972 International Business Machines Corporation.)

11.4.3 Transient-State Flow Systems

In the case of significant seasonal variations in the configuration of the water table, it will be necessary to consider a transient flow system analysis. This is typically the case when slope stability is threatened by a pronounced change in precipitation due to seasonal rains or snowmelt or tropical storms. Figure 11.14 illustrates the variation in the water table in semi-arid, seasonally frozen regions with soils containing macropores that allow rapid recharge. The first and last

schematics capture water-table fluctuations in regions that are not subject to freezing.

Figure 11.15 illustrates the development of a perched water table during the transient evolution of a flow pattern. The hydrostratigraphy, Figure 11.15(a), consists of a clay unit ($K_0 = 1$, $n = 0.60$) set within an unsaturated zone of higher saturated hydraulic conductivity. At time $t = 0$, the total hydraulic head, h, is shown in Figure 11.15(b). Recharge to the flow system occurs at an inflow rate of $I = 0.09 K_0$ in Figure 11.15(c), forming a perched water table above the clay layer and an increase in the regional water table after 310 hours. At 460 hours, the total hydraulic head indicates a perched water table that, were the clay extended to the valley slope, would produce a significant slope stability hazard; we shall return to this topic in Chapter 15. Figure 11.15(d) indicates that flow is draining the perched water table at its toe. Figure 11.15(e) illustrates how the development of the transient flow system affects the recharge path for water entering the flow system at two different locations in the recharge area. Note that it is both the topography and the contrasts in hydraulic conductivity that govern the flow pattern.

11.5 Aquifers and Aquitards

Aquifers and aquitards are the basic elements in any hydrostratigraphic model of a groundwater flow system. Any aquifer is a potential source of water supply or a hazard in foundation engineering that must be acknowledged. Groundwater flow through aquifers occurs by (i) intergranular flow by advection, (ii) flow through fractures or solution channels or (iii) some combination of both. Aquitards may act to protect underlying aquifers from surface contamination or may host landfills.

First, we consider granular, alluvial aquifer systems deposited by fluvial processes. Figure 1.6 shows alluvial gravels that constitute a very permeable aquifer irrespective of the fact that the deposit in the photograph is dry. That is, an aquifer is not necessarily saturated with groundwater at any particular time; rather, the word implies that it is capable of transmitting and yielding significant quantities of water irrespective of its current state of saturation. Second, we consider bedrock aquifers, which by definition exhibit lithification of sediments, and which are equally important sources of groundwater, e.g., the Chalk of southern England, shown in Figure 2.27. Igneous and metamorphic rocks may be used as groundwater supplies in some areas, where fracture zones within them may impart significant hydraulic conductivity that is important in mine dewatering. Finally,

BOX 11.3 | EXCAVATING INTO A CONFINED AQUIFER CAN BE HAZARDOUS TO YOUR PROJECT

The relationship between grain size and specific yield (Figure 11.3) is important when draining a water-table aquifer at a construction site in that it allows an estimate of the volume of water that may need to be recovered. However, when excavations expose a confined aquifer under an artesian head, it may be discovered that excavation has created a long-term problem that could have been avoided if the flow system hosting the aquifer had been identified beforehand.

Artesian aquifers are confined aquifers with a hydraulic head that exceeds ground surface. If the confining layer is removed during excavation for some building or transportation-link foundation, soil heave and groundwater discharge will occur. The city of Richmond Hill, north of Toronto, Ontario, Canada, is situated on the southern slope of the Oak Ridges Moraine aquifer, which is confined by a low-permeability glacial till cap. Figure 11.16 shows the flow system created by recharge on the moraine; Figure 8.12 shows the moraine exposed in an aggregate pit nearby.

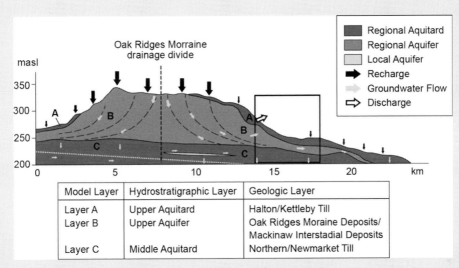

Figure 11.16 The groundwater flow system (top) in the Oak Ridges Moraine aquifer, Ontario, Canada (from Gerber, 1999). The box identifies the construction area around Richmond Hill, where the aquifer is under artesian head (from Di Biase et al., 2017). (Reproduced with permission of Rick Gerber and Steve Di Biase.)

In 1968, construction of an underpass beneath a railroad track caused the excavation of half ($\sim$5 m) of the till cap that confined the aquifer in the area shown in Figure 11.16. The excavation caused uncontrolled flow of soil and water. An initial drainage system was installed and an artificial groundwater-fed spring was established to lower the head in the excavation. This proved to be a temporary solution to a problem requiring a better-engineered dewatering system for the confined aquifer to remove groundwater that would otherwise compromise an expanded underpass beneath the railroad tracks.

In designing such a system, it was necessary to measure the hydraulic properties of the confined aquifer by hydraulic testing. Pumping tests at several nearby wells allowed engineers to estimate the transmissivity (T) and storativity (S) based on approximations to the Theis solution implicit in (11.21) and (11.22). This simplification, the Cooper–Jacob semi-approximate method of solution, is the basis of most pumping-test analyses and derives from the fact that u is small when r is small and t is large. This allows the transmissivity to be estimated by $T = 2.30Q/(4\pi\Delta(dh))$, where $\Delta(dh)$ is the drawdown difference per log cycle of time.

Transmissivity is estimated as shown in Figure 11.17(a), which is a distance–drawdown plot of a pumping test that shows the drawdown with respect to the radial distance (r) from the pumping well. Butler (1990) indicates that the Cooper–Jacob method provides meaningful estimates of the hydraulic properties of heterogeneous

BOX 11.3 (CONT.)

aquifers provided large-time data are employed. The 72-hour pumping test, which was conducted from a fully penetrating well at 8.7 L/s (see Figure 11.9), yielded a transmissivity of 165 m²/day and a storativity of 7×10^{-4} for the aquifer of 15 m thickness. These values allowed the design of the dewatering system to depressurize the aquifer by 10 m, as shown in Figure 11.17(b). Figure 11.5 illustrates the construction of the dewatering wells, which were pumped at rates between 4 and 10 L/s per well with overlapping cones of influence to ensure full dewatering.

(a)

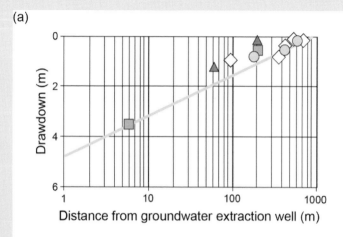

(b)

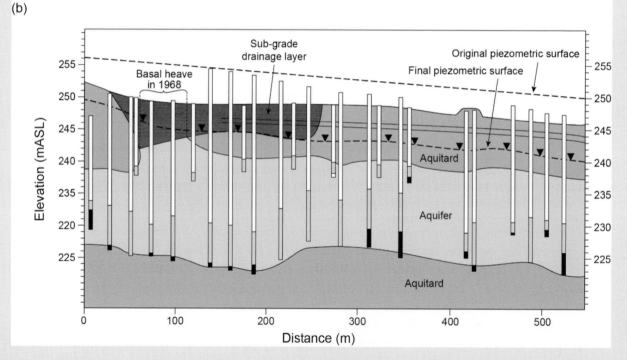

Figure 11.17 (a) Distance–drawdown plot of a pumping test in the Oak Ridges Moraine aquifer employing the semi-approximate Cooper–Jacobs method of analysis conducted. The symbols represent drawdowns (dh) at observation wells at various distances from the groundwater extraction well. The aquifer is approximately 15 m thick. (b) Layout of the aquifer dewatering system installed in the construction area of the Oak Ridges Moraine aquifer. A total of 40 wells were installed to lower the piezometric surface by 10 m. The excavated area of basal heave and groundwater discharge in 1968 is shown. The shallow monitoring wells were gauged daily and flow rates in all monitoring and dewatering wells were measured. (From Di Biase et al. (2017); reproduced with permission.)

Source: From Di Biase et al. (2017).

we consider the role of aquitards in hydrogeology that protect aquifers, influence groundwater flow patterns and provide attractive environments for hazardous waste storage.

11.5.1 Alluvial Aquifers

Figure 9.15 indicates how fluvial systems create different textural regimes that determine the productivity of alluvial aquifers. Figure 11.18 shows a cross-section through glaciofluvial sediments near Tacoma, Washington, that are resting on an interglacial silt. While not saturated with groundwater, it is nevertheless an aquifer because it can transmit groundwater and illustrates what alluvial aquifers look like in cross-section; frequently, many channels like this one are deposited adjacent to one another.

By far the greatest volumes of groundwater pumped in the USA are from alluvial aquifers (Johnston, 1997). The Central Valley aquifer system in California was pumped during 1961–77 at a rate of $Q = 466\,\mathrm{m^3/s}$, while the High Plains and Gulf Coast aquifers were pumped in the 1980s at a rate of $Q = 273\,\mathrm{m^3/s}$ and $Q = 382\,\mathrm{m^3/s}$, respectively. Quaternary alluvium is the largest groundwater source in Canada, but not in the UK, where bedrock aquifers are more important than are alluvial aquifers.

Figure 11.18 Glaciofluvial sand and gravel, Tacoma Harbor, Washington state. While this deposit is unsaturated, it illustrates the architecture of very permeable, semi-confined sand and gravel aquifers that are productive water-supply aquifers and vulnerable to contamination. The light-coloured Quaternary interglacial sediments at the top are remnant ash beds from 15–70 ka. Beneath them is the strongly oxidized glacial outwash gravel, Qpog or Quaternary pre-Olympia glacial, 70–780 ka. It overlies an oxidized silt that is partially eroded by the gravel channel. Geologic interpretation by Kathy Goetz Troost, University of Washington.

The Quaternary glaciations had a profound effect on the development of aquifers and aquitards throughout the world. Glacial erosion first created new channel networks and then meltwater flowed in huge volumes down the Pleistocene drainage network, often eroding new channels in the warm-weather months and depositing large volumes of sediment when meltwater receded. Advancing glaciers then deposited overlying till sheets. Today such buried-valley aquifers are vital groundwater resources, e.g., the Teays–Mahomet valley in the US Midwest and those beneath the Canadian Prairies (Figure 8.14).

The alluvial floodplain aquifers of the Mississippi, Missouri, Ohio and other rivers were also affected by the Quaternary glaciations, resulting in deep thicknesses of coarse alluvium overlain by overbank silts (Sharp, 1988). Both confined and water-table conditions occur with an average hydraulic conductivity of $K = 1 \times 10^{-3}$ m/s. Davis's (1988) review of alluvial valleys in the western USA includes the High Plains aquifer, which stretches from Texas to South Dakota, the intermontane basins of the Pacific coast and the alluvial valleys of the Basin and Range Province, e.g., Nevada and Arizona. These aquifers are often overdrawn by pumping due to the intense demands of population growth and irrigated agriculture and many have undergone subsidence (Chapter 13).

Prominent among alluvial aquifers are those associated with alluvial fans emerging from mountain fronts (see Figure 9.16). Neton et al. (1994) pointed out that these aquifers are very heterogeneous as a result of the wide range of structural, volcanic, tectonic and depositional processes that created such fans. Hydrofacies within these fans have mean hydraulic conductivity ranges of $K = 10^{-1.5}$ m/s for clast-supported gravels to $K = 10^{-4}$ m/s for sandy units to $K = 10^{-9}$ m/s for laminated fine-grained units.

11.5.2 Bedrock Aquifers

Many bedrock aquifers are vital water-supply aquifers, e.g., the Floridan aquifer, the Edwards–Trinity aquifer in Texas, and the carbonate and sandstone formations underlying much of England and the US Midwest (Johnston, 1997). For geotechnical engineers, these are probably not as important as alluvial aquifers because their greater depth removes them from infrastructure foundations.

Perhaps because they often displayed artesian heads at great distances from possible sources of recharge, bedrock aquifers have long been of great interest to hydrogeologists and engineers interested in water supply. The more famous bedrock

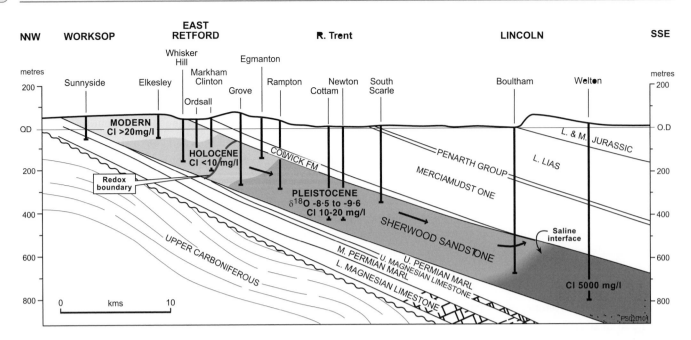

Figure 11.19 Sherwood sandstone aquifer, England, is an example of a stratiform aquifer with bounding upper and lower aquitards. The figure shows the spatial distribution of groundwater quality downdip with reference to chloride and stable oxygen isotope concentrations, which is discussed in Chapter 12. (From Smedley and Edmunds (2002); reproduced with the permission of John Wiley and Sons.)

aquifers, such as the Dakota sandstone and the dolomite and sandstones beneath the upper Midwest of the USA, were identified in the late 1800s as **stratiform aquifers** (Tolman, 1937) on account of the aquifer being bounded above and below by shale confining beds with discharge occurring in a downdip artesian basin. Stratiform cross-sections are still used to show patterns of flow and geochemical change in bedrock aquifers, although leakage to and from the main aquifer across the confining beds is clearly important from water balance considerations.

Figure 11.19 presents the Sherwood Sandstone of the English Midlands as a stratiform aquifer and its geochemical evolution. It is one of several important Permian–Triassic age sandstone aquifers of fluvial and eolian origin (**red beds**) that are major water-supply aquifers in Europe and North Africa. The English Permo-Triassic sandstones are similar to many other continental red-bed formations with flow through both fractures and the intergranular matrix (Tellam and Barker, 2006). The transmissivity varies from $0.017 \, \text{m}^2/\text{s}$ in the recharge area to 0.001–$0.010 \, \text{m}^2/\text{s}$ downdip in the confined aquifer. The average linear groundwater velocity varies between 1 and 9 m/yr (Smedley and Edmunds, 2002). The recharge area has modern groundwater with ages of less than 100 years old, although downdip the water is older, first of Holocene and then Pleistocene age.

Bedding-plane fractures in sandstones greatly enhance the permeability of the formation and are most likely to be found at contacts between rock types. The Tunnel City Sandstone is part of the major Cambrian–Ordovician aquifer that is an important water supply for the cities of southern Wisconsin and Illinois; total production from this aquifer across the Midwest is $34 \, \text{m}^3/\text{s}$. The shaded lines within Figure 11.20 show the presence of high-permeability and laterally continuous zones within the sandstone that were correlated with natural gamma logs. Swanson et al. (2006) present the flowmeter log for well WDNR-7 indicating a significant increase in flow at an elevation of about 240 m above mean sea level. Hydraulic testing of straddle-packer intervals (see Figure 5.14) indicates $K \sim 1 \times 10^{-3}$ m/s in the high-permeability zones, decreasing to 1×10^{-6} m/s elsewhere in the sandstone.

The Paskapoo bedrock aquifer of Alberta (Grasby et al., 2008) exhibits hydraulic conductivities as high as 1×10^{-3} m/s in thick, coarse channel sandstones with an open framework within the grains (see Figure 9.19), yielding a porosity range of 20–36%. Well yields are believed to depend on intergranular flow associated with the thickness and areal extent of the channel sandstones. Groundwater flow through fractured zones appears to be limited to (i) the northeast–southwest directional trend (Figure 3.17) aligned with the regional maximum horizontal stress field, which is independent of the channel geometry, and (ii) fractures within thin sandstone beds (Figure 3.16).

Carbonate aquifers, both limestone and dolostone, are regionally important water-supply aquifers in the upper

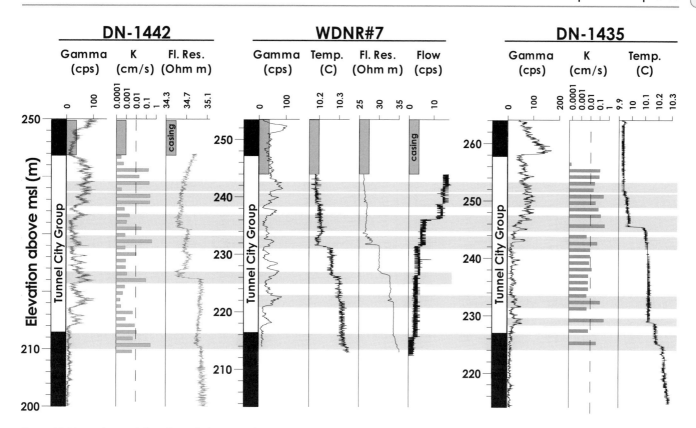

Figure 11.20 Preferential flow through the Tunnel City Sandstone, southern Wisconsin, shown in three borehole logs. Natural gamma, hydraulic conductivity *K*, fluid resistivity and flowmeter logs show laterally extensive, high-permeability fracture zones in this sandstone formation. (From Swanson et al. (2006); reproduced with the permission of Elsevier.)

Midwest of the USA ($Q \sim 34\,m^3/s$), Texas ($Q \sim 35\,m^3/s$), Florida and other southeastern states ($Q \sim 120\,m^3/s$) and northwest Europe. We discuss the evolution of groundwater quality in the Floridan aquifer in Chapter 12.

The Chalk is the most important of the English aquifers ($Q = 45\,m^3/s$). The porosity of the Chalk matrix is 25–45%; however, it is the fractures in the Chalk and the larger pores that store the extractable groundwater. The Chalk matrix has such small pore throats that much of this pore water is not readily extracted by pumping. The fracture regime is partly due to primary discontinuities in which the fracture spacing is ~10 cm, while the secondary permeability comes from solution enhancement of the fractures in which the spacing is ~5–10 m (Lawrence et al., 1990; MacDonald and Allen, 2001). Figure 2.27 shows the bedding-plane fractures in the Chalk, usually three orthogonal joint sets with an effective porosity of 0.1–2%.

11.5.3 Aquitards

In North America and northern Europe, aquitards are usually associated with surficial glacial deposits of till and lacustrine clays or deeper shales, siltstones and mudstones. Neuzil (1994) has summarized the porosity and hydraulic conductivity data for clays and shales, and van der Kamp (2001) has reviewed methods of estimating the hydraulic conductivity of shallow aquitards.

The migration of solutes through aquitards is by molecular diffusion (Ingebritsen et al., 2006). Therefore, it is the thermal-kinetic energy of the solute that drives the random motion, creates the concentration gradients and explains the temperature dependence of diffusion. Diffusion in such low-permeability environments leads to migration rates of a few millimetres per year. Thus, aquitards can act as archives of ancient waters that infiltrated thousands of years before. Remenda et al. (1994) used oxygen isotopes sampled in thick clay deposits in southern Manitoba and adjacent North Dakota to demonstrate that the isotope profile through the 20–30 m thick clay aquitard was likely due to the infiltration and slow migration of Pleistocene-aged water from glacial Lake Agassiz.

Post-depositional alteration of aquitards yields preferential pathways for groundwater and contamination to penetrate otherwise confined aquifers and thereby undermines

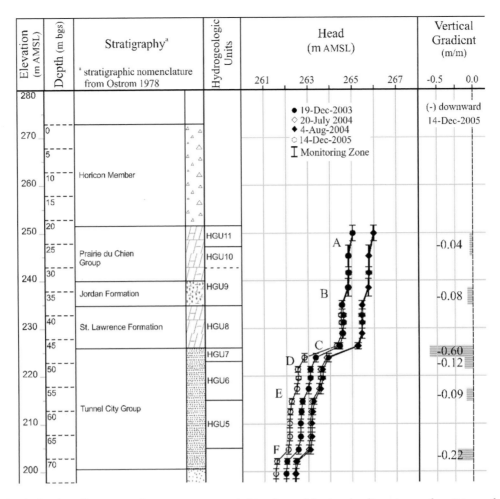

Figure 11.21 Hydraulic head profiles measured over a two-year period in a layered fractured sedimentary rock unit in southern Wisconsin. The inflections in head that occur between adjacent permeable units are most probably due to poor vertical connectivity of fracture sets rather than distinctly lower vertical hydraulic conductivity across the zone of head drop. That is, the fracture sets are not continuous throughout the vertical section shown and thus the largest head drops are localized. (From Meyer et al. (2008); reprinted by permission of Springer from Environmental Geology © 2007.)

the property of **aquitard integrity**, i.e., the absence of high-permeability "short circuits" in an aquitard (Cherry et al., 2004). Figure 11.1 shows fractured till from southern Manitoba described by Grisak and Cherry (1975), who determined by injection of tritium-containing water that these fractures were hydraulically conductive.

Figure 11.20 demonstrates the importance of bedding-plane fractures as groundwater flow conduits in the horizontal direction. In the vertical direction, the low-permeability layers of rock inhibit groundwater flow provided that they are not fractured and are continuous. Figure 11.21 shows the particular importance of very thin beds that resist vertical groundwater flow and cause the largest head drops. The inflections in the head profile shown are due to the discontinuous nature of the fractures throughout the vertical section shown. These two figures are a complementary pair, illustrating flow and resistance to flow in fractured, layered rocks.

11.6 Summary

Our goal in this chapter was to describe hydrogeological processes and parameters that the geotechnical and geoenvironmental engineer is likely to encounter in practice, i.e., those important in groundwater extraction, dewatering, slope stability, subsidence, etc. Consequently we began our discussion by identifying the conceptual structure on which

Darcy's law (1856) is founded, namely the porosity and hydraulic conductivity of the porous medium and the hydraulic head gradient across it. We further showed that the hydraulic head is derived from the fluid potential of Hubbert (1940). This allowed us to identify the relevant parameters in the groundwater flow equation, i.e., head, transmissivity, storativity and compressibility. The nature of water wells and hydraulic testing to determine these parameters (K, T, S) was then described.

We discussed the theory of gravitationally driven groundwater flow systems and recharge–discharge relationships. Fluid flow in the subsurface proceeds from locations of high hydraulic head, known as recharge areas, to locations of low hydraulic head, known as discharge areas; usually this means from topographically high to topographically low areas. The hydraulic gradient points in the general direction of flow but not the specific direction, which is influenced by the properties of the soil and rock through which flow occurs. We concluded with an outline of the nature of water-supply aquifers and the aquitards that protect them and affect groundwater flow patterns.

Above all, we have stressed how heterogeneities in the spatial distribution of hydraulic conductivity in soils and rocks are to be expected and make site-specific testing essential. These heterogeneities, whether present in small-scale formations or in large-scale groundwater flow systems, result in patterns of groundwater flow that differ greatly from any homogeneous system.

11.7 FURTHER READING

Bear, J., 1988, *Dynamics of Fluids in Porous Media*. New York, NY: Dover Books. — Originally published in 1972, the standard theoretical analysis of flow through porous media.

Bear, J., 2007, *Hydraulics of Groundwater*. New York, NY: Dover Books. — Originally published in 1979, a rigorous account of groundwater hydraulics now published inexpensively.

Freeze, R.A. and Cherry, J.A., 1979, *Groundwater*. Englewood Cliffs, NJ: Prentice-Hall. — This classic text has been in print for over 30 years because of its clarity of writing and thoughtful explanation of physical and chemical hydrogeological processes.

Ingebritsen, S., Sanford, W. and Neuzil, C., 2006, *Groundwater in Geologic Processes*, 2nd edn. Cambridge, UK: Cambridge University Press. — This book contains numerous thoughtful sections on hydrogeology and hydromechanical processes, including poroelastic theory from a hydrogeological perspective.

Jorgensen, D.G., 1980, *Relationships between Basic Soils: Engineering Equations and Basic Ground-water Flow Equations*. Water Supply Paper 2064, Washington, DC: US Geological Survey. — This report provides a helpful guide to negotiating the differences in geotechnical and hydrogeological terms.

Schwartz, F.W. and Zhang, H, 2003. *Fundamentals of Ground Water*. Hoboken, NJ: John Wiley and Sons. — A fine modern text on hydrogeology that has become the successor to Freeze and Cherry's textbook and is prepared with more attention to numerical simulation of groundwater systems.

Todd, D.K. and Mays, L.W., 2005, *Groundwater Hydrology*, 3rd edn. Hoboken, NJ: John Wiley and Sons. — The first edition was published in 1959 based upon Todd's work in California and established itself as the standard engineering text for groundwater hydrologists.

Tóth, J., 2009, *Gravitational Systems of Groundwater Flow: Theory, Evaluation, Utilization*. Cambridge, UK: Cambridge University Press. — A monograph on the nature of regional groundwater flow systems, especially in deep basins, by the founder of the topic.

11.8 Questions

Question 11.1. Develop a graph showing the variation in average linear groundwater velocity v_{gw} as a function of porosity for a sandstone with a hydraulic conductivity of $K = 1 \times 10^{-4}$ m/s, a total porosity of 25% and subject to a hydraulic gradient of 10^{-3}. Assume the total porosity is the maximum value of the effective porosity and estimate v_{gw} down to $n = 0.1\%$.

Question 11.2. If equation (11.4) has units of total potential energy per unit mass, then show why it has dimensions of length^2/time^2.

Question 11.3. Consider a sandstone aquifer of 100 m thickness (i.e., b), a porosity of 20% with $S = 0.0001$. What is the compressibility of this aquifer during groundwater extraction? What is the proportion of water produced by its expansion to that from the compression of the aquifer?

Question 11.4. In Figure 11.9 the total hydraulic head at the ground surface is 400 units but it is between 500 and 600 units after 460 hours. However, the water table is still well below ground surface and the ground surface elevation has not changed, so why has the total head increased so much?

Question 11.5. The fluid density of the Michigan Basin brine beneath Michigan is $\rho_f = 1.16\,\text{g/cm}^3$. The total hydraulic head in the St. Peter Sandstone is 100 m above the ground surface elevation of 300 m above mean sea level (amsl). Assuming hydrostatic conditions throughout the Basin, at what elevation (in metres amsl) is the pore pressure 39 MPa?

Question 11.6. You have a contract to drill a well into the St. Peter Sandstone at this same depth but need to tell your driller to order a blow-out preventer (BOP) because State of Michigan drilling regulations require the BOP to prevent possible explosions due to natural gas ignition. To specify the size of the BOP, you must determine the pressure gradient from the information in question 11.5. The driller works in traditional engineering units of psi and ft. What is the pressure gradient in the Michigan Basin in psi/ft?

12 Groundwater Quality and Contamination

In Chapter 7 we reviewed the geochemical controls on soils, sediments, rocks and associated natural waters. We return to this topic to discuss the properties of natural and contaminated groundwaters. We will employ the shorthand term GWQ to indicate groundwater quality. We begin with a discussion of GWQ in alluvial, glacial and bedrock aquifers – those used for water-supply purposes – and then proceed to discuss environmental isotopes, which provide critical tools to understanding groundwater flow systems. We then turn to the topic of groundwater contamination that may affect the geotechnical or geoenvironmental engineer through a need to understand (i) the origin of contamination of a municipal groundwater supply or construction site or (ii) the development of a plume of contaminants requiring remediation. This topic is followed by consideration of how one investigates site conditions to identify potential pathways of contamination. We conclude with consideration of salt-water intrusion into coastal aquifers and of general aquifer vulnerability.

12.1 Groundwater Quality

Groundwater quality (GWQ) is the sum of the chemical (and sometimes microbial) properties of a groundwater sample. In Box 12.1 we explain the various terms used to define GWQ in North America and Europe. Here we consider GWQ in alluvial, glacial and bedrock aquifers and relate the chemical properties to their position in a groundwater flow system and the mineralogy of the aquifer.

Note that, in the following discussion of GWQ, only old groundwaters tend to be entirely free of anthropogenic compounds such as tritium, nitrate or organic contaminants.

Also note that the driving forces for hydrochemical change are associated with the evolutionary concepts discussed in Chapter 7: the control of (i) pH by dissolved carbon dioxide (Figure 7.4) and (ii) Eh (or E_H) by the redox reactions involving the oxidation of dissolved organic carbon (DOC) by dissolved oxygen (DO), NO_3, $Mn(IV)$, $Fe(III)$ and SO_4 (Figure 7.5).

We begin our discussion by considering GWQ in alluvial aquifers, which provide groundwater supplies to many municipalities. In Europe, groundwater production from alluvium is often from bank-filtration wells that induce flow from nearby rivers as well as capture groundwater that would otherwise discharge to the river. Bank filtration was identified in Germany in the 1870s as a process by which water from the Rhine and Elbe rivers might be pretreated *in situ* to remove fecal bacteria.

Table 12.1 shows the distribution of major ions from alluvial aquifers adjacent to the Missouri, Mississippi, Red and Brazos Rivers. Sharp (1988) characterized this GWQ as being of a calcium–magnesium–bicarbonate type, i.e., a Ca–Mg–HCO_3 **hydrochemical facies**, with dissolved Fe and Mn sometimes present due to anoxic conditions causing the dissolution of the ferric and manganese oxide coatings on the sand grains (see section 7.6). pH values were typically in the range 7.0 to 8.0 in aquifers with potable groundwater, i.e., total dissolved solids TDS $\leq$ 1000 mg/L.

The glacial aquifer samples are from the Waterloo Moraine Aquifer of Ontario, Canada. The upper aquifer groundwaters have low TDS values and are also of a Ca–Mg–HCO_3 facies dictated by the presence of carbonate minerals. Agricultural seepage has affected the upper aquifer (U in Table 12.1) and has contributed ~12 mg/L of nitrate. As is often the case,

BOX 12.1 | OFFICIAL DEFINITIONS OF GROUNDWATER QUALITY

In Europe, the terminology is that of the European Commission in Brussels. Thus Edmunds and Shand (2008) refer to the natural (pristine) GWQ as the **baseline GWQ**, which is "derived from natural geological, biological, or atmospheric sources", with no influence on anthropogenic effects. They recognized that a different definition is required "when the present day status of groundwater may represent the starting point of monitoring", which they define as the **background** state. However, in North America the USGS definitions (Lee and Helsel, 2005) are used, which are the opposite of the European terms:

background GWQ describes the pristine GWQ derived from natural geological, biological, or atmospheric
    sources in the absence of identifiable anthropogenic influences; whereas

baseline GWQ describes the GWQ representing the starting point of monitoring, such as "pre-drilling" around
    natural gas wells, and which typically includes some influences of human activities; and

ambient GWQ is that measured at some time and place without any assumption being made as to
    anthropogenic influences.

Table 12.1 Groundwater quality in alluvial and bedrock aquifers.

Aquifer	TDS	pH	*Eh*	Ca	Mg	Na	Fe	HCO$_3$	Cl	SO$_4$
Missouri alluvium, USA	497	7.7	—	114	26	22	2.5	451	3.2	36
Mississippi alluvium, USA	353	7.6	—	73	21	15	5.2	360	7.5	9.8
Red River alluvium, USA	490	7.4	—	62	40	39	1.3	279	52	10
Brazos alluvium, USA	770	7.0	—	152	36	119	3.5	600	111	124
Waterloo Moraine, Canada, U	899	6.7	0.4	126	35	83	0.1	364	167	67
Waterloo Moraine, Canada, L	879	7.3	0.0	158	42	26.7	0.7	263	23	353
Lincolnshire Lst, UK, 1	630	7.1	0.4	150	7.1	15	—	256	34	110
Lincolnshire Lst, UK, 2	571	7.2	0.2	120	9.2	20	—	286	25	110
Lincolnshire Lst, UK, 3	1181	8.4	0.0	4.2	3.5	370	—	463	308	25
Sherwood Sst, UK, 1	505	7.8	0.3	48.3	28.2	72.8	0.04	149	138	65.7
Sherwood Sst, UK, 2	312	7.9	0.1	27.1	28.0	5.4	0.13	215	6.3	22.7
Sherwood Sst, UK, 3	355	7.6	0.1	46.4	24.2	10.3	0.08	189	10	70.4

Notes: All data in mg/L, except *Eh* (in volts) and pH. TDS = total dissolved solids; U = upper; L = lower. GWQ data for alluvium are median values from Sharp (1988), for Waterloo Moraine from Stotler et al. (2011), for the Lincolnshire Limestone are from boreholes 1, 10A and 17 from Edmunds and Walton (1983) and for the Sherwood Sandstone are from boreholes 9, 25 and 35 from Smedley and Edmunds (2002).

dissolved oxygen (DO) at ~2 mg/L is also present with nitrate, and dissolved iron is very low in such conditions. In the lower aquifer (L in Table 12.1), dissolved oxygen is less abundant, dissolved iron is more abundant and nitrate is ~1 mg/L. The measured *Eh* values reflect the presence or absence of DO.

The bedrock aquifer data are from two English aquifers – the Lincolnshire Limestone and the Sherwood Sandstone – that are heavily pumped for municipal and industrial purposes and are somewhat contaminated in their recharge areas. GWQ data from three production wells in each aquifer are listed in Table 12.1, with the first well in the recharge area and the second and third wells further downdip. The Triassic Sherwood Sandstone is shown in profile in Figure 11.19, in which the Jurassic Lincolnshire Limestone is indicated as "L. & M. JURASSIC" and outcropping to south-southeast.

The Sherwood Sandstone is divided into three zones (Smedley and Edmunds, 2002). In the unconfined outcrop area (Sherwood sample 1), where modern groundwaters (<100 years old) occur, the chloride concentrations exceed 20 mg/L but the groundwater is typically oxygenated such that dissolved iron is <1 mg/L. The pH is the result of dolomite dissolution. Downgradient in the unpolluted groundwater of Holocene age (Sherwood sample 2), the chloride concentrations are <10 mg/L. Dissolved oxygen becomes absent in these Holocene groundwaters and thus Smedley and Edmunds identify a **redox boundary**, downdip of which Fe and Mn concentrations rise. Sherwood sample 3 is from groundwaters of Pleistocene age in excess of 10 ka. This groundwater is anoxic, at equilibrium with carbonate minerals but undergoing dissolution of gypsum or anhydrite, causing an increase in calcium and sulfate concentrations.

The Lincolnshire Limestone is distinguished by two zones (Edmunds and Walton, 1983). The first 12 km of the aquifer (samples 1 and 2) is the oxidizing zone, in which $Eh \sim 400$ mV and the groundwater is at equilibrium with calcium carbonate (pH $\sim$ 7). The groundwater at the redox barrier (sample 2) is similar to that updip. However, downdip of the redox barrier (sample 3) sulfate reduction causes loss of SO_4 and ion exchange of Ca for Na occurs. Additionally mixing of the old groundwater with **connate** water, i.e., saline trapped during sedimentary diagenesis, becomes evident, causing large increases in Na, HCO_3 and Cl.

Figure 12.1 illustrates how the major ions comprising GWQ may be represented graphically. This Piper diagram is composed of two triangular plots that feed a diamond-shaped plot. Anion and cation data are in the form of meq/L (see equation (7.3)), with anions and cations being plotted separately in triangles as percentages of total equivalents of anions or cations. Each sample is then projected onto the diamond where the intersection of points from the cationic and anionic triangles defines the GWQ.

The patterns indicate hydrochemical facies so that the connate water (Lincolnshire Limestone sample 3) is distinct by having 0% Ca and $\sim$60% Cl. The alluvial samples are closely grouped in all three plots, while the other samples are more dispersed, reflecting the evolution of the different groundwaters and their age.

12.2 Environmental Isotopes

We have discussed isotopes in two contexts so far. In Chapter 7 we considered the sorption of radioactive strontium, and in Chapter 8 we used the stable oxygen isotope ratio ($\delta^{18}O$) as a proxy for climatic variations over Quaternary time (see Box 8.1). Here we consider the use of naturally occurring stable and radioactive isotopes in hydrogeology. Clark (2015) refers to these as **environmental isotopes**, i.e., "naturally occurring isotopes of the major elements that participate in processes at the earth's surface", e.g., ^{18}O and ^2H (i.e., deuterium, D).

12.2.1 Applications of Environmental Isotopes

A particular value of stable environmental isotopes lies in identifying the **provenance** or origin of the water sample. The Sherwood Sandstone shown in Figure 11.19 is an example of how natural groundwater quality can be identified on the basis of geologic age using ^{18}O. Another contribution made by stable isotopes such as ^{13}C and ^{34}S is their application in understanding **water–rock interaction**, i.e., the nature of mineral dissolution–precipitation reactions that yield GWQ.

Table 12.2 illustrates a third application of environmental isotopes, **diffusional profiling**, by which isotopes present in pore-water samples collected from aquitards are used to demonstrate that molecular diffusion – not groundwater flow – must be responsible for their extraordinarily slow rates of migration. Such information may be useful in siting a nuclear waste repository on the grounds that diffusional migration of isotopes is likely to allow many half-lives of the entombed radioactive waste to pass before any waste

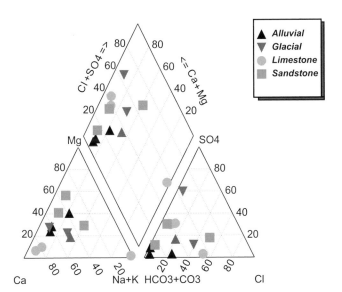

Figure 12.1 Piper plot of alluvial, glacial and bedrock groundwater quality data from Table 12.1.

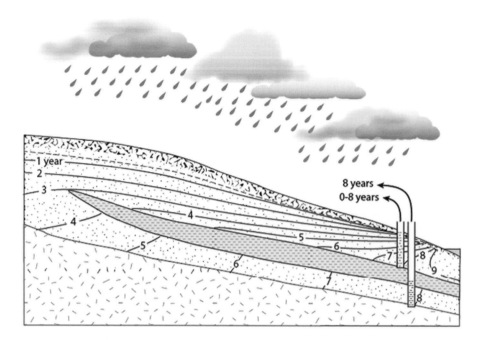

Figure 12.2 The reliability of an age date for groundwater very much depends on the degree to which the sample analyzed is not a mixture of groundwaters of various ages. In this figure Clark (2015) shows that the sample from the water-table aquifer is a composite of groundwater recharged in an open system, i.e., an extensive recharge area, resulting in a wide spectrum of potential ages, whereas that from the deeper confined aquifer is the result of advection in confined aquifer with little or no mixing with other waters. (Reproduced with the permission of Taylor and Francis Group LLC Books.)

radionuclides reach ground surface; Gautschi (2001) and Clark et al. (2013) provide examples of this application using helium, chloride and $\delta^{18}O$ profiles.

When advection controls solute transport, the application of environmental isotopes to the **calibration of groundwater flow models** has become of increasing importance. In this case, an age-dating isotope is used to estimate the mean residence time of the groundwater assuming that the isotope moves at approximately the same velocity as the groundwater volume in which it is dissolved; for ^{18}O and ^{3}H (tritium) this assumption is clearly satisfied. For example, the ^{18}O values shown in Figure 11.19 for the Sherwood Sandstone indicate that groundwater extracted from the Cottam and Newton wells is of Pleistocene not modern origin. Figure 12.2 illustrates how difficult it might be to arrive at a groundwater age when recharge to the flow system sampled occurs over an open system as opposed to the closed system of the Sherwood Sandstone and similar deep aquifers.

The age of the groundwater is a clear indication of **groundwater vulnerability** because the presence of young groundwater midway towards a discharge zone in an aquifer indicates rapid migration through the aquifer. An old groundwater that is being extracted at rates greater than it is being recharged is being mined.

12.2.2 Radioisotopes

The wide range of half-lives of radioisotopes shown in Table 12.2 allows their application to many hydrogeological issues, in particular model calibration and groundwater vulnerability assessment. An engineer's assessment of groundwater residence time in a flow system will dictate his or her choice of which radioisotope to choose for analysis for the particular problem. Thus, the rate of radioactive decay, as indicated by the radioisotope's half-life, will dictate the engineer's choices.

Radioactive decay is described by a first-order linear differential equation, in which the number of radionuclei (N) of any radioisotope that decays is given by the initial value problem:

$$\frac{dN}{dt} = -\lambda N, \quad t > 0, \tag{12.1}$$

$$N(0) = N_0, \tag{12.2}$$

where t is elapsed time after radioactive decay began (t_0), λ is the decay constant (units of time) and N_0 is the number of radionuclei present at time zero. The solution to these equations is

$$N(t) = N_0\, e^{-\lambda t} \tag{12.3}$$

Table 12.2 Some uses of environmental isotopes.

Stable isotope	Implied ratio	Applications
D	$^2H/^1H$	Temperature of groundwater at recharge; diffusional profiling
^3He	$^3He/^4He$	Diffusional profiling in clay and limestones
^{13}C	$^{13}C/^{12}C$	Water–rock interaction of carbonate species in groundwater
^{18}O	$^{18}O/^{16}O$	Temperature of rainout or recharge to groundwater
^{34}S	$^{34}S/^{32}S$	Water–rock interaction and fate of sulfide species
Radioisotope	Half-life (a)	Applications
^3H	12.4	Calibration of groundwater flow and transport models
^{14}C	5730	Age dating up to a maximum of 30 000 a
^{36}Cl	301 000	Calibration of groundwater flow and transport models

Table 12.3 Evolution of groundwater quality in the Floridan Aquifer with distance from the recharge area. (Data from Plummer and Sprinkle (2001).)

Well	Distance	SEC	pH	Ca	Mg	SO₄	SI$_{Cal}$	SI$_{Dol}$	SI$_{Gyp}$	Age
Polk City	0.0	303	7.76	38	8.1	12	0.08	−0.16	−2.66	1 400
Haines City	20	321	7.64	42	5.6	2.5	0.06	−0.41	−3.30	6 800
Canoe Creek	55	443	7.71	52	5.6	9.8	0.24	−0.15	−2.65	13 900
Roper Groves	74	1284	7.57	64	26	66	0.11	0.16	−1.90	27 500
Kempfer Grove	93	2168	7.68	67	30	83	0.09	0.17	−1.83	21 700
Melbourne	112	2185	7.55	103	50	147	0.10	0.19	−1.47	20 600
Ingraham	121	2770	7.43	130	70	170	0.10	0.29	−1.40	21 500

Note: Distance is in kilometres; specific electrical conductance (SEC) is in microsiemens per cm (μS/cm) and is a measure of TDS; Ca, Mg and SO₄ are in mg/L; saturation indices (SI) for calcite, dolomite and gypsum are in log units; age is in years. The groundwater is anoxic throughout the aquifer.

and the time elapsed since radioactive decay began is

$$t = \frac{1}{\lambda} \ln \frac{N_0}{N_t} \quad \text{or} \quad t = -\frac{1}{\lambda} \ln \frac{N_t}{N_0}. \tag{12.4}$$

The radioisotopes identified in Table 12.2 allow estimates of approximate groundwater ages – t in (12.4) – if the principal process by which $N(t)$ varies within a groundwater flow system is radioactive decay (Clark, 2015). In such cases, the computed age is the elapsed time since the radioisotope was recharged to the flow system and was no longer influenced by atmospheric processes. If other processes cause the radioisotope concentration to change, such as ^{14}C retention in an aquitard or groundwater mixing, then the computed age is an apparent one. Constraints such as these lead to a maximum age using ^{14}C ages in groundwater of 30 000 a, often much shorter. The application of the much longer-lived ^{36}Cl is constrained by mixing with chloride-rich formation waters

but it can sometimes be applied for age dating of 100 000 to 1 000 000 a.

The Floridan Aquifer supplies 57% of the fresh water of Florida and is continually threatened with contamination that seeps through the overlying aquitard. Plummer and Sprinkle (2001) demonstrate how the various geochemical reactions that complicate ^{14}C age dating of groundwater are accommodated into a hydrogeochemical reaction model, i.e., one that integrates water–rock interactions. These reactions include: (i) de-dolomitization, i.e., dolomite dissolution and calcite precipitation caused by the dissolution of gypsum or anhydrite (see question 12.1); (ii) sulfate reduction arising from oxidation of organic carbon; and (iii) mixing with saline groundwater near the coast.

The hydrochemical evolution of the groundwater along the 120 km flow path from the middle of the Floridan peninsular to the Atlantic coast is presented in Table 12.3. The integration of these reactions is by hydrogeochemical simulation and yields a ^{14}C age at the coast (near Cape Canaveral) of 21 500 years. Therefore much of the fresh groundwater in the Floridan Aquifer is the result of recharge during the last glaciation, indicating that care must be taken to protect and sustain this resource.

Tritium – often written HTO to indicate that only one hydrogen atom is radioactive in tritiated water – was produced by atmospheric bomb testing that peaked in 1963, which provided an event marker. Since that time, this "bomb tritium" has been rained out of the hydrologic cycle and what we measure today is usually natural tritium formed in the upper atmosphere by cosmic ray bombardment of nitrogen. It is reported in picocuries per litre (pCi/L) and often converted to "tritium units" (TU), where $3.19\,\mathrm{pCi/L} \equiv 1\,\mathrm{TU}$.

Tritium decay to helium producing ^{3}He extends tritium dating where the 1963 peak is poorly defined and provides independent travel times by which groundwater flow and transport models might be constrained (Solomon et al., 1992). Thus a ^{3}H–^{3}He age determined for a sampling location in a flow system reflects the true travel time of that sample volume from its origin at the water table in the recharge area. Murphy et al. (2011) illustrate how ^{3}H–^{3}He age dating was used to verify a groundwater transport model of trichloroethene in a deltaic aquifer.

12.3 Groundwater Contamination

We shall consider two general types of groundwater contamination: (i) miscible and (ii) immiscible. **Miscible contaminants** mix readily with groundwater because they are dissolved by infiltrating water, e.g., landfill leachate or agricultural seepage created by rainfall on fertilized fields, and are transported as solutes by groundwater flow. **Immiscible contaminants** are sparingly soluble in groundwater and must be treated very differently because their behaviour *in situ* requires consideration of interfacial processes and their migration is described by two-phase or multi-phase flow, not solute transport equations. These include non-aqueous phase liquids (NAPLs), such as gasoline, and fugitive natural gas that may have escaped from a leaky gas well. Dissolution of NAPLs generates plumes of aqueous contaminants that then behave as solutes. The dissolved components of NAPLs are subject to sorption, just like ^{90}Sr and other inorganic contaminants discussed in Chapter 7, as well as biodegradation, just as DO and nitrate are biotransformed with the oxidation of organic carbon.

Diffusion, advection and dispersion are the transport processes by which solutes, i.e., dissolved contaminants, migrate in the subsurface. Numerical simulation of dissolved contaminant migration is thus referred to as contaminant transport in that groundwater is doing the transporting. NAPLs are not typically transported by groundwater; NAPLs migrate due to their density difference with groundwater irrespective of whether they are liquids (e.g., gasoline, chlorinated solvents) or gases (e.g., methane). We will simplify Simpson's (1962) definitions of these transport terms.

- **Diffusion**: the spreading of molecules or ions within a fluid-filled porous medium in a direction tending to equalize concentrations in all parts of the system; it is understood to occur as a result of the thermal-kinetic energy of the molecules or ions in the absence of fluid advection.
- **Advection**: transport of molecules or ions by groundwater flow and measured as occurring at the average linear groundwater velocity ($v = n_e/q$).
- **Dispersion**: the spreading of molecules or ions within a fluid-filled porous medium in a direction tending to equalize concentrations in all parts of the system; it occurs as a result of fluid advection and accounts for transport at velocities other than the mean groundwater velocity.

12.3.1 The Geoenvironmental Perspective

The processes under discussion in this chapter affect the engineer in many ways in that he or she may be designing a waste-management system to control contaminant migration, such as a municipal landfill or an industrial wastewater lagoon or nuclear-waste repository. In this context, a desirable outcome is that it is diffusion not advection from the waste-management system that is the principal contaminant transport process, because diffusion is slow. Alternatively, the engineer may be contracted to identify a wellhead protection area for a water-supply well. If the remediation of a plume of contaminants migrating away from an industrial plant is the context, then remedial processes to capture the plume will require a full appreciation of the advection and dispersion of the contaminants and how remediation will be affected by them.

It is important to bear in mind how geologic features – both structural discontinuities and sedimentary facies – profoundly affect how both miscible and immiscible contamination spread in the subsurface. These result in preferential flow pathways – fracture systems in rock and high-permeability gravels in granular sediments – that strongly influence arrival times of contaminants at site boundaries, rivers and wells. Similarly, the migration of immiscible contaminants in fractured rock and granular sediments requires particular attention to a number of physical-chemical parameters, such as fluid viscosity and density and the interfacial tension (IFT) between groundwater and the immiscible phase.

With these thoughts in mind, we will define the solute transport and fugitive gas processes associated with contaminant migration in this section and then proceed to consider (i) NAPLs, (ii) the fate of organic contaminants, (iii) the characterization of contaminated sites requiring remediation, (iv) seawater intrusion into coastal aquifers, and (v) wellhead protection for municipal wellfields.

12.3.2 Solute Transport

Diffusion is the temperature-dependent spreading of ions or molecules in a three-dimensional (x, y, z) porous medium in the absence of advection and is described by Fick's first and second laws (Bear, 1972; Ingebritsen et al., 2006):

$$D_m \left(\frac{\partial^2 C}{\partial x^2} + \frac{\partial^2 C}{\partial y^2} + \frac{\partial^2 C}{\partial z^2} \right) = n_e \frac{\partial C}{\partial t}, \qquad (12.5)$$

where D_m is coefficient of molecular diffusion (m^2/s), C is the concentration of the molecule or ion, n_e is the effective porosity and t is time. Figure 12.3 illustrates a particular solution of this equation for one-dimensional diffusion with the results for two representative diffusion coefficients. The relatively short distances of travel over long periods of time indicate why isolation of nuclear and other toxic wastes suggests disposal within diffusion-dominated systems (see diffusional profiling discussion in section 12.2.1).

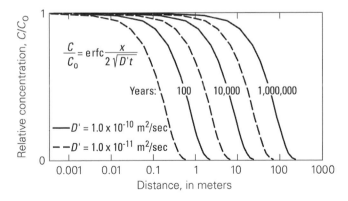

Figure 12.3 Rates of molecular diffusion illustrated by solutions to the one-dimensional diffusion equation with diffusion coefficients typical of geological media. (From Ingebritsen et al. (2006); reproduced with the permission of Cambridge University Press.)

Advection is the fluid flux of molecules or ions being transported at the average groundwater velocity according to Darcy's law, i.e., the fluid flux associated with the specific discharge:

$$\frac{\partial C}{\partial t} = -v_x \frac{\partial C}{\partial x}. \qquad (12.6)$$

Equation (12.6) is written for flow at velocity v_x in the x direction, with the negative sign indicating that the concentration is increasing in the x direction.

The advective flux of molecules and ions occurs at the average linear groundwater velocity; however, heterogeneities within a flow system will cause spreading because many molecules or ions are not transported at the average linear velocity but at slower or faster speeds in the same general direction. Assuming that the statistical velocity distribution of molecules and ions is normal, then dispersion is the process accounting for the transport of molecules and ions about the mean value.

Ingebritsen et al. (2006) show these effects in Figure 12.4 as different pathways leading to the spreading of molecules and ions, in which both diffusion and dispersion are combining to spread contaminants in a porous medium. These two processes are related by the coefficient of hydrodynamic dispersion, a tensor (D_{ij}) with units m^2/s, which, for an isotropic porous medium, is (Bear, 1972)

$$D_{ij} = \alpha_T v \delta_{ij} + (\alpha_L - \alpha_T) \frac{v_i v_j}{v} + \frac{D_m}{n_e} \delta_{ij}, \qquad (12.7)$$

where α_T and α_L are characteristic properties of the porous medium in units of length called **dispersivities**, v_i, v_j are the spatial components of the velocity field of average v, and δ_{ij} is the Kronecker delta function. Equation (12.7) allows us to write the general equation of advection and dispersion as

$$\frac{\partial C}{\partial t} = \frac{\partial}{\partial x_i} D_{ij} \frac{\partial C}{\partial x_j} - \frac{\partial v_i C}{\partial x_i}, \qquad (12.8)$$

where v_i is the average linear velocity.

The longitudinal coefficient of hydrodynamic dispersion is often written simply as $D_L = \alpha_L \cdot v$, in which α_L is the longitudinal dispersivity. Figure 12.5 presents Gelhar's (1986) illustration of the **scale effect** of dispersion, with α_L plotted against the distance between the monitoring well and the source of the contaminant into the flow system. As the scale increases, the dispersivity also increases logarithmically. Some of this implied contaminant spreading may be due to the dilution of a sample in long-screened monitoring wells suggesting *in situ* contaminant dilution due to dispersion, but part is without doubt due to contaminant dispersion induced by heterogeneities over large distances.

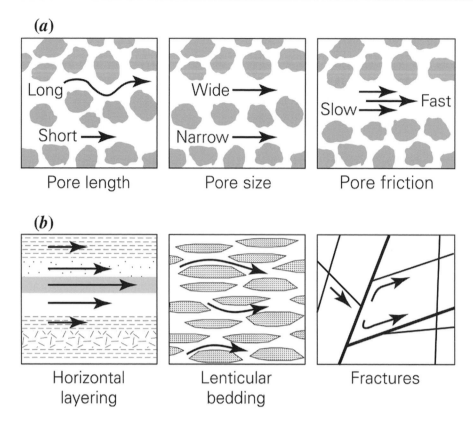

Figure 12.4 Physical mechanisms causing dispersion at (a) the pore scale and (b) the field scale. (From Ingebritsen et al. (2006); reproduced with the permission of Cambridge University Press.)

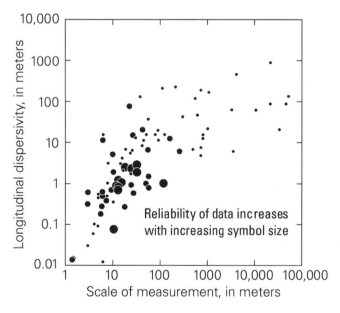

Figure 12.5 Values of longitudinal dispersivity as a function of the scale at which they were measured. (After Gelhar (1986), from Ingebritsen et al. (2006); reproduced with the permission of Cambridge University Press.)

Gillham et al. (1984) noted that this scale effect could be explained in heterogeneous sand and gravel aquifers by considering contaminant transport as an **advection–diffusion**

process. Contaminants are transported through high-permeability pathways with flow predominantly parallel to bedding with transverse diffusion into adjacent lower-permeability sediments. Advection–diffusion therefore results in slower contaminant transport in the high-permeability units because of loss of contaminants to the adjacent low-permeability units and thus the average contaminant velocity is slower than the average linear groundwater velocity, i.e. v in (12.6). Critically, this means that contaminant is stored in relatively low-permeability units from which its release by **back-diffusion** is slow, which process hinders aquifer remediation (Chapman and Parker, 2005).

12.4 Immiscible Contaminants

Gasoline, chlorinated solvents, creosote, coal tar and diesel fuel are all immiscible with groundwater and all have been widely used and spilled by industrial societies. Here we briefly consider their behaviour in the subsurface with particular reference to their migration in heterogeneous soils. Remediation of these oil-like contaminants has preoccupied geo-environmental engineers since the 1980s when it was realized

Table 12.4 Physical-chemical properties of some chlorinated solvents.

Chemical name Units	Symbol	Mol. wt. (g/mol)	Density (g/mL)	Viscosity (mPa s)	Aqueous solubility (mg/L)	K_{oc} (mL/g)
Carbon tetrachloride	CTET	153.8	1.59	0.97	785	439
1,2-Dichloroethane	EDC	99.0	1.26	0.84	8690	14
1,1,1-Trichloroethane	TCA	133.4	1.35	0.84	720	152
Trichloroethene	TCE	131.5	1.46	0.57	1100	126
Tetrachloroethene	PCE	165.8	1.63	0.90	200	364
Chlorofluorocarbon-113	CFC-113	187.4	1.56	0.68	158	372

Sources: Schwille (1988), Pankow and Cherry (1996) and, for CFC-113, Jackson et al. (1992).

Table 12.5 Physical-chemical properties of multi-component field NAPLs. (From Jackson et al. (2006).)

Chemical name Units	Density (g/mL)	Viscosity (mPa s)	IFT (mN/m)
Weathered gasoline	0.88	6.9	12
Weathered jet fuel	0.88	36	8
Creosote	1.12	41-67	17
Coal tar	1.02	19	22

Note: IFT = interfacial tension.

that widespread groundwater contamination was associated with their dissolution in the subsurface. This section begins with the concept of fluid saturations and interfacial properties in granular soils and sediments, and then addresses the migration of immiscible contaminants: denser-than-water non-aqueous-phase liquids (**DNAPLs**), such as chlorinated solvents; lighter-than-water hydrocarbon NAPLs (**LNAPLs**) and **viscous NAPLs**. We conclude with discussions of the generation of contaminant plumes by NAPL dissolution and the recent problem of fugitive gas migration. Table 12.4 lists some of the important parameters of chlorinated solvents in their initial state prior to industrial use; Table 12.5 presents similar information on multi-component LNAPLs and viscous NAPLs that were collected from field sites.

12.4.1 Fluid Saturations and Interfacial Behaviour

The saturation of one fluid in a pore space was introduced in section 11.4.2, where the hydraulic conductivity of an unsaturated soil was shown in Figure 11.12. We continue this discussion by replacing air with an NAPL. When this NAPL

has stopped draining through the soil, it is said to be trapped at **residual saturation** (S_{or}). If the measured saturation exceeds the residual saturation, then the NAPL is considered to be potentially mobile or **free phase** and can continue draining.

Values of S_{or} depend on the soil texture, the wetting state, which identifies the liquid phase (groundwater or NAPL) that adheres to the soil grains, and the duration that the NAPL has been trapped *in situ* and thus available for dissolution, volatilization or further drainage. Most importantly, residual saturations depend on whether the system is two-phase (water–NAPL) or, when dewatering occurs and the water table drops, three-phase (air–water–NAPL). In lab experiments, Wilson et al. (1990) showed two-phase saturations in sandy soils of ~27% and three-phase saturations of ~9%, because of the decrease in interfacial tension forces between water and NAPL (see question 12.2). Rather than residual saturations, Schwille (1988) reported his results for trapping of DNAPLs as retention capacities: three-phase values were $30\,L/m^3$ of sandy soil and two-phase values were $50\,L/m^3$. *In situ* estimates of saturations obtained by partitioning inter-well tracer tests for five Superfund sites across the USA (Jackson and Jin, 2005) indicated saturations of 0.4–4.7%, which suggests that the duration of NAPL residence *in situ* at these sites (probably 20–40 years) had allowed dissolution and volatilization to occur, yielding such low values relative to experimentally observed residual saturations.

The spatial distribution of NAPL at the pore scale is controlled by **wettability**, which refers to the tendency of one fluid to spread on or adhere to a mineral surface in the presence of another immiscible fluid. The contact angle between a fluid and a mineral surface is the standard measure of wettability. Surfaces are considered water wet if the contact angle (θ) measured through water is $\theta < 75°$, oil wet if $\theta = 105–180°$ and intermediate or mixed wet in between.

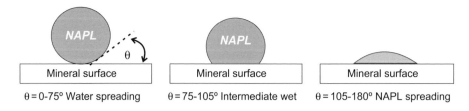

Figure 12.6 Contact angles and spreading phenomena by NAPL on pore surfaces.

Figure 12.6 shows how NAPLs might wet a pore surface. Initially mineral surfaces are water wet; however, with time, NAPL often wets those same surfaces, making NAPL displacement difficult hydraulically. In water-wet media, blobs of NAPL are trapped in the centre of large pores and fractures and can sometimes be displaced by water flooding, while in oil-wet media NAPL occupies smaller pores and is difficult to displace without altering the interfacial properties back to water wet. Intermediate or mixed-wet systems describe the condition where larger pores are organic wetting, while smaller pores are water wetting. When $\theta = 160°$ (see Figure 12.6), oil spreading begins; this phenomenon explains the appearance of oily sheens at air–water interfaces due to the balance of interfacial tensions between the three phases present (see Lake, 1989; Dwarakanath et al., 2002).

Soap bubbles in water exhibit a stress that leads to a contraction in their size (Dullien, 1979). The stress causing the size reduction is **interfacial tension** (IFT), referred to as surface tension in air–water systems. Let the pore space be described as a bundle of capillary tubes and the interface of the bubble have a mean radius of curvature, r. Then the balance of forces across the interface at equilibrium is described by the Laplace–Young equation, by which the capillary pressure (p_c) is defined as

$$p_c = p_{nw} - p_w = \frac{2\sigma}{r}\cos\theta, \qquad (12.9)$$

where p_{nw} and p_w are the non-wetting and wetting phase hydrostatic pressures, and σ is the interfacial tension in units of mN/m, i.e. force per unit length of the interface. The **entry pressure** (p_e) is the critical capillary pressure at which there is non-wetting phase penetration of pores. For cylindrical pores, (12.9) indicates that it is a function of pore-throat size, contact angle and interfacial tension (NRC, 2005).

Dwarakanath et al. (2002) showed that samples of various NAPLs recovered from waste-disposal sites tend towards NAPL-wet status and that their interfacial tensions were much lower than reference values for the principal NAPL present because of the incorporated oil, grease and detergents from use in industrial operations. Thus a DNAPL composed of

chlorinated solvents from a solvent disposal site had an interfacial tension of just 8.6 mN/m, a density of 1.37 g/mL and a viscosity of 0.8 mPa s, while freshly produced trichloroethene, the principal component of this DNAPL, has equivalent values of 34.5 mN/m, 1.46 g/mL and 0.57 mPa s. Similarly, studies at the Savannah River site in the USA (Dou et al., 2008) and with the Sherwood Sandstone in the UK (Harrold et al., 2001) indicate that NAPLs from waste-disposal sites display complex NAPL-wetting characteristics. Care therefore should be taken in using relationships based on simple water-wetting properties.

12.4.2 Migration of NAPLs

When $p_{nw} > p_e$, the capillary pressure of the non-wetting phase is sufficient to cause its entry into the water-wet capillary or pore throat. If we consider DNAPL so that we can temporarily avoid the issue of buoyancy, then DNAPL penetration leads to the idealization of its pressure–depth distribution as shown in (1) on Figure 12.7, in which S_{dnapl} and S_w are the DNAPL and water saturations such that they fill the pore space. Sufficient DNAPL has entered this soil column to cause it to pond and accumulate at a depth such that the entry pressure of the soil unit or rock at that depth is referred to as a **capillary barrier**, in which $p_{nw} < p_e$. Dewatering leads to DNAPL drainage and its mobilization, which is shown to increase the head of the DNAPL pool in (2) of Figure 12.7.

The flow of groundwater and mobile DNAPL in a porous medium is governed by their Darcy fluxes (Dake, 1994),

$$q_w = -\frac{k k_{rw}}{\mu_w} A \frac{\Delta p_w}{\Delta L}, \qquad (12.10)$$

$$q_d = -\frac{k k_{rd}}{\mu_d} A \frac{\Delta p_d}{\Delta L}, \qquad (12.11)$$

where k is the intrinsic permeability (m^2), k_{rw} and k_{rd} are dimensionless relative permeabilities of the groundwater and the DNAPL, μ_w and μ_d are the absolute viscosities of the groundwater and DNAPL, A is the cross-sectional area of flow through the porous medium, and $\Delta p_w/\Delta L$ and $\Delta p_d/\Delta L$

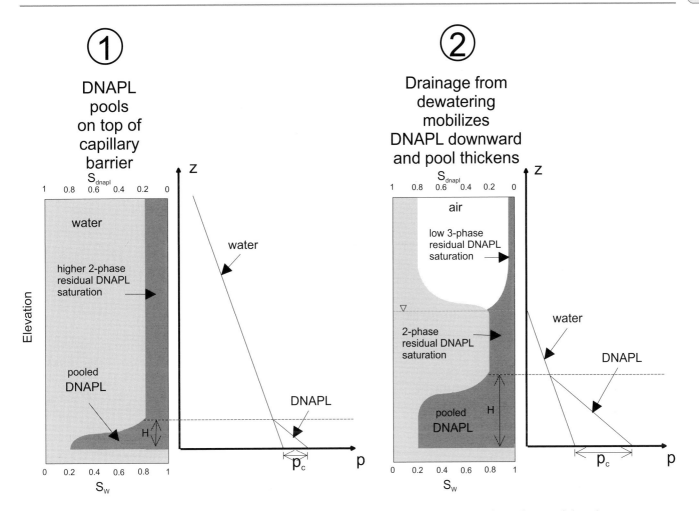

Figure 12.7 Conceptual model of DNAPL pool atop capillary barrier before and after dewatering. (After Jackson et al. (2006); © 2008 Canadian Science Publishing or its licensors. Reproduced with permission.)

are the respective pressure gradients. Figure 12.8 shows the **relative permeability** curves for the two phases that govern how the DNAPL flows to an extraction well; it will be limited by the coexistence in the porous medium of the aqueous phase. In the case of viscous DNAPLs, such as creosote, coal tar or diesel fuel, there is the additional problem posed by their high viscosities that cause them to migrate very slowly, reducing k_{rd} and lengthening their remediation (see Jackson et al., 2006).

The migration of DNAPL in the subsurface is strongly influenced by the volume released, the density and viscosity of the DNAPL and the capillary properties of the soil or rock encountered as migration occurs. Gerhard et al. (2007) provide a numerical assessment of these and other factors; they conclude that the DNAPL density and viscosity and the average soil permeability are the most important factors in DNAPL migration over time. Kueper et al. (1989) illustrated the effect of soil heterogeneity using a parallel-plate sand tank model with a heterogeneous distribution of sand units.

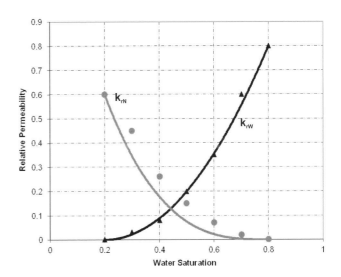

Figure 12.8 Relative permeability curves for NAPL (k_{rN}) and water (k_{rW}). (From Jackson et al. (2006); © 2008 Canadian Science Publishing or its licensors. Reproduced with permission.)

Table 12.6 Physical-chemical properties of the sand units in Figures 12.9 and 12.10.

Sand unit	Permeability (cm^2)	Grain size (mm)	P_e (cm water)	S_{or}
#16 Silica	5×10^{-6}	1.19	3.77	0.078
#25 Silica	2×10^{-6}	0.71	4.43	0.069
#50 Silica	5×10^{-7}	0.30	13.5	0.098
#70 Silica	8×10^{-8}	0.21	33.1	0.189

Table 12.6 lists the properties of the four sand units embedded in the sand box. Figure 12.9 presents the capillary-pressure curves for the four sands embedded in the sand box under DNAPL imbibition or water drainage. Note that, the finer the sand, the higher the entry pressure (p_e) must be to penetrate the water-wet sand and reduce its water saturation.

The critical head or DNAPL thickness (T_D) required so that the entry pressure of this particular capillary barrier is exceeded and penetration occurs is given by (McWhorter and Kueper, 1996)

$$T_D = 9.6 \left(\frac{\rho_w}{\Delta \rho} \right) \left(\frac{\sigma_{nw}}{\sigma_{aw}} \right) \left(\frac{K}{n} \right)^{-0.403}, \qquad (12.12)$$

where ρ_w is groundwater density, $\Delta \rho$ is the density difference between DNAPL and groundwater, σ_{nw} and σ_{aw} are the non-wetting and air–water interfacial tensions, respectively, K is the saturated hydraulic conductivity of the soil and n is its porosity.

The sand box is shown in Figure 12.10 for the condition occurring 310 s after tetrachloroethene (PCE) was allowed to enter the sand tank from the top, with water draining along each side of the sand box. The figure shows the numerical simulation of this experiment using the UTCHEM simulator (Pope et al., 1999) that matches the experimental results of Kueper et al. (1989) well but also allows estimation of the PCE saturations that were not measured in the experiment. A number of important observations may be made:

- The PCE was unable to penetrate unit 3 (#50) silica because it did not achieve the capillary entry pressure ($p_e = 13.5$ cm) required, although the sand unit is only 10 cm below the source inlet. The PCE moved laterally and bypassed it. Note on Figure 12.9 that this entry pressure is very close to 100% wetting-phase saturation for this sand.
- After 310 s the PCE DNAPL has invaded only the coarse sand units 1 and 2 and has ponded on the fine sand (#70) silica shown as unit 4.

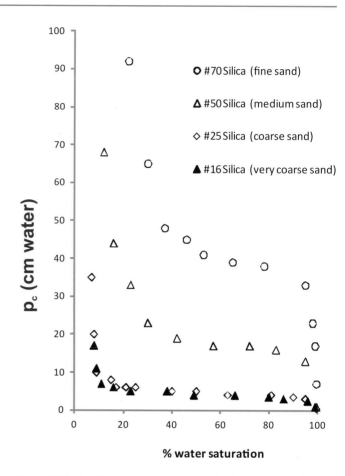

Figure 12.9 Capillary-pressure curves for the four sands used to construct the sand box model of Figure 12.10. (After Kueper et al. (1989); reproduced with permission from Elsevier.)

- The permeability of unit 2 (#25) silica is sufficiently less than that of unit 1 (#16) that there is some pooling (∼60%) saturation above the interface and lateral DNAPL flow within unit 2. Therefore there is "a continuity of pressure but not of saturation" (Kueper et al., 1989).
- Viscous DNAPLs, such as creosote (see Table 12.5), require 7 hours rather than 310 s to mimic the migration of PCE – see discussion of travel time in Gerhard et al. (2007).

We may use (12.9) for the capillary pressure behaviour of LNAPL following its release from fuel-storage tanks at the ground surface. The spatial distribution of petroleum hydrocarbons in the subsurface is affected by their buoyancy because they have specific gravities in the range of 0.78 for fresh gasoline up to 0.94 for diesel fuel (Brost and DeVaull, 2000). However, often what is discovered at a site is weathered LNAPL, as in Table 12.5.

The release of LNAPL to the subsurface, perhaps from a leaking underground storage tank, and its buoyant trapping at the water table is shown in Figure 12.11. In the case

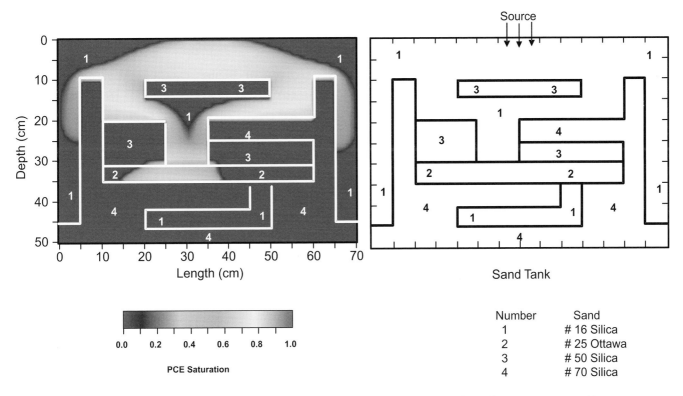

Figure 12.10 The Kueper sand-box experiment: simulation at 310 s after PCE release by M. Jin, formerly, INTERA, Austin, Texas. (Reproduced with permission from Elsevier.) See colour plate section.

Number	Sand
1	# 16 Silica
2	# 25 Ottawa
3	# 50 Silica
4	# 70 Silica

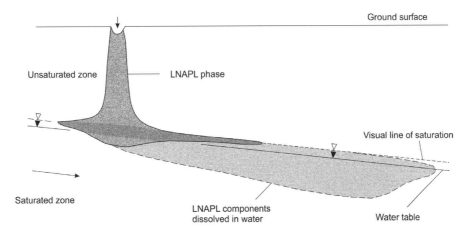

Figure 12.11 LNAPL release with penetration and buoyant trapping at the water table. (From Schwille (1975); with the permission of the IAHS.)

shown (Schwille, 1975), the volume of LNAPL infiltrating the soil exceeded its retention capacity and thus drained to the water table. Infiltrating water passing through the unsaturated zone and lateral groundwater flow generate a dissolved phase plume. The vertical zone of LNAPL contamination is referred to as the **smear zone** that will contain residual LNAPL.

As the water table drops, LNAPL will drain as it goes from two-phase to three-phase conditions. Thus it is common to find LNAPL pooling at depth at the end of a dry period. Figure 12.12 presents a conceptual model of LNAPL mobility with varying water-table elevations. Lundegard and Johnson (2006) describe the substantial natural attenuation of the LNAPL zone by volatilization, aerobic biodegradation and vapour transport in the unsaturated zone, which may cause indoor air pollution, as well as dissolution and further biodegradation in the saturated zone (see section 12.5).

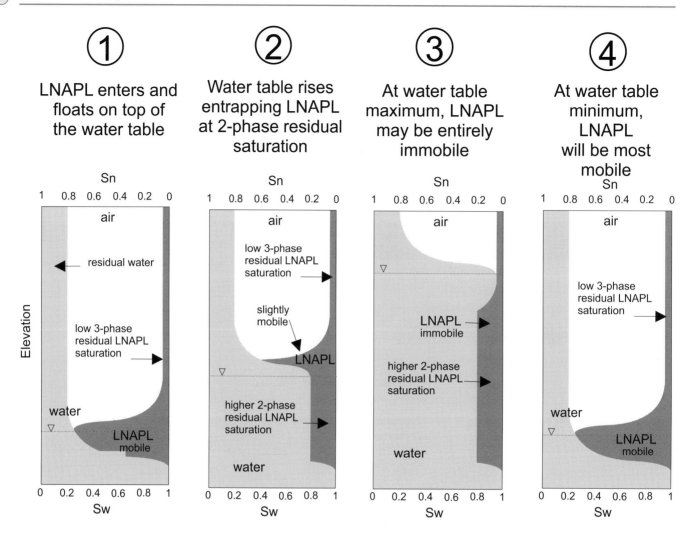

Figure 12.12 Conceptual model of LNAPL saturation profile with fluctuating water table, where S_n is the LNAPL saturation and S_W is the water saturation. (Drawn by J. Ewing, INTERA, Austin, Texas.)

12.4.3 Plume Generation by NAPLs

We may draw a similar schematic to Figure 12.11 but for a DNAPL zone generating a plume of dissolved compounds, perhaps a dissolved chlorinated solvent (e.g., TCE, TCA) leaking from a vapour degreasing system as in Figure 12.13. Note here that the inset pictures show the system as still being water wet, with the DNAPL occupying the interior of the pores; with time, this may change to DNAPL wet.

The DNAPL generates a dense vapour plume that contaminates the water-table aquifer; measured relative to air, vapour densities of the chlorinated solvents listed in Table 12.4 vary from 1.12 for PCE to 4.46 for CFC-113. The liquid density of the DNAPL causes its deep penetration, with ponding at the base of the water-table aquifer and subsequent lateral migration into the deep aquifer. Groundwater flowing left to right through the vertical zones of residual DNAPL

will ultimately cause contamination of the water-supply well shown. The length of such plumes may be very large if they go undetected in high-permeability aquifers over several years, and the dissolved solutes are sufficiently soluble and persistent in the redox conditions encountered.

Figure 12.14 shows a fully developed TCA plume that migrated from leaking underground storage tanks at the IBM plant in Silicon Valley, California (Jackson, 1998). The plume was detected in a water-supply well in San Jose, 6 km away, and its origin identified on the basis of the presence of both TCA and a chlorofluorocarbon solvent (CFC-113) then used in electronics manufacturing. Installation of the numerous monitoring wells shown led the investigators back to the plant. Mackay and Cherry (1989) calculated that the plume shown – later remediated at a cost of about $100 million – contained only 130 L of dissolved solvent within five billion litres of groundwater!

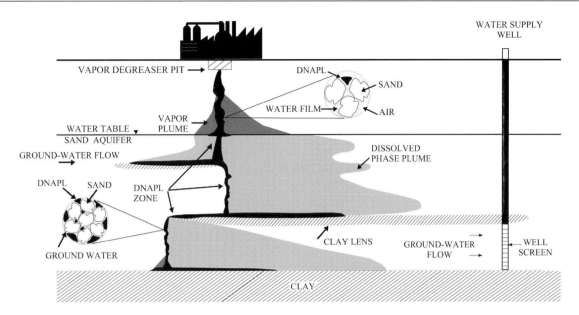

Figure 12.13 Migration of DNAPL in a sand aquifer and the development of DNAPL zones and dissolved-phase plumes, one of which is likely to contaminate a water-supply well. (From Jackson (1998); reprinted by permission of Springer from *Hydrogeology Journal* © 2003.)

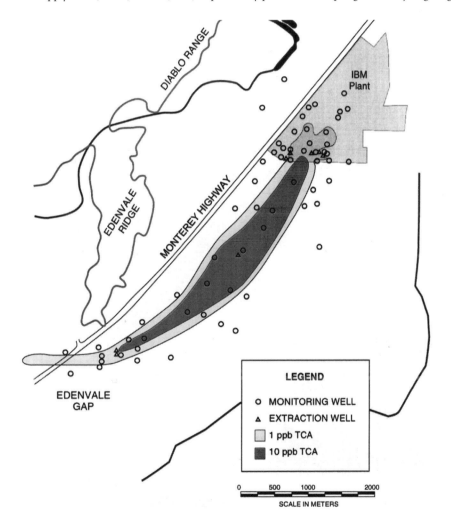

Figure 12.14 The TCA plume in South San Jose, California, in 1989 following its discovery by the contamination of the Tully well in San Jose. (From Jackson (1998); reprinted by permission of Springer from *Hydrogeology Journal* © 2003.)

The typical concentrations of TCA measured in the area of the DNAPL source zone at the IBM plant were approximately 100 µg/L, a small fraction of the aqueous solubility of TCA (720 mg/L, Table 12.4), although other compounds were much higher, e.g., CFC-113 at ~10 mg/L. It is most likely that this low concentration was the result of samples collected from monitoring wells with long well screens penetrating both clean and contaminated soils, so causing the sample from the contaminated zone to be diluted by clean groundwater. It is also possible that the monitoring wells were located too far from the DNAPL source zone to measure TCA concentrations in the mg/L range that indicate nearby DNAPL – see discussion in Kueper and Davies (2009).

However, most NAPLs are not single-component liquids and consequently the dissolved concentrations follow a form of **Raoult's law** (Bannerjee, 1984), which states that the **effective solubility** of an NAPL ($C_{e,i}$) is the product of the mole fraction (X_i) and its aqueous solubility ($C_{s,i}$) as listed in Table 12.4: $C_{e,i} = X_i C_{s,i}$. Therefore Raoult's law modifies what might be measured near a DNAPL source zone and reduces the measured concentrations of the various chlorinated solvents due to the presence within the NAPL of other solvents (e.g., CFC-113 in this case), solvent additives, oil and grease from degreasing operations as well as particulate matter (Jackson and Dwarakanath, 1999).

Similarly, Raoult's law applies to LNAPLs such as gasoline. Figure 12.15 shows a plume of gasoline-contaminated groundwater ten years after it was generated by groundwater flow through an LNAPL zone beneath a former underground storage tank. Chapelle et al. (2002) report that the maximum concentration of benzene measured in monitoring well LB-EX-1 four years after the LNAPL release was ~45 mg/L. Assuming this to be the effective solubility of benzene in the LNAPL, we can compute the mole fraction of the benzene (C_6H_6) remaining in the LNAPL given the aqueous solubility of benzene to be 1750 mg/L (Brost and DeVaull, 2000): $X_i = C_{e,i}/C_{s,i} = 45/1750 = 0.026$. This result is consistent with typical benzene concentrations in gasoline of 2–5%.

12.4.4 Fugitive Gas Migration

Before proceeding with a discussion of the fate of dissolved-phase plumes, we pause to consider the migration of fugitive natural gas released by leaking oil and gas wells. This concern developed very publicly in the USA with the application of hydraulic fracture stimulation of shale-gas reservoirs. In turn, this raised the possibility of the upward migration of natural gas towards shallow aquifers and surface waters, perhaps causing the accumulation of toxic and radioactive materials (derived from the reservoir) in soil and stream sediments near oil and gas wells. Rather than hydraulic fracturing, which can only rarely cause environmental problems at the ground surface (Dusseault and Jackson, 2014), concern soon became focused on gas migration within the wellbore annulus – the space between the well casing and the borehole rock wall that may or may not be fully completed with cement following drilling of a production well and with abandoned oil and gas wells (Jackson, 2014). Using noble-gas and hydrocarbon-gas tracers to identify gas migration pathways, Darrah et al. (2014) documented contamination of shallow domestic wells by natural gas due to a failure of annular cement seals.

The physical basis of fugitive gas migration is the development of free gas bubbles in the casing annulus adjacent to gas-rich zones that may not be the producing formation but which are likely to be under pressures of a few megapascals. These gas bubbles nucleate into gas slugs that have a strong tendency to rise in any microannulus of micrometre size that may form from cement shrinkage occurring during cement setting (Dusseault et al., 2000). Thus the microannulus may occur at the contact between cement sheath and borehole rock wall, allowing the gas slug to form a column and migrate upwards due to its buoyancy in formation waters, i.e., brine. The gas rises with a displacement pressure $p_d = zg(\rho_w - \rho_g)$, where z is the height of the gas column, g is the gravitational constant, and ρ_w and ρ_g are the densities of the brine and the natural gas.

As the gas column rises, it expands in length due to lower hydrostatic pressures encountered, i.e., ~10–11 kPa/m. The gas pressure–depth profile is much less (i.e., ~50 Pa/m). Therefore, upon nearing the ground surface, the gas column will likely have a pressure of hundreds of kilopascals (or even 1–2 MPa). Therefore the gas column will have the potential to easily exceed the capillary entry pressures of aquifers that are penetrated by the wellbore, for which p_e will be of the order of ~10 kPa or less in the case of open fractures. In this manner fugitive gas may enter shallow bedrock aquifers and migrate to a low-pressure zone, quite possibly a water well, where it will cause biodegradation of the GWQ.

12.5 Fate of Dissolved Contaminants

Dissolved contaminants are transported by groundwater and subjected to various **natural attenuation** processes, including dilution at production wells or other mixing processes, dispersion, volatilization, sorption and biodegradation. Both

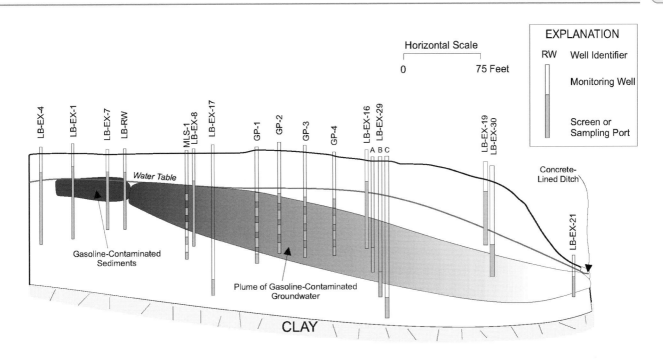

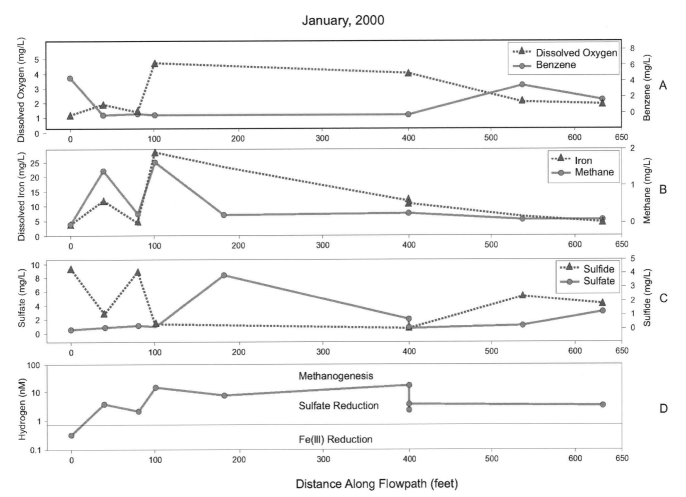

Figure 12.15 The generation of a benzene plume by LNAPL leakage from an underground storage tank into an underlying sand aquifer, Laurel Bay, South Carolina, USA. The lower panels, which are aligned with the cross-section, indicate the groundwater quality changes evident as various redox reactions occur sequentially. (From Chapelle et al. (2002); with the permission of John Wiley and Sons.)

fuel hydrocarbon (Figure 12.15) and chlorinated hydrocarbon plumes (Figure 12.14) are affected by these processes; inorganic contaminants are affected by dilution, dispersion and sorption. We shall consider the effects of sorption and biodegradation on contaminant concentrations and focus on organic contaminants, having discussed inorganic sorption in section 7.7.

12.5.1 Contaminant Sorption

In general, **hydrophobic compounds** are strongly sorbed, as their name suggests. These include high-molecular-weight compounds such as polyaromatic hydrocarbons (PAHs), which are commonly found in creosote and coal tar and which lack the functional groups like oxygen that promote dissolution in water. **Hydrophilic compounds** – those with high aqueous solubilities because they readily bond with the water molecule – tend to be poorly sorbed. These include additives in chlorinated solvents and gasoline, such as 1,4-dioxane used in TCA degreasing solvent, phenols found in creosote and coal tar, and methyl tert-butyl ether (MTBE), an oxygenate additive that was used in gasoline.

Sorption of groundwater contaminants and other solutes results in their retardation relative to the average linear groundwater velocity. The standard estimate of contaminant mobility is the retardation factor, R_f, given by

$$R_f = 1 + \frac{\rho_b K_D}{n} = \frac{v}{v_c}, \qquad (12.13)$$

where ρ_b is the bulk density of the porous medium (kg/m^3), K_D the distribution coefficient (m^3/kg) from section 7.7, n the porosity (dimensionless), v the groundwater velocity and v_c the contaminant velocity. The distribution coefficient can be obtained from batch or column tests in the laboratory or it can be estimated if the fractional organic carbon content (f_{oc}) of the porous medium is known. For this case, we use the organic-carbon partition coefficients K_{oc} for each compound, some of which are listed in Table 12.4, and compute the distribution coefficients from the relation $K_D = K_{oc} f_{oc}$. The variability in organic sorbent used to experimentally determine K_{oc} will lead to some uncertainty (Seth et al., 1999). Roy and Griffin (1985) provide K_{oc} values for 37 common organic solvents.

12.5.2 Contaminant Biodegradation

Dissolved hydrocarbons biodegrade in groundwater according to the **biodegradation reaction** (Borden, 2001):

$$HC + Ox + microbes + nutrients$$
$$\Rightarrow CO_2 + microbes + waste\ products$$

On the left-hand side of the arrow are the reactants for this biogeochemical reaction, in particular the dissolved hydrocarbon (HC) and an oxidant (Ox). The hydrocarbon is oxidized by an electron acceptor, such as dissolved oxygen or another less powerful oxidant, such as the oxygen bound into manganese (MnO_2) and ferric oxides (FeOOH) or in sulfate (SO_4); these reactions are illustrated in Figure 7.5. Such electron acceptors are reduced in the process of oxidizing the hydrocarbon and become waste products shown on the right-hand side of the reaction, in this case dissolved manganese, ferrous iron and sulfide. This redox reaction requires the catalysis by natural microorganisms that are unchanged by their role in the reaction but do require nutrients to sustain their function.

The arrow indicates that this reaction is not an equilibrium chemical reaction but one that goes to completion if there are sufficient electron acceptors, microbes and nutrients to complete the oxidation of the hydrocarbons to carbon dioxide. For example, we may write the oxidation of toluene as

$$C_7H_8 + 9O_2 \Rightarrow 7CO_2 + 4H_2O \qquad (12.14)$$

Borden (2001) states that the complete oxidation of toluene and many other hydrocarbons requires ∼3 mg DO/L for each 1 mg HC/L. However, the aqueous solubility of DO at 25°C is only 8.2 mg/L; consequently fully oxygen-saturated groundwater can only biotransform a limited amount of hydrocarbon contaminant, after which time other oxidants would function as electron acceptors to maintain biotransformation of the hydrocarbon. Full oxidation to carbon dioxide is not always achieved. In many cases, some intermediate contaminant is produced rather than CO_2. Examples include the dechlorination reaction that produces vinyl chloride or 1,1-dichlorethene from PCE or TCE, or that which converts MTBE to TBA. Both TBA and vinyl chloride are more toxic than their parent compounds.

The waste products of such reactions include indicators of bioremediation, e.g., dissolved Mn, Fe, H_2S and methane (CH_4). The biodegradation reaction shown above will convert at least some of the hydrocarbons to carbon dioxide, which hydrolyzes to form carbonic acid and subsequently bicarbonate in groundwater (see equation (7.28)). One product of this reaction is bicarbonate, which is usually measured as total alkalinity (see equation (7.29)). The proton produced by this reaction acidifies the groundwater and will cause the dissolution of carbonate minerals and even feldspars within the aquifer (Bennett et al., 1993), thus adding to the total dissolved solids (TDS) in groundwater. Therefore the waste products of the biodegradation reaction include many inorganic species as well as organic ones.

Figure 12.15 illustrates the redox reactions associated with benzene biodegradation at the LNAPL site discussed earlier. The lower panels show that the LNAPL zone itself (i.e., distance 0–80 m) remains anoxic with dissolved Fe, H_2S and CH_4 present. Downgradient of the LNAPL zone, the dissolved hydrogen gas concentrations are used to identify a large zone where methane is being generated (100–400 m), beyond which sulfate reduction is occurring (this zone was partially remediated by the addition of an oxygen-release compound). The pattern shown in the figure took ten years to develop with Fe(III) reduction occurring in the first four years, followed by methanogenesis and sulfate reduction (Chapelle et al., 2002).

12.5.3 Landfill Leachate

A particularly common problem for geoenvironmental engineers is the management of landfill leachate, which is generated by infiltration of precipitation through the municipal (and perhaps industrial) wastes. Should there be any failure of the soil or clay liner on which the waste is disposed, whether due to slope instability (Mitchell et al., 1990; Koerner and Soong, 2000) or other mechanism, leachate may seep from the landfill. Older landfills that lacked liners, such as that in Oklahoma studied by the US Geological Survey (Cozzarelli

et al., 2011), have released significant leachate plumes to the underlying soils, which have provided insight into leachate reactions in sandy aquifers beneath landfills.

Christensen et al. (2001) concluded that many contaminants in leachate plumes are naturally attenuated during plume migration and that leachate plumes seldom travel further than 1 km from the landfill. Most notable is the presence of elevated concentrations of dissolved organic carbon (30–29 000 mg/L) causing the groundwater to become anoxic, yielding high concentrations of Fe, Mn and H_2S following the sequence of reduction reactions shown in Figure 7.5. Christensen et al. (2001) note that these concentrations of redox-sensitive compounds depend on whether the waste is in the early acid phase or the later methanogenic phase of biodegradation. Also strongly elevated are those ions that do not undergo redox reactions, e.g., chloride (~2000 mg/L), K, Na.

Table 12.7 illustrates changes that occur to groundwater in the sand aquifer beneath a landfill in Oklahoma, USA. The specific electrical conductance (SEC) of the groundwater increases sixfold as it passes beneath the landfill and carries downgradient a very significant TDS load, mainly Na–Cl. Not recorded by the SEC measurements is the nonvolatile DOC increase from 3 to over 200 mg/L. Methane was oxidized anaerobically, probably by sulfate reduction, but was generally well below aqueous saturation (~30 mg/L) that might indicate gas-phase migration. Heavy metals (Ni, Cr) were probably mobilized by complexation with DOC to form soluble

Table 12.7 Median concentrations of groundwater quality parameters measured at the Norman, Oklahoma, landfill. (From Cozzarelli et al. (2011).)

Parameter	Units	Background GWQ	Landfill source	Downgradient GWQ
Specific conductance	μS/cm	1461	8832	5025
Nonvolatile DOC	mg/L	3.3	227	87
pH		7.1	7.0	6.8
HCO_3	mg/L	604	3987	2107
Ammonium	mg/L	2.1	532	151
Calcium	mg/L	132	104	245
CH_4	mg/L	0.06	8.2	3.7
DO	mg/L	0.14	0.1	0.1
Iron	mg/L	2.7	5.1	10.5
Arsenic	μg/L	2.2	13.5	8.0
Chromium	μg/L	2.0	10.5	5.5
Nickel	μg/L	<0.1	26.6	13.0

compounds. Arsenic was released from the landfill in concentrations close to the drinking-water limit (10 µg/L).

12.6 Characterization of Contaminated Sites

It has been wisely remarked that successful site remediation requires successful site characterization. Experience has shown the necessary importance of high-resolution sampling of plumes using multi-level sampling wells (Figure 12.16) to understand the nature of the source and how efficient remedial designs can be developed. The general methods

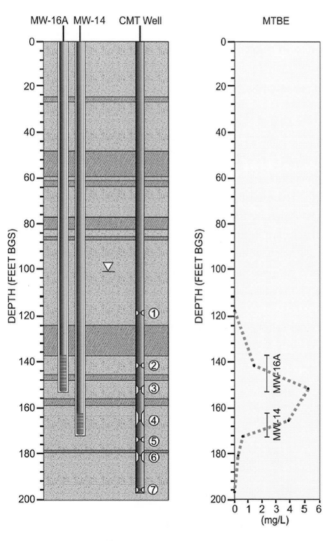

Figure 12.16 High-resolution of groundwater quality became possible in the 1980s with the development of multi-level sampling (MLS) wells. MLS wells, such as this continuous multi-channel tubing (CMT) well, allow the contaminant concentrations in individual sedimentary layers or fractures to be sampled without incurring the sample dilution that is inherent in the use of long-screened monitoring wells shown at the left. (From Einarson and Cherry (2002); with the permission of John Wiley and Sons.)

and tools employed are summarized in Table 12.8 and Box 12.2, with special emphasis on the application of geophysical tools.

The US National Research Council (NRC, 2005) identified four principal objectives in site characterization:

1. *Understanding the nature of the source.* What is the source producing the groundwater contamination, e.g., DNAPL, LNAPL, landfill leachate, and what is the anticipated behaviour of the individual components of the source?

2. *Characterization of the site hydrogeology.* What are the geological units that comprise the site and what are their hydrogeological properties? What are the preferential pathways leading off-site? Can the flow system be described quantitatively and simulated by mathematical modelling?

3. *Determining the geometry of the source zone and its dissolution rate.* How is the source material spatially distributed with respect to the geological units present? If NAPLs are present, what is their chemical composition and fluid saturations – residual as well as mobile NAPL? What is their potential for future migration off-site? If soluble mineral phases constitute the source – e.g., fly ash or radioactive waste – what is their rate of dissolution?

4. *Understanding natural attenuation processes.* What roles do transport and biodegradation processes play in the natural attenuation of the plume and the source?

12.7 Seawater Intrusion

Fresh groundwater typically discharges along coastlines with a chloride concentration of ≤ 20 mg/L. However, in coastal aquifers, a salt-water interface exists, seaward of which is salt water with a chloride concentration of 19 000 mg/L (TDS = 35 000 mg/L). Figure 12.17 shows the chloride concentration profile beneath Miami, Florida, where the Biscayne Aquifer is contaminated by seawater intrusion.

Hubbert (1940) showed that, above some datum, e.g., the base of the Biscayne Aquifer, the fresh-water (h_f) and salt-water (h_s) heads could be defined by

$$h_f = \frac{p}{\rho_f g} + z \quad \text{and} \quad h_s = \frac{p}{\rho_s g} + z,$$

where ρ_f is the density of fresh water, ρ_s is the density of salt water, p is the pore pressure at that point of measurement and z is the vertical location above the datum on the interface. To locate the interface between fresh and salt water, Hubbert reasoned that continuity of fluid pressure meant that these

Table 12.8 Methods and tools for site characterization. (Modified after NRC (2005).)

Method	Tools	Purpose
Historical information		Establish source composition and spatial distribution (air photos)
Regional geology		Establish site conceptual model and hydrogeology
Geophysical methods	Seismic refraction	Establish 3D hydrostratigraphic map, e.g., Watkins & Spieker (1971)
	Seismic reflection	High-definition lateral stratigraphy, e.g., Sharpe et al. (2003)
	Electrical resistivity	Identify water table, buried waste, conductive plumes and karst features
	Electromagnetic tools	Identify shallow hydrostratigraphy, conductive plumes, buried wastes, e.g., Greenhouse and Slaine (1986)
	Ground-penetrating radar	Establish shallow site stratigraphy, e.g., Aspirion and Aigner (1999)
	Magnetic tools	Location of wastes in steel drums
Direct push tools	Cone penetrometer (CPT) and rotary hammer	Characterization of shallow (<35 m) unconsolidated subsurface including pore pressure, GWQ samples, e.g., Pitkin et al. (1999)
Core retrieval and analysis	Hollow-stem augers, sonic drilling and hollow-stem or piston tubes	Measurement of spatial distribution of soil contamination, e.g., Munch and Killey (1985), McElwee et al. (1991)
Downhole methods	Video, flow metering, borehole geophysics, etc.	Measurement of formation properties, and flow regimes
Monitoring wells	Single-screen wells, and multi-level systems	Measurement of head and collection of GWQ samples, e.g. Einarson (2006)
Hydraulic testing	Slug tests and inter-well pump tests	Establish hydraulic properties and interconnections in aquifer system
Groundwater analysis	Monitoring wells	Delineation of source zones and plumes

---

BOX 12.2 | WORKFLOW FOR CHARACTERIZATION OF A GRANULAR AQUIFER BENEATH A LANDFILL

The objective of site characterization is usually to inform the decision-making process that leads to site remediation. This is invariably an iterative process lasting one or even many years depending on the complexity of the site and the availability of funding. An impressive example of such characterization is presented by Tremblay et al. (2014) in their development of a conceptual model for a sand aquifer beneath a landfill near Québec City in Canada. A workflow was developed in the following sequence:

1. *Continuous 2D geophysical surveys.* Initially both ground-penetrating radar (GPR) and electrical resistance tomography (ERT) surveys were conducted. Some 28 km of GPR survey was conducted and was complemented by 8 km of ERT, which was used where the GPR penetration depth was limited by high electrical conductivities. This testing defined the boundaries of the flow system and the hydrostratigraphy,
2. *Direct push testing.* The hydrostratigraphy of the flow system was investigated using cone penetrometer testing (CPT) and soil moisture resistivity (SMR) mapping. CPT samples permitted textural analysis, while dielectric measurements from SMR surveys indicated porosity and water content.
3. *Monitoring wells.* Fully screened monitoring wells (5 cm diameter) were installed in locations identified by the direct push testing. These permitted the collection and analysis of groundwater samples and a series of flowmeter and slug tests.
4. *Conceptual site model development.* Once the field data had been acquired, the flow system was defined in terms of its hydrofacies and hydraulic conductivity distribution, leading to the development of a flow and transport model. The groundwater quality data allowed an assessment of the natural attenuation of the leachate emanating from the landfill.

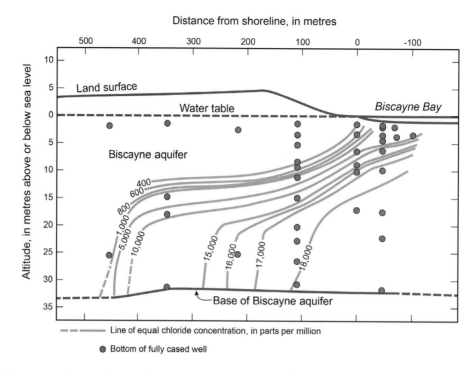

Distance from shoreline, in metres

Figure 12.17 Chloride concentrations in the transition zone in the Biscayne Aquifer near Miami, Florida (shown by the lines of equal chloride concentration), in which seawater contains 19 000 mg/L. The dots represent the well screens of fully cased wells at which depth the sample was collected. (Modified from Kohout (1964); courtesy of the US Geological Survey.)

two heads must be equal at the interface and therefore the elevation of the interface, z, is given by

$$z = \frac{\rho_f}{\rho_f - \rho_s} h_f - \frac{\rho_s}{\rho_f - \rho_s} h_s. \qquad (12.15)$$

Figure 12.17 indicates that in reality there is not a sharp interface between salt water and fresh groundwater because groundwater discharge and tidal fluctuations create a dynamic situation. Rather, a zone of transition occurs reflecting a number of factors, including fresh-water over-extraction by municipalities, decrease in surface-water infiltration due to channelization, artificial recharge and Pleistocene sea-level variations.

To obtain high-resolution interface information, in terms of hydraulic heads and salinity, the US Geological Survey has installed multi-level monitoring wells in southern California. Hanson et al. (2009) concluded that seawater intrusion in the coastal aquifers of southern California is associated with advancement of the salt-water plume through coarse-grain units at the base of fining-upward sequences of marine sediments, a finding similar to that of Neilson-Welch and Smith (2001) for the Fraser estuary of British Columbia near Vancouver. Edwards et al. (2009) describe how the structural geology of the Los Angeles–Long Beach area created pathways by which salt water intruded into the Dominguez Gap

area caused by overpumping after the 1920s. Further south in San Diego, Anders et al. (2014) identified pre-modern seawater intrusion into regional groundwater that Sengebush et al. (2015) ascribed to seawater inundation of the southern Californian coastline 120 ka during the last interglacial when the coastline was much lower than it is today. Barlow and Reichard (2010) have reviewed seawater intrusion studies along North American coasts.

In northwestern Europe, the Chalk of England, Sweden and France requires careful groundwater extraction so that intrusion is minimized, especially given the extensive network of **fissures** in the Chalk that allows inland seepage of seawater. In Brighton, England, seawater intrusion into the Chalk is managed by seasonally adjusting pumping rates so that the high winter groundwater flow is captured near the coast while peak summer demands are met by inland wells (Downing, 1998). The management of the Lincolnshire Chalk around Grimsby (Hutchinson et al., 2012), which is threatened by intrusion from the Humber Estuary, has been accomplished using a simple fresh-water model without resort to variable-density simulators developed by the US Geological Survey and necessarily used there – e.g., Nishikawa et al. (2009) in southern California and Hughes et al. (2009) in Florida. The effects of sea-level change on coastal aquifers is an issue of current concern that is addressed using variable-density models (e.g., Green and MacQuarrie, 2014) because of the complexity of other controlling factors, e.g., changing recharge and groundwater-extraction rates.

12.8 Wellhead Protection

Public-supply, domestic or farm wells may become contaminated by (1) short circuits to the wells due to well-construction failures or contaminated groundwater finding short flow paths to well screens or (2) migration of contaminants from more distant sources. The issue of short circuits to drinking-water wells has been addressed by Landon et al. (2010) for public-supply wells and for domestic or farm wells by Jackson and Heagle (2015). Here we focus on the concept of wellhead protection, which technique was adopted in Germany in the 1970s by identifying a groundwater catchment area for a particular public-supply well (Kinzelbach et al., 1992).

All groundwater flow lines within a catchment area would lead to the water well. Therefore the recharge area (A) of the catchment is a function of the extraction rate of the well (Q) and the recharge rate (R): $A = Q/R$. The catchment area is

the assemblage of streamtubes captured by and discharging to the water well. Thus the catchment area of the water well that was contaminated by the TCA plume in Figure 12.14 would include the area of the IBM plant, where solvent releases occurred. The larger the ratio of streamtubes with clean groundwater to those with contaminated groundwater, the greater the dilution of the contaminant will be at the water well (Jackson and Mariner, 1994).

In its simplest form, the determination of **wellhead protection areas** usually is based on a two-dimensional advective groundwater flow model of the aquifer containing the vulnerable well. The model neglects dispersion, sorption and biodegradation, as well as the third (vertical) dimension. This advective approach allows travel times to be computed. More sophisticated approaches are of course possible with more complex groundwater models, such as the **aquifer vulnerability** concept (Frind et al., 2006) involving particle back-tracking.

Figure 12.18 shows well vulnerability maps for the Greenbrook wellfield in Kitchener, Ontario (Canada). Figure 12.18(a) shows the catchment area and the maximum

(a) Max. conc. expected at the well field (C/Co)

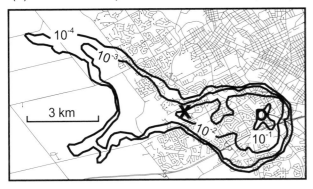

(b) Time to reach maximum concentration, years

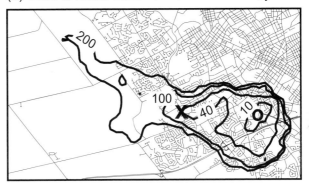

(c) Time taken to breach threshold, years

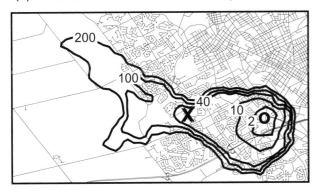

(d) Exposure time to concentrations above threshold, years

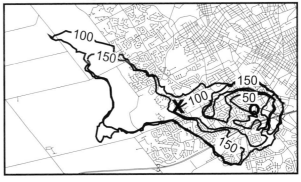

X reference point   **O** Greenbrook well field

Figure 12.18 Wellfield vulnerability maps, Greenbrook well, Kitchener, Ontario, Canada. (From Frind et al. (2006); with the permission of John Wiley and Sons.)

concentration that might be expected to discharge from the Greenbrook well, shown at bottom right, in normalized concentration units, i.e., maximum $C/C_0 = 1.0$. Figure 12.18(b) indicates times of travel from a reference point marked **X**, i.e., ~40 years. Figure 12.18(c) presents the time required to exceed a threshold value of 10^{-4} relative to the source concentration, i.e., ~35 years. Finally, Figure 12.18(d) indicates that the threshold value between the reference point and the well is exceeded for ~100 years, leading to well decommissioning after ~35 years.

12.9 Summary

Using principles learned in Chapter 7, we have examined the evolution of groundwater quality in alluvial and bedrock aquifers in England and North America. The application of stable and radioactive isotopes to problems of groundwater quality interpretation has been reviewed. The larger part of this chapter has been devoted to the issue of groundwater contamination that may affect the engineer through his or her need to manage sanitary landfills, protect groundwater-supply wells and remediate plumes of dissolved chemicals. In this context we have described how NAPLs migrate in the subsurface and how they generate dissolved-phase plumes, which are subject to attenuation by sorption and biodegradation. We have also introduced the issues of contamination by oil and gas wells, landfill leachate contamination, seawater intrusion into coastal aquifers and the establishment of wellhead protection areas.

12.10 FURTHER READING

Charbeneau, R.J., 2000, *Groundwater Hydraulics and Pollutant Transport*, 2nd edn. Upper Saddle River, NJ: Prentice-Hall. — This remains an excellent summary of the fluid mechanics and hydrologic principles necessary to understand groundwater contamination, especially LNAPL problems. Includes useful spreadsheet computer programs.

Clark, I.D., 2015, *Groundwater Geochemistry and Isotopes*. Boca Raton, FL: CRC Press. — A modern and comprehensive introduction to understanding groundwater quality issues with a geochemist's treatment of contamination and environmental isotopes. The discussion of groundwater dating in chapter 8 is particularly helpful.

Dullien, F.A.L., 1992, *Porous Media: Fluid Transport and Pore Structure*, 2nd edn. San Diego, CA: Academic Press. — A fine introduction to transport processes, pore structure and interfacial phenomena in a very readable format.

Ingebritsen, S., Sanford, W. and Neuzil, C., 2006, *Groundwater in Geologic Processes*, 2nd edn. Cambridge, UK: Cambridge University Press. — Chapter 3 is a fine review of solute transport processes and their influence on GWQ.

Kueper, B.H., Wealthall, G.P., Smith, J.W.N., Leharne, S.A. and Lerner, D.N., 2003, *An Illustrated Handbook of DNAPL Transport and Fate in the Subsurface*. Bristol, UK: Environment Agency. — A lucid and comprehensive introduction to the migration and fate of DNAPLs in soil and rock, with specific reference to conditions in the UK.

NRC (National Research Council), 2005, *Contaminants in the Subsurface: Source Zone Assessment and Remediation*. Washington, DC: National Academies Press. — A readable account of the science and management of source-zone characterization and remediation by a distinguished panel of experts.

Pankow, J.F. and Cherry, J.A. (eds), 1996, *Dense Chlorinated Solvents and Other DNAPLs in Groundwater: History, Behavior and Remediation*. Portland, OR: Waterloo Press. — The standard reference for those concerned with DNAPL issues. Clearly written and exhaustive.

Wiedemeier, T.H., Rifai, H.S., Newell, C.J. and Wilson, J.T., 1999, *Natural Attenuation of Fuels and Chlorinated Solvents in the Subsurface*. New York, NY: John Wiley and Sons. — A comprehensive account of natural attenuation reactions.

12.11 Questions

Question 12.1. Consider Table 12.3 on the Floridan Aquifer. De-dolomitization is the result of concurrent gypsum dissolution causing calcite precipitation as noted in the text. Complete the geochemical reaction that describes de-dolomitization:

$$1.8\ CaSO_4 + 0.8CaMg(CO_3)_2 \Rightarrow ??$$

and show it is consistent with the data in Table 12.3.

Question 12.2. Compute DNAPL drainage due to dewatering. Assume a soil zone of one cubic metre (porosity = 35%) with the water table at the top surface. Under these two-phase conditions, the residual saturation in the saturated zone (S_{ors}) is 27%. Let the water table drain to the lower surface so that the one cubic metre soil zone drains DNAPL to the residual saturation of the vadose zone (S_{orv}) of 9%. What is the volume of DNAPL released downwards by dewatering?

Question 12.3. Prior to the development of equation (12.15) describing the salt water–fresh water interface, it was customary to use the Ghyben–Herzberg formula to locate the interface:

$$z_{GH} = \frac{\rho_f}{\rho_s - \rho_f} h_{WT},$$

where z_{GH} is the depth below sea level to a point on the interface and h_{WT} is the height of the water table above sea level at that same location. Assume that the density of fresh water and salt water are 1.00 kg/L and 1.025 kg/L, respectively, and compute how much the salt water–fresh water interface will rise should the water table be lowered one metre by pumping?

Question 12.4. Estimate the entry pressure for silica sand 2 (#25 Ottawa) in Figure 12.10 in cm of PCE.

Question 12.5. Figure 7.8 shows two adsorption isotherms for ^{90}Sr in a sand aquifer. Assume the groundwater velocity is 30 m/yr, the bulk density of the sand is 1.7 g/cm^3 and its porosity is 0.38 mL/cm^3, i.e., millilitres of groundwater per cubic centimetre of porous medium. Using equation (12.13), how much farther in one year does a ^{90}Sr plume in the unweathered sand travel compared with that in the weathered sand?

PART

IV Geological Hazards

Part IV addresses those natural hazards that are predominantly geological in their occurrence. As such, engineers are engaged in preventing damage from these hazards to infrastructure or mitigating the effects of previous events. In either case, the engineer is obliged to understand why certain hazardous events occur in various geological environments and what the risk is of their occurrence or re-occurrence.

We have introduced the hazards associated with volcanic activity in Chapter 2 with a discussion of Mount St. Helens (Figure 2.4), which last erupted catastrophically in 1980, and of the various volcanic hazards (Figure 2.5) that afflict communities situated near volcanoes. We return to this topic with a discussion of a particular volcanic hazard in Chapter 15, where we discuss lahars, a type of debris flow initiated by volcanic activity.

The chapters that comprise Part IV do not explicitly address hazards arising from **climate change**. However, the role of climate change in this context is to exacerbate various processes through drought, intense rainfall events and storm surges onto erodible shorelines. Therefore, climate-change-induced drought will cause many communities to rely more heavily on groundwater supplies, which in turn will likely result in increased land subsidence in sedimentary basins and karst terrain. Alternatively, more intense rainfall will promote flooding and stream channel erosion, karst dissolution and landslides. Tropical cyclones – as well as those in temperate regions – will cause storm surges from the sea onto erodible shorelines that are experiencing worldwide sea-level rise. Therefore, while the geological processes described in the last four chapters of this book have occurred in times long before climate change was evident, the effect of climate change is one of increasing the hazards that the engineer must address.

In Chapter 13 the role of groundwater, mineral and oil and gas extraction on land subsidence is considered. Karst phenomena are a constant source of public concern in regions of carbonate and sulfate bedrock through the development of solution channels and sinkholes. We shall show that many of these unstable ground conditions are principally associated with the action of groundwater on the solid skeleton of soil or rock. Thus, whether the concern is subsidence due to pumping groundwater or petroleum or ground failure due to dissolution of soluble rocks, an understanding of the geotechnical and hydrogeological processes is essential in interpreting existing conditions. However, subsidence is in many cases due to the extraction of minerals, particularly coal, which we address by considering subsidence in US and European coalfields.

In Chapter 14, we consider seismic faulting, seismic waves, active tectonics and the generation of earthquakes. Tectonic forces are those associated with the effects of plate tectonics, earthquakes resulting from plate movements and the subsequent stress redistribution in the Earth's crust. It is common knowledge that earthquakes cause damage to infrastructure. We consider the effects of the 1989 earthquake near Loma Prieta, California, which caused the collapse of highway bridges and other infrastructure in that part of California, and the hazard posed by even larger subduction-zone earthquakes to the North American region of Cascadia – Oregon, Washington and British Columbia. We consider the nature of faults that arise from the brittle failure of the Earth's crust, the detection of active faults, the nature of ground motion in general and liquefaction in particular.

Landslides, the subject of Chapter 15, are triggered by many processes, including earthquakes. Frequently groundwater, snowmelt or rainfall – or some combination of these – is involved in triggering a landslide. Figure 15.2 shows a still-moving landslide near Slumgullion, Colorado, with the main scarp that marks the edge of the slide shown on the horizon and the slide itself with its distorted trees in the foreground. We consider the characterization of landslide motion using inclinometers and piezometers and the mechanisms by which slope failure occurs, both in soil and in rock. We complete this chapter by reviewing an in-depth analysis of the worst landslide in recent European history – the Vaiont landslide disaster in northern Italy – that took the lives of nearly 2000 people in 1963 and was caused by incomplete geotechnical characterization.

Finally, Chapter 16 describes the hazards experienced along coastlines and in estuaries arising from storms and wave action, tsunamis caused by earthquakes, subsidence due to groundwater and oil and gas production, landslides and the steady rise of sea levels throughout the world. In an age that is deeply concerned with climate change, some of the most profound geomorphic changes will occur along coastlines and in heavily populated estuaries. In particular, we consider the erosion of shorelines along the US Gulf and Atlantic coasts from storm surges associated with tropical cyclones, to the failure of coastal bluffs along the Great Lakes and to cliff erosion along the Californian and English coastlines from wave action.

CHAPTER

13 Land Subsidence and Karst

Among geological hazards important to construction and infrastructure maintenance, the issue of land subsidence is a continuing technical challenge made more difficult by the social and economic context in which it occurs, especially the supply of water to cities in arid climates and of irrigation water to successful agricultural economies. We treat subsidence in this part of this book as a hazard of hydrogeological origin, although it may in fact be promoted by mineral or oil and gas extraction rather than groundwater extraction or groundwater dissolution of soluble rocks causing karst.

We begin this chapter by considering the relationship between groundwater extraction and land subsidence that results in a loss of groundwater storage and, in some instances, the formation of earth fissures that may erode into gullies channelling flash floods. Then we consider land subsidence caused by the oxidation of organic soils during agricultural development. Mining-induced subsidence, which is well established in the mining regions of Europe and elsewhere, is then discussed. Finally we consider the critical issue of karst-related problems such as sinkhole development, cavity detection in areas of construction, and aquifer and dam vulnerability arising from dissolution of soluble rocks.

13.1 Groundwater-Extraction-Induced Subsidence

The sedimentary basins of the world are host to large urban and rural populations that use the groundwaters of the basins as a major water supply for municipal, industrial and, particularly, agricultural uses. Not all sedimentary basins are subject to land subsidence induced by groundwater extraction. If the

sediments have been lithified into sedimentary rocks, groundwater extraction may cause elastic responses to groundwater extraction without causing compaction of the sediments.

However, if the sediments are both unlithified and contain significant thicknesses of compressible fine-grained deposits (principally clays and silts), there is a strong probability that these sediments will undergo inelastic compaction with prolonged groundwater extraction. This phenomenon has been noted throughout the US Southwest and in the Texas Gulf Coast, Mexico, China, Taiwan, Japan, Vietnam, Thailand, Italy, Spain, Iran, Indonesia and London (UK). Sedimentary compaction will cause differential subsidence across the basin – and at small scales – sometimes resulting in the kind of failure shown in Figure 13.1. The same effect is noticed with the extraction of oil and gas rather than groundwater. Our particular concern is with alluvial and lacustrine sediments deposited in the intermontane basins – often called **basin-fill aquifers** – and deltas across the world.

Subsidence of coastal areas around Long Beach Harbor in California were observed in the 1930s when production began from the Wilmington oilfield. In 1948, Karl Terzaghi, acting as a consulting engineer, indicated that the production of oil, gas and brine was the cause of the subsidence. By this time, the US Navy and the City of Long Beach were involved in multi-million-dollar mitigation efforts to protect the port facilities from rising sea levels and to repair streets, water pipelines and railroad tracks deformed by the subsidence (~9 m by 1969). In 1949, the subsidence was associated with the failure of the casings of 200 oil wells because of a 33 cm displacement of a local fault. Terzaghi indicated in his report that fluid extraction from the oilfield had reduced the pore pressure and increased the effective stress in the basin sediments, especially

Figure 13.1 Damage to a house in Pahrump, Nevada, USA, caused by subsidence induced by groundwater extraction.

the fine-grained units that became compressed, thus leading to land subsidence and the relative increase in sea levels within Long Beach Harbor (Goodman, 1998). Subsidence was halted in the Long Beach area only by seawater injection into the petroleum reservoirs to prevent further subsidence.

The US Geological Survey has identified the range of subsidence hazards that develop from groundwater depletion:

- changes in elevation and slope of streams, canals and drains;
- damage to bridges, roads, railroads, storm drains, sanitary sewers, pipelines, canals and levees;
- damage to private and public buildings (as in Figure 13.1);
- enhanced coastal and river flooding;
- failure of well casings from forces generated by compaction of fine-grained materials in aquifers; and
- development of earth fissures that intercept surface-water drainage and cause gully erosion.

Subsidence can be halted by careful groundwater management that typically involves regulatory intervention, as in the cases of Houston, Texas, and the Santa Clara Valley, California (see Galloway et al., 1999).

In this section, we consider the principle of effective stress (introduced in section 1.3.4) in the context of land subsidence. We illustrate how fine-grained hydrogeologic units within alluvial-basin sediments (**aquitards**) respond by drainage and compaction in response to a reduction in pore pressure. The same principles are involved in soil consolidation; however, in consolidation theory, an increase in the total vertical load on sediments is involved rather than an *in situ* reduction in pore pressure. Consequently we will use the geological term

"**compaction**" rather than the geotechnical term "consolidation" to emphasize the difference in context. Then we will consider how the aquifer changes in response to prolonged groundwater extraction and how earth fissures develop and groundwater storage is lost. Finally, we outline the various methods of measuring land subsidence currently in use.

13.1.1 Compaction

The arrival of a freight train at a railroad station overlying an elastic artesian aquifer causes an increase in the pore pressure within the aquifer (Ferris et al., 1962). The progressive transfer of this increased load to the grains of the aquifer skeleton causes the water level to rise in a monitoring well installed within the aquifer. Similar effects are caused by barometric and tidal loading of artesian aquifers. These effects are elastic in nature and do not permanently deform the aquifer structure or cause loss of the pore volume that in part constitutes the capacity of the aquifer to store groundwater.

The hydrogeological and geomechanical responses differ in the case of heterogeneous coarse- and fine-grained unconsolidated sediments, such as those that constitute the aquifer systems underlying deltas and intermontane basins, in which case the aquifer system loses pore volume through compaction of the relatively more compressible aquitards (confining units and interbeds) within the aquifer system with prolonged groundwater extraction. These low-permeability interbeds are deposited in the intermontane basins of the world by sheetfloods and debris flows that cross alluvial fans (see Figure 9.16) and as lacustrine deposits, which may have formed as pluvial lakes during the Quaternary. Blair and McPherson (1994) note the presence of mud from sheetfloods in the centres of the alluvial basins, with coarser materials present on the flanks of the fans.

In regional subsidence cases, hydrogeologists typically treat land subsidence in response to groundwater extraction as a one-dimensional consolidation of the sedimentary column involving both elastic and inelastic deformation following Terzaghi's classical theory of consolidation of a finite-thickness layer of saturated soil (see Coduto, 1999; Wang, 2000). This deformation is usually dominated by compaction of the fine-grained sediments, as is shown in Figure 13.2(B). When pumping occurs, groundwater flows towards the extraction well (not shown in Figure 13.2B) and there is a slight movement of coarse-grained sediments towards the well, resulting in silts and sands entering the filter pack around the well screen and perhaps the well itself.

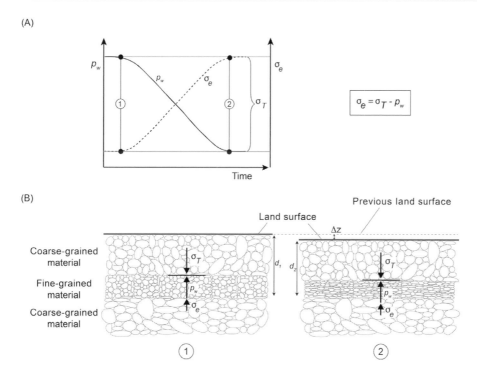

Figure 13.2 (A) The principle of effective stress is illustrated for the case of land subsidence caused by groundwater extraction leading to the drainage and compaction of fine-grained materials comprising aquitards. (B) Deformation of the aquitard materials occurs as the effective stress (s_e) within the fine-grained material increases in direct response to the overall reduction in pore pressure (p_w). The measured vertical displacement of the land surface ($\Delta z = d_1 - d_2$) constitutes land subsidence. (From Sneed and Galloway (2000); courtesy of the US Geological Survey.)

However, a larger vertical displacement occurs as the fine-grained aquitard material, which is interbedded within or forming confining units between the coarse-grained aquifer units, responds to the lowered pore pressure in the aquifer units. Deformation of the aquitard materials occurs as the **effective stress** (σ_e) within the fine-grained material increases in direct response to the overall reduction in **pore pressure** (p_w). The measured vertical displacement of the land surface (Δz) constitutes **land subsidence**.

The total vertical stress on the aquitard – σ_T in Figure 13.2 – is counterbalanced by the upward stresses exerted by the pore pressure and the effective stress. With continued pumping of groundwater, the proportion of the total stress supported by the pore pressure necessarily declines and the effective stress within the aquitard increases (see Figure 13.2A). The consequence is aquitard drainage, as groundwater generally flows vertically from the aquitard to the adjacent aquifer, with accompanying compaction. As a rule of thumb, hydrostatic considerations indicate that a 1 m drop in the water level will result in a $\sim$10 kPa increase in the effective stress within a clay unit beneath the water table.

Terzaghi's theory of consolidation was adopted by hydrogeologists (Riley, 1969) to interpret compaction and drainage

of aquitards within intermontane basins, such as those of the Basin and Range Province of the US Southwest, or within deltaic deposits. Among its assumptions are that (1) the aquifer–aquitard system is fully saturated, (2) the layer being compacted is horizontal with negligible solid-grain compressibility, (3) the vertical drainage of groundwater from the aquitard is governed by Darcy's law, and (4) there is a linear reduction in void ratio with each increase in effective stress.

13.1.2 Aquitard Drainage

Previously in Chapter 11 we had defined the **specific storage** (s_s) of an aquifer volume in (11.12) as $s_s = \rho g(\alpha + n\beta)$, where ρ is the specific weight of water, g is the acceleration of gravity, α is the matrix (vertical) compressibility of the porous medium, n is the porosity and β is the compressibility of water. In the context of aquitard drainage, the specific storage is the volume of water that a unit volume of aquifer releases from storage under a unit decline in hydraulic head. It is the sum of two terms – compressibility of the aquifer and of the groundwater.

In the conceptual diagram of aquitard compaction shown in Figure 13.2(B), we ascribe the permanent compaction to the fine-grained aquitard units within the aquifer because the inelastic compressibility of sands and gravels is usually insignificant by comparison with that of silts and especially clays (only when the aquifer units contain micaceous sands is the aquifer response to groundwater pumping inelastic). During compaction, the aquitard – denoted here by primed terms with $'$ – drains by one-dimensional flow following (11.18) as (Galloway and Burbey, 2011)

$$\frac{\partial^2 h}{\partial z^2} = \frac{s'_{sk}}{K'_z}\frac{\partial h}{\partial t}, \qquad (13.1)$$

where s'_{sk} is the skeletal specific storage of the compacting aquitard in m^{-1} and K'_z is its vertical hydraulic conductivity (m/s). We neglect the specific storage term associated with the compressibility of water (β) because it is much less than the compressibility of clays (see Table 11.2) and define s'_{sk} by

$$s'_{sk} = \rho g \alpha'_k, \qquad (13.2)$$

where α'_k is the skeletal compressibility of the aquitard. A drained consolidation test in an oedometer provides an estimate of α'_k, which is reported in units of m^2/kN.

The analysis and simulation of aquifer–aquitard system compaction requires that the preconsolidation stress of the system first be estimated. The **preconsolidation stress** is defined as the maximum previous effective stress to which the system has been subjected and is estimated by a drained test of the fine-grained aquitard material in an oedometer (section 10.3.3). The ratio of the preconsolidation stress to the present effective stress is the overconsolidation ratio from soil mechanics. Sneed and Galloway (2000) indicate that, in a hydrogeological context, the preconsolidation stress may be considered as having developed during the previous lowest hydraulic head measured in wells after allowing sufficient time for fluid pressures to equilibrate throughout the aquifer system. We may expect overconsolidation in deltaic or alluvial intermontane basin aquifers for a variety of reasons, including erosion or geological unloading, prehistoric groundwater-level declines, desiccation, diagenesis and weathering.

Once stresses are imposed on an aquifer–aquitard system in excess of the preconsolidation stress, compaction of fine-grained sediments occurs as shown in Figure 13.2(B), while the coarse sand and gravel units remain essentially unaltered. The amount of compaction of the aquitard (Δz) is estimated by $\Delta z = s'_{sk} z_0 \Delta h$, where z_0 is the original thickness of the aquitard and Δh is the change in hydraulic head adjacent to it.

Terzaghi's theory of consolidation due to compaction of clay layers indicated that the drainage period would be lengthy because of their low vertical hydraulic conductivity, which might be of the order of 10^{-12} to 10^{-7} m/s. If the compacting aquitard drains both vertically upwards and downwards into adjacent aquifers as in Figure 13.2 – i.e., a doubly draining aquitard – the required time for $>$90% of ultimate compaction is (Galloway and Burbey, 2011)

$$\tau' = \frac{s'_s (z_0/2)^2}{K'}, \qquad (13.3)$$

where s'_s is the specific storage of the aquitard ($\cong s'_{sk}$), z_0 is the aquitard thickness and K'_z is the vertical hydraulic conductivity of the aquitard.

In the US Geological Survey's MODFLOW groundwater flow simulator, the skeletal specific storage is divided into elastic and inelastic components and specified separately for the aquitard and the coarse-grained aquifer units (see Leake and Galloway, 2010). The aquifer units undergo insignificant compaction compared with the clay-rich aquitards and we may expect the inelastic component of aquitard specific storage to be between 30 and several hundred times larger than the aquitard's elastic component. For regional subsidence studies (on the order of hundreds of km^2 or more), the simple one-dimensional Terzaghi approach is suitable. Where local two- or three-dimensional deformation of the pore volume occurs, such as around a pumping well in unconsolidated sediments, Biot's theory of poroelasticity may be used to simulate the hydromechanical effects (Burbey, 2006; Galloway and Burbey, 2011). In this case, the skeletal specific storage for the aquifer–aquitard system is estimated from the shear modulus and Poisson's ratio and the compaction is estimated as above.

The effect of delayed aquitard drainage is that subsidence is itself delayed and progressive. Figure 13.3 shows the hydrologic response over a 90-year period of the basin aquifer in the Antelope Valley, north of Los Angeles, California (Sneed and Galloway, 2000), which was originally artesian where confined. While the water levels in the aquifer units have declined progressively over time, the heads in the two thick aquitards have lagged these levels. The shallow thick aquitard at ~60 m is laterally continuous and drained upward to about 1948, then after 1958 was double draining until about 1968 when the water table receded to the top of this aquitard. The deep thick aquitard at 100 m depth, which is laterally discontinuous, continues to be double draining.

Figure 13.4(A) illustrates a hydrograph of a monitoring well in the Antelope Valley subject to seasonally varying rates of groundwater extraction that began 80 years earlier. Such monitoring wells are screened in and measure hydraulic heads in

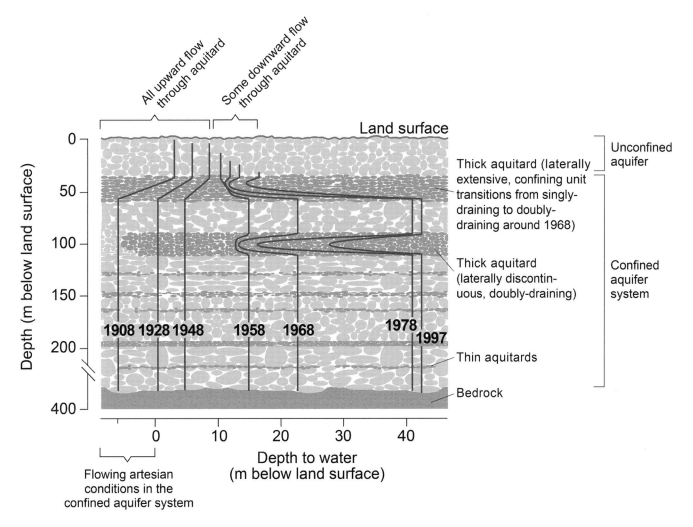

Figure 13.3 Hydraulic heads in the aquifer beneath Antelope Valley, California, have progressively decreased with time as groundwater development occurred, starting in 1908 when the confined aquifer was artesian. Aquitard drainage lagged behind the heads in the aquifer because of the low vertical hydraulic conductivities in the aquitards that were draining during compaction. (From Galloway et al. (1999); courtesy of the US Geological Survey.)

the aquifer units rather than reflecting heads in the aquitards. During 1993–96, a period of groundwater level recovery, the seasonal low water levels increased and their amplitudes also increased. Despite the inter-annual recovery trend and the seasonal recovery of the water level, compaction progressively increased, indicative of the hydrodynamic lag of head changes in the aquitard with respect to changing heads in the adjacent aquifers. Furthermore, the rate of subsidence appeared to be constant. Simulation of aquitard drainage, which is illustrated in Figure 13.4(B), indicates that subsidence began in the late 1950s after water levels had fallen 20 m. The thin aquitards drained and compacted first, followed by drainage of the thick aquitards. Although groundwater levels stabilized in the late 1990s, the goodness-of-fit of the simulations to the measured compaction indicate continued compaction beyond 1997.

In Figure 13.2(B), the compression of the fine-grained aquitard layer produces a clay structure aligned perpendicular to the effective stress with lower porosity than before. This is consistent with studies of clays undergoing standard consolidation testing. Delage and Lefebvre (1984) observed that this anisotropic structure develops with one-dimensional consolidation of Champlain Sea clays and it must be expected that compaction by groundwater extraction in compressible aquifer–aquitard systems will reduce the vertical hydraulic conductivity of the compressed aquitard (state 2) compared with its original state (state 1). Rudolph and Frind (1991) concluded that K' and s'_s were indeed being reduced during compaction of clays near Mexico City as groundwater extraction progressed. They noted that these decreased in vertical hydraulic conductivity, and specific storage caused

(A)

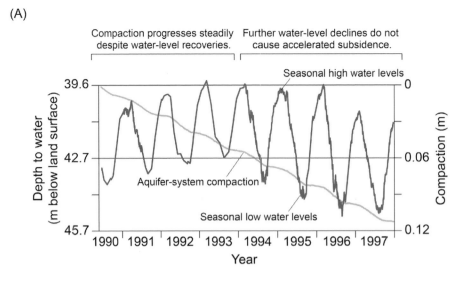

(B)

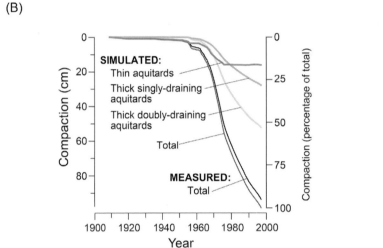

Figure 13.4 (A) A hydrograph records the seasonal water-level fluctuations from groundwater extraction in a monitoring well in Antelope Valley, California. Aquifer system compaction reflects the hydrodynamic lag between aquifers and aquitards. Compaction is progressive but delayed compared with the minimum water levels. (B) Measured and simulated compaction for the period 1908–97 during which groundwater extraction occurred. (From Galloway et al. (1999); courtesy of the US Geological Survey.)

a reduction in the aquitard leakage flux to adjacent aquifer units. Consequently, producing wells had to increase rates and expand the hydraulic-head drawdown cone of influence to maintain a steady supply rate.

13.1.3 Fissure Development and Growth

A particular feature of aquifer–aquitard compaction is the development of long fissures that may develop quite suddenly; one in Arizona is 11 km long. The progressive development of fissures in Nevada from groundwater depletion is illustrated by Figure 13.5. The causes of earth fissuring have been explored by Sheng et al. (2003) and are believed to derive from the hydraulic forces driving aquifer material and groundwater

horizontally during pumping (Helm, 1994) and the effects that these forces have on sediments and geological structures within the basin.

These authors considered several conceptual models that produce earth fissures, including dipping faults, bedrock ridges and thinning of aquitards, and evaluated the sensitivity of various parameters in producing tensile failures. They concluded that earth fissuring is a multi-step process that begins with the movement of the aquifer solids and groundwater towards the pumping well, thus initiating a zone of tensile stress at depth. Differential horizontal and vertical displacements then develop along weakness planes as tensile failure occurs. The net result of these differential movements is the formation of a tensile failure zone at depth that can

Figure 13.5 Progressive development of fissures, Pahrump Valley, Nevada. (a) An incipient fissure photographed behind the house shown in Figure 13.1; the aperture of the fissure is approximately 1 cm as shown by the scale. (b) A well-developed fissure. (c) A fissure that has been widened into a gully by flash flood erosion.

migrate to the ground surface and evolve into a fissure. The earth fissure becomes a zone of enhanced porosity that is vulnerable to the infiltration of contaminated surface waters into the aquifer.

Field examples of the role of faults and heterogeneous sedimentary structures are particularly instructive. In the intermontane basin aquifer of Queretaro, Mexico, groundwater depletion caused differential land subsidence of the order of 10–60 mm/yr over a 40-year period. The drawdown resulted in the slippage of high-dip-angle faults and low-dip-angle stratigraphic joints, resulting in the shearing of water-well casings (Carreón-Freyre et al., 2016). Burbey (2002) investigated subsidence in the basin-fill aquifer of Las Vegas, Nevada, and concluded that vertical differential subsidence is associated with the variations in thickness of aquitard units crossing vertical faults. His analysis identified five criteria that, when coupled with groundwater extraction, generate earth fissures:

1. An arid or semi-arid climate with a relatively deep unsaturated zone overlying a deforming aquifer–aquitard system that is being subjected to hydraulic stress from groundwater extraction, e.g., the basin-fill intermontane aquifers of the US Southwest.
2. Long-term depletion of the aquifer system by groundwater extraction causing deformation.
3. Unconsolidated and fully saturated basin-fill sediments containing aquitard materials exhibiting significant vertical compressibility.

4. The presence of hydrogeological discontinuities, e.g., high-angle Quaternary faults in basin sediments, causing abrupt changes in horizontal hydraulic conductivity (and thus transmissivity) leading to tensile failure of the brittle unsaturated zone and fissuring.
5. Differential subsidence caused by the variable thicknesses of aquitard units leading to bending of the brittle unsaturated zone, which has little cohesive strength by virtue of its low moisture content.

13.1.4 Measuring Subsidence

Table 13.1 presents estimates of subsidence rates from areas across the world indicating that there are several viable approaches to measuring land subsidence. Sometimes it is quite obviously displayed when an anchored structure, such as a water well, protrudes above the sunken ground surface yielding a local estimate of ground-surface displacement. The original method of measuring land subsidence was spirit levelling by traditional survey methods using a theodolite and, more recently, geodimeters with laser range finding complemented by satellite-based systems (Galloway et al., 1999). Table 13.2 summarizes the capabilities of the various measurement approaches.

The US Geological Survey introduced the use of **extensometers** in the 1950s and these have provided estimates of both vertical and horizontal displacements in subsiding

Table 13.1 Rates of subsidence from groundwater extraction over specified periods of time. (From Galloway and Burbey (2011).)

Location	Rate of subsidence (mm/yr)	Period	Measurement method
Bangkok, Thailand	30	2006	Levelling
Bologna, Italy	40	2002–2006	Differential interferometry
Houston–Galveston, Texas, USA	40	1996–1998	Differential interferometry
Jakarta, Indonesia	250	1997–2008	GPS
Mexico City, Mexico	300	2004–2006	Differential interferometry
Murcia, Spain	35	2008–2009	Differential interferometry
Tehran Basin, Iran	205–250	2004–2008	Differential interferometry
Yunlin, China	100	2002–2007	Levelling

Table 13.2 Some methods of measurement of land subsidence and aquifer–aquitard system compaction. (From Galloway and Hoffmann (2007).)

Method	Sense of displacement	Resolution (mm)	Spatial density (samples/survey)
Spirit level	Vertical	0.1–1	10–100
Geodimeter	Horizontal	1	10–100
Borehole extensometer	Vertical	0.01–0.1	1–3
Horizontal tape extensometer	Horizontal	0.3	1–10
Horizontal invar-wire extensometer	Horizontal	0.0001	1
GNSS	Vertical	20	10–100
InSAR	Range	1–10	$10^5 - 10^7$

basins. Figure 13.6 shows a two-stage, counterweighted pipe extensometer developed by the USGS. With compaction of the aquifer–aquitard system, the reference surface subsides in elevation with the land surface, and the deeper extensometer pipe appears to rise in elevation with respect to the shallow extensometer pipe because the deeper pipe is anchored in sediments that are not deforming, i.e., ideally below the zone of aquifer-system compaction. In this manner the vertical subsidence is measured over the depth interval between the shallow seated piers supporting the reference surface and the bottom of the deep extensometer pipe. Horizontal extensometers use tapes or wires to measure displacement along fixed distances of perhaps 10 m.

Since the 1980s, **global navigational satellite systems** (GNSS, or GPS when the US satellite system is referred to) using Earth-orbiting satellites have become the standard method of conducting land surveys because of their high precision over distances of several to tens of kilometres. Networks of GPS stations have been established to monitor

subsidence, including one for the Antelope Valley in California and others for basins where land subsidence is attributed to groundwater extraction.

Most recently, radar interferometry has been adopted to measure **subsidence bowls** with extraordinary spatial detail and have demonstrated the development of subsidence bowls over compacting aquifer–aquitard systems. By repeated satellite measurements, **interferometric synthetic aperture radar** (InSAR) is able to map deformation in the ground surface with spatial resolutions of the order of a few metres, with the size of a picture element (pixel) being roughly 80 m². InSAR is used to compile a time series of radar line-of-sight images from a satellite to yield measurements of ground displacement.

Hydrogeological applications of InSAR include the identification of groundwater flow barriers, such as fault zones and textural changes in alluvial-basin sediments, and constraining numerical models of aquifer–aquitard compaction (Galloway and Hoffmann, 2007). These authors provide useful examples

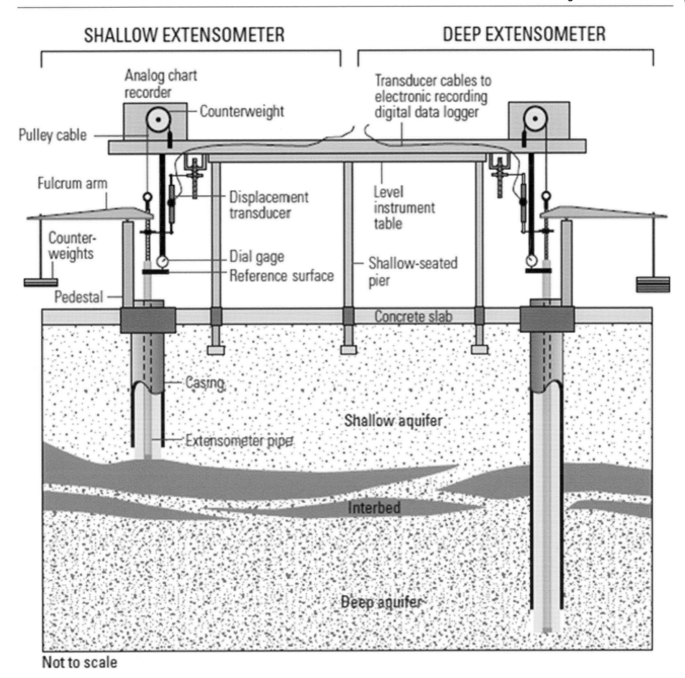

Figure 13.6 Schematic of a borehole pipe extensometer used by the US Geological Survey in Antelope Valley, California. The extensometer measures vertical displacement at two depth intervals in the aquifer. The displacement is measured as a movement of the pipe relative to the reference surface and indicates shortening (subsidence) or lengthening (uplift) of the distance between the shallow-seated piers and the anchor depths of the extensometer pipes. (From Galloway et al. (1999); courtesy of the US Geological Survey.)

of its hydrogeological application in areas of California and Nevada, while Chaussard et al. (2014) describe how it has allowed determination of subsidence rates of 5–30 cm/yr beneath many Mexican cities, which occupy intermontane basins with significant deposits of lacustrine clays as well as alluvial sands and gravels. Thus, InSAR provides a regional subsidence map that can be complemented by the use of

local measurements of higher resolution using levelling, geodimeters, GNSS or extensometers.

13.2 Mining-Induced Subsidence

Two hundred years of coal, gold and salt mining have caused land subsidence features across the industrial world that

BOX 13.1 | SUBSIDENCE FROM DRAINAGE OF ORGANIC SOILS

Tidal marshlands exist in deltas and coastal plains across the world and host agricultural developments that have caused the oxidation of the peat-rich soils followed by their subsidence, e.g., the Dutch coastal plain, the Sacramento–San Joaquin delta of northern California and the Zennare Basin, south of the Venice Lagoon in the Po delta of Italy. Consequently, sea levels threaten their operation with flooding. These marshlands may contain delicate ecosystems and wildlife that are protected under law, which has led to the restoration of the marshlands.

The sediments of coastal plains and deltas are composed of clastic sediments, such as sand, silts and clays as in Figure 9.17, and organic matter, which is formed at sea level. These **peatlands** are sustained by frequent flooding, sediment deposition and their biological productivity. Human agricultural cultivation typically prevents annual inundation by the building of levees or dykes and lowers the water table to allow air entry into the root zone of crops. In this manner, peat formation ceases and oxidation of the organic matter proceeds, leading to contraction of the plant fibres and pores. The peat thus shrinks with oxidation and carbon is lost through evolution of carbon dioxide. The peat thus loses its buoyancy and this new load then compacts the underlying peat-rich sediments causing further subsidence.

It has been estimated that the Dutch coastal plain, consisting of up to 20 m of sediment, has lost $\sim$20 km^3 of peat volume over the past 1000 years of cultivation. Much of the peat was lost to surface mining for fuel until the late nineteenth century that led to the creation of lakes up to 6 m deep. Peat mining and biodegradation have caused land subsidence, with approximately 25% of the Netherlands now situated below mean sea level. Most of the subsidence (>70%) is thought to be due to peat oxidation.

The case of the Sacramento–San Joaquin delta is particularly instructive. The farms in the delta exist in 55 discrete islands or tracts of land that are protected from flooding by levees, i.e., berms composed of dredged sediments that have been built up continually over the period 1860–1930. Inside the levees, continual drainage is needed to maintain the water table sufficiently deep (1–2 m) to prevent waterlogging of the crops that require aeration. Over time, this drainage has exposed the organic soils, composed of peat and mineral sediments, to oxidation as air entered the now-depressed unsaturated zone. The peat decomposes and is converted to carbon dioxide that vents to the atmosphere leading to land subsidence. Thus the fields, enclosed behind the levees, subside over time at rates of 2–7 cm/yr, with some fields now more than 5 m below sea level.

The management of agriculture and wildlife in the Sacramento–San Joaquin delta is made much more complex by virtue of the fact that the delta is a critical part of the water-transfer system by which fresh water is released from dams in the foothills of the Sierra Nevada Mountains and routed through the delta and on by aqueduct to southern California and the Santa Clara (Silicon) and San Joaquin valleys. About 50% of all streamflow in California passes through the delta, with about 25% of this flow pumped southwards through the aqueducts of state and federal water systems. In this manner, the municipal water supplies of the Los Angeles and San Diego areas are largely met and irrigation water supplied to Californian farmers. Levee failure is a constant source of worry because of spring flooding or, far more seriously, from earthquakes or liquefaction associated with seismic shaking that could affect many levees.

Near Venice, the Zennare Basin has subsided 1.5–2.0 m over the past 70 years, leading to a flood hazard from storm surges of the Adriatic Sea and to the threat of seawater intrusion from nearby rivers and the sea. The drainage system that permits continued farming – ditches, levees and pumping stations – requires continued maintenance and rebuilding to maintain the water table at depth, as subsidence has depressed the ground surface to as much as 4 m below sea level. Consequently, peat oxidation continues and the land subsides at 2–5 cm/yr.

Sources: From Ingebritsen and Ikehara (1999), Gambolati et al. (2006) and Erkens et al. (2016).

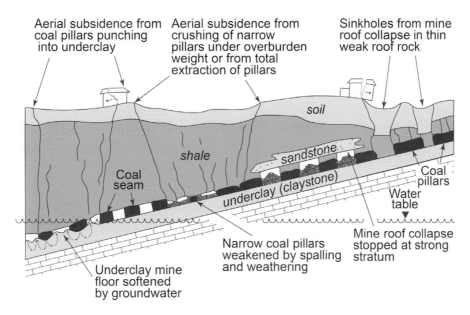

Aerial subsidence from coal pillars punching into underclay

Aerial subsidence from crushing of narrow pillars under overburden weight or from total extraction of pillars

Sinkholes from mine roof collapse in thin weak roof rock

soil

shale

sandstone

Coal seam

underclay (claystone)

Coal pillars

Water table

Narrow coal pillars weakened by spalling and weathering

Mine roof collapse stopped at strong stratum

Underclay mine floor softened by groundwater

Figure 13.7 Modes of subsidence caused by coal mining in Appalachia, USA. Note the vertical fracturing in the overlying shale that contributes to the generation of mine water, often of very poor quality from acid-rock drainage. (Modified from Turka and Gray (2005).)

increasingly pose hazards to infrastructure in the twenty-first century. Dewatering for mining is obviously necessary and this can also add hydrogeological consequences. In the Ruhr coalfields of Germany, 25 m of subsidence has been measured (Harnischmacher and Zepp, 2014) associated with the famous Zollverein mine in Essen. Coal mining in Dortmund nearby caused as much as 9 m of subsidence, producing a lake 200 m wide and 450 m long as the subsided ground intersected the local water table (Bell et al., 2000). While these are extreme examples, even small amounts of subsidence due to mineral extraction can have costly implications that may be delayed in effect by many years.

Subsidence caused by old mine workings – and these may date back hundreds of years to the Middle Ages – are often associated with tabular seams of coal, gypsum or other minerals that outcropped at the surface. These ores were mined by a room-and-pillar method by which the minerals were extracted, leaving behind rooms supported – at least temporarily – by pillars of the ore itself. The pillars over time might have been mined or suffered progressive failure because of the increased overburden load transferred to them. Room collapse and subsidence of the land surface might occur many years later. Consequently, there are many towns and cities now faced with subsidence hazards, in particular in the former coal-mining areas of the UK, the Ruhr in Germany, South Africa and in Appalachia in the USA. Alternatively, early mining may have proceeded by the progressive excavation of vertical shafts of perhaps 10 m depth to the ore body followed by a bell-shaped excavation of the ore.

The subsidence sinkholes that develop above the mined cavities are collapse features caused by either **pillar collapses** or, when the roof material fails, **crown holes**. These collapse features fill with broken rock from the overburden and may create **chimneys** through the overburden that may grow upwards over one hundred metres to the ground surface. Figure 13.7 illustrates the modes of subsidence observed in the coal-mining areas of Appalachia, USA, which are described by Turka and Gray (2005) and Mabry (2009). The use of geophysical tools to detect mining-related voids is actively pursued in the USA (Gardner, 2009) and by the British Geological Survey (e.g., Gunn et al., 2009). Box 13.2 provides practical advice on investigating subsidence in coal-mining regions.

Price (2009) indicates that the maximum height (Z) to which a collapse feature (or void) might propagate upwards can be estimated by the expression $Z = z/(B_f - 1)$, where z is the thickness of the tabular ore body, e.g., a coal seam, and B_f is the bulking factor, which is the ratio of the volume of the rock rubble filling the collapse feature to the volume occupied by the same weight of rock when undisturbed. Values of B_f range from 1.2 for soft shales and mudstones to 1.5 for sandstones.

Since the nineteenth century, subsurface coal mining has been by **longwall mining**, in which a cutting machine works along the length of a coalface (the longwall) in a supported area known as a panel (Brady and Brown, 1993; Donnelly et al., 2008; Price, 2009). The width of the panel is 30–300 m and perhaps 1 km in length. Figure 13.8(a) illustrates longwall

---

BOX 13.2 | SUBSIDENCE FROM COAL MINING

R.E. Gray's recommendations (published by the American Society of Civil Engineers for the geotechnical engineering community) to geotechnical engineers concerning mining-induced subsidence are based on long experience in Appalachia where there are believed to be 300 000 abandoned mines. These recommendations, which also relate to other mineral-mining operations, can be summarized as follows:

- Know the geological structure of the region in which you are working.
- Determine if there is a historical record of regional problems associated with coal mining.
- While larger overburden thicknesses above a mine do provide some reduction in likelihood of subsidence, no overburden thickness really provides safety from experiencing subsidence.
- Subsidence may occur long after mining has ended, although total extraction by longwall mining or excavation of pillars will shorten the after-effects.

Source: From Gray (2009).

---

mining of a horizontal coal seam with the development of a **subsidence trough** at ground surface (Figure 13.8b). This feature is outlined by the angle of draw, values for which are 8–45°. As the working face advances, subsidence occurs within about two years above the collapsed worked panel while there may be a slight rise in the ground surface ahead of the working face from tensional forces due to stress redistribution (Figure 13.8c).

Consequently, the surficial effects of longwall mining are typically soon apparent; however, this is not the case when longwall mining reactivates faults. The nature of fault reactivation caused by mining-induced subsidence is complex. Quite obviously, the prevailing and pre-existing stress distribution and the geological history of the fault are important. Also important are the frictional properties of the fault, e.g., Mohr–Coulomb parameters, the hydraulic heads within the fault zone and various mining factors, e.g. depth and rate of extraction. Donnelly et al. (2008) note that simple normal faults are most prone to slip downdip when their dip is in excess of 70° and the fault-zone infill material does not provide high frictional resistance.

With the abandonment of much coal mining in the twenty-first century because of cheaper abundant natural-gas supplies, groundwater that had previously been prevented from invading coal mines began to resaturate abandoned mines. This **mine water** is distinctive not only because it exhibits all the undesirable properties of acid-rock drainage (section 7.8), having come into contact with exposed sulfide minerals in and above the ore body (e.g., Younger, 2002), but also because of its effect in increasing pore pressures in fault zones and thus promoting fault slip through shear-

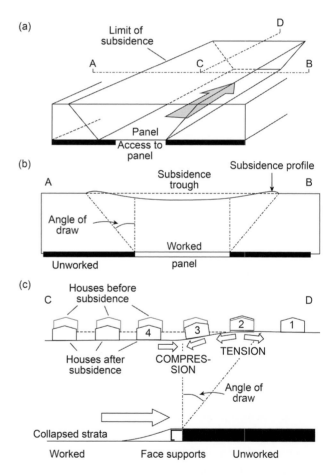

Figure 13.8 Subsidence effects for a uniformly level landscape caused by longwall mining, i.e., the total extraction of the coal seam. (From Price (2009); reprinted by permission from Springer © 2009.)

strength reduction. In this context, Booth (2007) has noted that the shallow fissures and bedding-plane discontinuities that develop with subsidence following the onset of longwall

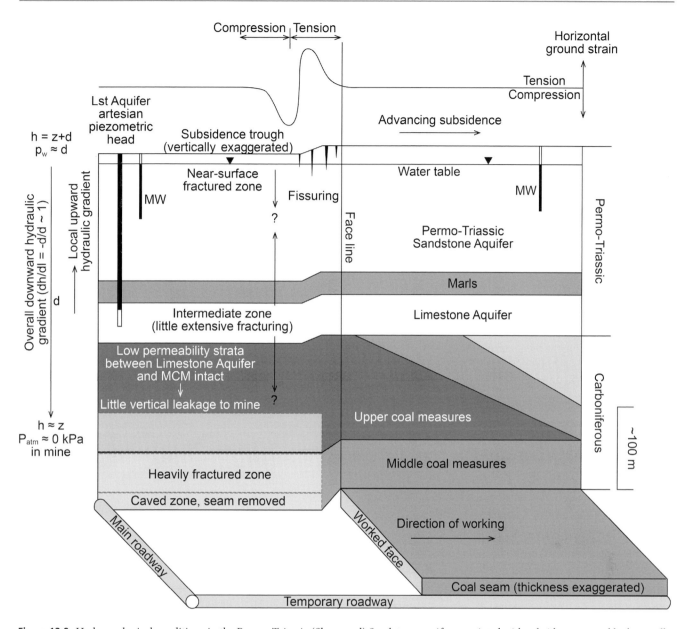

Figure 13.9 Hydrogeological conditions in the Permo-Triassic (Sherwood) Sandstone aquifer associated with subsidence caused by longwall mining in the Nottinghamshire coalfield, UK. (From Shepley et al. (2008); with permission of The Geological Society of London.)

coal mining can allow a confined aquifer to begin to behave as an unconfined aquifer. Through-going fracturing of the confining unit, such as the shale in Figure 13.7, will cause sulfide-mineral oxidation within the former confining unit and, with recharge from this shale unit, groundwater quality degradation of the aquifer below.

In the East Midlands of the UK, the Permo-Triassic Sandstone aquifer is an important groundwater supply; it is also known as the Sherwood Sandstone, which was introduced in Chapter 11 as an example of a stratiform aquifer. In Figure 13.9, Shepley et al. (2008) consider the hydrogeological effects of longwall mining on the outcrop area of the aquifer as shown in Figure 11.19, applying a conceptual model used

by Booth et al. (2000) and Booth (2006) for coalfields in Appalachia and Illinois:

1. A lower heavily fractured zone of greatly increased hydraulic conductivity drains directly into the mine. The thickness of this zone is ~30–40 times the thickness of the extracted ore body.

2. An intermediate zone subsides coherently with little through-going fracturing. In many cases this is an aquitard, but in this case this zone includes an artesian limestone aquifer shown as the Magnesian Limestone in Figure 11.19, which is part of the Permian Zechstein Group.

3. An uppermost zone consists of shallow strata subject to extensional stress and fissuring but which is unaffected by dewatering operations in the coal mine.

The shallow fissuring in the sandstone aquifer caused by mining-induced subsidence has increased its transmissivity but complicated the dynamics of flow between local streams and the sandstone aquifer due to large-scale fracturing of the streambeds. In some upstream areas, flows have decreased and exposed the aquifer to surface-water pollution, while flows have increased in some downstream areas because of subsidence.

13.3 Karst

We conclude this chapter on land subsidence by turning to the dissolution of soluble rocks – carbonates and evaporites – that pose engineering problems where these rocks are close to the ground surface and thus subject to enhanced dissolution by the infiltration of groundwater undersaturated with respect to carbonate, sulfate and chloride minerals.

Karst terrain is identifiable by its abundance of sinkholes that may have been inundated by the water table to form lakes, such as the mantled karst of Florida. But karst is not always found where it might be expected. Figure 13.10(a) shows ponds formed in sinkholes in the San Juan Mountains of Colorado. Karst terrain is characterized by a number of common features: sinkholes, disappearing streams, springs and limestone pavements with clint-and-grike features as in Figure 13.10(b). The standard English-language reference for karst geoscience is that of Ford and Williams (2007).

Initially we consider the evolution of karst terrain in its geochemical and hydrogeological context and then turn to how it evolves through moderate karst formation in temperate environments to extreme karst in tropical regions. Then we discuss the geotechnical problems that karst creates and site characterization in karst terrains, in particular the use of geophysical tools in the detection of karst.

13.3.1 Dissolution of Soluble Rocks

Karst may form in any soluble rock such as limestone, dolomite, gypsum, halite or sylvite. The term **soluble rock** indicates that recharge of typically acidic rainfall (pH $\sim$ 5.7, see section 7.5) will cause significant dissolution of the rock mass (at least in temperate climates), creating such geotechnical hazards as sinkholes and hydrogeological conditions allowing turbulent flow in the channel network.

Figure 13.10 Karst terrain. (a) Ponds occupying sinkholes in the Leadville Limestone, San Juan Mountains, Colorado, USA. The limestone is Mississippian period but was not karstified until the Pennsylvanian. (b) Clint-and-grike feature that has developed on the exposed Amabel dolomite, Manitoulin Island, Lake Huron, Ontario, Canada. Rock hammer for scale is shown bridging the fissure, i.e., the "grike".

The driving force for dissolution of such rocks is the geochemical undersaturation of infiltrating precipitation (see section 7.4) with respect to the minerals present. If we consider limestone as a suitable example of karst development, then the acidity of rainwater infiltration will cause carbonate mineral dissolution and the increase in pH, calcium and bicarbonate ions in groundwater (i.e., equation (7.30) and Figure 7.4). With carbonate rocks, the presence of dissolved carbon dioxide in the infiltrating water drives the dissolution of carbonate minerals; these groundwaters are considered "aggressive". But carbonate groundwaters may remain undersaturated with respect to calcite (i.e., SI < 0.0), particularly when pH < 7.5. This is not the case with the evaporite minerals, for which simple congruent dissolution occurs rapidly.

The evolution of groundwater in carbonate aquifers and streams draining carbonate rocks depends on the rate of surface reaction of calcium and hydrogen ions at the surface of calcite minerals; it is suspected that mineral impurities within the calcite, e.g., aluminosilicates, may inhibit the dissolution of calcite crystals. Above pH 6, the rate of calcite dissolution in carbonate aquifers slows significantly and equilibrium with calcite (i.e., SI = 0.0, equation (7.26)) may take many years (Plummer and Wigley, 1976; Eisenlohr et al., 1999).

It should also be appreciated that the solubility of carbonate minerals is strongly temperature-dependent. In alpine areas, such as is shown in Figure 13.10(a), the solubility product of calcite at $0°C$ is $\log K_{sp} = -8.38$, whereas at $30°C$ it is -8.51. For $P_{CO_2} = 10^{-1.5}$ or 3%, calcite solubility at $10°C$ is $127\,mg/L$, whereas at $25°C$ it is $100\,mg/L$ (see Langmuir, 1997). This decrease in solubility is reflected in calcite precipitation in warm climates promoting carbonate reefs in warm seas and **caliche** in the soils of the US Southwest and other semi-arid climates.

While the groundwater will not necessarily become super-saturated with these minerals, there is a curious situation with carbonate minerals that allows further dissolution of carbonate minerals when mixing of two calcite-saturated groundwaters occurs because they converge from separate flow channels in karst. This is known as **corrosion mixing** and can cause undersaturation of the mixed groundwater, thus promoting further carbonate-mineral dissolution and karstification of carbonate aquifers.

Why this should be so is explained in Figure 13.11, in which the solubility curve for Ca^{2+} is an approximate function of the cube root of the pressure of CO_2. Consequently, when two calcite-saturated waters with differing gas pressures mix, the result is an undersaturated solution that only achieves equilibrium with calcite by dissolving more calcite at the expense of its dissolved (aqueous) carbon dioxide concentration, which typically means its bicarbonate ions (see Langmuir, 1997; Appelo and Postma, 2005). Mixing corrosion therefore provides additional dissolution of carbonate minerals in aquifers that display karst features, which prompts us to enquire as to the nature of such aquifers.

13.3.2 Groundwater Flow in Carbonate Aquifers

There are many carbonate aquifers in the world that are not considered karstic, i.e., they do not exhibit obvious features such as caves and large solution channels or conduits that

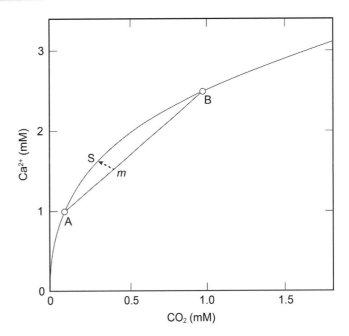

Figure 13.11 The solubility of calcite in pure water at $25°C$ and 1 atm total pressure showing calcium and aqueous CO_2 concentrations. The curved line represents solution equilibrium with calcite; therefore groundwaters plotting below the line are undersaturated with respect to calcite, while those plotting above the line are supersaturated. A and B represent calcite-saturated groundwaters that mix in karst channels to form solution "m" that only reaches solubility equilibrium with calcite by dissolving additional calcite to reach a calcium concentration given by S. (Figure reproduced from Langmuir (1984), Physical and chemical characteristics of carbonate water. In P.E. Lamoreaux, B.M. Wilson and B.A. Memeon, eds, *Guide to the Hydrology of Carbonate Rocks*, © UNESCO 1984. Used by permission.)

exhibit turbulent flow. Over geologic time, the pre-existing bedding-plane fractures (see Figure 5.9) will enlarge but not to the extent that the aquifers might be considered karstic; to do so would be to obscure the geotechnical and hydrogeological importance of karst.

However, when carbonate rocks are present at or close to the ground surface in a climate in which recharge is abundant and soil-zone CO_2 production is significant, there is a strong likelihood that karst features will develop. Infiltration will bring low-pH groundwaters into contact with carbonate rocks and promote dissolution-enlarged pathways within the carbonate aquifer that are dendritic in geometry and form self-organized networks of drainage channels in the subsurface, which are illustrated in Figure 13.12 (Worthington, 2009). The evolution of the flow pattern shown involves enlargement of pre-existing fractures, likely bedding-plane discontinuities with variable apertures, causing a positive feedback loop in which increasing amounts of flow travel along channels

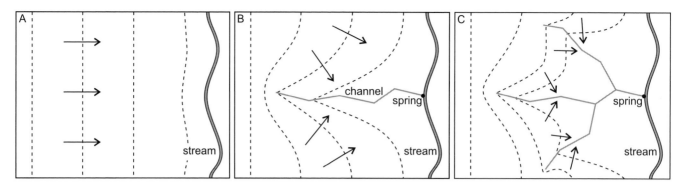

Figure 13.12 The evolution of self-organized drainage channels within carbonate aquifers leading to the increasingly large hydraulic conductivities indicative of karst conditions. (A) An initial flow pattern within a homogeneous flow field draining to a stream. (B) Once the first channel has developed, the flow system evolves into one in which further channel development is organized around the first channel. (C) Finally, other channels develop and create the dendritic pattern of karstic aquifers. (From Worthington (2009); reproduced with the permission of John Wiley and Sons.)

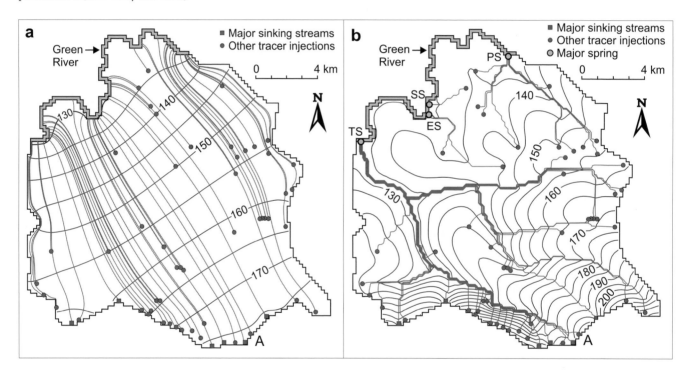

Figure 13.13 Simulated patterns of hydraulic head distribution in the karst aquifer at Mammoth Cave, Kentucky, USA, based on measured heads (metres above sea level) and spring-flow discharges and 58 tracer tests. (a) The distribution of head in a homogeneous porous-medium aquifer with $K = 1.1 \times 10^{-3}$ m/s. (b) A channel network within the karst aquifer that reproduces the principal features of the head distribution as interpreted by the field evidence. The hydraulic conductivity distribution in this simulation varies from 4×10^{-5} to 7 m/s, which was the simulated K value for the final 6.6 km of channel to Turnhole Spring, marked TS. (Figure from Worthington (2009); reprinted by permission of Springer from *Hydrogeology Journal* © 2009.)

or conduits in which dissolution enlargement of fissures is occurring because of preferential flow. Once channels along their full length are enlarged sufficiently, a kinetic break-through is thought to occur and flow rates increase dramatically as flow becomes turbulent and caves develop (Dreybrodt, 1996). Generally, areal recharge favours the development of many small dissolution-enlarged fissures; however, sinking streams are associated with the development of caves.

Consequently, the groundwater flow pattern of subsurface drainage within a karst aquifer of any particular mean hydraulic conductivity is very different from that of a homogeneous porous-medium aquifer with a similar mean hydraulic conductivity. Figure 13.13 illustrates the development of high-permeability channel networks in the karst aquifer at Mammoth Cave, Kentucky, USA, which has been intensively investigated over the years, including numerous tracer tests, spring-flow surveys and head measurements in 1500 wells. Hydraulic conductivity measurements by field testing yield values in the range 2×10^{-4} m/s (pumping test) to 6×10^{-6} m/s (geometric mean of six slug tests), none of which tests were conducted in channels partially mapped by tracer testing. The head distribution in the aquifer was known to exhibit troughs associated with the channel network.

The two simulations by Worthington (2009) shown in Figure 13.13 assume either a homogeneous K field or one that is the result of the development of a channel network. The simulated average discharge from the porous-medium aquifer to the springs along the Green River is 620 L/s, which is much less than the measured discharge of 5800 L/s, which was closely approximated by the channel-network model. The porous-medium aquifer was simulated with an isotropic hydraulic conductivity of 1.1×10^{-3} m/s, whereas the channel-network model required a range of values from 4×10^{-5} to 7 m/s to achieve a satisfactory fit to the head pattern measured in the field.

The general range of hydraulic conductivity and porosity values for karst aquifers is presented in Figure 13.14. It illustrates that high-porosity carbonate aquifers tend to have relatively low hydraulic conductivities indicative of flow through the matrix. Carbonate aquifers with low porosities ($\sim$0.001) are associated with fractures and channel networks. Such fractures and channels provide rapid transport for tracers and contaminants in carbonate aquifers, while flow through the matrix is much slower. This point was made earlier in section 11.1.2 in discussing the dual-porosity nature of the Upper Floridan aquifer in the Old Tampa wellfield as revealed by a tracer test.

Channel networks, fissures associated with clint-and-grike landforms (Figure 13.10b) and dissolution-enlarged fractures are known to develop in all unconfined carbonate aquifers, thus making them susceptible to contamination. Concerns about the protection of a very large wellfield near Miami, Florida, threatened by open-pit mining led the US Geological Survey (Renken et al., 2005) to conduct a tracer test to evaluate possible pathogen transport in the triple-porosity Biscayne aquifer. The three-porosity aquifer system is composed of (1) the low-permeability matrix, (2) large voids known as vugs that are interconnected as stratiform pathways, and (3) channels formed by solution pipes, bedding-plane vugs and caverns. An injected tracer took only 6.5 hours to travel 100 m through the channels and vugs to the production well, which was pumping at 0.49 m^3/s. Figure 13.15 illustrates some vertical and horizontal karst features that are found in the Upper Floridan aquifer, which provide infiltration and

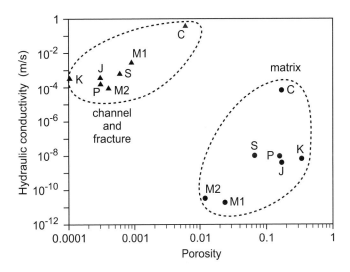

Figure 13.14 Hydraulic conductivity and porosity values for the matrix, fractures and channels measured in various carbonate rocks: C, Cenozoic limestone, Mexico; J, Jurassic limestone, England; K, Cretaceous limestone, England; M1, Mississippian limestone, Kentucky; M2, Mississippian limestone, England; P, Permian limestone, England; and S, Ontario dolomite. (From Worthington (2009); reproduced with the permission of John Wiley and Sons.)

transport pathways for groundwater flow and contaminant migration.

Worthington and Smart (2017) discuss an actual case of the migration of pathogens to municipal wells situated in a dual-porosity carbonate aquifer that resulted in seven deaths in Ontario. The fractures in this aquifer are a few millimetres to a few centimetres in aperture ($n = 0.0005$, $K = 10^{-4}$ m/s), allowing groundwater velocities of the order of $\sim$100 m/day. Therefore, irrespective of whether a carbonate aquifer has the typical channel or conduit features of a karst or simply is a mature fractured carbonate aquifer, carbonate aquifers are susceptible to rapid groundwater flow transporting contaminants to water wells.

13.3.3 Evolution of Karst Terrain

The terminology of karst is extensive, perhaps because it is observed and described worldwide by geoscientists, who have identified particular features that require naming. Many of the terms are of Balkan origin, having been identified in Serbia, Croatia and Slovenia. Some features, such as **sinkholes**, are better known by their engineering term but have an equivalent European name, i.e., dolines, which is the preferred geomorphological term. Sinkholes are the characteristic geomorphological feature of karst terrains and may be subdivided into

Figure 13.15 Vertical and horizontal solution features, Haile Quarry, Florida. (a) A cover-collapse sinkhole. (b) The channels that drain the sinkholes and create the high hydraulic conductivities measured in karst aquifers. The photo scale indicates that the central aperture channel is approximately 10 cm wide.

various types (see Waltham et al., 2005), some of which, i.e., dissolution, collapse and caprock sinkholes, form within the exposed rock and are considered relatively stable but must be connected to an underlying drainage system that may pose a hazard. Others, such as dropout and suffosion sinkholes, form

beneath a soil cover that Waltham and Fookes (2003) indicate produce a greater engineering hazard. These are illustrated in Figures 13.16 and 13.17, and discussed in Box 13.3.

It is helpful to divide karst features into surficial dissolution features known as **epikarst**, and deep dissolution features known as **hypogenic karst**, both of which are characterized by cavity development and the potential for sudden sinkhole collapse. We have previously described thermokarst in section 8.5 as that related to melting of ground ice. Additionally, the term **paleokarst** is used to identify solution surfaces or cavities that are inert with respect to further growth in void volume because of deep burial in a sedimentary column and the groundwater has become saturated with respect to carbonate and sulfate minerals. These ancient void features may be filled with sediment and thus easily missed in borehole logs or they may persist as unoccupied cavities.

An engineering-oriented classification of karst terrain has been developed by Waltham and Fookes (2003), which is shown in Figure 13.16. They define the various stages of karst development in five classes that are useful in identifying local issues that may arise with karst terrain and approaches to their investigation and mitigation:

kI.   This is juvenile karst that may be observed in desert and periglacial environments. Sinkholes are rare, the top-of-rock or rockhead is uniform and only small caves exist.

kII. Youthful karst exhibits an irregular, fissured bedrock surface with many small caves (<3 m) and few sinkholes. It is likely to be the minimum karst state in any temperate region, such as the Great Lakes region of North America. The clint-and-grike terrain shown in Figure 13.10(b) is an example.

kIII. Mature karst is observed in temperate climates and some wet tropical regions. The rockhead is extensively fissured and there is extensive cave development. The Yorkshire Dales and areas of the USA such as Kentucky and Missouri are examples.

kIV. This is complex karst typical of tropical regions, e.g., Malaysia, Indonesia and southern China, with large sinkholes, non-uniform rockhead and large cave openings (>5 m) presenting significant foundation problems.

kV.   Extreme karst occurs only in humid tropical areas and is characterized by high topographical relief, large sinkholes and cave systems with dimensions of >15 m across and a terrain distinguished by pinnacle and tower features.

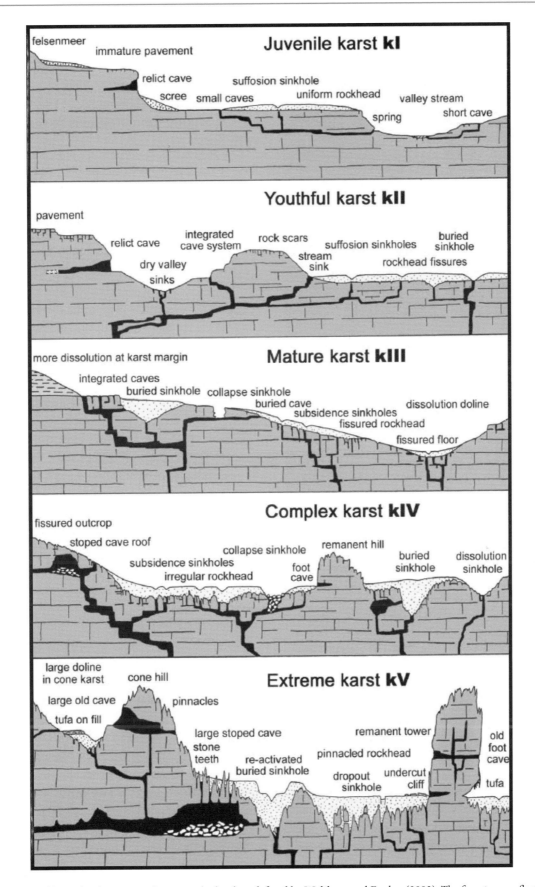

Figure 13.16 Stages of karst development in limestone bedrock as defined by Waltham and Fookes (2003). The five stages reflect increasing amounts of recharge of atmospheric precipitation, which is itself slightly acidic, thus promoting carbonate mineral dissolution. The panels identify a variety of sinkholes, some of which are covered by soil, others are in exposed bedrock. The limestone is presumed to be horizontally bedded with bedding-plane discontinuities. (From Waltham and Fookes (2003); with permission of The Geological Society of London.)

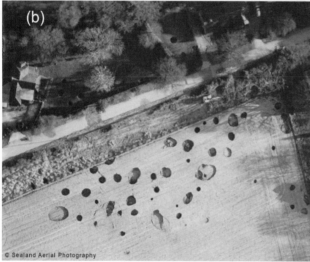

Figure 13.17 Sinkhole hazards: (a) Sinkhole development in Pennsylvania, USA. (Photo courtesy of Jim Lolcama, KCF Groundwater Inc.) (b) Sinkholes may be created by the failure of water pipes that promote rapid dissolution of soluble rocks. This air photo shows the development of sinkholes on the Chalk due to a burst water pipe in Sussex, England. The uneven rockhead beneath the field is caused by Chalk dissolution. Dissolution pipes within the Chalk are infilled with soft clay and sand from the overlying strata, which creates potential problems for foundation design. See Matthews et al. (2000) for discussion of the causes. (Photo used by courtesy of Sealand Aerial Photography.)

Waltham and Fookes (2003) also consider the stability of cave roofs in limestone terrain with respect to the magnitude of the imposed load that can be safely supported by the rock when it is assumed that its unconfined compressive strength UCS $\sim$ 80 MPa. Figure 13.18 illustrates the cave roof thicknesses that will sustain stresses (referred to as safe bearing pressures, SBP) of 2 and 4 MPa assuming a factor of safety of 3. The predicted curves indicate the failure loads for jointed beds and those with bed spacings of 0.5 and 1 m thickness. These

authors note that, for the rock specimen studied, a roof-cover thickness of intact rock that is 70% of the cave width provides stability to the cave.

13.3.4 Engineering Problems Associated with Karst

Karst terrain is defined by sinkholes, caves and complex top-of-rock topography (i.e., rockhead), all of which pose considerable hazards to construction and infrastructure maintenance (see Waltham et al., 2005):

- *Bridge foundations*: Bridge foundations require careful locating in karst terrain, in particular where the rockhead is known to be irregular. Fookes et al. (2015) discuss the belated discovery of a cave in the foundation of a bridge pier in the UK that was overlooked during site investigation of class kII terrain. Because most caves are <10 m in width, site investigation should respond accordingly in order to identify rockhead and void features.

- *Dam and navigation-lock foundations*: These present an equally critical site investigation issue because the high hydraulic gradients, which are created by the impounded water, increase the rates of solution widening of channels and bedding-plane fissures, leading to failure of constructed dams or locks. Goodman (1993) has summarized some of the problems experienced by the Tennessee Valley Authority in dam construction, some of which required 20 000 m^3 of grout to seal caves and other voids that were identified during site excavation. Lienhart (2013) identifies the primary geological causes of dam and lock failure in this same area of the USA as (i) valley stress relief, by which valley-bottom rocks become heavily fractured with removal of their overburden whether by fluvial or glacial scouring of the valley, and (ii) subsequent dissolution widening of the fracture network into karst. Numerical simulations of this issue for both limestone and gypsum dam foundations illustrate how leakage can reach excessive rates of flow involving turbulence in periods of 25 years for limestone foundation rock and only five years for gypsum rocks (Dreybrodt et al., 2002). In Chapter 15, we address the disaster at the Vaiont Dam in Italy in 1963 when a massive landslide associated with karst caused heavy loss of life.

- *Tunnelling*: A 16 km railroad tunnel in South Korea encountered construction problems of groundwater inflow, sinkholes and ground subsidence along a fault lineament (Song et al., 2012). Unfortunately, the pre-construction site characterization had failed to identify the presence of these

BOX 13.3 | THE SUDDEN APPEARANCE OF SINKHOLES IN FLORIDA

The mild winter climate of Florida has made the state the largest US producer of winter strawberries. The growing area is situated directly above the Upper Floridan aquifer. In January 2010, following nine nights of freezing temperatures over an 11-day period, there was a sudden surge of complaints from 760 well owners that their wells had gone dry, and the equally sudden appearance of 140 soil-cover collapse sinkholes in the growing area. Why should freezing temperatures be highly correlated with such phenomena in this same area?

Sprinkler irrigation is widely used to protect high-value crops like strawberries during freezing periods. A layer of ice sprayed on fruit and other crops provides a level of protection against freeze damage and loss of the crop. Consequently, when cold air plunged southwards across the USA in January 2010, strawberry growers in Florida turned on their irrigation pumps and the water table in the Floridan aquifer declined (~15 m) beneath the base of the confining layer separating the surficial aquifer and the Upper Floridan aquifer. Over that 11-day period the pumping rate in the growing area was 140 000 m^3/s.

Those water wells with submersible pumps set at relatively shallow depths would have been most likely to go dry, i.e., the pump was not sufficiently submerged beneath the water table to function properly or was suspended above the depressed water table, in which cases its operation may have been affected by overheating, cavitation, etc. Most of the dry-well complaints occurred at the time of maximum drawdown or shortly after water-table recovery began. Well and pump repairs and replacements cost $1 200 000.

The surficial aquifer and the confining unit are composed of sandy deposits overlying the Floridan aquifer. As vertical solution pipes develop (e.g., Figure 13.15a), the underlying voids and channels drain infiltrating water and soil debris away, leaving behind a soil arch in the confining layer supported partly by the buoyancy of the groundwater. By lowering the water table, the buoyant force supporting the soil cap is removed and the soil arch is downwashed or ravelled into a new cover-collapse or cover-subsidence sinkhole.

Source: After Peterson and Rumbaugh (2012) and Aurit et al. (2013).

karst features; consequently it was necessary to undertake a major programme of electrical resistivity surveying and borehole coring to allow the tunnel to be completed. The groundwater inflow caused failure of shotcrete used to support the tunnel and introduced a brown mud into the tunnel. The source of the inflow was traced to a stream situated 235 m above the tunnel with inflow travelling a 300 m pathway into the tunnel. The tunnel roof face was sealed by injecting grout 5–14 m into the roof. A waterproof membrane was installed beneath the stream and 15 m of grout was injected beneath the stream channel to prevent further loss of streamflow to the karst network that had inundated the tunnel. A monitoring programme for ground settlement and hydraulic heads was established in the area of the sinkhole. Similarly, three deep, long tunnels excavated in the Silurian dolomite aquifer beneath Milwaukee, Wisconsin, USA, were subject to inflows of 50–220 L/s (Day, 2004). It appears that geoscientists had missed the potential for karst development beneath Milwaukee on the grounds that glacial sediments had alleviated the possibility of karst development; however, this karst developed prior to the Quaternary.

- *Highways*: The karst of eastern Tennessee is evident at ground surface in the form of sinkholes, cave entrances, sinking or disappearing streams and outcrops of weathered limestone. The principal problems that face the state highways department arise from sinkhole development in roadside ditches that are induced by runoff infiltrating through the ditch and causing sinkholes to develop by soil-cover collapse (Moore, 2006). Furthermore, highway spills of fuels can readily contaminate aquifers and nearby water wells. The highways department determined that a number of proactive measures could be undertaken to minimize or avoid this problem, including relocation of a highway, lining drainage ditches and curbs for embankments sections.

These site histories point to the need for engineering teams to properly characterize carbonate-rock terrain prior to construction. We now turn to methods of site investigation in karst terrain.

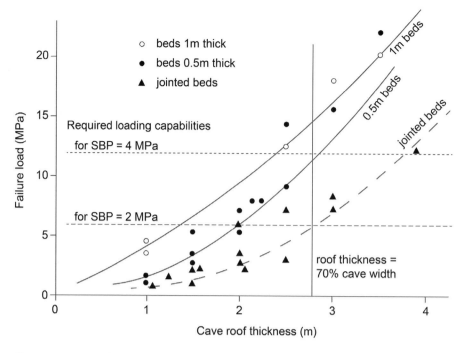

Figure 13.18 Stability of limestone cave roofs with respect to the failure load versus the cave-roof thickness for rock with $Q = 4$–10 in the NGI system and a geomechanics classification of RMR = 40–60 with UCS = 80 MPa. If the roof thickness is 70% of the cave width, the roof is considered stable provided the load does not exceed 2 MPa. The safe bearing pressure (SBP) is indicated with a factor of safety of 3. (From Waltham and Fookes (2003); with permission of The Geological Society of London.)

13.3.5 Site Investigation in Karst Terrain

Any site investigation in carbonate or evaporite terrain should begin with a review of the regional technical literature and of aerial photographs. These should identify the propensity for sinkhole development and the existence of other karstic phenomena. A simple phone call to the appropriate highways department or geological survey may provide much useful guidance. Goodman (1993), Waltham et al. (2005) and Fookes et al. (2015) provide general overviews of site investigation in soluble rocks. Tracer tests using dyes or microspheres are commonly used to trace karst channels, particularly when groundwater-supply wells are under threat, such as the studies by the US Geological Survey in Florida (Robinson, 1995; Renken et al. 2005). Drilling and coring are typically expensive and time-consuming and may miss important dissolution features; however, they are likely to be necessary to evaluate karst hazards prior to construction if infrastructure failure is to be avoided.

Consequently, geophysical methods, conducted by an expert geophysicist, are widely used to undertake a preliminary assessment. Useful guidance in the geophysical detection of karst features is given by Matthews et al. (2000) for the English Chalk, Yuhr (2009) for the Floridan limestone and, generally, Chalikakis et al. (2011). In the USA, **ground-penetrating radar** (GPR), **electromagnetics** (EM) and **electrical resistivity imaging** (ERI) are all widely used. Carpenter et al. (1998) provide a particularly helpful case history of sinkhole detection at the Oak Ridge site in Tennessee using GPR complemented by EM and resistivity measurements.

In Florida (see Yuhr, 2009), GPR has the highest resolution within the upper 10 m for identifying evidence of karst, such as the soil pipes shown in Figure 13.15(a) and other epikarst features. This allows precise information on potential drilling locations to confirm dissolution and subsidence features. **Microgravity** is particularly useful in detecting conduits and caves of 10–15 m in diameter at depths of 30–100 m because they present low-density contrasts to the surrounding rock mass and thus stand out as gravity anomalies at such depths. **Multi-channel analysis of surface waves** (MASW) has proven useful in identifying the weathered surface of karst and soil ravelling and/or cavities in weathered limestone.

Seismic reflection will provide high-resolution images of karst structures, whether they are of limestone, gypsum or salt. A seismic reflection survey of the Floridan Aquifer for wellfield development by Odum et al. (1999) identified what are believed to be solution pipes (Figure 13.15a), collapse sinkholes (Figure 13.17a), isolated cavities and narrow fracture zones. Sargent and Goulty (2009) determined by seismic reflection that subsidence around Darlington, UK, was indicated by subsidence in a limestone formation at 40–50 m below ground surface (bgs) believed to be caused by dissolution in the underlying gypsum beds at 70 m bgs. The mining of salt beds in Kansas (Croxton and Henthorne, 2003), both solution and rock-salt mining, as well as natural dissolution, has created sinkholes and subsidence bowls near highways and railroads around the city of Hutchinson. These have been investigated by the Kansas Geological Survey using seismic reflection because of the considerable depth to the Hutchinson Salt Member at 200–400 m. While seismic reflection is costly compared with GPR and EM, it may be the preferred geophysical tool when information on deep-seated problems is required.

13.4 Summary

Land subsidence may occur because of (i) the subsurface extraction of groundwater, oil and gas, coal or other minerals, (ii) the oxidation of organic soils or (iii) the natural or enhanced dissolution of carbonate or evaporite rocks. Irrespective of the cause, subsidence presents the geotechnical engineer with a design challenge as well as a continuing problem of monitoring and mitigation of infrastructure. Such subsidence is typically irreversible and the public may be unaware of the immediate hazards or the long-term challenge and expense.

In this chapter, we have considered the effect of groundwater depletion on regional subsidence of basins caused by the dewatering of clay-rich aquitards with the attendant loss of groundwater storage. Similarly, oil and gas production can cause subsidence, as occurred around Los Angeles and Houston. The principle of effective stress is again central to our understanding of land subsidence when pore pressures are reduced by fluid extraction. Lowering of the water table in organic soils, as in the Netherlands and California, leads to the oxidation of the soils and their subsidence, which in coastal areas creates a potential risk of marine invasion of the lowered coastline. The extraction of coal and other minerals has caused mining-induced subsidence across the former industrial heartlands of Europe, North America and Asia.

We conclude with a lengthy discussion of karst phenomena as they affect carbonate and evaporite formations that undergo natural or enhanced dissolution. Karst produces many potential geotechnical hazards, in particular the development of sinkholes and shallow cavities that may cause failure of structures, such as bridges, dams, locks, tunnels and highways, because of the enhanced flow of groundwater through the karst. Furthermore, rapid flow of contaminated groundwater through karst creates a significant hazard to the many groundwater supplies developed in carbonate aquifers.

13.5 FURTHER READING

Brady, E.H.G. and Brown, E.T., 1993, *Rock Mechanics for Underground Mining*, 2nd edn. London, UK: Chapman and Hall. — This textbook contains helpful discussions of the geomechanical aspects of longwall mining and of mining-induced surface subsidence.

Ford, D. and Williams, P., 2007, *Karst Hydrogeology and Geomorphology*. Hoboken, NJ: John Wiley and Sons. — A comprehensive account of karst phenomena, in particular karst hydrogeology.

Galloway, D., Jones, D.R. and Ingebritsen, S.E., 1999, *Land Subsidence in the United States*. Circular 1182. Reston, VA: US Geological Survey. — This freely available report on all the aspects considered in this chapter is an excellent guide to land subsidence.

Goodman, R.E., 1993, *Engineering Geology: Rock in Engineering Construction*. New York, NY: John Wiley and Sons. — Chapter 5 is an excellent introduction to the engineering geology of karstic rocks and the mitigation efforts required to safely construct dams in their presence.

Kratzch, H., 1983, *Mining Subsurface Engineering*. Berlin, Germany: Springer. — This textbook is the standard reference to subsidence issues arising from coal mining in Europe and is recommended by US experts on the subject.

Waltham, T., Bell, F. and Culshaw, M., 2005, *Sinkholes and Subsidence: Karst and Cavernous Rocks in Engineering and Construction*. Chichester, UK: Springer-Praxis. — This monograph, now available online, covers all engineering aspects of karst by three very knowledgeable engineering geologists.

Wang, H.F., 2000, *Theory of Linear Poroelasticity with Applications to Geomechanics and Hydrogeology*, Princeton, NJ: Princeton University Press. — A helpful guide to the geomechanical implications of fluid addition to or removal from porous media.

Wyllie, D.C., 1999, *Foundations on Rock*, 2nd edn. London, UK: Spon, Taylor and Francis. — A section of this textbook is devoted to foundation engineering in karst rocks, including spread footings and driven piles.

13.6 Questions

Question 13.1. What are the units of s'_{sk}? Do a dimensional analysis.

Question 13.2. Discuss in specific terms why a preconsolidation stress should differ from any present effective stress within an aquifer–aquitard system, such as is shown in Figure 13.2.

Question 13.3. The compressibility of soils and rocks (α) is related to Young's modulus of elasticity (E), which measures stiffness. Estimate the compressibility of the clays listed in Table 4.1 based on the relation $\alpha = 3(1 - 2\nu)/E$, where ν is Poisson's ratio, and compare with Table 11.2.

Question 13.4. Consider an elastic, confined aquifer 60 m thick with storativity = 5×10^{-4} and a porosity of 0.3. What is the (elastic) compression of this aquifer?

Question 13.5. Consider a 12 m thick compressible clay aquitard, such as that shown in Figure 13.2, with overlying and underlying aquifers. Assume no flow across this aquitard as the hydraulic head is lowered 15 m regionally, i.e., in both aquifers. Let the specific storage of the aquitard (s'_{sk}) be 1.0×10^{-3} m^{-1}. What is the vertical subsidence?

Question 13.6. Assume that a doubly draining 12 m thick aquitard, similar to that shown in Figure 13.2, has a specific storage = 1.0×10^{-3} m^{-1}, and the head is lowered sufficiently to dewater the aquitard regionally. What is the required time for >90% of ultimate aquitard compaction if $K' = 10^{-8}$ m/s?

Question 13.7. Consider Figure 13.7 showing subsidence in an area of longwall coal mining. What geophysical tools would you employ to investigate a similar abandoned area to determine (i) the degree of fissuring in the overlying shale and (ii) the location of voids in the worked coal seam beneath it?

Question 13.8. Does dissolution of limestone occur below the water table?

Question 13.9. Given the discussion of the difficulties of tunnelling through karst in section 13.3.4, what specifications would you recommend be defined for a site investigation where a tunnel is to be excavated in limestone at a depth of 20–100 m?

Question 13.10. If you are to construct a groundwater-supply wellfield in a carbonate aquifer suspected of being karstic, why would you recommend undertaking an inter-well tracer test?

14 Seismicity and Earthquakes

Some regions are renowned because of the earthquakes they have experienced this century, e.g., China, Japan, New Zealand, Sumatra and California, all of which are located along plate margins, as shown in Figure 2.2. Others have experienced a quiet seismic history in modern times, but major earthquakes are known to have occurred during the Quaternary and will again occur, such as in Cascadia, which is the Pacific Northwest of the USA and British Columbia in Canada. In this chapter we discuss why this might be so and how it affects the role of geotechnical engineering in the twenty-first century.

The occurrence and consequences of the next major earthquake in any country with a history of seismic activity is a matter of life and death. Japan is perhaps the best-informed society, having suffered two recent massive earthquakes in 16 years: the 1995 Kobe and 2011 Fukushima disasters. In the USA, journalists and science writers have described just how serious the "Next Big One" might be for California and Cascadia. Without hyperbole, geoscientists like Hough and Bilham (2006) have outlined the hazardous future of the greater Los Angeles region, beneath which are a dozen major faults, while Clague et al. (2006) and Schulz (2015) describe the ominous seismic threat posed by plate subduction in Cascadia.

In earlier chapters of this book we considered issues such as subduction of lithospheric plates at plate boundaries (section 2.1), the nature of faults and shear zones (section 3.3), stress regimes (section 4.2) and the deformation of rocks (section 4.3). Here we return to these topics to illustrate their role in modern earthquake science. Then we proceed with a discussion of the types of seismic waves and of seismometry before turning to the nature of seismically active faults and their

detection. We conclude with sections that address ground motion, liquefaction and seismic hazard analysis.

14.1 Introduction to Seismic Faulting

Modern earthquake science began with a series of field studies by Grove Karl Gilbert in California in 1872, Alexander McKay in New Zealand in 1888 and Bunjiro Koto in Japan in 1891. Each of these geologists mapped field scarps from recent earthquakes. These early geologists identified fault scarps in the landscape of Owens Valley, California, the South Island of New Zealand and Honshu, Japan, all of which were the result of recent earthquakes. Figure 14.1 shows the scarps of such active normal faults along the Wasatch Front in Utah, USA.

Complementing this fieldwork, land surveys in the 1890s in Sumatra and India attributed the change in the angles measured between survey monuments to ground movements caused by recent earthquakes. These findings indicated that earthquakes produced not just vertical movement of the land but also horizontal movement. Most importantly, it became clear during the early twentieth century that, irrespective of whether or not an earthquake produced a surface fault such as those shown Figure 14.1, all earthquakes were the result of slippage of geological faults, i.e., **seismic faulting** (NRC, 2003). This was the conclusion of Professor Koto of the Imperial University in Tokyo in 1891 when he described a 100 km surface fault across southern Japan and deduced that its movement was the result of the earthquake.

Faulting may occur as illustrated in Figure 3.8. In particular, geomorphological mapping and seismic interpretation of earthquakes has identified the importance of (i) **strike-slip**

Figure 14.1 Fault scarps, Little Cottonwood Canyon, near Salt Lake City, Utah. The three fault scarps shown (arrowed) are the result of normal faulting during the late Pleistocene and Holocene. The individual scarps are 10 m or more high, with an estimated slip rate of 0.9 mm/yr for 16 000 years (Hanson and Schwartz, 1982). (Photo courtesy of the Utah Geological Survey.)

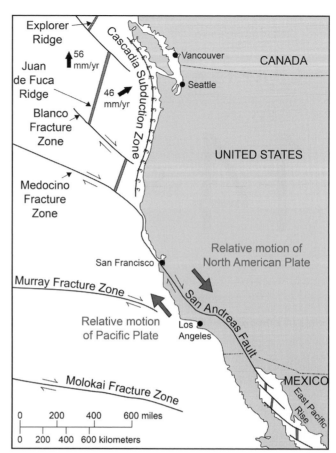

Figure 14.2 Schematic representation of major fault, ridge and subduction zones along the west coast of North America. (Modified from Kious and Tilling (undated); courtesy of the US Geological Survey; with plate-migration rates from Clague et al. (2006).)

faults, sometimes known as wrench faults, (ii) **normal faults** in which the down-dropped block had slipped downdip, and (iii) **thrust faults**, which are reverse faults with a dip <45°. Combinations of these types are possible. Anderson's 1905 theory of faulting (section 4.2.2) provided a coherent theory to understand the field observations of faults based on a comparison of principal stresses, one of which was always the vertical stress (σ_v). The fault slips when the shear strength of the fault surface is exceeded by the applied stresses, i.e., the Mohr–Coulomb strength criterion (equation (1.8)) is exceeded. If the shear stress on a fault plane is reduced by elevated pore pressure, as in equation (1.12), then reverse faulting is possible. Overpressuring (section 4.2.4) makes many faults weak, including the San Andreas in California.

On the west coast of North America (Figure 14.2), the two principal types of damaging earthquakes are strike-slip movements along transform-fault boundaries and quakes associated with subduction-zone thrust faults. Figure 2.7 shows how both of these tectonic processes have shaped the west coast. Note that at this time we are omitting those earthquakes that occur within the boundaries of a continental plate, which are known as **intraplate earthquakes**. Here we focus on the San Andreas and neighbouring strike-slip faults and the mechanisms involved in thrust faulting along Cascadia's subduction zone.

The San Andreas Fault (SAF) is a transform fault of the kind illustrated in Figure 2.6 and is 1300 km long and as much as 10 km wide in places. Most transform faults occur on the

ocean floor offsetting two spreading ridges; however, the SAF is one of a few that occur mainly onshore connecting the East Pacific Rise in Mexico with a series of spreading ridges off the Pacific Northwest. Figure 14.3 traces its path through southern California to the Mexican border, where it connects with the Brawley Seismic Zone and, according to Yeats (2012), displacement shifts to the Imperial Fault.

Research into past ruptures of the SAF indicates that it has failed numerous times over the past 1500 years, with events occurring quite frequently over periods of 50–200 years, followed by quiescent intervals of 200–300 years (Hough, 2002). The 1906 San Francisco earthquake destroyed 28 000 buildings, killed at least 3000 people and produced a rupture of 450 km in length with an average right-lateral slip of 4.5 m. The 1989 Loma Prieta earthquake near Santa Cruz, California, caused 63 deaths and $6.8 billion in property damage.

The 1906 San Francisco earthquake prompted an intensive investigation that stimulated earthquake science in the USA (Zoback, 2006). This investigation of the 1906 quake revealed

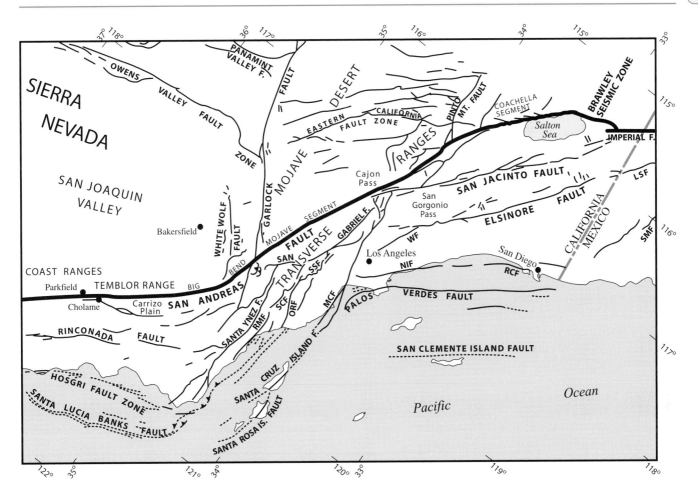

Figure 14.3 Active fault map of southern California. RCF refers to the Rose Canyon Fault, NIF is the Newport–Inglewood Fault. Dotted faults are covered by water or sediments but may be active. This map should not be considered definitive because fault traces are subject to continual revision. (Modified by Yeats (2012) based on the original by Wallace (1990). Reproduced with the permission of Cambridge University Press.)

that the SAF could be traced throughout most of California. Triangulation surveying in 1906–07 near San Francisco determined the horizontal displacements of the ground surface that occurred during the earthquake, i.e., **coseismic displacement**, and indicated that fence lines on either side of the SAF moved in opposite directions with respect to the fault trace, as shown in Figure 14.4.

The Commission of Inquiry into the 1906 disaster concluded that the coseismic displacement was the result of the accumulation of elastic strain prior to the rupture of the fault, and that the fault slip itself indicated the release of external elastic forces, likely applied from considerable distance. This theory, developed by H.F. Reid of Johns Hopkins University in 1910, became known as the **elastic rebound model** and suggested that the slip was the final stage of an earthquake cycle along strike-slip faults that would begin again with the accumulation of strain on parts of the fault that were temporarily locked in place. Geomorphic evidence

of previous SAF ruptures indicated that repetition was likely. Perhaps most importantly, the Commission concluded that geologic conditions together with the design and construction of the buildings of San Francisco were the primary factors controlling the property damage and loss of life caused by the earthquake. It so happened that much of the city was destroyed not by the earthquake but by the fires that began shortly thereafter.

Reid's theory had the accumulating **interseismic elastic strain**, i.e., that which accumulates between earthquakes, dissipated by the coseismic strain release and displacement in an equal and opposite manner. The horizontal displacement shown in Figure 14.4 is associated with a shear strain that causes a preseismic flexure to a distance d on each side of the fault trace. This flexure can now be monitored by **GNSS** or GPS. This distance d is also the approximate depth at which the fault is locked, i.e., is not exhibiting creep from tectonic forces. With coseismic displacement, an offset of the two fault

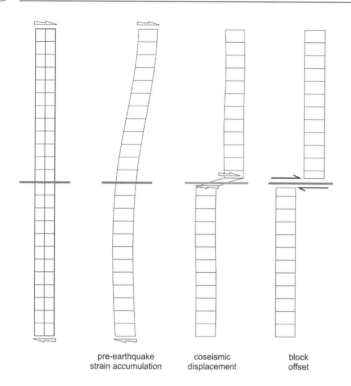

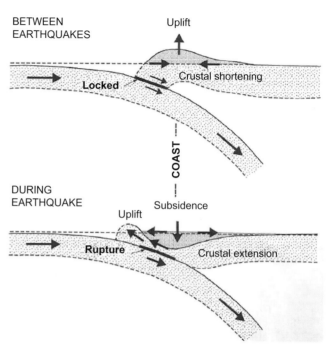

Simplified earthquake cycle

Figure 14.4 The elastic rebound hypothesis that was developed following the 1906 San Francisco earthquake. A fence crosses the fault and displays pre-earthquake accumulation of strain. When the fault slipped in a right-lateral sense, coseismic displacement separated the fence segments across the trace of the fault. Elastic rebound causes the offset of the two blocks separated by the fault. The fence line indicates a structure crossing the fault and extending more than 100 km in either direction. (From Zoback (2006); courtesy of the Geological Society of America.)

Figure 14.5 The elastic dislocation model of a subduction zone earthquake as in Cascadia in which the subducting plate is locked to the continental plate in the fault zone. Prior to an earthquake, stress accumulates and an uplift develops as a flexural bulge inland of the zone. With rupture of the locked fault, stored elastic energy is released as an earthquake and the marine edge of the continental plate rebounds upward and the bulge collapses. The cycle is repeated by relocking of the fault zone and by the subsequent re-accumulation of strain. (From Clague (1997); reproduced with the permission of John Wiley and Sons.)

blocks occurred, resulting in the condition of zero net strain. In fact, we now know 100 years later that, while Reid's general principle is correct, the loading of the fault is more complex and not fully elastic with the steady accumulation of strain. The tectonic forces are of course those produced by plate tectonics, a process of which Reid had only the slimmest of understanding in 1910 (Scholz, 2002).

Earthquakes associated with subduction zones, such as Cascadia, also exhibit an earthquake cycle. Figure 14.5 illustrates the development of stress caused by the locked fault-zone boundary between the continental crust and the subducting plate. Inland of the plate boundary, a flexural bulge causes uplift and crustal shortening. Also inland are the volcanic arcs that characterize the volcanic terrain associated with subduction zones, e.g., Mount St. Helens (Figures 2.3 and 2.4). Sudden rupture of the fault releases the accumulated elastic strain energy as an earthquake. This is referred to as an **elastic dislocation model**. The last earthquake in Cascadia occurred in 1700 and ruptured the entire length (1000 km)

of the subduction zone. Similar giant subduction earthquakes are believed to have occurred in Chile in 1960, Alaska in 1964 and Sumatra in 2004 (Clague, 1997).

Strike-slip earthquakes are the result of rupture and slippage along a fault surface typically at depths of 6–20 km at a rate of ~1 m/s; however, the focal depth for subduction thrust earthquakes is substantially deeper, although the 2011 Tohoku earthquake had a focal depth of 29 km. (Note that the US Geological Survey considers "shallow focal depths" as being at <70 km, with "deep" earthquakes having a focal depth >300 km.) Furthermore, the earthquake is the result of the accumulation of strain energy caused by tectonic forces within the Earth's crust and slip typically lasts 1–10 s. Each slippage begins the earthquake cycle of strain accumulation and slippage again – known as **stick–slip** to engineers familiar with friction and sliding. Finally, large earthquakes usually leave a geomorphic trace that may be identifiable by surface faults, such as shown in Figure 14.1 (Sibson, 2011).

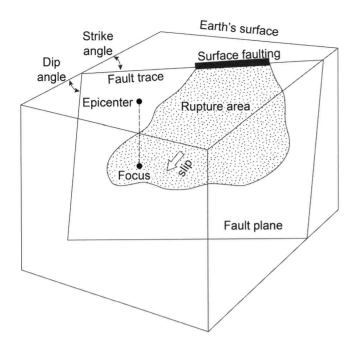

Figure 14.6 Model of a typical earthquake source. The source or focus is often referred to as the hypocentre. The depth below the epicentre to the focus is referred to as the focal depth. The displacement shown in this schematic is of a normal fault slip, but could be in other directions. The surface fault is projected above ground surface as is shown in the exposures in Figure 14.1. (Modified from Ziony and Yerkes (1985); courtesy of the US Geological Survey.)

14.2 Seismic Waves and Seismometry

We may think of an earthquake source as similar to that shown in Figure 14.6, in which the fault ruptures at a **focus** or **hypocentre** 6–20 km below ground surface and, in this case, is presumed to have slipped downdip as a normal fault, i.e., $\sigma_v = \sigma_1$. Obviously, if the principal stresses were of different magnitude, the slip might be in a different direction. We identify the location on the ground surface immediately above the focus as the **epicentre** of the earthquake. If the fault has a surface expression, and most with $M > 6$ do, then the fault may have created a scarp at the surface, three of which are shown in Figure 14.1. These fault scarps provide evidence of active faulting.

Seismic energy radiating from the earthquake is emitted as seismic waves that are of two kinds: body waves, which are acoustic waves, and surface waves. Primary or **P waves** and secondary or **S waves** travel through the body of the Earth, being reflected and refracted by the core and mantle of the Earth. Figure 14.7 illustrates the passage of P and S waves. P and S waves are the fastest seismic waves and the first to appear on a seismogram; these are also used for index

testing of rock specimens to detect fracturing (section 5.3.1). P waves are compressive in nature, whereas S waves, being shear waves, are transmitted through fluids without causing deformation. Because of anisotropy in rock properties, S waves may decompose into a horizontal component, S_H, and a vertical component, S_V.

The elasticity of an isotropic, homogeneous rock of density ρ is a function of its compressibility or bulk modulus κ and its shear or rigidity modulus μ. The velocities of P and S waves are given by

$$v_P = \sqrt{\frac{\kappa + \frac{4}{3}\mu}{\rho}}, \tag{14.1}$$

$$v_S = \sqrt{\frac{\mu}{\rho}}. \tag{14.2}$$

Primary velocities are of the order of 4–7 km/s for limestone, 2–5.5 km/s for sandstone, 1.8–4 km/s for shale and 5–6 km/s for granite. The ratio $v_P/v_S \sim 1.5$–2.5 and is strongly dependent on the porosity and fluid saturation of the rock (Guéguen and Palciauskas, 1994).

Surface waves, such as Rayleigh and Love waves (Figure 14.7), travel relatively slowly and only across the Earth's surface, but are often associated with the coseismic shaking damage. The modified Mercalli intensity scale listed in Box 14.1 gives semi-quantitative information on these earthquake effects. Surface waves are similar to waves crossing an ocean surface carrying up and down any floating object. In a Rayleigh wave, the motion is that of a vertical ellipse, the amplitude of which ellipse quickly decreases with depth. Love waves, which are somewhat faster than Rayleigh waves but slower than S waves, are similar to a horizontal S wave with motion perpendicular to the direction of wave propagation.

Both body and surface waves are shown in Figure 14.8, which illustrates the acquisition of three seismograms at a station with seismometers oriented in different directions to acquire both horizontal (E–W and N–S) as well as vertical seismic signals. These signals may have been recorded on separate seismometers, or a single combined unit, with electronic storage and processing of the data.

The Mercalli scale (see Box 14.1) is useful for public communication purposes only. In 1935, Charles Richter of the California Institute of Technology developed a magnitude scale based on the distance from the seismometer to the earthquake focus, the type of seismograph, the maximum amplitude of the seismic wave on the seismograph and local conditions (NRC, 2003). The Richter magnitude, denoted by M_L, is still cited by news media following an earthquake, but,

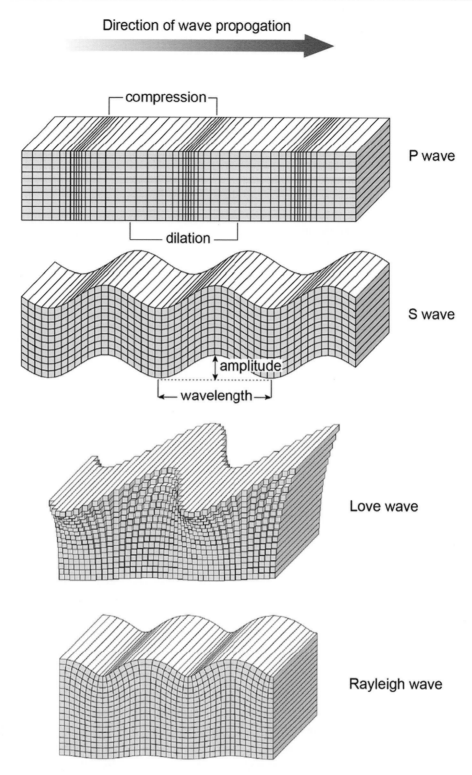

Figure 14.7 Seismic waveforms. Elastic responses of rocks to the passage of P and S waves are shown with parameters defined. The velocity of wave travel is equal to the wavelength divided by its period (T), which is the time interval between successive wave crests. The frequency of a sinusoidal wave is $1/T$ and is reported in hertz, i.e., cycles per second. Love and Rayleigh surface waves are shown in exaggerated detail with horizontal propagation. Love waves exhibit purely transverse motion, whereas Rayleigh waves contain both vertical and radial motion. In both cases, the wave amplitude decays strongly with depth. (Modified from Shearer (2009); reproduced with the permission of Cambridge University Press.)

BOX 14.1 | MODIFIED MERCALLI INTENSITY SCALE

The standard measure of earthquake intensity is the **modified Mercalli intensity scale** (MM scale), which was developed in 1931. It is composed of 12 increasing levels of intensity, ranging from barely sensible shaking to catastrophic destruction.

It is used to communicate to the public those seismic effects that might be felt or witnessed in a particular location. The lower values indicate sensible effects, i.e., those felt by people, while the higher numbers address structural damage and require the input of structural engineers to assign intensity values of VIII or above. Magnitudes (M) listed are those given by the US Geological Survey; average peak accelerations ($g = 9.8$ m/s^2) are from Bolt (1993). The 12 levels are as follows.

I. Not felt except by a very few under especially favourable circumstances. $M1$–$M3$

II. Felt only by a few persons at rest, especially on upper floors of buildings. Delicately suspended objects may swing. $M3.0$–$M3.9$

III. Felt quite noticeably by persons indoors, especially on upper floors of buildings. Many people do not recognize it as an earthquake. Standing automobiles may rock slightly. Vibration similar to the passing of a truck. Duration estimated. $M3.0$–$M3.9$

IV. Felt indoors by many and outdoors by few during the day. At night, some awakened. Dishes, windows, doors disturbed; walls make cracking sound. Sensation like a heavy truck striking the building. Standing automobiles rocked noticeably. $M4.0$–$M4.9$; $0.015g$–$0.02g$

V. Felt by nearly everyone; many awakened. Some dishes and windows broken. Unstable objects overturned. Pendulum clocks may stop. $M4.0$–$M4.9$; $0.03g$–$0.04g$

VI. Felt by all; many frightened. Some heavy furniture moved; a few instances of fallen wall plaster. Damage slight. $M5.0$–$M5.9$; $0.06g$–$0.07g$

VII. Damage negligible in buildings of good design and construction; slight to moderate in well-built ordinary structures; considerable damage in poorly built or badly designed structures; some chimneys broken. $M5.0$–$M5.9$; $0.10g$–$0.15g$

VIII. Damage slight in specially designed structures; considerable damage in ordinary substantial buildings with partial collapse. Damage great in poorly built structures. Fall of chimneys, factory stacks, columns, monuments, walls. Heavy furniture overturned. $\geq M7$; $0.25g$–$0.30g$

IX. Damage considerable in specially designed structures; well-designed frame structures thrown out of alignment. Damage great in substantial buildings, with partial collapse. Buildings shifted off foundations. Underground pipes broken. $\geq M7$; $0.50g$–$0.55g$

X. Some well-built wooden structures destroyed; most masonry and frame structures destroyed with foundations. Rails bent. $\geq M7$; $> 0.6g$

XI. Few, if any, masonry structures remain standing. Bridges destroyed. Rails bent greatly. $\geq M7$

XII. Damage total. Lines of sight and level are distorted. Objects thrown into the air. $\geq M7$

Source: From US Geological Survey (1997).

because it is dependent on the seismic instrument and local conditions, it is unsatisfactory as a quantitative measure of an earthquake's kinetic energy. Seismological practice now uses other magnitude scales that are based on the mechanics of fault slip, such that when $M_L > 6$ the Richter magnitude is unreliable.

Prior to the rupture of a fault, shear forces on either side of the fault trace constitute a double-couple force system shown in Figure 14.9, in which the shear forces (F) are separated from the fault trace by a distance b; therefore the moment of the couple is $F \cdot 2b$. In Figure 14.6 we show a rupture area as being the source of the earthquake; consequently the moment of the couple must reflect the area A of the rupture plane shown in Figure 14.9, the elastic shear or rigidity modulus (μ) and the strain produced by the earthquake, i.e., its slip or displacement d. This product is called the **seismic moment**

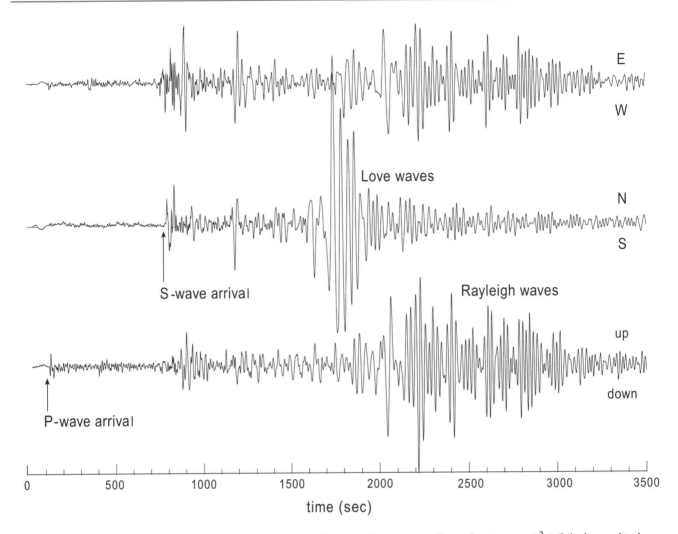

Figure 14.8 Seismograms showing surface waves. The amplitude of the waveforms is typically acceleration in m/s^2. While the amplitudes of the surface waves – Love and Rayleigh waves – are larger than the S-wave amplitudes, it is not necessarily the case that the surface waves will cause more damage than the S waves. (From Mussett and Khan (2000). Reproduced with the permission of Cambridge University Press.)

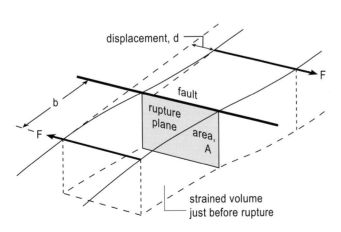

Figure 14.9 Definition of the seismic moment. (From Mussett and Khan (2000). Reproduced with the permission of Cambridge University Press.)

(M_0) and is estimated in newton metres (N m) by $M_0 = A\mu d$ (Mussett and Khan, 2000; Scholz, 2002). The elastic shear modulus $\mu = E/(2 + 2\nu)$ is a measure of the shear resistance of the rock in which the fault is located; it is estimated from the S-wave velocity.

The standard measure of earthquake magnitude now used is the **moment magnitude** (M_W), which matches the Richter magnitude at lower magnitudes but more accurately estimates the size of large earthquakes (Shearer, 2009):

$$M_W = \tfrac{2}{3}(\log_{10} M_0 - 9.1). \qquad (14.3)$$

An additional measure is the surface-wave magnitude, M_S, which is a function of the large amplitudes of surface waves, as illustrated by Figure 14.8. Some values of M_S and M_W are listed in Table 14.1. The moment magnitude (M_W) is sometimes given simply as M and is an empirical dimensionless measure.

Table 14.1 Measurements of the magnitude of some recent earthquakes (From NRC (2003), except for the four most recent entries, which are from the Caltech Tectonics Observatory website, for which the seismic moments were estimated from equation (14.3).)

Date	Location	M_S	M_W	M_0 (10^{18} J)
1906	San Francisco, USA	8.25	8.0	1 000
1960	Chile	8.3	9.5	200 000
1964	Alaska, USA	8.4	9.2	82 000
1985	Mexico City, Mexico	8.1	8.0	1 100
1989	Loma Prieta, USA	7.1	6.9	27
1994	Northridge, USA	6.6	6.7	12
1999	Izmit, Turkey	7.8	7.4	242
1999	Chi-Chi, Taiwan	7.7	7.6	340
2001	Bhuj, India	8.0	7.6	340
2004	Sumatra, Indonesia	8.5	9.2	70 000
2008	Sichuan, China	8.0	7.9	800
2010	Haiti	—	7.0	35
2011	Tohoku, Japan	—	9.0	36 000

From Table 14.1 it is apparent that, while strike-slip earthquakes, such as the 1906 San Francisco earthquake, cause significant damage and loss of life, they are in fact an order of magnitude smaller than the major subduction-type earthquakes that occurred in 1960 in Chile, 1964 in Alaska and 2004 in Sumatra. According to the Caltech Tectonics Observatory, the Sumatran fault ruptured over a distance of 1500 km, lasting 10 minutes, while the 1994 Northridge quake ruptured over only 20 km and lasted just 15 seconds. As Table 14.1 shows, M_S and M_W differ; this is because they are measured over different bandwidths of the seismic signal. The reader is referred to Scholz (2002) for further information on earthquake magnitude scales.

Table 14.1 hints at some important lessons, e.g., very large earthquakes occur infrequently and only on big faults. Strain energy must build over time before rupture and slip occur because plate tectonics produces slow migration of the plates that cause the strain energy to build. The larger the fault, the larger the seismic moment might be. Figure 14.10 makes this conclusion explicit. The estimation of a maximum magnitude for a fault is not straightforward because our seismic records are short compared with the length of many earthquake cycles. The maximum magnitude of a San Andreas Fault event appears to be $M8$, i.e., similar to the 1906 San Francisco ($M_W 8.0$) and 1857 Fort Tejon ($M_W 7.9$) events. This

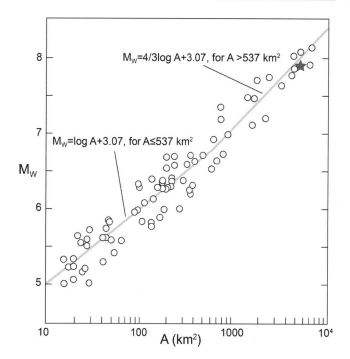

Figure 14.10 Moment magnitude plotted against source area for strike-slip earthquakes. The star (top right) indicates the 2002 Denali earthquake in Alaska that produced a 340 km surface rupture on the Denali Fault and two adjacent faults. The fault ruptured beneath the Trans-Alaska Oil Pipeline, which withstood 5.5 m of lateral offset and 1 m of vertical offset without damage. (Modified from Haeussler et al. (2004). Courtesy of the US Geological Survey.)

magnitude is better constrained than the estimate for other southern California faults.

The number of small earthquakes for a particular region is typically related to the number of larger ones by the **b-value curve**, which is a simple plot of the logarithm of the number of earthquakes per year (N) greater than a given magnitude, thus (Shearer, 2009) $\log_{10} N = a - bM$, where a is a constant related to regional seismicity and $b \sim 1.0$, which is also known as the Gutenberg–Richter recurrence law (Gutenberg and Richter, 1944). While there are many small earthquakes, the energy they expend is small compared to larger-magnitude if infrequent quakes. Therefore, globally, one annual event with magnitude 8 implies 10 000 events of magnitude 4 annually. For $b = 1.0$, the constant a is simply the magnitude of the average annual earthquake for the region.

14.3 Friction and Faults

The work required to make a fault rupture and slip is provided by the strain energy released by an earthquake:

$$E_s = \frac{\Delta\sigma}{2\mu} M_0, \qquad (14.4)$$

where E_s is the seismic energy (joules) and $\Delta\sigma$ is the shear stress drop that occurs during fault rupture, its mean being $\Delta\sigma \sim 3\,\text{MPa}$ with a range of 0.3–50 MPa. Equation (14.4) allows us to identify the magnitude of an earthquake based on the seismic energy radiated:

$$\log_{10} E_s = 1.5 M_S + 4.8. \tag{14.5}$$

The implication of (14.5) is that a unit increase in earthquake magnitude results in a $\sim$32-fold increase in kinetic energy released (Scholz, 2002; NRC, 2003).

Because (1) $\Delta\sigma \sim 3\,\text{MPa}$ for typical earthquakes and (2) earthquakes typically display an average slip (d) to fault length (L) ratio of $10^{-4} < d/L < 10^{-5}$ (Fagereng and Toy, 2011), it is possible to scale the results shown in Figure 14.10 to develop the schematic representation of relative earthquake size shown in Figure 14.11. The fault geometry shown assumes an elliptical rupture area and applies to earthquakes from $M_W 2$ to $M_W 9$. It illustrates a particular geological mapping problem when suspected active faults are to be investigated: a mapped outcrop of 20 m is perhaps meaningful for a $M_W 4$ event but would have little relevance to an $M_W 6$ event. A typical geological map area of 25 km^2 is equally small relative to an $M_W 9$ event.

Geological mapping has indicated that some horizontal movement during earthquakes can be accommodated without faulting by ductile folding of sedimentary rock or by plastic deformation of igneous or metamorphic rock, as shown in Figure 4.9. However, often stresses cause faulting with deformation by ductile shear of shallow rocks (<6 km) involving **granular flow** (see Figure 4.8a).

In the **seismogenic zone** at depths of 6–12 km, where frequent earthquakes occur, much of the deformation is by **cataclastic flow**, as in Figure 4.8(b). Cataclasis occurs when a rock being deformed under confining stress undergoes brittle crushing of the mineral grains together with microfracturing, frictional sliding and grain rotation (Fossen, 2016). This permanent deformation may begin elastically; however, above 40–50% of the peak shear strength, the rock undergoes progressive microcracking within a brittle–ductile regime. With increased stress, the microcracks coalesce, the crack density increases, the shear strength of the deforming rock is reduced with continuing slip (i.e., slip weakening) and a macroscopic shear-fracture surface develops (Ohnaka, 2013). Brittle microfracturing and plastic deformation at greater depths produces mylonite, as shown in Figure 3.12.

Irrespective of the deformational mechanism, fault rupture occurs when the frictional strength of the fault is exceeded

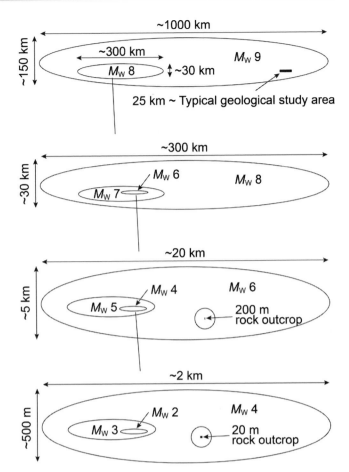

Figure 14.11 Relative size of the rupture areas for earthquakes of $M_W 2$–9 based on the empirical relationships shown in Figure 14.10. The dimensions shown were adapted to present elliptical rupture areas by Fagereng and Toy (2011) for simplicity of presentation. (From Fagereng and Toy (2011). Reproduced with the permission of The Geological Society of London.)

by the shear stresses at failure. Thus, we might incorporate the Mohr–Coulomb strength criterion (equation (1.8)) within Amonton's law (equation (1.12)) as

$$\tau = \tau_0 + \mu(\sigma_n - p_w), \tag{14.6}$$

where τ is the shear strength at failure, τ_0 is the cohesion, μ is the friction coefficient and $\sigma_n - p_w$ is the effective normal stress acting on the fault. The progression of events that lead to rupture along a pre-existing fault has been regarded as a **stick–slip** process representing frictional instability since the work of Gilbert and Reid. When the fault slips, the asperities that constitute the sawtooth edges of faults are crushed and form a gouge zone.

Consequently, the nature of the friction coefficient has attracted much experimental investigation based on the expression (Scholz, 2002) $\tau = \mu_0 \sigma_n^m$, where μ_0, n and m are constants. The base friction coefficient (μ_0) is measured by standard laboratory test methods as shown in Figure 4.10. Measurements of various rock types lead to **Byerlee's law**:

$$\tau = 0.85\sigma_n, \qquad 10\,\text{MPa} \leq \sigma_n \leq 200\,\text{MPa}, \qquad (14.7)$$

$$\tau = 50 + 0.6\sigma_n, \quad 200\,\text{MPa} \leq \sigma_n \leq 1500\,\text{MPa}. \qquad (14.8)$$

Quite remarkably, these conditions hold for the majority of carbonate and silicate rocks with μ_0 confined to a range 0.6–0.8. Above 350°C, which exceeds the temperature of the seismogenic zone, stick–slip behaviour transitions into ductile creep, i.e., plastic flow in Figure 4.9. Therefore, Byerlee's law provides a first approximation of the strength of faults.

Should the fluid pressure ($p_w = \rho_w gz$) in (14.6) exceed the hydrostatic gradient ($\sim$10 MPa/km, depending on fluid density) and approach the total lithostatic stress ($\sigma_T = \rho gz$), then the frictional resistance to sliding of thin rock formations between other stable formations disappears. These low-angle reverse or **thrust faults** can exceed 10 km in length. The requirement that high fluid pressures are present indicates that shale caprocks may confine these fluids following their generation by compaction of sediments. This class of faults was not identified by Anderson (section 4.2.2).

It follows that, when fluid pressures are sufficiently raised by intentional fluid injection, earthquakes may be induced. The injection of wastewater into the fractured Precambrian bedrock beneath the Rocky Mountain Arsenal near Denver, Colorado, in the 1960s triggered a number of earthquakes, which were not recognized as such immediately. The formation pressures reached a maximum increase exceeding 3 MPa beneath the epicentre of the earthquakes (Ingebritsen et al., 2006).

Similarly, 50 years later, it became apparent that fluid injection for oil and gas production using hydraulic fracture stimulation near pre-existing faults and, particularly, of wastewaters from oil and gas production were responsible for numerous earthquakes (typically $M \geq 3$, maximum $M_W 5.7$) in the mid-continent of the USA (Ellsworth, 2013), i.e., they were typically large enough to be felt but were not generally damaging. Similar experiences have been documented in Europe (Grigoli et al., 2017). These indicate that many faults may be **critically stressed**, meaning that they are in a state of incipient frictional failure (Zoback, 2010). Actual failure may be triggered by a minor increase in fluid pressure or by an increase in the shear stress or by a decrease in the normal stress on the fault.

14.4 Detection of Active Faults

The state of California adopted the Alquist–Priolo Earthquake Fault Zoning Act in 1972 to mitigate the hazard of surface faulting to structures for human occupancy and has continually updated it to accommodate new findings from earthquake science. The Act requires the identification of "potentially and recently active traces of the San Andreas … and other faults, or segments thereof, that the State Geologist determines to be sufficiently active and well-defined to constitute a potential hazard to structures from surface faulting or fault creep".

An **active fault** is defined in California as one that has "had surface displacement within Holocene time (about the last 11 000 years)". Bryant and Hart (2007) provide guidelines for those wishing to investigate the hazards of surface faulting, including stereoscopic interpretation of aerial photographs, surface mapping, trenching, boreholes, test pits, cone penetrometer testing (CPT) and various geophysical investigations, in particular high-resolution seismic reflection and ground-penetrating radar (GPR). This is the domain of paleoseismologists.

Paleoseismology is defined as the study of prehistoric earthquakes, in particular their location, timing and size (McCalpin, 2009). Alquist–Priolo has had a significant influence on the development of paleoseismology, in that it made demands on building developers to have geologists define the presence of active faults prior to construction. With a similar motive, Hanson et al. (1999) have summarized the process by which seismogenic faults of $M > 5$ might be identified to guide siting of nuclear reactors, in which paleoseismic investigations have played a significant role in seismic hazard assessments (Meehan, 1984; Serva and Slemmons, 1995).

The benefit that paleoseismology provides is that it permits an extrapolation from the few hundred or thousand years exposed in a trench to tens or hundreds of thousand years by allowing geoscientists to use well-established empirical regression relationships (e.g., Wells and Coppersmith, 1994) between earthquake magnitude and surface-fault rupture dimensions, for which measurements in excavations are essential.

Treiman (2010) has provided a helpful guide to those who wish to undertake an investigation as required by Alquist–Priolo while stressing the difficulty of identifying where and how a particular fault might rupture at ground surface and estimating the amount of slip. Among the many issues that complicate the intent of the Alquist–Priolo Act, Treiman noted that rupture on primary strike-slip faults is often very

Figure 14.12 Trenches in Coachella Valley, California. (Courtesy of Roy Shlemon.)

Figure 14.13 Wallace Creek in the Carrizo Plain of central California. The Creek discharges from the Tremblor Mountains and has incised its way through alluvial-fan deposits before crossing the San Andreas Fault (SAF) in the foreground. The drainage offset shown by the double-headed arrow is approximately 128 m, which displacement occurred over 3587 years. Right-lateral slip has caused the stream channel to become "beheaded". (Photo courtesy of the US Geological Survey.)

complex, such that fault traces splinter into arrays, known as *en echelon* patterns, characterized by groups of sub-parallel fault traces. These complicate the siting and extent of trenches to uncover the faults because geologic maps may show only a fraction of the secondary fault traces that become evident with recurrence of fault slip. Slemmons (1995) endorsed this concern but in the context of normal faulting in the Basin and Range Province of the western USA, where he noted that complex fault patterns were particularly evident when $M > 6.5$–7.5.

The detection of active faults using trenches has become a standard tool of paleoseismology. Figure 14.12 shows a trench excavated near a suspected strike-slip fault trace, which is the principal focus of such trenches in California. The trench shown was excavated to a depth of 5–10 m; however, that depth is only sufficient to expose soils of ~4–6 ka, not 11 ka. This is because of rapid land subsidence associated with groundwater extraction (see Chapter 13).

Once a trench is established, the mapping of the exposed soil profile may require lengthy assessment and sufficient soil sampling to define the stratigraphy and to age date the profile. Borchardt (2010) recommends that two ages be defined using radiocarbon analysis of charcoal samples: (1) t_0, the age when soil formation and/or aggradation began (units of ka); and (2) t_b, the age when the soil or the soil stratum was buried. The difference ($t_0 - t_b$) is the duration of time of soil development and/or aggradation. Shlemon (1985) used soil stratigraphy as a complement to geotechnical site studies at a liquefied natural-gas plant and a nuclear reactor to date the last fault-slip displacement and its recurrence interval.

An engineering geologist will consider geodetic or geomorphic evidence of active tectonics, which are well described by Keller and Pinter (2002). Geodetic techniques attempt to measure strain along the fault zone as it develops, employing very-long-baseline interferometry, satellite laser ranging, GNSS/GPS, InSAR and strain meters, as discussed in section 13.1.4 and by NRC (2003).

Geomorphic evidence may involve (i) fluvial responses to tectonic activity, such as uplift that leaves the former floodplain as a fluvial terrace or a change in the drainage pattern, (ii) coastal uplift exposing a now-inactive shoreline, (iii) the development of fold belts associated with buried (hidden) faults, and (iv) various secondary effects of fault rupture, e.g., landslides, tsunami flooding (see Box 14.2), and sand boils and other liquefaction phenomena.

Perhaps the classic example of the advantages of paleoseismic investigations was the determination of the slip rate of the San Andreas Fault by examination of offset stream drainage patterns at Wallace Creek in the Carrizo Plain of central California, which is shown in Figure 14.13. The offset of Wallace Creek shown caused Sieh and Jahns (1984) to excavate eight trenches across and adjacent to the fault trace in the area shown in the figure and two other trenches to the left of the photo to inspect an abandoned channel of Wallace

BOX 14.2 | IDENTIFYING THE LAST GREAT EARTHQUAKE IN THE CASCADIA SUBDUCTION ZONE

One of the great achievements of paleoseismology has been its elucidation of the threat of a great earthquake ($M \geq 8$) associated with the Cascadia subduction zone in the Pacific Northwest of the USA and British Columbia in neighbouring Canada (see Figure 14.5).

Back in the 1970s a team of experts had concluded that Cascadia was aseismic, that is, it was not subject to major earthquakes. This threat emerged in the geoscience community around 1990 when there was sufficient evidence that Cascadia was experiencing "the calm before the quake". Thomas Heaton of the US Geological Survey summarized the evidence for a great Cascadia earthquake, listing elevated marine terraces west of Seattle, tree-ring data from dead trees in an intertidal zone, landslides crashing into Lake Washington in Seattle – all in the last 1100 years. However, to use Heaton's words, "the most compelling evidence that great subduction earthquakes" have occurred in this same time period was the description of ancient sand sheets in Willapa Bay, Washington state, by Brian Atwater of the USGS and their interpretation as tsunami deposits from enormous waves generated by a great subduction earthquake.

In the years since Heaton's alert, the scientific evidence for a future great Cascadia subduction zone earthquake has reached a consensus such that buildings in Seattle and Vancouver have been retrofitted to withstand seismic shaking and some threatened coastal communities have organized tsunami evacuation plans and drills (others have dithered).

Atwater's discovery is illustrated in Figure 14.14. Similar tsunami-deposited sands were later discovered on Vancouver Island in British Columbia by John Clague of the Geological Survey of Canada. The sequence of events is: (1) a great earthquake causes land subsidence (see Figure 14.5), with flooding by the tsunami and deposition of sand sheets; (2) the lowered land surface becomes a marsh, and any trees are killed by salt water in the intertidal zone; (3) uplift elevates the area as a beach, and soils and vegetation re-establish themselves, but the dead trees remain in place, buried by a metre of tidal mud and sand as a "ghost forest" to be investigated later by geoscientists; and (4) the cycle repeats itself.

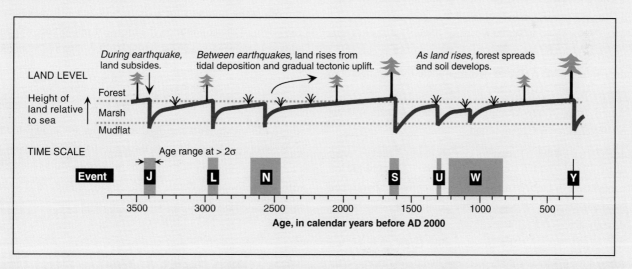

Figure 14.14 The evidence for seven earthquakes over the past 3500 years in the sediments of Willapa Bay, southwest Washington state, USA. The events in the time scale are shaded to indicate uncertainty in the ages identified; however, the most recent event (AD 1700) is known with certainty from Japanese records. Note that the recurrence intervals vary from 1000 to 300 years. The line recording land level was developed with microfossil data. The vertical drop in this line indicates the severity of the particular seismic event. (From Yeats (2012), based on the USGS report of Atwater and Hemphill-Haley (1997). Reproduced with the permission of Cambridge University Press.)

Creek that is symmetrical with the offset of the modern channel shown in the figure. The locations of the channel offsets act as piercing points, i.e., a feature that is offset by the fault and is apparent in two or more excavations dug parallel to the fault. The abandoned channel has been displaced 380 m, while the modern channel had been displaced 128 m.

With radiocarbon dates from charcoal samples to fix the geomorphic timeline, Sieh and Jahns (1984) described the late Pleistocene–Holocene evolution of Wallace Creek as follows:

1. The older alluvial fan deposited sediments that buried small fault scarps along the trace of the San Andreas Fault during the period including 19 ka.
2. At 13 ka, erosion of the older alluvial fan became inactive because of gully erosion (see Figure 14.13) that caused deposition of the eroded sediments in the younger fan alluvium.
3. About 10 ka, a new channel cut across the fault and the earlier channel was abandoned.
4. The new channel was the active channel of Wallace Creek until 3.7 ka by which time it had slipped ~250 m to the left (not shown in Figure 14.13) and was eventually abandoned by Wallace Creek, which then incised another new channel directly across the fault.
5. Between 3.7 ka and the present, this latest channel has slipped by 128 m as shown, i.e., a right-lateral offset.

The average slip rate since 3.7 ka has been 34 mm/yr at Wallace Creek based on the 128 m displacement. More recently Liu-Zeng et al. (2006) have analyzed a second site a few hundred metres further southeast along the fault and identified six sequential offsets that also produced an average slip rate of 34 mm/yr between AD 1210 and AD 1857,

which was the date of the nearby Fort Tejon earthquake now estimated as $M_W 7.9$. Therefore, along this 220 km reach of the San Andreas Fault, paleoseismic studies have provided a unique history of slip of five earthquake cycles, while most paleoseismic studies account for only two cycles.

Long-term **slip rates** may be estimated on the basis of the recurrence rate (Frankel et al., 1996)

$$\text{rate} = \frac{\mu \dot{u} L W}{M_0}, \qquad (14.9)$$

where μ is the shear modulus, $\dot{u}$ is the long-term slip rate, L and W are the length and width of the rupture zone of the fault and M_0 is the moment for a characteristic earthquake event. The effect of the slip rate on the moment magnitude has been estimated by Anderson et al. (1996) as

$$M_W = (5.12 \pm 0.12) + (1.16 \pm 0.07) \log L_{SR}$$
$$- (0.20 \pm 0.04) \log \dot{u}, \qquad (14.10)$$

where the means and standard deviations are shown for the surface rupture length (L_{SR}, in km) and slip rate ($\dot{u}$, in mm/yr). This regression equation is based on analysis for earthquakes with a seismogenic depth of 15–20 km; therefore it would exclude most subduction-thrust events.

14.5 Seismic Hazard Analysis

Many modern textbooks on geotechnical engineering, such as Briaud (2013), address seismic hazard analysis. Consequently, here we will confine our discussion to the geoscientific basis on which such geotechnical issues are considered and seek to link the geoscientific to the geotechnical themes. Certain topics are of particular importance in this context: the nature of strong ground motions, the development of seismic hazard maps and the occurrence of liquefaction. Two other related issues of great concern, seismically induced landslides and tsunamis, are considered in the following chapters.

14.5.1 Ground Motion

Ground motions are the movements of the Earth's surface induced by earthquakes. Measurements of ground motion may be presented as peak ground accelerations as a percentage of the Earth's gravity or as peak ground velocity or as peak displacements.

However, peak measurements are no longer the sole parameters in seismic hazard assessment and are complemented by spectral analysis of strong-motion records collected by observatories, in which the ground-motion accelerations are measured as a function of their frequency spectrum. Attenuation or amplification of seismic waves by soil and rock are then considered to determine seismic hazard. These factors are required for building codes used in the development of structural designs.

Seismic signals, such as the acceleration time series shown in Figure 14.8, are described as a probability density function of the energy radiated as a function of its frequency (or its period), i.e., a Fourier acceleration spectrum. The magnitudes of the low-frequency part of the spectrum are directly proportional to $\log M_0$ and therefore, by (14.3), to $\log M_W$ (Boore, 1983, 2003).

Seismic waves may be attenuated or amplified by three-dimensional geologic structures such as fault blocks or sedimentary basins. Of particular concern is amplification in soils, which occurs as S waves are refracted at the bedrock–soil boundary followed by amplification in the low-velocity soils and trapping within the shallow soil layers, leading to structural damage; similarly, surface waves cause destructive shaking.

Consequently, the average shear-wave velocity in the upper $\sim$30 m is critical in defining the seismic hazard for particular sites, e.g., hard rock or soft soil, and is used to derive seismic hazard levels. It was noted that the resonance of seismic waves in shallow lacustrine soils in Mexico City during the 1985 M_W 8.0 earthquake was 8–50 times that of hard bedrock sites (NRC, 2003; Kavazanjian et al., 2011).

Figure 14.15 illustrates the effects of the M_W 6.9 Loma Prieta earthquake in 1989. The epicentre of this earthquake was near Santa Cruz, California; however, serious damage and 60 fatalities occurred 100 km to the northeast, mostly along the Cypress viaduct in Oakland, partly because of the

way the S waves were propagated. The focus or hypocentre of the earthquake was at a depth of $\sim$15 km and the fault rupture was $\sim$40 km long. At the epicentre, the modified Mercalli intensities were lower than in San Francisco and Oakland (9 versus 7–8). Loma Prieta, which released just 3% of the energy of the 1906 earthquake, was one of the larger earthquakes in California since 1906. It has been extensively documented with respect to its geotechnical effects by the US Geological Survey (see the further reading at the end of the chapter).

Monitoring of strong ground motions for the design of safe foundations requires the siting of instruments known as accelerometers, which measure the peak accelerations imposed on structures by the seismic waves and the duration of shaking. While both vertical and horizontal accelerations from an earthquake acted upon the Cypress viaduct shown in Figure 14.15, the structural design would have considered the self-weight of the viaduct as constraining any significant vertical effects.

Consequently, it is the **peak horizontal acceleration** (PHA) that is more critical in causing damage to infrastructure through the side-to-side motion induced (Bolt, 1993), although the peak ground velocity and peak ground displacement are also employed by engineers in this context. The attenuation of the PHA for the Loma Prieta quake was analyzed for alluvial sites to the north and northwest of the epicentre such as in Oakland by Campbell (1998). The attenuation relationship decays with distance. At radial distances (d) greater than 79 km from the zone of seismogenic rupture, the peak horizontal acceleration was estimated as $\ln \mathrm{PHA} = 3.2 - 1.24 \ln(d + 7.8)$.

Similar ground-motion prediction equations have been developed for a variety of regions and fault types, with parameters that specify shear-wave velocities, such as Atkinson and Boore (2006) prediction equations for eastern North America. The general form of these equations is

$$\ln \mathrm{PHA} = b_1 + b_2(M_W - 6) + b_3(M_W - 6)^2 \\ + b_5 \ln d + b_v \ln \frac{\overline{v}_s}{v_A}, \qquad (14.11)$$

where the regression coefficients b_1, b_v, etc., identify regional properties, d is the distance from a specific site to the earthquake focus, $\overline{v}_s$ is the average shear-wave velocity and v_A is a shear velocity estimate for a type of site, e.g., soil or rock (Boore et al., 1997).

Next Generation Attenuation Models (https://ngawest2.berkeley.edu/) is software used to estimate ground motions that can affect any infrastructure site based on historical data.

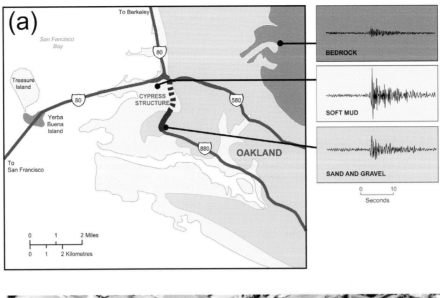

Figure 14.15 (a) Seismic shaking following the 1989 Loma Prieta earthquake, showing the amplification of the ground velocity during an aftershock by the soft mud compared with that in the bedrock in the Oakland hills. (b) Collapsed section of the Cypress viaduct of Interstate 880 that was built on the soft mud. That section built on sand and gravel remained standing despite severe shaking as shown above. (Photo and figures prepared by the US Geological Survey.)

The report by Boore and Atkinson (2007) contains ground-motion prediction equations for a subset of horizontal ground motions as a function of earthquake mechanism, distance from source to site, local average shear-wave velocity and fault type.

Predictions of strong ground motions are based on either a deterministic or a stochastic approach, which are reviewed by Baker (2016). In the deterministic approach, both a reference seismic event and a reference ground motion must be identified. In practice, worst-case events are chosen as the reference cases; however, this leads to the problem of defining what should be such a worst-case event given the typical shortness of the historical record.

Figure 14.16(a) is a map of two adjacent active faults that produce different response spectra in Figure 14.16(b). Note that the maximum spectral acceleration is produced by fault A in the high-frequency part of the spectrum and fault B in the low-frequency part. Consequently, there is not a single worst-case event producing the largest spectral acceleration amplitude for all periods. Here we are dealing with only two active faults with known locations; in earthquake-prone regions, more complex source patterns may be required.

BOX 14.3 | SEISMIC WAVE FREQUENCY AND GROUND ACCELERATION

The displacement (y) of a sinusoidal wave, such as that shown in Figure 14.7, at any time t at a location x is

$$y = A \sin\left(\frac{2\pi}{\lambda}(x - vt)\right),\qquad(14.12)$$

where the travel or phase velocity of this wave is its wavelength (λ) divided by the period of the wave (T), which is the time required for the wave to travel one wavelength, i.e., $v = \lambda/T = f \cdot \lambda$. The angular frequency of the wavelength is written as $\omega = 2\pi/T$ (radians per second, rad/s).

The lateral displacement of the point $x = 0$, such as might be caused by seismic vibration, can be written as

$$y = A \sin\left(\frac{2\pi}{T}t\right) = A \sin(\omega t).\qquad(14.13)$$

Differentiating twice with respect to time, we obtain the acceleration ($\ddot{y}$):

$$\ddot{y} = \frac{4\pi^2 A}{T^2} \sin\left(\frac{2\pi}{T}t\right) = -\omega^2 A \sin(\omega t) = -\omega^2 y.\qquad(14.14)$$

Equation (14.14) indicates that any particular ground acceleration, usually expressed relative to the acceleration of gravity ($g = 9.81 \text{ m/s}^2$), can be due to various combinations of displacement, i.e., ground-motion amplitude, and the wave frequency of this particular event.

Therefore, a displacement of 1.0 cm may be caused by a wave with a frequency of 10 Hz with an acceleration of $0.4g$ or by a wave of frequency 1 Hz with an acceleration of $0.04g$. Long-period waves tend to have small accelerations and relatively large displacements, whereas short-period waves, particularly those with frequencies greater than 10 Hz, have high accelerations and small amplitudes of displacement.

Sources: Resnick and Halliday (1960) and Bolt (2003).

A further problem is that the response spectra of Figure 14.16(b) are obtained from predictions by empirical models based on measured ground motions. Figure 14.16(b) actually shows the predicted median response spectral acceleration at 1 s developed from a ground-motion prediction equation, such as (14.11), with a Gaussian distribution of ground motions around the curves. Thus, 50% of the measured ground motions exceed the curves.

Figure 14.17 illustrates how the recorded ground motions for the 1999 Chi-Chi earthquake in Taiwan are distributed around the mean spectral acceleration at 1 s. At a distance of 1–10 km, the spectral accelerations vary between $0.15g$ and $1.0g$. Even if a "mean plus one standard deviation" approach is used, some 16% of recorded ground motions will be ignored. Therefore, what is estimated in deterministic seismic hazard analysis is really the "maximum considered earthquake" rather than a worst-case earthquake.

The probabilistic approach, **probabilistic seismic hazard analysis** (PSHA), tries to accommodate the variability in the recorded ground motions shown in Figure 14.17 – and to allow for ground motions that are not recorded but

might be reasonably predicted to occur in such a situation. PSHA involves five steps that are illustrated in Figure 14.18 (Baker, 2016):

(a) Identification of all earthquake sources that are capable of producing damaging ground motions at the site of interest. In many cases, the location of a fault is well known (e.g., Figure 14.6) and the fault plane may be defined as a line source. In some cases, such as with intraplate sources in eastern North America, the sources may be poorly defined and an area source is used.

(b) Characterization of the rates at which earthquakes of various magnitudes are expected to occur. The Gutenberg–Richter recurrence law or b-value curve, introduced earlier as $\log_{10} N = a - bM$, yields the rate of earthquakes (N) of magnitude M, and the coefficients a and b provide information from the historical record.

(c) Characterization of the distances of sources to sites associated with the potential earthquakes. It is generally assumed that there is an equal probability of an earthquake source being uniformly distributed along a fault.

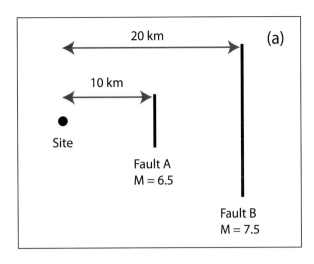

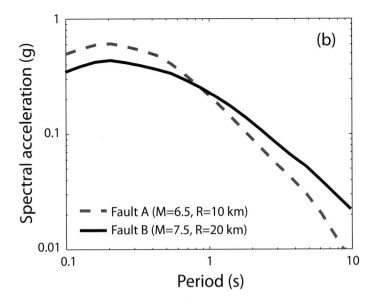

Figure 14.16 (a) Map view of a hypothetical site with two nearby active faults. (b) The predicted median response spectra from the two earthquakes shown in (a), illustrating that the event producing the maximum response spectrum may vary depending upon the period of interest. (From Baker (2016), reprinted with permission from *Applied Geology in California*, Anderson and Ferriz (eds), copyright 2016 by Star Publishing Company, Inc., Belmont, California, USA.)

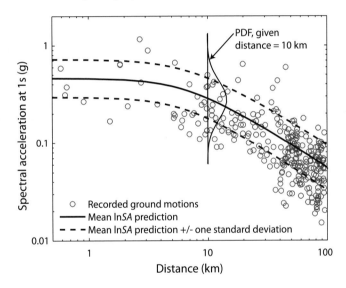

Figure 14.17 The observed spectral accelerations (at 1 s) from the 1999 Chi-Chi earthquake in Taiwan illustrating variability in ground-motion intensity within 100 km of the fault. The ground-motion prediction equation used to develop the mean-value curve and standard-deviation bracket may provide values of the natural logarithm of the spectral acceleration (shown here) or of the peak ground acceleration. This figure indicates a normally distributed set of ln SA values. (From Baker (2016), reprinted with permission from *Applied Geology in California*, Anderson and Ferriz (eds), copyright 2016 by Star Publishing Company, Inc., Belmont, California, USA.)

(d) Prediction of the distribution of ground-motion intensity as a function of earthquake magnitudes, distances and other relevant parameters. Ground-motion prediction

equations are used similar to (14.9), which are based on thousands of observed ground-motion intensities from numerous historic earthquakes. The plotted mean and standard deviation shown in Figure 14.17 represent one such prediction equation.

(e) Combine these uncertainties in earthquake size, its location and ground motion to compute the annual rate of exceeding a given ground-motion intensity. The result, shown in Figure 14.18(e), is known as a hazard curve for a single-source site.

14.5.2 Seismic Hazard Maps

Seismic hazard maps present the probability of seismic events, such as (i) one or more $M6.7$ earthquakes over 30 years for a particular region, as in Figure 14.19, or (ii) regional or national maps of peak ground accelerations (PGA) defined in terms of a PGA with a 10% chance of exceedance in 50 years. The development of such maps poses a number of issues concerning earthquake geology and seismology.

One is to identify the active fault that might rupture and to estimate the paleoearthquake magnitude and its recurrence interval that should be used in estimating the seismic hazard. Hopefully, the scarps of any surface ruptures have not been eroded and the surface rupture length can be accurately measured. The faults of southern California are relatively well defined (see Figure 14.3); however, this is not typically the case for intraplate earthquakes across much of

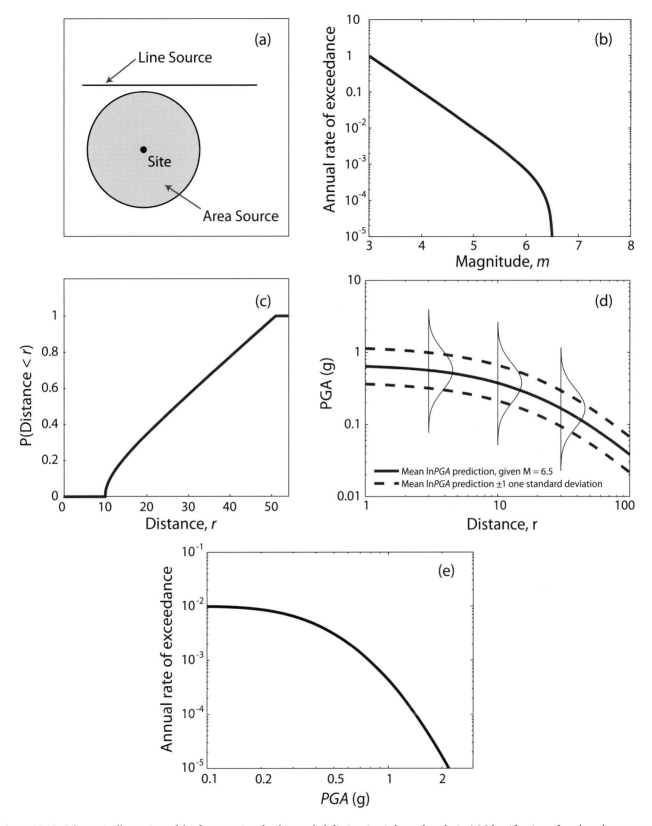

Figure 14.18 Schematic illustration of the five steps involved in probabilistic seismic hazard analysis. (a) Identification of earthquake sources. (b) Characterization of the distribution of earthquake magnitudes from each source. (c) Characterization of the distribution of source-to-site distances from each source. (d) Prediction of the resulting distribution of ground-motion intensity. (e) Combination of the first four steps to compute the annual rate of exceeding a given ground-motion intensity. (From Baker (2016), reprinted with permission from *Applied Geology in California*, Anderson and Ferriz (eds), copyright 2016 by Star Publishing Company, Inc., Belmont, California, USA.)

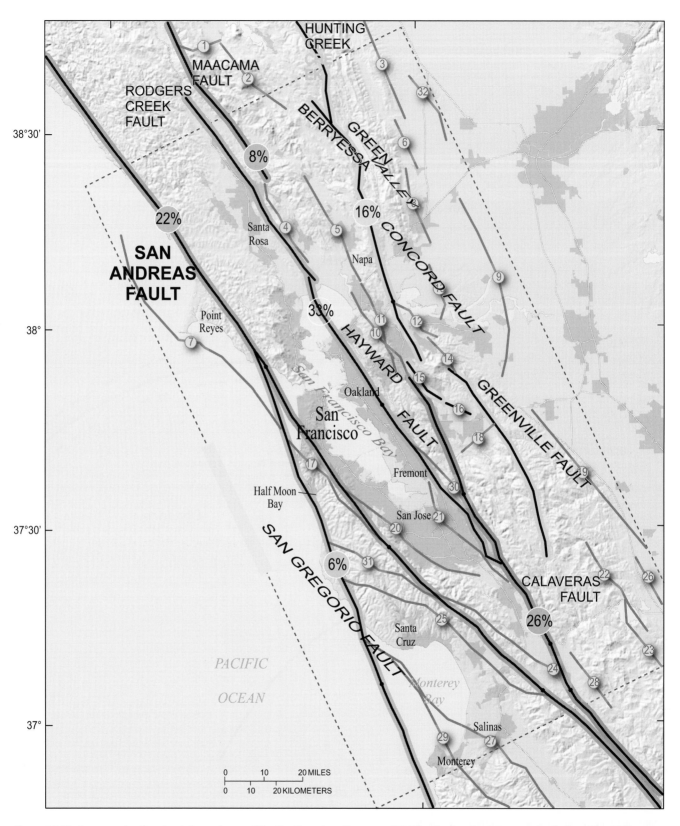

Figure 14.19 An example of a seismic hazard map of the San Francisco Bay area of California showing known active faults with a 72% probability of one or more *M*6.7 or greater earthquakes from 2014 to 2043. Yellow traces indicate 32 minor faults. (Courtesy of the US Geological Survey.) See colour plate section.

BOX 14.4 | **THE RECURRENCE INTERVAL FOR THE ROSE CANYON FAULT, SAN DIEGO, CALIFORNIA**

The City of San Diego is set in an area of great beauty, ironically created by seismic events in southern California, including effects caused by the Rose Canyon Fault (Figure 14.3). Trenches dug by Thomas Rockwell of San Diego State University indicated a cluster of five surface ruptures in the early Holocene, i.e., between 5000 and 9300 ka, and an additional rupture around AD 1650 ± 120 years, in which the surface displacement was as much as 3 m.

He noted that the construction of the first San Diego Mission in 1769 by the Spaniards precludes a large earthquake after that time "as a surface rupture in San Diego would assuredly have destroyed the Mission, and there is no record of such an event". The displacements observed in trenches along the Rose Canyon Fault suggest earthquakes of at least $M7$ and, based upon the regression equations of Wells and Coppersmith (1994), Rockwell estimates that the average value of these displacements indicates an earthquake with a moment magnitude $M_W7.3$. Earlier, Grant and Shearer (2004) of the University of California in San Diego estimated that the rupture of the full Newport–Inglewood and Rose Canyon fault zone would yield an $M7.4$ earthquake.

The **recurrence interval** of an $M7$ earthquake is the average time between recurring $M7$ events. For such an $M7$ event, Rockwell (2010) pointed out that, if the recurrence interval is calculated for the six earthquakes that have occurred during the Holocene epoch, then the average recurrence interval would be ~1500–3000 years, suggesting that San Diego would be safe for the near future. However, if the earthquakes on the Rose Canyon fault occur as clusters and the AD 1650 earthquake was the first of such a cluster, then Rockwell indicates that an average recurrence interval of ~800 years is reasonable.

Active faults are always being reinterpreted and their seismic hazard is subject to continual re-evaluation. In the case of the Rose Canyon Fault, new seismic data indicate that the Rose Canyon Fault is actually a continuation of the Newport–Inglewood Fault shown in Figure 14.3 (Sahakian et al., 2017). The Newport–Inglewood/Rose Canyon (NIRC) Fault comprises four principal fault segments separated by three short zones (<3 km) where faulting is absent, known as **stepovers**. By comparing this NIRC fault with the North Anatolian fault zone in Turkey, which is a particularly active strike-slip fault with numerous twentieth-century earthquakes of $M7$, these authors conclude that the short length of the stepovers gives a likelihood of an end-to-end rupture (89 km) of the NIRC >30–40% with a maximum magnitude of $M7.4$.

When San Diego's football stadium and fuel-storage tank farm were built in the 1960s in Mission Valley just 5 km from the Rose Canyon Fault, there was no perception that the fault might be later identified as being an active fault. Subsequently, this area of Mission Valley was identified as susceptible to liquefaction in an earthquake. Liquefaction-induced failures would raise the possibility of damage to water and fuel pipelines in Mission Valley. The stadium was subsequently reinforced to withstand the design earthquake.

Sources: Grant and Shearer (2004), Rockwell (2010) and Sahakian et al. (2017).

North America and Western Europe. Yeats' (2012) review of active faults is an appropriate initial reference, but even such a recent source may not be fully up to date; therefore the appropriate seismic hazard institution, e.g., the US Geological Survey or the California State Seismic Commission, should be consulted.

Many tasks associated with seismic hazard mapping – at least in the USA – are now conducted using standard software packages that are undergoing continual revision and refinement. *UCERF* (http://www.wgcep.org/; Field et al., 2013)

provides long-term estimates of seismic hazard in terms of magnitude, location and time-average frequency of potentially damaging earthquakes needed to establish building codes and earthquake insurance rates in California. The USGS' *ShakeMap* by Worden and Wald (2016) generates a map showing the distribution and severity of ground motions for a particular region. Accompanying the map are estimates of peak ground-motion parameters and the associated potential damage (https://earthquake.usgs.gov/data/shakemap/).

14.5.3 Liquefaction

Liquefaction is the loss of shear strength of a cohesionless soil during seismic (cyclic) loading. This dynamic loading scenario is considered cyclic because of the periodic nature of seismic waves, as shown in Figure 14.8, such that the sandy soil behaves as a fluid, causing individual grains to be suspended by the increase in pore pressure and re-deposited following ground motion. Liquefaction occurs in saturated cohesionless soils under undrained conditions subject to horizontal shear stress.

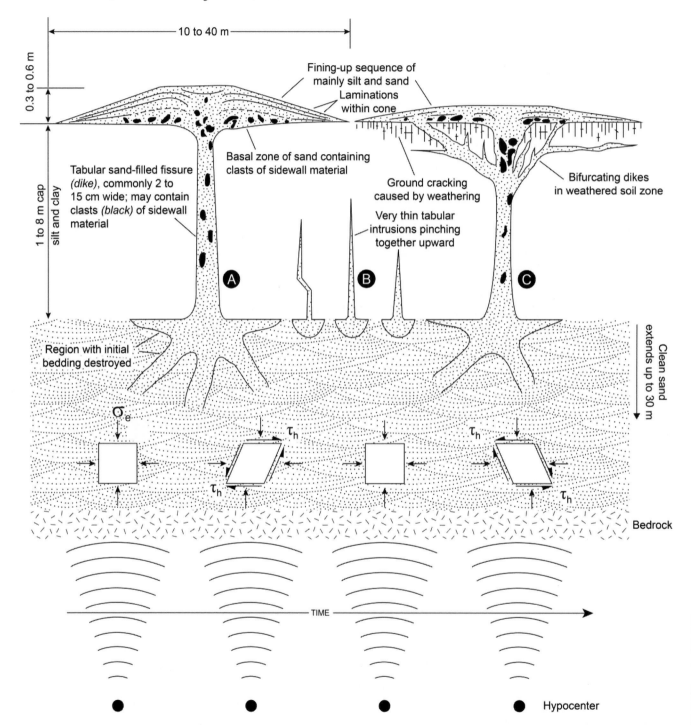

Figure 14.20 (A) Clastic dike penetrating fine-grained soils depositing sand boils. (B) Dikes that pinch together during ascent but do not break through cap. (C) Dikes that develop in plastic clays. Cyclic loading of a soil volume is illustrated in the lower part, with pre-liquefaction on the left, where σ_e is the initial vertical effective stress and τ_h is the horizontal cyclic shear stress. (From Obermeier et al. (2005). Figure courtesy of the US Geological Survey.)

The implications of liquefaction include the appearance of **sand boils** or geysers at the ground surface and the flow of sandy soils into gullies and rivers known as **lateral spreading** (Obermeier et al., 2005; Obermeier, 2009). These should not be confused with sand boils that form on the opposite side of a levee from a river in flood caused by piping of sand and its discharge on the landward side of the levee (Li et al., 1996).

Liquefaction-induced sand boils might be 10–40 m in diameter and represent the discharge of deep sand through a **clastic dike**, as shown in Figure 14.20.

Lateral spreading is the horizontal displacement of surficial blocks of soil comprising a fine-grained cap overlying liquefiable sand, shown in Figure 15.1(j), caused by liquefaction. It is detected in gently sloping (0.3–5%) ground adjacent to

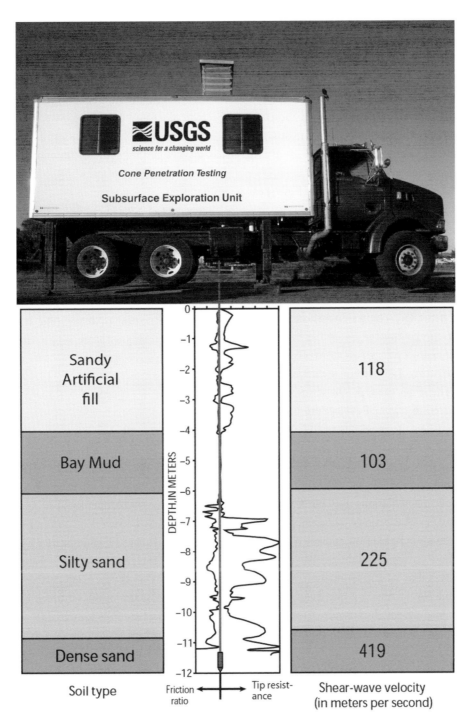

Figure 14.21 A cone penetrometer test log of the friction ratio and relative tip resistance with inferred geological units and their shear-wave velocities. See section 10.1.4 concerning CPT methods. (From US Geological Survey Fact Sheet 028-03, courtesy of T.L. Holzer, USGS.)

streams or gullies in which hydraulic fracturing of the soil cap breaks up the surface layer through which intrude sand boils. Rauch and Martin (2000) used case histories of lateral spreads to develop a predictive model based on seismic shaking, site topography and soil conditions. Holzer and Bennett (2007) investigated eight lateral spreads and concluded that the failure boundaries of the spreads coincided with abrupt lateral changes of sedimentary facies.

Liquefaction implies a sudden loss of shear stiffness during cyclic loading (Idriss and Boulanger, 2008). As ground shaking continues, the overburden load on the cohesionless soil is increasingly transferred from the soil skeleton to the pore water, causing an increase in pore pressure (Δp_w) and a decrease in the effective stress (σ_e). Consequently, the effective cohesion of the soil (c') in the Mohr–Coulomb strength criterion (equation (1.13), $s = c' + \sigma_e \tan \phi'$) goes to zero and, because $\sigma_e = \sigma_n - p_w$ (equation (1.11), for $\sigma_T = \sigma_n$), we may write the condition of liquefaction in terms of the shear strength of the soil as

$$s = \sigma_e \tan \phi' = (\sigma_n - p_w - \Delta p_w) \tan \phi', \qquad (14.15)$$

where $\tan \phi'$ is the effective friction angle and Δp_w is referred to as the **excess pore pressure** that ultimately dissipates, causing densification of the soil. The residual shear strength of this soil is a small fraction of its pre-liquefied strength.

Additionally, we define the **cyclic stress ratio** (CSR) as the uniform cyclic shear stress (τ_{cyc}) divided by the initial vertical effective stress ($\sigma_{e,v}$). The larger the CSR, the fewer numbers of seismic cycles required to trigger liquefaction, whereas a small value of CSR indicates the requirement for numerous cycles to cause failure for any particular relative density of cohesionless soil.

The loss of shear strength produces much greater concern because of the damage to foundations and the floating of underground utilities such as fuel-storage tanks and pipelines made buoyant by the excess pore pressure indicated by Δp_w. Liquefaction is well known to cause building collapse caused by the bearing capacity failure of the foundation. Cohesive soils, such as clays and plastic silts, are also susceptible to strain caused by cyclic loading, which is usually referred to as cyclic softening to distinguish it from liquefaction (see Idriss and Boulanger, 2008).

The focus of this section is the nature of liquefaction phenomena and their recognition in the field rather than the geotechnical theory that has developed to explain and predict liquefaction, which may be found in geotechnical texts (e.g., Coduto, 1999; Briaud, 2013) and the monograph

of Idriss and Boulanger (2008). Because sample disturbance in the field during the collection of cohesionless soils for laboratory testing has resulted in unreliable test results, site characterization of liquefiable soils, including the estimation of the CSR, now increasingly depends on field investigation using cone penetrometer testing (Lunne et al., 1997), as in Figure 14.21.

In their development of maps of anticipated liquefaction-induced failure, Youd and Perkins (1978) compiled information on the susceptibility of sediments to strong seismic shaking, which is summarized in Table 14.2. This table indicates that the **liquefaction susceptibility** of cohesionless sediments is substantially reduced with the age of the sediment because older and pre-Pleistocene sediments are likely to be lithified by diagenetic cements and thus are no longer cohesionless. Note that some of the most damaging liquefaction is associated with hydraulic sandy fills placed into bays and estuaries.

Holzer et al. (2011) conducted a probabilistic assessment of the sediments considered in Table 14.2 to estimate the probability ranges for the Youd and Perkins (1978) sedimentary classes – see Table 14.3. This table is based on water-table depths of 1.5 and 5.0 m below ground surface for earthquakes of magnitude $M6.0$ or $M7.5$, assuming a peak ground acceleration of $0.25g$. The liquefaction probability index (LPI) is defined over a 20 m cumulative thickness of liquefiable layers as

$$\text{LPI} = \int_0^{20\,\text{m}} Fw(z)\,\text{d}z, \qquad (14.16)$$

where $F = 1 - \text{FS}$ for $\text{FS} \leq 1$ and $F = 0$ for $\text{FS} > 1$, and $w(z) = 10 - 0.5z$, where FS is the liquefaction factor of safety and z is the depth in metres for which LPI is estimated. The factor of safety exceeds 1 for cohesionless soils above the saturated zone, consequently $F = 0$. LPI values can range from 0 to 100; however, in areas of sand boils and lateral spreading, median LPI values were 5 and 12, respectively.

Although Table 14.3 specifies a water-table depth of either 1.5 or 5.0 m, this should conservatively be considered to be the depth to the top of the saturated zone, i.e., the surface at which the pores are fully saturated ($\theta = n$), which defines the top of the capillary fringe (see section 11.4.2). During ground shaking it is reasonable expect that liquefaction will deform sediments in the capillary fringe as well, although in many cases it might be relatively thin and therefore of no practical significance. Lohman (1979) lists capillary rise for sediments with $n = 0.41$ as 6.5 cm in very coarse sand (1–2 mm) to 42.8 cm in fine sand (0.1–0.2 mm).

Table 14.2 Liquefaction susceptibility: the likelihood of saturated cohesionless sediments to be susceptible to liquefaction based on their Quaternary age. (From Youd and Perkins (1978).)

Sediment type	Spatial distribution of cohesionless sediments within sedimentary deposits	Recent <500 yr	Holocene	Pleistocene
Continental sediments				
River channel	Locally variable	Very high	High	Low
Floodplain	Locally variable	High	Moderate	Low
Alluvial fans	Widespread	Moderate	Low	Low
Marine terraces	Widespread	—	Low	Very
Lacustrine	Variable	High	Moderate	Low
Colluvium	Variable	High	Moderate	Low
Loess	Variable	High	High	High
Glacial till	Variable	Low	Low	Very low
Residual soils	Rare	Low	Low	Very low
Coastal zone				
Delta	Widespread	Very high	High	Low
Estuarine	Locally variable	High	Moderate	Low
Beach				
high-wave energy	Widespread	Moderate	Low	Very low
low-wave energy	Widespread	High	Moderate	Low
Lagoonal	Locally variable	High	Moderate	Low
Artificial or made ground				
Uncompacted fill	Variable	Very high	—	—
Compacted fill	Variable	Low	—	—

Furthermore, an appreciation of the nature of shallow groundwater flow systems (section 11.4.2; Figures 11.13 and 11.15) will likely have impressed upon the reader that the liquefaction susceptibilities shown in Table 11.2 reflect the fact that shallow water tables are a permanent feature of groundwater discharge areas, where the water table is kept shallow by upward groundwater flow driven by heads propagated through the flow system. Consequently, river channels, floodplains and deltas are susceptible to liquefaction if the sediments present are cohesionless and saturated.

Because of historic damage from the 1811–12 New Madrid **intraplate earthquakes** (see Stein and Mazzotti, 2007), Chung

and Rogers (2013) investigated how variable groundwater depths along the floodplain of the Missouri River might affect liquefaction susceptibility in the case of an earthquake in the St. Louis, Missouri, area. They ascribed LPI values of >5 and <15 to a moderate risk of liquefaction, for which groundwater was 3.2 to 4.5 m depth, and >15 for a severe risk (~1 m depth) and developed hazard maps for the floodplain of the Missouri River as it passes St. Louis.

The investigations following the 1989 Loma Prieta earthquake indicated that damaging liquefaction was identifiable approximately 98 km from the epicentre or 84 km from the end of the seismic source zone (Holzer, 1998), such as the

Table 14.3 Probability ranges for the liquefaction susceptibilities in Table 14.2. (From Holzer et al. (2011).)

Susceptibility	Probability range		
	M6.0 PGA = 0.25g WT = 1.5 m	M7.5 PGA = 0.25g WT = 1.5 m	M7.5 PGA = 0.25g WT = 5.0 m
Very low	0	0	0
Low	<0.08	<0.04	<0.02
Moderate	0.08–0.30	0.04–0.14	0.02–0.07
High	0.30–0.62	0.14–0.20	0.07–0.14
Very high	>0.62	>0.20	>0.14

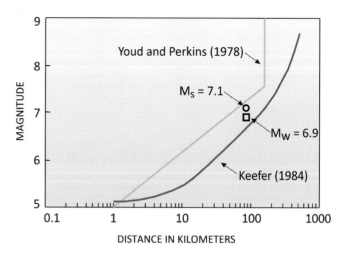

Figure 14.22 Earthquake magnitude versus maximum distance to damaging liquefaction measured from the seismic source zone of the 1989 Loma Prieta earthquake ($M_S = 7.1$; $M_0 = 6.9$). The correlation curves displayed are those of Youd and Perkins (1978), who used the surface-wave magnitude (M_S) while Keefer (1984) used the moment magnitude (M_W). The damage is illustrated in Figure 14.15. (Reproduced from Holzer (1998), courtesy of the US Geological Survey.)

damage to the highway structure shown in Figure 14.15 considered by Field et al. (1994). This liquefaction was attributed to local amplification of seismic waves by soft sediments and reinforcing reflections from the Mohorovičić discontinuity (see Figure 2.1). Figure 14.22 illustrates how far from the seismic source zone damaging liquefaction might occur and allows engineers to assess whether critical infrastructure, such as major highways or fuel-storage facilities, are threatened by a nearby active fault.

14.6 Summary

We began the chapter by identifying seismic faulting as the cause of earthquakes following the accumulation of strain along the fault caused by plate movements. The observations of coseismic displacement during the 1906 San Francisco earthquake led to the development of the elastic rebound model in which the stick–slip process of strain accumulation and subsequent slippage re-occurs in the earthquake cycle. Seismic waves – P, S and surface waves – have been described and the particular threat to infrastructure posed by shear (S) waves has been stressed.

The seismic moment (M_0) was defined and used to estimate the moment magnitude (M_W), which has replaced the Richter or local magnitude (M_L) as the preferred magnitude for reporting the seismic source. The strain energy released by an earthquake was estimated from the seismic moment, and the relationship between the shear strength at failure of a fault and the normal stress acting on the fault was described by Byerlee's law.

The detection of active faults has spurred the development of paleoseismology – the geologic study of prehistoric earthquakes. This work has provided insight into the major earthquakes along various fault segments throughout the world. We concluded with a discussion of the elements of seismic hazard assessment, notably the estimation of ground motions at a particular site on the ground surface, the occurrence of liquefaction and the characterization of the seismic source in terms of maximum earthquake magnitude and its recurrence interval. Consideration of seismically induced landslides and tsunamis follows in the next two chapters.

14.7 FURTHER READING

Anderson, R.S. and Anderson, S.P., 2010, *Geomorphology: The Mechanics and Chemistry of Landscapes*. Cambridge, UK: Cambridge University Press. — This excellent textbook has a review of tectonic geomorphology that will be helpful to those interested in surface faulting and other geomorphic indicators of earthquakes.

Bolt, B.A., 2003, *Earthquakes*, 5th edn. New York, NY: W.H. Freeman. — A concise overview of earthquakes and seismology written at the undergraduate level.

Hough, S.E., 2002, *Earthshaking Science: What We Know (and Don't Know) about Earthquakes*. Princeton, NJ: Princeton University Press. — A very readable popular scientific book for the interested lay-person explaining earthquake processes and effects.

Idriss, I.D. and Boulanger, R.W., 2008, *Soil Liquefaction during Earthquakes*. Oakland, CA: Earthquake Engineering Research Institute. — An updated monograph on all aspects of liquefaction investigations: behaviour, triggering, consequences and mitigation in sandy sediments, and cyclic softening of clays and plastic silts.

Keller, E.A. and Pinter, N., 2002, *Active Tectonics: Earthquakes, Uplift and Landscape*, 2nd edn. Upper Saddle River, NJ: Prentice-Hall. — This is a most useful short guide to tectonic geomorphology and will assist the engineer to appreciate the surficial features that characterize earthquake country.

Kramer, S.L., 1996, *Geotechnical Earthquake Engineering*. Upper Saddle River, NJ: Prentice-Hall. — Chapter 4 contains a lengthy discussion of probabilistic seismic hazard analysis.

Meehan, R.L., 1984, *The Atom and the Fault*. Cambridge, MA: MIT Press. — An account of conflict among geologists and engineers over the siting of a nuclear reactor in earthquake country.

Ohnaka, M., 2013, *The Physics of Rock Failure and Earthquakes*. Cambridge, UK: Cambridge University Press. — A rigorous physics-based account of rock failure mechanics, seismic-source generation, large earthquake cycles and their seismic activity.

Scholz, C.H., 2002, *The Mechanics of Earthquakes and Faulting*, 2nd edn. Cambridge, UK: Cambridge University Press. — This text remains the basic geomechanical reference for the nature of faulting and the generation of earthquakes.

Shearer, P.M., 2009, *Introduction to Seismology*, 2nd edn. Cambridge, UK: Cambridge University Press. — A quantitative introduction to the topic suitable for engineers studying geotechnical earthquake engineering.

Stein, S. and Mazzotti, S. (eds), 2007, *Continental Intraplate Earthquakes: Science, Hazard, and Policy Issues*. GSA Special Paper 425. Boulder, CO: Geological Society of America. — A useful source for those interested in earthquakes occurring within the interiors of tectonic plates, such as those in eastern North America.

US Geological Survey, 1992–98, *The Loma Prieta, California, Earthquake of October 17, 1989*. USGS Professional Papers 1550–1552. Washington, DC: US Geological Survey. — This series of reports by the US Geological Survey describes investigations into all aspects of this earthquake, including geotechnical and hydrologic issues.

US Nuclear Regulatory Commission, 1997, *Recommendations for Probabilistic Seismic Hazard Analysis: Guidance on Uncertainty and Use of Experts: Main Report*. NUREG/CR-6372, vol. 1. Washington, DC: US Nuclear Regulatory Commission. — This volume can be downloaded without charge from the website of the US Nuclear Regulatory Commission. It addresses seismic hazard analysis for nuclear reactors in the USA.

Yeats, R., 2012, *Active Faults of the World*. Cambridge, UK: Cambridge University Press. — A compilation of information on surface faults and seismological and geodetic data providing basic information for regional seismic hazard assessment for regions across the world.

14.8 Questions

Question 14.1. The 1906 San Francisco earthquake is estimated to have had a magnitude of $M_W 8.0$. Approximately what was the size of the source area involved?

Question 14.2. What was the direction of the maximum principal stress for the 1906 San Francisco earthquake: σ_v or σ_{Hmax}?

Question 14.3. What was the direction of the maximum principal stress for the fault ruptures that produced the scarps in Figure 14.1: σ_v or σ_{Hmax}?

Question 14.4. What are the approximate velocities of shear waves in alluvium if $v_P \sim 2000$ m/s and in mud if $v_P \sim 1000$ m/s?

Question 14.5. The seismic moment for the 1906 San Francisco earthquake is estimated as 1000×10^{18} J. Use the answer from question 14.1 to estimate the slip or displacement.

Question 14.6. If we assume that the slip rate of the San Andreas Fault near San Francisco in 1906 was the same as that measured for the Carrizo Plain at Wallace Creek, what would be the surface rupture length of the 1906 earthquake?

Question 14.7. Fuel-storage tank farms and oil refineries are susceptible to earthquake-induced damage under certain circumstances. Discuss.

Question 14.8. Search for a seismic hazard map for the region in which you live. Identify the active faults that contribute to seismic hazard and the soil conditions and sedimentary basins that might cause amplification of shear waves in populated areas. Define areas subject to damage from liquefaction and landslides.

CHAPTER

15 Landslides

In terms of loss of life, landslides are one of the most serious geological hazards, particularly in tropical countries where rainfall from tropical cyclones falling on steep hillsides causes massive debris flows into heavily populated lowlands. In America, the US Geological Survey (Highland, 2004) estimates that each year landslides cause $3.5 billion damage and kill 25–50 people. In Europe, human catastrophes in the 1960s, such as the Vaiont Reservoir disaster in Italy in 1963 and the failure of coal-mine tailings deposits in Wales (UK) in 1966, caused significant changes in geotechnical engineering practice throughout the world.

In the US Pacific Northwest and in western Canada, where the transportation corridors often pass through mountainous terrain, landslides threaten highways, railroad tracks and pipelines. The Oso landslide in Washington state in March 2014 claimed 43 lives and buried a state highway. Where active volcanoes occupy the mountain slopes above urban areas (e.g., the Seattle–Tacoma area of Washington state, in Indonesia and central America), mudflows known as **lahars** have been mapped as having flowed 100 km from the volcanic slopes and deposited cubic kilometres of volcanic debris. Hong Kong experienced ~16 000 landslides between 1924 and 2003, an average of 0.6 landslides per km^2 per year, and has responded by developing risk assessment practices based on detailed engineering geological investigations (Geotechnical Engineering Office of Hong Kong, 2007; Parry, 2016).

We introduced the reader to slope movements earlier in this book. Rock-slope failure was the subject of section 4.6, in which the failure modes of rock slopes were discussed. The major geomorphic features of landslides, e.g., the **main scarp**, the **toe**, the **failure or slip surface**, and their association with rock weathering were considered in section 6.3. In section 10.3.2, the laboratory measurement of shear strength of soil specimens was described.

We begin this chapter with a discussion of further terminology, because landslides come in many shapes and sizes, then turn to the triggers of slope instability and the characterization and monitoring of landslides or potential landslides. We follow this introductory material with a discussion of the failure of soil and rock slopes and view unstable slopes as being part of a groundwater flow system. The chapter concludes with discussions of, first, the predisposing factors that make a slope susceptible to failure and, second, a case history of slope failure in the Italian Alps in 1963, which has been the subject of much analysis by geotechnical engineers and engineering geologists. We shall also review other case histories to understand landslide processes.

15.1 Definitions, Types and Processes

Two nouns are used to produce a term that identifies the material involved in a landslide and the sense of movement, e.g., debris flow and rock topple. This classification is attributed to the work of David Varnes of the US Geological Survey (Cruden and Varnes, 1996; Cruden and Couture, 2010). The materials involved are divided into three classes – rock, debris and earth – while the motions involved are fivefold – flows, falls, slides, spreads and topples.

Table 15.1 indicates the combinations, some of which modes are much more common than others, e.g., rock fall, debris flow and earth slide. Furthermore, the phrases can be combined to indicate that more than one type of movement

Table 15.1 Abbreviated classification of slope movements. (From Cruden and Varnes (1996) and Highland (2004).)

Type of movement	Type of material		
	Bedrock	Coarse soil	Fine soil
Falls	Rock fall	Debris fall	Earth fall
Topples	Rock topple	Debris topple	Earth topple
Slides (rotational or translational)	Rock slide	Debris slide	Earth slide
Lateral spreads	Rock spread	Debris spread	Earth spread
Flows	Rock flow	Debris flow	Earth flow

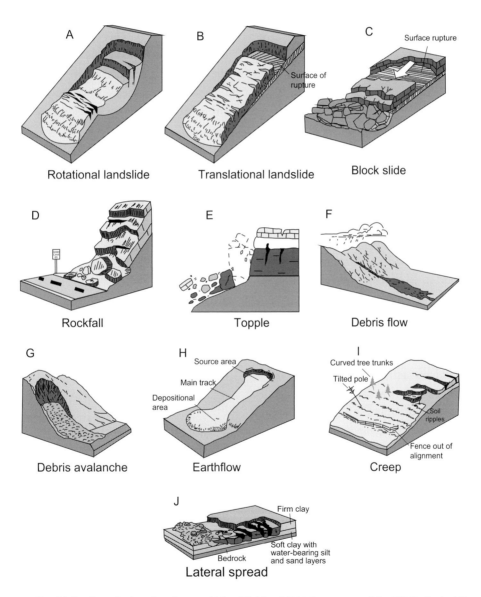

Figure 15.1 Major types of landslides described in this chapter. (After Highland (2004); courtesy of the US Geological Survey.)

and/or material was involved, e.g., rock fall–debris flow, which tells the reader that in different parts of the landslide differently sized materials were moving by different modes. The descriptor *rock* indicates that the displaced material is a hard, firm mass that was intact and in place prior to its movement downslope. *Debris* indicates a material that consists of 20–80% grains > 2 mm, while *earth* indicates material with 80% or more of the grains < 2 mm, which is the upper size limit of sand grains. To these descriptors we add terms to define the moisture content – *dry*, *moist* and *wet* – and the activity of the displaced material – *active*, *reactivated* and *inactive*.

Common modes of slope failure are shown in Figure 15.1. Rotational and translational slides (Figure 15.1A and 15.1B) are differentiated by the geometry of their surfaces. A **rotational slide** has a curved surface – concave upwards – with some backward tilting of the uppermost displaced material. **Translational slides** move along a planar or undulating surface; should only a single block of material be displaced, the slide is referred to as a block slide (Figure 15.1C). Translational slides are typically relatively shallow in depth compared with rotational slides: the ratio of depth of the failure surface to its length is 0.15–0.33 for rotational slides and <0.1 for translational slides. **Wedge slides** are types of translational slides caused by slope failure along the intersection of two sets of discontinuities, e.g., Figure 4.19. The detached materials in a **rock fall** (Figure 15.1D) travel predominantly by falling, rolling or bouncing through air at a very rapid rate, with separation occurring at discontinuities such as joints and bedding planes. **Topples** (Figure 15.1E) are due to the forward rotation out of the plane of the slope of rock due to gravity and, often, the effects of freeze–thaw cycles weakening attachment points at joints between the detached and in-place rock masses.

Figure 15.1F and 15.1G show two types of debris slides. A **debris flow** is composed of <50% fine-grained particles and rapidly moves downslope as a result of intense rainfall or snowmelt. A debris avalanche is also a very rapid event, which is defined as moving between 3 m/min and 5 m/s, i.e., the speed of a running person. **Earth flows** (Figure 15.1H) display an hourglass shape and are most frequently observed in clay-rich soils and rocks on fully saturated moderate slopes. **Creeps** are defined by their slowness of movement, perhaps only a metre or two a year (Figure 15.1I), and are characterized by their distinctive bent tree trunks. Creeps are often seasonal in nature caused by variations in soil temperature and moisture. Finally, **lateral spreads** (Figure 15.1J) are features of almost flat slopes and are associated with fine-grained soils that undergo liquefaction, perhaps caused by an earthquake or

sudden failure within the body of displaced materials. Failure typically begins in a small part of the landslide and then rapidly spreads.

15.2 Landslide Triggering Mechanisms

In Chapter 1, we considered Amonton's law of friction. The frictional stress resisting the movement of a block of soil or rock down an inclined plane is a function of its mass, gravity, slope angle and area of contact between block and planar surface (equation (1.3)). This relationship may be modified by the pore pressure within the discontinuity separating the block from the planar surface when groundwater is present (equation (1.12)). Therefore, we have already identified a number of **predisposing factors** in slope stability: the mass of the displaced material, gravity modified by slope angle, and the presence or absence of groundwater affecting the normal stress creating the frictional resistance to movement. But what might set the block of soil or rock in motion downslope? We shall consider three mechanisms: hydrologic, seismic and volcanic.

15.2.1 Hydrologic Triggering

The most frequent of causes of slope instability is the seepage of rainfall and/or snowmelt into the material comprising the slope, i.e., into the surface of rupture. An example of this was given by Coe et al. (2003) for the Slumgullion earth flow in southwestern Colorado, which is shown in Figure 15.2. This earth flow is estimated to have been active for about 300 years

Figure 15.2 The Slumgullion earth flow, southwestern Colorado. (From Coe et al. (2003); courtesy of the US Geological Survey.)

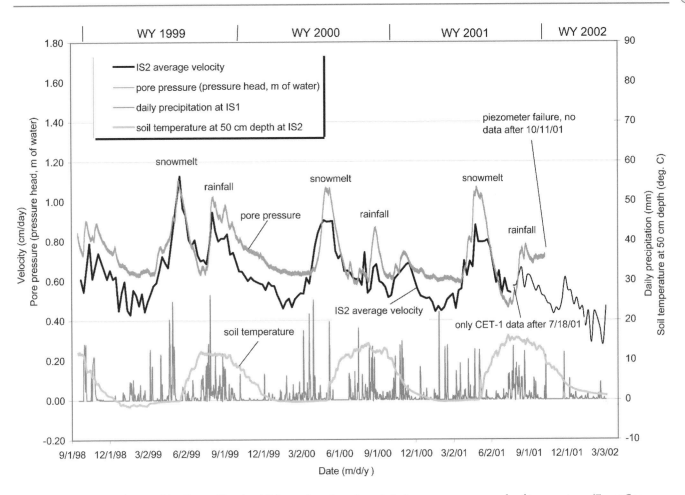

Figure 15.3 Average velocity of the Slumgullion landslide as a function of precipitation, pore pressure and soil temperature. (From Coe et al. (2003); courtesy of the US Geological Survey.)

(Varnes and Savage, 1996; Coe et al., 2003); an older inactive part sits beneath the active slide. The main scarp is clearly visible at the top of the photo and the active toe is in the foreground where some trees are tilted by the movement of the earth flow. The active part of the landslide is thought to be 12–30 m thick, and comprises an estimated volume of 20×10^6 m^3 and is 3.9 km long.

Figure 15.3 shows the variation in landslide velocity, soil temperature, pore pressure and precipitation (i.e., rainfall plus snowmelt as measured as station IS1 shown in Figure 15.2) over four water years, which are defined as the period from October 1 to September 30. The velocity of the earth flow at station IS2 (shown on Figure 15.2) on the toe of the slide is strongly correlated with snowmelt and rainfall events over the four-year duration of record. Landslide velocity is lowest in mid-winter when infiltration is at a minimum but peaks with snowmelt infiltration. Coe et al. (2003) recommend that pore pressures should be measured if at all possible at numerous locations at the base of the landslide. While this is most desirable, it is often quite difficult to achieve because of the slopes involved, their accessibility and their potential to fail. We may conclude that the Slumgullion earth flow accelerates each spring with a decrease in effective stress caused by the increased pore pressures at the failure surface. More recently Schulz et al. (2009) have identified the triggering effect of atmospheric tides on the movement of the Slumgullion earth flow; tidal changes in air pressure are transmitted throughout the earth flow causing changes in pore pressure and hence effective stresses leading to "the nearly perpetual movement of the landslide".

More complex hydrologic triggering involves toe erosion by a river at the base of a slope, as shown in Figure 15.4, or by mining or construction excavation. The removal of the toe of a slope quite obviously causes an imbalance in the sliding and resisting forces. In the case of toe erosion by a river, the triggering is often a complex and interacting suite of processes integrating pore pressure, mass of eroded slope and river stage that act to promote slope instability. An excellent example of this is the case of the Thompson River slides of British Columbia, Canada, described by Eshraghian et al. (2008).

Figure 15.4 Toe erosion along the Athabasca River, Alberta, Canada.

However, it needs to be remembered that the state of water saturation of the soil or rock forming the slope is of critical importance in determining the probability of slope failure. Figure 11.12 introduced the reader to the concept of a hydraulic conductivity that is dependent on the state of water saturation. Where a soil slope remains unsaturated, the negative pore pressure or **matric suction** yields a type of cohesion that strengthens the soil, thus inhibiting slope failure. Therefore the intensity of precipitation and the rate of infiltration become critical factors in slope stability.

The Oso landslide in Washington state (Figure 15.5a) occurred following an intense three-week period of rainfall (perhaps >760 mm), which probably triggered the fatal landslide. The investigation of this landslide encountered a familiar problem – the new landslide occurred in the same locality as earlier landslides that affected the strength of the materials involved in the 2014 event. However, the 2006 landslide at Oso only travelled 100 m, known as the **runout distance**, whereas the 2014 event involved 8×10^6 m^3 of material and travelled approximately 1 km in length across the floodplain of the nearby river, thus damming it (Keaton et al., 2014; Wartman et al., 2016).

Various lines of evidence point to **liquefaction** of wet sediment (Figure 15.5b) at the base of the Oso landslide that caused it to act as a debris flow or **debris–avalanche flow**. Numerical simulations of the event conducted by the US Geological Survey (Iverson et al., 2015) indicate that the high mobility of this landslide was due to the initial conditions of high porosity and water content in the basal sediment. When this loosely packed soil failed by shearing, it contracted with a concomitant increase in pore-water pressure

and decrease in shear resistance that was likely maintained during flow.

Iverson (2000) had earlier shown that experimentally triggered landslides at a debris-flow flume in Oregon (USA) varied in their rate and behaviour with the initial soil porosity. Where the porosity of sandy loam soil was >0.5, an external pore-pressure increase could accelerate the soil to speeds >1 m/s, whereas the same soil at an initial porosity of 0.4 moved downslope at a rate of only 0.002 m/s. The high-porosity soil contracted during slope failure, then partially liquefied but maintained high pore pressures. The low-porosity soil dilated upon failure and then slid downslope intermittently, with the episodes of sliding slowed or halted by decreases in pore-water pressures.

Debris flows may be considered to be part of a continuum, shown in Figure 15.6(a), in which the sediment concentration of a flow dictates the rheological properties of that flow. Typical floods in rivers contain limited sediment concentrations, typically <10 wt.% of fine-grained sediments, and have no shear strength, i.e., the flood is a Newtonian fluid, therefore the applied shear stress (τ) is a function of the dynamic viscosity (μ) and the rate of shear strain (dv/dy) or $\tau = \mu \cdot$ dv/dy, where v is fluid velocity over depth y (Pierson and Scott, 1985; Pierson, 2005).

With increasing sediment concentration, the flow regime undergoes a transition to a non-Newtonian fluid with low shear strength known as **hyperconcentrated flow**. Debris flows involve high-shear-strength fluid behaviour, in which case the **yield strength** (τ_y) referred to in Figure 15.6 is that of a non-Newtonian fluid known as a Bingham plastic, which acts like a solid at low shear stress and as a fluid at

Figure 15.5 The Oso landslide, Washington state, USA, March 22, 2014. (a) View of the Oso landslide and geologic exposures in the scarps. Timber harvesting had occurred in years previous to the landslide above the scarp and may have permitted increased groundwater recharge to the glaciofluvial outwash. Active seepage was observed at the outwash/till contact by the GEER team two months after the event. (b) Liquefied lacustrine sediments on the flank of the landslide. Note the rafted block set in the flow material. (Courtesy GEER, Geotechnical Extreme Events Reconnaissance, Jeff Keaton and Joseph Wartman, Team Leaders; Keaton et al. (2014).)

elevated shear stress (see Middleton and Wilcock, 1994). It is defined by

$$\tau = \tau_y + \mu_p \cdot dv/dy, \qquad (15.1)$$

where τ_y is the **yield strength** of the fluid and μ_p is the plastic viscosity. Figure 15.6(b) shows how texture and mineralogical composition affect the yield strength as sediment concentration increases.

15.2.2 Seismic Triggering

Earthquakes can cause strong ground motions and thus cause rock falls on steep slopes and block slides on lesser slopes.

Keefer (1984) analyzed 40 historic earthquakes to determine the maximum distance from the epicentre of the quake to where a landslide might be expected for any particular magnitude of quake. The resulting curve is shown in Figure 15.7 – compiled from Keefer and Manson (1998) – with the subsequent Loma Prieta earthquake of 1989 ($M_w = 6.9$) plotting on the magnitude–distance curve; this earthquake triggered several thousand landslides over an area of 15 000 km^2.

The issue of seismic triggering of landslides, i.e., **coseismic displacement**, is of particular concern to those responsible for the design of dams and the safety of existing dams because of the potential for a catastrophic dam break and release of the stored water. The classic treatment of this issue was undertaken by Newmark (1965) of the University of Illinois. Newmark considered the case of a rigid friction block sliding on an inclined plane when subjected to an acceleration caused by an earthquake. The displacement of the landslide is estimated from a double integration of that portion of the acceleration–time history (see Figure 14.12) that exceeds the threshold acceleration required to overcome frictional resistance and commence sliding. This displacement is known as a **Newmark displacement** (D_N) (in cm) and provides a preliminary analysis of coseismic displacement of slopes.

Jibson (2007) has developed various regression equations to estimate the Newmark displacement that can be used for regional but not site-specific analysis. Using over 2000 strong motion records from 30 earthquakes, mostly from California, he determined that the Newmark displacement can be represented by

$$\log(D_N) = 0.561 \log(I_a) - 3.833 \log\left(\frac{a_c}{a_{max}}\right) - 1.474 \pm 0.616, \qquad (15.2)$$

where the Arias intensity (I_a, in m/s) is defined in order to characterize strong ground shaking, a_c is the critical or yield acceleration in terms of g, the acceleration due to gravity, a_{max} is the peak ground acceleration and the final term is the standard deviation of the regression equation with $R^2 = 0.75$:

$$a_c = (FS - 1)g \sin \alpha, \qquad (15.3)$$

$$I_a = \frac{\pi}{2g} \int_0^d [a(t)^2] \, dt, \qquad (15.4)$$

where FS is the factor of safety (see section 15.4.2), d is the duration of strong ground motion, a is the ground acceleration and t is time. Sepúlveda et al. (2005) used the Newmark method to demonstrate that the rock-slope failures caused by the $M_w = 6.7$ Northridge earthquake of 1994 were influenced

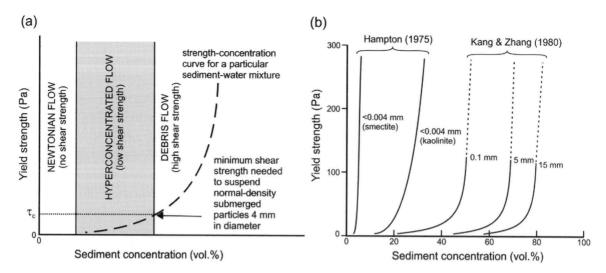

Figure 15.6 The yield strength of sediment–water mixtures with respect to the concentration of suspended sediment. (a) Flow regimes defined for an idealized yield-strength curve of a poorly sorted sediment–water mixture. (b) Range of measured yield strength–concentration curves, i.e., rheological properties, for various sediment–water mixtures illustrating the effect of texture and mineralogy. (From Pierson and Scott (1985), reproduced from Manville et al. (2013), with permission of Cambridge University Press.)

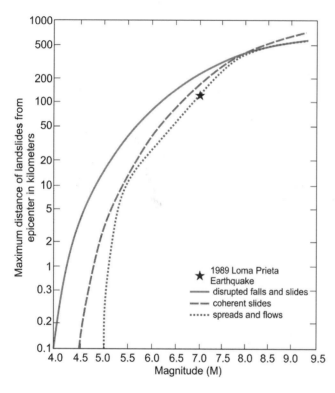

Figure 15.7 Maximum distance from the epicentre of an earthquake to a landslide. (From Keefer (1984); courtesy of the US Geological Survey.)

Figure 15.8 Earth flow at Notre-Dame-de-la-Salette, Québec, triggered by an M_w5.0 earthquake on June 23, 2010 with an epicentre located about 14 km away. (Photo courtesy of Charles O'Dale, used with permission.)

by the topographic amplification of the ground motion by the steep canyon walls.

Seismic triggering is especially important in the context of sensitive glaciomarine clays of the kind that are common in eastern Canada and Scandinavia (see section 8.4.2). Failure of these clays occurs following peak axial deformation in standard drained triaxial tests and likely is initiated by the elevation of the water table and/or toe erosion by a river. Retrogressive failures are quite common. Because of the homogeneity of the Champlain Sea deposits, a circular failure surface is often assumed and Bishop's modified method (see Box 15.2) is used for slope stability analysis (Lefebvre, 1996). Figure 15.8 (and Figure 8.11) shows a landslide in Québec that was seismically triggered by an M_w5 earthquake in 2010.

Brooks (2014) has associated these slope failures in the Ottawa Valley of eastern Canada with paleoseismicity that radiocarbon dating of woody debris identifies as occurring in clusters around 1000 and 5150 ^{14}C years before present. The initial disturbance within the clay causes liquefaction that in turn creates an earth flow of viscous soil that may flow great distances at very low angles. Iverson et al. (2015) indicate that debris flows transform into mobile, high-speed flows when slopes exceed 20°, although the average slope at Oso was <20°.

15.2.3 Volcanic Triggering

Vulcanism can release large volumes of lava rubble, loose ash and cinders that combine with melted snow to form volcanic debris flows, known as **lahars**. These pose serious risks in many parts of the world. Oppenheimer (2011) reports that a modest eruption of a volcano in Colombia, South America, in 1985 created a lahar that killed over 20 000 people. Iverson

et al. (1998b) report that lahars typically inundate areas 20 times larger than rock avalanches of a similar volume because the mobility of lahars is enhanced by the interaction of the solid particles with the transporting fluid (i.e., melted snow) that is absent in rock avalanches. While the return periods for lahars may be of the order of thousands of years, the consequences of such lahars can be truly enormous, even in the USA, as more than 150 000 people live on the surface of former lahars associated with Mount Rainier in Washington state. Table 15.2 presents information on the six major volcanoes of the Cascade Range of Washington and Oregon.

The lahars associated with the Mount Rainier volcano are of two textural varieties. One is defined as relatively clay-rich and contains more than 3–5% clay-sized particles and produces cohesive debris flows. The other contains <3–5% fines and develops non-cohesive debris flows. The clay-sized content of lahars is a distinctive feature of the behaviour of these debris flows throughout the Pacific Rim and is composed of a wide variety of clay minerals (Scott et al., 1995).

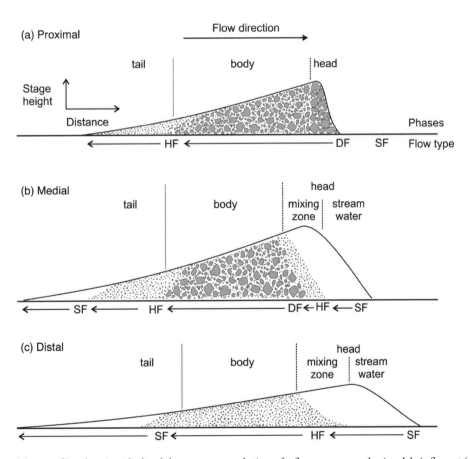

Figure 15.9 Schematic lahar profiles showing idealized downstream evolution of a flow as a non-cohesive debris flow at (a) proximal, (b) medial and (c) distal locations. The profiles show longitudinal separation of sediment, mixing with other streamflow in the same channel or from a tributary and the reversible transformation of flow regimes. SF indicates streamflow; HF, hyperconcentrated flow; DF, debris flow. (From Manville et al. (2013), with permission of Cambridge University Press.)

Table 15.2 Major volcanoes of the Cascades, Pacific Northwest, USA. (From Pringle (1994).)

Volcano	Eruption type	Latest activity	Holocene activity
Mount Baker	Ash, lava	1975 steam emission	Lahars in three valleys
Glacier Peak	Ash	Before 1800	Lahars extending > 100 km
Mount Rainier	Ash, lava	1820–1850 tephra	Lahars in three valleys
Mount St. Helens	Ash, lava, dome	1980 to present	Eruptions and lahars
Mount Adams	Lava, ash	3500 yr BP	Lahars
Mount Hood	Ash, dome	1865 + late 1700s	Lahars down two valleys

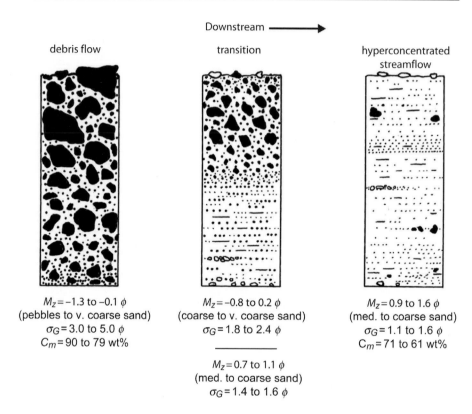

Downstream ⟶

debris flow

transition

hyperconcentrated streamflow

$M_z = -1.3$ to -0.1 ϕ
(pebbles to v. coarse sand)
$\sigma_G = 3.0$ to 5.0 ϕ
$C_m = 90$ to 79 wt%

$M_z = -0.8$ to 0.2 ϕ
(coarse to v. coarse sand)
$\sigma_G = 1.8$ to 2.4 ϕ

$M_z = 0.9$ to 1.6 ϕ
(med. to coarse sand)
$\sigma_G = 1.1$ to 1.6 ϕ
$C_m = 71$ to 61 wt%

$M_z = 0.7$ to 1.1 ϕ
(med. to coarse sand)
$\sigma_G = 1.4$ to 1.6 ϕ

Figure 15.10 Schematic representation of sedimentary facies of the March 1982 snowmelt-triggered lahar at Mount St. Helens, USA, showing the distal transformation from a debris-flow lahar to hyperconcentrated-streamflow lahar. The ranges in mean grain size (M_s), sorting (σ_G) and sediment mass concentration (C_m) are indicated. All units are in the ϕ scale (see section 10.2.1). See Figure 15.9 for those parts of lahar profiles producing these facies. (From Pierson and Scott (1985), reproduced from Manville et al. (2013), with permission of Cambridge University Press.)

The lahars associated with the Mount St. Helens eruption of 1982 (subsequent to that shown in Figure 2.4) were caused by the melting of 10^7 m^3 of snow and ice in the volcanic crater (Pierson and Scott, 1985). Much of this meltwater discharged from the crater lake into a gulley system that was readily eroded and created a lahar by incorporating volcanic debris. Figure 15.9 shows the migration of a non-cohesive lahar, in which the flow regimes are identified. Progressive downstream dilution involved a transition, 27 km from the crater, from a debris flow to hyperconcentrated flow to stream

water. The various sediment facies following deposition are shown in Figure 15.10.

15.3 Characterization and Monitoring of Landslides

The purpose of characterization is to anticipate landslides, to quantify the geometry of the suspect slope and to measure various material properties that affect slope stability; see the standard reference on landslides by Turner and Schuster (1996). Given the engineer's professional requirement to protect life where geological hazards exist, the scope of work should identify a clear and sufficient purpose and have an adequate budget for the characterization. This should begin with a simple desktop survey of available information on the formations involved and their geomorphology, followed by aerial photograph interpretation, i.e., terrain analysis (see section 6.5) and a site reconnaissance by walkover.

Aerial photographs at approximately 1 : 50 000 will likely indicate areas of unstable ground by revealing paleolandslide scars. It is important during the characterization process to view photographs not just of the immediate vicinity under study, but of nearby areas with similar terrain to identify other landslide scars. Airborne LiDAR applications to landslides have greatly improved our ability to read the landscape (e.g., Haugerud et al., 2003; Lan et al., 2010b). In addition, terrestrial digital photogrammetry and terrestrial laser scanning have been recently applied to the analysis of retrogressive landslides in sensitive Champlain Sea clays in Québec (Canada) and of rotational landslides along Swiss river banks (Jaboyedoff et al., 2009).

Typically these studies would be followed by surface mapping (see Box 15.1), preliminary drilling and sampling of soil and/or rock, initial laboratory testing of samples and a preliminary report to the client. Subsequently, the potential of a serious failure involving possible loss of life or infrastructure may indicate the need for additional site characterization through an iterative process. This would likely include the installation of **in-place inclinometers**, to measure lateral displacement, and piezometers, to measure pore pressures.

BOX 15.1 | GEOLOGICAL MAPPING OF THE JOHNSON CREEK LANDSLIDE, OREGON

The Johnson Creek landslide in Oregon, USA, is a 200 m long translational landslide that has been carefully mapped and monitored since 2002 by engineering geologists with the Oregon Department of Geology and Mineral Industries. Figure 15.11 is an engineering geology map (upper figure) and associated cross-section (lower figure) that present the important details of the landslide threatening to undermine the coastal highway (US 101) along the Pacific Ocean.

The cross-section of the landslide indicates boreholes completed with inclinometers (LT-1, LT-2 and LT-3) straddling the surface of rupture and piezometers, some (LT-1p, LT-2p and LT-3p) with vibrating-wire transducers installed to allow continuous monitoring of pore pressures. These transducers were set in sand packs and the screened intervals isolated with bentonite. Other piezometers were simple boreholes referred to as "sand-packed piezometers". Unfortunately the hydraulic conductivities of the formations were not measured by slug testing the piezometers; however, the basal shear zone was likely of the order of 10^{-5} m/s or higher.

The regional bedrock formation is the Tertiary Astoria Formation that dips seaward at approximately 17° west and is exposed along the coastline. This formation is predominantly a siltstone but contains interbeds of tuffaceous mudstone (shown as the dark band near the top of the formation along the coastal terrace) and sandstone. The surface of rupture, the basal shear zone, daylights at the beach. Mean sea level elevations are shown. A test pit was excavated at the beach and showed that the toe of the slide had rotated backwards by approximately 60°. Wave erosion of the toe clearly is a significant factor in triggering displacements.

Above the bedrock are 5–7 m of well-sorted marine terrace sand of Pleistocene age that form an upper recharge zone; however, the underlying bedrock is relatively impermeable. Figure 15.11 shows normal block faulting of both the bedrock and the marine terrace sand between the beach and inclinometer LT-1. A tectonic fault, unrelated to the landslide, is required to explain the offset of marker beds; this is shown dipping 60° to the northnorthwest. Through the centre of the map runs US 101, which is shown by the "Fill" outline.

The map and the associated cross-section clearly identify the directional movement of the blocks that constitute the landslide. These are steeply dipping to the west whereas the toe block dips back to the east. The main scarp is shown by the closely spaced topographical contours along the eastern edge of the map. Due west of the scarp is a headwall graben with colluvial sediments and a sag pond, which we previously (Chapter 14) associated as a geomorphic indicator of faulting. Above the scarp, the marine sands form a terrace.

BOX 15.1 | (CONT.)

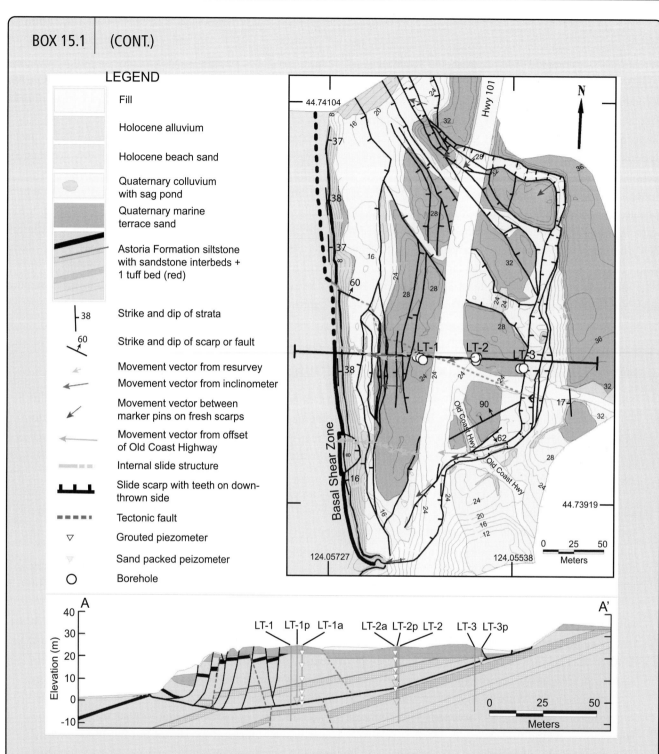

Figure 15.11 Johnson Creek landslide, Oregon. (From Priest et al. (2011); with the permission of the Association of Environmental and Engineering Geologists.)

The measured pore pressure and inclinometer readings indicate that the landslide begins to move as a single mass in response to rainfall once the hydraulic head in LT-2 is ~10 m above the elevation of the surface of rupture. Pore-pressure transmission from the headwall graben through the basal shear zone occurs at a rate of 1–6 m/h, which may be compared with the rate of infiltration through the marine terrace sands of ~0.05 m/h.

Source: Priest et al. (2011).

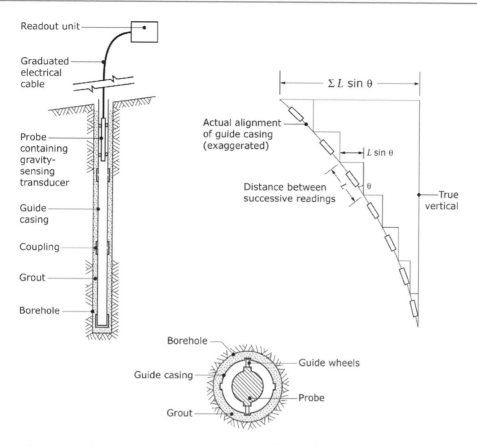

Figure 15.12 In-place inclinometer and principle of operation. (From Dunnicliff (1993); courtesy John Wiley and Sons.)

The inclinometer shown in Figure 15.12 and associated piezometers (e.g., Figure 11.4) are the two most critical elements of a landslide monitoring program. Other instruments such as extensometers, which measure displacement relative to a fixed point, e.g., beneath the surface of rupture, and time-domain reflectometry, which monitors the deformation of a cable by slope movement, are also commonly used. In-place inclinometers are installed after the surface of rupture is clearly identified and may consist of a single probe drawn up the guide casing, as in Figure 15.12, or multiple sensors. The inclinometer casing is grooved to allow the guide wheels of the inclinometer to descend the casing with ease; the grooves are aligned in the expected direction of displacement. The transducer in the sensors detects lateral displacement by monitoring the Earth's gravity field as the borehole casing is deformed by the slope movement. The sensors are connected to a data acquisition system and can be monitored remotely. In future, GPS/GNSS monitoring of receivers on slopes may develop into a real-time monitoring system; however, at present only pilot schemes are being assessed (Ning and Fish, 2017).

Landslide warning systems are created to provide timely notification of an impending hazard. The city of Orting,

Washington state, is partly built on old lahar flows from Mount Rainier and has a siren warning system to alert residents to evacuate to high ground when seismometers detect evidence of volcanic activity and acoustic flow monitors detect lahar migration (LaHusen, 2008). A possible repeat of the 1903 rock avalanche that killed 70 in the town of Frank, Alberta, is monitored by continuous measurement of fractures on Turtle Mountain, on the south margin of the 1903 slide (Froese et al, 2009). Displacements are monitored by use of photogrammetric targets that can be regularly resurveyed, electronic distance measurements and passive microseismic stations, each with a geophone, data recorder, power and telemetry to transmit real-time data. Eberhardt (2012) has reviewed the various techniques that may be used to monitor slope movements and points to the importance of both investigative monitoring to understand the nature of a particular slope over time and predictive monitoring to provide a warning of change of slope behaviour.

Heavily populated areas of intense rainfall and hilly terrain require special monitoring networks of rain gauges rather than a single rain gauge as at the Johnson Creek in Oregon. The San Francisco Bay area, Hong Kong and Rio de Janiero, all with populations of 6–10 million and subject to intense storms falling on hilly terrain, have developed sophisticated landslide warning systems. These integrate rainfall forecasts with real-time rainfall measurements from a network of rain gauges and an algorithm for hydrologic triggering of landslides. Hong Kong's system correlates the slope failure rate with the maximum 24 hour rainfall in each of 700 grid cells that each represent an area of 1.5 km × 1.2 km (Chan and Pun, 2004).

15.4 Slope Failure in Soils

The earth flows at the Slumgullion landslide in Colorado (Figure 15.2) and at Notre-Dame-de-la-Salette in Québec (Figure 15.8) involved predominantly fine-grained silts and clays. Traditionally, geotechnical engineers regard such soil materials as governed by the Coulomb equation (1.13), where the shear strength (s) is expressed in terms of the effective normal stress on a plane (i.e., σ_e), i.e., $s = c' + \sigma_e \tan \phi'$, where s is in units of kPa, c' is the effective cohesion, and $\tan \phi'$ is the effective friction angle. When the soil is only partially saturated, we may write (Barbour, 1998; Vanapalli, 2009)

$$s = c' + (\sigma_n - p_a) \tan \phi' + (p_a - p_w) \tan \phi^b, \quad (15.5)$$

where p_w is the pore-water pressure, p_a is the pore-air pressure, $(\sigma_n - p_a)$ is the net normal stress, $(p_a - p_w)$ is the matric suction and $\tan \phi^b$ is the shear strength contribution to matric suction. Vanapalli (2009) has provided a helpful review of the historical development of concepts of the shear strength of unsaturated soils, and several textbooks on unsaturated zone soil mechanics, e.g., Fredlund and Rahardjo (1993), are available.

15.4.1 Measurement of the Critical Parameters

The determination of parameters for the solution of these equations is commonly undertaken by laboratory testing of core samples or direct *in situ* measurements or preferably both. Lab testing might involve a simple shear-box test or a triaxial cell (see section 10.3.2). The nature of the tests will depend partly on the type of soil being investigated and the state of the core sample available for testing. *In situ* tests for measuring the undrained shear strength (s_u) in fine-grained soils include the field vane test and a cone penetration test (CPT). At Slumgullion (Figure 15.2), a field vane test indicated a maximum $s_u = 38$ kPa and minimum values of 4–5 kPa with continued rotation of the vane after the maximum value

had been reached (Varnes and Savage, 1996). CPT yields the following empirical relationship (Lunne et al., 1997):

$$s_u = \frac{(q_c - \sigma_{v0})}{N_k}, \quad (15.6)$$

where q_c is the measured cone resistance, σ_{v0} is the total vertical *in situ* stress and N_k is an empirical constant reflecting the overconsolidation ratio and the plasticity index and is of the order of 10–20.

The independent variables in (15.5) vary with soil type. The effective cohesion of a saturated sand or gravel, i.e., a *cohesionless* soil, is zero or close to zero when a diagenetic cement is present. The value of ϕ' will vary from around 30° for silty sands (SM) to 40° for poorly sorted gravels (GW). That is, the effective friction angle varies proportionally with the dry bulk density of the soil material. In the case of fine-grained soils, Lefebvre (1996) indicates that we may expect the effective cohesion for the sensitive Champlain Sea clays of Figure 15.8 to be approximately 7 kPa and the effective friction angle to vary from 28° at a preconsolidation stress of 100 kPa to 44° at 400 kPa. Higgins and Modeer (1996) tested eolian silts – loess deposits (Figure 1.7) – and found that the cohesion of these materials varied from 21 to 69 kPa and the effective friction angle was between 9° and 21°.

15.4.2 Soil Slope Stability Analysis

To conduct a slope stability analysis in today's professional environment means that the geotechnical engineer and engineering geologist have access to software that allows rapid simulation involving finite-element analysis and a sophisticated graphical user interface to simplify problem solving, e.g., SEEP/W (Geo-Slope, Calgary, Alberta) or SVSlope (Soil Vision, Saskatoon, Saskatchewan). An engineer might wish to begin the analysis of a soil slope stability problem by using slope stability charts (e.g., Duncan, 1996) for slopes in homogeneous soils; these provide useful screening tools for preliminary analysis.

But if it is necessary to model complex hydrologic processes, such as transient pore pressures, and complicated soil stratigraphy, then computational software will be required. The case histories of Eshraghian et al. (2008) in British Columbia and Hughes et al. (2016) in Northern Ireland provide excellent examples of the use of computational software to interpret detailed field geotechnical and hydrogeological observations based on landslides that have cut transportation routes.

In Box 15.2 we introduce the basic method of slices by which a failure or slip surface is first identified and then

BOX 15.2 | BISHOP'S MODIFIED METHOD

The slope of Figure 15.13 is subdivided into $N = 10$ slices and estimates of the numerator $\Sigma(N_2)$ and the denominator $\Sigma(N_1)$ of the factor-of-safety equation are computed according to the following equations:

$$N_1 = W \sin \alpha, \tag{15.7}$$

$$N_2 = \frac{\left\{ \left[\dfrac{W}{\cos \alpha} - p_w l \right] \tan \phi + cl \right\}}{\left[1 + \dfrac{\tan \alpha \tan \phi}{FS_a} \right]}, \tag{15.8}$$

$$FS_c = \frac{\Sigma(N_2)}{\Sigma(N_1)}. \tag{15.9}$$

where W is the weight of the slice (kN/m), ϕ is the friction angle, c is the cohesion (kPa), p_w is the pore-water pressure (kN/m^2), α is the angle between the base of the slice and the horizontal, l is the length (m) of the slip surface segments measured along the base of the slice, and FS_a and FS_c are the initially assumed and subsequently computed factors of safety.

Table 15.3 presents a worked example related to Figure 15.13. Each slice is populated with the parameter values for (15.6) and (15.7), from which N_1 is computed and then N_2, using an initial estimate of FS_a to yield FS_c. This value of FS_c is then used iteratively until there is no change in the computed factor of safety. Note that the values that are changing with iteration are in the slices towards the top of the hypothesized slip surface and that the pore-water pressure is zero in this example. It is not particularly surprising that the FS exceeds unity and the slope is stable.

Table 15.3 Worksheet for calculation of Bishop's modified method. (From Duncan (1996).)

								$FS_a =$ 1.43	1.51	1.52
Slice	W	l	α	c	ϕ	p_w	N_1	N_2	N_2	N_2
1	112	5.3	−32	35.9	0.0	0	−60	192	192	192
2	297	4.9	−22	35.9	0.0	0	−111	177	177	177
3	499	4.7	−13	35.9	0.0	0	−112	170	170	170
4	726	4.6	−4	35.9	0.0	0	−51	165	165	165
5	903	4.6	4	35.9	0.0	0	63	165	165	165
6	1028	4.7	13	35.9	0.0	0	231	170	170	170
7	1003	4.9	22	35.9	0.0	0	376	177	177	177
8	818	5.3	32	35.9	0.0	0	433	192	192	192
9	587	6.7	43	4.8	35	0	400	408	415	416
10	128	5.6	55	4.8	35	0	105	108	111	111
						$\Sigma =$	1275	1925	1934	1935
						$FS_c =$		1.51	1.52	1.52

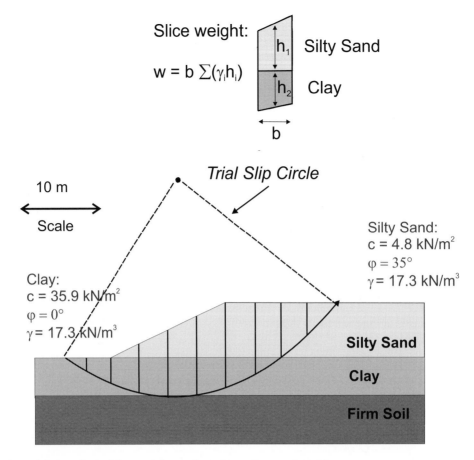

Figure 15.13 Bishop's modified method of slope stability analysis. (From Duncan (1996).)

the potentially unstable soil mass is subdivided so that the equilibrium of each slice of this soil mass is evaluated. Thus, as in our introduction to sliding demonstrated by Amonton's law (Figure 1.9), we consider a soil mass on the verge of failing and compute the shear stresses promoting slope movement and the shear strength of the soil mass restraining it. The ratio of the first to the second is referred to as the **factor of safety**, FS = s/τ, where, as before, s is the strength of the soil and τ is the shear stress required to maintain equilibrium. Where the shear strength of the soil exceeds the shear stress along any failure surface, the slope is stable. Once the shear stress exceeds the shear strength, the slope is expected to fail.

We shall conduct a limit equilibrium analysis of the slope shown in Figure 15.13. The factor of safety will vary along the failure surface but FS, as defined here, is simply the result of (15.6). Having identified the potential failure surface most likely to give the minimum FS, the soil slope is divided into N vertical slices as shown in Figure 15.13. We assume that we are dealing with a circular failure surface, which is a common assumption for relatively homogeneous soil materials. The

lower surface of each slice is considered to approximate a straight line in order to allow a computation of forces. Duncan (1996) and Briaud (2013) consider the commonly used methods of limit equilibrium analysis and we shall use Duncan's example of Bishop's (1955) modified method in Box 15.2.

15.4.3 Unsaturated Soils

We have noted that hydrologic triggering of landslides is the most likely cause of slope instability; therefore, the soil hydrological processes that create saturated conditions leading to slope failure are of great practical importance. This is the realm of unsaturated soil mechanics, which was mentioned earlier when we considered the effect of partially saturated soil conditions on the Mohr–Coulomb failure criterion, i.e., equation (15.4). The reader can find detailed discussions of this issue by Fredlund and Rahardjo (1993), Barbour (1998), Fredlund and Houston (2009) and Vanapalli (2009). In this section we examine such unsaturated conditions and how they effect slope failure.

Equation (15.4) demonstrates how unsaturated soil parameters affect the shear strength of the soil material within a slope. In Chapter 11, we discussed how the pressure head might be negative, in which case we refer to it as ψ and $p_w = \rho g \psi$. If this negative pressure can be maintained as a suction throughout the slope, then the likelihood of slope failure is greatly reduced (Zhang et al., 2004).

In Figure 11.12 we showed the relationship between the fractional moisture content, θ, and the pressure head, ψ. In unsaturated soil mechanics this relationship is typically represented by the water saturation, i.e., $S_w = \theta/n$, rather than θ on the ordinate and the matric suction or negative pressure, i.e., $p_w - p_a$ on the abscissa. This representation of Figure 11.12 is referred to as the **soil-water characteristic curve** (SWCC) and therefore defines the relationship between the degree of saturation and the suction. As Figure 11.12 demonstrates, these curves are not unique to a particular soil, but depend upon the initial conditions and how they are wetted and dried; thus they display hysteresis.

Figure 15.14 shows a typical SWCC in the upper schematic and, in the lower schematic, how the shear strength of the soil depends on the matric suction. The critical parts of

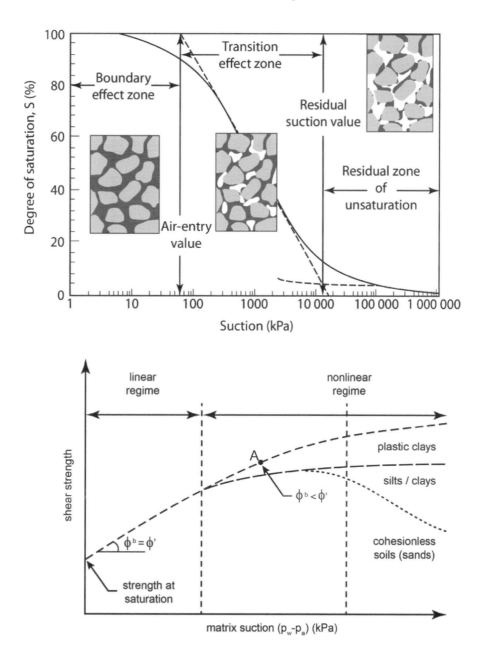

Figure 15.14 (top) Soil-water characteristic curve (top) showing different zones indicating decreasing water content with increasing suction. (bottom) The variation of shear strength of unsaturated soils in various zones of unsaturation for different soil types. (After Vanapalli (2009).)

this curve are the inflection at high saturations indicating air entry, referred to as the **air-entry value** (AEV), and the asymptote indicating irreducible water saturation, which is usually referred to as the **residual saturation**. Barbour (1998) describes how the SWCC varies as a function of grain size, consolidation and compaction, and Fredlund and Houston (2009) discuss the methods for the assessment of unsaturated soil properties in geotechnical engineering practice.

Vanapalli (2009) identified three zones of suction in Figure 15.14. In the *boundary effect zone*, the soil remains saturated until the AEV is met, after which the desaturation leads to dewatering of the soil. (Hydrogeologists refer to this zone of saturated soils that are under suction and exist just above the water table as the capillary fringe.) The *transition effect zone* represents the zone of decreased water saturation. Eventually the *residual saturation zone* is reached at very low suctions.

In Figure 15.14, we note that the shear strength increases linearly with matric suction until the AEV is reached and $\phi^b = \phi'$ in (15.4). Once desaturation occurs, there is a non-linear increase in shear strength. At point A on Figure 15.14 the shear strength contribution due to suction is less than the angle of shearing resistance, i.e., $\phi^b < \phi'$. In the residual zone of unsaturation, the texture of the soil determines the shear strength of the soil and, while clays might increase in shear strength, cohesionless soils do not.

Thus we can see that the factor of safety is very much improved by maintaining unsaturated soil conditions in fine-grained soils because the shear strength of the soil is increased over that at fully saturated conditions, and $s > \tau$ throughout the increase in suction shown in Figure 15.14 and the factor of safety will remain greater than unity. Quite obviously, drainage of slopes that are thought to be unstable promotes unsaturated conditions.

15.5 Slope Failure in Rock Masses

The concept of a sliding block on a tilted surface that we examined in Chapter 1 to illustrate Amonton's law is applicable to hard rocks containing bedding planes and joints, which are potential zones of weakness. The rock mass in Figure 1.9 is stable at increasing dip angles (θ) until it reaches the friction angle (ϕ), at which inclination the rock mass becomes unstable and slides downslope under its own weight. The coefficient of friction (μ) depends on the nature of the surface, e.g., a discontinuity, a jointed rock mass or weathered rock with residual soil. The block will remain stable as long as

the forces preventing sliding are greater than the gravitational component ($mg \sin \theta$). The penetration of groundwater into this surface will tend to lower the factor of safety by increasing the pore pressure and decreasing the shear strength, leading to slope failure (Figure 1.13).

In section 4.6 we discussed the principal modes of slope failures in rock masses, i.e., plane failure, wedge failure, toppling and circular failure, as shown in Figure 4.17. Plane failures involve a fracture surface that daylights the slope face, whereas a wedge failure daylights the line of intersection of two intersecting fracture planes. Toppling failures occur in hard rocks in which rock masses are defined by their discontinuities, e.g., bedding planes, that dip into the rock slope. Circular failures occur in waste rock or rock so completely fractured that there is no remaining structural pattern.

The presence or absence of discontinuities, such as joints and bedding planes, determines how we approach the potential for slope failure in rock masses; Figure 15.15 illustrates the workflow (Norrish and Wyllie, 1996; Wyllie, 2018). In the presence of discontinuities, it is important to undertake a stereographic analysis of the discontinuities (Figure 3.23) to determine if the slope has the potential to fail and then to conduct a limit equilibrium analysis. In the absence of discontinuities, the structural geology of the rock slope will not determine the potential for slope failure and the methods of soil mechanics (as in section 15.4) are used.

15.5.1 Characterizing the Strength of Rock Slopes

Given our analysis of slope stability in soils, the reader would rightly expect that the friction angle and cohesion are at the heart of our analysis of slope failure in rocks, i.e., we are dealing with a rock whose properties are considered representable by a Mohr–Coulomb strength analysis. This characterization of rock-slope strength is shown in Figure 15.15 and illustrates the importance of proper mapping of discontinuities. In cases involving plane and wedge failures, the shear strength of the discontinuity itself must be estimated, whereas, when there is no evident fracture surface that might cause slope failure, it is the shear strength of the rock mass itself that must be estimated. Figure 15.16 shows five scenarios of failure surfaces represented as Mohr–Coulomb materials, for which Wyllie and Norrish (1996) summarize the following properties:

- *Infilled fracture.* The fracture-infill material may exhibit cohesion by virtue of the presence of a strong calcite infill that has healed the fracture surface. If the material is weak

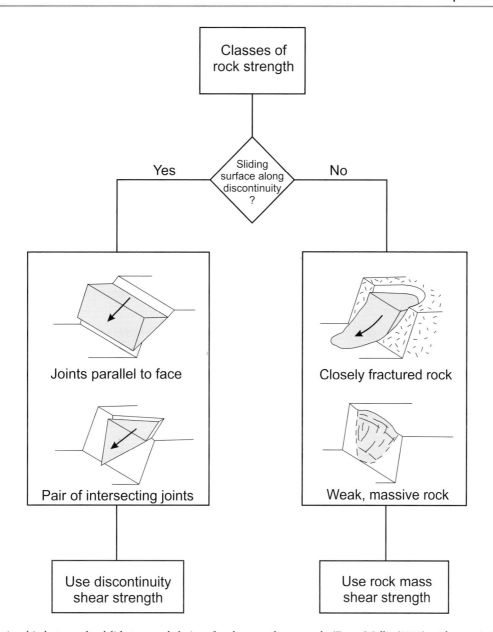

Figure 15.15 Relationship between landslide type and choice of rock strength approach. (From Wyllie (2018); with permission.)

clay or fault gouge (see Figure 3.11), the friction angle will likely be low, with perhaps some cohesion if the infilling material is undisturbed.

- *Smooth, clean fracture.* Figure 15.16 shows this fracture surface as having zero cohesion and the friction angle defines the interface (ϕ). The friction angle of the rock itself is related to grain size and mineral content.
- *Rough, clean fracture.* This fracture surface will also have zero cohesion, while the effective friction angle comprises two components: (a) that of the rock itself (ϕ), and (b) that due to i, the angle of inclination of the rough surface (see discussion of Patton's law in the following subsection), and the ratio between rock strength and the normal stress. With

an increase in the normal stress, the asperities or saw-cut edges that have an angle of inclination i are sheared off and the total friction angle is diminished to a residual value.

- *Fractured, strong rock.* The Mohr–Coulomb strength envelope for the shear strength of a fractured but strong rock mass, in which the failure surface partially lies on fracture surfaces and partially within the intact rock, is curved rather than straight as with the discontinuities. At low normal stress, for which there is little confinement of the fractured rock and the individual fragments may move and rotate, the cohesion is low; however, the friction angle is high because the sliding surface is effectively rough. At higher normal stresses, crushing of the rock fragments

Table 15.4 Ranges of friction angles for various rock types. (From Wyllie (2018); with permission.)

Rock class	Friction angle range (°)	Typical rock types
Low friction	20–27	Schists (rich in mica), shale, marl
Medium friction	27–34	Sandstone, siltstone, chalk, gneiss, slate
High friction	34–40	Basalt, granite, limestone, conglomerate

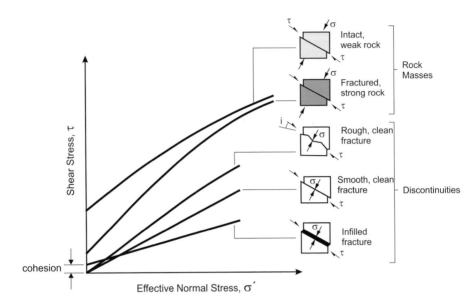

Figure 15.16 Relationships between shear and normal stresses on the rupture surface for five different structural conditions. (From Wyllie (2018); with permission.)

commences, with the result that the friction angle diminishes. The shape of the Mohr–Coulomb strength envelope is governed by the degree of fracturing and the strength of the intact rock.

- *Intact, weak rock*. Rocks such as tuff, shale and marl are fine-grained and have a low friction angle. However, because this rock contains no fractures, the cohesion can be higher than that of a strong but closely fractured rock.

Consequently, field testing must guide us to how to approach the characterization of slope stability. If surface mapping of rock outcrops or diamond-bit drilling identify the presence of discontinuities that indicate the likelihood of planar or wedge failures, then it will be necessary to determine the shear strength of the discontinuity itself. However, if no discontinuity is identified, then it is the shear strength of the rock mass itself that must be determined.

15.5.2 The Shear Strength of Discontinuities

Table 15.4 shows the variation of friction angle with rock type. In the case of a planar discontinuity, i.e., a clean fracture surface or joint, in which the shearing failure is parallel to the direction of shear stress, cohesion will be zero and the shear strength of the rock will directly depend on its friction angle. If that rock contains micaceous minerals, e.g., talc, chlorite or mica, or is composed of fine-grained materials, it will display a low friction angle, whereas coarser-grained rocks will yield a larger angle.

However, it is often the case that the surface of a joint is not planar and clean, but rough and clean, as shown in the middle of Figure 15.16. Patton (1966) investigated this possibility using plaster of Paris models of such fractures in which the fracture surface had a sawtooth ramp pattern. In such a case, the maximum strength envelope measured in the laboratory with a shear box for the specimen will be $S = N \tan(\phi + i)$, where S is the shear force applied to the fracture as in Figure 4.10(c), N is the total normal load on the specimen and i is the angle of inclination of the failure surface with respect to the direction of application of the shearing force, i.e., it is the inclination of the sawtooth above the horizontal. The angle increment i can be used in a shear strength (Mohr) graph such as Figure 15.16 to show the increase in the total friction angle compared with that of a smooth, clean fracture.

The effect of this surface roughness is to increase the effective friction angle. Converting S and N by dividing through by the respective surface areas to which the shear and normal forces were applied yields **Patton's principle** (see Patton, 1966; Wyllie, 2018):

$$\tau = \sigma \tan(\phi + i), \qquad (15.10)$$

where, as before, τ is the shear stress and σ is the normal stress applied to the surface, ϕ is the friction angle and i is the angle of inclination of the rough surface (see the rough, clean fracture in Figure 15.16). Note in Figure 15.16 that, as the fracture surface becomes more heterogeneous, the maximum strength envelope exhibits curvature. This can be explained by the presence of numerous irregular sawtooth "teeth", each of a different angle of inclination, e.g., $i_1, i_2, \ldots, i_n$, etc.

15.5.3 The Shear Strength of Rock Masses

For the case in which the rock is strongly fractured and there is no identifiable structural pattern or we are dealing with waste rock piles, Wyllie (2018) advises that empirical methods must be used. Firstly we may resort to a forensic or back-analysis procedure that involves the investigation of a failed rock slope or one in the process of failing. Secondly we may employ the geological strength index method introduced in Chapter 5. Irrespective of approach, we will need to estimate the strength of the intact rock.

The back-analysis procedure was discussed by Hoek and Bray (1981), who attempted to establish a relationship between friction angles and cohesion for a wide variety of geological materials from soils through soft rock to undisturbed hard rock masses. They found no simple linear relationship of the one to the other. As we have seen, soils such as sensitive glaciomarine clays may have friction angles of $28°$ to $44°$ and very low cohesive strength, i.e., <10 kPa; similarly weathered granites from Cornwall, England, to Hong Kong also have low cohesion. Weathered soft rocks such as limestone and sandstone can be expected to have higher cohesion values and friction angles of the order of $20°$ to $30°$. We may expect that undisturbed hard rock masses – granites and the like – will exhibit a cohesion of >200 kPa and friction angles $>30°$.

The geological strength index (GSI) method was introduced in section 5.4.3 to link rock strength to geological properties. It has become commonplace to directly obtain friction angles and cohesion values using the Hoek–Brown strength model. A curved shear strength envelope is developed from a measurement of the uniaxial compressive strength of a sample of intact rock and two GSI constants: m_i, the material constant for the intact rock, and s, a measure of the degree of fracturing. Two examples of the use of the GSI approach to rock-slope failure are those of Nichol et al. (2002), who studied toppling failures in British Columbia, and Benko and Stead (1998), who determined Young's modulus and the Poisson ratio using the GSI approach to study the catastrophic failure of the Frank slide in Alberta in 1903.

15.5.4 Rock-Slope Stability Analysis

As with the analysis of soil slopes, the slope stability analysis of rock masses increasingly employs the use of numerical simulators. Lorig and Stead (2018) list the following reasons for using numerical models of slopes:

- If other modes of failure can be shown to be unlikely, they can be extrapolated with confidence beyond the observed and measured phenomena in comparison to other methods (e.g., limit equilibrium methods) in which the failure mode is explicitly defined.
- They can incorporate key geologic features such as faults and hydraulic heads providing more realistic approximations of the behaviour of real slopes than can analytical models. By contrast, analytic or limit equilibrium models may oversimplify the important conditions.
- They can help explain observed phenomena.
- They can evaluate multiple variations of geological models, failure modes and design options.

Complementing such numerical models, which solve the constitutive equations for elastic or more complicated nonlinear materials, are less sophisticated but rigorously defined simulators that provide guidance to geotechnical engineers and engineering geologists responsible for the safety of transportation routes. The Colorado Rockfall Simulation Program (Jones et al., 2000) was developed at the Colorado School of Mines to assist in the construction of Interstate Highway I-70 through Glenwood Canyon, Colorado. The variables that determine rock-fall behaviour are shown in Table 15.5.

The acceleration or deceleration of **rock fall** are controlled in particular by the slope inclination, length and roughness. The simulator models the roughness of the surface of the slope by varying the slope angle randomly between limiting values of rock size and surface roughness. The slope material properties provide the means by which the rock-fall collisions and rebound are separated into normal and tangential components of rock velocity. The rock friction is calculated

Table 15.5 Parameters determining rock-fall behaviour. (From Jones et al. (2000).)

Factor	Parameter
Slope geometry	Slope inclination Slope length Surface roughness Lateral variability
Slope material properties	Slope coefficients Rock coefficients
Rock geometry	Rock size Rock shape
Rock material properties	Rock durability Rock mass

and the energy lost during bouncing is computed, producing a new velocity. Results using the simulator indicate that the behaviour of rock fall is most influenced by slope angle followed by surface roughness. The material coefficients are less important but determine the energy absorbed by impact. Where conditions along transportation routes warrant, commercial rock-fall drapes can be employed to restrain falling rocks; field testing has shown that rocks with energies of 3000 kJ can be restrained by such rock-fall barriers.

15.6 Case History: The Vaiont Reservoir Disaster

It is fitting to end this chapter with a cautionary story for geotechnical engineers involving enormous loss of life and infrastructure. This disaster was caused by our limited ability at the time (1963) to understand the coupling of hydrogeological, geotechnical and geomorphic processes that created a massive landslide in the Italian Dolomites. A decade after this disaster, the theoretical basis for understanding this coupling was outlined and the concept that unstable slopes were part of a local or regional groundwater flow system was quantified using the newly developed groundwater flow models.

The original idea was presented in the context of open-pit mine dewatering by Patton and Deere (1971). They noted that, if the open pit was located in a groundwater discharge area, there was "a greater possibility that excess pore-water pressures may be found in the walls and beneath the floor of the mine. In such cases the slope stability problems are likely to be aggravated. Furthermore, without some knowledge of the regional flow pattern erroneous conclusions can be drawn with respect to the suitability of drainage facilities and other

remedial measures." They concluded that "knowledge of the regional flow system is the starting point for understanding the role of fluid pressures in a mine pit."

Subsequently, this hydrogeological perspective of slope stability was addressed by numerical simulation of flow systems by Hodge and Freeze (1977) for dipping layered sedimentary rock formations and by others, including Lefebvre (1986) for sensitive Champlain Sea clays and the rate of valley deepening as erosion occurred. Since that time the use of groundwater flow models coupled with estimates of shear strength to study slope stability have become more common, leading to such sophisticated models as those presented by Iverson (2000), who stressed the importance of transient pore-water pressure transmission during and following storms, by Jiao et al. (2005) for the steep slopes of Hong Kong, by Eshraghian et al. (2008) for slope failure in British Columbia and by Hughes et al. (2016) for slope failure in an excavated drumlin in Northern Ireland. But this was not the case in the late 1950s and early 1960s, which was an age of building great dams (we seem to have moved on to the age of great tunnels). Picarelli et al. (2012) have provided a critical review of the hydrogeology of slopes, in which the slope is seen as a component of a groundwater flow system of uncertain stability.

The Vaiont Dam is a double-curved, thin-arch concrete dam that was completed in 1960. It was 3.4 m wide at the top and 22.7 m at its base and stood 265.5 m high (Kiersch, 1964). It is important to understand that this dam did not fail; rather, the Vaiont catastrophe was a failure to anticipate the effects of a massive landslide that would displace the water of the reservoir behind the dam over its crest and into the valley below. In the 1960s, studies of the stability of reservoir slopes were often not included in the planning of dam construction projects and this was indeed the case at Vaiont (Semenza and Ghirotti, 2000). The Vaiont reservoir disaster focused worldwide geotechnical attention on reservoir slopes, leading to the seminal study of Hendron and Patton (1985), commissioned by the US Army Corps of Engineers, which recognized the general nature of the problem that the Vaiont disaster posed to dams and their reservoirs in the USA.

The Vaiont Valley is shown in Figure 15.17 in cross-section. On the northern side of the Vaiont Valley is Mt. Salta (2039 m amsl), which is a peak in the Borgà Massif thrust upwards by the nappe of which it is an eroded remnant. On the southern side of the Valley stands Mt. Toc (1921 m amsl), forming the recharge area of the Vaiont Limestone (Middle Jurassic) aquifer, which discharges to the Vaiont Valley where the dam was erected. The bedding planes of this limestone dip at angles of 13° to 45° to the north. Beneath this limestone aquifer

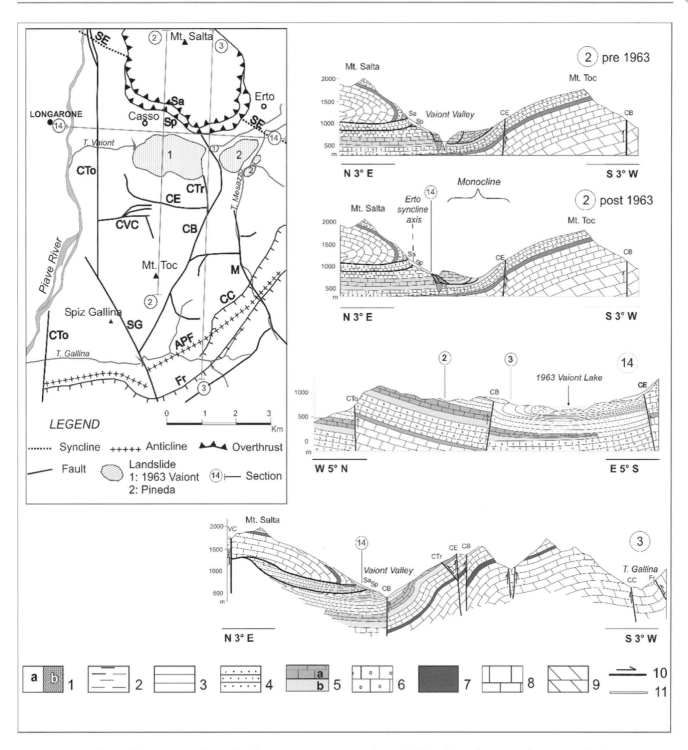

Figure 15.17 Geology of the Vaiont Valley, Italy. The Vaiont Dam is situated south of the village of Casso on the map, in which 1 represents the 1963 landslide. Cross-sections 2, 3 and 14 are indicated on the map. Faults are indicated by CE, CB, Fr, etc. Note the overthrust fold beneath Mt. Salta and the anticline APF. Legend: 1a, Quaternary; 1b, stratified alluvial gravels; 2, Eocene Flysch Fm.; 3, Paleocene marl; 4, Upper Cretaceous Scaglia Rosa Fm.; 5a, Socchèr Limestone Fm.; 5b, Ammonitico Rosso and Fonzaso Fms.; 6, Jurassic Vaiont Limestone Fm.; 7, Igne Fm.; 8, Soverzene Fm.; 9, Triassic Dolomia Principale Fm.; 10, faults and overthrusts; 11, failure surface of the 1963 landslide. (From Ghirotti (2012), reproduced with the permission of Cambridge University Press.)

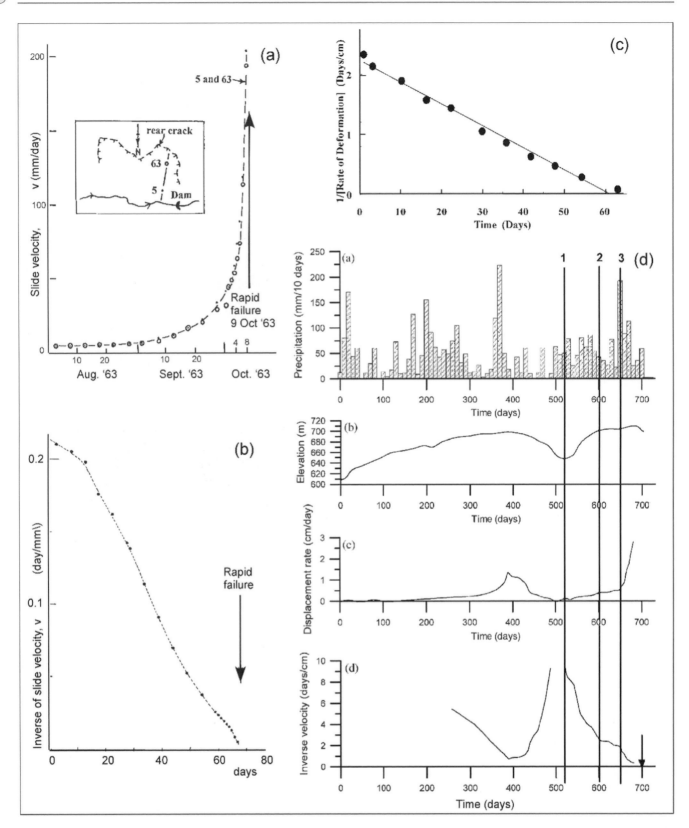

Figure 15.18 Landslide triggering, Vaiont, 1963. (a) Displacement rate on the north side valley wall in the months preceding the landslide. (b) The inverse of the slide velocity. (c) Inverse trend in pre-failure acceleration. (d) Time series of measurements in the two years prior to the landslide: 10-day precipitation, elevation of the reservoir water-level surface, displacement velocity, and inverse velocity of displacement. Vertical lines 1–3 indicate changes in trend of reservoir filling and precipitation. Failure occurred on October 9, 1963 (arrow). (From Ghirotti (2012), reproduced with the permission of Cambridge University Press.)

lie formations of the Lower Jurassic and the Upper Triassic Dolomite, after which the Italian Dolomites are named. Note how the Borgà nappe overrode the younger (Upper Cretaceous and Paleocene) formation, thus placing older rocks above younger rocks.

In the Vaiont Valley itself are shown a number of additional formations in the upper part of Figure 15.17. Unit 5 is the Socchèr formation, which grades down into the Fonzaso clay, its lowermost part. The Fonzaso (montmorillonite) clay is separated from the Socchèr by the failure surface of a paleolandslide, which is the feature that Semenza (2010) identified in 1959, four years before the catastrophic landslide. Above the failure surface are zones of both intact and disturbed Socchèr formation in which the intact zone was displaced rotationally while the rubbly zone underwent folding and faulting during sliding. Semenza and Ghirotti (2000) concluded that, subsequent to the paleolandslide, cementation of the rotated intact zone had taken place such that the rock did not look disturbed and thus earlier geological surveys missed the evidence for the earlier landslide. The original post-glacial Vaiont Valley was filled by glaciofluvial gravels and a new channel was cut into the Fonzaso clay immediately to the south during the Holocene.

The Vaiont reservoir was intended to rise to an elevation of 772.5 m amsl upon its completion in September 1960. Filling of the reservoir began in February 1960 and by May it had reached 595 m. Even this increase in the hydraulic head of the reservoir caused the development of a 1 m wide and 2.5 km long fissure, which developed at an elevation of 1100 m amsl (see Figure 15.17), and revealed the geometry of the paleolandslide. In November 1960, when the reservoir stood at only 650 m, a relatively small landslide of 700 000 m³ produced a 2 m high wave that broke against the dam. This event lead to the lowering of the reservoir and the building of a bypass tunnel beneath the northern slope of the Valley so that any further slide masses that split the reservoir into two halves would cause the two halves to be connected hydraulically.

Figure 15.18 presents the displacement rates in the valley slope above the dam (Figure 15.18a,b,c) and various time series of measurements of precipitation, reservoir water level and slope displacement in the months before the landslide (Figure 15.18d). By October 4, 1963, the reservoir level was over 700 m elevation and the slope movements had reached 50 mm/day (Figure 15.18a).

On October 9, 1963, the International Committee of Large Dams (ICOLD) was meeting in Prague and discussing Terzaghi's recently published argument that it was inappropriate

Figure 15.19 The Vaiont Dam and landslide, photographed in 2001. (Photograph courtesy of Paolo Semenza.)

to construct reservoirs in deeply incised valleys where paleolandslides had been detected (Semenza, 2010). At 10:39 pm that evening a mass of 270 million m³ slid into the reservoir at a very high velocity (20–30 m/s), sending a 200 m high wave to overtop the newly completed Vaiont Dam. The wave caused the loss of nearly 2000 lives, mostly in the small town of Longarone. The lower part of Figure 15.17 shows the displacement of the slide, while Figure 15.19 shows the landslide in 2001 with the dam still intact.

Hendron and Patton (1985) examined the causes of the slide and worked with the local geologist Semenza. They determined that the slope failure was not caused by a continual wetting of slope materials from the rise of the reservoir, as had been proposed by the consulting geotechnical engineer on the project. This trigger, according to Hendron and Patton, would only reduce the factor of safety by ∼12%. Rather, the presence of the thick (1–50 mm; dip, 30–36°N) Fonzaso clays that formed an aquitard and the uplift pressures from the Vaiont aquifer beneath on the basal surface of the clays were the responsible causes. Hendron and Patton's report provides much photographic evidence of this failure surface. Kiersch (1964) commented that swelling of the clay created additional uplift and contributed to the motion of the paleolandslide. Since this time, numerous additional studies have been undertaken, which have been summarized by Ghirotti (2012).

Later Patton (2006) wrote: "the accumulated evidence suggested that high fluid pressures were present beneath the Vaiont Slide and that the slide could have been prevented by drainage". But such drainage could only be correctly emplaced if one understood the hydrogeological regime that became evident from the collaboration of Semenza, Hendron and Patton. That is, it would have been inadequate to drain the paleolandslide itself without draining the underlying Vaiont limestone aquifer to reduce the upward hydraulic gradient on the clays. Thus the paleolandslide and the clays (drained residual $\phi = 6$–$10°$) were the predisposing factors, while the rise in hydraulic head (Figure 15.18b) within the paleolandslide material, which was caused by heavy October precipitation on the karstic recharge area of Mt. Toc, triggered the slope failure.

Of course, there was no reservoir present when the paleolandslide made its initial descent some time after deglaciation. That could only be explained by the high hydraulic heads present in the limestone following recharge by heavy rains or snowmelt on the karstic area high above the reservoir.

Paronuzzi et al.'s (2013) more recent analysis differs from that of Hendron and Patton's (1985) original assessment. They conclude, based on numerical simulation, that the cycles of filling and drawing down the reservoir before the slope failure induced cyclic pore-pressure changes within the low-strength shear zone at the base of the paleolandslide (unit 3, Figure 15.17) and that these changes destabilized the slope. Thus, the two analyses point to different interpretations of the hydrogeological regime triggering the slope failure.

In her review of the Vaiont disaster, Ghirotti (2012) indicates that two professional factors stand out: (i) the failure of the project engineers to realize the threat posed by the unstable rock slopes above the reservoir and their subsequent failure to employ adequate countermeasures and (ii) the limited knowledge among engineering geologists of rock-slope instability in the mid-twentieth century.

15.7 Summary

We have identified a standard classification for landslides that considers the type of material displaced as rock, debris (coarse soil) or earth (fine soil) and the nature of the slope movement. We have noted that the trigger of slope movement is usually hydrologic in nature, but sometimes volcanic or seismic. The displacement itself is monitored by piezometers, inclinometers, benchmarks and test pits, which have been discussed for a number of important landslides.

The mechanics of slope failure have been examined for soils and rocks and the concept of the factor of safety introduced and estimated using Bishop's modified method. We have noted that slope failure in rock masses requires consideration of the structural geology of the discontinuities within the rock mass. Slope failure along such discontinuities dictates that we must determine the shear strength of the fracture or joint or bedding plane itself. When failure is not along a discontinuity, then the shear strength of the rock mass itself is determined. The critical parameters are the Mohr–Coulomb variables: the friction angle and the cohesion.

Furthermore, we defined predisposing factors as critical in establishing the conditions necessary for the triggering of slope failure along its surface of rupture. The predisposing factors we identified were soil saturation and the nature of the soil-water characteristic curve, the weathering of clays and shales, and the weathering of igneous rocks. We concluded with an examination of the Vaiont reservoir disaster, in which the limits of geotechnical practice were revealed in the 1960s resulting in much loss of life. Given that the engineer's responsibility is first and foremost to public safety, this event set the stage for significant improvement in combining geotechnical, geomorphic and hydrogeological methods to study the stability of slopes surrounding reservoirs.

15.8 FURTHER READING

Clague, J.J. and Stead, D., 2012, *Landslides: Types, Mechanisms and Modeling*. Cambridge, UK: Cambridge University Press. — An advanced review of the literature featuring 32 papers reviewing various aspects of landslides, including triggers, geological materials, monitoring, simulation and case histories.

Geotechnical Engineering Office of Hong Kong, 2007, *Engineering Geology Practice in Hong Kong.* GEO Publication No. 1, The Government of the Hong Kong Special Administrative Region (https://www.cedd.gov.hk/eng/publications/geo/geo_p107.html). — An impressive discussion of the role of engineering geology in an area with acute geological hazards, particularly landslides from intense rainfall events. The intended reader is "an experienced geotechnical practitioner".

Highland, L. and Bobrowsky, P., 2008, *The Landslide Handbook: A Guide to Understanding Landslides.* USGS Circular 1325, Reston, VA: US Geological Survey, US Department of the Interior. — This colourful guide to landslides may be downloaded from the USGS website without cost. It provides excellent photographs and schematics to help the reader understand the nature of landslides and their management.

Jakob, M. and Hungr, O. (eds), 2005, *Debris-flow Hazards and Related Phenomena.* Berlin, Germany: Springer. — A detailed review of knowledge concerning debris flows, debris avalanches and debris floods with particular attention paid to predictive techniques.

Keaton, J.R., Wartman, J., Anderson, S.C., Benoît, J., deLaChapelle, J., Gilbert, R. and Montgomery, D.R., 2014, *The 22 March 2014 Oso Landslide, Snohomish County, Washington.* GEER (Geotechnical Extreme Events Reconnaissance webpage address: http://www.geerassociation.org). — A forensic analysis of the landslide that took 43 lives, which is summarized in Wartman et al. (2016). The analysis by Iverson et al. (2015) on the landslide's mobility is also essential reading to understand this event.

Machan, G. and Bennett, V.G., 2008, *Use of Inclinometers for Geotechnical Instrumentation on Transportation Projects: State of the Practice.* Transportation Research E-Circular (E-C129). Washington, DC: Transportation Research Board (http://www.trb.org/Publications/Blurbs/160335.aspx). — Strongly recommended reading for anyone intending to use an inclinometer.

Semenza, E., 2010, *The Story of Vaiont Told by the Geologist Who Discovered the Landslide.* Ferrara, Italy: K-flash. — Originally published in Italian in 2001 and written by the geologist who identified a paleolandslide before the 1963 tragedy. The preface is written by "Skip" Hendron and Frank Patton, themselves eminent geotechnical and geological engineers, who argue that this book should be required reading for all geotechnical engineers and engineering geologists. Their work, with E. Semenza, established the causes of the 1963 landslide.

Sidle, R.C. and Ochiai, H., 2006, *Landslides: Processes, Prediction, and Land Use.* Water Resources Monograph 18. Washington, DC: American Geophysical Union. — A particularly useful guide for those working in forested catchments that are landslide-prone.

Turner, A.K. and Schuster, R.L. (eds), 1996, *Landslides: Investigation and Mitigation.* Special Report 247, Transportation Research Board, National Research Council. Washington, DC: National Academy Press. — The authoritative treatment of the subject prepared by many contributors. Becoming a little outdated with the widespread use of new technologies, e.g., LiDAR, but a very comprehensive analysis of all basic mechanisms.

Turner, A.K. and Schuster, R.L. (eds), 2012, *Rockfall: Characterization and Control.* Transportation Research Board, National Research Council. Washington, DC: National Academy Press. — A comprehensive review of the methods of analyzing rock-fall phenomena and selecting rock-fall mitigation and maintenance options.

Wyllie, D.C., 2018, *Rock Slope Engineering. Civil Applications,* 5th edn. Boca Raton, FL: CRC Press. — The fifth edition of the classic by Hoek and Bray (1981) that brings the text up to date with respect to recent advances in rock-slope engineering and numerical simulation and over 30 years of research and practice.

15.9 Questions

Question 15.1. Determine the Mohr–Coulomb strength parameters for the following pairs of data: $\sigma_n = 12$, $\tau = 6$; $\sigma_n = 22$, $\tau = 10$; $\sigma_n = 21$, $\tau = 13$; $\sigma_n = 25$, $\tau = 11$; $\sigma_n = 26$, $\tau = 11$; $\sigma_n = 29$, $\tau = 12$; $\sigma_n = 27$, $\tau = 14$; $\sigma_n = 37$, $\tau = 17$; and $\sigma_n = 46$, $\tau = 21$. Assume the data represent results from drained shear-box tests; therefore the estimated Mohr–Coulomb parameters are effective cohesion and effective friction angle.

Question 15.2. Many early reports of Mohr–Coulomb shear strength parameters give results in non-SI units. For example, compute the cohesion in kPa for a weathered rock given as 2000 g/cm^2, i.e., force per unit area.

Question 15.3. What is the minimum rate of erosion in mm/yr of the London Clay, which was deposited in the early Eocene, if ≥ 150 m has been eroded?

Question 15.4. Develop a spreadsheet computer program to determine the factor of safety using Bishop's modified method of slope stability (see Table 15.3). First, reproduce the results of Table 15.3 and relate the spreadsheet rows to the slices in Figure 15.13; slices 9 and 10 are on the right-hand side of the trial slip circle. Assume that it is discovered that the ground surface to the right of the slip circle, which happens to slope sharply upwards just out of view, is to be irrigated for crop growth. It is decided to install piezometers to monitor the depth of the potential failure surface in the silty sand in slices 9 and 10. The measured hydraulic head after some months of irrigation in slice 9 is 6 m and that in slice 10 is 4 m. Compute the new factor of safety for the slope.

Question 15.5. Assume that the clay in Figure 15.13 is retested and it is discovered that the cohesive strength is not 35.9 but only 24 kN/m^3. Recompute the FS.

Question 15.6. Name the type of landslide motions that caused the Vaiont reservoir disaster.

CHAPTER

16 Coastal Hazards

In this final chapter we consider coastal processes, in particular coastal erosion from storm surges, which is a topic of increasing concern as the global climate changes and sea levels rise at over 3 mm/a. For countries with long coastlines along which a large proportion of the national population lives – such as the USA, Chile, Britain, the Netherlands, Japan, Bangladesh, China, Australia and New Zealand – the threat of a rising sea level and damaging storm surges will mean huge infrastructure costs in the near future. Table 16.1 illustrates some of the costs sustained by the USA in the past 120 years.

Storm surges cause seawater inundation of low-lying coastal areas, sometimes with great loss of life. In 1953, an unexpected storm surge inundated areas of England and the Netherlands, in particular, causing over 2000 fatalities. Hurricane Katrina in 2005 induced events leading to the failure of levees in New Orleans, Louisana, with a similar loss of life and damage to the city in excess of $100 billion dollars.

In developed nations, such catastrophes provoke intense post-mortem investigations and large expenditures on new flood-damage infrastructure. Developing nations rebuild as best they can.

Coastal erosion of beaches leads to expensive beach nourishment projects, often conducted in the USA by the US Army's Corps of Engineers at the urging of local politicians. Many of these beaches require continual replenishment with dredged sand (at a typical cost of $1–5 million per kilometre) that is subsequently eroded by tides and surges. All beaches are areas of active geomorphological change that adjust to waves, tides and surges continually. Where some beaches may be protected by groins or rock jetties, adjacent beaches will likely suffer increased erosion.

Figure 16.1 illustrates the net erosional and accretionary rates of shoreline change along the US Northeast coast as measured over long-term and short-term intervals by the US Geological Survey (Hapke et al., 2013). On the US West

Table 16.1 Some major US coastal storms since 1900. Storm category is given by the Saffir–Simpson scale (see Table 16.2) in which an intensity of 1 indicates a minimal storm surge of ~1 m to an intensity of 5, a catastrophic storm with a surge of >6 m. (From the US National Academy of Sciences (2014).)

Storm	Date	Landfall location	Cost of damage (billions US$)	Fatalities	Category
Hurricane Sandy	2012	Florida to Maine	66	159	2
Hurricane Ike	2008	Galveston, Texas	32	112	2
Hurricane Katrina	2005	Louisiana	149	1833	3
Hurricane Andrew	1992	Florida and Gulf Coast	45	61	4
Hurricane Hazel	1954	The Carolinas	2	95	4
Galveston Hurricane	1900	Galveston, Texas	<1	6000–8000	4

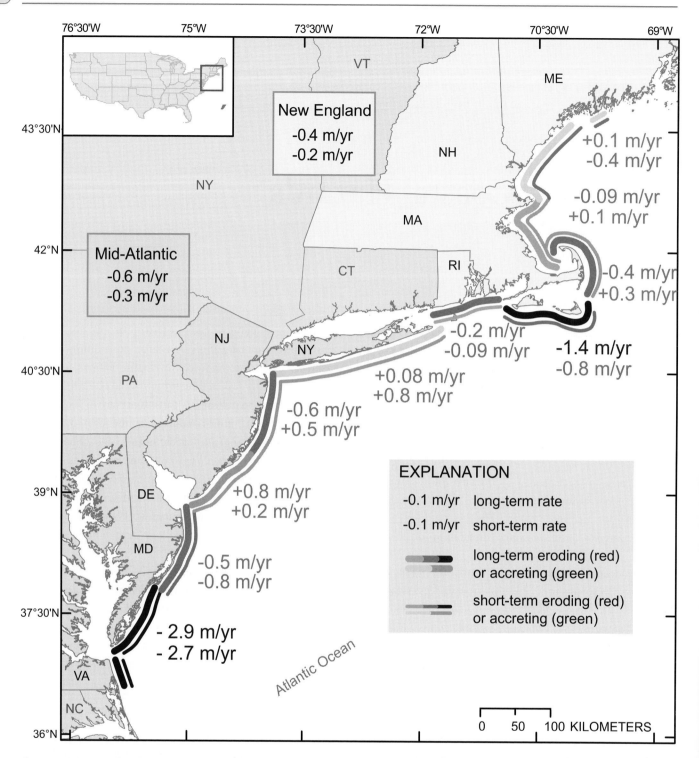

Figure 16.1 Erosion (−ve) and accretion (+ve) rates of shoreline change along the US northeast coast. Long-term rates were calculated over a period of approximately 150 years prior to year 2000, while short-term rates were calculated over a period of 1960–70 through year 2000. The values shown indicate regionally averaged net rates of shoreline change with about 60% of the shoreline indicating net erosion during the recent period. (From Hapke et al. (2013), courtesy of the US Geological Survey.) See colour plate section.

Figure 16.2 Stress relief fractures in coastal cliffs following loss of horizontal confinement with cliff retreat, San Diego County, California. These are Eocene sandstones of the Torrey Formation.

coast, weakly cemented sandstone cliffs like those shown near San Diego, California, in Figure 16.2 erode at rates of 3 cm/a in well-lithified Cretaceous sandstones to 43 cm/a in Pleistocene sands (Benumof et al., 2000), because of which one-third of the coastline of southern California is now armoured. In southern England, the white chalk cliffs along the English Channel appear to be eroding at rates ten times faster than they have eroded over the past 7000 years (Hurst et al., 2016).

Coastal landforms are the first topic of consideration in this chapter, followed by a discussion of waves, tides and sea levels. We then address the origin and nature of storm surges and examine how recent hurricanes have damaged the US eastern and Gulf of Mexico coasts, in particular barrier islands along these coasts. Next, we return to the nature of tsunamis that have their origin in seismic faulting on the ocean floor or to submarine landslides and to land subsidence along the US Gulf Coast, apparently induced by oil and gas extraction. We then consider how the strength of sediments and rocks affects

coastal cliff erosion using examples from the Great Lakes region of North America, California and England. Finally, we examine offshore hazards to infrastructure, such as posed by turbidity currents, gassy sediments and iceberg scouring to pipelines, drilling rigs and under-sea cables.

16.1 Coastal Landforms

The principal features of a beach profile are illustrated in Figure 16.3. This cross-section, shown perpendicular to the shoreline, consists of (i) an offshore zone, (ii) the nearshore zone, (iii) the beach or shore, and (iv) the coast. Waves transport sand from the offshore zone and break in the nearshore zone. The foreshore zone is alternatively wet or dry with wave action (the swash zone), while the backshore zone consists of flat-lying berms and scarps from earlier high water levels.

Coastal landforms take many forms. Picturesque coasts include headland bays, where rocky headlands project into the sea, and other rocky coasts and fjords. We shall consider the erosion of weakly lithified rocky coastlines later in this chapter. The deposition of stream sediments in deltas forms critical ecosystems and cities, such as have grown up around the Nile, Mississippi, Ganges and Fraser deltas (see section 9.2.3). Carbonate reefs form tropical coral reefs along some warm water coastlines such as that of the Central American Caribbean and Queensland (Australia) coasts. Permafrost coasts in Arctic regions, introduced in section 8.5.3, are subject to severe wave erosion that is exacerbated by the longer periods of ice-free seas that climate change has caused. Coastal plains are common in many regions and we will discuss coastal erosion along the coastal plains of the US Atlantic and Gulf coasts for which much research has been conducted.

Figure 16.4 presents some coastal landforms in plan view, based on common features in the Gulf of Mexico described by Morton et al. (2004). Barrier islands may be created either by the drowning of dunes, or by longshore drift of sand along the coastline, or by the upward growth of a submerged sandbar. Tidal inlets separate barrier islands and are maintained either by ebb and flood currents under natural regimes or by rock jetties where shipping channels have been created. Behind the inlets are coastal lakes or lagoons that form the outlets to rivers discharging fresh water. Therefore, these bays exhibit variable salinity regimes that may result in the flocculation of fine-grained sediments and the formation of wetlands, salt marshes and swamps. Coastal bluffs may be covered by vegetation, which is essential in stabilizing the bluffs. The

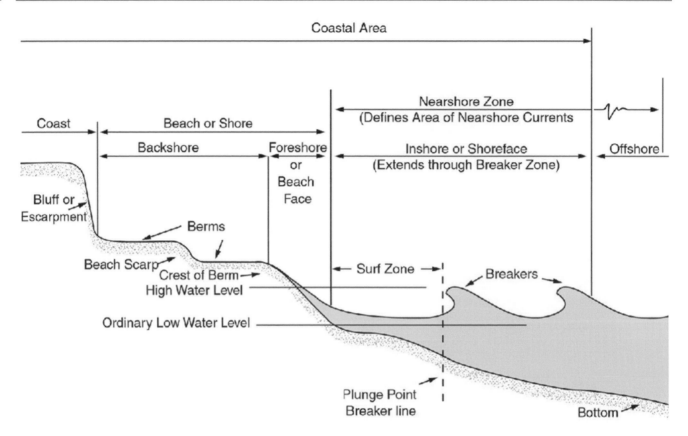

Figure 16.3 The standard terminology of the beach profile. (From US Army Corps of Engineers (1984).)

coastline may be distinguished by dunes that protect the landward areas from inundation. These features are common to both marine and inland areas such as the Great Lakes of North America.

As is indicated above, natural processes are often overridden by the imposition of engineered structures that alter the pattern of erosion and deposition. These structures might include rock jetties lining tidal inlets or groins that protrude into the nearshore zone to trap sediment in place. This will cause an expansion of the beach on the up-current side of the groin, where sand accretion occurs, and erosion and beach loss on the down-current side of the groin from longshore drift of sediments as shown in Figure 16.5. Similarly, sea walls will confine the wave energy at the wall and cause increased erosion in the narrow zone along the wall. Eventually, the beach on which the wall was constructed may disappear and the wall itself threatened by undermining.

16.2 Waves: Form and Energy

Coastal landforms undergo geomorphic change from wave action caused by wind-induced effects on the sea surface or, to a lesser extent, by water displacement by boats and ships.

Waves are produced by air flow over a smooth water surface that develops into a sinusoidally shaped waveform as in Figure 16.6. Minor air pressure differences over the water surface are subsequently magnified by increasing wind speed, creating wave crests (low pressure) and troughs (high pressure). A rise or fall of air pressure of 100 Pa (1 millibar) results in a fall or rise, respectively, of 1 cm in the sea level, i.e., an inverse barometric response. The periods of these waves are from a few seconds to ~15 s.

As waves from offshore approach a coastline, the sea-bottom topography begins to exert an influence on them, causing refraction of the wave fronts and changes in their height and velocity, as shown in Figure 16.4. As the wave height increases, the wave breaks in turbulent flow in the surf zone of the shoreline, causing sediment transport and rock erosion. The maximum wave height can be approximated by $H_{max}/h \approx 0.75$ (Holthuijsen, 2007), where h is the water depth. Wave energy therefore plays a significant role in creating and eroding beaches, cliffs and bluffs, sorting sediments on the shoreface and transporting sediments into and along the nearshore zone.

Figure 16.6 illustrates the principal components of a water wave, in which L is the wavelength and $\eta(x, t)$ is the

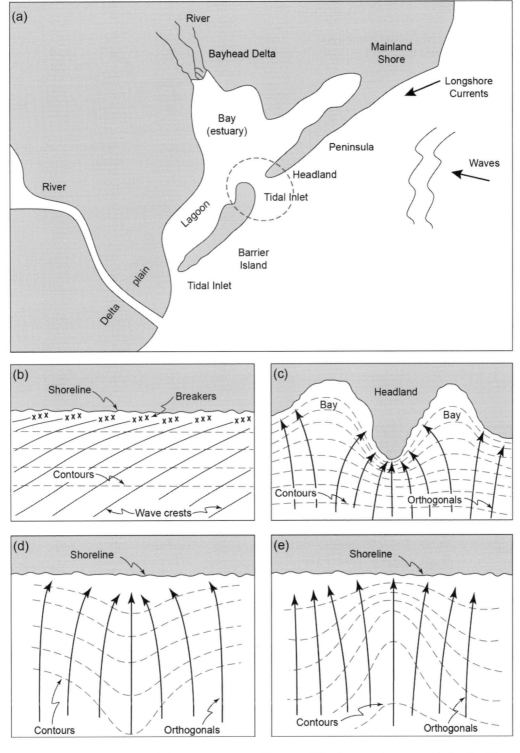

Figure 16.4 (a) Landforms common to the Gulf of Mexico coastline of the USA. (b) Refraction along a straight beach with parallel bottom contours. (c) Refraction along an irregularly shaped shoreline. (d) Refraction by a submarine ridge. (e) Refraction by a submarine trough or canyon. ((a) From Morton et al. (2004).(b)–(e) From US Army Corps of Engineers (1984).)

displacement of the water surface from the mean sea level (Dean and Dalrymple, 2002):

$$\eta(x,t) = \frac{H}{2}\cos[k(x-Ct)] = \frac{H}{2}\cos[(kx-\sigma t)], \quad (16.1)$$

where H is the wave height, k is the wavenumber defined as $k = 2\pi/L$, C is the celerity of the travelling wave, i.e., $C = L/T$, where T is the wave period, and σ is the angular frequency of the wave. Equation (16.1) is the fundamental

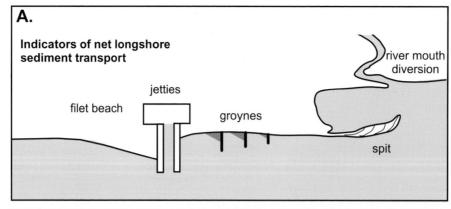

Figure 16.5 Longshore sediment transport. Sediment is trapped on the up-current side of a barrier, such as (A) jetties or groynes or (B) a harbour breakwater. The net effect is a widening of the up-current beach and a depletion of the down-current beach, or lengthening of spits and diversion of river mouths. (From Davidson-Arnott (2010); reproduced courtesy of Cambridge University Press.)

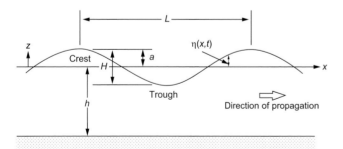

Figure 16.6 Characteristics of waves. (From US Army Corps of Engineers (1984).)

formula for wave motion in elementary wave theory, otherwise known as the small-amplitude or linear theory, developed by Airy in 1845 (US Army Corps of Engineers, 1984; Dean and Dalrymple, 2002). Wave motion is assumed to have

a periodicity of T seconds, repeating identically every period in the wave direction.

The **celerity** is the time for two successive wave crests to pass a particular point. However, wavelength (L) is dependent on water depth (h in Figure 16.6) and celerity is $C = L/T = \sigma/k$, where the angular frequency (σ) is written $\sigma^2 = gk \tanh(kh)$. This is known as the dispersion relationship; it accounts for the frequency dispersion of waves of different periods in that longer-period waves travel faster than shorter-period waves.

This equation for angular dispersion may be rewritten in terms of the wavelength as (Dean and Dalrymple, 2002)

$$L = \frac{g}{2\pi} T^2 \tanh(kh) = L_0 \tanh(kh), \qquad (16.2)$$

where L_0 is the deep-water wavelength, in which case kh is large and $\tanh h$ approaches unity. Therefore, we may write $L_0 = gT^2/2\pi$ and the celerity of a deep-water wave is $C_0 = gT/2\pi$. This dispersion equation is not simply solved because L appears on both sides of the equation, i.e., $k = 2\pi/L$, requiring an iterative solution or the use of a good approximation (see Dean and Dalrymple, 2002).

In shallow water, where $h/L < 1/25$, i.e., small kh, $\tanh h = kh$ and (16.2) yields $C_s = \sqrt{gh}$ and $L_s = T\sqrt{gh} = TC$. A summary of expressions for Airy's linear wave theory is provided in the *Shore Protection Manual* of the US Army Corps of Engineers (1984), including equations for wave celerity, wavelength, group velocity and subsurface pressure in shallow, transitional and deep waters. Davidson-Arnott (2010) provides a succinct summary of wave theory and dynamics.

Wave height increases approximately with the square of the wind speed and with the distance over which the wind has blown, known as the **fetch**. The energy of a wave is a function of its height:

$$E = \tfrac{1}{8}\rho g H^2, \tag{16.3}$$

where E is the **wave energy** per unit surface area in N m/m^2, i.e., J/m^2, ρ is the water density and g is the acceleration due to gravity. This energy term is sometimes referred to as the specific energy or energy density and is the sum of the kinetic and potential energies. The kinetic energy is that due to the velocity of individual water particles associated with wave motion, while the potential energy is that associated with the position of the water particles relative to the wave trough. If we assume that the density of seawater is $\sim$1025 kg/m^3, then a 1 m high wave would exhibit a total energy of 1257 J/m^2.

Because the energy of waves migrates at the speed of groups or packets of waves, not at the speed of individual waves, we write the group celerity as C_g and this accounts for waves of differing frequency. Furthermore, because wave energy per unit surface area increases with the square of wave height, we may write the energy flux in watts per metre of wave, or **wave power**, as

$$\omega = EC_g = \tfrac{1}{8}\rho g H^2 \sqrt{gh}. \tag{16.4}$$

Davidson-Arnott (2010) refers to this expression as the energy flux or the rate at which energy is transmitted in the direction of wave propagation. In deep water, $C_g = 0.5C$, while in shallow water $C_g = C$ (Dean and Dalrymple, 2002). Consequently, large waves contribute the most energy to a shoreface and thus are the principal geomorphic agents of sediment transport and coastal erosion (see Anderson and Anderson, 2010).

The frequency profile of waves is important in the design of coastal and offshore infrastructure; critical frequencies that might cause structural vibrations should be identified and the resonance effects inhibited. Figure 16.7 illustrates the various frequencies and their associated periods attributable to various types of waves. Trans-tidal waves are long waves generated by fluctuations in the Earth's crust and atmosphere. Tides and storm surges are relatively infrequent wave forms, with the former produced by Earth–Moon–Sun gravitational interactions and the latter by low-atmospheric-pressure weather systems and the associated winds. Both of these are considered in the next section. Higher-frequency waves include (i) tsunamis, most often induced by seafloor earthquakes, (ii) seiches, which are resonant standing waves observed in

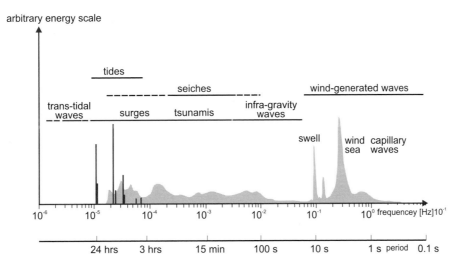

Figure 16.7 Frequency structure of waves in the ocean, their periods and their relative energies. (From Davidson-Arnott (2010) from a figure by Holthuijsen (2007) based on work by W.H. Munk of Scripps Institution of Oceanography in La Jolla, California; reproduced courtesy of Cambridge University Press.)

bays or harbours caused by wind or some other natural cause, and (iii) high-frequency wind-generated waves (see Holthuijsen, 2007; Davidson-Arnott; 2010; Pugh and Wood-worth, 2014).

16.3 Sea-Level Change

At any particular moment, the measured sea level $X(t)$ is the sum of three time series:

$$X(t) = Z_0(t) + T(t) + R(t), \qquad (16.5)$$

where $Z_0(t)$ is the **mean sea level** (MSL), $T(t)$ is the component of sea level associated with the astronomical tide, and $R(t)$ is a residual that reflects meteorological and other effects. How these components interact over various time periods and in different locations affects the protection of coastal infrastructure. Here we consider, in order, (i) astronomical tides, (ii) the observed and predicted changes to the MSL, and (iii) the residual effects represented by $R(t)$, e.g., storm surges associated with hurricanes and typhoons, tsunamis caused by seismic faulting or a landslide beneath sea level and coastal subsidence. The reader is referred to Pugh and Woodworth (2014) for further reading.

16.3.1 Tides

The Moon and the Sun generate astronomical tides in the ocean, from which the tides spread as waves to coastal waters. Because gravitational attraction between Earth and Moon and between Earth and Sun is the cause of tidal responses of the ocean, tides are often referred to as gravitational tides.

Newton's law of gravitation states that all particles in the Universe are attracted to other particles with a force that is proportional to the product of their masses and inversely proportional to their distance apart squared. The force of gravitational attraction between celestial bodies (Earth, Moon, Sun, etc.) is the vector sum of the constituent parts of each body, and the total forces can be calculated assuming the attraction emanates from a single point within that body, i.e. the centre of the sphere, which is Gauss' law of gravitation. Therefore, the force of attraction between the Earth and the Moon, with masses m_1 and m_2 and separated by a distance R is: force $= Gm_1m_2/R^2$, where G is the universal gravitational constant (6.67×10^{-11} N m^2 kg^{-2}). A few calculations indicate that the Moon exerts less than 1% of the gravitational attraction on the Earth as does the Sun.

However, the tidal effect of a distant body is a function of $1/R^3$ not $1/R^2$. This is because particles on the Earth's surface, i.e., nearer the Moon, are subjected to a stronger gravitational force by the Moon than particles at the Earth's centre. The difference between the forces necessary for the Moon's orbit and those experienced at the Earth's surface are responsible for generating tides. Pugh and Woodworth (2014) show that the solar tidal forces are 0.46 weaker than the lunar ones because of the extreme distance of the Sun, i.e., the Moon has roughly twice the tidal effect of the Sun.

Tides display not only the vertical rise and fall in the sea level along shorelines, but also a horizontal ebb and flow. The **tidal range** is the difference between low and high tidal water levels and is equal to twice the tidal amplitude. The **tidal period** is the time between a tidal high water level and the next high, or between successive low water levels. Figure 16.8 presents five different tidal regimes representing locations scattered across the world. It illustrates the fact that no two tides are precisely alike, even along the same coastline, because the coastal geomorphology has an effect on tidal response. Tidal regimes in sites located adjacent to oceans, such as those of Bermuda on the Atlantic Ocean and Mombasa, Kenya, on the Indian Ocean, are dominated by the **semi-diurnal tide** with each tidal cycle taking an average of 12 hours and 25 minutes, or roughly two cycles per day.

The range of the semi-diurnal tide increases and decreases over a 14 day period, with the maximum range known as the spring tide, which occurs a few days after the alignment of the Earth, Moon and Sun (see Figure 16.9). The minimum range is referred to as the neap tide; it occurs shortly after the first and last quarters of the Moon when the longitude of the Moon differs by 90° from the longitude of the Sun. The maximum semi-diurnal tides occur in bays that are funnel-shaped, such as the Bay of Fundy in the Canadian Maritime provinces or in the Bristol Channel and Severn estuary of England, where the mean spring tidal ranges exceed 12 m.

Diurnal tides, with a daily tidal cycle, are prominent in certain gulf locations such as Karumba on the Gulf of Carpentaria in northern Queensland, Australia, the Gulf of Mexico and the South China Sea. These increase with the lunar or solar declination north and south of the equator. The tidal regime of San Francisco exhibits a **mixed tide** composed of both diurnal and semi-diurnal tides. The distortion of tides in very shallow waters on continental shelves yields unusual tidal regimes, such as those of Courtown on the Irish Sea and Southampton on the English Channel, with small tidal ranges.

Tidal channel currents occur where narrow channels connect two seas, such as the Singapore Strait or the Strait of Messina in the Mediterranean or the Cape Cod Canal in New England, USA. These currents reflect the different tidal

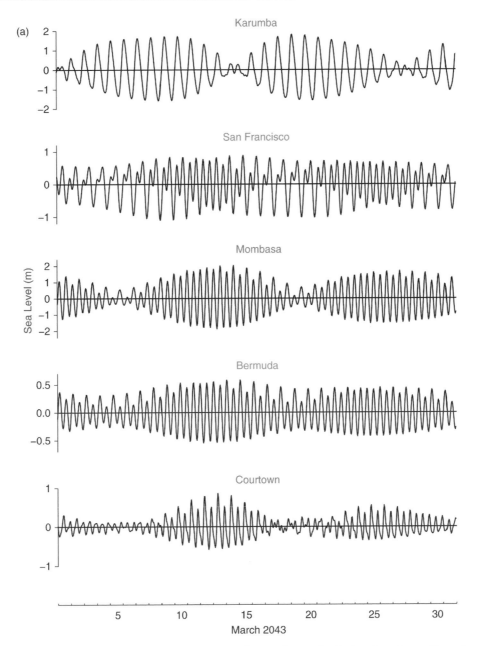

Figure 16.8 Tidal patterns across the world. These are tidal predictions for March 2043 for the locations shown: Karumba in the Gulf of Carpentaria in Australia; San Francisco, California, on the west coast of the USA; Mombasa on the Indian Ocean coast of Kenya; the island of Bermuda in the western North Atlantic Ocean; and Courtown, Ireland, on the Irish Sea. (From Pugh and Woodworth (2014), with permission of Cambridge University Press.)

ranges and phases at the opposite ends of the channels. Tidal waves or bores differ from channel currents in that these are single waves with a steep leading edge of 1–3 m that propagate up confined narrow channels with no outlet and occur only where large tidal ranges exist.

16.3.2 The Rise in Mean Sea Level

The mean sea level (MSL) has varied substantially over the course of the Quaternary and continues to change but at a much decreased rate. The huge change in MSL occurred between 20 and 6 ka with the melting of continental glaciers; the resulting **sea-level rise** (SLR) followed the 120 m decrease during the Last Glacial Maximum (LGM; 18 ka) as discussed in section 8.1. Recent MSL changes are of the order of millimetres per year, which nevertheless are extremely worrying because of the need to protect coastal cities.

MSL data are acquired from various types of tidal gauges, some of which are shown in Figure 16.10. Tide gauges are obviously important in that they represent the historically

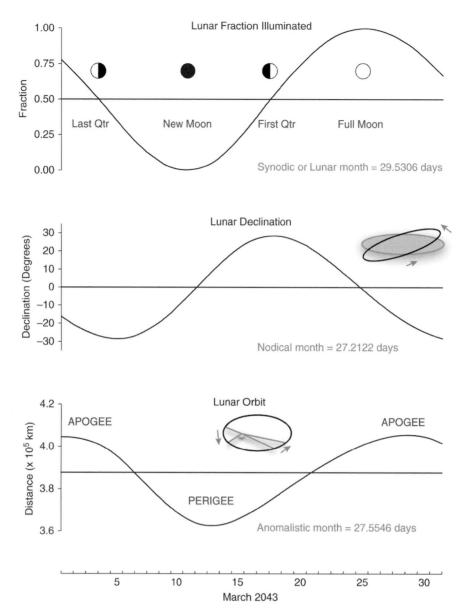

Figure 16.9 Lunar characteristics responsible for spring and neap tides. Every 14.76 days, solar and lunar tide-producing forces combine at new and full moon to give large spring tidal ranges. Neap tides occur twice monthly when the Moon's longitude differs from that of the Sun by 90°. The lunar declination north and south of the equator varies over a period of 27.21 days and the lunar orbit varies over a period of 27.55 days. The maximum diurnal tidal range occurs when the lunar declination is at its maximum; correspondingly, when the declination is at a minimum, the range becomes small. When the Moon is at its maximum distance from the Earth, the lunar apogee, semi-diurnal tidal ranges are less than when it is at its perigee or its closest approach. (From Pugh and Woodworth (2014), with permission of Cambridge University Press.)

archived record of MSL; however, their records must be corrected for changes in the elevation of the land surface adjacent to the tide gauge. In particular, because of the relatively small changes in MSL presently being calculated, variations of similar size caused by vertical land movements associated with glacial isostatic adjustment (GIA) must be accounted for.

The tidal record is corrected using the absolute gravimeter and GPS/GNSS receivers that correlate land movements with sea-level measurements. GIA has caused the MSL in Stockholm, Sweden, to fall ~0.3 m since 1900 as the Scandinavian Shield rebounded during the Holocene at ~9 mm/a (see Figure 2.8). Unfortunately, some areas are experiencing SLR from the collapse of the bulge that was created ahead of the ice sheet; this issue of coastal subsidence is discussed below.

Complicating an estimation of long-term SLR are several short-term changes in MSL. One is the seasonal cycle of MSL in coastal areas caused by discharge from major rivers such as the Ganges River and the St. Lawrence River. The Ganges

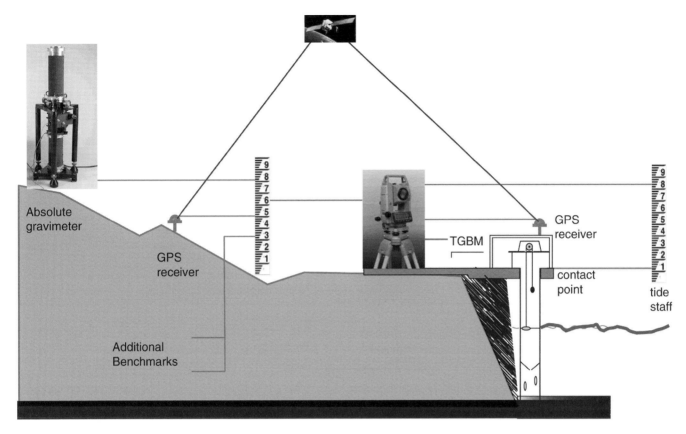

Figure 16.10 The precise measurement of sea level requires a network of datum levelling to reliably define the tide gauge datum. The absolute gravimeter provides information on the vertical land movements, while the GPS/GNSS receivers correlate land movements with sea-level measurements. The tidal gauge bench mark (TGBM) is a stable surface adjacent to a tidal gauge, to which the gauge zero is referred. (From Pugh and Woodworth (2014), with permission of Cambridge University Press.)

drains the runoff from the annual monsoon into the Bay of Bengal, while the St. Lawrence drains snowmelt from the Great Lakes Basin into the Gulf of St. Lawrence. A second source of short-term variability in MSL occurs with inter-annual patterns such as the El Niño–Southern Oscillation of the Pacific Ocean. In normal years, the easterly trade winds cause upwelling of warm waters in the western Pacific; however, El Niño events occurring over 3–7 year cycles are triggered by a relaxation of the trade winds and currents elevate the MSL on the South American side of the Pacific.

However, it is the long-term change in sea level from glacial melting and oceanic warming that is cause for current concerns related to climate change. Engineers likely remember their experiences in the chemistry laboratory with volumetric flasks or graduated cylinders that are calibrated to yield a fixed volume of water at a temperature of 20°C. Were these containers warmed above this temperature, the meniscus would rise, reflecting volumetric thermal expansion. Similarly, the warming of the atmosphere causes volumetric expansion of the oceans. Melting of glaciers adds to the oceanic volume.

The Intergovernmental Panel on Climate Change (IPCC), an international group of scientific experts, has issued a series of reports on SLR since 1990, with the latest (at the time of writing) released in 2014. This final report noted (IPCC, 2014) that the observed rate of SLR during 1971–2010 of 2 mm/a was likely to increase during the twenty-first century to 8–16 mm/a by 2081–2100 "with medium confidence". This indicates an absolute SLR of 28–97 cm by 2100 depending on the atmosphere–ocean scenario modelled. Current (2017) rates of SLR are >3 mm/a, e.g., the US National Aeronautics and Space Administration (https://sealevel.nasa.gov/), but increasing quite rapidly from recent substantial contributions from the melting of the Greenland Ice Sheet (Chen et al., 2017).

However, in southeast Florida, SLR is occurring at ∼ 9 ± 4 mm/a, approximately triple the rate prior to 2006 (Wdowinski et al., 2016). Miami Beach, which is built on a barrier island off the city of Miami, has been experiencing increasing rates of flooding from rainfall and from tidal events. These authors recommend that engineering works, particularly

Table 16.2 Saffir–Simpson hurricane wind scale. (From Liu (2007) and US National Hurricane Center, NOAA (2012).)

Category	Sustained wind speeds (km/h)	Surge (m)	Types of damage associated with hurricane winds and central pressure of hurricane
1	119–153	1.2–1.5	Very dangerous winds will produce some damage to roofs and trees; electrical power outages (>98.0 kPa)
2	154–177	1.8–2.4	Extremely dangerous winds will cause extensive damage; near-total electrical power loss (96.5–97.9 kPa)
3	178–208	2.7–3.7	Devastating damage; roads blocked from uprooted trees; well-built frame houses may incur major damage; electricity and water supplies suspended for days (94.5–96.4 kPa)
4	209–251	4.0–5.5	Catastrophic damage; houses severely damaged; most trees damaged or uprooted; electrical power poles blown down; power outages for weeks or months (92.0–94.4 kPa)
5	>252	>5.5	Catastrophic damage will occur; high percentage of framed houses destroyed; fallen trees and power poles isolate residential areas; areas no longer habitable (<92.0 kPa)

storm drainage systems and sea walls, be designed on the basis of regional estimates of SLR not global ones. In November 2016, the *New York Times* reported on the increasing wariness of home buyers along such exposed coastlines because of high flood-insurance rates and the concern that a collapse in the market for waterfront properties will cause immense financial damage to the US economy.

16.3.3 Storm Surges

Large-scale beach and dune erosion and the widespread destruction of coastal structures are caused by **storm surges**, typically produced by tropical cyclones, which are low-pressure atmospheric systems that may be 200–2000 km across and originate in warm equatorial waters. Tropical cyclones are not frontal systems; rather they announce their arrival with a sudden and significant drop in atmospheric (barometric) pressure and generate high sustained winds. In the northern hemisphere, cyclones swirl around a low-pressure centre in a counterclockwise direction dictated by the Coriolis force; in the southern hemisphere, cyclones rotate in a clockwise fashion. Tropical cyclones with sustained maximum surface wind speeds of at least 33 m/s are known as hurricanes, or as typhoons in East Asia. A typical hurricane may rise 10–20 km into the atmosphere and release prodigious amounts of rainfall at sea and over land, particularly when encountering hilly or mountainous terrain or becoming stalled over coastal regions. Table 16.2 summarizes the Saffir–Simpson hurricane wind scale in use in North America.

Storm surges have their origin in the abnormally low pressures within the centre of tropical cyclones that cause the

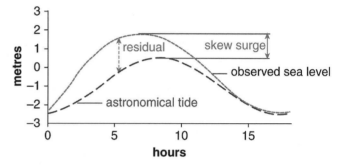

Figure 16.11 The skew surge is defined as the maximum observed sea level during a storm surge event minus the maximum predicted astronomical tide. The two maxima would refer to the same high tide but could be at slightly different times to reflect meteorological forcing of the storm surge. It reflects the mismatch between predicted high tide and observed sea level. (From Pugh and Woodworth (2014), with permission of Cambridge University Press.)

sea level to mound upwards, producing a barometric tide. This mounding is greatly exaggerated by horizontal winds that enhance the elevated sea level. The height of the mound is proportional to the square of the wind speed and to the fetch, and inversely proportional to the average water depth. Because tropical cyclones are sustained by the warm surface waters, which are typically of the order of 20–50 m depth, the bathymetric depth to the seabed is important.

Emanuel (2005) identified a number of factors that influence the size of a surge for hurricanes approaching in a direction perpendicular to the coastline:

- the barometric pressure in the eye or centre of the hurricane is well below the normal barometric pressure at sea level of ~101 kPa, perhaps <95 kPa;
- the onshore wind is large and sustained at ≥33 m/s;

- the hurricane approaches the coast rapidly;
- where the seafloor produces a shoaling effect, in which the shoal is present as an offshore submerged ridge or sandbar, the vertical mixing of colder deeper water with the warm surface layer is reduced while the high heat transfer to the hurricane is sustained; and
- where the coastline is indented by bays and estuaries, the confinement of the surge may amplify its height.

Where the storm approaches the coastline at an oblique angle, the surge will be higher with a coastline to the right

of the hurricane in the Northern Hemisphere and to the left in the Southern Hemisphere. Most critically and irrespective of the direction of approach, when the surge arrives with the high tide, the damage from the storm surge will most likely be maximized. Figure 16.11 illustrates the definition of the surge, or skew surge as it is referred to in storm-surge modelling, which is the residual sea-level elevation that, added to the astronomical tide, yields the observed sea level.

Figure 16.12 shows the unexpected storm surge from the North Sea of 1953 that devastated parts of Britain, the Netherlands and Belgium with over 2500 fatalities. Since that era,

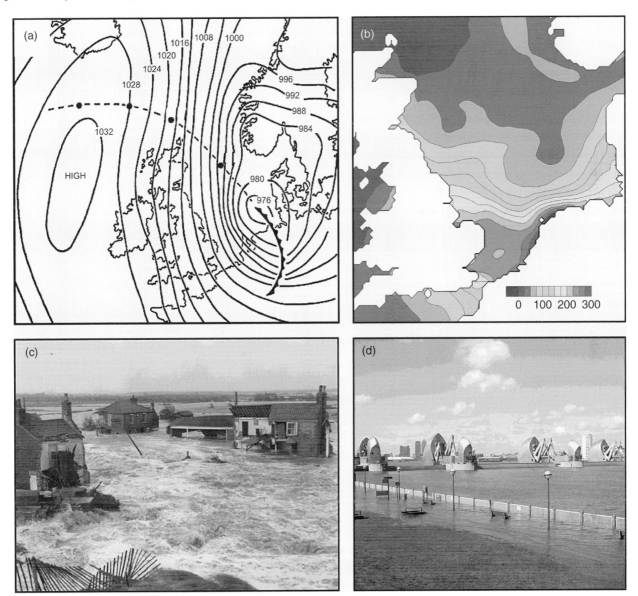

Figure 16.12 A 1953 storm surge in the North Sea inundated parts of Britain, Belgium and the Netherlands, which suffered the greatest loss of life. (a) The track of the storm centre from January 30 (00:00 h) to February 1 (00:00 h), with the black dots indicating the centre in 12 h intervals. (b) The maximum computed surge throughout the affected area in centimetres. (c) Flooding on the Norfolk coast of England. (d) The Thames Barrier that was completed in 1984 to protect central London from a similar storm surge. (From Pugh and Woodworth (2014), with permission of Cambridge University Press.)

Table 16.3 The USGS storm impact scale for barrier islands. Here η_{98} is the extreme high water level attained during a storm, defined as the 98% exceedance level, and η_{50} is the storm-induced mean water level, defined by only storm surge, tide and wave setup. (From Sallenger (2000) and Birchler et al. (2014).)

Impact level and regime	Range of parameters	Predictions of beach changes
1: Swash	$\eta_{98} < Z_t$	Runup is confined to the foreshore of the beach. Moderate conditions. During storms, the foreshore is eroded and sand is transported offshore. Following the storm, sand is transported onshore gradually, over weeks to months; thus eroded sand is replaced and little net change to beach.
2: Collision	$\eta_{98} > Z_t$	Runup collides with the base of the foredune ridge. Collision erodes the dune and transports sand offshore. Net erosion occurs because sand is not readily returned to the dune.
3: Overwash	$\eta_{98} > Z_c$	Runup exceeds the elevation of the dune crest. Sand is transported landward tens or hundreds of metres. Erosion of the beach foreshore and deposition landward of dune.
4: Inundation	$\eta_{50} > Z_c$	Storm surge exceeds and continuously submerges dune or berm crest. Net landward transport of sand. On narrow barrier islands, inundation allows strong currents to cross the island and carve breaches.

satellite tracking of tropical cyclones has provided monitoring of the approach of such storms, and coastal communities can prepare for sustained high winds, intense rainfall and storm surges; however, they cannot expect to avoid damage to the coast and the infrastructure built along it.

Clearly a need developed to identify the vulnerability of coastlines in terms of the probability for coastal change when threatened by storm surges. The US Geological Survey (Sallenger, 2000; Stockdon et al., 2007; Birchler et al., 2014) developed a storm impact scale, Table 16.3, to classify storm-induced patterns and magnitudes of net erosion and accretion on barrier islands along the US Gulf of Mexico and southeast Atlantic coasts. This scaling model identifies four storm impact levels related to the variables identified in Figure 16.13, with each impact level being associated with a particular storm impact regime; the storm may or may not be a tropical cyclone. Figure 16.14 illustrates the three more severe impact levels: collision, overwash and inundation.

We employ this approach to discuss potential storm-surge effects on vulnerable coastlines, keeping in mind that it is designed for particular coastlines in the USA. The storm effect is considered to comprise (i) the astronomical tide, (ii) the storm surge and (iii) the wave runup on the beach, which includes the wave setup and the swash. **Wave runup** refers to the maximum height reached by a wave on a beach above a stillwater level, i.e., the elevation of the sea when there are no waves present. The **wave setup** is the increase in mean sea level above the stillwater level, which is defined as the sum of the tide, surge and any seasonal or inter-annual component of mean sea level as discussed above. The setup is generated by the momentum transfer to the water column by waves that are breaking or otherwise dissipating their energy (FEMA, 2005).

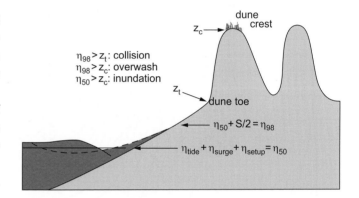

Figure 16.13 Sketch defining geomorphological and hydrodynamic parameters used in scaling storm impacts on barrier islands. Here η_{98} is the extreme high water level attained during a storm, defined as the 98% exceedance level; η_{50} is the storm-induced mean water level, defined by only storm surge, tide and wave setup; and $S/2$ is one-half the swash rise. (From Stockdon et al. (2007), courtesy of the US Geological Survey.)

The **swash** is the temporary ebb and flow of the waves on the beach.

Storm surges cause a significant part of the damage associated with hurricanes. Additionally, the intense winds damage buildings and project roof members, trees and other objects in whirlwinds through communities, killing residents and further damaging structures, while the accompanying heavy rains cause flooding. Over four days in August 2017, Hurricane Harvey released over 100 cm of rain on the Houston–Beaumont area of Texas; this represented approximately 19 times the daily discharge of the Mississippi River according to the Texas state climatologist. Of particular geotechnical concern is the potential for storm surges to undermine flood protection structures (see Box 16.1).

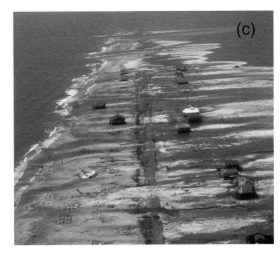

Figure 16.14 Examples of dune response along the Gulf of Mexico coastline described by the USGS storm-scaling model: (a) collision response, Orange Beach, Alabama; (b) overwash response, Dauphin Island, Alabama; and (c) inundation response, Dauphin Island, Alabama. (Photos courtesy of the US Geological Survey and US Army Corps of Engineers.)

BOX 16.1 | THE FAILURE OF THE NEW ORLEANS FLOODWALLS DURING HURRICANE KATRINA

The investigation of catastrophes caused by failures in engineering practice is the means by which the engineering profession improves its technical performance and regains the trust of the public. Canadian engineers wear an iron ring to remind themselves of their responsibility for public safety and ethical practice. Legend has it that the iron rings are made from the wreckage of the Québec Bridge in Canada, which collapsed in 1907 during construction, killing 75 workers. The failure of flood-control structures in and around New Orleans during the storm surge of Hurricane Katrina in 2005 caused ~1000 fatalities and was judged by a panel of independent experts (Seed et al., 2006) as a historic failure of American engineering.

The city of New Orleans has been flooded by hurricanes six times over the past century: in 1915, 1940, 1947, 1965, 1969 and by Hurricane Katrina in 2005. While Hurricane Katrina was the cause of the storm surge, the disaster is more appropriately identified as a civil engineering failure according to the independent engineering experts who analyzed what happened and why it happened in the months following the catastrophe.

New Orleans is located on deltaic and marsh deposits sandwiched between the Mississippi River to its south and to the north Lake Pontchartrain, which is connected to the Gulf of Mexico. In August 2005, the counterclockwise winds of Hurricane Katrina struck the city and its eastern suburbs from the north and east with surges of 4–5 m.

BOX 16.1 (CONT.)

The communities to the east of the city were soon flooded as levees (i.e., floodwalls) were overtopped. Despite 40 years of planning and construction, the protective ring of floodwalls was incomplete in 2005 before this category 3 hurricane reached landfall.

The greatest death toll during Katrina occurred in downtown New Orleans, and it was caused principally by three breaches of floodwalls along the 17th Street and London Avenue canals. It is important to appreciate that these failures were not due to overtopping of the levees, although this had been made more likely by the subsidence of New Orleans over the years. Surge levels in the drainage canals were well below the design levels and the tops of the floodwalls.

Figure 16.15 illustrates the failure of a portion of the 17th Street Canal floodwall. This levee failed by lateral translation because of the low residual shear strength of a sensitive clay layer (1.9–2.5 kPa) and the high pore pressure within this layer caused by the surge level within the canal. The footwall sheet pile had been installed too shallow to be anchored into the lacustrine clay that was the target depth because in this section of the canal the contact between the marsh materials, containing the sensitive clay layer, was much deeper than had been detected during site investigation of the canal in 1980–81.

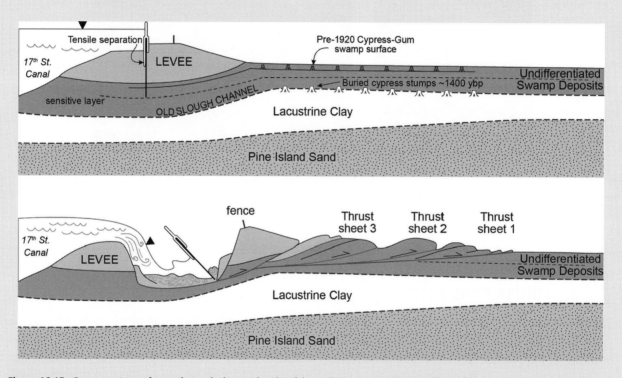

Figure 16.15 Cross-sections of strata beneath the north side of the 17th Street Canal breach, New Orleans, Louisiana, USA, before (top) and after (bottom) its failure during Hurricane Katrina, August 2005. (Modified from Rogers et al. (2008); with permission from ASCE.)

By spacing boreholes along the canal at separation distances of 500 ft (i.e., 152 m) apart, the contact between the marsh sediments and the underlying clay was misplaced. The contact was actually much deeper over a distance of 100 m; however, this old slough channel where the failure occurred was within the 152 m separation of investigation boreholes. Much of the translational movement that undermined the sheet pile of the floodwall was caused by slippage of the organic silty clay of ~25 mm thickness. The failure produced three separate thrust sheets of marsh sediments that included displaced zones of the organic silty clay.

The failure of the three canal floodwalls was responsible for ~80% of the floodwater that inundated downtown New Orleans and displaced 450 000 people. Although there are few dams in the USA whose failure would threaten as many people as lived in New Orleans in 2005, the design levels of safety and reliability used for the floodwalls were those developed for unpopulated agricultural areas and were inappropriate for a large urban population.

Sources: Seed et al. (2006) and Rogers et al. (2008, 2015).

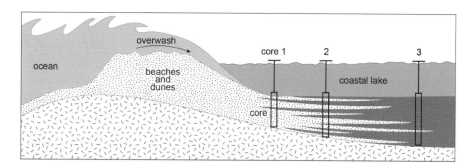

Figure 16.16 Coastal lakes or lagoons may contain a sedimentary record of hurricane-related overwash. The sand layers in these lakes represent beach sediments that have been transported over the dune barriers and deposited in the lee of the dunes. The sand thickness is greater closer to the dunes and decreases away from them, where mud dominates the lacustrine sediments. (After Liu (2007).)

Furthermore, geotechnical engineers need to know the recurrence interval of major hurricanes, i.e., category 3 or greater. When storm surges overwash or inundate barrier islands, the sandy sediments deposited in coastal lakes and marshes behind the dunes provide evidence of great prehistoric hurricanes, such as the overwash fans found in New Jersey 11 years before Hurricane Sandy devastated the New Jersey coast (Donnelly et al., 2001). The study of these sediments has been called **paleotempestology** (Emanuel, 2005). These investigations have a similar purpose to paleoseismic projects conducted to establish the recurrence intervals of large earthquakes (section 14.4).

Figure 16.16 illustrates the nature of such investigations. It requires evidence that the sandy sediments are indeed from overwash or inundation deposits by showing that: (i) the thickness of the sand zones decreases with distance from the beach and dunes, as would be expected of an ocean source; (ii) microfossils in the sand deposits are typical marine species; and (iii) the sandy interlayers and muddy, organic sediments that comprise the deposits in these coastal lakes indicate that the lakes have always been in close proximity to the coastal shoreline throughout the hundreds or thousands of years that ^{14}C dating of organic debris reveals (Liu, 2007).

However, paleotempestology is controversial because there is concern that only the largest storm events are preserved and that overwash deposits may represent incomplete sedimentary records (Hippensteel et al., 2013), thus yielding uncertain or even misleading information about return periods of hurricanes. This is an area of active research.

16.3.4 Tsunamis

Tsunamis are a series of long waves that are propagated across the ocean as a result of seismic faulting or a landslide on the seabed or into the sea or are due to some meteorological cause. In the Pacific Ocean basin, over 80% of tsunamis are caused by seabed seismic faulting, sometimes involving $M9$ subduction earthquakes. The remainder are associated with submarine volcanic eruptions or submarine landslides triggered by earthquakes, volcanoes or meteorological causes associated with the inverse barometric effect (see Pugh and Woodworth, 2014).

Tsunamis, when generated by earthquakes, can damage infrastructure far more seriously than the earthquake itself. The tsunami that struck the Fukushima nuclear reactor on the Pacific coast of Japan was ~8 m higher than the reactor's protective sea wall. Along the coastline of Japan, approximately 20 000 people died from the tsunami, although none apparently died from the immediate release of radiation that arose from the damage to the nuclear reactors.

In Chapter 14, we discussed (Box 14.2) the evidence for the tsunami associated with the 1700 Cascadia paleoearthquake

(Atwater and Hemphill-Haley, 1997). Regional coseismic subsidence from plate-boundary earthquakes was recorded in sand sheets laid down on top of soil layers in coastal marshlands along the Pacific Northwest of the USA and in adjacent British Columbia (see Figure 14.14). These sand/soil couples provide a means by which the recurrence intervals of large tsunamis may be estimated by radiocarbon dating of organic matter in the buried soil layers.

16.3.5 Coastal Subsidence

Coastal subsidence relative to sea levels is most often associated with the extraction of groundwater, oil or gas, as was discussed in Chapter 13. Morton et al. (2006) have argued that prolonged petroleum reservoir pressure depletion has caused both land subsidence and fault reactivation along the Louisiana coastline. Figure 16.17 illustrates the strong correlation between oil and gas extraction and land subsidence and coastal wetland loss. The reduced rate of subsidence since the mid-1990s corresponds to a reduction in petroleum

production along this coastline as production has shifted further offshore (Morton and Bernier, 2010).

However, sometimes other mechanisms are at play, including shallow Holocene sediment compaction, neotectonic activity and diminished sediment replenishment by rivers. It is these processes, when coupled with the worldwide increase in sea levels associated with glacial melting and oceanic warming, that are believed responsible for the relative rise in sea level versus the land surface in the densely populated Nile delta coast of Egypt (Stanley and Clemente, 2017). By examining drill cores obtained landward of the coast, these authors determined average rates of **sediment compaction** for sections across the delta from ~3.7 mm/a to ~8.4 mm/a, which were caused by overburden loading.

In the Nile delta, **neotectonic effects**, including earthquakes and other fault movements, have occurred since the Tertiary and have continued to the present; these are the result of ongoing interactions among the various plates that intersect in the eastern Mediterranean (see Figure 2.2). **Sediment replenishment** by the Nile River has been drastically reduced

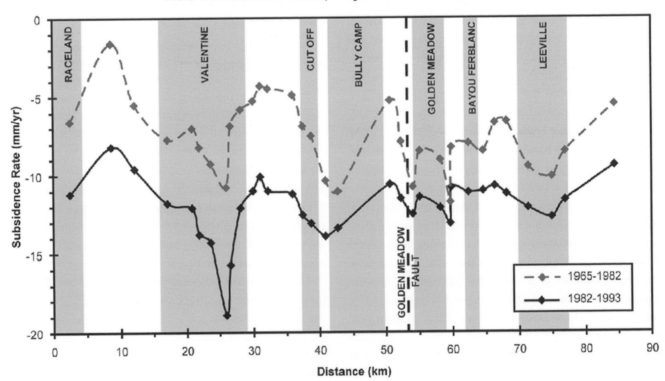

Figure 16.17 Historical subsidence rates calculated from re-levelling of seven benchmarks by the US National Geodetic Survey (NGS). The oil- and gas-producing areas are shown in grey. The two plots indicate a strong correlation between oil and gas production, the projection of the Golden Meadow fault, and rates of land subsidence and wetland loss in Bayou Lafourche, Louisiana, USA. Note that the rates of subsidence increased over the two time periods monitored by the re-levelling. (From Morton et al. (2006).)

by the construction of major dams over the last 100 years. Such replenishment would have minimized or moderated the problems of coastal submergence by worldwide sea-level rise and land subsidence triggered by neotectonics and/or sediment compaction. Consequently, the worldwide SLR > 3 mm/a from glacial melting and oceanic warming explains less than half the SLR measured along the Nile delta (Stanley and Clemente, 2017).

Tectonic activity in the form of earthquakes has caused coastal subsidence elsewhere. The coastal areas of northeastern Japan subsided over 1 m as a result of the 2011 M_W9.0 Tohuku earthquake; and a series of moderate ($M_W = 5.3-7.1$) earthquakes in Christchurch, New Zealand, during 2010–11 caused floodplain subsidence of 0.5–1.0 m adjacent to the Pacific Ocean. This increased the long-term flood hazard to the city both from upstream storms draining through the city, caused by waterway deformation, and from storm surges from the sea (Hughes et al., 2015). In 2014, following the floodplain subsidence, heavy rains caused widespread flooding in the city's suburbs.

In Figure 2.8, we noted that an ice sheet produces a forebulge that progressively collapses with time once the ice sheet has melted. This **glacial isostatic adjustment** (GIA) means that estimates of relative SLR must account for changes in the elevation of the land surface, as shown in Figure 16.10. The melting of the Laurentide Ice Sheet (Figure 8.1), which covered much of eastern Canada and the northeastern USA until 10–20 ka, has resulted in the progressive collapse of the forebulge, which poses SLR problems along the US Atlantic coastline.

A hinge line separating GIA-induced uplift from GIA-induced forebulge collapse occurs off the Maine coast, and subsidence rates of <0.8 mm/a are observed in Maine, increasing to rates of 1.7 mm/a in Delaware, and to rates <0.9 mm/a in the Carolinas. Norfolk, Virginia, with its huge US Navy base, is plagued by flooding because of its location and has experienced an SLR rate of 3.9 mm/a, measured from about 1950 through 2006 (Eggleston and Pope, 2013). Most of this relative SLR is due to land subsidence, most of which is caused by aquifer–aquitard compaction (Chapter 13); however, GIA-induced land subsidence further increases the flood risk to tidewater Virginia.

16.4 Stability of Coastal Cliffs and Bluffs

Coastal cliffs are created along sea coasts by wave action, the rise of sea levels or landslides. We refer to these same geomorphic features along lakeshores as **coastal bluffs**. These are dynamic geomorphic features that typically erode landward ("retreat") either progressively or episodically.

Because coastal erosion threatens urban development and transportation links on the top of cliffs and bluffs, as well as on their slopes and base, the stability of cliffs and bluffs is of geotechnical concern. Erosion of coastal cliffs or bluffs may shut highway or rail routes and cause the loss or abandonment of suburban developments built without consideration of the hazards of coastal retreat; these imply either replacement costs or loss of a municipal tax base (Hampton and Griggs, 2004).

Igneous and metamorphic rocks may form essentially stable coasts with little measurable retreat. However, weak sedimentary rocks and both cohesive and granular sediments are subject to erosion, such as the cliff shown in Figure 16.2 experiencing progressive failure by the collapse of small blocks. Coastal erosion is also important in that it is the source of sediments that form beaches or mud-rich tidal flats or salt marshes, which are of ecological significance. Therefore, if cliff or bluff erosion is inhibited by revetments, walls or other coastal protection measures, the supply of sediment for nearby beaches or marshes may be eliminated and a costly program of beach nourishment may be required.

In this section – and in order of increasing strength of geological materials – we will consider erosion of (i) bluffs along the shoreline of the Great Lakes of North America, (ii) cliffs along the Californian coast, and (iii) the Chalk cliffs of southern England and northwest Europe.

16.4.1 Lake Erie Bluffs

Bluff erosion along the shoreline of the Great Lakes has several causes. Mickelson et al. (2004) concluded that the principal causes were (i) toe erosion by wave action, (ii) wave erosion of the nearshore lake bed, (iii) the abrasive effect of entrained sand and gravel grains in a wave on the nearshore lake bottom, known as nearshore downcutting, (iv) sheet-wash and rill (small gully) erosion of the slopes during runoff and (v) groundwater discharge. Ice may protect the shoreline from wave erosion, although it may gouge the lake bed.

Figure 15.4 shows the effect of toe erosion promoting slope failure in a fluvial setting. The bluffs of the Great Lakes shoreline are typically cohesive glacial till. The base of bluffs that are continually undergoing wave attack will slump and transport the detritus away from the base of the bluff. As noted earlier, waves tend to break when $H_{max}/h \approx 0.75$; consequently, for any particular wave height, a shallow nearshore slope will tend to dissipate wave energy farther away from the bluff.

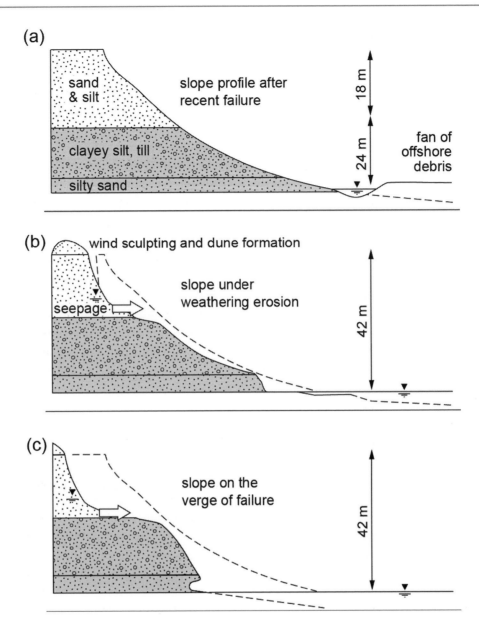

Figure 16.18 Erosion sequence leading to bluff failure along the north shore of Lake Erie, Ontario: (a) slope profile after recent failure; (b) slope under weathering erosion; and (c) slope on the verge of failure. (From Quigley et al. (1977). © 2008 Canadian Science Publishing or its licensors. Reproduced with permission.)

Kamphuis (1987) determined that the recession rate (R) for a glacial till bluff could be written in terms of the breaking wave height (H_b) and the wave power in watts per metre of shoreline (P_b):

$$R = K_1 H_b^{3.5} \qquad (16.6)$$

and

$$R = K_2 P_b^{1.4}, \qquad (16.7)$$

where K_1 and K_2 are constants. Equation (16.6) indicates that a 3 m wave height will produce a recession rate 500 times that

of a 0.5 m wave. Kamphuis concluded that the effect of small wave heights on the erosion of till bluffs was negligible.

Quigley et al. (1977) investigated several till bluffs along the shore of Lake Erie to determine the factors governing their retreat, which they estimated at 0.5–3.0 m/a over the last 150 years. They determined that retreat at this and the other sites was cyclic, involving (i) toe erosion by wave attack, (ii) cliff steepening by toppling and slaking and (iii) landslide initiation returning the slope of the bluff face to an angle of <39°. Figure 16.18 shows one of these sites, in which $R = 2.8$ m/a and $P_b^{1.4} \sim 2$ kW/m. The till and the silt at this site had undrained shear strengths of 250 and 110 kPa, respectively, with friction angles of 28° and 34°, respectively.

The cyclic process begins once the offshore debris apron from the previous landslide, which is shown in Figure 16.18(a), is eroded and removed by wave action. At a slope angle of 39°, the factors of safety computed by Bishop's modified method (see Box 15.2) are all less than one when groundwater discharge is occurring from the mid-slope area of the bluff face shown in Figure 16.18. Therefore, the cyclic failure of this site is promoted by (a) groundwater seepage at the contact between the overlying sand and silt with the basal till midway up the bluff face and (b) toe erosion causing bluff steepening that is unsustainable given the shear stresses that develop in the weakened bluff face.

Studies of till erosion in a 100 m long wave tank indicate that plunging breakers were erosive of fixed till beds, particularly so when sand was entrained in the waves. When a thin layer (10 mm) of sand was applied to the top of the till surface, the wave conditions employed did not scour the sand, which acted to protect the till. However, a thin layer of moving and discontinuous sand resulted in the suspension of the sand, leading to a high erosion rate (0.5 mm/h) of the till bed through its abrasion (Skafel and Bishop, 1994).

16.4.2 Californian Cliffs

In our introduction to coastal processes, we noted that the erosional retreat of the cliffs along the coastline of San Diego County, California, shown in Figure 16.2 is of the order of 3 cm/a in sandstones to 43 cm/a in unconsolidated sands. Such rates pose severe problems to cliff-top communities. Furthermore, about two-thirds of the California shoreline is threatened by high erosion rates (Griggs and Patsch, 2004), which are associated with strong storms arising from the El Niño weather effect when periods of high rainfall and large waves promote cliff erosion and retreat (Hapke and Green, 2006).

Benumof and Griggs (1999) concluded that the principal factors associated with cliff stability along the coastline of San Diego County were the intact rock strength and joint structure rather than wave action. Highly weathered rocks with continuous joint systems with spacings of 50–300 mm between joints and apertures of 5–20 mm were considered weak; groundwater discharge was a significant hazard in such cliff faces. These are additional examples of the **weak rocks** defined in section 5.4.4.

More recently, Collins and Sitar (2008) undertook a study of coastal cliff stability, employing high-resolution laser scanning of weakly and moderately cemented sandstone cliffs along the northern Californian coast, which are shown in Figures 16.19 and 16.20. The first figure shows the cyclic failure of weakly cemented cliffs caused by wave action; the second shows a similar failure that is caused by rainfall and subsequent seepage through coarser sand layers within the sandstone.

Both sands have a uniform fine-grained texture but the weakly cemented sand has only 1.4% fines, whereas the moderately cemented sand has 12.1% fines and this difference appears to dictate their respective strengths. The geotechnical properties of these sandstones are summarized in Table 16.4 (Collins and Sitar, 2009). The water content is critical to the behaviour of these specimens in that wetting the weakly cemented sand causes it to disaggregate, whereas the moderately cemented sand loses much tensile strength and its unconfined compressive strength (UCS) decreases by 60%. Using LiDAR images and a limit equilibrium analysis, Collins and Sitar (2011) concluded that the weakly cemented sand failed through shear but the moderately cemented sand underwent tensile failure.

16.4.3 The English Chalk Cliffs

The Chalk cliffs extend discontinuously along the English Channel from Devon in southwest England to Yorkshire in the northeast. They are late Cretaceous in age and are generally considered weak rocks (as in section 5.4.4), i.e., UCS = 2–8 MPa, of relatively low density in southeast England but higher elsewhere (see Bowden et al., 2002). The engineering properties of the Chalk (see Mortimore, 2012) are of great importance because the Chalk is present in foundations and excavations in many parts of northwestern Europe, e.g., the tunnel beneath the English Channel and the high-speed train route linking Paris with London. Here we consider two styles of cliff failure in the Chalk along the coast of southern England described by Mortimore et al. (2004), in which rock structure rather than wave action is the more important factor in cliff retreat.

The first style is shown in Figure 16.21, which involves a relatively simple vertical failure of the cliff face along a tension crack. The rock is the Seaford Chalk, which is shown in close-up in Figure 2.27. At some time, the vertical stress at the critical point exceeds the shear strength of the joint and the cliff face collapses in a **runout** but retains its stratigraphic integrity with the topsoil still intact. The collapse begins as a **toppling**

Table 16.4 Geotechnical properties of weakly and moderately cemented sand, northern California. The first two rows show the sandstones at *in situ* water contents, whereas the final row shows results of the sandstone following wetting. (From Collins and Sitar (2009).)

Sand material	UCS (kPa)	Friction angle ϕ (°)	Cohesion c (kPa)	Tensile strength τ_0 (kPa)	Water content (%)
Weakly cemented	13	39	6	0	8.9
Moderately cemented	340	46	69	32	12.6
Moderately cemented	124	47	34	6	22.1

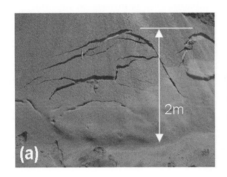
(a) 2m

(b) 10m

(c) 24m

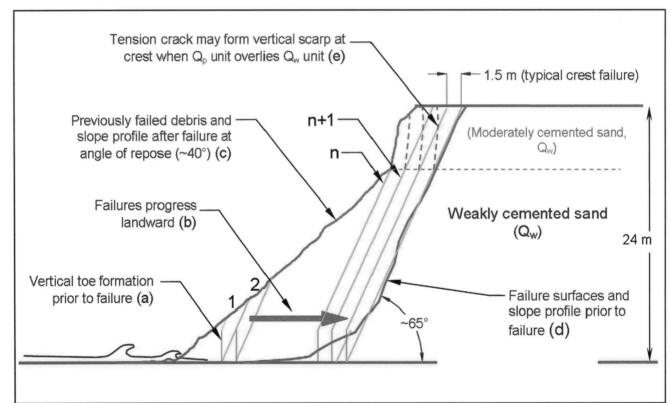

Figure 16.19 A geomorphic model of bluff failure caused by wave action on weakly cemented sandstones, i.e., UCS = 5–30 kPa. (a) Wave action on previously failed debris causes failure of small slabs less than 2 m high. (b) Shear failures increase landward with height as shown in the main panel, i.e., 1, 2, . . . , n, but without the retreat of the crest. (c) Wave action on previously failed debris with an angle of repose of 40°. The erosion cycle begins again once the intact slope fails and a new debris deposit is created. (From Collins and Sitar (2008), courtesy of the US Geological Survey.)

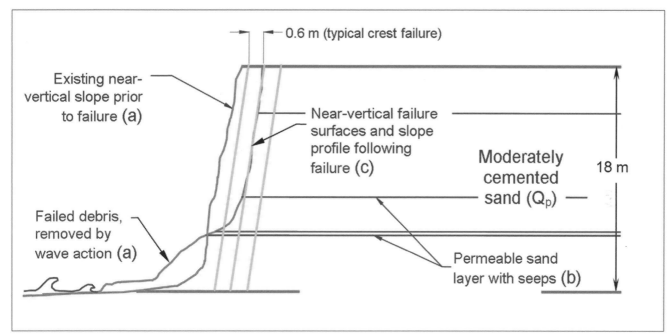

Figure 16.20 A geomorphic model of the failure of a sandstone bluff caused by groundwater flow through permeable layers within the sandstone of an otherwise moderately cemented sandstone, i.e., UCS ∼ 340 kPa. Wave action removes the failed debris while loss of tensile strength causes slab failure and retreat of the bluff. (From Collins and Sitar (2008), courtesy of the US Geological Survey.)

failure (Figure 4.17) but then becomes a **plane failure** such that translated basal beds are located furthest from the base of the cliff.

The second style, illustrated by Figure 16.22, is more complicated and involves **plane** and **wedge** block failures (Figure 4.17) that may initiate cliff failure from any elevation in the cliff face, i.e., because of the interlocking nature of the blocks, cliff failure may proceed from the top down or the bottom up. The fractures in the Chalk are steeply dipping (60–70°). The Newhaven Chalk joints are filled with flint (chert) sheets, whereas the Holywell Chalk exhibits clay-filled polished surfaces.

Chalk forms the northern coast of France, where its failures have been described by Duperret et al. (2004). They indicate that failure of the French chalk was often triggered by rainfall. In one case, a prolonged period of rain preceded by about one month the failure of less fractured chalk, whereas an intense

rainfall following a dry period would be sufficient to cause failure of cliffs that have karst features, such as pipes.

16.5 Nearshore Hazards

Finally, we consider a number of nearshore hazards that affect geotechnical engineering practice: (i) iceberg scour, (ii) gassy sediments and (iii) submarine landslides. Each of these poses problems to offshore infrastructure such as communication cables, offshore platforms and hydrocarbon pipelines.

16.5.1 Iceberg Scouring

Iceberg scouring of coastal seabeds leaves pronounced ridge-and-groove topographies at water depths of 100–230 m. On the seabed of the continental shelf of the Labrador Sea

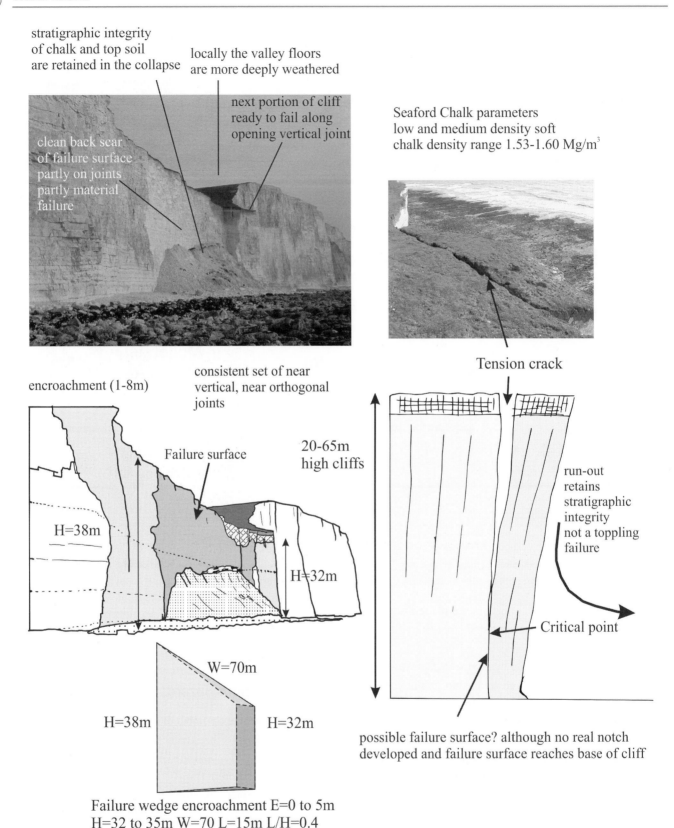

Figure 16.21 Styles of cliff failure in the Chalk of Sussex. The Seven Sisters are the icons of the English Chalk cliffs along the English Channel. The failure occurs along tension cracks that develop from vertical joint sets with a limited runout of debris at the base of the cliff. A failure in May 2016 caused one of these cliffs to retreat 10 m. (From Mortimore et al. (2004). Reproduced with permission of The Geological Society of London.)

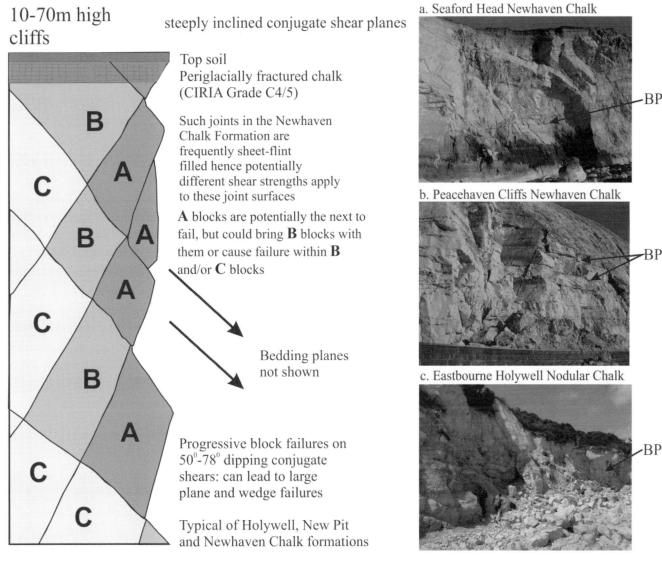

10-70m high cliffs

steeply inclined conjugate shear planes

Top soil
Periglacially fractured chalk
(CIRIA Grade C4/5)

Such joints in the Newhaven
Chalk Formation are
frequently sheet-flint
filled hence potentially
different shear strengths apply
to these joint surfaces

A blocks are potentially the next to
fail, but could bring **B** blocks with
them or cause failure within **B**
and/or **C** blocks

Bedding planes
not shown

Progressive block failures on
50^0-78^0 dipping conjugate
shears: can lead to large
plane and wedge failures

Typical of Holywell, New Pit
and Newhaven Chalk formations

a. Seaford Head Newhaven Chalk — BP

b. Peacehaven Cliffs Newhaven Chalk — BP

c. Eastbourne Holywell Nodular Chalk — BP

Figure 16.22 Styles of cliff failure in the Chalk of Sussex. In this case, block failure originates with the presence of steeply inclined conjugate shear fractures that have little to do with undermining by energetic waves. BP indicates bedding planes, which are usually marl seams. Progressive failure occurs first with A, then B, then C blocks. Sliding occurs along failure planes that have undergone weathering and thus have lost shear strength. (From Mortimore et al. (2004). Reproduced with permission of The Geological Society of London.)

between northeastern Canada and Greenland, icebergs incise grooves to an average depth of 1–2 m and width of 30–40 m (Woodworth-Lynas et al., 1991). Such features clearly indicate a hazard to any pipelines that might be laid in Arctic areas of Alaska, Canada, Norway and Russia, where hydrocarbons are being produced.

These scours are also evident in lake beds that were once adjacent to ice sheets, such as Glacial Lake Agassiz in Manitoba (Dredge, 1982) and Lake Ontario, Canada (Eden and Eyles, 2002). In the glaciolacustrine clayey silt of Lake Ontario, near Toronto, Ontario, a scour of 10 m width and 4 m depth formed at a water depth of <150 m and occurred 45 ka, prior to the Last Glacial Maximum. In northern Manitoba, Glacial Lake Agassiz left criss-crossing linear scours of 300–1800 m in length.

Palmer (1997) suggested safe depths of pipeline burial should be based on the assumption that, even below the depth of ice scour and pipeline rupture, there exists a zone in which the ice would cause any pipeline to deform plastically. In this shallow zone, the seabed sediment would be remoulded to its critical state, which is a function of the sediment's effective stress and volumetric packing. Below this remoulded zone is one in which elastic deformation of the pipeline would occur under safe conditions.

16.5.2 Gassy Sediments

Gassy sediments are distinguished by the presence of a significant part of the pore volume being filled with gas that exceeds the solubility of the gas in the aqueous phase that fills the balance of the pore volume. This occluded gas alters the shear strength, compressibility and hydraulic conductivity of these sediments in ways that lead to submarine slope failures through volume change and **liquefaction**, thus complicating the design of offshore foundations (Chillarige et al., 1997; Grozic et al., 1999).

Methane is often present as the gas phase. It is derived from biogenic and thermogenic processes that produce gas within the Earth. The biogenic processes are similar to those discussed in section 12.5.2, in which dissolved organic matter is converted into methane under reducing conditions at shallow depths, whereas thermogenic gas generation occurs at temperatures $>120°C$ indicating considerable depth. Methane may become trapped in the pores of marine sediments and may be either biogenic or thermogenic, which can be determined by stable ^{13}C analysis (Clark, 2015).

Particularly in cold regions, but also in sediments along continental margins worldwide, methane may occur in the form of a **gas hydrate**, a crystalline compound in which the gas is trapped within the ice lattice of water (Hunt, 1979). Methane hydrates dissociate in the pores of marine sediments through an increase in temperature or a decrease in pressure causing free gas to accumulate in pores. This inhibits drainage from pores because the presence of a gas phase reduces the relative permeability of pore water (see Figure 12.8) and acts as a non-aqueous phase blocking drainage. This, in turn, would tend to create an excess pore pressure and thus promote slope instability (Locat and Lee, 2002).

Chillarige et al. (1997) suggested that submarine slope failures in the Fraser delta (Figures 9.17 and 9.18) occurred during tidal drawdown; in a geotechnical context, this constitutes a situation of undrained unloading with respect to marine sediments. Amaratunga and Grozic (2009) demonstrated the validity of this hypothesis experimentally using sands fully saturated ($S_w = 1.0$; $S_g = 0.0$) with water containing dissolved gas. As the total stress was reduced to simulate tidal drawdown, gas dissolved from the pore water causing the pore pressure to increase and forcing a decrease in the effective stress. Eventually, the specimen collapsed from strain softening, which occurs in these (Ottawa) sands when $S_w = 0.9$ according to Grozic et al. (1999).

In deltaic slopes, such situations cause **flow liquefaction** first, as the gravitational shear stresses exceed the strength of the marine sediments, and then the initiation of flow slides. On level seabeds, a reduction in bearing capacity of the sediments will occur.

16.5.3 Submarine Landslides

The example of flow liquefaction in gassy sands is just one mechanism that might cause a submarine landslide. Such gravity-induced flow slides occur when the shear strength of the sediment (τ_{fs}) is exceeded by the imposed shear stress, i.e., $\tau_{fs} < \tau$ in equation (1.3). Locat and Lee (2002) indicate that submarine landslides are typically initiated by earthquakes, tidal changes, sedimentation and wave loading in shallow water. Submarine landslides are typically much larger than terrestrial ones, with **runout** distances from several kilometres to over 100 km, such as the Storegga slide (see Box 14.2).

Wave loading is one mechanism that obviously does not have a terrestrial counterpart. Cyclic wave loading was suspected of causing slope failures in the Gulf of Mexico that destroyed an oil platform during a hurricane (Bea et al., 1983). In the delta of the Yellow River off China, circular depressions were observed in shallow water depths (4–8 m) on the upper deltaic sediments. Xu et al. (2009) concluded that these were caused by loss of strength by the marine silts that underwent deformation induced through wave action by the to-and-fro movement of a bowl-shaped sediment mass that subsequently consolidated.

The general form of a submarine slope failure is that of a rotational slide (Figure 15.1A), although translational slides may result from failure along bedding planes (Figure 15.1B). Following initiation, there may be a transition into a debris flow (see Figure 15.1F) and subsequent formation of a turbidity current and then migration on the seafloor until final deposition occurs.

Figure 16.23 illustrates the generation of a turbidity current on a submarine slope. The minimum thickness for the dense-flow phase to migrate (H_c in m) is given by Locat and Lee (2002) as

$$H_c = \frac{2.42 I_L^{-3.13}}{(\gamma - \gamma_w)\sin\beta}, \qquad (16.8)$$

where I_L is the Atterberg liquidity index $I_L = (w - w_p)/I_P$ relating the water content to the plasticity index, γ and γ_w are the total unit weight of sediment and water, respectively, and β is the angle of the submarine slope. The liquidity index is used to reflect the importance of the yield strength of the fluid (see equation (15.1)), which is the dominant factor in the resistance to flow that the dense fluid encounters.

BOX 16.2 | THE STOREGGA SUBMARINE LANDSLIDE OF 8150 KA

The sizes of submarine landslides are difficult to imagine. Several are estimated to have runout lengths of over 100 km and displaced sediment volumes of 10^{10} m^3. One of these is the Storegga landslide, which is believed to have been initiated by gas-hydrate dissociation.

Bondevik et al. (2003) discovered tsunami deposits in the Shetland Islands located midway between Scotland and Norway. They excavated a prominent sand layer 30–40 cm thick at the shoreline that they traced over 150 m as it thinned to an elevation of 9.2 m above high-tide levels. This sand contained woody debris and clasts of peat that had been ripped from below the sand.

The radiocarbon date on the peat indicated an age of approximately 7300 ^{14}C years or 8150 ka, at which time the sea level was 10–15 m lower on Shetland than today. The runup of the tsunami was therefore ~20 m at a location 150–200 km from its source on the Norwegian continental shelf.

Sources: Locat and Lee (2002) and Bondevik et al. (2003).

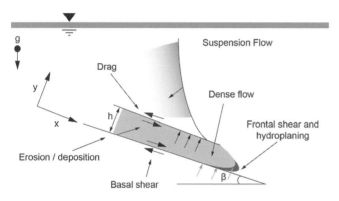

Figure 16.23 Schematic of a turbidity current illustrating the suspension and dense-flow regimes, drag forces, basal and frontal shear producing erosion. For the dense-flow phase to become mobile, its thickness (h) must exceed H_c in equation (16.8). (From Locat and Lee (2002) © 2008 Canadian Science Publishing or its licensors. Reproduced with permission.)

Therefore, the dense-flow phase requires a minimum thickness before it becomes mobile and then is subject to drag forces on its surface, basal and frontal shear and uplift from hydroplaning. The basal shear will produce both erosional and depositional surfaces. The dense fluid remains in an undrained state during flow and may migrate along submarine canyons or other deep topographic features.

Because the *in situ* investigation of submarine landslides poses problems of accessibility, the nature of slope failures in the nearshore environment is often uncertain. Recently, evidence has accumulated that the origin of these failures rests in the presence of a **weak layer** within marine sediments, which is defined by Locat et al. (2014) as a layer or band of sediment (or rock) that has a strength (either potentially or actually) that is sufficiently lower than the adjacent units. This

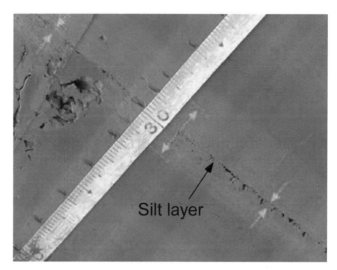

Figure 16.24 A thin silt weak layer, approximately 1 mm thick, in marine clay that failed in testing after remoulding with ~10 cm displacement. The red arrows indicate the failure-induced softening on either side of the silt layer. A Mohr–Coulomb failure analysis (equation (1.13)) of the silt would be quite different from that of the adjacent clay. This silt layer would be of lower effective cohesion and thus lower effective shear strength than the clay. (From Locat et al. (2014). Reprinted by permission of Springer, © 2014.) See colour plate section.

strength contrast acts as a potential zone of weakness along which a failure surface may develop. The weak layer may be either inherited – in which case it was of low strength prior to the initiation of a landslide, perhaps caused by high pore pressure – or induced by strain softening or an earthquake. Figure 16.24 illustrates a weak layer in marine clay. Locat et al. (2014) summarize the evidence for such weak layers in a variety of marine sediments.

16.6 Summary

We have considered the nature of coastal landforms, with particular emphasis on those of the Atlantic and Gulf coasts of the USA because of their detailed analysis and similarity to conditions on other coasts across the world. A general beach profile was defined as were the form and energy of waves. Wave power was shown to be strongly dependent on the height of the wave, which is believed to be the principal agent of coastal erosion.

The measured sea level is composed of a mean sea level plus tidal effects and a residual component. Tides are generated by gravitational forces exerted by the Moon and, to a lesser extent, the Sun. These cause the common semi-diurnal tide as well as diurnal and mixed tides. Because of their nature, they can be predicted with considerable accuracy. Meteorological factors often dictate the magnitude of the sea-level residual in the form of storm surges and lesser wave runups. Sea-level rise (SLR) was shown to be a mix of several factors, in particular glacial melting and ocean warming. However, there are seasonal variations in SLR associated with regional runoff from major rivers and other increases or decreases in sea level that are associated with glacial isostatic adjustment. Global SLR is currently (2017) estimated to be >3 mm/a.

Coastal erosion has initiated many multi-million-dollar geotechnical projects, including beach nourishment and coastal protection structures, as well as forcing administrative mitigation measures. The storm surges that damage coastal cities, beaches and cliffs are driven by cyclonic storms, including hurricanes and typhoons, that cause their own temporary but damaging SLR effect through barometric effects. We considered the erosion of lakeshore bluffs and coastal cliffs along the Great Lakes, in California and in England. The strength of the geological materials forming these slopes is critical to their stability, although all are in various degrees of retreat.

We concluded with a discussion of nearshore hazards, which might threaten communications cables, pipelines or other structures with foundations on the seabed or lake beds. These hazards include iceberg scour of the seabed or lake beds; the exsolution of gas from sediments or the dissociation of gas hydrates contained within those sediments; and submarine landslides caused by earthquakes, gassy sediments, tidal changes, sedimentation or wave action.

16.7 FURTHER READING

Davidson-Arnott, R., 2010, *Introduction to Coastal Processes and Geomorphology*. Cambridge, UK: Cambridge University Press. — The best introductory source of information on elementary wave theory, beach processes and the geomorphology of coasts.

Davis, R.A. and Fitzgerald, D.M., 2004, *Beaches and Coasts*. Malden, MA: Blackwell Science. — Less technical than Davidson-Arnott, an introduction to processes and issues along coasts and on beaches.

Dean, R.G. and Dalrymple, R.A., 2002, *Coastal Processes with Engineering Applications*. Cambridge, UK: Cambridge University Press. — The standard text on coastal engineering written for graduate students and researchers.

Emanuel, K., 2005, *Divine Wind: The History and Science of Hurricanes*. New York, NY: Oxford University Press. — A most interesting discussion of the origin of hurricanes and their history, written for the technically minded individual.

Hampton, M.A. and Griggs, G.B. (eds), 2004, *Formation, Evolution, and Stability of Coastal Cliffs: Status and Trends*. USGS Professional Paper 1693. Denver, CO: US Geological Survey. — A survey of articles on regional issues concerning the coastal cliffs and lakeshore bluffs of the USA.

Pugh, D. and Woodworth, P., 2014, *Sea-Level Science: Understanding Tides, Surges, Tsunamis and Mean Sea-Level Changes*. Cambridge, UK: Cambridge University Press. — A detailed account of all aspects of sea-level processes and their measurement.

US National Academy of Sciences, 2014, *Reducing Coastal Risk on the East and Gulf Coasts*. Washington, DC: National Academies Press. — A recent review of the origin and nature of coastal risks along the US East and Gulf coasts and how they might be mitigated when so many people wish to reside by the seashore.

16.8 Questions

Question 16.1. A procession of waves of similar height is known as a wave train. For a wave train with a period of 5 s in water of depth $h = 3.05$ m, the exact solution of equation (16.2) by Newton–Raphson iteration is $L_s = 25.1$ m. (i) What would be the deep-water wavelength, L_0, for this wave train? (ii) What is the error obtained if one uses the simplified expression for the shallow-water wavelength, $L_s = T\sqrt{gh}$?

Question 16.2. What is the wave power in W/m of each wave in the wave train of question 16.1 if successive waves have a height (i) 0.2 m and (ii) 2.0 m?

Question 16.3. List some conditions under which a storm surge is most likely to overtop a sea wall built to prevent such overtopping?

Question 16.4. The term "relative sea-level rise" refers to the observed change of sea level relative to the land surface at a particular point. What are the factors that affect the measurement of relative sea-level rise?

Question 16.5. Rate the erodibility of these rocks and sediments as geological materials during storm surges based on what you have read in this chapter: (i) siltstones, such as shown in Figure 1.5; (ii) dune sediments, as shown in Figure 16.14(a); (iii) glacial till along the shoreline of Lake Michigan in Wisconsin; (iv) European chalk cliffs with pronounced joint patterns (e.g., Figure 2.27); and (v) red-bed sandstones, such as those shown in Figure 2.23. Refer to Figures 4.13 and 9.7.

GLOSSARY

Common geoscientific terms that are printed in **bold** in the text should be memorized: many are defined here in the glossary. The definitions of some of these terms listed here are taken from Klein and Philpotts (2013) and Fossen (2016) and are used here with the permission of Tony Philpotts, Haakon Fossen and Cambridge University Press. Many of the terms in **bold** in the text that are not defined here may be found in Bates and Jackson (1984).

abrasion (glacial) – subglacial erosion of bedrock leaving a smoothed surface, sometimes with striations

acid-rock or acid-mine drainage – the aqueous dissolution of sulfide minerals in waste tailings and rock exposed by mining and the subsequent production of acidic sulfate- and metal-rich waters

active fault – defined in California as a fault that has had surface displacement within Holocene time (about the last 11 000 years)

active layer (of permafrost) – the shallow regolith or rock zone that experiences seasonal freezing and thawing

activity (ionic) – the chemical potential of an ion; pure solids and liquids have activities of unity; in natural and contaminated waters, the activity of an ionic or neutral species i is a_i and is related to the analytical concentration m_i by $a_i = y_i \cdot m_i$, where y_i is the activity coefficient

adsorption edge – the critical pH range over which sorption of cations and oxyanions occurs in porous media

advection – the transport of molecules or ions by groundwater flow occurring at the average linear groundwater velocity according to Darcy's law

advection–diffusion – the transport of contaminants through high-permeability pathways with flow predominantly parallel to bedding with transverse diffusion into adjacent lower-permeability sediments

aggrading – stream channels actively depositing sediment are considered aggrading

aggregate – in the construction industry, it is a loose mixture of sand, gravel, pebbles or crushed stone; it can be used by itself as road and railway ballast, or to stabilize slopes; it is also mixed with approximately one-third Portland cement to form concrete

air-entry value – the soil-moisture suction at which air enters water-saturated pores

alluvial deposits – stream-laid sediments of various grain size

alluvial fans – alluvial sediments deposited on a gently sloping landform along a mountain front in arid zones that appear as a fan when viewed on a map with its apex at the mouth of a valley issuing from the mountain front

amphibole – a major rock-forming silicate group with infinitely extending double tetrahedral chains: hydrous with Ca, Mg, Fe, Na, Al and Si as major constituents

Anderson's theory of faulting – the theory that the nature of faulting, i.e., normal, strike-slip, reverse, is determined by how the magnitude of the vertical stress is related to the magnitude of the other principal stresses

andesite – a volcanic rock that characterizes composite volcanoes formed along island arcs above subduction zones; it is composed of approximately equal proportions of pyroxene and plagioclase

angle of internal friction – in the Mohr–Coulomb failure criterion, the angle of inclination of a surface within a block of rock at which sliding of the overlying part of the block occurs

angular unconformity – an unconformity in which the sedimentary beds below the plane of unconformity are truncated and have a different orientation than beds above the unconformity

anhydrite – an evaporite mineral composed of anhydrous calcium sulfate

annual exceedance probability (AEP) – the probability of a particular event, such as a flood, being exceeded annually; a 0.2 AEP flood has a 20% chance of occurring in any given year, and this corresponds to a five-year recurrence-interval flood; AEP terminology implies that the occurrence of a rare flood or other event in one year does not reduce the chances of another similar rare flood or event following within a short time period; an AEP is always a fraction of one

aquifer – a formation, group of formations, or part of a formation that contains sufficient saturated permeable material to yield significant quantities of water to wells and springs; the permeable materials will comprise a porous medium, in which the pore volume may be composed of fractures or dissolution features or granular interstices

aquitard – a formation, group of formations, or part of a formation of low hydraulic conductivity compared with adjacent formations

architectural elements – groups of sedimentary facies that describe distinctive shapes, dominant textures and internal geometries of fluvial deposits

arenite – a sedimentary rock composed of sand-size grains

argillaceous rocks – rocks having a significant proportion of clay minerals, such as shale or argillaceous limestone

arkose – a sandstone containing more than 25% feldspar

arsenopyrite – FeAsS, a sulfide mineral often found in veins of crystalline rock

augers – screw-shaped drilling tools that are rotated into the soil or sediment with applied force in order to recover core or to excavate a hole within which a well may be installed

avulsion – the natural process by which flow diverts out of an established river channel into a new permanent course on the adjacent floodplain

_b_-value curve – also known as the Gutenberg–Richter recurrence law, the _b_-value curve is a plot of the logarithm of the number of earthquakes per year (N) greater than a given magnitude, $\log_{10} N = a - bM$, where a is a constant related to regional seismicity and $b \sim 1.0$

back-diffusion – molecular diffusion of ions or solutes that have become stored in relatively low-permeability units back into the high-permeability sediments or fractures in which they originated, causing an increase in concentration in the high-permeability units

bars – sedimentary ridges, typically sand and gravel, created by streams and currents often restricting navigation on rivers and coasts

basal slip – the sliding of the ice over the glacier bed material facilitated by meltwater lubrication and implying the deformation of the bed

basalt – a fine-grained igneous rock composed predominantly of equal proportions of plagioclase feldspar and pyroxene (or olivine); it is the most common igneous rock and forms most of the ocean floor

baseflow – groundwater discharge sustaining streamflow between periods of storm runoff

basement – the crust of the Earth beneath the sedimentary rock cover; Precambrian Shield terrains form the basement rocks of the continents

base-metal ore deposits – deposits of Cu, Pb, Zn and other metal sulfide minerals transported through the crust by high-temperature fluids

batholiths – an extremely large intrusive igneous body with an outcrop area exceeding 100 km^2; they are usually composed of granite and granodiorite

Becker drill – percussion drilling method especially suited for penetrating coarse sediments, e.g., cobbles and boulders in alluvium or in glacial till

bed – the basic sedimentary unit of a particular exposure of rock and synonymous with strata or layer

bedding-plane fractures – usually horizontal discontinuities evident at contacts between rock types or texturally different strata

bedforms – large-scale sedimentary features observable in an alluvial channel, which are recognizable at a distance, e.g., ripples or dunes

bed material – sediment that forms the bed and lower banks of an alluvial channel

bed shear stress – the stress required to entrain a sediment particle of maximum diameter, D, by a particular flow

biochemical sediments – sediments deposited by marine organisms building calcite or silica skeletons

biodegradation reaction – a microbially catalyzed chemical reaction that oxidizes a hydrocarbon molecule to yield carbon dioxide and waste products

bioturbation – the burrowing activities of worms, crayfish or mammals in soil that causes increased infiltration of rain and snowmelt

Bishop's modified method – a limit-equilibrium method of slope stability analysis, in which the slope is partitioned into slices with forces on each slice determined

borehole breakouts – spalling or caving of opposite walls of a borehole indicating the orientation of σ_{Hmin}

borehole geophysics – the use of geophysical probes lowered down a borehole by a wireline to develop a profile of the lithology, porosity, bulk density or other property of rock penetrated by the borehole

breccia – a rock composed of angular rock fragments produced by brittle fracturing; they can be of sedimentary, igneous, tectonic or meteorite impact origin

bridge scour – the failure of bridge foundations and piers by fluvial erosion, particularly during flood events

brittle deformation – those mechanical processes causing the sample to yield and then fail, including tensile failure, granular flow and cataclastic flow

Byerlee's law – the relationship between the critical shear stress on a fracture and the related normal stress across it; because the normal stress reflects the depth of the crust, this law models the critical shear strength through the frictional upper crust

cable-tool drilling – a percussion drilling method driving a chisel-type bit and casing into soils or poorly consolidated rock

calcite – the mineral calcium carbonate

calcrete (or caliche) – a mineral deposit formed by the precipitation of calcite from water as it evaporates from arid and semi-arid soils

capillary barrier – soil or rock, which is in contact with a non-wetting fluid, having an entry pressure exceeding the non-wetting pressure of the contact fluid and thus acting to prevent the penetration of that fluid into the soil or rock

capillary fringe – capillarity causes the pore volume immediately above the water table to become saturated but under tension or negative pressure

carbonate minerals – minerals in which the anionic group is carbonate as in calcite, $CaCO_3$

cataclastic flow – flow of rock during deformation by means of cataclasis, involving brittle crushing of grains accompanied by frictional sliding and rotation, occurring at a scale making the deformation appear continuous and distributed over a zone

celerity – the time for two successive wave crests to pass a particular point, where $C = L/T$ in which L is the wavelength and T is the wave period

cementation – a stage during the diagenesis of sediments into rock during which there is grain overgrowth by cementation, usually involving calcium carbonate, iron oxide or silica cements that precipitate from pore waters flowing through the sediment and create mineral bonds between grains

chalk – a carbonate mudstone (usually white) formed from the skeletal remains of plankton and algae and typically containing flint nodules and clay minerals

channels (alluvial) – an architectural element indicating the deepest part of a stream along which flow is concentrated

chemical grains – grains exhibiting interlocking crystal growth fabric acquired during diagenesis

chert – a sedimentary rock composed primarily of extremely fine-grained quartz; flint is synonymous with chert but is used more commonly in archeological literature

chimneys – a subsidence feature caused by underground mining or karst dissolution in which a cavity fills with broken rock or soil debris from the overburden

circular failure of rock slopes – rock-slope failure occurring in extremely weak rock as a result of numerous randomly oriented joint planes

clastic – refers to a rock or sediment composed principally of broken fragments that are derived from pre-existing rocks or minerals

clasts – clasts are large sedimentary particles, such as pebbles in a conglomerate, that have been transported by water

clast-supported – an adjective indicating that sedimentary detrital particles, such as pebbles in a conglomerate, touch one another and form an interconnected framework; *see* matrix-supported

clay minerals – hydrous aluminum silicate minerals with layered structures resulting from the weathering of feldspars

cleavage – the breaking of a mineral along crystallographic planes

clinometer – a device for measuring the angle of tilt of a planar surface; the required bearing is measured with a compass; the two measurements provide the means by which the orientation of a geological planar surface is recorded

coal – a sedimentary rock composed predominantly of fossilized plant remains

cohesion – in Mohr–Coulomb materials, cohesion is the shear strength of a sample of the material in the absence of either compressive or tensile normal stresses

compaction, mechanical – the process by which sediment particles are gravitationally compressed by some overburden load during burial while interstitial liquid is expelled upwards

competence – the ability of a stream to entrain and transport sediment of particular size

compressibility – the fractional change in bulk volume of a porous medium with a unit change in confining stress; a saturated porous medium of sand may be compressed through the compression of (i) the water, (ii) the sand grains and (iii) the porous medium through grain slippage and the rearrangement of the sand grains into a lower-porosity packing geometry

compressive strength – in unconfined compression testing, it is the ratio of peak load to the cross-sectional area of the specimen

cone resistance – the ratio of total force acting on a cone penetration testing (CPT) cone to the cross-sectional area of the cone

conglomerate – a sedimentary rock composed of gravel-size particles

congruent dissolution – aqueous dissolution of a mineral to yield groundwater with the same ionic composition as the mineral

consolidation – in geological terms, consolidation indicates compression, cementation and lithification of sediments; in soil mechanics, consolidation refers to a change in volume as a result of a stress change

contact metamorphism – metamorphism caused by heat from a cooling body of magma

core disking – the splitting of drill core into disk-shaped pieces upon recovery from a borehole due to stress relief

corestones – unweathered or partially weathered blocks of parent rock in a soil profile

corrosion mixing – when two calcite-saturated groundwaters mix, through the stream-tube convergence of separate karst flow channels, the resulting groundwater may be undersaturated with calcium carbonate and therefore cause additional dissolution of the karst bedrock

coseismic displacement – a seismically induced displacement of ground occurring simultaneously as a nearby earthquake

crest – the highest elevation on the upper surface of a particular bed in a vertical cross-section of an anticline

critically stressed faults – a fault that is stressed to the limit of its strength such that any changes in confining or stress or pore pressure may result in frictional failure

cross-bedding – bedding that is at a significant angle to the main bedding in a sedimentary rock; it can be formed, for example, by erosion and deposition in stream channels, deltas and sand dunes, and by the swash and backwash of waves on a beach

crown holes – subsidence sinkholes caused when the roof material fails above mined cavities

cubic law – the hydraulic conductivity of a fracture in a rock mass as a function of the apparent width of the fracture aperture and the aqueous density and viscosity

cut bank – the steep bank on the exterior of a meander bend where active erosion occurs

cutting and filling – the process by which streams erode in one place and then deposit at a nearby downstream location until the surfaces of erosion and deposition are continuous

cyclic stress ratio (CSR) – the ratio of the uniform cyclic shear stress to the initial vertical effective stress

debris or debris-avalanche flow – a landslide involving relatively coarse-grained material referred to as debris, which is generally larger than 2 mm

deformation modulus – the ratio of Young's modulus to the unconfined compressive strength of a rock mass

degrading – stream channels that are undergoing a net sediment loss due to erosion

deltaic deposits – the triangular- or fan-shaped alluvial sediments that occur where rivers discharge to the surface waters

DEM – a digital elevation model of a landscape

detrital sediments – a term that applies to any particle resulting from the weathering of a pre-existing rock

diabase – a dike rock of basaltic composition; referred to as dolerite in the UK

diagenesis – chemical, physical and biological changes that affect sediment or sedimentary rock after initial deposition but excluding weathering or metamorphism

diamicton – an unsorted mixture of glacial sediments indicating an uncertain depositional process

diffusion – the spreading of molecules or ions within a fluid-filled porous medium in a direction tending to equalize concentrations in all parts of the system caused by the thermal-kinetic energy of the molecules and ions

diffusional profiling – the depth profile of isotopes or ions present in pore-water samples collected from aquitards in a stratigraphic column where the slow migration of the isotopes or ions is the result of molecular diffusion rather than groundwater flow

dikes – an igneous sheet-like body that cuts across the layering in the surrounding rocks; most dikes have steep dips

diorite – a plutonic igneous rock composed predominantly of pyroxene and plagioclase; andesite is its volcanic equivalent

dip – the dip of a bed is the angle, relative to the horizontal, at which the bed slopes downwards

disconformity – an unconformity in which beds above and below the unconformity are parallel

discontinuity – a joint in otherwise intact rock capable of transmitting groundwater

disequilibrium compaction – the rapid burial of sediment within a sedimentary basin without compensating drainage at depth causing overpressure

dispersal trains – linear patterns of detrital rocks known as glacial erratics deposited along the migration path of glaciers during the Pleistocene and before

dispersion – the spreading of molecules or ions within a fluid-filled porous medium in a direction tending to equalize concentrations in all parts of the system caused by fluid advection

dissolution – during diagenesis, geothermal heating and sediment loading cause elevated temperatures and pressures that result in mineral dissolution; at ambient temperatures and pressures, dissolution of minerals occurs within groundwater provided the groundwater is undersaturated with respect to a particular mineral

diurnal tide – a tide causing a once-daily variation in sea level

DNAPL – dense non-aqueous phase liquids; an immiscible liquid that is sparingly soluble in water

DO – dissolved oxygen

dolerite – a dike rock of basaltic composition; referred to as diabase in the USA

dolomite – a mineral formed by the replacement process whereby half of the calcium in the calcite structure is replaced by magnesium, thus converting calcite into dolomite; also a sedimentary rock of which >50% of the rock is composed of dolomite; this rock is sometimes called dolostone

downstream accretion – an architectural element indicating sedimentary deposits located within river channels that are actively eroding, transporting and depositing sand and gravel to produce bars and islands in the middle of the channel

drawdown cone – the cone of depression of hydraulic head spreading radially outwards from a pumping well

drift – an old and now-discouraged term to indicate rock transported and deposited by glaciers

drilling mud – a carefully formulated heavy suspension, usually in water but sometimes in oil, used in rotary drilling; it commonly consists of bentonitic clays, chemical additives and weighting materials such as barite

dual porosity – a porous medium of rock or soil displaying two separate pore networks, such as intergranular voids, macropores from bioturbation or a fracture system

ductile deformation – rock that can sustain deformation before fracturing or faulting

earth, as in earthflow – landslide slope material that is at least 80% composed of sand-, silt- and clay-sized particles

effective porosity – the interconnected porosity within a porous medium (rock, sediment or soil) per unit volume implying that is the pore volume responsible for flow

effective pressure – the effective stress at the base of a glacier

effective solubility – the actual solubility of a component in a multi-component NAPL; according to Raoult's law, it is the product of the mole fraction and the aqueous solubility of a component of the NAPL

effective stress – the actual stress experienced by a soil or rock volume measured as the total stress minus the pore pressure within the volume

elastic dislocation model – the process by which the sudden rupture of a fault in a subduction zone releases the accumulated elastic strain energy as an earthquake, following which the fault re-locks and strain begins to accumulate again

elastic rebound model – process describing an earthquake cycle along strike-slip faults that begins with the accumulation of strain on parts of the fault that are temporarily locked in place followed by the rupture of the fault and further accumulation of strain energy until the cycle is repeated

electroneutrality – the requirement that the sum of the cations and the sum of the anions in an aqueous sample are equal in terms of equivalents per litre

eluviation – the light-coloured soil horizon that is distinguished by the removal of soluble minerals and ions through leaching leaving quartz as a residue

EM (electromagnetic) survey – geophysical survey method involving the measurement of electrical and magnetic fields induced in the subsurface but detected at the ground surface; used in mineral exploration, hydrogeological and salinity studies

entry pressure – in a two-phase system, the critical capillary pressure at which there is non-wetting phase penetration of water-wet pores

eolian deflation – the effect of wind action by which desert pavements develop from the erosion and transport of fine-grained soils

eolian deposits – sediments, mainly silt and fine sand, transported by suspension in air flows and then deposited as loess

epicentre – the location on the ground surface immediately above the focus or hypocentre of an earthquake

epikarst – surficial karst dissolution features

equilibrium reactions – at chemical equilibrium in the reversible system $A + B = C + D$, the forward reaction rate by which the products C and D are created is equal to the backward reaction rate in which the reactants A and B are re-created; in chemical thermodynamics, equilibrium indicates that no further change or reaction is needed because the temperature, pressure and chemical potential are everywhere the same throughout the system, and the Gibbs free energy is at a minimum

equivalent rock mass – the volume of rock containing numerous small aperture discontinuities yielding hydraulic-test properties equivalent to that of rock with a single well-defined aperture

erodibility – as defined by Briaud (2008), the ratio of the rate of erosion in m/s to the shear stress in the horizontal direction causing erosion

ERT (electrical resistance tomography) – a geophysical imaging technique that investigates the patterns of electrical resistivity by measuring the voltages arising from the application of an electrical current to the ground

esker – a subglacial sedimentary deposit comprising a linear or sinuous ridge of coarse materials

evaporite – a nonclastic sedimentary rock composed primarily of minerals that resulted from the extensive or total evaporation of a saline liquid

extensometer – an instrument to measure vertical and horizontal displacements in subsiding basins

extrusive rock – a rock formed by eruption of magma onto the Earth's surface

facies – term used in sedimentology and related fields to describe a sedimentary feature on the basis of texture or origin or lithology; also used in metamorphic petrology

factor of safety (FS) – in slope stability investigations, FS measures the ratio of the strength of the soil to the shear stress required to maintain equilibrium; thus where the shear strength exceeds the shear stress of a surface, the slope is stable and FS > 1.0

failure or strength envelope – that envelope of coordinates of σ and τ defined by experimentally obtained Mohr's circles below which conditions are stable and above which conditions lead to failure

fault – surface or narrow planar zone with displacement parallel to the surface, generally used for brittle materials

fault block – the rock mass along a fault that slipped because of rupture

fault zone – a series of closely spaced sub-parallel faults

feldspar – a major rock-forming silicate mineral group with compositions rich in K, Na, Ca, Al and Si

felsic rocks – light-coloured igneous rocks that contain feldspars and silica

ferric oxides – a family of oxides composed of ferric iron (Fe^{3+}), such as hematite and goethite

fissility – the property of shale to split into thin sheets

fissures – fluid-filled extension fractures that can be up to hundreds of metres in length but with no visible lateral displacement

flint – what archeologists call chert

fluid potential – the total head at a location in a flow system expressed as energy per unit mass

flysch – thick accumulations of alternating marine sandstones and mudstones

focus or hypocentre – the location in the subsurface where the fault initially ruptures

folds – a geologic structure formed by ductile deformation of essentially planar beds into curved surfaces

foliation – the property of a rock to break into thin sheets that are bounded by planes along which typically platy minerals are aligned; this alignment may result from sedimentation of clay minerals or from the growth of micas in a preferred orientation in metamorphic rocks

Foraminifera (forams) – single-celled microfossils of micrometre (micron) or centimetre size that are abundant in marine sediments and used for stratigraphic and radiometric dating

formation – a series of beds, perhaps of various rock types, that has been similarly deposited and which is defined as the smallest mappable unit

fossilized stresses – stresses created by either glacial loading or uplift and/or erosion of sediments from a sedimentary basin

fractionation – the small change in the isotopic ratios of an element, particularly that due to a change in temperature, caused by the greater vibrational energy of the lighter isotope

fracture – a sharp, narrow planar discontinuity that will conduct fluids, e.g., shear fracture, extension fracture

fracture (of mineral grains) – the irregularity of a mineral surface produced by breakage

fracture pressure – the fluid pressure at which hydraulic fracturing of a rock occurs

free phase (re NAPL) – the condition exhibited by a non-aqueous-phase liquid (NAPL) in which its fluid saturation exceeds the residual saturation of the pore volume; therefore the NAPL is potentially mobile and thus can drain vertically or migrate laterally under a suitable gradient

frost cracking – the action of freezing water expanding within microcracks in rocks causing sufficient strain to fracture the rock

frost heave – when freezing temperatures create expansion of soil water as ice, a suction potential is also created in soil such that water migrates to the horizontal ice lens being formed, thus promoting segregation of ice in silts and fine sand

frozen debris lobes – frozen masses of soil, rock, vegetation and ice that move down unstable slopes in permafrost terrain

fugitive gas – the upward migration of natural gas, released by leaking oil and gas wells, within a wellbore or a vertical fracture caused by its buoyancy with respect to water

gabbro – a plutonic igneous rock composed of approximately equal amounts of pyroxene (and olivine) and plagioclase; its volcanic equivalent is basalt

gangue minerals – the valueless rock or mineral aggregates in an ore that are not economically desirable but cannot be avoided in mining

gas hydrate – a crystalline compound in which the gas is trapped within the ice lattice of water

geological map – presentation in plan view of rock types that form outcrops or would be exposed as outcrops were surficial unconsolidated sediments and soils removed; the orientation of geological structures is indicated by symbols, topographic contours are shown and various geological features are identified such as strikes and dips of formations, faults, mine workings, etc.

geological strength index (GSI) – a rock-mass classification system for estimating the reduction in rock-mass strength for different geological conditions as identified by field observations

geomechanics classification system / rock-mass rating (RMR) – a classification system to characterize rock-mass deformability

geomodel – a conceptual model of technical information pertaining to the geological structure and geomorphology of a site for engineering investigation or construction purposes

geomorphic hazards – hazards to people and infrastructure arising from unstable slopes, particularly where associated with volcanic and seismic activity, heavy rains, snow and ice

geophysical processes – processes that cause a change in the Earth's crust through tectonic or geomorphic activity

geotechnical scour number – a measurement of fluvial erosion in terms of scour depth per unit stream power

glacial isostatic adjustment (GIA) – depression of the (asthenospheric) mantle and the crust of the Earth under the load imposed by ice sheets

glaciofluvial deposits – glacially derived sediments deposited in contact with or beneath ice sheets or by meltwater streams draining glaciers

glaciolacustrine deposits – glacially derived sediments deposited in proglacial lakes developed from ponded meltwater, i.e., adjacent to ice sheets

glaciomarine deposits – glacially derived sediments deposited on the continental shelf either adjacent to an ice sheet (proximal) or distant to an ice margin (distal)

glaciotectonic deformation – deformation caused by glacial migration producing both proglacial structures that are characterized by shear and compressional forces and subglacial shearing and extensional features

global meteoric water line – the relationship of the stable isotopic ratios of oxygen ($^{18}O/^{16}O$) to hydrogen ($^{2}H/^{1}H$) in water samples

gneiss – a metamorphic rock in which prominent layers are produced by variations in the abundance of minerals; often platy minerals, such as micas, alternate with more granular minerals, such as quartz and feldspar

gouge – a fine-grained material in fault zones formed by the grinding of rocks together along the fault

GPR (ground-penetrating radar) – a geophysical method of electromagnetic mapping, sometimes referred to as "georadar", that sends short pulses of radar signals generated by a transmitter mounted on a sledge or wheeled vehicle into shallow soil or rock and the reflections of which are detected on the same antenna as used in transmitting

GPS (global positioning system) – an American geodetic positioning system informed by an array of satellites providing radiolocation and radionavigation information; international systems are referred to as GNSS (global navigation satellite system)

graded – a layer in which mineral grains gradually change size and or density from bottom to top of the layer; in sedimentary rocks, graded beds result from settling of grains in turbidity currents

granodiorite – a plutonic igneous rock containing approximately one-third quartz and two-thirds feldspar, of which plagioclase is more abundant than alkali feldspar; it contains a higher percentage of mafic minerals than does granite and hence is slightly darker than granite

granular flow – the frictional sliding and rolling of particles that occur during deformation of rock, sand or soil

gravel bar – an architectural element indicating a fluvial deposit of predominantly gravel-size particles comprising sediment stored indefinitely in a stream channel

gravity (geophysical) method – field acquisition and interpretation of measured variations in the Earth's gravitational field arising from differences in the density of rocks and sediments

greywacke – sandstone containing more than 15% muddy matrix

ground model – a commonly used term of British origin to indicate a conceptual model of the geological structure and geomorphology of a site

ground motion – movements of the Earth's surface induced by earthquakes and measured as peak ground accelerations, velocities or displacements

groundwater discharge – seepage of groundwater to a stream or wetlands

groundwater flow model – the solution of the groundwater flow equation using a numerical simulator, such as the USGS's MODFLOW, to simulate the hydraulic-head distribution within a flow system given the distribution of hydraulic conductivity within that flow system

groundwater flow system – the flow net of hydraulic head in a groundwater basin illustrating the location of recharge zones in upland areas receiving precipitation input (high hydraulic head), transition zones and discharge zones, where seepage of groundwater to streams and wetlands occurs under conditions of low hydraulic head relative to the recharge zone; small-scale local flow systems may exist within larger-scale regional systems; the flow system will become altered by the presence of extraction wells that will create local discharge areas

groundwater recharge – infiltration of rainfall or snowmelt into a flow system

groundwater vulnerability – a measure of the potential for groundwater to become contaminated by infiltration of wastes or recharge of poor-quality groundwater

group – an assemblage of geological formations with common features that are named for purposes of identification

GSI – *see* geological strength index

GWQ – shorthand term used in this textbook for groundwater quality, irrespective of its state of contamination

gypsum – a white mineral ($CaSO_4 \cdot 2H_2O$) that is precipitated from evaporating seawater

halite – a white evaporite mineral ($NaCl$)

hand specimen – a piece of mineral or rock that can be held in the hand for evaluation of its macroscopic properties

hardness – the resistance of a mineral to scratching by a sharp point, edge, or other mineral; a hard mineral is one with a hardness greater than 7

hard rocks – a common, if rather unscientific, term for igneous and metamorphic rocks

hiatus – an interruption in a stratigraphic sequence indicating the cessation of deposition or the erosion of rock such that the stratigraphic record is discontinuous

hornfels – a contact metamorphic rock lacking foliation

hue – soil colour

humus – decomposed organic matter in soil

hydraulic conductivity – if a porous medium is isotropic and the fluid is homogeneous, the hydraulic conductivity of the medium is the volume of water at the existing kinematic viscosity that will move in unit time under a unit hydraulic gradient through a unit area measured at right angles to the direction of flow (USGS)

hydraulic diffusivity – the hydraulic conductivity of the saturated porous medium when the unit volume of water moving is that involved in changing the head a unit amount in a unit volume of medium, i.e., T/S or K/s_s

hydraulic fracturing – a natural or human-induced process by which the fluid pressure in a finite zone of rock causes a fracture to be induced perpendicular to the minimum principal stress

hydraulic geometry – a set of empirical equations relating the geometry of a stream channel (width, depth, slope) and flow velocity with its downstream discharge

hydraulic gradient – the change in static head per unit of distance in a given direction

hydraulic head – the total head of water at a given point, being the sum of elevation head, pressure head and velocity head

hydraulic testing – analysis of volumes of soil and rock under controlled conditions to determine hydraulic parameters, e.g., K, T, S

hydrochemical facies – the major ions measured in groundwaters from certain formations (or zones in formations) identifying a particular groundwater quality

hydrofacies – a sedimentary feature identified on the basis of its limited ranges of hydraulic conductivity and grain size

hydrophilic compounds – ions or molecules with high aqueous solubility because their polarity allows them to readily bond with the water molecule

hydrophobic compounds – ions or molecules with low aqueous solubility because they lack functional groups permitting them to bond with the water molecule

hydrostratigraphy – the characterization of a stratigraphic column by which zones of high and low hydraulic conductivity are identified

hydrothermal brine – a hot aqueous fluid of high salinity generated with magma that is commonly the source of mineral deposits

hydrothermal ore deposits – ore deposits formed by hydrothermal brines either in areas of magmatic activity, such as at plate boundaries, or in regions undergoing magmatic and tectonic activity, or in sedimentary basins that have been invaded by brines after migrating over great distances

hypogenic karst – deep karst dissolution features

hysteresis – in the context of soil physics, where a porous medium exhibits different wetting and drying pathways such that the hydraulic conductivity and water content are not single-valued functions of pressure head (suction)

iceberg scouring – the erosion and furrowing of seafloor or lake-bottom sediments by the keels of icebergs that are taller than the depth of water into which they have drifted

igneous rocks – rocks formed by the solidification of molten rock (i.e., magma)

ignimbrites – rock formed from ash flows

immiscible contaminants – liquids that do not dissolve into the solution they come into contact with, resulting in an interface; in many cases of groundwater contamination, the dissolution of an immiscible contaminant is slight but may exceed allowable drinking-water concentrations

inclinometers – instruments emplaced on slopes to measure lateral displacement and therefore indicate slope stability

index tests – a set of basic geomechanical tests that measure the strength and durability of rock specimens

INSAR – interferometric synthetic aperture radar, which is used to measure displacement in the ground surface from repeat mapping

intact rock – rock that contains no through-going fractures

interfacial tension – tension exerted on a surface, which acts perpendicular to and inwards from the surface, tending to cause the area of the interface to decrease

interglacial event – a period of time, such as today, when ice sheets have receded from their maximum extent and are confined to polar and mountainous regions

interseismic elastic strain – the strain deformation that accumulates between earthquakes and which is dissipated by the coseismic strain release and displacement in an equal and opposite manner

interstadial event – a period of time during which there is somewhat greater warmth than during a glacial cycle but not as significant as that occurring during an interglacial

intraplate earthquakes – earthquakes that occur within the boundaries of a continental plate rather than at plate boundaries

intrusive rock – the intrusion of magma or salt into an existing rock and its subsequent lithification

ion activity product – the product of the activity of two ions that may be compared to the solubility product to determine whether the compound is saturated or unsaturated in the aqueous solution

ionic strength – in chemical thermodynamics, the ionic strength is the measure of the total effective ionic concentration in a solution

ironstone – an iron-rich sedimentary rock

irreversible reactions – chemical reactions that, at standard temperatures and pressures, are not equilibrium reactions; consequently the products are not convertible back into the reagents

isomorphous substitution – refers to the extent of atomic (or ionic) substitution in a specific site of a crystal structure

isostasy – the state of buoyant equilibrium between masses of rock with different densities; lower-density continental rocks, for example, float higher than denser oceanic rocks in the asthenosphere

joint – a rock fracture with a small displacement or opening normal to the fracture plane but no visible displacement along the fracture plane

jökulhlaups – an Icelandic term for the sudden and calamitous outburst of water from a glacier during volcanic eruption; more generally, it is now used to indicate the sudden outflow of glacial meltwater from an ice-dammed lake or subglacial reservoir causing severe flooding

kame – a stratified meltwater deposit formed in a river flowing along the side of a glacier or in a crevasse

kaolinite – a simple clay mineral having a single tetrahedral and single octahedral layer and which exhibits neither plastic nor expandable properties

known unknowns – those processes that can be reasonably anticipated but not quantified based on past experience as exemplified by case histories

lahar – a debris flow or mudflow initiated by volcanic eruption

laminated sand – an architectural element of sandy sheet deposits in which the laminations may be due to textural differences or to the presence of alternating bands of dark and light minerals and are usually <1 cm

landfill leachate – contaminated water produced by the dissolution of components within municipal solid waste exposed to rainfall or snowmelt

Last Glacial Maximum (LGM) – the time of occurrence of the most recent maximum geographic extent of the continental ice sheets

lateral accretion – a fluvial sedimentary deposit characterized by low-angle cross-bedding of sand

lateral spread – the flow of sandy soils into gullies and rivers induced by liquefaction during an earthquake

laterite – a reddish-brown soil rich in hydrous Fe, Mn and Al oxides derived from the intense weathering of granite in humid tropical climates

lateritic weathering profiles – the intense weathering of Al-rich source rocks, such as granites, to produce laterites

law of faunal succession – fossil organisms (fauna and flora) succeed one another in a definite and recognizable order, such that the relative age of rocks can be determined from their fossil content

law of mass action – the basic principle of aqueous chemical equilibrium by which the rate of a reaction is directly proportional to the concentration of each reacting substance

law of original continuity – strata deposited by water flow continue laterally in all directions until they thin or pinch out as a result of lack of sediment

law of original horizontality – water-laid sediments are deposited in strata that are horizontal and parallel to the Earth's surface

law of superposition – in any sequence of sedimentary strata that has not been overturned, the youngest stratum is at the top and the oldest is at the bottom; each stratum or bed is younger than the bed beneath it but older than the bed above it

LiDAR – light detection and ranging, i.e., terrestrial laser scanning

limestone – sedimentary rock composed principally of calcium carbonate with minor amounts of clay minerals, chert, quartz, pyrite and dolomite

liquefaction – the loss of shear strength of a cohesionless soil during seismic (cyclic) loading

liquefaction susceptibility – the likelihood of saturated cohesionless sediments to be susceptible to liquefaction based on their Quaternary age

lithic grains – particle fragments in sediments derived from previously formed rocks

lithic rocks – sedimentary rocks containing fragments of previously formed rocks

lithification – the process by which unconsolidated sediment is converted into sedimentary rock

lithified – the compaction, cementation and recrystallization of unconsolidated sediment as it is converted into rock

lithofacies – the characterization of a rock on the basis of its textural properties

LNAPL – light non-aqueous-phase liquid, such as hydrocarbon fuels

load–deformation curve – the geomechanical response of a rock specimen by deformation (strain) from an imposed load (stress)

longwall mining – the mining process by which a cutting machine works along the length of a coal face (the longwall) in a supported area known as a panel

macropores – large interconnected pore systems associated with fractures or root holes but comprising only a small fraction of the total porosity

mafic rocks – dark-coloured igneous rocks that contain magnesium and ferric iron minerals

magma – molten rock; most magma is not totally molten but contains some solids that are carried along in the liquid

magnetic method – a geophysical method used to identify ferrous metal objects such as abandoned oil and gas wells or subsurface pipes and in mineral exploration and structural geology

mantle – that part of the Earth lying between the Mohorovičić discontinuity and the core–mantle boundary; it is divided into the upper and lower mantle by the seismic discontinuity at a depth of ∼660 km

marble – a rock formed from the metamorphism of limestone and composed predominantly of calcite, but it may also contain dolomite

marl – a muddy limestone containing a significant proportion of clay

massive ground ice – metre-thick ice zones within shallow permafrost regolith and soils

MASW – multi-channel analysis of surface waves; a geophysical method for investigating the shallow subsurface (down to a few tens of metres) through the analysis of low-frequency (1–30 Hz) seismic surface waves

matrix-supported – an adjective indicating that sedimentary detrital particles, such as pebbles in a conglomerate, do not touch one another but are supported by a finer-grained matrix that surrounds them; *see* clast-supported

member – a formal rock unit identified within a formation that has specific characteristics

metamorphic rocks – any rock that has been changed in texture, mineralogy or chemical composition as a result of changes in environmental factors such as temperature, pressure, directed stress or shear, and composition of fluids

metamorphism – the change that takes place in rocks either in texture, mineralogy or chemical composition as a result of changes in temperature, pressure, or composition of fluids in the environment

meteoric water – a misleading term that has nothing to do with meteors, but rather indicates water that has been part of the hydrologic cycle in the recent past

mica – a family of sheet silicate minerals, such as biotite and muscovite

micro-granite – granitic rock with crystals that can be seen only with the aid of a microscope

microgravity – a geophysical tool used in karst to identify voids

microstructure – structures that range in size from the atomic scale to that of grain aggregates, observable with the aid of a microscope

mid-ocean ridge – the topographic high produced by the creation of new hot oceanic crust at a divergent plate boundary; ridges are often located near the centre of an ocean, as in the case of the Mid-Atlantic Ridge, but they need not be, as in the case of the East Pacific Rise

migmatite – a mixed metamorphic and igneous rock formed when the rock was raised to a temperature at which partial melting occurred; the melt tends to be granitic in composition and forms light-coloured layers in the darker refractory metamorphic rock

Milankovitch cycles – cyclical climate changes caused by variations in the Earth's orbital eccentricity about the Sun (~100 000 years), and the tilt (~40 000 years) and precession (19 000–23 000 years) of its axis

mineralogy – the scientific study of naturally occurring solids with an ordered atomic arrangement and a definite (but commonly not fixed) chemical composition, and of inorganic origin

mine water – groundwater that has invaded coal mines and acquired deleterious chemical characteristics associated with sulfide mineral oxidation

miscible contaminants – compounds that readily dissolve into the solution they come into contact with

mixed tide – a tidal regime where both the diurnal and semi-diurnal components are significant

modified Mercalli intensity scale – a standard measure of earthquake intensity

Mohorovičić discontinuity – the seismic discontinuity defining the base of the Earth's crust; the sharp increase in seismic velocities across this boundary is believed to result from a change from crustal rocks to mantle peridotite (also known as Moho or M discontinuity)

Mohr–Coulomb strength criterion – variation of peak shear strength of a rock specimen under confining conditions described by a linear relationship between normal and shear stresses

Moh's scale of hardness – a relative hardness scale of 1–10, based on the hardness of ten common minerals used to evaluate a mineral's hardness

moment magnitude – the standard measure of earthquake magnitude, which is a function of the seismic moment and is dimensionless

moraine – ridge or mound of glacial debris, largely till or possibly of lacustrine, fluvial or marine origin

MSL – mean sea level

mylonite – an extremely fine-grained metamorphic rock formed by recrystallization of rock in a fault zone

mudstone – a calcareous sedimentary rock composed of more than 90% mud-size particles (<0.0625 mm)

NAPL – non-aqueous-phase liquid, i.e., an immiscible liquid

neotectonics – a poorly defined term indicating tectonic processes occurring in recent geologic history

Newmark displacement – a method to compute the coseismic displacement of a landslide estimated from that part of the acceleration–time history of a design earthquake that exceeds the threshold acceleration required to overcome frictional resistance and commence sliding

nonconformity – an unconformity where sedimentary beds were deposited on basement rocks

normal fault – a fault in which the hanging wall (upper block) moves downwards relative to the footwall; normal faults result from crustal extension

obsidian – a glassy volcanic rock, usually of granitic composition

offset beds – beds that are displaced across a strike-slip fault

olivine – a green silicate mineral of igneous origin rich in magnesium and iron

ore deposits – a concentration of ore minerals that can be extracted and sold for a profit

ore mineral – that part of an ore, or ore deposit, usually metallic, that is economically desirable, as contrasted with the gangue

orthoquartzite – a clastic sedimentary rock composed of pure quartz sandstone

outcrop – part of geologic formation that is exposed and visible at the ground surface

overbank fines – an architectural element of silts and clays that have been transported in suspension from a stream channel into the adjacent flood plain by overbank flow during flood events

overcoring – strain relaxation method by isolating a drill core section *in situ*, the expansion of which is then measured to estimate stress

overland flow – rainfall runoff that occurs under conditions when the underlying soil is fully saturated or infiltration is otherwise inhibited, e.g., by trapped air

overpressured systems – fluid pressure profiles with depth in sedimentary rocks that exhibit pore pressures greater than hydrostatic

oxidation reaction – the removal of an electron from an atom or group of atoms resulting in an increase in the valence of the oxidized species

oxyanions – anions containing an oxygen functional group responsible for the negative charge

P waves – primary seismic waves that are transmitted in the direction of wave propagation as body waves by compressional and dilational effects on adjacent rock particles

paleokarst – surfaces or cavities in soluble rocks that are inert with respect to further growth in void volume because of deep burial in a sedimentary column and the

groundwater has become saturated with respect to carbonate and/or sulfate minerals

paleoseismology – the study of prehistoric earthquakes, in particular their location, timing and size

paleostage indicator – a historically high stage of a river whose evidence is recorded by one or more geomorphic indicators

paleotempestology – the study of the sandy sediments deposited in coastal lakes and marshes behind the dunes that might provide evidence of great prehistoric hurricanes

Patton's principle – the effect of an increase in the friction angle of a rough surface over that of an equivalent clean fracture surface on the shear strength of that surface

peak horizontal acceleration – a measure of ground motion of the Earth's surface induced by an earthquake indicating the maximum horizontal acceleration at a station

peatland – wetland composed of clastic sediments and organic matter, in particular those formed at sea level in coastal plains and deltas

pegmatite – an exceptionally coarse-grained igneous rock, usually of granitic composition and forming irregular dikes and lenses

percussion drilling – drilling with a pneumatic hammer tool, typically used for hard rock and boulder-rich sediments

peridotite – a plutonic igneous rock composed essentially of olivine and pyroxene

periglacial – geomorphological features indicating freeze–thaw processes in non-glacial alpine areas and at the margins of glaciers, which may have long retreated from the area

permafrost – soil or rock that has remained frozen ($\leq 0^\circ$C) for at least two years

permeability – the relative ease with which a porous medium can transmit a liquid under a potential gradient and is a property of the medium alone and is independent of the nature of the liquid and of the force field causing movement; also referred to as intrinsic or absolute permeability; secondary permeability refers to post-depositional changes in sediments and sedimentary rocks that enhance permeability or, in the case of hard rocks, fracturing that enhances permeability

petrography – the description and classification of rocks by examination of thin sections of rock with a petrographic microscope

petrology – the origin, occurrence, structure and history of rocks

petrophysical testing – measurements typically conducted for the oil industry that measure porosity, permeability and fluid saturations in reservoir rocks and cap rocks

phenocryst – crystals in igneous rocks that are significantly larger than most of the crystals in the rock; igneous rocks containing phenocrysts are described as porphyritic

phyllite – a metamorphic rock with a prominent foliation; the foliation surfaces are less regular than in slate and have a sheen; individual mineral grains on the foliation plane are not visible to the unaided eye

pillar collapse – pillar collapse occurs in mines when they no longer are capable of providing the bearing capacity of the excavation

pitchstone – a volcanic glass similar to metallurgical slag and containing numerous phenocrysts

planar cleavage – the breakage of a mineral along a single planar direction, as shown by mica

plane failure – failure of a rock slope occurring when a discontinuity has a strike that is parallel to the rock slope and a dip angle that is greater than the angle of friction of the discontinuity

plate tectonics – the deformation of the Earth as a result of the motion of lithospheric plates

ploughing – erosion of cavities in till produced by glacial abrasion; in the case of bedrock, the abrasion is referred to as quarrying

plucking – the erosion of a streambed by hydraulic forces; also quarrying

plunge – the inclination of the crest of a fold

plutonism – the process by which an igneous body or rock is formed at depth in the Earth; most plutonic rocks are relatively coarse-grained

point bar – fluvial deposit formed on the inside of a growing meander bend by lateral accretion

point-load strength testing – a simple strength test by which a rock specimen is compressed to indicate the tensile stress at failure

Poisson effect – the stress readjustment resulting from the erosion of a vertical component of rock

Poisson's ratio – the ratio of lateral to axial strain in an elastic medium

pop-ups – anticlinal pop-ups occurring on the floor of a quarry caused by unloading of the overlying rock or blast-induced fractures

pore ice – ice present in the pores of soils

pore pressure – the hydrostatic stress exerted in all directions by water present in soil pores or rock fractures

porosity – the fraction of void space within a porous medium (rock, sediment or soil) per unit volume

preconsolidation stress – the maximum previous effective stress that unconsolidated sediments have been subject to resulting in compaction

predisposing factors – factors that cause slopes to become more susceptible to failure by landslide

probable maximum flood – the peak stream discharge from a probable maximum rainfall event

provenance – the source area in which a particular material originates, irrespective of whether the material is sediment derived from distant bedrock outcrops or water that infiltrated into a groundwater flow system in a distant recharge area

pyrite – iron sulfide (FeS_2), the oxidation of which generates acid waters

pyroxene – a major rock-forming silicate mineral group with infinitely extending chains of tetrahedral $(SiO_4)^{4-}$ groups and compositions rich in Mg, Fe, Ca, Al and Si

Q method – a rock-mass classification system developed to describe the peak shear strength of tension fractures, particularly used in tunnel support applications; it is also known as the NGI system or the rock-mass quality rating system

quarrying – erosion of till or bedrock by glacial or hydraulic processes

quartz – crystalline silica (SiO_2), a hard and chemically resistant mineral

Quaternary period – the period of the Cenozoic Era beginning 2.6 Ma comprising the Pleistocene and Holocene epochs

quick clays – sensitive clays that quickly lose their shear strength when remoulded

radiometric methods – methods of age dating based on the radioactive decay of natural isotopes found in certain minerals that comprise these rocks and which allow dating through the growth of daughter isotopes, in particular the decay of uranium isotopes to stable lead isotopes

rainout – the isotopic depletion of an atmospheric vapour mass through condensation such that the resulting rainfall is enriched in ^{18}O and deuterium

Raoult's law – in terms of gaseous systems, the partial pressure of solvent vapour in equilibrium with a dilute solution is the product of the vapour pressure of the pure solvent times the mole fraction of the solvent in the solution; similarly, the effective solubility of an NAPL

$(C_{e,i})$ is the product of the mole fraction (X_i) and its aqueous solubility $(C_{s,i})$

recurrence interval – the average number of years within which a given event will be equalled or exceeded (of an earthquake, flood, etc.); also known as the return period

redox boundary – that part of a groundwater flow system where a particular redox reaction creates a change in groundwater quality, e.g., reduction of dissolved oxygen

redox reaction – a chemical reaction in which an oxidant consumes one or more electrons to become reduced while simultaneously a reductant is oxidized by removal of one or more electrons; each redox reaction requires that one species is oxidized while another is reduced

reduction reaction – the addition of electrons to an atom or group of atoms resulting in a decrease in the valence of the reduced species

regional metamorphism – metamorphism that takes place on a regional scale as a result of increasing temperature and pressure on a regional scale; it is usually associated with deformation related to tectonic plate convergence

regolith – the unconsolidated material above bedrock

relative permeability – the reduced permeability of one phase within a porous medium relative to the (absolute) permeability because of the presence of one or two other phases (gas, oil, water) within the pores

relict structural features – those features in weathered rock that are inherited from the unweathered bedrock, e.g., faults, foliation, bedding planes and joints

reopening pressure – the fluid pressure exerted within an isolated section of borehole that causes a previously induced hydraulic fracture to reopen but which is less than the fluid pressure required for fracturing

representative elementary volume (REV) – the volumetric range of soil or rock that will provide a reliable average parameter estimate such that this average at a point within this range should remain, more or less, constant

reserves and resources – reserves of ore or oil and gas are those volumes of rock or hydrocarbon, respectively, that are extractable on the basis of current economic, engineering and regulatory conditions; resources are ore bodies or estimates of oil and gas volumes that have the potential for future extraction

reservoir rock – a sedimentary rock with high porosity in which oil and natural gas can accumulate and is sufficiently permeable to allow economic extraction

residual saturation – the percentage or fraction of pore space containing a particular liquid after it has been allowed to drain; the liquid is thus retained by capillary forces

reverse fault – a fault in which the hanging wall moves up the fault plane relative to the footwall

Reynolds number (*Re*) – a numerical ratio between inertial forces and viscous forces in a fluid, which allows one to predict whether flow will be laminar or turbulent

rhyolite – a volcanic rock of granitic composition; it is often glassy, and when completely glassy is called obsidian

rippability – the mechanical ability to use machines to excavate rock without having to use explosives to fragment the rock beforehand

rock fall – materials detached from an elevated rock face because of its failure and deposited beneath the outcrop

rock-forming mineral – a mineral that is a common constituent of various rock types found in the Earth's crust

rockhead – the top elevation of unweathered or only slightly weathered rock above which is soil-like material, i.e., the "top-of-rock"

rock mass – the volume of rock that is the subject of engineering investigation, design and construction and which comprises fractures and intact rock

rock-mass rating (RMR) – a rock-mass or geomechanics classification system to characterize rock-mass deformability

rock material – intact rock that may be used in laboratory testing to determine geomechanical properties but which cannot represent the effects of fractures on those properties when they are present in the rock mass

rotary drilling – drilling by rotary action of a drill bit with the aid of a drilling fluid ("mud") that recovers the milled rock particles from downhole and which is recycled into and out of the borehole via a mud pit

rotational slide – slope failure in which the failure surface is curved and concave upwards and the uppermost displaced material undergoes some backward tilting during sliding

runout distance – the distance of travel of displaced slope material from a landslide as measured along the receiving surface at the base of the slope

S waves – shear or secondary seismic waves that are transmitted as body waves with the adjacent rock particles being displaced perpendicular to the direction of wave propagation

sand bedform – an architectural element of fluvial sediments that includes channel fills and crevasse splay deposits

sand boils – sand geysers observed at the ground surface indicating liquefaction events

sandstone – a sedimentary rock composed predominantly of sand-sized grains (0.0625–2.0 mm), which are usually made of quartz

saprock – weathered rock that retains the original rock fabric, but crumbles by hand to individual grains often referred to as grus

saprolite – highly weathered rock in which more than 20% of weatherable minerals (biotite, feldspars, olivine, hornblende, etc.) have altered to clay minerals, while the original rock fabric is retained

saturated hydraulic conductivity – a term used in soil physics and unsaturated-zone soil mechanics, where flow under unsaturated conditions is considered, to indicate that the porous medium is fully water-saturated; therefore, the saturated hydraulic conductivity is the hydraulic conductivity under conditions of atmospheric pressure and an absence of soil suction

saturation index – the indicator of chemical equilibrium between a mineral and its pore water in which SI = 0.0; SI > 0.0 for oversaturated solutions, and SI < 0.0 for undersaturated solutions

scale – the quantitative relationship between distance on a map with actual distance on the surface of level ground

scale effect – the variation in a parameter as the scale of measurement increases

scan line – a reference trace line on a surface of interest allowing measurements to be taken of features along that axis

scarp, main – a steep surface along which slope failure has occurred by displacement downslope and which indicates ground above which there has been no displacement

schist and schistosity – a metamorphic rock in which medium- to coarse-grained platy minerals, such as micas, or needlelike minerals, such as amphiboles, create a prominent foliation or schistosity

scour critical – that scour depth which, if subjected to the design flood, would sufficiently erode the foundation of a bridge pier so that the design load of the bridge would not be supported

seafloor spreading – spreading of the oceanic crust associated with divergence from a mid-ocean ridge

sea-level rise (SLR) – the rise in the mean sea level associated with long-term phenomena, such as climatic warming from the melting of continental ice sheets and the warming of oceans

seawater intrusion – the intrusion of seawater into coastal aquifers whether under natural conditions or induced by groundwater extraction near the coastline

sedimentary rocks – a sedimentary layer that is distinguishable from layers above and below it on the basis of rock type, grain size or physical property

sediment gravity flow – an architectural element indicating sands and gravels deposit by gravity displacement

sediment replenishment – replenishment of deltas by sediment deposited on the delta by discharging rivers

segregation ice – ice that formed as thin horizontal lenses in soils and which are associated with frost heave

seismic faulting – the rupture and slippage of faults that is associated with seismic events

seismic hazard analysis – probabilistic analysis of those coseismic effects posing a hazard to humans and infrastructure, e.g., tsunamis, landslides, liquefaction and ground shaking

seismic moment – the product of area of faulting times the rigidity modulus of the rock times the displacement of the fault in newton metres

seismic refraction and reflection – geophysical test methods using induced seismic waves to determine geological structural and stratigraphic features, rippability, rock strength, etc.

seismogenic zone – that depth in the Earth's crust of 6–12 km where earthquakes originate most commonly

seismology – the science of earthquake generation and seismic wave propagation through the Earth

semi-diurnal tide – the tidal change of sea level twice in a lunar day

sensitivity (of clays) – the property of clays indicating their loss of shear strength when remoulded, defined by the ratio of the specimen's undisturbed to remoulded shear strengths

serpentinite – a rock composed primarily of the mineral serpentine formed by the hydrothermal alteration of ultramafic rocks

settling velocity – the rate by which a sediment particle is deposited in still clear water at a specified temperature

shale – an extremely fine-grained (<0.004 mm) sedimentary rock that is characterized by a prominent foliation or fissility parallel to the bedding

shear fracture – discontinuity in which there is displacement of one side relative to the other along the trace of the fracture caused by shear

shear strength – the peak shear stress of a rock or soil specimen

shear testing – measurement of shear strength of a material using a shear box, in which the normal stress is maintained constant

shear zone – the tabular deformed zone of rock associated with shear strain and ductile deformation exhibiting such features as parallel fractures and crushed rock

Shields parameter – the critical shear stress at which the bed-load of a stream begins to be displaced and suspension of sediment occurs

siderite – the mineral ferrous carbonate: $FeCO_3$

silicate mineral – mineral with the anionic $(SiO_4)^{4-}$ tetrahedral group, as in olivine, $(Mg,Fe)_2SiO_4$

siliciclastic sediment – sediment formed from the solid weathering products of rocks, which consist primarily of silicate minerals

sill – a sheet-like intrusive igneous body that parallels the layering in the intruded rock; most large sills have shallow dips

sinkhole – a circular depression in soluble carbonate or sulfate rock that has been created by dissolution and collapse processes and may be filled with water

site characterization model – a preliminary model of the geological structure and geomorphology of a site based on reconnaissance mapping and data interpretation; *see also* ground model and geomodel

slackwater flood deposits – the primary indicator of a historically high stage of a river assumed to represent deposits formed in eddies at the mouths of tributaries of rivers that caused sedimentation of sands and silts

slake durability testing – destructive tests conducted on weathered or clay-rich rocks to determine their rate of disaggregation and hence their durability

slaking – the failure of shale or mudstone during weathering when the rock undergoes wetting-induced disintegration

slate – a fine-grained metamorphic rock characterized by a prominent foliation resulting from the preferred orientation of platy minerals such as muscovite and chlorite

slaty cleavage – the extremely planar cleavage that is characteristic of slate and is caused by the preferred alignment of micaceous minerals; the cleavage typically parallels the axial plane of folds

slickensides – the polished and striated surface of a fault plane caused by frictional sliding

slip (of a fault) – the shear motion along a fault

slip or failure surface – the surface of rupture of a landslide that forms the lower boundary of displaced material below the original ground surface

slip rate (of a fault) – the rate at which offset occurs along a fault

slug tests – a hydraulic test of soil or rock to determine the local hydraulic conductivity adjacent to a well screen by injection of a slug of water, air or a metal cylinder below the water level within the well

smectite – a clay mineral with a 2 : 1 layer structure, i.e. two tetrahedral layers and a single octahedral layer

soft rocks – a common, if rather unscientific, term for sedimentary rocks

softcopy photogrammetry – a technique of terrain analysis employing photogrammetric imaging and specialized 3D glasses to view terrain images on computer monitors

soil – in engineering geology, soil is all unconsolidated material above the bedrock; in soil science, it refers to the natural medium for growth of land plants

soil behaviour type index – an index used in interpreting cone penetrometer test (CPT) results to indicate the soil texture

solifluction – the slow downslope sludging over a frozen substratum of the surface soil of arctic and alpine regions, rendered liquid during the annual thaw by its high water content and disaggregated clay particles

solubility (mineral) – the concentration of a solute in an aqueous solution containing a second phase, such as a solid or a gas, in equilibrium with the solution; it is understood that the second phase is the source of the solute

soluble rock – rock that readily dissolves in contact with water, e.g., carbonate, sulfate and halide rocks

solutes – ions or molecules that have dissolved in a solution such as water

sonic drilling – a modern drilling technology in which a borehole is drilled and cased by rotating and vibrating the drill rods and casing at sonic resonance frequencies causing the drill bit to vibrate up and down and rotate

sorption, sorbent, sorbate – sorption is the removal of dissolved ions or molecules from an aqueous or gaseous system onto a mineral or organic solid-phase sorbent; the sorbate is the dissolved ion or organic molecule that is sorbed by the sorbent; the terms imply nothing about the reversibility of the reaction

sorting – the degree of uniformity of grain size in a sediment sample; grains that have a small range of particle size are well sorted or, in soil mechanics terms, poorly graded

specific discharge – the rate of discharge of groundwater per unit area of the porous medium measured at right angles to the direction of flow

specific retention – the ratio of (i) the volume of water which the rock or soil, having been saturated beforehand, will retain following gravity drainage to (ii) the volume of rock or soil

specific storage – the volume of water released from or taken into storage per unit volume of the porous medium per unit change in head

specific yield – the ratio of (i) the volume of water which the rock or soil, having been saturated beforehand, will release by gravity drainage to (ii) the volume of rock or soil

spherical projection – the first step in the process of stereographic projection

standard penetration test – the number of hammer blows required to drive a drill rod with an attached soil sampler 150 mm or 6 inches into soil (or weak rock); this N value yields information on the geotechnical engineering properties of the soil

stereogram – a circular diagram that depicts the projected positions of perpendiculars to planes

stereographic projection – a projection procedure in which point locations in the upper (and lower) hemisphere are projected onto the equatorial plane by viewing from the South Pole (or North Pole)

stick–slip process – a description of an earthquake cycle dominated by friction and sliding with each cycle beginning with slippage followed by strain accumulation and then further slippage

stiffness – the measure of a material's resistance to strain as indicated by the ratio of the force applied to a specimen to its extension and is usually expressed by Young's modulus

stillwater level – the sum of the tide, surge and any seasonal or inter-annual component of mean sea level

stochastic method – with respect to the prediction of ground motion induced by seismicity, the stochastic method describes a probabilistic description of a seismic time series, such as peak horizontal accelerations at a station, and an estimation of the ground motions that have not been recorded but might be reasonably predicted to occur

stockwork – a three-dimensional lattice-like network of closely spaced veins containing secondary minerals often associated with mineral deposits

storativity – the volume of water an aquifer releases or takes into storage per unit surface area of the aquifer per unit change in head; also known as the storage coefficient

storm surge – the changes in sea level in coastal waters caused by air pressures and winds acting on the sea surface

strain – in geomechanics, it is the deformation of rock relative to its original dimension caused by rigid-body translation or rotation or by changes in the internal configuration of the rock; in structural geology, strain is limited to the third cause, i.e., particles within a rock change positions relative to each other

strata-bound ore deposits – deposits in volcanic and sedimentary rocks that originated as hydrothermal fluids and often migrated as high-salinity waters

stratum (*plural* **strata**) – a layer or bed of a particular sedimentary formation distinguishable from adjacent layers on account of texture, colour or other attribute

stream power – the rate at which sediment is eroded or transported, i.e., the rate at which fluvial work is performed

strength – the stress necessary to cause an undisturbed sample of soil or rock to fail

strike – the azimuth, or compass reading relative to due north, of a horizontal line on a dipping surface

strike-slip fault – a fault with slip that has occurred along the direction of the fault strike

structural geology – the branch of geology that is concerned with the geometry, spatial distribution and formation of geologic structures and their deformation, in particular faults and folds in rocks; the scale studied is smaller than tectonics

structure – in describing rocks, it is used for features that are on a larger scale than the relations between individual mineral grains, i.e., texture

stylolite – a dark irregular line in limestone formed where insoluble residue has accumulated on a surface where solution of the rock has taken place

subduction – the process by which a lithospheric plate is consumed by its descent beneath another (typically continental) plate

subglacial deposits – deposits formed beneath a glacier, including till and glaciofluvial deposits, such as eskers

subsidence (coastal) – subsidence of the coastline relative to sea level caused by (i) groundwater or hydrocarbon extraction or (ii) sediment compaction without compensating replenishment or (iii) glacial isostatic adjustment

subsidence troughs (or bowls) – topographic depressions observed at the ground surface caused by mining of groundwater, hydrocarbons or ore beneath the depression

subsurface storm flow – precipitation that infiltrates and migrates towards streams by flowing principally through the unsaturated zone, although flow itself may be along pathways that are temporarily saturated

supergene ore deposits – mineral deposits formed at the surface under oxidizing weathering conditions

supraglacial deposits – melt-out and flow tills from sediment entrained by the glacier as well as kames and moraines deposited at the glacier's margin

surface of rupture – the surface of rupture of a landslide that forms the lower boundary of displaced material below the original ground surface; also referred to as the slip surface

surface waves – Rayleigh and Love coseismic waves that are confined to travel along the Earth's surface and are associated with much shaking damage

suspended load – sediment entrainment in the flowing water (or air)

suspended bed-material load – bed sediments that become entrained in flowing water

swash – the temporary ebb and flow of the waves on a beach

syenite – a plutonic igneous rock composed primarily of alkali feldspar

taliks – unfrozen zones within permafrost terrain

tectonic forces – regional-scale processes that create a set of geological structures and deformation patterns

tectonics – the Earth's regional-scale structural and deformational features and evolution and the processes that developed them

tensile failure – the occurrence of extension fractures formed under tension

tensile strength – the magnitude of tensile stress that a rock can withstand before failing

terrain analysis – the interpretation of stereographic images of landscapes to undertake site assessment or to determine the potential for damage arising from natural and man-made geohazards

terrigeneous sediments – detrital or clastic mineral grains derived from the weathering of continental rocks

thermokarst – the hummocky terrain produced by subsidence following the melting of ground ice

thrust fault – a reverse fault with a dip $<45°$

tidal period – the time between a tidal high water level and the next high, or between successive low water levels

tidal range – the difference between low and high tidal water levels; equal to twice the tidal amplitude

till – a dense, overconsolidated sediment deposited beneath a glacier and often exhibiting an internal fabric aligned in the direction of ice movement

toe of landslide – the furthest extent of the displaced slide material

toppling – a mode of failure of rock slopes that occurs when two sets of fracture planes dip into the slope of the rock mass

transform fault – a fault occurring where two plates slide past one another

translational slide – a landslide that migrates along a planar or undulating surface

transmissivity – the rate at which water at prevailing kinematic viscosity is transmitted through a unit width of the aquifer under a unit hydraulic gradient

travertine – a rock formed where springs discharge warm waters that are saturated with calcium carbonate

triaxial compression test – a test of compressive strength of a specimen when the axial stress applied is increased to failure as the horizontal stresses are maintained constant

triggering mechanism – the natural process, whether hydrologic, volcanic or seismic, that initiates a landslide

trough – the lowest elevation on the lower surface of a particular bed in a vertical cross-section of a syncline

tsunami – a series of long waves propagated across an ocean as a result of seismic faulting or a landslide on the seabed or into the sea or to some meteorological effect

tuff – a volcanic rock formed from the solidification of tephra or ash

tunnel support – the reinforcement of an underground excavation with supporting materials such as concrete, shotcrete or rock bolts

turbidites – a sedimentary rock formed from sediment deposited from a turbidity current

Udden–Wentworth scale – a scale for measuring the grain size of sedimentary particles

ultramafic rock – an igneous rock composed of more than 90% olivine and pyroxene

ultrasonic pulse velocities – used as index tests of rock specimens to measure primary (P) or shear (S) waves and thus to estimate the dynamic elastic constants of the rock, e.g., Young's modulus, Poisson's ratio, the shear modulus and the bulk modulus

unconfined compression test – a test of a rock's uniaxial compressive strength (UCS) with no horizontal confinement

unconformity – a boundary in a sequence of rocks formed by either a period of erosion or a lack of deposition before overlying sedimentary beds were deposited

underpressured systems – fluid pressure profiles with depth in sedimentary rocks that exhibit pore pressures less than hydrostatic

uniaxial compressive strength – a rock's compressive strength measured with an unconfined compression test

Unified Engineering Geology Mapping System (UEGMS) – a system for preparing engineering geology maps based on a set of mapping symbols identifying the genesis, lithology and various qualifying attributes of the map area

uniformitarian principle – geological processes observed today are no different from those that have operated throughout Earth history and are responsible for structural and geomorphic features we may observe, i.e., the present is the key to the past

unknown unknowns – these are unexpected or unforeseeable conditions, which pose a potentially greater risk to life and infrastructure simply because they cannot be anticipated based on past experience or investigation

Unified Soil Classification System (USCS) – classifies soils as coarse- or fine-grained while noting their degree of grading (sorting) and the type of clay if present; commonly used to log soils recovered from boreholes

varves – sedimentary layers whose rhythmic banding results from annual fluctuations in the rate of deposition

velocity (average linear) – the average interstitial groundwater velocity assuming direct linear flow through a porous medium without deviation because of pore structure

vermiculite – the clay mineral with a 2 : 1 layer structure, i.e. two tetrahedral layers and a single octahedral layer, formed by biotite oxidation

vesicle – a gas bubble in a volcanic or hypabyssal rock

volcanic hazard – volcanic hazards may exist proximal to the volcano, e.g., volcanic gases and projectiles in the form of a pyroclastic density current, or may occur distal to it, e.g., increased sediment loading and deposition in nearby rivers affected by fallout and lahars

volcanic rock – igneous rock formed on the Earth's surface by the eruption of magma mainly from a central vent

vug – a small cavity in a rock, or in a vein, usually lined with crystals

wall – the block of rock adjacent to the fault plane

wash – suspended sediment

water–rock interaction – the mineral dissolution and precipitation reactions and associated redox reactions that produce a distinct groundwater quality

wave energy – the sum of the kinetic and potential energies of a wave expressed in terms of its height and the specific weight of the water column

wave runup – the maximum height reached by a wave on a beach above a stillwater level

wave setup – the increase in mean sea level above the stillwater level

weak layer – a layer or band of sediment (or rock) that has a strength (either potentially or actually) that is sufficiently lower than the adjacent units and is perceived to be the cause or potential cause of a landslide

weak rock – a consolidated earth material possessing an unusual degree of bedding or foliation separation, fissility, fracturing, weathering and/or alteration products, and a significant content of clay materials, altogether having the appearance of a rock, yet behaving partially as a soil, and often exhibiting a potential to swell or slake, with the addition of water

weathering – the processes by which rocks are decomposed or broken down chemically and physically by reaction with the atmosphere, water and ice

wedge failure – a mode of failure of rock slopes that occurs when two discontinuity planes intersect such that the dip of the slope exceeds the dip of the line of intersection of the two discontinuity planes associated with the potentially unstable wedge

well – a borehole usually completed with a casing to prevent collapse of soil or rock into the borehole and a screen in contact with a target formation to allow fluid to move into and out of the well for monitoring, testing or production purposes

well completion – the process by which a borehole is completed to function as a well for a particular purpose, whether it be adding a well screen for production of a fluid or hydraulically fracturing a formation penetrated by the well to allow such production

wellhead protection areas – the catchment area of a water-supply well identified by the assemblage of stream tubes captured by the well

wettability – the tendency of one immiscible fluid to spread on or adhere to a mineral surface in the presence of another immiscible fluid

whole rock analysis – an analytical technique that measures the geochemical composition of a rock specimen

Wisconsin glacial stage – the most recent glaciation of eastern North America, which began approximately 90 ka with the radial expansion of the Laurentide Ice Sheet

yielding – stress-induced failure of a rock

Young's modulus – a measure of the stiffness of the rock, i.e., the ratio of stress and strain of an elastic rock, otherwise known as the modulus of elasticity

zeolites – a group of hydrous framework silicates (tectosilicates) with generally large voids as part of the framework structure; these voids may house H_2O, Na, Ca and/or K atoms

REFERENCES

Adams, J.L., 1991, *Flying Buttresses, Entropy, and O-Rings: The World of an Engineer*. Cambridge, MA: Harvard University Press.

Amaratunga, A. and Grozic, J.L.H., 2009, On the undrained unloading behaviour of gassy sands. *Canadian Geotechnical Journal* 46: 1267–1276.

Anand, R.R., 1998, *Regolith–Landform Evolution and Geochemical Dispersion from the Bodington Gold Deposit, Western Australia*. CRC LEME Open File Report 3, Canberra, ACT, Australia.

Anand, R.R. and Paine, M., 2002, Regolith geology of the Yilgarn Craton, Western Australia: implications for exploration. *Australian Journal of Earth Sciences* 49(1): 3–162.

Anders, R., Mendez, G.O., Futa, K. and Danskin, W.R., 2014, A geochemical approach to determine sources and movement of saline groundwater in a coastal aquifer. *Groundwater* 52(5): 756–768.

Anderson, D.E., Goudie, A.S. and Parker, A.G., 2007, *Global Environments through the Quaternary: Exploring Environmental Change*. Oxford, UK: Oxford University Press.

Anderson, E.M., 1951, *The Dynamics of Faulting*. Edinburgh, UK: Oliver and Boyd.

Anderson, J.G., Wesnousky, S.G. and Stirling, M.W., 1996, Earthquake size as a function of fault slip rate. *Bulletin of the Seismological Society of America* 86: 683–690.

Anderson, M.P., Aiken, J.S., Webb, E.K. and Mickelson, D.M., 1999, Sedimentology and hydrogeology of two braided stream deposits. *Sedimentary Geology* 129: 187–199.

Anderson, R.O., 1984, *Fundamentals of the Petroleum Industry*. Norman, OK: University of Oklahoma Press.

Anderson, R.S. and Anderson, S.P., 2010, *Geomorphology: The Mechanics and Chemistry of Landscapes*. Cambridge, UK: Cambridge University Press.

Annandale, G.W., 2000, Prediction of scour at bridge pier foundations founded on rock and other earth materials. *Journal of the Transportation Research Board* 1696: 67–70.

APEGBC (Association of Professional Engineers and Geoscientists of British Columbia), 2016, *Site Characterization for Dam Foundations in BC*. APEGBC Professional Practice Guidelines, Version 1.0. Burnaby, BC: APEGBC.

Appelo, C.A.J. and Postma, D., 2005, *Geochemistry, Groundwater and Pollution*, 2nd edn. Leiden, The Netherlands: Balkema.

Armstrong, D.K. and Carter, T.R., 2010, *The Subsurface Paleozoic Stratigraphy of Southern Ontario*, Special Volume 7. Sudbury, ON: Ontario Geological Survey.

ASCE (American Society of Civil Engineers), 2008, *Civil Engineering Body of Knowledge for the 21st Century: Preparing the Civil Engineer for the Future*, 2nd edn. Reston, VA: ASCE.

Aspirion, U. and Aigner, T., 1999, Towards realistic aquifer models: three-dimensional georadar surveys of Quaternary gravel deltas (Singen Basin, SW Germany). *Sedimentary Geology* 129: 281–297.

ASTM (American Society for Testing and Materials), 1995, *Using Rock-Mass Classification Systems for Engineering Purposes*, D5878-95. Philadelphia, PA: ASTM.

ASTM (American Society for Testing and Materials), 2004, *Standard Test Method for Slake Durability of Shales and Similar Weak Rocks*, D4644-04. Philadelphia, PA: ASTM.

ASTM (American Society for Testing and Materials), 2005a, *Standard Test Method for Determination of the Point Load Strength Index of Rock*, D5731-05. Philadelphia, PA: ASTM.

ASTM (American Society for Testing and Materials), 2005b, *Laboratory Determination of Pulse Velocities and Ultrasonic Elastic Constants for Rock*, D2845-05. Philadelphia, PA: ASTM.

ASTM (American Society for Testing and Materials), 2005c, *Standard Test Method for Splitting Tensile Strength of Intact Rock Core Specimens*, D3967-05. Philadelphia, PA: ASTM.

ASTM (American Society for Testing and Materials), 2007, *Standard Test Method for Compressive Strength and Elastic Moduli of Intact Rock Core Specimens under Varying States of Stress and Temperatures*, D7012-07. Philadelphia, PA: ASTM.

ASTM (American Society for Testing and Materials), 2008, *Standard Test Method for Performing Laboratory Direct Shear Strength Tests of Rock Specimens under Constant Normal Force*, D5607-08. Philadelphia, PA: ASTM.

ASTM (American Society for Testing and Materials), 2017, *Standard Practice for Classification of Soils for Engineering Purposes (Unified Soil Classification System)*, D2487-17. West Conshohocken, PA: ASTM International.

Atkinson, G.M. and Boore, D.M., 2006, Earthquake ground-motion prediction equations for eastern North America. *Bulletin of the Seismological Society of America* 96(6): 2181–2205.

Atkinson, J.H., Fookes, P.G., Miglio, B.F. and Pettifer, G.S., 2003, Destructuring and disaggregation of Mercia Mudstone during full-face tunnelling. *Quarterly Journal of Engineering Geology and Hydrogeology* 36(4): 293–303.

Atwater, B.F. and Hemphill-Haley, E., 1997, *Recurrence Intervals for Great Earthquakes of the Past 3,500 Years at Northeastern Willapa Bay, Washington*. USGS Professional Paper 1576. Denver, CO: US Geological Survey.

Aurit, M.D., Peterson, R.O. and Blanford, J.I., 2013, A GIS analysis of the relationship between sinkholes, dry-well complaints and groundwater pumping for frost-freeze protection of winter strawberry production in Florida. *PLoS ONE* 8(1): e53832.

Bahr, J.M., Moline, G.R. and Nadon, G.C., 1994, Anomalous pressures in the deep Michigan basin. In *Basin Compartments and Seals* (ed. P. Ortoleva), AAPG Memoir 61, pp. 153–165.

Bai, T. and Pollard, D.D., 2000, Fracture spacing in layered rocks: a new explanation based on the stress transition. *Journal of Structural Geology* 22: 43–57.

Bair, E.S. and Lahm, T.D., 1996, Variations in capture-zone geometry of a partially penetrating pumping well in an unconfined aquifer. *Groundwater* 34(5): 842–852.

Baker, G.S. and Jol, H.M. (eds), 2007, *Stratigraphic Analysis Using GPR*. Special Paper 432. Boulder, CO: Geological Society of America.

Baker, J.W., 2016, An introduction to probabilistic seismic hazard analysis (PSHA). In *Applied Geology in California* (eds R. Anderson and H. Ferriz). AEG Special Publication 26, pp. 943–965. Belmont, CA: Star Publishing.

Ball, S., 2005, Hard-hat tour of Hoover Dam, North Shore of Lake Mead, and Valley of Fire. *Field Trip 2*, Association of Engineering Geologists, 2005 Annual Meeting, Las Vegas, NV.

Bannerjee, S., 1984, Solubility of organic mixtures in water. *Environmental Science and Technology* 18(8): 587–591.

Barbour, S.L., 1998, The soil-water characteristic curve: a historical perspective. *Canadian Geotechnical Journal* 35: 873–894.

Barlow, P.M. and Moench, A.F., 1999, *WTAQ – A Computer Program for Calculating Drawdowns and Estimating Hydraulic Properties for Confined and Water-Table Aquifers*. Water Resources Investigations Report 99-4225. Northborough, MA: US Geological Survey.

Barlow, P.M. and Reichard, E.G., 2010, Saltwater intrusion in coastal regions of North America. *Hydrogeology Journal* 18(1): 247–260.

Barton, N., 2002, Some new Q-value correlations to assist in site characterization and tunnel design. *International Journal of Rock Mechanics and Mining Sciences* 39: 185–216.

Barton, N., Lien, R. and Lunde, J., 1974, Engineering classification of rock masses for the design of tunnel support. *Rock Mechanics* 6: 189–236.

Barton, N., Bandis, S. and Bakhtar, K., 1985, Strength, deformation and conductivity coupling of rock joints. *International Journal of Rock Mechanics and Mining Sciences & Geomechanics Abstracts* 22: 121–140.

Batenipour, H., Alfaro, M., Kurz, D. and Graham, J., 2014, Deformations and ground temperatures at a road embankment in northern Canada. *Canadian Geotechnical Journal* 51: 260–271.

Bates, R.L. and Jackson, J.A. (eds), 1984, *Dictionary of Geological Terms*, 3rd edn. New York, NY: Anchor Books.

Bea, R.G., Wright, S.G., Sircar, P. and Niedoroda, A.W., 1983, Wave-induced slides in South Pass Block 70, Mississippi Delta. *ASCE Journal of Geotechnical Engineering* 109(4): 619–644.

Bear, J., 1972, *Dynamics of Fluids in Porous Media*. New York, NY: Elsevier.

Bear, J., 2018, *Modeling Phenomena of Flow and Transport in Porous Media*. Heidelberg, Germany: Springer.

Beard, D.C. and Weyl, P.K., 1973, Influence of texture on porosity and permeability of unconsolidated sand. *AAPG Bulletin* 57(2): 349–369.

Bedrossian T.R., Hayhurst, C.A., Short, W.R. and Lancaster, J.T., 2014, Surficial geologic mapping and associated GIS databases for identification of alluvial fans. *Environmental and Engineering Geoscience* 20(4): 335–348.

Belitz, K. and Bredehoeft, J.D., 1988, Hydrodynamics of Denver Basin: explanation of subnormal fluid pressures. *AAPG Bulletin* 72(11): 1334–1359.

Bell, F.G., Stacey, T.R. and Genske, D.D., 2000, Mining subsidence and its effect on the environment: some differing examples. *Environmental Geology* 49(1–2): 135–152.

Bell, J.S., 1996, Petro geoscience 1. In situ sedimentary rocks (Part 1): Measurement techniques. *Geoscience Canada* 23(2): 85–100.

Benko, B. and Stead, D., 1998, The Frank slide: a reexamination of the failure mechanism. *Canadian Geotechnical Journal* 35: 299–311.

Bennett, P.C., Siegel, D.E., Baedecker, M.J. and Hult, M.F., 1993, Crude oil in a shallow sand and gravel aquifer – 1: Hydrogeology and inorganic geochemistry. *Applied Geochemistry* 8(6): 529–549.

Bennett, S.J., Rhoton, F.E. and Dunbar, J.A., 2005, Texture, spatial distribution, and rate of reservoir sedimentation within a highly erosive, cultivated watershed: Grenada Lake, Mississippi. *Water Resources Research* 41, W01005.

Benumof, B.T. and Griggs, G.B., 1999, The dependence of seacliff erosion rates on cliff material properties and physical processes: San Diego County, California. *Shore and Beach* 67(4): 29–41.

Benumof, B.T., Storlazzi, C.D., Seymour, R.J. and Griggs, G.B., 2000, The relationship between incident wave energy and seacliff erosion rates: San Diego County, California. *Journal of Coastal Research* 16(4): 1162–1178.

Berner, R.A., 1981, A new geochemical classification of sedimentary environments. *Journal of Sedimentary Petrology* 51(2): 359–365.

Bernknopf R.L., Brookshire, D.S., Soller, D.R., et al., 1993, *Societal Value of Geologic Maps*. USGS Circular 1111. Denver, CO: US Geological Survey.

Bérubé, M.-A., Locat, J., Gélinas, P. and Changnon, J.-Y., 1986, Black shale heaving at Sainte-Foy, Quebec, Canada. *Canadian Journal of Earth Sciences* 23: 1774–1781.

Bethke, C.M., 2008, *Geochemical and Biogeochemical Reaction Modeling*, 2nd edn. Cambridge, UK: Cambridge University Press.

Bevan, M.J., Endres, A.L., Rudolph, D.L. and Parkin, G., 2005, A field scale study of pumping-induced drainage and recovery in an unconfined aquifer. *Journal of Hydrology* 315: 52–70.

Bieniawski, Z.T., 1978, Determining rock mass deformability: experience from case histories. *International Journal of Rock Mechanics and Mining Sciences & Geomechanics Abstracts* 15: 237–247.

Bierkens, M.F.P., 1996, Modeling hydraulic conductivity of a complex confining layer at various spatial scales. *Water Resources Research* 32(8): 2369–2382.

Birchler, J.J., Stockdon, H.F., Doran, K.S. and Thompson, D.M., 2014, *National Assessment of Hurricane-Induced Coastal Erosion Hazards – Northeast Atlantic Coast*. USGS Open-File Report 2014-1243, 34pp. Reston, VA: US Geological Survey.

Bishop, A.C., Woolley, A.R. and Hamilton, W.R., 2001, *Cambridge Guide to Minerals, Rocks and Fossils*. Cambridge, UK: Cambridge University Press.

Bishop, A.W., 1948, A new sampling tool for use in cohesionless sands below ground water level. *Géotechnique* 1(1): 125–131.

Bishop, A.W., 1955, The use of the slip circle in the stability analysis of slopes. *Géotechnique* 5(1): 7–17.

Bishop, A.W., 1959, Norwegian Geotechnical Institute lecture 1955, Oslo, Norway.

Bjørlykke, K., 1989, *Sedimentary and Petroleum Geology*. Berlin, Germany: Springer.

Bjornstad, B., Last, G. and Fecht, K., 2002, *Ice-Age Floods Features in the Vicinity of the Pasco Basin and the Hanford Reach National Monument*. Field trip notes prepared by the Ice-Age Floods Institute and the Columbia River Exhibition of History, Science and Technology, Kennewick, WA, USA.

Black, J.H., 2010, The practical reasons why slug tests (including falling and rising head tests) often yield the wrong value of hydraulic conductivity. *Quarterly Journal of Engineering Geology and Hydrogeology* 43: 345–358.

Black, P., Riddols, B. and Horrey, P., 2010, Petrographic determination of clay content within East Coast Bays Formation for mechanized tunnelling projects. In *Geologically Active* (eds A.L. Williams, G.M. Pinches, C.Y. Chin, T.J. McMorran and C.I. Massey), Proceedings of IAEG2010, 11th Congress of the International Association for Engineering Geology and the Environment, Auckland, New Zealand. Boca Raton, FL: CRC Press.

Blackbourn, G.A., 2009, *Cores and Core Logging for Geoscientists*. Caithness, UK: Whittles.

Blair, S.C. and Cook, N.G.W., 1998, Analysis of compressive fracture in rock using statistical techniques: Part II. Effect of microscale heterogeneity on macroscopic deformation. *International Journal of Rock Mechanics and Mining Sciences* 35(7): 849–861.

Blair, T.C. and McPherson, J.G., 1994, Alluvial fans and their natural distinction from rivers based on morphology, hydraulic processes, sedimentary processes, and facies assemblages. *Journal of Sedimentary Research* A64(3): 450–489.

Blair, T.C. and McPherson, J.G., 1998, Recent debris-flow processes and resultant form and facies of the dolomite alluvial fan, Owens Valley, California. *Journal of Sedimentary Research* 68(5): 800–818.

Blair, T.C. and McPherson, J.G., 1999, Grain-size and textural classification of coarse sedimentary particles. *Journal of Sedimentary Research* 69(1): 6–19.

Blair, T.C. and McPherson, J.G., 2009, Processes and forms of alluvial fans. In *Geomorphology of Desert Environments*, 2nd edn (eds A.J. Parsons and A.D. Abrahams). Berlin, Germany: Springer.

Blowes, D.W., Ptacek, C.J. and Jurjovec, J., 2003, Mill tailings: hydrogeology and geochemistry. In *Environmental Aspects of Mine Wastes* (eds J.L. Jambor, D.W. Blowes and A.I.M. Ritchie), ch. 5, pp. 95–116. Ottawa, ON: Mineralogical Association of Canada,

Blum, M.D. and Roberts, H.H., 2012, The Mississippi Delta region: past, present, and future. *Annual Reviews of Earth and Planetary Science* 40: 655–683.

Boart Longyear, 2014, *Diamond Products Catalog*. [Available from http://www.boartlongyear.com/performance-tooling/performance-tooling-catalogs.]

Bock, H., 2004, *Report of the IAEG Joint European Working Group: Professional Tasks, Responsibilities and Co-operation in Ground Engineering*. [Available on the website of the International Association of Engineering Geology and the Environment, www.iaeg.info.]

Bock, H., 2006, Common ground in engineering geology, soil mechanics and rock mechanics: past, present and future. *Bulletin of Engineering Geology and the Environment* 65: 209–216.

Bolt, B.A., 1993, *Earthquakes and Geological Discovery*. New York, NY: Scientific American Library.

Bolt, B.A., 2003, *Earthquakes*, 5th edn. New York, NY: W.H. Freeman.

Bondevik, S., Mangerud, J., Dawson, S., Dawson, A. and Lohne, Ø., 2003, Record-breaking height for 8000-year-old tsunami in the North Atlantic. *EOS, Transactions of the American Geophysical Union* 84(31): 289–293.

Boore, D.M., 1983, Stochastic simulation of high-frequency ground motions based on seismological models of the radiated spectra. *Bulletin of the Seismological Society of America* 73(6): 1865–1894.

Boore, D.M., 2003, Simulation of ground motion using the stochastic method. *Pure and Applied Geophysics* 160: 635–676.

Boore, D.M. and Atkinson, G.M., 2007, *Boore–Atkinson NGA Ground Motion Relations for the Geometric Mean Horizontal Component of Peak and Spectral Ground Motion Parameters*. PEER Report 2007/01. Pacific Earthquake Engineering Research Center, University of California Berkeley, USA.

Boore, D.M., Joyner, W.B. and Fumal, T.E., 1997, Equations for estimating horizontal response spectra and peak acceleration from western North American earthquakes: a summary of recent work. *Seismological Research Letters* 68: 128–153.

Booth, C.J., 2006, Groundwater as an environmental constraint of longwall coal mining. *Environmental Geology* 49: 796–803.

Booth, C.J., 2007, Confined–unconfined changes above longwall coal mining due to increases in fracture porosity. *Environmental and Engineering Geoscience* 13(4): 355–367.

Booth, C.J., Curtiss, A.M., Demaris, P.J. and Bauer, R.A., 2000, Site-specific variation in the potentiometric response to subsidence above active longwall mining. *Environmental and Engineering Geoscience* 6(4): 383–394.

Borchardt, G., 2010, Determining relative age of faulting by using soil stratigraphy: problems and misconceptions. *Environmental and Engineering Geoscience* 16(1): 31–39.

Borden, R.C., 2001, Natural bioattenuation of anaerobic hydrocarbons and chlorinated solvents in groundwater. In *Groundwater Contamination by Organic Pollutants: Analysis and Remediation* (ed. J.J. Kaluarachchi). ASCE Manuals and Reports on

Engineering Practice No. 100, pp. 121–151. Reston, VA: American Society of Civil Engineers.

Boultbee, N., Stead, D., Schwab, J. and Geertsma, M., 2006, The Zymoetz River rock avalanche, June 2002, British Columbia, Canada. *Engineering Geology* 83: 76–93.

Bowden, A.J., Spink, T.W. and Mortimore, R.N., 2002, The engineering description of chalk: its strength, hardness and density. *Quarterly Journal of Engineering Geology and Hydrogeology* 35: 355–361.

Brady, E.H.G. and Brown, E.T., 1993, *Rock Mechanics for Underground Mining*, 2nd edn. London, UK: Chapman and Hall.

Brennand, T.A. and Shaw, J., 1994, Tunnel channels and associated landforms, south-central Ontario: their implications for ice-sheet hydrology. *Canadian Journal of Earth Sciences* 31(3): 505–522.

Briaud, J.-L., 2006, Bridge scour. *Geotechnical News* September, pp. 54–56.

Briaud, J.-L., 2008, Case histories in soil and rock erosion: Woodrow Wilson Bridge, Brazos River meander, Normandy Cliffs, and New Orleans levees. *ASCE Journal of Geotechnical and Geoenvironmental Engineering* 134(10): 1425–1447.

Briaud, J.-L., 2013, *Geotechnical Engineering: Unsaturated and Saturated Soils*. Hoboken, NJ: John Wiley and Sons.

Briaud, J.-L., Ting, F.C.K., Chen, H.C., Han, S.W. and Kwak, K.W., 2001, Erosion function apparatus for scour rate predictions. *ASCE Journal of Geotechnical and Geoenvironmental Engineering* 127(2): 105–113.

Briaud, J.-L., Chen, H.-C., Govindasamy, A.V. and Storesund, R., 2008, Levee erosion by overtopping in New Orleans during the Katrina Hurricane. *ASCE Journal of Geotechnical and Geoenvironmental Engineering* 134(5): 618–632.

Bridge, J.S., 2003, *Rivers and Floodplains: Forms, Processes, and Sedimentary Record*. Oxford, UK: Blackwell Science.

Bridge, J.S. and Demicco, R.V., 2008, *Earth Surface Processes, Landforms and Sediment Deposits*. Cambridge, UK: Cambridge University Press.

Bristow, C.S. and Jol, H.M. (eds), 2003, *Ground Penetrating Radar in Sediments*. Special Publication 211. London, UK: Geological Society of London.

Brooks, G.R., 2014, Prehistoric sensitive clay landslides and paleoseismicity in the Ottawa valley, Canada. In *Landslides in Sensitive Clays: from Geosciences to Risk Management* (eds J.-S. L'Heureux et al.). Advances in Natural and Technological Hazards Research, vol. 36, pp. 119–131. Dordrecht, The Netherlands: Springer.

Brost, E.J. and DeVault, G.E., 2000, *Non-Aqueous Phase Liquid (NAPL) Mobility Limits in Soil*. Soil and Groundwater Research Bulletin No. 9, American Petroleum Institute.

Brown, E.T. and Hoek, E., 1978, Trends in relationships between measured in-situ stresses and depth. *International Journal of Rock Mechanics and Mining Sciences & Geomechanics Abstracts* 15: 211–215.

Bryant, W.A. and Hart, E.W., 2007, *Fault-Rupture Hazard Zones in California: Alquist–Priolo Earthquake Fault Zoning Act with Index to Earthquake Fault Zone Maps*. Sacramento, CA: California Geological Survey.

BSI (British Standards Institution), 2018, *Geotechnical Investigation and Testing. Identification and Classification of Soil. Principles for a Classification*, BS EN ISO 14688-2:2018. London, UK: British Standards Institution.

Bull, W.B., 1977, The alluvial-fan environment. *Progress in Physical Geography* 1(2): 222–270.

Burbey, T.J., 2002, The influence of faults in basin-fill deposits on land subsidence, Las Vegas Valley, Nevada, USA. *Hydrogeology Journal* 10(5): 525–538.

Burbey, 2006, Three-dimensional deformation and strain induced by municipal pumping, Part 2: Numerical analysis. *Journal of Hydrology* 330: 422–434.

Burland, J., 2007, Terzaghi: back to the future. *Bulletin of Engineering Geology and the Environment* 66: 29–33.

Burland, J., 2012a, A brief history of the development of geotechnical engineering. In *ICE Manual of Geotechnical Engineering*, ch. 3. London, UK: Institution of Civil Engineers.

Burland, J., 2012b, The geotechnical triangle. In *ICE Manual of Geotechnical Engineering*, ch. 4. London, UK: Institution of Civil Engineers.

Burt, T.P. and Williams, P.J., 1976, Hydraulic conductivity in frozen soils. *Earth Surface Processes* 1(4): 349–360.

Butler, J.J., 1990, The role of pumping tests in site characterization: some theoretical considerations. *Ground Water* 28(3): 395–402.

Butler, J.J., 2005, Hydrogeological methods for estimation of spatial variations in hydraulic conductivity. In *Hydrogeophysics* (eds Y. Rubin and S. Hubbard). Dordrecht, The Netherlands: Springer.

Butler, J.J., Garnett, E.J. and Healey, J.M. 2003, Analysis of slug tests in formations of high hydraulic conductivity. *Ground Water* 41(5): 620–630. [This paper presents results from a freeware spreadsheet "Simple procedures for analysis of slug tests in formations of high hydraulic conductivity using spreadsheet and scientific graphics software". Kansas Geological Survey Open-File Report 2000_40. Available from http://www.kgs.ku.edu/Hydro/Publications/OFR00_40/index.html.]

Campbell, K.W., 1998, Empirical analysis of peak horizontal acceleration, peak horizontal velocity, and modified Mercalli intensity. In *The Loma Prieta, California, Earthquake of October 17, 1989 – Earth Structures and Engineering Characterization of Ground Motion* (ed. T.L. Holzer). USGS Professional Paper 1552-D. Washington, DC: US Geological Survey.

Campbell, S.D.G. and Parry, S., 2002, *Report on the Investigation of Kaolin-Rich Zones in Weathered Rocks in Hong Kong*. GEO Report No. 312. Hong Kong, China: Geotechnical Engineering Office.

Carey, A.E., Lyons, W.B. and Owen, J.S., 2005, Significance of landscape age, uplift, and weathering rates to ecosystem development. *Aquatic Geochemistry* 11: 215–239.

Carpenter, P.J., Doll, W.E. and Kaufman, R.D., 1998, Geophysical character of buried sinkholes on the Oak Ridge Reservation, Tennessee. *Journal of Environmental and Engineering Geophysics* 3(3): 133–145.

Carreón-Freyre, D., Cerca, M., Ochoa-González, G., Teatini, P. and Zuñiga, F.R., 2016, Shearing along faults and stratigraphic joints controlled by land subsidence in the Valley of Queretaro, Mexico. *Hydrogeology Journal* 24(3): 657–674.

CCME (Canadian Council of Ministers of the Environment), 1984, *Subsurface Assessment Handbook for Contaminated Sites*. Prepared for the Canadian Council of Ministers of the Environment by the Waterloo Centre for Groundwater Research, University of Waterloo, Waterloo, Ontario, Canada.

Cey, E.E. and Rudolph, D.L., 2009, Field study of macropore flow processes using tension infiltration of a dye tracer in partially saturated soils. *Hydrological Processes* 23: 1768–1779.

Chalikakis, K., Plagnes, V., Guerin, R., Valois, R. and Bosch, F.P., 2011, Contribution of geophysical methods to karst-system exploration: an overview. *Hydrogeology Journal* 19: 1169–1180.

Chan, R.K.S. and Pun, W.K., 2004, Landslip warning system in Hong Kong. *Geotechnical News* December, pp. 33–35.

Chandler, R.J., 2000, Clay sediments in depositional basins: the geotechnical cycle. *Quarterly Journal of Engineering Geology and Hydrogeology* 33: 7–39.

Chandler, R.J. and Apted, J.P., 1988, The effect of weathering on the strength of London Clay. *Quarterly Journal of Engineering Geology* 21: 59–68.

Chanson, H., 2004, *Environmental Hydraulics of Open Channel Flows*. Burlington, MA: Elsevier Butterworth-Heinemann.

Chapelle, F.H. and McMahon, P.B., 1991, Geochemistry of dissolved inorganic carbon in a coastal plain aquifer. 1. Sulfate from confining beds as an oxidant in microbial CO_2 production. *Journal of Hydrology* 127: 85–108.

Chapelle, F.H., Bradley, P.M., Lovely, D.R., O'Neill, K. and Landmeyer, J.E., 2002, Rapid evolution of redox processes in a petroleum-contaminated aquifer. *Ground Water* 40(4): 353–360.

Chapman, D., Metje, N. and Stärk, A., 2010, *Introduction to Tunnel Construction*. London, UK: Spon, Taylor and Francis.

Chapman, S.W. and Parker, B.L., 2005, Plume persistence due to aquitard back diffusion following dense nonaqueous phase liquid source removal or isolation. *Water Resources Research* 41(12): W12411.

Chapuis, R.P., 2004, Predicting the saturated hydraulic conductivity of sand and gravel using effective diameter and void ratio. *Canadian Geotechnical Journal* 41: 787–795.

Chapuis, R.P., 2008, Predicting the saturated hydraulic conductivity of natural soils. *Geotechnical News* 26(2): 47–50.

Chapuis, R.P. and Aubertin, M., 2003, On the use of the Kozeny–Carman equation to predict the hydraulic conductivity of soils. *Canadian Geotechnical Journal* 40: 616–628.

Chaussard, E., Wdowinski, S., Cabral-Cano, E. and Amelung, F., 2014, Land subsidence in central Mexico detected by ALOS InSAR time-series. *Remote Sensing of Environment* 140: 94–106.

Chen, X., Zhang, X., Church, J.A., et al., 2017, The increasing rate of global mean sea-level rise during 1993–2014. *Nature Climate Change* 7: 492–496.

Cherry, J.A., Parker, B.L., Bradbury, K.R., et al., 2004, *Role of Aquitards in the Protection of Aquifers from Contamination: A "State of the Science" Report*. Denver, CO: AWWA Research Foundation.

Chigira, M. and Oyama, T., 1999, Mechanism and effect of chemical weathering of sedimentary rocks. *Engineering Geology* 55: 3–14.

Chillarige, A.V., Morgenstern, N.R., Robertson, P.K. and Christian, H.A., 1997, Seabed instability due to flow liquefaction in the Fraser River delta. *Canadian Geotechnical Journal* 34(4): 520–533.

Christensen, T.H., Kjeldsen, P., Bjerg, P.L., et al., 2001, Biogeochemistry of landfill leachate plumes. *Applied Geochemistry* 16: 659–718.

Christiansen, E.A. and Sauer, E.K., 1998, Geotechnique of Saskatoon and surrounding area, Saskatchewan, Canada. In *Urban Geology of Canadian Cities* (eds P.F. Karrow and O.L. White). Special Paper 42, pp. 117–145. St. John's, NL: Geological Association of Canada.

Chuhan, F.A., Kjelstad, A., Bjorlykke, K. and Hoeg, K., 2003, Experimental compression of loose sands: relevance to porosity reduction during burial in sedimentary basins. *Canadian Geotechnical Journal* 40: 995–1011.

Chung, J.-W. and Rogers, J.D., 2013, Influence of assumed groundwater depth on mapping liquefaction potential. *Environmental and Engineering Geoscience* 19(4): 377–389.

Church, M., 1992, Channel morphology and typology. In *The Rivers Handbook* (eds P. Calow and G.E. Petts), vol. 1, pp. 126–143. Oxford, UK. Blackwell Science.

Church, M., 2002, Geomorphic thresholds in riverine landscapes, *Freshwater Biology* 47: 541–557.

Church, M., 2003, Grain size and shape. In *Encyclopedia of Sediments and Sedimentary Rocks* (ed. G.V. Middleton), p. 339. Dordrecht, The Netherlands: Kluwer.

Church, M., 2006, Bed material transport and the morphology of alluvial river channels. *Annual Reviews of Earth and Planetary Science* 34: 325–354.

Clague, J.J. (ed.) 1996, *Paleoseismology and Seismic Hazards, Southwestern British Columbia*. GSC Bulletin 494. Ottawa, ON: Geological Survey of Canada.

Clague, J.J., 1997, Evidence for large earthquakes at the Cascadia Subduction Zone. *Reviews of Geophysics* 35(4): 439–460.

Clague, J.J., Lutenauer, J.L., Pullan, S.E. and Hunter, J.A., 1991, Postglacial deltaic sediments, southern Fraser River delta, British Columbia. *Canadian Journal of Earth Sciences* 28(9): 1386–1393.

Clague, J.J., Lutenauer, J.L., Monahan, P.A., et al., 1998, Quaternary stratigraphy and evolution of the Fraser delta. In *Geology and Natural Hazards of the Fraser River Delta, British Columbia* (eds J.J. Clague, J.L. Lutenauer and D.C. Mosher). GSC Bulletin 525, pp. 57–90. Ottawa, ON: Geological Survey of Canada.

Clague, J.J., Yorath, C., Franklin, R. and Turner, B., 2006, *At Risk: Earthquakes and Tsunamis on the West Coast*. Vancouver, BC: Tricouni Press.

Clark, I.D., 2015, *Groundwater Geochemistry and Isotopes*. Boca Raton, FL: CRC Press.

Clark, I.D. and Fritz, P., 1997, *Environmental Isotopes in Hydrogeology*. Boca Raton, FL: CRC Press.

Clark, I.D., Al, T., Jensen, M., et al., 2013, Paleozoic-aged brine and authigenic helium preserved in an Ordovician shale aquiclude. *Geology* 41(9): 951–954.

Coduto, D.P., 1999, *Geotechnical Engineering: Principles and Practices*. Upper Saddle River, NJ: Prentice-Hall.

Coduto, D.P., Yeung, M.R. and Kitch, W.A., 2011, *Geotechnical Engineering: Principles and Practices*, 2nd edn. Upper Saddle River, NJ: Prentice-Hall.

Coe, J.A., Ellis, W.L., Godt, J.W., et al., 2003, Seasonal movement of the Slumgullion landslide determined from Global Positioning System surveys and field instrumentation, July 1998–March 2002. *Engineering Geology* 68: 67–101.

Colgan, P.M., Mickelson, D.M. and Cutler, P.M., 2003, Ice-marginal terrestrial landsystems: southern Laurentide ice sheet margin. In *Glacial Landsystems* (ed. D.J.A. Evans), pp. 111–142. London, UK: Hodder Arnold.

Collier, M., Webb, R.H. and Schmidt, J.C., 2000, *Dams and Rivers: A Primer on the Downstream Effects of Dams*. USGS Circular 1126. Denver, CO: US Geological Survey.

Collins, B.D. and Sitar, N., 2008, Processes of coastal bluff erosion in weakly lithified sands, Pacifica, California, USA. *Geomorphology* 97: 483–501.

Collins, B.D. and Sitar, N., 2009, Geotechnical properties of cemented sands in steep slopes. *ASCE Journal of Geotechnical and Geoenvironmental Engineering* 135(10): 1359–1366.

Collins, B.D. and Sitar, N., 2011. Stability of steep slopes in cemented sands. *ASCE Journal of Geotechnical and Geoenvironmental Engineering* 137(1): 43–51.

Cozzarelli, I.M., Böhlke, J.K., Masoner, J., et al., 2011, Biogeochemical evolution of a landfill leachate plume, Norman, Oklahoma. *Ground Water* 49(5): 663–687.

Crosby, C.J., Arrowsmith, J.R., Nandigam, V. and Baru, C., 2011, Online access and processing of LiDAR topography data. In *Geoinformatics: Cyberinfrastructure for the Solid Earth Sciences* (eds G.R. Keller and C. Baru), ch. 16. Cambridge, UK: Cambridge University Press.

Croxton, N.M. and Henthorne, R.W., 2003, High resolution seismic reflection investigation of salt dissolution features in Kansas. In *Proc. 3rd Int. Conf. on Applied Geophysics*, Orlando, FL, December.

Cruden, D.M. and Couture, R., 2010, More comprehensive characterization of landslides: review and additions. In *Geologically Active* (eds A.L. Williams, G.M. Pinches, C.Y. Chin, T.J. McMorran and C.I. Massey), Proceedings of IAEG2010, 11th Congress of the International Association for Engineering Geology and the Environment, Auckland, New Zealand, pp. 1033–1041. London, UK: Taylor and Francis.

Cruden, D.M. and Varnes, D.J., 1996, Landslide types and processes. In *Landslides: Investigation and Mitigation* (eds A.K. Turner and R.L. Schuster). Transportation Research Board Special Report 247, pp. 36–75. Washington, DC: National Academies Press.

Cuffey, K.M. and Paterson, W.S.B., 2010, *The Physics of Glaciers*, 4th edn. New York, NY: Elsevier.

Cummings, D.I., Gorrell, G., Guilbault, J.-P., et al., 2011, Sequence stratigraphy of a glaciated basin fill, with a focus on esker sedimentation. *Geological Society of America Bulletin* 123(7–8): 1478–1496.

Cummings, D.I., Russell, H.A.J. and Sharpe, D.R., 2012, Buried-valley aquifers in the Canadian Prairies: geology, hydrogeology, and origin. *Canadian Journal of Earth Sciences* 49: 987–1004.

Dake, L.P., 1994, *The Practice of Reservoir Engineering*. New York, NY: Elsevier.

Dapples, E.C., 1959, *Basic Geology for Science and Engineering*. New York, NY: John Wiley and Sons.

D'Appolonia, E., 1979, Statement of Evidence of E. D'Appolonia, D'Appolonia Consulting Engineers, Pittsburgh, Pennsylvania. In *Proceedings of the British Columbia Royal Commission of Inquiry into Uranium Mining, Phase V: Waste Disposal*, p. 9. [Cited by Shields, D.H., 2015, Giving credit where credit is due. *Geotechnical News* 33(4): 33–34.]

Darcy, H., 1856, *The Public Fountains of the City of Dijon* (transl. P. Bobeck). Paris, France: Victor Dalmont. [Republished, 2004, in English, by Kendall/Hunt, Dubuque, Iowa.]

Darrah, T.H., Vengosh, A., Jackson, R.B., Warner, N.R. and Poreda, R.J., 2014, Noble gases identify the mechanisms of fugitive gas migration in drinking-water wells overlying the Marcellus and Barnett Shales. *Proceedings of the National Academy of Sciences* 111(39): 14 076–14 081.

Darrow, M.M., Bray, M.T. and Huang, S.L., 2012, Analysis of a deep-seated landslide in permafrost, Richardson Highway, south-central Alaska. *Environmental and Engineering Geoscience* 18(3): 261–280.

Davidson-Arnott, R., 2010, *Introduction to Coastal Processes and Geomorphology*. Cambridge, UK: Cambridge University Press.

Davis, G.H., 1988, Western alluvial valleys and the high plains. In *The Geology of North America*, Vol. O-2, *Hydrogeology* (eds W. Back, J.S. Rosenshein and P.R. Seaber), ch. 34. Boulder, CO: Geological Society of America.

Davison, C.C., Chan, T., Brown, A., et al., 1994, *The Disposal of Canada's Nuclear Fuel Waste: Site Screening and Site Evaluation Technology*. Report AECL-10713, COG-93-3. Pinawa, MB: Atomic Energy of Canada Ltd.

Day, J., Diederichs, M. and Hutchinson, J., 2015, Common core: core logging procedures for characterization of complex rock masses as input into geomechanical analysis for tunnel design. *Tunnels and Tunnelling* 2015(1): 26–32.

Day, M.J., 2004, Karstic problems in the construction of Milwaukee's deep tunnels. *Environmental Geology* 45: 859–863.

Dean, R.G. and Dalrymple, R.A., 2002, *Coastal Processes with Engineering Applications*. Cambridge, UK: Cambridge University Press.

Dearman, W.R., 1991, *Engineering Geological Mapping*. Oxford, UK: Butterworth-Heinemann.

Deere, D.U. and Deere, D.W., 1988, The rock quality designation (RQD) index in practice. In *Rock Classification Systems for Engineering Purposes* (ed. L. Kirkaldie), ASTM STP 984, pp. 91–101. Philadelphia, PA: American Society for Testing Materials.

Deere, D.U. and Patton, F.D., 1971, Slope stability in residual soils. In *Proc. 4th Panamerican Conf. of Soil Mechanics and Foundation Engineering*, San Juan, Puerto Rico, vol. 1. Reston, VA: American Society of Civil Engineers.

de Freitas, M.H., 2009, Geology: its principles, practice and potential for geotechnics. *Quarterly Journal of Engineering Geology and Hydrogeology* 42: 397–441.

de Grandpré, I., Fortier, D. and Stephani, E., 2012, Degradation of permafrost beneath a road embankment enhanced by heat advected in groundwater. *Canadian Journal of Earth Sciences* 49: 953–962.

Delage, P. and Lefebvre, G., 1984, Study of the structure of a sensitive Champlain clay and of its evolution during consolidation. *Canadian Geotechnical Journal* 21: 21–35.

de Marsily, G., 1986, *Quantitative Hydrogeology Groundwater Hydrology for Engineers*. Orlando, FL: Academic Press.

Deng, J.H., Lee, C.F. and Ge, X.R., 2001, Characterization of the disturbed zone in a large rock excavation for the Three Gorges Project. *Canadian Geotechnical Journal* 38: 55–106.

De Weist, R.J.M., 1965, *Geohydrology*. New York, NY: John Wiley and Sons.

Di Biase, S., Rochetta, M. and Nadarajah, A., 2017, Maintaining stable excavation base by pressurizing an underlying confined artesian aquifer: a case study in Richmond Hill, Ontario. In *Proc. 70th Canadian Geotechnical Conf. and 12th Joint CGS/IAH-CNC Groundwater Conf.*, Ottawa, Ontario, Canada. Richmond, BC: Canadian Geotechnical Society.

Dobbs, M.R., Culshaw, M.G., Northmore, K.J., Reeves, H.J. and Entwhistle, D.C., 2012, Methodology for creating national engineering geological maps of the UK. *Quarterly Journal of Engineering Geology and Hydrogeology* 45: 335–347.

Doe, T.W., Zieger, M., Enachescu, C. and Böhner, 2006, In-situ stress measurements in exploratory boreholes. *Felsbau* 24(4): 39–47.

Domenico, P.A., 1972, *Concepts and Models in Groundwater Hydrology*. New York, NY: McGraw-Hill.

Donnelly, J.P., Roll, S., Wengren, M., et al., 2001, Sedimentary evidence of intense hurricane strikes from New Jersey. *Geology* 29(7): 615–618.

Donnelly, L.J., Culshaw, M.G. and Bell, F.G., 2008, Longwall mining-induced fault reactivation and delayed subsidence ground movement in British coalfields. *Quarterly Journal of Engineering Geology and Hydrogeology* 41: 301–314.

Dou, W., Omran, K., Grimberg, S.J., Denham, M. and Powers, S.E., 2008, Characterization of DNAPL from the U.S. DOE Savannah River Site. *Journal of Contaminant Hydrology* 97: 75–86.

Downing, R.A., 1998, *Groundwater: Our Hidden Asset*. Keyworth, UK: British Geological Survey.

Dredge, L.A., 1982, Relict ice-scour marks and late phases of Lake Agassiz in northernmost Manitoba. *Canadian Journal of Earth Sciences* 19(5): 1079–1087.

Dreybrodt, W., 1996, Principles of early development of karst conduits under natural and man-made conditions revealed by mathematical analysis of numerical models. *Water Resources Research* 32(9): 2923–2935.

Dreybrodt, W., Romanov, D. and Gabrovsek, F., 2002, Karstification below dam sites: a model of increasing leakage from reservoirs. *Environmental Geology* 42: 518–524.

Dullien, F.A.L., 1979, *Porous Media: Fluid Transport and Pore Structure*. San Diego, CA: Academic Press. [2nd edn., 1992.]

Duncan, J.M., 1996, Soil slope stability analysis. In *Landslides: Investigation and Mitigation* (eds A.K. Turner and R.L. Schuster). Special Report 247, ch. 13, pp. 337–371. Washington, DC: Transportation Research Board, National Research Council.

Dunne, T. and Black, R.D., 1970, Partial area contributions to storm runoff in a small New England watershed. *Water Resources Research* 6(5): 1296–1311.

Dunnicliff, J., 1993, *Geotechnical Instrumentation for Monitoring Field Performance*. New York, NY: John Wiley and Sons.

Duperret, A., Genter, A., Martinez, A. and Mortimore, R.N., 2004, Coastal chalk cliff instability in NW France: role of lithology, fracture pattern and rainfall. In *Coastal Chalk Cliff Instability* (eds R.N. Mortimore and A. Duperret). Engineering Geology Special Publication 20, pp. 32–55. London, UK: Geological Society of London.

Dusseault, M.B. and Jackson, R.E., 2014, Seepage pathway assessment for natural gas to shallow groundwater during well stimulation, in production, and after abandonment. *Environmental Geosciences* 21(3): 107–126.

Dusseault, M.B., Gray, M.N. and Nawrocki, P., 2000, *Why Oilwells Leak: Cement Behaviour and Long-Term Consequences*. SPE Paper 64733, pp. 1–8. Richardson, TX: Society of Petroleum Engineers.

Dwarakanath, V., Jackson, R.E. and Pope, G.A., 2002, Influence of wettability on the recovery of NAPLs from alluvium. *Environmental Science and Technology* 36(2): 227–231.

Dyck, J.H., Keys, W.S. and Meneley, W.A., 1972, Application of geophysical logging to groundwater studies in southeastern Saskatchewan. *Canadian Journal of Earth Sciences* 9: 78–94.

Eberhardt, E., 2012, Landslide monitoring: the role of investigative monitoring to improve understanding and early warning of failure. In *Landslides: Types, Mechanisms and Modeling* (eds J.J. Clague and D. Stead), pp. 222–234. Cambridge, UK: Cambridge University Press.

Eden, D.J. and Eyles, N., 2002, Case study of a relict iceberg scour exposed at Scarborough Bluffs, Toronto, Ontario: implications for pipeline engineering. *Canadian Geotechnical Journal* 39: 519–534.

Edmunds, W.M. and Shand, P., 2008, Groundwater baseline quality. In *Natural Groundwater Quality* (eds W.M. Edmunds and P. Shand), pp. 1–21. Malden, MA: Blackwell.

Edmunds, W.M. and Walton, N.R.G., 1983, The Lincolnshire limestone – hydrogeochemical evolution over a ten-year period. *Journal of Hydrology* 61: 201–211.

Edwards, B.D., Ehman, K.D., Ponti, D.J., et al., 2009, Stratigraphic controls on saltwater intrusion in the Dominguez Gap area of coastal Los Angeles. In *Earth Science in the Urban Ocean: The Southern California Continental Borderland* (eds H.J. Lee and W.R. Normark). Special Paper 454, pp. 375–395. Boulder, CO: Geological Society of America.

Eggleston, J. and Pope, J., 2013, *Land Subsidence and Relative Sea-Level Rise in the Southern Chesapeake Bay Region.* USGS Circular 1392. Reston, VA: US Geological Survey.

Einarson, M., 2006, Multilevel ground-water monitoring. In *Practical Handbook of Environmental Site Characterization and Ground-Water Monitoring* (ed. D.M. Nielsen), pp. 808–848. Boca Raton, FL: CRC, Taylor and Francis.

Einarson, M.D. and Cherry, J.A., 2002, A new multilevel ground water monitoring system using multichannel tubing. *Ground Water Monitoring and Remediation* 22(4): 52–65.

Eisenlohr, L., Meteva, K., Gabrovsek, F. and Dreybrodt, W., 1999, The inhibiting action of intrinsic impurities in natural calcium carbonate minerals to their dissolution kinetics in aqueous H_2O–CO_2 solutions. *Geochimica et Cosmochimica Acta* 63(7/8): 989–1002.

Ellsworth, W.L., 2013, Injection-induced earthquakes. *Science* 341: 142–149.

Emanuel, K., 2005, *Divine Wind: The History and Science of Hurricanes.* New York, NY: Oxford University Press.

England, P., Molnar, P. and Richter, F., 2007, John Perry's neglected critique of Kelvin's age for the Earth: a missed opportunity in geodynamics. *GSA Today* 17(1): 4–9.

Erkens, G., van der Muelen, M.J. and Middelkoop, H., 2016, Double trouble: subsidence and CO_2 respiration due to 1,000 years of Dutch coastal peatlands cultivation. *Hydrogeology Journal* 24: 551–568.

Eshraghian, A., Martin, C.D. and Morgenstern, N.R., 2008, Movement triggers and mechanisms of two earth slides in the Thompson River Valley, British Columbia, Canada. *Canadian Geotechnical Journal* 45: 1189–1209.

Evans, D.J.A., Phillips, E.R., Hiemstra, J.F. and Auton, C.A., 2006, Subglacial till: formation, sedimentary characteristics and classification. *Earth Science Reviews* 78: 115–176.

Eyles, N. and Eyles, C.H., 1992, Glacial depositional systems. In *Facies Models: Response to Sea Level Change* (eds R.G. Walker and N.P. James). St. John's, NL: Geological Association of Canada.

Eyles, N. and Sladen, J.A., 1981, Stratigraphy and geotechnical properties of weathered lodgement till in Northumberland, England. *Quarterly Journal of Engineering Geology* 14: 129–141.

Fagereng, A. and Toy, V.G., 2011, Geology of the earthquake source: an introduction. In *Geology of the Earthquake Source: A Volume in Honour of Rick Sibson* (eds A. Fagereng, V.G. Toy and J.V. Rowland), Special Publication 359, pp. 1–16. London, UK: Geological Society of London.

Fairhurst, C., 2003, Stress estimation in rock: a brief history and review. *International Journal of Rock Mechanics and Mining Sciences*, 40: 957–973.

Fasullo, J.T. and Trenberth, K.E., 2008, The annual cycle of the energy budget: Part II: Meridional structure and poleward transports. *Journal of Climate* 21: 2314–2326.

FEMA (Federal Emergency Management Agency), 2005, *Wave Runup and Overtopping: FEMA Coastal Flood Hazard Analysis and Mapping Guidelines Focused Study Report.* Washington, DC: US Department of Homeland Security.

Ferguson, G. and Jasechko, S., 2015, The isotopic composition of the Laurentide Ice Sheet and fossil groundwater. *Geophysical Research Letters* 42(12): 4856–4861.

Ferguson, H.F. and Hamel, J.V., 1981, Valley stress relief in flat-lying sedimentary rocks. In *Proc. Int. Symp. on Weak Rock*, Tokyo, Japan, September, pp. 1235–1240.

Ferris, J.G., Knowles, D.B., Brown, R.H. and Stallman, R.W., 1962, *Theory of Aquifer Tests.* Water-Supply Paper 1536-E. Washington, DC: US Geological Survey.

Field, E.H., Hough, S.E., Jacob, K.H. and Friberg, P.A., 1994, Site response in Oakland, California near the failed section of the Nimitz Freeway. In *The Loma Prieta, California, Earthquake of October 17, 1989 – Strong Ground Motion* (ed. T.L. Holzer). USGS Professional Paper 1551-A, pp. 169–179. Washington, DC: US Geological Survey.

Field, E.H., Biasi, G.P., Bird, P., et al., 2013, *Uniform California Earthquake Rupture Forecast*, Version 3 (UCERF3) – *The Time-Independent Model.* US Geological Survey Open-File Report 2013-1165; California Geological Survey Special Report 228; and Southern California Earthquake Center Publication 1792. [Available from http://pubs.usgs.gov/of/2013/1165/.]

Flint, R.F., 1955, *Pleistocene Geology of Eastern South Dakota.* USGS Professional Paper 262. Washington, DC: US Government Printing Office.

Folk, R.L., 1980, *Petrology of Sedimentary Rocks.* Austin, TX: Hemphill.

Fookes, P., Pettifer, G. and Waltham, T., 2015, *Geomodels in Engineering Geology – An Introduction.* Caithness, UK: Whittles.

Ford, D. and Williams, P., 2007, *Karst Hydrogeology and Geomorphology.* Hoboken, NJ: John Wiley and Sons.

Fossen, H., 2016, *Structural Geology*, 2nd edn. Cambridge, UK: Cambridge University Press.

Francis, R.M., 1989. *Hydrogeology of the Winter River Basin, Prince Edward Island.* Department of the Environment, Prince Edward Island, Canada.

Frankel, A., Mueller, C., Barnhard, T., et al, 1996, *National Seismic-Hazard Maps: Documentation June 1996.* USGS Open-File Report 96-532. Denver, CO: US Geological Survey.

Franklin, J.A. and Dusseault, M.B., 1989, *Rock Engineering.* New York, NY: McGraw-Hill.

Fredlund, D.G. and Houston, S.L., 2009, Protocol for the assessment of unsaturated soil properties in geotechnical engineering practice. *Canadian Geotechnical Journal* 46: 694–707.

Fredlund, D.G. and Rahardjo, H., 1993, *Soil Mechanics for Unsaturated Soils.* New York, NY: John Wiley and Sons.

Freeze, R.A., 1972, Subsurface hydrology at waste disposal sites. *IBM Journal of Research and Development* 16(2): 117–129.

Freeze, R.A., 1974, Streamflow generation. *Review of Geophysics and Space Physics* 12(4): 627–647.

Freeze, R.A. and Cherry, J.A., 1979, *Groundwater.* Englewood Cliffs, NJ: Prentice-Hall.

French, H.M., 1996, *The Periglacial Environment*, 2nd edn. Harlow, UK: Longman.

Frind, E.O., Molson, J.W. and Rudolph, D.L., 2006, Well vulnerability: a quantitative approach for source water protection. *Ground Water* 44(5): 732–742.

Froese, C.R., Moreno, F., Jaboyedoff, M. and Cruden, D.M., 2009, 25 years of movement monitoring on South Peak, Turtle Mountain: understanding the hazard. *Canadian Geotechnical Journal* 46: 256–269.

Galloway, D.L. and Burbey, T.J., 2011, Review: Regional land subsidence accompanying groundwater extraction. *Hydrogeology Journal* 19: 1459–1486.

Galloway, D.L. and Hoffmann, J., 2007, The application of satellite differential SAR interferometry-derived ground displacements in hydrogeology. *Hydrogeology Journal* 15: 133–154.

Galloway, D.L., Jones, D.R. and Ingebritsen, S.E., 1999, *Land Subsidence in the United States*. Circular 1182. Reston, VA: US Geological Survey.

Gambolati, G., Putti, M., Teatini, P. and Gasparetto Stori, G., 2006, Subsidence due to peat oxidation and impact on drainage infrastructures in a farmland catchment south of the Venice Lagoon. *Environmental Geology* 49: 814–820.

Gardner, G.H., 2009, Mine voids: what's down there anyway? Characterizing mine voids for civil construction. *Geo-Strata* 13(1): 24–27.

Gautschi, A., 2001, Hydrogeology of a fractured shale (Opalinus Clay): implications for deep geological disposal of radioactive waste. *Hydrogeology Journal* 9: 97–107.

Geertsema, M. and Clague, J.J., 2011, Pipeline routing in landslide-prone terrain. *Innovation*, July/August, pp. 17–21.

Gelhar, L.W., 1986, Stochastic subsurface hydrology from theory to applications. *Water Resources Research* 22: 135S–145S.

Geological Society of London's Engineering Group Working Party, 1977, The description of rock mass for engineering purposes. *Quarterly Journal of Engineering Geology* 10(4): 355–388.

Geological Society of London's Engineering Group Working Party, 1990, Tropical residual soils. *Quarterly Journal of Engineering Geology* 23: 1–101.

Geotechnical Engineering Office of Hong Kong, 2007, *Engineering Geology Practice in Hong Kong*. GEO Publication No. 1. Government of the Hong Kong Special Administrative Region.

Gerber, R.E., 1999, *Hydrogeologic Behaviour of the Northern Till Aquitard near Toronto, Ontario*. Ph.D. Thesis, Department of Geology, University of Toronto, Ontario, Canada.

Gerber, R.E. and Howard, K., 2000, Recharge through a regional till aquitard: three-dimensional flow model water balance approach. *Ground Water* 38(3): 410–422.

Gerhard, J., Pang, T. and Kueper, B.H., 2007, Time scales of DNAPL migration in sandy aquifers via numerical simulation. *Ground Water* 45(2): 147–157.

Ghazvinian, E., Diederichs, M. and Archibald, J., 2011, Challenges related to standardized detection of crack initiation thresholds for lower-bound or ultra-long-term strength prediction of rock. In *Proc. 2011 Pan-Am Canadian Geotechnical Society Geotechnical Conf.*, Toronto, Ontario, Canada.

Ghirotti, M., 2012, The 1963 Vaiont landslide, Italy. In *Landslides: Types, Mechanisms and Modeling* (eds J.J. Clague and D. Stead), pp. 359–372. Cambridge, UK: Cambridge University Press.

Gibbard, P.L., Head, M.J., Walker, M.J.C. and the Subcommission on Quaternary Stratigraphy, 2009, Formal ratification of the Quaternary System/Period and the Pleistocene/Epoch with a base at 2.58 Ma. *Journal of Quaternary Science* 25(2): 96–102.

Gilbert, R. (comp.), 1994, *A Field Guide to the Glacial and Postglacial Landscape of South-Eastern Ontario and Part of Québec*. GSC Bulletin 453. Ottawa, ON: Geological Survey of Canada.

Gillham, R.W., Sudicky, E.A., Cherry, J.A. and Frind, E.O., 1984, An advection–diffusion concept for solute transport in heterogeneous unconsolidated geologic deposits. *Water Resources Research* 20(3): 369–378.

Goldthwait, R.P., 1991, The Teays Valley problem; a historical perspective. In *Geology and Hydrogeology of the Teays–Mahomet Bedrock Valley Systems* (eds W.N. Melhorn and J.P. Kempton). Special Paper 258, pp. 3–8. Boulder, CO: Geological Society of America.

Goodman, R.E., 1989, *Introduction to Rock Mechanics*, 2nd edn. New York, NY: John Wiley and Sons.

Goodman, R.E., 1993, *Engineering Geology: Rock in Engineering Construction*. New York, NY: John Wiley and Sons.

Goodman, R.E., 1998, *Karl Terzaghi: The Engineer as Artist*. Reston, VA: ASCE Press.

Goodman, R.E. and Ahlgren, C., 2000, Evaluating safety of concrete gravity dam on weak rock: Scott Dam. *Journal of Geotechnical and Geoenvironmental Engineering* 126(5): 429–442.

Goodman, R.E. and Bray, J.W., 1977, Toppling of rock slopes. In *Proc. Specialty Conf. on Rock Engineering for Foundations and Slopes*, vol. 2, pp. 201–234. New York, NY: ASCE.

Goudie, A.S. and Viles, H.A., 2016, *Geomorphology in the Anthropocene*. Cambridge, UK: Cambridge University Press.

Graf, W.L., 1983, Variability of sediment removal in a semiarid watershed. *Water Resources Research* 19(3): 643–652.

Graf, W.L., 2001, Damage control: restoring the physical integrity of America's rivers. *Annals of the Association of American Geographers* 91(1): 1027.

Graham, J. and Teller, J.T., 1984, Leda clay from deep boreholes at Hawkesbury, Ontario. Part I: Geology and geotechnique: Discussion. *Canadian Geotechnical Journal* 21: 733–734.

Graham, R.C., Rossi, A.M. and Hubbert, K.R., 2010, Rock to regolith conversion: producing hospitable substrates for terrestrial ecosystems. *GSA Today* 20(2): 4–9.

Grant, L.B. and Shearer, P.M., 2004, Activity of the off-shore Newport–Inglewood Rose Canyon fault zone, coastal South California, from relocated microseismicity. *Bulletin of the Seismological Society of America* 94(2): 747–752.

Grasby, S.E., Chen, Z., Hamblin, A.P., Wozniak, P.R.J. and Sweet, A.R., 2008, Regional characterization of the Paskapoo bedrock aquifer system, southern Alberta. *Canadian Journal of Earth Science* 45: 1501–1516.

Gray, R.E., 2009, Commentary. *Geo-Strata* 13(1): 12.

Green, N.R. and MacQuarrie, K.T.B., 2014, An evaluation of the relative importance of the effects of climate change and groundwater extraction on seawater intrusion in coastal aquifers in Atlantic Canada. *Hydrogeology Journal* 22: 609–623.

Greenhouse, J.P. and Slaine, D.D., 1986, Geophysical modelling and mapping of contaminated groundwater around three waste disposal sites in southern Ontario. *Canadian Geotechnical Journal* 23(3): 372–384.

Griffiths, J.S. (ed.), 2001, *Land Surface Evaluation for Engineering Practice.* Engineering Geology Special Publication 18, pp. 39–42. London, UK: Geological Society of London.

Griffiths, J.S., 2002, *Mapping in Engineering Geology.* London, UK: Geological Society.

Griffiths, J.S., 2017, Technical note: Terrain evaluation in engineering geology. *Quarterly Journal of Engineering Geology and Hydrogeology* 50: 3–11.

Griffiths, J.S. and Stokes, M., 2008, Engineering geomorphological input to ground models: an approach based on earth systems. *Quarterly Journal of Engineering Geology and Hydrogeology* 41: 73–91.

Griffiths, J.S., Mather, A.E. and Stokes, M., 2015, Mapping landslides at different scales. *Quarterly Journal of Engineering Geology and Hydrogeology* 48: 29–40.

Griggs, G.B. and Patsch, K.B., 2004, California's coastal cliffs and bluffs. In *Formation, Evolution and Stability of Coastal Cliffs – Status and Trends* (eds M.A. Hampton and G.B. Griggs). USGS Professional Paper 1693, pp. 53–64. Denver, CO: US Geological Survey.

Grigoli, F., Cesca, S., Priolo, E., et al., 2017, Current challenges in monitoring, discrimination, and management of induced seismicity related to underground industrial activities: a European perspective. *Reviews of Geophysics* 55: 310–340.

Grisak, G.E., 1975, The fracture porosity of glacial till. *Canadian Journal of Earth Sciences* 12(3): 513–515.

Grisak, G.E. and Cherry, J.A., 1975, Hydrologic characteristics and response of fractured till and clay confining a shallow aquifer. *Canadian Geotechnical Journal* 12(1): 23–43.

Grisak, G.E., Cherry, J.A., Vonhof, J.A. and Blumele, J.P., 1976, Hydrogeologic and hydrochemical properties of fractured till in the interior plains region. In *Glacial Till* (ed. R.F. Leggett), pp. 304–355. Ottawa, ON: Royal Society of Canada.

Grozic, J.L., Robertson, P.K. and Morganstern, N.R., 1999, The behavior of loose gassy sand. *Canadian Geotechnical Journal* 36(3): 482–492.

Gudmundsson, A., 2011, *Rock Fractures in Geological Processes.* Cambridge, UK: Cambridge University Press.

Guéguen, Y. and Palciauskas, V., 1994, *Introduction to the Physics of Rocks.* Princeton, NJ: Princeton University Press.

Gunn, D.A., Dyer, B., Nelder, L.M., et al., 2009, 3D seismic tomography survey of a mineshaft at Pewfall, St. Helens. In *Engineering Geology for Tomorrow's Cities* (eds M.G. Culshaw, H.J. Reeves, I. Jefferson and T.W. Spink). Engineering Geology Special Publication 22. IAEG2006 Paper no. 302. London, UK: Geological Society of London.

Gutenberg, B. and Richter, C.F., 1944, Frequency of earthquakes in California. *Bulletin of the Seismological Society of America* 34(4): 185–188.

Haeussler, P.J., Schwartz, D.P., Dawson, T.E., et al., 2004, Surface rupture of the 2002 Denali Fault, Alaska, earthquake and comparison with other strike-slip ruptures. *Earthquake Spectra* 20(3): 565–578.

Halford, K.J. and Kuniansky, E.L., 2002, *Documentation of Spreadsheets for the Analysis of Aquifer-Test and Slug-Test Data.* Open-File Report 02-197. Carson City, NV: US Geological Survey.

Hampton, M.A. and Griggs, G.B. (eds), 2004, *Formation, Evolution, and Stability of Coastal Cliffs – Status and Trends.* USGS Professional Paper 1693. Denver, CO: US Geological Survey.

Haneberg, W.C., 2008, Using close range terrestrial digital photogrammetry for 3-D rock slope modeling and discontinuity mapping in the United States. *Bulletin of Engineering Geology and the Environment* 67(4): 57–469.

Hanor, J.S., 1993, Effective hydraulic conductivity of fractured clay beds at a hazardous waste landfill, Louisiana Gulf Coast. *Water Resources Research* 29(11): 3691–3698.

Hanson, K.L. and Schwartz D.P. (eds), 1982, *Guidebook to Late Pleistocene and Holocene Faulting along the Wasatch Front and Vicinity: Little Cottonwood Canyon to Scipio, Utah.* AGU Chapman Conference on Fault Behavior and the Earthquake Generation Process, Snowbird, Utah.

Hanson, K.L., Kelson, K.I., Angell, M.A. and Lettis, W.R., 1999, *Techniques for Identifying Faults and Determining their Origins.* NUREG/CR-5503. Washington, DC: US Nuclear Regulatory Commission.

Hanson, R.T., Izbicki, J.A., Reichard, E.G., et al., 2009, Comparison of groundwater flow in Southern California coastal aquifers. In *Earth Science in the Urban Ocean: The Southern California Continental Borderland* (eds H.J. Lee and W.R. Normark). Special Paper 454, pp. 345–373. Boulder, CO: Geological Society of America.

Hapke, C.J. and Green, K.R., 2006, Coastal landslide material loss rates associated with severe climatic effects. *Geology* 34(12): 1077–1080.

Hapke, C.J., Kratzmann, M.G. and Himmelstoss, E.A., 2013, Geomorphic and human influence on large-scale coastal change. *Geomorphology* 190: 160–170.

Harbaugh, A.W., 2005, *MODFLOW-2005, The U.S. Geological Survey Modular Ground-Water Model – the Ground-Water Flow Process.* U.S. Geological Survey Techniques and Methods 6-A16 (Book 6, Modeling techniques, Section A, Ground water, ch. 16). Reston, VA: US Geological Survey.

Harnischmacher, S. and Zepp, H., 2014, Mining and its impact on the earth surface in the Ruhr District (Germany). *Zeitschrift für Geomorphologie* 58(3): 3–22.

Harris, D.J. and Feldman, A.D. (eds), 2002, *Proc. Workshop on Hydrologic Research Needs for Dam Safety*, November 2001. US Army Corps of Engineers.

Harrold, G., Gooddy, D.C., Lerner, D.N. and Leharne, S.A., 2001, Wettability changes in trichloroethylene-contaminated sandstone. *Environmental Science and Technology* 35(7): 1504–1510.

Hart, A.B., Griffiths, J.S. and Mather, A.E., 2009, Some limitations in the interpretation of vertical stereo photographic images for a landslide investigation. *Quarterly Journal of Engineering Geology and Hydrogeology* 42: 21–30.

Hart, M.W., Shaller, P.J. and Farrand, G.T., 2012, When landslides are misinterpreted as faults: case studies from the Western

United States. *Environmental and Engineering Geoscience* 18(4): 313–325.

Harvey, J.C., 1982, *Geology for Geotechnical Engineers*. Cambridge, UK: Cambridge University Press.

Hatheway, A.W., 1999, Origins and formation of weak-rock masses: a guide to field work. In *Characterization of Weak and Weathered Rock Masses* (eds P.M. Santi and A. Shakoor). Special Publication No. 9, pp. 23–35. College Station, TX: Association of Engineering Geologists: Texas A&M University.

Hatheway, A.W., 2005, What we should expect from undergraduate training of engineers practicing in the geosciences. *The Professional Geologist*, September/October.

Haugerud, R.A., Harding, D.J., Johnson, S.Y., et al., 2003, High-resolution lidar topography of the Puget Lowland, Washington – a bonanza for earth science. *GSA Today* 13(6): 4–10.

Head, K.H., 2006, *Manual of Soil Laboratory Testing*, vol. I, *Soil Classification and Compaction Tests*, 3rd edn. Caithness, UK: Whittles.

Head, K.H. and Epps, R.J., 2011, *Manual of Soil Laboratory Testing*, vol. II, *Permeability, Shear Strength and Compressibility Tests*, 3rd edn. Caithness, UK: Whittles.

Head, K.H. and Epps, R.J., 2014, *Manual of Soil Laboratory Testing*, vol. III, *Effective Stress Tests*, 3rd edn. Caithness, UK: Whittles.

Healy, D., Sibson, R.H., Shipton, Z. and Butler, R., 2012, Stress, faulting, fracturing and seismicity: the legacy of Ernest Masson Anderson. In *Faulting, Fracturing and Igneous Intrusion in the Earth's Crust* (eds D. Healy, R.W.H. Butler, Z. Shipton and R.H. Sibson). Special Publication 367, pp. 1–6. London, UK: Geological Society of London.

Heaton, T.H., 1990, The calm before the quake? *Nature* 343: 511–512.

Hecht, L., Thuro, K., Plinninger, R.J. and Cuney, M., 1999, Mineralogical and geochemical characterization of hydrothermal alteration and episyenitization in the Königshain granites, northern Bohemian Massif, Germany. *International Journal of Earth Sciences* 88: 236–252.

Heinz, J. and Aigner, T., 2003, Hierarchical dynamic stratigraphy in various Quaternary gravel deposits, Rhine glacier area (SW-Germany): implications for hydrostratigraphy. *International Journal of Earth Sciences* 92: 923–938.

Heinz, J., Kleineidam, S., Teutsch, G. and Aigner, T., 2003, Heterogeneity patterns of Quaternary glaciofluvial gravel bodies (SW-Germany): application to hydrogeology. *Sedimentary Geology* 158: 1–23.

Helm, D.C., 1994, Hydraulic forces that play a role in generating fissures at depth. *Bulletin of the Association of Engineering Geologists* 31(3): 293–304.

Hencher, S.R., 2012, *Practical Engineering Geology*. London, UK: Spon, Taylor and Francis.

Hencher, S.R. and Lee, S.G., 2010, Landslide mechanisms in Hong Kong. In *Weathering as a Predisposing Factor to Slope Movements* (eds D. Calcaterra and M. Parise). Engineering Geology Special Publication 23, pp. 77–103. London, UK: Geological Society of London.

Hendron, Jr., A.J. and Patton, F.D., 1985, *The Vaiont Slide, A Geotechnical Analysis Based on New Geologic Observations of the Failure Surface*. Technical Report GL-85-5. Washington, DC: US Army Corps of Engineers.

Herget, G., 1988, *Stresses in Rock*. Rotterdam, The Netherlands: A.A. Balkema.

Higgins, J.D. and Modeer, Jr., V.A., 1996, Loess. In *Landslides: Investigation and Mitigation* (eds A.K. Turner and R.L. Schuster). Special Report 247, ch. 23, pp. 585–606. Washington, DC: Transportation Research Board, National Research Council.

Highland, L., 2004, *Landslide Types and Processes*. Fact Sheet 2004-3072. Denver, CO: US Geological Survey. [Available from http://pbs.usgs.gov/fs/2004/3072/.]

Highland, L. and Bobrowsky, P., 2008, *The Landslide Handbook: A Guide to Understanding Landslides*. USGS Circular 1325. Reston, VA: US Geological Survey, US Department of the Interior.

Hildes, D.H.D., Clarke, G.K.C., Flowers, G.E. and Marshall, S.J., 2004, Subglacial erosion and englacial sediment transport modelled for North American ice sheets. *Quaternary Science Reviews* 23: 409–430.

Hippensteel, S.P., Eastin, M.D. and Garcia, W.J., 2013, The geological legacy of Hurricane Irene: implications for the fidelity of the paleo-storm record. *GSA Today* 23(12): 4–10.

Hodge, R.A. and Freeze, R.A., 1977, Groundwater flow systems and slope stability. *Canadian Geotechnical Journal* 14(4): 466–476.

Hoek, E., 1999, Putting numbers to geology – an engineer's viewpoint. *Quarterly Journal of Engineering Geology* 32: 1–19.

Hoek, E. and Bray, J.W., 1981, *Rock Slope Engineering*, rev. 3rd edn. London, UK: Spon.

Hoek, E. and Brown, E.T., 1997, Practical estimates of rock mass strength. *International Journal of Rock Mechanics and Mining Sciences & Geomechanics Abstracts* 34(8): 1165–1186.

Holt, R.M., 1997, *Conceptual Model for Transport Processes in the Culebra Dolomite Member, Rustler Formation*. Report SAND97-0194. Albuquerque, NM: Sandia National Laboratories.

Holthuijsen, L.H., 2007, *Waves in Oceanic and Coastal Waters*. Cambridge, UK: Cambridge University Press.

Holzer, T.L., 1998, Introduction. In *The Loma Prieta, California, Earthquake of October 17, 1989 – Liquefaction* (ed. T.L. Holzer). USGS Professional Paper 1551-B. Washington, DC: US Geological Survey.

Holzer, T.L. and Bennett, M.J., 2007, Geologic and hydrogeologic controls of boundaries of lateral spreads: lessons from USGS liquefaction case histories. In *Conf. Presentations, 1st North American Landslide Conf.*, Vail, CO (eds V.R. Schaefer, R.L. Schuster and A.K. Turner). Zanesville, OH: Association of Environmental and Engineering Geologists.

Holzer, T.L., Noce, T.E. and Bennett, M.J., 2011, Liquefaction probability curves for surficial geologic deposits. *Environmental and Engineering Geoscience* 17(1): 1–21.

Hooke, R. LeB., 2005, *Principles of Glacier Mechanics*, 2nd edn. Cambridge, UK: Cambridge University Press.

Hoque, M.A. and Pollard, W.H., 2009, Arctic coastal retreat through block failure. *Canadian Geotechnical Journal* 46: 1103–1115.

Hough, S.E., 2002, *Earthshaking Science: What We Know (and Don't Know) about Earthquakes*. Princeton, NJ: Princeton University Press.

Hough, S.E. and Bilham, R.G., 2006, *After the Earth Quakes: Elastic Rebound on an Urban Planet*. New York, NY: Oxford University Press.

House, P.K., Webb, R.H., Baker, V.R. and Levish, D.R. (eds), 2002, *Ancient Floods, Modern Hazards: Principles and Applications of Paleoflood Hydrology*. Washington, DC: American Geophysical Union.

Hozik, M.J., Parrott, Jr., W.R. and Talkington, R.W., 2003, Geologic structures, maps and block diagrams. In *Laboratory Manual in Physical Geology*, 6th edn (ed. R.M. Busch), pp. 169–180. For American Geological Institute and the National Association of Geoscience Teachers. Upper Saddle River, NJ: Pearson Education.

Hsieh, P.A. and Bredehoeft, J.D., 1981, A reservoir analysis of the Denver earthquakes: a case of induced seismicity. *Journal of Geophysical Research* 82(B2): 903–920.

Hubbert, M.K., 1940, The theory of ground-water motion. *Journal of Geology* 48(8): 785–944.

Hudson, J.A., 1989, *Rock Mechanics Principles in Engineering Practice*. London, UK: Butterworths.

Hudson, J.A. and Harrison, J.P., 1997, *Engineering Rock Mechanics: An Introduction to the Principles*. Oxford, UK: Pergamon.

Hughes, D.A.B., Clarke, G.R.T., Harley, R.M.G. and Barbour, S.L., 2016, The impact of hydrogeology on the instability of a road cutting through a drumlin in Northern Ireland. *Quarterly Journal of Engineering Geology and Hydrogeology* 49: 92–104.

Hughes, J.D., Vacher, H.L. and Sanford, W.E., 2009, Temporal response of hydraulic head, temperature and chloride concentrations to sea-level changes, Floridan aquifer system, USA. *Hydrogeology Journal* 17(4): 793–815.

Hughes, M.W., Quigley, M.C., van Ballegooy, S., et al., 2015, The sinking city: earthquakes increase flood hazard in Christchurch, New Zealand. *GSA Today* 25(3–4): 4–10.

Hungr, O., 2001. *Task Force on the Promotion of Geological Engineering and Engineering Geology in Canada*, Preliminary Report. May 16. [Available from http://1ss.gsc.nrcan.gc.ca/egd/TaskForceE.html.]

Hunt, J.M., 1979, *Petroleum Geochemistry and Geology*. San Francisco, CA: W.H. Freeman.

Hurst, M.D., Rood, D.H., Ellis, M.A., Anderson, R.S. and Dornbusch, U., 2016, Recent acceleration in coastal cliff retreat rates on the south coast of Great Britain. *Proceedings of the National Academy of Sciences* 113(47): 13 336–13 341.

Hutchinson, M.J., Ingram, R.G.S., Grout, M.W. and Hayes, P.J., 2012, A successful model: 30 years of the Lincolnshire Chalk model. In *Groundwater Resources Modelling: A Case Study from the UK* (eds M.G. Shepley, M.I. Whiteman, P.J. Hulme and M.W. Grout). Special Publication 364, pp. 173–191. London, UK: Geological Society of London.

Idriss, I.D. and Boulanger, R.W., 2008. *Soil Liquefaction during Earthquakes*. Oakland, CA: Earthquake Engineering Research Institute.

Ingebritsen, S.E. and Ikehara, M.E., 1999, Sacramento–San Joaquin delta. In *Land Subsidence in the United States* (eds D.L. Galloway, D.R. Jones and S.E. Ingebritsen). Circular 1182. Reston, VA: US Geological Survey.

Ingebritsen, S.E., Sanford, W. and Neuzil, C., 2006, *Groundwater in Geologic Processes*, 2nd edn. Cambridge, UK: Cambridge University Press.

IPCC (Intergovernmental Panel on Climate Change), 2014, *Climate Change 2014: Synthesis Report*. Cambridge, UK: Cambridge University Press.

International Society for Rock Mechanics (ISRM) Commission on Standardization of Laboratory and Field Tests, 1978, Suggested methods for the quantitative description of discontinuities in rock masses. *International Journal of Rock Mechanics and Mining Sciences & Geomechanics Abstracts* 15(6): 319–368. [Also available in *Rock Characterization Testing and Rock Monitoring: ISRM Suggested Methods* (ed. E.T. Brown).]

Ireson, A.M., van der Kamp, G., Ferguson, G., Nachshon, U. and Wheater, H.S., 2013, Hydrogeological processes in seasonally frozen northern latitudes: understanding, gaps and challenges. *Hydrogeology Journal* 21: 53–66.

Iverson, N.R., 2010, Shear resistance and continuity of subglacial till: hydrology rules. *Journal of Glaciology* 56(200): 1104–1114.

Iverson, N.R., Hooyer, T.S. and Baker, R.W., 1998a, Ring-shear studies of till deformation: Coulomb–plastic behavior and distributed strain in glacier beds. *Journal of Glaciology* 44(148): 634–642.

Iverson, R.M., 2000, Landslide triggering by rain infiltration. *Water Resources Research* 36(7): 1897–1910.

Iverson, R.M., Schilling, S.P. and Vallance, J.W., 1998b, Objective delineation of lahar-inundation hazard zones. *Geological Society of America Bulletin* 110(8): 972–984.

Iverson, R.M., Reid, M.E., Iverson, N.R., et al., 2000, Acute sensitivity of landslide rates to initial soil porosity. *Science* 29: 513–516.

Iverson, R.M., George, D.L., Allstadt, K., et al., 2015, Landslide mobility and hazards: implications of the 2014 Oso disaster. *Earth and Planetary Science Letters* 412: 197–208.

Izzo, D., 2004, Reengineering the Mississippi. *Civil Engineering* 74(7): 38–45.

Jaboyedoff, M., Demers, D., Locat, J., et al., 2009, Use of terrestrial laser scanning for the characterization of retrogressive landslides in sensitive clay and rotational landslides in river banks. *Canadian Geotechnical Journal* 46: 1379–1390.

Jackson, R.B., 2014, The integrity of oil and gas wells. *Proceedings of the National Academy of Sciences* 111(30): 10 902–10 903.

Jackson, R.E., 1998, The migration, dissolution, and fate of chlorinated solvents in the urbanized alluvial valleys of the southwestern USA. *Hydrogeology Journal* 6: 144–155.

Jackson, R.E. and Dwarakanath, V., 1999, Chlorinated degreasing solvents: physical-chemical properties affecting aquifer contamination and remediation. *Ground Water Monitoring and Remediation* 19(4): 102–110.

Jackson, R.E. and Heagle, D.J., 2015, Sampling domestic/farm wells for baseline groundwater quality and fugitive gas. *Hydrogeology Journal* 24: 269–272.

Jackson, R.E. and Inch, K.J., 1983, Partitioning of strontium-90 among aqueous and mineral species in a contaminated aquifer. *Environmental Science and Technology* 17(4): 231–237.

Jackson, R.E. and Inch, K.J., 1989, The in-situ adsorption of ^{90}Sr in a sand aquifer at the Chalk River Nuclear Laboratories. *Journal of Contaminant Hydrology* 4: 27–50.

Jackson, R.E. and Jin, M., 2005, The measurement of DNAPL in low-permeability lenses within alluvial aquifers by partitioning tracers. *Environmental and Engineering Geoscience* 11(4): 405–412.

Jackson, R.E. and Mariner, P.E., 1994, Estimating DNAPL composition and VOC dilution from extraction well data. *Ground Water* 33(3): 407–414.

Jackson, R.E. and Patterson, R.J., 1982, Interpretation of pH and Eh trends in a fluival-sand aquifer system. *Water Resources Research* 18(4): 1255–1268.

Jackson, R.E., Patterson, R.J., Graham, B.W., et al., 1985, *Contaminant Hydrogeology of Toxic Organic Chemicals at a Disposal Site, Gloucester, Ontario. 1: Chemical Concepts and Site Assessment*. Ottawa, ON: National Hydrology Research Institute, Inland Waters Directorate.

Jackson, R.E., Lesage, S. and Priddle, M.W., 1992, Estimating the fate and mobility of CFC-113 in groundwater: results from the Gloucester Landfill project. In *Groundwater Contamination and Analysis at Hazardous Waste Sites* (eds R.E. Jackson and S. Lesage), pp. 511–526. New York, NY: Marcel Dekker.

Jackson, R.E., Dwarakanath, V., Ewing, J.E. and Avis, J., 2006, Migration of viscous non-aqueous phase liquids (NAPLs) in alluvium, Fraser River lowlands, British Columbia. *Canadian Geotechnical Journal* 43(7): 694–703.

Jackson, R.G., 1976, Depositional model of point bars in the Lower Wabash River. *Journal of Sedimentary Petrology* 46(3): 579–594.

Jambor, J.L., 2003, Mine-waste mineralogy and mineralogical perspectives of acid–base accounting. In *Environmental Aspects of Mine Wastes* (eds J.L. Jambor, D.W. Blowes and A.I.M. Ritchie), ch. 6, pp. 117–145. Ottawa, ON: Mineralogical Association of Canada.

Jiao, J.J., Wang, X.-S. and Nandy, S., 2005, Confined groundwater zone and slope instability in weathered igneous rocks in Hong Kong. *Engineering Geology* 80: 71–92.

Jibson, R.W., 2007. Regression models for estimating coseismic landslide displacement. *Engineering Geology* 91: 209–218.

John, B.S. and Jackson, Jr., L.E., 2009, Stonehenge's mysterious stones. *Earth* January, pp. 36–43.

Johnson, A.I., 1967, *Specific Yield – Compilation of Specific Yields for Various Materials*. Water Supply Paper 1662-D. Washington, DC: US Geological Survey.

Johnston, R.H., 1997, Source of water supplying pumpage from regional aquifer systems of the United States. *Hydrogeology Journal* 5(2): 54–63.

Jones, B.M., Arp, C.D., Jorgenson, M.T., et al., 2009, Increase in the rate and uniformity of coastline erosion in Arctic Alaska. *Geophysical Research Letters* 36, L03503.

Jones, C.L., Higgins, J.D. and Andrew, R.D., 2000, *Colorado Rockfall Simulation Program*, Version 4.0 (for Windows). Denver, CO: Colorado Department of Transportation.

Jordan, D.W. and Pryor, W.A., 1992, Hierarchical levels of heterogeneity in a Mississippi River meander belt and application to reservoir systems. *AAPG Bulletin* 76(10): 1601–1624.

Jorgensen, D.G., Gogel, T. and Signor, D.C., 1982, Determination of flow in aquifers containing variable density water. *Groundwater Monitoring Review* 2(2): 40–45.

Jorgenson, M.T., Shur, Y.L. and Pullman, E.R., 2006, Abrupt increase in permafrost degradation in Arctic Alaska. *Geophysical Research Letters* 33, L02705.

Julien, P., 2002, *River Mechanics*. Cambridge, UK: Cambridge University Press.

Kamb, B., Raymond, C.F., Harrison, W.D., et al., 1985, Glacier surge mechanism: 1982–1983 surge of Variegated Glacier, Alaska. *Science* 227(4686): 469–479.

Kamphuis, J.W., 1987, Recession rate of glacial till bluffs. *ASCE Journal of Waterway, Port, Coastal and Ocean Engineering* 113(1): 60–73.

Karrow, P.F., Dreimanis, A. and Barnett, P.F., 2000, A proposed diachronic revision of Late Quaternary time-stratigraphic classification in the Eastern and Northern Great Lakes area. *Quaternary International* 54: 1–12.

Kavazanjian, E., Wang, J.-N.J., Martin, G.R., et al., 2011, *LRFD Seismic Analysis and Design of Transportation Geotechnical Features and Structural Foundations Reference Manual*. Geotechnical Engineering Circular no. 3, FHWA-NHI-11-032. Washington, DC: Federal Highways Administration.

Keaton, J.R., 2013, Estimating erodible rock durability and geotechnical parameters for scour analysis. *Environmental and Engineering Geoscience* 19(4): 319–343.

Keaton, J.R. and DeGraff, J.V., 1996, Surface observation and geologic mapping. In *Landslides: Investigation and Mitigation* (eds A.K. Turner and R.L. Schuster). Special Report 247, ch. 9, pp. 178–230. Washington, DC: Transportation Research Board, National Research Council.

Keaton, J.R. and Rinne, R., 2002, Engineering-geology mapping of slopes and landslides. In *Geoenvironmental Mapping: Methods, Theory and Practice* (ed. P.T. Bobrowsky), pp. 9–27. Lisse, The Netherlands: A.A. Balkema.

Keaton, J.R., Mishra, S.K. and Clopper, P.E., 2012, *Scour at Bridge Foundations on Rock*. National Cooperative Highway Research Program (NCHRP) Report 717. Washington, DC: Transportation Research Board.

Keaton, J.R., Wartman, J., Anderson, S.C., et al., 2014, *The 22 March 2014 Oso Landslide, Snohomish County, Washington*. GEER Association. [Available from Geotechnical Extreme Events Reconnaissance webpage at http://www.geerassociation.org.]

Keefer, D.K., 1984, Landslides caused by earthquakes. *Geological Society of America Bulletin* 95(4): 406–421.

Keefer, D.K. and Manson, M.W., 1998, Regional distribution and characteristics of landslides generated by the earthquake. In *The Loma Prieta, California, Earthquake of October 17, 1989 – Landslides* (ed. D.K. Keefer). US Geological Survey Professional

Paper 1551-C, pp. C7–C32. Washington, DC: US Government Printing Office.

Keller, C.K., van der Kamp, G. and Cherry, J.A., 1989, A multiscale study of the permeability of a thick clayey till. *Water Resources Research* 25(11): 2299–2317.

Keller, E.A. and Pinter, N., 2002, *Active Tectonics: Earthquakes, Uplift and Landscape*. Upper Saddle River, NJ: Prentice-Hall.

Kemeny, J. and Turner, K., 2008, *Ground-Based LiDAR: Rock Slope Mapping and Assessment*. FHWA-CFL/TD-08-006. Lakewood, CO: US Department of Transportation, Federal Highways Administration.

Kennett, J.P., 1995, Latest Quaternary benthic oxygen and carbon isotopic stratigraphy: hole 893A, Santa Barbara Basin, California. In *Proceedings of the Ocean Drilling Program, Scientific Results*, Leg 146, Part 2 (eds J. Kennett, J.G. Baldauf and M. Lyle). Alexandria, VA: National Science Foundation.

Kesler, S.E. and Simon, A.C., 2015, *Mineral Resources, Economics and the Environment*. Cambridge, UK: Cambridge University Press.

Kiersch, G.A., 1964, Vaiont reservoir disaster. *Civil Engineering* 34(3): 32–39.

Kinzelbach, W., Marburger, M. and Chiang, W.-H., 1992, Determination of groundwater catchment areas in two and three spatial dimensions. *Journal of Hydrology* 134(1–4): 221–246.

Kious, W.J. and Tilling, R.I., undated, *This Dynamic Earth: The Story of Plate Tectonics*. Washington, DC: US Geological Survey.

Kirk, P.A., Campbell, S.D.G., Fletcher, C.J.N. and Merriman, R.J., 1997, The significance of primary volcanic fabrics and clay distribution in landslides in Hong Kong. *Journal of the Geological Society, London* 154: 1009–1019.

Klein, C. and Philpotts, A.R., 2013, *Earth Materials: Introduction to Mineralogy and Petrology*. Cambridge, UK: Cambridge University Press.

Klein, C. and Philpotts, A.R., 2016, *Earth Materials: Introduction to Mineralogy and Petrology*, 2nd edn. Cambridge, UK: Cambridge University Press.

Klint, K.E., Nilsson, B., Troldborg, L. and Jakobsen, P.R., 2013, A poly morphological landform approach for hydrogeological applications in heterogeneous glacial sediments. *Hydrogeology Journal* 21: 1247–1264.

Knill, J., 2003, Core values: the first Hans Cloos lecture. *Bulletin of Engineering Geology and the Environment* 62: 1–34.

Knill, J., Cratchley, C.R., Early, K.R., et al., 1970, The logging of rock cores for engineering purposes. *Quarterly Journal of Engineering Geology* 3(1): 1–24.

Koerner, R.M. and Soong, T.-Y., 2000, Stability assessment of ten large landfill failures. In *Advances in Transportation and Geoenvironmental Systems Using Geosynthetics*, pp. 1–38. Reston, VA: American Society of Civil Engineers.

Kohout, F.A., 1964, *The Flow of Fresh Water and Salt Water in the Biscayne Aquifer of the Miami Area, Florida*. US Geological Survey Water-Supply Paper 1613-C, pp. 12–32. Washington, DC: US Government Printing Office.

Koltermann, C.E. and Gorelick, S.M., 1996, Heterogeneity in sedimentary deposits: a review of structure-imitating, process-imitating, and descriptive approaches. *Water Resources Research* 32(9): 2617–2658.

Konrad, J.-M., 1994, Sixteenth Canadian Geotechnical Colloquium: Frost heave in soils: concepts and engineering. *Canadian Geotechnical Journal* 31: 223–245.

Kueper, B.H. and Davies, K.L., 2009, *Assessment and Delineation of DNAPL Source Zones at Hazardous Waste Sites*. EPA/600/R-09/119. Cincinnati, OH: US Environmental Protection Agency.

Kueper, B.H., Abbott, W. and Farquhar, G., 1989, Experimental observations of multiphase flow in heterogeneous porous media. *Journal of Contaminant Hydrology* 5: 83–95.

Lacelle, D., Juneau, V., Pellerin, A., Lauriol, B. and Clark, I.D., 2008, Weathering regime and geochemical conditions in a polar desert environment, Haughton impact structure region, Devon Island, Canada. *Canadian Journal of Earth Science* 45(10): 1139–1157.

LaHusen, R., 2008, Debris flow and lahar management in the Pacific Northwest. In *GeoEdmonton'08, Conf. Proc., 61st Canadian Geotechnical Conf.*, Edmonton, Alberta, pp. 1069–1070.

Lake, L.W., 1989, *Enhanced Oil Recovery*. Englewood Cliffs, NJ: Prentice-Hall.

Lamb, S. and Sington, D., 1998, *Earth Story: The Forces that Have Shaped Our Planet*. Princeton, NJ: Princeton University Press.

Lan, H., Martin, C.D. and Hu, B., 2010a, Effect of heterogeneity of brittle rock on micromechanical extensile behavior during compression loading. *Journal of Geophysical Research* 115, B011202.

Lan, H., Martin, C.D., Zhou, C. and Lim, C.H., 2010b, Rockfall hazard analysis using LiDAR and spatial modeling. *Geomorphology* 118(1–2): 213–223.

Landon, M.K., Jurgens, B.C., Katz, B.G., et al., 2010, Depth-dependent sampling to identify short-circuit pathways to public-supply wells in multiple aquifer settings in the United States. *Hydrogeology Journal* 18: 577–593.

Langer, W.H. and Knepper, D.H., 1995, *Geologic Characterization of Natural Aggregate*. USGS Open-File Report 95-582. Denver, CO: US Geological Survey.

Langmuir, D., 1984, Physical and chemical characteristics of carbonate water. In *Guide to the Hydrology of Carbonate Rocks* (eds P.E. Lamoreaux, B.M. Wilson and B.A. Memeon), pp. 60–130. Paris, France: UNESCO,

Langmuir, D., 1997, *Aqueous Environmental Geochemistry*. Upper Saddle River, NJ: Prentice-Hall.

Larsen, N.J. and Piotrowski, J.A., 2003, Fabric pattern in a basal till succession and its significance for reconstructing subglacial processes. *Journal of Sedimentary Research* 73(5): 725–734.

Lato, M., Hutchinson, J., Diederichs, M., Ball, D. and Harrap, R., 2009, Engineering monitoring of rockfall hazards along transportation corridors: using mobile terrestrial LiDAR. *Natural Hazards and Earth System Sciences* 9: 935–946.

Lattman, L.A. and Parizek, R.R., 1964, Relationship between fracture traces and the occurrence of ground water in carbonate rocks. *Journal of Hydrology* 2: 73–91.

Lawrence, A.R., Chilton, P.J., Barron, R.J. and Thomas, W.M., 1990, A method for determining volatile organic solvents in Chalk pore waters (southern and eastern England) and its relevance to the evaluation of ground water contamination. *Journal of Contaminant Hydrology* 6: 377–386.

Leake, S.A. and Galloway, D.L., 2010, Use of the SUB-WT package for MODFLOW to simulate aquifer-system compaction in Antelope Valley, California, USA. In *Land Subsidence, Associated Hazards and the Role of Natural Resources Development* (eds D. Carreón, M. Cerca and D.L. Galloway). IAHS Publication 339, pp. 61–67. Wallingford, UK: IAHS Press.

LeBlanc, A.-M., Fortier, R., Allard, M., Cosma, C. and Buteau, S., 2004, Seismic cone penetration test and seismic tomography in permafrost. *Canadian Geotechnical Journal* 41(5): 796–813.

Lee, J.R. and Phillips, E., 2013, Glacitectonics – a key approach to examining ice dynamics, substrate rheology and ice-bed coupling. *Proceedings of the Geologists' Association* 124: 731–737.

Lee, L. and Helsel D., 2005, Baseline models of trace elements in major aquifers of the United States, *Applied Geochemistry* 20(8): 1560–1570.

Lee, S.G., 1987, *Weathering and Geotechnical Characterization of Korean Granites*. Ph.D. Thesis, London University.

Leeder, M., 1999, *Sedimentology and Sedimentary Basins: from Turbulence to Tectonics*. Oxford, UK: Blackwell Science.

Lefebvre, G., 1986, Slope instability and valley formation in Canadian soft clay deposits. *Canadian Geotechnical Journal* 23: 261–270.

Lefebvre, G., 1996, Soft sensitive clays. In *Landslides: Investigation and Mitigation* (eds A.K. Turner and R.L. Schuster), Special Report 247, ch. 24, pp. 607–619. Washington, DC: Transportation Research Board, National Research Council.

Lemieux, J.-M., Kirkwood, D. and Therrien, R., 2009, Fracture network analysis of the St-Eustache quarry, Quebec, Canada, for groundwater resources management. *Canadian Geotechnical Journal* 46: 828–841.

Leopold, L. and Maddock, Jr., T., 1953, *The Hydraulic Geometry of Stream Channels and Some Physiographic Interpretations*. US Geological Survey Professional Paper 252. Washington, DC: US Government Printing Office.

Levish, D.R., 2002, Paleohydrologic bounds: non-exceedance information for flood hazard assessment. In *Ancient Floods, Modern Hazards: Principles and Applications of Paleoflood Hydrology* (eds P.K. House, R.H. Webb, V.R. Baker and D.R. Levish), pp. 175–190. Washington, DC: American Geophysical Union.

Lewis, C.F.M., Taylor, B.B., Stea, R.R., et al., 1998, Earth science and engineering: urban development in the Metropolitan Halifax Region. In *Urban Geology of Canadian Cities* (eds P.F. Karrow and O.L. White). Special Paper 42, pp. 409–444. St. John's, NL: Geological Association of Canada.

Li, Y., Craven, J., Schweig, E.S. and Obermeier, S.F., 1996, Sand boils induced by the 1993 Mississippi River flood: could they one day be misinterpreted as earthquake-induced liquefaction? *Geology* 24(2): 171–174.

Lienhart, D.A., 2013, Long-term geological challenges of dam construction in a carbonate terrane. *Environmental and Engineering Geoscience* 19(1): 1–25.

Lim, S.S. and Martin, C.D., 2010, Core disking and its relationship with stress magnitude for Lac du Bonnet granite. *International Journal of Rock Mechanics and Mining Sciences* 47: 254–264.

Lingley, W.S. and Jazdzewski, S.P., 1994, Aspects of growth management planning for mineral resource lands. *Washington Geology* 22(2): 36–45.

Lisle, R.J. and Leyshon, P.R., 2004, *Stereographic Projection Techniques for Geologists and Civil Engineers*, 2nd edn. Cambridge, UK: Cambridge University Press.

Lisle, R.J., Brabham, P. and Barnes, J., 2011, *Basic Geological Mapping*, 5th edn. Chichester, UK: Wiley-Blackwell.

Liu, K.-B., 2007, Uncovering prehistoric hurricane activity. *American Scientist* 95: 126–133.

Liu-Zeng, J., Klinger, Y., Sieh, K., Rubin, C. and Seitz, G., 2006, Serial ruptures of the San Andreas fault, Carrizo Plain, California, revealed by three-dimensional excavations. *Journal of Geophysical Research* 111: B02306.

Lo, K.Y., Wai, R.S.C., Palmer, J.H.L. and Quigley, R.M., 1978, Time-dependent deformation of shaly rocks in southern Ontario. *Canadian Geotechnical Journal* 15: 537–547.

Locat, J. and Lee, H.J., 2002, Submarine landslides: advances and challenges. *Canadian Geotechnical Journal* 39: 193–212.

Locat, J., Lefebvre, G. and Ballivy, G., 1984, Mineralogy, chemistry and physical properties interrelationships of some sensitive clays from Eastern Canada. *Canadian Geotechnical Journal* 21: 530–540.

Locat, J., Leroueil, S., Locat, A. and Lee, H., 2014, Weak layers: their definition and classification from a geotechnical perspective. In *Submarine Mass Movements and their Consequences* (eds S. Krastel, J.-H. Behrmann, D. Völker, et al.), pp. 3–12. Cham, Switzerland: Springer International.

Loew, S., Barla, G. and Diederichs, M., 2010, Engineering geology of Alpine tunnels: past, present and future. In *Geologically Active* (eds A.L. Williams, G.M. Pinches, C.Y. Chin, T.J. McMorran and C.I. Massey), Proceedings of IAEG2010, 11th Congress of the International Association for Engineering Geology and the Environment, Auckland, New Zealand. Boca Raton, FL: CRC Press.

Logan, C., Russell, H.A.J., Sharpe, D.R. and Kenny, F.M., 2006, The role of GIS and expert knowledge in 3-D modelling, Oak Ridges Moraine, Southern Ontario. In *GIS for the Earth Sciences* (ed. J.R. Harris). GAC Special Paper 44. St. John's, NL: Geological Association of Canada, Memorial University of Newfoundland.

Lohman, S.W., 1979, *Ground-Water Hydraulics*. USGS Professional Paper 708. Washington, DC: US Geological Survey.

Lohman, S.W., Bennett, R.R., Brown, R.H., et al., 1972, *Definitions of Selected Ground-Water Terms – Revisions and Conceptual Refinements*. USGS Water-Supply Paper 1988. Washington, DC: US Geological Survey.

Lorig, L. and Stead, D., 2018, Numerical analysis. In *Rock Slope Engineering: Civil Applications*, 5th edn (ed. D.C. Wyllie), ch. 12, pp. 319–348. Boca Raton, FL: CRC Press.

Lucia, F.J., 1999, *Carbonate Reservoir Characterization*. New York, NY: Springer.

Lucius, J.E., Langer, W.H. and Ellefsen, K.J., 2007, *An Introduction to Using Surface Geophysics to Characterize Sand and Gravel Deposits*. USGS Circular 1310. Reston, VA: US Geological Survey.

Lundegard, P.D. and Johnson, P.C., 2006, Source zone natural attenuation at petroleum hydrocarbon spill sites – II: Application to a former oil field. *Groundwater Monitoring and Remediation* 26(4): 93–106.

Lunne, T., Robertson, P.K. and Powell, J.J.M., 1997, *Cone Penetration Testing in Geotechnical Practice*. London, UK: Spon.

Mabry, R.E., 2009, Uncertainties in evaluating mine subsidence. *Geo-Strata* 13(1): 20–22.

MacDonald, A.M. and Allen, D.J., 2001, Aquifer properties of the Chalk of England. *Quarterly Journal of Engineering Geology and Hydrogeology* 34: 371–384.

Mackay, D.M. and Cherry, J.A., 1989, Groundwater contamination: pump-and-treat remediation. *Environmental Science and Technology* 23(6): 630–636.

Mackay, J.R., 1970, Disturbances to the tundra and forest tundra environment of the western Arctic. *Canadian Geotechnical Journal* 7: 420–432.

Maerz, N.H., Youssef, A.M., Otoo, J.N., Kassebaum, T.J. and Duan, Y., 2013, A simple method for measuring discontinuity orientations from terrestrial LiDAR data. *Environmental and Engineering Geoscience* 19(2): 185–194.

Manville, V., Major, J.J. and Fagents, S.A., 2013, Modeling lahar behavior and hazards. In *Modeling Volcanic Processes: The Physics and Mathematics of Volcanism* (eds S.A. Fagents, T.K.P. Gregg and R.M.C. Lopes), pp. 300–330. Cambridge, UK: Cambridge University Press.

Marinos, P., Hoek, E. and Marinos, V., 2006, Variability of the engineering properties of rock masses quantified by the geological strength index: the case of ophiolites with special emphasis on tunnelling. *Bulletin of Engineering Geology and the Environment* 65: 129–142.

Martin, C.D., Kaiser, P.K. and McCreath, D.R., 1999, Hoek–Brown parameters for predicting the depth of brittle failure around tunnels. *Canadian Geotechnical Journal* 36: 136–151.

Martin, C.D., Kaiser, P.K. and Christiansson, R., 2003, Stress, instability and design of underground excavations. *International Journal of Rock Mechanics and Mining Sciences* 40: 1027–1047.

Masset, O. and Loew, S., 2010, Hydraulic conductivity distribution in crystalline rocks, derived from inflows to tunnels and galleries in the Central Alps, Switzerland. *Hydrogeology Journal* 18: 863–891.

Matheson, G.D. and Quigley, P., 2016, Evaluating pyrite-induced swelling in Dublin mudrocks. *Quarterly Journal of Engineering Geology and Hydrogeology* 49: 47–66.

Mathez, E.A. and Webster, J.D., 2004, *The Earth Machine: The Science of a Dynamic Planet*. New York, NY: Columbia University Press.

Matthews, M.C., Clayton, C.R.I. and Rigby-Jones, J., 2000, Locating dissolution features in Chalk. *Quarterly Journal of Engineering Geology and Hydrogeology* 33: 125–140.

Matula, M. (chair.), 1981, Recommended symbols for engineering geology mapping. *Bulletin of the International Association of Engineering Geology* 24: 227–234.

Mavko, G., Mukerji, T. and Dvorkin, J., 2009, *The Rock Physics Handbook*, 2nd edn. Cambridge, UK: Cambridge University Press.

Mayne, P.W., Christopher, B.R. and DeJong, J., 2001, *Manual of Subsurface Investigations*. National Highway Institute FHWA NHI-01-031. Washington, DC: Federal Highways Administration.

Mazzotti, S., Lambert, A., Van der Kooij, M. and Mainville, A., 2009, Impact of anthropogenic subsidence on relative sea-level rise in the Fraser River delta. *Geology* 37(9): 771–774.

McCalpin, J.P. (ed.), 2009, *Paleoseismology*. International Geophysics, vol. 95. Amsterdam, The Netherlands: Academic Press, Elsevier.

McCarthy, D.F., 1998, *Essentials of Soil Mechanics and Foundations: Basic Geotechnics*. Upper Saddle River, NJ: Prentice-Hall.

McElwee, C.D., Butler, J.J. and Healey, J.M., 1991, A new sampling system for obtaining relatively undisturbed samples of unconsolidated sand and gravel. *Groundwater Monitoring and Remediation* 11: 182–191.

McLean, D.G., Church, M. and Tassone, B., 1999, Sediment transport along lower Fraser River: 1. Measurements and hydraulic computations. *Water Resources Research* 35(8): 2533–2548.

McWhorter, D.B. and Kueper, B.H., 1996, Mechanics and mathematics of the movement of dense non-aqueous phase liquids (DNAPLs) in porous media. In *Dense Chlorinated Solvents and Other DNAPLs in Groundwater* (eds J. Pankow and J.A. Cherry). Portland, OR: Waterloo Press.

Meehan, R., 1984, *The Atom and the Fault: Experts, Earthquakes and Nuclear Power*. Cambridge, MA: MIT Press.

Meinardus, H.W., Jasek, N.A., Grisak, G.E. and Saulnier, Jr., G.J., 1993, Offset shotholes in West Texas: evidence for neotectonics, stress relief, or blasting phenomena? *Bulletin of the Association of Engineering Geologists* 30(4): 427–442.

Meinzer, O.E., 1922, *The Occurrence of Ground Water in the United States, with a Discussion of Principles*. Ph.D. Thesis, University of Chicago.

Melhorn, W.N. and Kempton, J.P. (eds), 1991, *Geology and Hydrogeology of the Teays–Mahomet Bedrock Valley Systems*. Special Paper 258. Boulder, CO: Geological Society of America.

Meyer, J.R., Parker, B.L. and Cherry, J.A., 2008, Detailed hydraulic head profiles as essential data for defining hydrogeologic units in layered fractured sedimentary rock. *Environmental Geology* 56(1): 27–44.

Miall, A.D., 1992, Alluvial deposits. In *Facies Models: Response to Sea Level Change* (eds R.G. Walker and N.P. James), pp. 119–142. St. John's, NL: Geological Association of Canada.

Micheletti, N., Chandler, J.H. and Lane, S.N., 2015, Investigating the geomorphological potential of freely available and accessible structure-from-motion photogrammetry using a smartphone. *Earth Surface Processes and Landforms* 40(4): 473–486.

Mickelson, D.M., Edil, T.B. and Guy, D.E., 2004, Erosion of coastal bluffs in the Great Lakes. In *Formation, Evolution, and Stability of Coastal Cliffs – Status and Trends* (eds M.A. Hampton and G.B. Griggs). USGS Professional Paper 1693, pp. 107–129. Denver, CO: US Geological Survey.

Middleton, G.V. and Wilcock, P.R., 1994, *Mechanics in the Earth and Environmental Sciences*. Cambridge, UK: Cambridge University Press.

Migon, P., 2010, Mass movement and landscape evolution in weathered granite and gneiss terrains. In *Weathering as a Predisposing Factor to Slope Movements* (eds D. Calcaterra and M. Parise). Engineering Geology Special Publication 23, pp. 33–45. London, UK: Geological Society of London.

Milsom, J. and Eriksen, A., 2011, *Field Geophysics*, 4th edn. Chichester, UK: John Wiley and Sons.

Mitchell, J.K. and Soga, K., 2005, *Fundamentals of Soil Behavior*, 3rd edn. Hoboken, NJ: John Wiley and Sons.

Mitchell, J.K., Seed, R.B. and Seed, H.B., 1990, Kettleman Hills Waste Landfill slope failure: I: Liner-system properties. *Journal of Geotechnical Engineering* 116(4): 647–668.

Moeller, T. and O'Connor, R., 1971, *Ions in Aqueous Systems: An Introduction to Chemical Equilibrium and Solution Chemistry*. New York, NY: McGraw-Hill.

Mollard, J.D., 1988, Fracture lineament research and applications on the western Canadian plains. *Canadian Geotechnical Journal* 25: 749–767.

Montgomery, D.R., 2002, Valley formation by fluvial and glacial erosion. *Geology* 30(11): 1047–1050.

Moore, H.L., 2006, A proactive approach to planning and designing highways in east Tennessee karst. *Environmental and Engineering Geoscience* 12(2): 147–160.

Moore, P.L. and Iverson, N.R., 2002, Slow episodic shear of granular materials regulated by dilatant strengthening. *Geology* 30(9): 843–846.

Moore, R.D., Fleming, S.W., Menounos, B., et al., 2009, Glacier change in western North America: influences on hydrology, geomorphic hazards and water quality. *Hydrological Processes* 23(1): 42–61.

Morgenstern, N.R., 2000, Common ground. In *Proc. GeoEng 2000, 1st Int. Conf. on Geotechnical and Geological Engineering*, pp. 1–30. Melbourne, Australia: Australian Geomechanics Society.

Mortimore, R.N., 2012, Making sense of Chalk: a total-rock approach to its engineering geology. *Quarterly Journal of Engineering Geology and Hydrogeology* 45: 252–334.

Mortimore, R.N., Lawrence, J., Pope, D., Duperret, A. and Genter, A., 2004, Coastal cliff geohazards in weak rock: the UK Chalk cliffs of Sussex. In *Coastal Chalk Cliff Instability* (eds R.N. Mortimore and A. Duperret). Engineering Geology Special Publication 20, pp. 3–31. London, UK: Geological Society of London.

Morton, R.A. and Bernier, J.C., 2010, Recent subsidence-rate reductions in the Mississippi Delta and their geological implications. *Journal of Coastal Research* 26(3): 555–561.

Morton, R.A., Miller, T.L. and Moore, L.J., 2004, *National Assessment of Shoreline Change: Part 1. Historical Shoreline Changes and Associated Coastal Land Loss along the U.S. Gulf of Mexico*. Open File Report 2004-1043. St. Petersburg, FL: US Geological Survey.

Morton, R.A., Bernier, J.C. and Barras, J.A., 2006, Evidence of regional subsidence and associated interior wetland loss induced by hydrocarbon production, Gulf Coast region, USA. *Environmental Geology* 50: 261–274.

Mount Polley Independent Expert Investigation and Review Panel, 2015, *Report on Mount Polley Tailings Storage Facility Breach, Appendix C: Surface Investigations*. [Available from https://www.mountpolleyreviewpanel.ca/.]

Muir Wood, D., 2009, *Soil Mechanics: A One-Dimensional Introduction*. Cambridge, UK: Cambridge University Press.

Munch, J.H. and Killey, R.W.D., 1985, Equipment and methodology for sampling and testing cohesionless sediments. *Ground Water Monitoring Review* 5(1): 38–42.

Murphy, F. and Herkelrath, W.N., 1996, A sample-freezing drive shoe for a wire line piston core sampler. *Ground Water Monitoring and Remediation* 16(3): 86–90.

Murphy, S., Ouellon, T., Ballard, J.-M., Lefebvre, R. and Clark, I.D., 2011, Tritium–helium groundwater age used to constrain a groundwater flow model of a valley-fill aquifer contaminated with trichloroethylene (Quebec, Canada). *Hydrogeology Journal* 19: 195–207.

Mussett, A.E. and Khan, M.A., 2000, *Looking into the Earth: An Introduction to Geological Geophysics*. Cambridge, UK: Cambridge University Press.

Nagra, 1992, *Grimsel Test Site: The Radionuclide Migration Experiment: Overview of Investigations 1985–1990*. Nagra Technical Report NTB 91-04. Wettingen, Switzerland: Nagra.

National Academies (National Academies of Sciences, Engineering, and Medicine), 2017, *Volcanic Eruptions and their Repose, Unrest, Precursors, and Timing*. Washington, DC: The National Academies Press.

Neilson-Welch, L. and Smith, L., 2001, Saline water intrusion adjacent to the Fraser River, Richmond, British Columbia. *Canadian Geotechnical Journal* 38(1): 67–82.

Neton, M.J., Dorsch, J., Olson, C.D. and Young, S.C., 1994, Architecture and directional scales of heterogeneity in alluvial-fan aquifers. *Journal of Sedimentary Research* 64(2): 245–257.

Neuzil, C.E., 1993, Low fluid pressure within the Pierre Shale: a transient response to erosion. *Water Resources Research* 29(7): 2007–2020.

Neuzil, C.E., 1994, How permeable are clays and shales? *Water Resources Research* 30(2): 145–150.

Newmark, N.M., 1965, Effects of earthquakes on dams and embankments. *Géotechnique* 15(2): 139–160.

Nichol, S.L., Hungr, O. and Evans, S.G., 2002, Large-scale brittle and ductile toppling of rock slopes. *Canadian Geotechnical Journal* 39: 773–788.

Nickman, M., Spaun, O. and Thuro, K., 2006, Engineering geological classification of weak rocks. In *Engineering Geology for Tomorrow's Cities*. Proc. 10th Int. Association of Engineering Geologists International Congress, Nottingham, UK. London, UK: Geological Society of London.

Ning, Z. and Fish, M., 2017, A case study of Global Navigation Satellite System (GNSS) in landslide ground movement monitoring. *Geotechnical News* 35(3): 23–27.

Nishikawa, T., Siade, A.J., Reichard, E.G., et al., 2009, Stratigraphic controls on seawater intrusion and implications for groundwater management, Dominguez Gap area of Los Angeles, California, USA. *Hydrogeology Journal* 17(7): 1699–1725.

Noe, D.C., Higgins, J.D. and Olsen, H.W., 2007, Steeply dipping heaving bedrock, Colorado. *Environmental and Engineering Geoscience* 13(4): 289–344.

Norbury, D., 2010, *Soil and Rock Description in Engineering Practice*. Caithness, UK: Whittles.

Norman, D.K., Cederholm, C.J. and Lingley, W.S., 1998, Flood plains, salmon habitat, and sand and gravel mining. *Washington Geology* 26(2/3): 3–28.

Norrish, N.I. and Wyllie, D.C., 1996, Rock slope stability analysis. In *Landslides: Investigation and Mitigation* (eds A.K. Turner and R.L. Schuster). Special Report 247, ch. 15, pp. 391–425. Washington, DC: Transportation Research Board, National Research Council.

Novakowsi, K. and Lapcevic, P., 2004, Site characterization for the Smithville Phase IV bedrock remediation program. In *Proc. GéoQuébec 2004, 5th Joint CGS/IAH Conf.*, Québec City, Canada.

NRC (National Research Council), 2003, *Living on an Active Earth: Perspectives in Earthquake Science*. Washington, DC: National Academies Press.

NRC (National Research Council), 2005, *Contaminants in the Subsurface: Source Zone Assessment and Remediation*. Washington, DC: National Academies Press.

Obermeier, S.F., 2009, Using liquefaction-induced and other soft-sediment features for paleoseismic analysis. *International Geophysics* 95: 497–564.

Obermeier, S.F., Olson, S.M. and Green, R.A., 2005. Field occurrences of liquefaction-induced features: a primer for engineering geologic analysis of paleoseismic shaking. *Engineering Geology* 76(3–4): 209–234.

O'Connell, D.R.H., Ostenaa, D.A., Levish, D.R. and Klinger, R.E., 2002, Bayesian flood frequency analysis with paleohydrologic bound data. *Water Resources Research* 38(5): doi:10.1029/2000WR000028.

O'Connor, J.E. and Baker, V.R., 1992, Magnitudes and implications of peak discharges from glacial Lake Missoula. *Geological Society of America Bulletin* 104(3): 267–279.

Odum, J.K., Stephenson, W.J., Williams, R.A., et al., 1999, Shallow high-resolution seismic-reflection imaging of karst structures within the Floridan Aquifer System, Northeastern Florida. *Journal of Environmental and Engineering Geophysics* 4(4): 251–261.

Oelkers, E.H., 1996, Physical and chemical properties of rocks and fluids for chemical mass transport calculations. *Reviews in Mineralogy and Geochemistry* 34(1): 131–191.

Ohnaka, M., 2013, *The Physics of Rock Failure and Earthquakes*. Cambridge, UK: Cambridge University Press.

O'Leary, D. and Isidoro, A., 2016, The value of using softcopy mapping tools for siting of engineered tailing impoundments. *Geotechnical News* 34(4): 33–36.

Ollier, C.D., 2010, Very deep weathering and related landslides. In *Weathering as a Predisposing Factor to Slope Movements* (eds D. Calcaterra and M. Parise). Engineering Geology Special Publication 23, pp. 5–14. London, UK: Geological Society of London.

Oppenheimer, C., 2011, *Eruptions that Shook the World*. Cambridge, UK: Cambridge University Press.

Ostenaa, D.A. and O'Connell, D.R.H., 2005, *Big Lost River Flood Hazard Study, Idaho National Laboratory: Summary Document*. Report 2005-2, Bureau of Reclamation. Denver, CO: US Department of the Interior.

Osterkamp, T.E. and Jorgenson, M.T., 2009, Permafrost conditions and processes. In *Geological Monitoring* (eds R. Young and L. Norby), pp. 141–162. Boulder, CO: Geological Society of America.

Overduin, P.P., Strzelecki, M.C., Grigoriev, M.N., et al., 2014, *Coastal Changes in the Arctic*. Special Publication 388(1), pp. 103–129. London, UK: Geological Society of London.

Oyama, T. and Chigira, M., 1999, Weathering rate of mudstone and tuff on old unlined tunnel walls. *Engineering Geology* 55: 15–27.

Palmer, A., 1997, Geotechnical evidence of ice scour as a guide to pipeline burial depth. *Canadian Geotechnical Journal* 34(6): 1002–1003.

Palmstrom, A. and Berthelsen, O., 1988, The significance of weakness zones in rock tunnelling. In *ISRM Symposium on Rock Mechanics and Power Plants*, Madrid, Spain, pp. 381–388. Lisbon, Portugal: International Society for Rock Mechanics.

Pankow, J.F. and Cherry, J.A. (eds), 1996, *Dense Chlorinated Solvents and Other DNAPLs in Groundwater: History, Behavior and Remediation*. Portland, OR: Waterloo Press.

Park, C.B., Miller, R.D., Xia, J. and Ivanov, J., 2007, Multichannel analysis of surface waves (MASW) – active and passive methods. *The Leading Edge* 26(1): 60–64.

Paronuzzi, P., Rigo, E. and Bolla, A., 2013, Influence of filling–drawdown cycles of the Vajont reservoir on Mt. Toc slope stability. *Geomorphology* 191: 75–93.

Parry, S., 2016, Landslide hazard assessments: problems and limitations. Examples from Hong Kong. In *Developments in Engineering Geology* (eds M.J. Eggers, J.S. Griffiths, S. Parry and M.G. Culshaw). Engineering Geology Special Publication 27, pp. 135–145. London, UK: Geological Society of London.

Parsons, P.J., 1960, *Movement of Radioactive Wastes through Soil. Part I: Soil and Ground-water Investigations in Lower Perch Lake Basin*. Atomic Energy of Canada Ltd., Chalk River, Report 1038.

Pâsse, T., 2004, *The Amount of Glacial Erosion of the Bedrock*. SKB Technical Report TR-04-25. Stockholm, Sweden: Swedish Nuclear Fuel and Waste Management Co.

Patton, F.D., 1966, Multiple modes of shear failure in rock. In *Proc. 1st Int. Congress on Rock Mechanics*, Lisbon, Portugal, pp. 509–513.

Patton, F.D. and Deere, D.U., 1971, Significant geological factors in rock slope stability. In *Planning Open Pit Mines*, pp. 143–151. Johannesburg, South Africa: South African Institute of Mining and Metallurgy.

Patton, F.D., 2006, The role of the Downie Slide in the development of 3D groundwater instrumentation. In *Proc. 59th Canadian Geotechnical Conf. and 7th Joint CGS/IAH-CNC*

Groundwater Specialty Conf., Vancouver, BC, Canada, pp. 1411–1418.

Pearson, F.J., 1999, What is the porosity of a mudrock? In *Muds and Mudstones: Physical and Fluid Flow Problems* (eds A.C. Aplin, A.J. Fleet, J.H.S. Macquaker). Special Publication 158, pp. 9–21. London, UK: Geological Society of London.

Peck, R.B., 1969, Advantages and limitations of the observation method in applied soil mechanics, *Géotechnique* 19(2): 171–187.

Peck, R.B., 1991, *Engineering Judgement*. DVD. Albuquerque, NM, and Richmond, BC: BiTech Publishers.

Pehme, P.E., Parker, B.L., Cherry, J.A., Molson, J.W. and Greenhouse, J.P., 2013, Enhanced detection of hydraulically active fractures by temperature profiling. *Journal of Hydrology* 484: 1–15.

Peterson, R.O. and Rumbaugh, J.O., 2012, *Hydrogeologic Impacts Observed during the January 2010 Freeze Event in Dover/Plant City, Hillsborough County, Florida*. Brooksville, FL: Southwest Florida Water Management District.

Petley, D., 2012, Remote sensing techniques and landslides. In *Landslides: Types, Mechanisms and Modeling* (eds J.J. Clague and D. Stead). Cambridge, UK: Cambridge University Press.

Pettijohn, F.J., Potter, P.E. and Siever, R., 1987, *Sand and Sandstone*, 2nd edn. New York, NY: Springer.

Picarelli, L. and Di Maio, C., 2010, Deterioration processes of hard clays and clay shales. In *Weathering as a Predisposing Factor to Slope Movements* (eds D. Calcaterra and M. Parise). Engineering Geology Special Publication 23, pp. 15–32. London, UK: Geological Society of London.

Picarelli, L., Urciuoli, G. and Russo, C., 2004, Effect of groundwater regime on the behavior of clayey slopes. *Canadian Geotechnical Journal* 41: 467–484.

Picarelli, L., Leroueil, S., Olivares, L., et al., 2012, Groundwater in slopes. In *Landslides: Types, Mechanisms and Modeling* (eds J.J. Clague and D. Stead), pp. 235–251. Cambridge, UK: Cambridge University Press.

Pierson, T.C., 2005, Hyperconcentrated flow – transitional process between water flow and debris flow. In *Debris-Flow Hazards and Related Phenomena* (eds M. Jakob and O. Hungr), pp. 159–202. Berlin, Germany: Springer.

Pierson, T.C. and Scott, K.M., 1985, Downstream dilution of a lahar: transition from debris flow to hyperconcentrated streamflow. *Water Resources Research* 21(10): 1511–1524.

Piotrowski, J.A., 1997, Subglacial hydrology in north-western Germany during the last glaciation: groundwater flow, tunnel valleys and hydrological cycles. *Quaternary Science Reviews* 16: 169–185.

Pitkin, S.E., Cherry, J.A., Ingleton, R.A. and Broholm, M., 1999, Field demonstrations using the Waterloo ground water profiler. *Ground Water Monitoring and Remediation* 19(2): 122–131.

Plummer, L.N. and Sprinkle, C.L., 2001, Radiocarbon dating of dissolved inorganic carbon in groundwater from confined parts of the Upper Floridan aquifer, Florida, USA. *Hydrogeology Journal* 9: 127–150.

Plummer, L.N. and Wigley, T.M.L., 1976, The dissolution of calcite in CO_2-saturated solutions at 25° and 1 atmosphere total pressure. *Geochimica et Cosmochimica Acta* 40: 191–202.

Pollard, D.D. and Fletcher, R.C., 2005, *Fundamentals of Structural Geology*. Cambridge, UK: Cambridge University Press.

Pollard, W.H. and French, H.M., 1980, A first approximation of the volume of ground ice, Richards Island, Pleistocene Mackenzie delta, Northwest Territories, Canada. *Canadian Geotechnical Journal* 17: 509–516.

Pope, G.A., Sepehrnoori, K., Sharma, M.K., et al., 1999, *Three-Dimensional NAPL Fate and Transport Model*. Report EPA/600/R-99/011. Cincinnati, OH: US Environmental Protection Agency.

Post, V.E. and von Asmuth, J.R., 2013, Hydraulic head measurements – new technologies, classic pitfalls. *Hydrogeology Journal* 21(4): 737–750.

Poston, S.W. and Berg, R.R., 1997, *Overpressured Gas Reservoirs*. Richardson, TX: Society of Petroleum Engineers.

Price, D.G. (ed./comp. by M.H. de Freitas), 2009, *Engineering Geology: Principles and Practice*. Berlin: Springer.

Price, N.J. and Cosgrove, J.W., 2005, *Analysis of Geological Structures*. Cambridge, UK: Cambridge University Press.

Price, W.A., 2003, Challenges posed by metal leaching and acid rock drainage, and approaches to address them. In *Environmental Aspects of Mine Wastes* (eds J.L. Jambor, D.W. Blowes and A.I.M. Ritchie), ch. 1, pp. 1–10. Ottawa, ON: Mineralogical Association of Canada.

Priest, G.R., Schulz, W.H., Ellis, W.L., et al., 2011, Landslide stability: role of rainfall-induced, laterally propagating, pore-pressure waves. *Environmental and Engineering Geoscience* 17(4): 315–335.

Pringle, P., 1994, Volcanic hazards in Washington – a growth management perspective. *Washington Geology* 22(2): 25–33.

Proctor, R.J., 1981, Let's teach geology to the civil engineering student. *Engineering Issues – Journal of Professional Activities* 107(EI1): 61–63.

Pryor, W.A., 1973, Permeability–porosity patterns and variations in some Holocene sand bodies. *American Association of Petroleum Geologists Bulletin* 57(1): 162–189.

Pugh, D. and Woodworth, P., 2014, *Sea-Level Science: Understanding Tides, Surges, Tsunamis and Mean Sea-Level Changes*. Cambridge, UK: Cambridge University Press.

Pusch, R., 1995, *Rock Mechanics on a Geological Base*. Amsterdam, The Netherlands: Elsevier Science.

Pye, K. and Miller, J.A., 1990, Chemical and biochemical weathering of pyritic mudrocks in a shale embankment. *Quarterly Journal of Engineering Geology* 23: 365–381.

Quigley, R.M., Zajic, J.E., McKyes, E. and Yong, R.N., 1973, Biochemical alteration and heave of black shale: detailed observations and interpretations. *Canadian Journal of Earth Sciences* 10: 1005–1015.

Quigley, R.M., Gelinas, P.J., Bou, W.T. and Packer, R.W., 1977, Cyclic erosion–instability relationships: Lake Erie north shore bluffs. *Canadian Geotechnical Journal* 14: 310–323.

Quigley, R.M., Gwyn, Q.H.J., White, O.L., et al., 1983, Leda clay from deep boreholes at Hawkesbury, Ontario. Part I: Geology and geotechnique. *Canadian Geotechnical Journal* 20: 288–298.

Quinn, P., Cherry, J.A. and Parker, B.L., 2012, Hydraulic testing using a versatile straddle packer system for improved transmissivity estimation in fractured-rock boreholes. *Hydrogeology Journal* 20: 1529–1547.

Raaen, A.M., Horsrud, P., Kjørholt, H. and Økland, D., 2006, Improved routine estimation of the minimum horizontal stress component from extended leak-off tests. *International Journal of Rock Mechanics and Mining Sciences* 43(1): 37–48.

Rauch, A.F. and Martin, J.R., 2000, EPOLLS model for predicting average displacements on lateral spreads. *Journal of Geotechnical and Geoenvironmental Engineering* 126(4): 360–371.

Raudsepp, M. and Pani, E., 2003, Application of Rietveld analysis to environmental mineralogy. In *Environmental Aspects of Mine Wastes* (eds J.L. Jambor, D.W. Blowes and A.I.M. Ritchie), ch. 8, pp. 165–180. Ottawa, ON: Mineralogical Association of Canada.

Rauh, F., Spaun, G. and Thuro, K., 2006, Assessment of the swelling potential of anhydrite in tunnelling projects. In *IAEG 2006: Engineering Geology for Tomorrow's Cities*. Paper 473. London, UK: Geological Society of London.

Raven, K.G. and Gale, J.E., 1985, Water flow in a natural rock fracture as a function of stress and sample size. *International Journal of Rock Mechanics and Mining Sciences* 22(4): 251–261.

Raven, K.G., Lafleur, D.W. and Sweezey, R.A., 1990, Monitoring well into abandoned deep-well disposal formations, Sarnia, Ontario. *Canadian Geotechnical Journal* 27: 105–118.

Raven, K.G., Novakowski, K.S., Yager, R.M. and Heystee, R.J., 1992, Supernormal fluid pressures in sedimentary rocks of southern Ontario–western New York State. *Canadian Geotechnical Journal* 29(1): 80–93.

Reed, S.J.B., 2005, *Electron Microprobe Analysis and Scanning Electron Microscopy in Geology*, 2nd edn. Cambridge, UK: Cambridge University Press.

Remenda, V.H., Cherry, J.A. and Edwards, T.W.D., 1994, Isotopic composition of old ground water from Lake Agassiz: implications for late Pleistocene climate. *Science* 266: 1975–1978.

Renken, R.A., Cunningham, K.J., Zygnerski, M.R., et al., 2005, Assessing the vulnerability of a municipal well field to contamination in a karst aquifer. *Environmental and Engineering Geoscience* 11(4): 319–331.

Resnick, R. and Halliday, D., 1960, *Physics for Students of Science and Engineering*. New York, NY: John Wiley and Sons.

Reynolds, J.M., 2011, *An Introduction to Applied and Environmental Geophysics*, 2nd edn. Chichester, UK: John Wiley and Sons.

Ridley, J., 2013, *Ore Deposit Geology*. Cambridge, UK: Cambridge University Press.

Riley, F.S., 1969, Analysis of borehole extensometer data from central California. In *Land Subsidence*, vol. 2 (ed. L.J. Tison). IASH Publication 89, pp. 423–431. Gentbrugge, Belgium: International Association of Scientific Hydrology.

Rivera, A., 2014. Groundwater basics. In *Canada's Groundwater Resources* (ed. A. Rivera), pp. 22–61. Markham, ON: Fitzhenry & Whiteside.

Roberson, J.A. and Crowe, C.T., 1993, *Engineering Fluid Mechanics*, 5th edn. Boston, MA: Houghton Mifflin.

Robert, A., 2003, *River Processes: An Introduction to Fluvial Dynamics*. London, UK: Arnold.

Robertson, P.K., 2009, Interpretation of cone penetration tests – a unified approach. *Canadian Geotechnical Journal* 46: 1337–1355.

Robertson, P.K., 2010, Soil behavior type from the CPT: an update. In *CPT'10, Second Int. Symp. on CPT*, Huntingdon Beach, CA.

Robinson, J.L., 1995, *Hydrogeology and Results of Tracer Tests at the Old Tampa Well Field in Hillsborough County, with Implications for Wellhead-Protection Strategies in West-Central Florida*, USGS Water-Resource Investigation Report 93-4171. Tallahassee, FL: US Geological Survey.

Rockwell, T., 2010, The Rose Canyon fault zone in San Diego. In *Fifth Int. Conf. on Recent Advances in Geotechnical Earthquake Engineering and Soil Dynamics*, San Diego, California, Paper 7.06c.

Rogers, J.D., 1992, Reassessment of the St. Francis Dam failure. In *Engineering Geology Practice in Southern California* (eds B.W. Pipkin and R.J. Proctor). AEG Special Publication 4, pp. 639–666. Belmont, CA: Star Publishing.

Rogers, J.D., 2006, Subsurface exploration using the standard penetration test and the cone penetrometer test. *Environmental and Engineering Geoscience* 12(2): 161–179.

Rogers, J.D., Boutwell, G.P., Schmitz, D.W., et al., 2008, Geologic conditions underlying the 2005 17th Street Canal levee failure in New Orleans. *Journal of Geotechnical and Geoenvironmental Engineering*, 134(5): 583–601.

Rogers, J.D., Kemp, P.G., Bosworth, H.J. and Seed, R.B., 2015, Interaction between the US Army Corps of Engineers and the Orleans Levee Board preceding the drainage canal wall failures and catastrophic flooding of New Orleans in 2005. *Water Policy* 17: 707–723.

Ross, M., Campbell, J.E., Parent, M. and Adama, R.S., 2009, Paleo-ice streams and the subglacial landscape mosaic of the North American mid-continent prairies. *Boreas* 38(3): 421–439.

Rowland, J.C. and Coon, E.T., 2016, From documentation to prediction: raising the bar for thermokarst research. *Hydrogeology Journal* 24: 645–648.

Roy, W.R. and Griffin, R.A., 1985, Mobility of organic solvents in water-saturated soil materials. *Environmental Geology and Water Science* 7(4): 241–247.

Rudolph, D.L. and Frind, E.O., 1991, Hydraulic response of highly compressible aquitards during consolidation. *Water Resources Research* 27(1): 17–30.

Ruland, W., Cherry, J.A. and Feenstra, S., 1991, The depth of fractures and active ground-water flow in a clayey till plain in southwestern Ontario. *Ground Water* 29(3): 405–417.

Russell, H.A.J., Sharpe, D.R. and Cummings, D.I., 2007, A framework for buried-valley aquifers in southern Ontario. In *OttawaGeo2007, Proc. 8th Joint CGS/IAH-CNC Groundwater Conf.*, Ottawa, ON, pp. 386–393.

Rust, B.R., 1977, Mass flow deposits in a Quaternary succession near Ottawa, Canada: diagnostic criteria for subaqueous outwash. *Canadian Journal of Earth Sciences* 14: 175–184.

Rutledge, D.R. and Meyerholtz, S.Z., 2005, Using the global positioning system (GPS) to monitor the performance of dams. *Geotechnical News* 23(4): 25–28.

Ruxton, B.P. and Berry, L., 1957, Weathering of granite and associated erosional features in Hong Kong. *Geological Society of America Bulletin* 68: 1263–1282.

Sahakian, V., Bormann, J., Driscoll, N., et al., 2017, Seismic constraints on the architecture of the Newport–Inglewood/Rose Canyon fault: implications for the length and magnitude of future earthquake ruptures. *Journal of Geophysical Research: Solid Earth* 122(3), 2085–2105.

Sallenger, Jr., A.H., 2000, Storm impact scale for barrier islands. *Journal of Coastal Research* 16(3): 890–895.

Samsonov, S.V., d'Oreye, N., González, P.J., et al., 2014, Rapidly accelerating subsidence in the Greater Vancouver region from two decades of ERS–ENVISAT–RADARSAT-2 DInSAR measurements. *Remote Sensing of Environment* 143: 180–191.

Santi, P.M., 2006, Field methods for characterizing weak rock for engineering. *Environmental and Engineering Geoscience* 12(1): 1–11.

Santi, P.M. and Gregg, J.M., 2002, Engineering stratigraphic columns. *Environmental and Engineering Geoscience* 8(3): 237–241.

Sargent, C. and Goulty, N.R., 2009, Seismic reflection survey for investigation of gypsum dissolution and subsidence at Hell Kettles, Darlington, UK. *Quarterly Journal of Engineering Geology and Hydrogeology* 42: 31–38.

Sass, I. and Burbaum, U., 2009, A method for assessing adhesion of clays to tunneling machines. *Bulletin of Engineering Geology and the Environment* 68: 27–34.

Sauer, E.K., 1978, The engineering significance of glacier ice-thrusting. *Canadian Geotechnical Journal* 15(4): 457–472.

Sauer, E.K., Egeland, A.K. and Christiansen, E.A., 1993, Preconsolidation of tills and intertill clays by glacial loading in southern Saskatchewan, Canada. *Canadian Journal of Earth Sciences* 30: 420–433.

Saunders, A.D., 2005, Large igneous provinces: origin and environmental consequences. *Elements* 1: 259–263.

Schaetzl, R.J. and Thompson, M.L., 2015, *Soils: Genesis and Geomorphology*, 2nd edn. New York, NY: Cambridge University Press.

Schandl, E.S., 2001, *A Petrographic Report: Churchill Falls – West Tailrace Tunnel*. Toronto, ON: Geoconsult.

Scholz, C.H., 2002, *The Mechanics of Earthquakes and Faulting*, 2nd edn. Cambridge, UK: Cambridge University Press.

Schulz, K., 2015, The really big one. *New Yorker* July 20.

Schulz, W.H., Kean, J.W. and Wang, G., 2009, Landslide movement in southwest Colorado triggered by atmospheric tides. *Nature Geoscience* 2: 863–866.

Schwille, F., 1975, Groundwater pollution by mineral oil products. In *Groundwater Pollution Symposium*, Proceedings of the Moscow Symposium, August 1971. IAHS-AISH Publication 103, pp. 226–240. Wallingford, UK: IAHS Press.

Schwille, F., 1988, *Dense Chlorinated Solvents in Porous and Fractured Media: Model Experiments* (transl. J.F. Pankow). Chelsea, MI: Lewis.

Scott, J.S., 1971, *Surficial Geology of the Rosetown Map-Area, Saskatchewan*. GSC Bulletin 190. Ottawa, ON: Geological Survey of Canada.

Scott, J.S., 2003, *A Review of the Geology and Geotechnical Characteristics of Champlain Sea Clays of the Ottawa River Valley with Reference to Slope Failures*. Open File Report 4475. Ottawa, ON: Geological Survey of Canada.

Scott, K.M., Vallance, J.W. and Pringle, P.T., 1995, *Sedimentology, Behavior and Hazards of Debris Flows at Mount Ranier, Washington*. US Geological Survey Professional Paper 1547. Washington, DC: US Government Printing Office.

Seager, R., 2006, The source of Europe's mild climate. *American Scientist* 94: 334–341.

Seal, R.R. and Hammarstrom, J.M., 2003, Geoenvironmental models of mineral deposits: examples from massive sulfide and gold deposits. In *Environmental Aspects of Mine Wastes* (eds J.L. Jambor, D.W. Blowes and A.I.M. Ritchie), ch. 2, pp. 11–50. Ottawa, ON: Mineralogical Association of Canada.

Seed, R.B., Bea, R.G., Abdelmalak, R.I., et al., 2006, *Investigation of the Performance of the New Orleans Flood Protection Systems in Hurricane Katrina on August 29, 2005*. Report of the Independent Levee Investigation Team. Berkeley, CA: University of California.

Semenza, E., 2010, *The Story of Vaiont Told by the Geologist who Discovered the Landslide*. Ferrara, Italy: K-Flash.

Semenza, E. and Ghirotti, M., 2000, History of the 1963 Vaiont slide: the importance of geological factors. *Bulletin of Engineering Geology and the Environment* 59: 87–97.

Sen, M.A. and Abbott, M.A.W., 1991, Hydrogeological investigation of a fault in clay. *Quarterly Journal of Engineering Geology* 24: 413–426.

Sengebush, R.M., Heagle, D.J. and Jackson, R.E., 2015, The late Quaternary history and groundwater quality of a coastal aquifer, San Diego, California. *Environmental and Engineering Geoscience* 21(4): 249–275.

Senger, R.K. and Fogg, G.E., 1987, Regional underpressuring in deep brine aquifers, Palo Duro Basin, Texas. 1. Effects of hydrostratigraphy and topography. *Water Resources Research* 23(8): 1481–1493.

Sepúlveda, S.A., Murphy, W., Jibson, R.W. and Petley, D.N., 2005, Seismically induced rock slope failures resulting from topographic amplification of strong ground motions: the case of Pacoima Canyon, California. *Engineering Geology* 80: 336–348.

Serva, L. and Slemmons, D.B. (eds), 1995, *Perspectives in Paleoseismology*. Special Publication 6. Zanesville, OH: Association of Engineering Geologists.

Seth, R., Mackay, D. and Muncke, J., 1999, Estimating the organic carbon partition coefficient and its variability for hydrophobic chemicals. *Environmental Science and Technology* 33: 2390–2394.

Shakoor, A. and Erguler, Z.A., 2009, Characterizing the slaking behavior of clay-bearing rocks. *GeoStrata* 10(2): 39–41.

Sharp, Jr., J.M., 1988, Alluvial aquifers along major rivers. In *The Geology of North America*, vol. O-2, *Hydrogeology* (eds W. Back, J.S. Rosenshein and P.R. Seaber), ch. 33. Boulder, CO: Geological Society of America.

Sharpe, D.R., 1994, Peterborough drumlin field. In *A Field Guide to the Glacial and Postglacial Landscape of Southeastern Ontario and Part of Québec* (ed. R. Gilbert). GSC Bulletin 453, pp. 66–69. Ottawa, ON: Natural Resources Canada.

Sharpe, D.R., Pugin, A., Pullan, S.E. and Gorrell, G., 2003, Application of seismic stratigraphy and sedimentology to regional hydrogeological investigations: an example from Oak Ridges Moraine, southern Ontario, Canada. *Canadian Geotechnical Journal* 40: 711–730.

Sharpe, D.R., Pugin, A., Pullan, S.E. and Shaw, J., 2004, Regional unconformities and the sedimentary architecture of the Oak Ridge Moraine area, southern Ontario. *Canadian Journal of Earth Sciences* 41: 183–198.

Sharpe, D.R., Pugin, A. and Russell, H.A.J., 2013, *The Significance of Buried Valleys to Groundwater Systems in the Oak Ridges Moraine Region, Ontario: Extent, Architecture, Sedimentary Facies and Origin of Valley Settings in the ORM Region.* Open File Report 6980. Ottawa, ON: Geological Survey of Canada.

Shearer, P.M., 2009, *Introduction to Seismology*, 2nd edn. Cambridge, UK: Cambridge University Press.

Sheng, Z., Helm, D.C. and Li, J., 2003, Mechanisms of earth fissuring caused by groundwater withdrawal. *Environmental and Engineering Geoscience* 9(4): 351–362.

Shepley, M.G., Pearson, A.D., Smith, G.D. and Banton, C.J., 2008, The impact of coal mining subsidence on groundwater resources management of the East Midlands Permo-Triassic Sandstone aquifer, England. *Quarterly Journal of Engineering Geology and Hydrogeology* 41: 425–438.

Sheorey, P.K., 1994, A theory for in situ stresses in isotropic and transversely isotropic rock. *International Journal of Rock Mechanics and Mining Sciences & Geomechanics Abstracts* 31: 23–34.

Shields, D., 2015, Giving credit where credit is due. *Geotechnical News* 33(4): 33–34.

Shlemon, R.J., 1985, Application of soil-stratigraphic techniques to engineering geology. *Bulletin of the Association of Engineering Geologists* 22(2): 129–142.

Sibson, R.H., 2011, The scope of earthquake geology. In *Geology of the Earthquake Source: A Volume in Honour of Rick Sibson* (eds A. Fagereng, V.G. Toy and J.V. Rowland). Special Publication 359, pp. 319–331. London, UK: Geological Society of London.

Sidle, R.C. and Ochiai, H., 2006, *Landslides: Processes, Prediction, and Land Use.* Water Resources Monograph 18. Washington, DC: American Geophysical Union.

Sieh, K.E. and Jahns, R.H., 1984, Holocene activity of the San Andreas fault at Wallace Creek, California. *Geological Society of America Bulletin* 95: 883–896.

Simpson, E.S., 1962, *Transverse Dispersion in Liquid Flow Through Porous Media.* US Geological Survey Professional Paper 411-C. Washington, DC: US Government Printing Office.

Simpson, J.M., Darrow, M.M., Huang, S.L., Daanen, R.P. and Hubbard, T.D., 2016, Investigating movement and characteristics of a frozen debris lobe, south-central Brooks Range, Alaska. *Environmental and Engineering Geoscience* 22(3): 259–277.

Sims, J.E., Elsworth, D. and Cherry, J.A., 1996, Stress-dependent flow through fractured clay till: a laboratory study. *Canadian Geotechnical Journal*, 33(3): 449–457.

Sitar, N., 1985, Goals for basic research in engineering geology. *Bulletin of the Association of Engineering Geologists* 22(4): 435–443.

Skafel, M.G. and Bishop, C.T., 1994, Flume experiments on the erosion of till shores by waves. *Coastal Engineering* 23: 329–348.

Slemmons, D.B., 1995, Complications in making paleoseismic evaluations in the Basin and Range Province, Western United States. In *Perspectives in Paleoseismology* (eds L. Serva and D.B. Slemmons). Special Publication 6, pp. 19–33. Zanesville, OH: Association of Engineering Geologists.

Slingerland, R. and Smith, N.D., 2004, River avulsions and their deposits. *Annual Review of Earth and Planetary Sciences* 32: 255–283.

Smedley, P.L. and Edmunds, W.M., 2002, Redox patterns and trace-element behavior in the East Midlands Triassic sandstone aquifer, U.K. *Ground Water* 40(1): 44–58.

Smith, A. and Ellison, R.A., 1999, Applied geological maps for planning and development: a review of examples from England and Wales, 1983–1996. *Quarterly Journal of Engineering Geology* 32: S1–S44.

Smith, K., 2007, Strategies to predict metal mobility in surficial mining environments. In *Understanding and Responding to Hazardous Substances at Mines Sites in the Western United States* (ed. J.V. DeGraaf). Reviews in Engineering Geology, XVII, pp. 25–45. Boulder, CO: Geological Society of America.

Smith, M.R. and Collis, L., 2001, *Aggregates: Sand, Gravel and Crushed Rock Aggregates for Construction Purposes*, 3rd edn (rev. by P.G. Fookes, J. Lay, I. Sims, M.R. Smith and G. West). Engineering Geology Special Publication 17. London: Geological Society of London.

Smith, N.D., Slingerland, R.L., Pérez-Arlucea, M. and Morozova, G., 1998, The 1870s avulsion of the Saskatchewan River. *Canadian Journal of Earth Sciences* 35: 453–466.

Smith, S.L., Romanowsky, V.E., Lewkowicz, A.G., et al., 2010, Thermal state of permafrost in North America: a contribution to the International Polar Year. *Permafrost and Periglacial Processes* 21: 117–135.

Sneed, M. and Galloway, D.L., 2000, *Aquifer-System Compaction: Analyses and Simulations – the Holly Site, Edwards Air Force Base, Antelope Valley, California.* U.S. Geological Survey Water-Resources Investigations Report 00-4015.

Snyder, N.P., Rubin, D.M., Alpers, C.N., et al., 2004, Estimating accumulation rates and physical properties of sediment behind a dam: Englebright Lake, Yuba River, northern California. *Water Resources Research* 40, W11301.

Solomon, D.K., Poreda, R.J., Schiff, S.L. and Cherry, J.A., 1992, Tritium and helium 3 as groundwater age tracers in the Borden Aquifer. *Water Resources Research* 28(3): 741–755.

Song, K.-I., Cho, G.-C. and Chang, S.-B., 2012, Identification, remediation, and analysis of karst sinkholes in the longest railroad tunnel in South Korea. *Engineering Geology* 135–136: 92–105.

Spagnoli, G, Feinendegen, M. and Fernandez-Steeger, T., 2011, Influence of salt solutions on the undrained shear strength and clogging of smectite–quartz mixtures. *Environmental and Engineering Geoscience* 17(3): 293–305.

Stanley, J.D. and Clemente, P.L., 2017, Increased land subsidence and sea-level rise are submerging Egypt's Nile Delta coastal margin. *GSA Today* 27(5): 4–11.

Stein, S. and Mazzotti, S. (eds), 2007, *Continental Intraplate Earthquakes: Science, Hazard, and Policy Issues*. Special Paper 425. Boulder, CO: Geological Society of America.

Stevenson, G.W., Lukajic, B., Smith, W.R. and Hynes, G., 2004, Rehabilitation of the West Tailrace Tunnel, Churchill Falls, Labrador. In *18th National Tunnelling Association of Canada Conf.*, Edmonton, AB.

Stockdon, H.F., Sallenger Jr., A.H., Holman, R.A. and Howd, P.A., 2007, A simple model for the spatially-variable coastal response to hurricanes. *Marine Geology* 238: 1–20.

Stotler, R.L, Fraper, S.K., El Mugammar, H.T., et al., 2011, Geochemical heterogeneity in a small stratigraphically complex moraine aquifer system (Ontario, Canada): interpretation of flow and recharge using multiple geochemical tracers. *Hydrogeology Journal* 19: 101–115.

Stumm, W. and Morgan, J.J., 1981, *Aquatic Chemistry: An Introduction Emphasizing Chemical Equilibria in Natural Waters*, 2nd edn. New York, NY: John Wiley and Sons. [1st edn, 1970; 3rd edn, 1996.]

Stumpf, A., Malet, J.P., Allemand, P., Pierrot-Deseilligny, M. and Skupinski, G., 2015, Ground-based multi-view photogrammetry for the monitoring of landslide deformation and erosion. *Geomorphology* 231: 130–145.

Sutphin, D.M., Drew, L.J., Fowler, B.K. and Goldsmith, R., 2002, *Techniques for Assessing Sand and Gravel Resources in Glaciofluvial Deposits – an Example Using the Surficial Geologic Map of the Loudon Quadrangle, Merrimack and Belknap Counties, New Hampshire*. Professional Paper 1627. Reston, VA: US Geological Survey.

Swanson, S.K., Bahr, J.M., Bradbury, K.R. and Anderson, K.M., 2006, Evidence for preferential flow through sandstone aquifers in southern Wisconsin. *Sedimentary Geology* 184: 331–342.

Swarbrick, R.E., Osborne, M.J. and Yardley, G.S., 2002, Comparison of over-pressure magnitude resulting from the main generating mechanisms. In *Pressure Regimes in Sedimentary Basins and their Prediction* (eds A.R. Huffman and G.L. Bowers). AAPG Memoir 76, pp. 1–12.

Tellam, J.H. and Barker, R.D., 2006, *Towards Prediction of Saturated-Zone Pollutant Movement in Groundwaters in Fractured Permeable-Matrix Aquifers: the Case of the UK Permo-Triassic Sandstones*. Special Publication 263(1), pp. 1–48. London, UK: Geological Society of London.

Terzaghi, K, 1961, Engineering geology on the job and in the classroom. *Journal of the Boston Society of Civil Engineers* 48: 97–109.

Testa, S.M., 2016, From aggregate availability to sustainability in California. In *Applied Geology in California* (eds R. Anderson and H. Ferriz). AEG Special Publication 26, pp. 379–391. Belmont, CA: Star Publishing.

Thewes, M. and Burger, W., 2005, Clogging of TBM drives in clay – identification and mitigation of risks. In *Underground Space Use: Analysis of the Past and Lessons for the Future* (eds Y. Erdem and T. Solak), pp. 737–742. London: Taylor and Francis.

Thorstenson, D.C., Fisher, D.W. and Croft, M.G., 1979, The geochemistry of the Fox Hills–Basal Hell Creek aquifer in southwestern North Dakota and northwestern South Dakota. *Water Resources Research* 15(6): 1479–1498.

Thuro, K. and Scholz, M., 2004, Deep weathering and alteration in granites – a product of coupled processes. In *Coupled Thermo-Hydro-Mechanical-Chemical Processes in Geosystems – Fundamentals, Modeling, Experiments and Applications* (ed. O. Stephanson). Geo-Engineering Book Series 2, pp. 785–790. Amsterdam, The Netherlands: Elsevier.

Tiab, D. and Donaldson, E.C., 2004, *Petrophysics: Theory and Practice of Measuring Reservoir Rock and Fluid Transport Properties*, 2nd edn. Oxford, UK: Elsevier.

Tingay, M., Reinecker, J. and Müller, B., 2008, *Borehole Breakout and Drilling-Induced Fracture Analysis from Image Logs*. World Stress Map Project. [Available from http://www.world-stress-map.org/.]

Todd, D.K. and Mays, L.W., 2005, *Groundwater Hydrology*, 3rd edn. Hoboken, NJ: John Wiley and Sons.

Tolman, C.F., 1937, *Ground Water*. New York, NY: McGraw-Hill.

Törnqvist, T.E., Bick, S.J., van der Borg, K. and de Jong, A.F.M., 2006, How stable is the Mississippi Delta? *Geology* 34(8): 697–700.

Tóth, J., 2009, *Gravitational Systems of Groundwater Flow*. Cambridge, UK: Cambridge University Press.

Treiman, J.A., 2010, Fault rupture and surface deformation. *Environmental and Engineering Geoscience* 16(1): 19–30.

Tremblay, L., Lefebvre, R., Paradis, D. and Gloaguen, E., 2014, Conceptual model of leachate migration in a granular aquifer derived from the integration of multi-source characterization data. *Hydrogeology Journal* 22: 587–608.

Trenberth, K.E., 2005, The impact of climate change and variability on heavy precipitation, floods and droughts. In *Encyclopedia of Hydrological Sciences* (ed. M.G. Anderson). New York, NY: John Wiley and Sons.

Trenberth, K.E., Dai, A., Rasmussen, R.M. and Parsons, D.B., 2003, The changing character of precipitation. *Bulletin of the American Meteorological Society* 84: 1205–1218.

Trenberth, K.E., Smith, L., Qian, T., Dai, Q. and Fasullo, J., 2007, Estimates of the global water budget and its annual cycle using observational and model data. *Journal of Hydrometeorology* 8: 758–769.

Trenberth, K.E., Fasullo, J.T. and Kiehl, J.T., 2009, Earth's global energy budget. *Bulletin of the American Meteorological Society* 90: 311–324.

Turcotte, D.L. and Schubert, G., 2002, *Geodynamics*, 2nd edn. Cambridge UK: Cambridge University Press.

Turekian, K.K., 1972, *Chemistry of the Earth*. New York, NY: Holt, Rinehart and Winston.

Turka, R.J. and Gray, R.E., 2005, Impacts of coal mining. In *Humans as Geologic Agents* (eds J. Ehlen, W.C. Hanneberg and R.A.

Larson). Reviews in Engineering Geology XVI, pp. 79–86. Boulder, CO: Geological Society of America.

Turner, A.K. and Schuster, R.L. (eds), 1996, *Landslides: Investigation and Mitigation*. Special Report 247, Transportation Research Board, National Research Council. Washington, DC: National Academy Press.

US Army Corps of Engineers, 1984, *Shore Protection Manual*, 4th edn. Vicksburg, MS: Waterways Experiment Station, Coastal Engineering Research Center.

US Bureau of Reclamation, 1998, *Engineering Geology Field Manual*, 2nd edn. Washington, DC: US Bureau of Reclamation.

US Geological Survey, 1997, *The Severity of an Earthquake*, USGS Report 1997-421-530. Washington, DC: US Government Printing Office. [Available from https://pubs.usgs.gov/gip/earthq4/severitygip.html.]

US Geological Survey, 2010, *Divisions of Geologic Time – Major Chronostratigraphic and Geochronologic Units*. USGS Fact Sheet 2010-3059. Washington, DC: US Geological Survey.

US National Academy of Sciences, 1996, *Alluvial Fan Flooding*. Washington, DC: National Academy Press.

US National Academy of Sciences, 2014, *Reducing Coastal Risk on the East and Gulf Coasts*. Washington, DC: National Academies Press.

US National Hurricane Center, NOAA, 2012, Saffir–Simpson hurricane wind scale. [Available from https://www.nhc.noaa.gov/aboutsshws.php/.]

Vanapalli, S.K., 2009, Shear strength of unsaturated soils and its applications in geotechnical practice. In *Unsaturated Soils: Experimental Studies in Unsaturated Soils and Expansive Soils* (eds O. Buzzi, S. Fityus and D. Sheng), pp. 579–598. Leiden, The Netherlands: CRC Press and Balkema.

van der Kamp, G., 2001, Methods for determining the in situ hydraulic conductivity of shallow aquitards – an overview. *Hydrogeology Journal* 9: 5–16.

van der Kamp, G. and Maathuis, H., 2012, The unusual and large drawdown response of buried-valley aquifers to pumping. *Groundwater* 50(2): 207–215.

van der Pluijm, B.A and Marshak, S., 1997, *Earth Structure: An Introduction to Structural Geology and Tectonics*. New York, NY: WCB/McGraw-Hill.

Van der Wateren, F.M., 2005, Ice-marginal terrestrial landsystems: southern Scandinavian ice sheet margin. In *Glacial Landsystems* (ed. D.J.A. Evans), pp. 166–203. London, UK: Hodder Arnold.

van Everdingen, R.O., 1987, The importance of permafrost in the hydrological regime. In *Canadian Aquatic Resources, Canadian Bulletin of Fisheries and Aquatic Sciences 215* (eds M.C. Healy and R.R. Wallace), pp. 243–276. Ottawa, ON: Department of Fisheries and Oceans.

Vargas, C. and Ortega-Guerrero, A., 2004, Fracture hydraulic conductivity in the Mexico City clayey aquitard: field piezometer rising-head tests. *Hydrogeology Journal* 12: 336–344.

Varnes, D.J. and Savage, W.Z. (eds), 1996, *The Slumgullion Earth Flow: A Large-Scale Natural Laboratory*. US Geological Survey Bulletin 2130. Washington, DC: US Government Printing Office.

Verhoogen, J., Turner, F.J., Wiess, L.E., Wahrhaftig, C. and Fyfe, W.S., 1970, *The Earth: An Introduction to Physical Geology*. New York, NY: Holt, Rinehart and Winston.

Wahlstrom, E.E., 1974, *Dams, Dam Foundations and Reservoir Sites*. Developments in Geotechnical Engineering, vol. 6. New York, NY: Elsevier.

Wald, J.A., Graham, R.C. and Schoeneberger, P.J., 2013, Distribution and properties of soft weathered bedrock at ≤1 m depth in the contiguous United States. *Earth Surface Processes and Landforms* 38: 614–626.

Walker, A.S., 2000, *Deserts: Geology and Resources*. Denver, CO: US Geological Survey.

Walker, J.D. and Cohen, H.A., 2009, *The Geoscience Handbook: AGI Data Sheets*, 4th edn, rev. Alexandria, VA: American Geological Institute.

Wallace, R.E. (ed.), 1990, *The San Andreas Fault System, California*. Professional Paper 1515. Washington, DC: US Geological Survey.

Wallach, J.L., Mohajer, A.A. and Thomas, R.L., 1998, Linear zones, seismicity, and the possibility of a major earthquake in the intraplate western Lake Ontario area of eastern North America. *Canadian Journal of Earth Sciences* 35: 762–786.

Waltham, A.C. and Fookes, P.G., 2003, Engineering classification of karst ground conditions. *Quarterly Journal of Engineering Geology and Hydrogeology* 36: 101–118.

Waltham, T., Bell, F. and Culshaw, M., 2005, *Sinkholes and Subsidence: Karst and Cavernous Rocks in Engineering and Construction*. Chichester, UK: Springer-Praxis.

Wang, B. and Saad, B., 2007, In-situ measurements of ground response to heat penetration induced by removal of organic cover in fine-grained permafrost soils. In *1st North American Landslide Conf.*, Vail, Colorado (eds V.R. Schaefer, R.L. Schuster and A.K. Turner), pp. 1595–1604.

Wang, B., Paudel, B. and Li, H., 2009, Retrogression characteristics of landslides in fine-grained permafrost soils, Mackenzie Valley, Canada. *Landslides* 6: 121.

Wang, H.F., 2000, *Theory of Linear Poroelasticity with Applications to Geomechanics and Hydrogeology*, Princeton, NJ: Princeton University Press.

Wartman, J., Montgomery, D.R., Anderson, S.A., et al., 2016, The 22 March 2014 Oso landslide, Washington, USA. *Geomorphology* 253: 275–288.

Waterloo Centre for Groundwater Research, 1994, *Subsurface Assessment Handbook for Contaminated Sites*. Report for Canadian Council of Ministers of the Environment, CCME-EPC-NCSRP-48E, Winnipeg, Manitoba.

Watkins, J.S. and Spieker, A.M., 1971, *Seismic Refraction Survey of Pleistocene Drainage Channels in the Lower Great Miami River Valley, Ohio*. US Geological Survey Professional Paper 605-B. Washington, DC: US Government Printing Office.

Wdowinski, S., Bray, R., Kirtman, B.P. and Wu, Z., 2016, Increasing flooding hazard in coastal communities due to rising sea level: case study of Miami Beach, Florida. *Ocean and Coastal Management* 126: 1–8.

Wells, D.L. and Coppersmith, K.J., 1994, New empirical relationships among magnitude, rupture length, rupture width, rupture area and surface displacement. *Bulletin of the Seismological Society of America* 84(4): 974–1002.

Wenk, H.-R. and Bulakh, A., 2004, *Minerals: Their Constitution and Origin*. Cambridge, UK: Cambridge University Press.

West, A.C.F., Novakowski, K.S. and Gazor, S., 2005, Usefulness of core logging for the identification of conductive fractures in bedrock. *Water Resources Research* 41, W03018.

Wilson, J.L., Conrad, S.H., Mason, W.R., Peplinski, W. and Hagan, E., 1990, *Laboratory Investigation of Residual Liquid Organics from Spills, Leaks, and the Disposal of Hazardous Wastes in Groundwater*. EPA/600/6-90/004. Ada, OK: US Environmental Protection Agency.

Wilson, J.T., Mandell, W.A., Paillet, F.L., et al., 2001, *An Evaluation of Borehole Flowmeters Used to Measure Horizontal Ground-Water Flow in Limestones of Indiana, Kentucky and Tennessee, 1999*. Water-Resources Investigations Report 01-4139. Indianapolis, IN: US Geological Survey.

Witherspoon, P.A., Wang, J.S.Y., Iwai, K. and Gale, J.E., 1980, Validity of the cubic law for fluid flow in a deformable rock fracture. *Water Resources Research* 16(6): 1016–1024.

Wohl, E.E. and Cenderelli, D.A., 2000, Sediment deposition and transport patterns following a reservoir sediment release. *Water Resources Research* 36(1): 319–333.

Woodhouse, C.A., Meko, D.M., MacDonald, G.M., Stahle, D.W. and Cook, E.R., 2010, A 1200-year perspective of 21st century drought in southwestern North America. *Proceedings of the National Academy of Sciences* 107(50): 21 283–21 288.

Woodworth-Lynas, C.M.T., Josenhans, H.W., Barrie, J.V., Lewis, C.F.M. and Parrott, D.R., 1991, The physical processes of seabed disturbance during iceberg grounding and scouring. *Continental Shelf Research* 11(8–10): 939–961.

Worden, C.B. and Wald, D.J., 2016, *ShakeMap Manual Online: Technical Manual, User's Guide, and Software Guide*. US Geological Survey. [Available from usgs.github.io/shakemap.]

Worthington, S.R.H., 2009, Diagnostic hydrogeologic characteristics of a karst aquifer. *Hydrogeology Journal* 17: 1665–1678.

Worthington, S.R.H. and Ford, D.C., 2009, Self-organized permeability in carbonate aquifers. *Ground Water* 47(3): 326–336.

Worthington, S.R.H. and Smart, C.C., 2017, Transient bacterial contamination of the dual-porosity aquifer at Walkerton, Ontario, Canada. *Hydrogeology Journal* 25: 1003–1016.

Wright, M., Dillon, P., Pavelic, P., Peter, P. and Nefiodovas, A., 2002, Measurement of 3-D hydraulic conductivity in aquifer cores at in situ effective stresses. *Ground Water* 40(5): 509–517.

Wyllie, D.C., 1999, *Foundations on Rock*, 2nd edn. London, UK: Spon, Taylor and Francis.

Wyllie, D.C., 2018, *Rock Slope Engineering: Civil Applications*, 5th edn. Boca Raton, FL: CRC Press.

Wyllie, D.C. and Norrish, N.I., 1996, Rock strength properties and their measurement. In *Landslides: Investigation and Mitigation* (eds A.K. Turner and R.L. Schuster). Special Report 247, ch. 14,

pp. 372–390. Washington, DC: Transportation Research Board, National Research Council.

Xu, G., Sun, Y., Wang, X., Hu, G. and Song, Y., 2009, Wave-induced shallow slides and their features on the subaqueous Yellow River delta. *Canadian Geotechnical Journal* 46(12): 1406–1417.

Yang, S.L., Zhang, J. and Xu, X.J., 2007, Influence of the Three Gorges Dam on downstream delivery of sediment and its environmental implications, Yangtze River. *Geophysical Research Letters* 34: L10401.

Yeats, R., 2012, *Active Faults of the World*. Cambridge, UK: Cambridge University Press.

Youd, T.L. and Perkins, D.M., 1978, Mapping liquefaction-induced ground failure potential. *Journal of the Soil Mechanics and Foundations Division* 104(4): 433–446.

Younger, P.L., 2002, The importance of pyritic roof strata in aquatic pollutant release from abandoned mines in a major, oolitic, berthierine–chamosite–siderite iron ore field, Cleveland, UK. In *Mine Water Hydrogeology and Geochemistry* (eds P.L. Younger and N.S. Robins). Special Publication 198, pp. 251-266. London, UK: Geological Society of London.

Yuhr, L.B., 2009, Site characterization in karst using surface geophysics. *Geo-Strata* 10(2): 34–38.

Zangerl, C., Eberhardt, E. and Loew, S., 2003, Ground settlements above tunnels in fractured crystalline rock: numerical analysis of coupled hydromechanical mechanisms. *Hydrogeology Journal* 11: 162–173.

Zapico, M.M., Vales, S.E. and Cherry, J.A., 1987, A wireline piston core barrel for sampling cohesionless sand and gravel below the water table, *Ground Water Monitoring Review* 7(3): 74–82.

Zhang, L.L., Fredlund, D.G., Zhang, L.M. and Tang, W.H., 2004, Numerical study of soil conditions under which matric suction can be maintained. *Canadian Geotechnical Journal* 41: 569–582.

Zhu, C. and Anderson, G., 2002, *Environmental Applications of Geochemical Modeling*. Cambridge, UK: Cambridge University Press.

Zimbelman, D., Watters, R.J., Bowman, S. and Firth, I., 2003, Quantifying hazards and risk assessments at active volcanoes. *EOS, Transactions, American Geophysical Union* 84(23): 213, 216–217.

Zimbelman, D., Watters, R.J., Firth, I., Breit, G.N. and Carrasco-Nunez, G., 2004, Stratovolcano stability assessment methods and results from Citlaltépetl, Mexico. *Bulletin of Volcanology* 66: 66–79.

Ziony, J.I. and Yerkes, R.F., 1985, Evaluating earthquake and surface-faulting potential. In *Evaluating Earthquake Hazards in the Los Angeles Region – An Earth-Science Perspective* (ed. J.I. Ziony). US Geological Survey Professional Paper 1360. Washington, DC: US Geological Survey.

Zoback, M.D., 2010, *Reservoir Geomechanics*. Cambridge, UK: Cambridge University Press.

Zoback, M.L., 2006, The 1906 earthquake and a century of progress in understanding earthquakes and their hazards. *GSA Today* 16(4/5): 4–11.

INDEX

ablation 175
abrasion (glacial) 181
acid-rock or acid-mine drainage 4, 33, 164
active fault 337
active layer (of permafrost) 190
activity (ionic) 155
adsorption edges 162
advection (of groundwater) 280
advection–diffusion process 282
aggrading (stream) 139
aggregate 30, 51, 101–106
air-entry value 370
alkalinity 157–158
alluvial deposits 204–210
alluvial fans 210–212
amphibole 32
Anderson's theory of faulting 60
andesite 41
angle of internal friction 12, 373
angular unconformity 66
anhydrite 33
annual exceedance probability 142
aquifer 245, 257–259, 266–271
    alluvial 269
    basin-fill 303
    bedrock 269–271
    vulnerability 297
    water table vs. confined 256
aquitard (integrity, drainage) 245, 271–272,
    304–308
architectural elements 204
arenite 44
argillaceous rocks 44
arkose 44
arsenopyrite 33
asthenosphere 22
augers 233
avulsion 140

b-value curve 335
back-diffusion 282
bars (gravel) 204
basal slip 178
basalt 42
baseflow 126, 265
basement 28
base-metal ore deposits 164
batholiths 38
Becker drill 222

bed 53
bedding-plane fractures 86, 270
bedforms 204
bed load and material 139, 197
bed shear stress 137
biochemical sediments 38
biodegradation reaction 292–293
bioturbation 214, 250
Bishop's modified method 367–368
borehole breakouts 78
borehole geophysics 99, 231
bottom-set beds 212
breccia 42, 60, 62
bridge scour 200
brittle deformation 84
Byerlee's law 337

cable-tool drilling 222
calcite 33
calcrete (or caliche) 131, 317
cap rock 47
capillary barrier 284
capillary fringe 261
carbonate minerals 33, 151
cataclastic flow 84, 336
celerity 385–386
cementation 216
chalk 48
channel (alluvial) 207
chemical grains and sediments 38, 231
chert 47
chimneys 313
circular failure of rock slopes 95
clastic dike 349
clasts and clastic sediments 38, 231
clay minerals 32, 167–173
clay weathering 134
cleavage 31, 36
climate change 301
clinometer 57
coal and coalification 48
coastal cliffs and bluffs 193, 399–405
cohesion 12, 15, 373
compaction (mechanical) 215, 304
compass 57
competence 138, 199
compressibility 251–252
compressive strength 88
connate water 277

cone penetrometer testing (CPT) 227–228, 295,
    366
conglomerate 42
congruent dissolution 151
consolidation 30, 240–241
contact 56 (between strata), 127 (lithic)
contact metamorphism 39, 48
core (of the Earth) 21
core disking 78
core logs 101–103 (rock), 226–227 (soil)
corestones 127
corrosion mixing 317
coseismic displacement 329, 359, 398
creeps 356
crest (of an anticline) 56
critically stressed faults 337
cross-bedding 44
crown holes 313
crust (of the Earth) 21
cubic law 91, 108
cut bank 206
cutting and filling 207
cyclic stress ratio (CSR) 350
cyclothem 48

debris or debris-avalanche flow (landslide) 133,
    356, 358
deformation modulus 83
degrading (stream) 139
deltaic deposits 212–214
DEM (digital elevation models) 144
detrital sediments 38, 231
diabase 41
diagenesis 42, 214–216
diamicton 184
diffusion 280
diffusional profiling 277
dikes 38
diorite 41
dip (true and apparent) 55
disconformity 67
discontinuities 6, 7, 86, 108
disequilibrium compaction 81
dispersal trains 183
dispersion 280
dissolution (during diagenesis) 216
diurnal tide 388
DNAPLs (dense non-aqueous-phase liquids)
    283

DO (dissolved oxygen) 159
dolerite 41
dolomite 33, 44
downstream accretion 207
drawdown cone 257
drift (glacial) 184
drill cuttings 222
drilling mud 100
dual porosity 249
ductile deformation 86

earthflow (landslide slope material) 133, 356
effective pressure 179
effective solubility 290
effective stress 15, 305, 348
elastic dislocation model 330
elastic rebound model 329
electroneutrality 151
eluviation 154
EM (electromagnetic surveys) 242, 295, 324
entry pressure (for NAPLs) 284
environmental isotopes 277–280
eolian deflation 131
eolian deposits 184
epicentre 331
epikarst 320
equilibrium reactions 150
equivalent rock mass 120
erodibility 200
ERT (electrical resistance tomography or imaging) 99, 229, 242, 295, 324
evaporite 33, 47
extensometers 309, 311
extrusive rock 38

facies 204
factor of safety 368
failure envelope 12
fault block 60
fault zone 60
faults 6, 54, 60, 86
feldspars 32, 151–152
felsic rocks 38
ferric oxides 33, 152
fissility 44
fissures 64, 87, 296, 308–309
flint 47
fluid potential 251
flysch 45
focus or hypocentre 331
folds 6, 54
foliation 48
foraminifera (forams) 176
foreset-beds 212
formation 54
fossilized stresses 76
fracking (see hydraulic fracturing)
fractionation 177
fracture (of mineral grains) 36

fracture pressure 115
fractures 6, 64, 102 (roughness), 369–372 (re rock mass strength)
free phase (NAPL) 283
free swell testing 113
frost cracking 130
frost heave 191
frozen debris lobes 194
fugitive gas 290

gabbro 41
gangue minerals 166
gas hydrate 406
geological map 55, 67
geological strength index (GSI) 110–112
geomechanics classification system (rock-mass rating RMR) 109
geomodel 220
geomorphic hazards 175
geophysical processes 9
georadar see GPR 99, 242
geotechnical scour number (m/(W/m)) 202
glacial isostatic adjustment (GIA) 26–27, 399
glaciofluvial deposits 184, 187–190
glaciolacustrine deposits 184
glaciomarine deposits 184, 186–187
glaciotectonic deformation 178–180
global meteoric water line 177
gneiss 49
GNSS (global navigation satellite system) see GPS 117, 310
GPR (ground-penetrating radar) 99, 229, 295, 324
GPS (global positioning system) 117
graded (texture) 43
granite 40
granodiorite 40
granular flow 84, 336
gravel bar 207
gravity (geophysical) method 99
greywacke 44
ground model 220
ground motion 341–344
groundwater contamination 280–294
groundwater discharge 126
groundwater flow models 278
groundwater flow system 126, 258–266
groundwater recharge 126
groundwater velocity 249
groundwater vulnerability 278
group (of formations) 54
GSI (geological strength index) 110–112, 373
GWQ (groundwater quality) 295
gypsum 33, 151

halite 33, 151
hardness 36, 106
hard rocks 29
hiatus 65
hornfels 34–35
hue (soil colour) 154

humus 154
hydraulic conductivity 15, 238 (lab), 248–249; saturated vs. unsaturated 250, 261
hydraulic diffusivity 255
hydraulic fracturing 75, 77
hydraulic geometry 138
hydraulic gradient 17
hydraulic head 16, 17
    errors in measurement 252
    in variable TDS waters 263
hydraulic testing 113–115 (rocks), 256–258
hydrochemical facies 275
hydrofacies 210
hydrophilic compounds 292
hydrophobic compounds 292
hydrostratigraphy 265
hydrothermal brine 50
hydrothermal ore deposits 164
hyperconcentrated flow 358
hypogenic karst 320
hysteresis (soil-water) 261

ion-activity product 156
iceberg scouring 403–405
igneous rocks 5
ignimbrites 41
immiscible contaminants 280
inclinometers 363, 365
index tests 106
INSAR (interferometric synthetic aperture radar) 310
intact rock 6
interfacial tension 284
interglacial event 176
interseismic elastic strain 329
interstadial event 176
intraplate earthquakes 328, 351
intrusive rock 38
ion activity product 156
ionic strength 155
ironstone 47
irreversible reactions 150
isomorphous substitution 173
isostasy 26

jökulhlaups 183
joint 64, 87, 136 (relict)

kaolinite 152, 168, 173 (photo)
kaolinization 40
karst 316–325
known unknowns 98

lahars 24, 361–362
laminated sand 207
landfill leachate 293
large igneous provinces 38
Last Glacial Maximum (LGM) 175
lateral accretion 207
lateral spread 349, 356
laterite 40

lateritic weathering profiles 165
law of faunal succession 29
law of mass action 154
law of original continuity 29
law of original horizontality 29
law of superposition 29
LiDAR (light detection and ranging) 116, 144–146
limestone 47, 48
liquefaction 348–352, 358, 406
liquefaction susceptibility 350
lithic grains 45
lithic rocks 60
lithification 39
lithified 29
lithofacies 208, 210
lithosphere 21
LNAPLs (light non-aqueous-phase liquids) 283
load–deformation curve 82–83
longwall mining 313–316

macropores 128, 250
mafic rocks 38, 60
magma 38
magnetic (geophysical) method 99, 295
mantle 21
marl 44
massive ground ice 191
MASW (multi-channel analysis of surface waves) 231, 324
matric suction 358, 366
member 54
metamorphic rocks 5
meteoric water 177
micas 32
micro-granite 40
microgravity 324
microstructure 83
mid-ocean ridges 24
migmatite 49
mineral 30
mineralogy 4
mine water 314
miscible contaminants 280
mixed tide 388
modified Mercalli intensity scale 331, 333
Mohr–Coulomb criterion 10, 13
Moh's scale of hardness 102
moment magnitude 334
MSL (mean sea level) 388
mylonite 50, 62
mudstones 44

NAPLs (non-aqueous phase liquids) 283
natural attenuation 290
neotectonics 398
Newmark displacement 359
nonconformity 67
normal fault 60, 76, 328

obsidian 41
offset beds 60

olivine 32
oolitic limestone 48
ore deposits 50
orthoquartzite 44
outcrop 6, 54
overbank fines 207
overcoring 77
overland flow 126
overpressured systems 81
oxidation reaction 158
oxyanions 162

P (primary) waves 331–332
paleokarst 320
paleoseismology 337–340
paleostage indicator 142
paleotempestology 397
Patton's principle 373
peak horizontal acceleration 341
peatlands 312
pegmatite 40, 102
percussion drilling 100
peridotite 42
periglacial 190, 191–193
permafrost 190–194
permeability 15, 107–108, 249
  secondary 250
petrography 4
petrology 4
petrophysical testing 108
phenocrysts 40
phyllite 49
pillar collapses 313
pitchstone 41
plane failure of rock slopes 94, 403
plate tectonics 23
ploughing (glacial) 180
plucking (hydraulic) 202
plunge (of a fold) 56
plutonism 50
point bars 206
Poisson effect 82
Poisson's ratio 14, 79
point-load strength testing 85, 106–107
pop-ups 78
pore ice 191
pore pressure 14–15, 305, 350, 366
porosity (total) 15, 248
  effective 108, 249
preconsolidation stress 306
predisposing factors (landslides) 356
principal stress 11
probable maximum flood 143
provenance (origin) 231, 277
pyrite 33, 152, 166–167, 170 (oxidation)
pyroxene 32

Q (or NGI) system 109
quarrying (glacial or hydraulic) 181, 202

quartz 32
Quaternary period 29, 175
quick clays 187

radioisotopes 278–280
radiometric methods 21
rainout 177
Raoult's law 290
recurrence interval (of an earthquake)
red beds 33, 270
redox boundary 277
redox reaction 153, 158–161
reduction reaction 161
regional metamorphism 39, 48
regolith 127, 172
relative permeability 285
relict structural features 133
reopening pressure 116
reserves and resources 165
residual saturation 283, 370
REV (representative elementary volume) 8
reverse fault 60, 77
reworking (by bioturbation) 214
rhyolite 40, 41
rippability 44, 99
rock fall 93, 356, 373–374
rock-forming minerals 38
rockhead (top of rock) 40, 49, 127, 222
rock mass 6, 75, 83
rock-mass rating (RMR) 109
rock material 75
rock quality designation (RQD) index 101
rotary drilling 100, 223
rotational slide 356
runout distance 358, 406

S (shear) waves 331–332
Saffir–Simpson hurricane wind scale 381, 392
sand bed form 207
sand boils 349
sandstone 43
saprock 127
saprolite 127, 172
saturation index 156
scale (of a map) 67
scale effect (of dispersion) 281–282
scan line 108
scarp, main 133, 354
schist and schistosity 49
scour, critical 200
seafloor spreading 24
sea-level rise (SLR) 389
seawater intrusion 294
sediment replenishment
sediment gravity flow 207
sedimentary rocks 5
sedimentation in reservoirs 202–203
segregation ice 191
seismic faulting 327

seismic hazard analysis 343–345
seismic moment 333–334
seismic refraction and reflection 99, 230, 242, 295, 324–325
seismogenic zone 336
seismology 23
semi-diurnal tide 388
sensitivity (of clays) 187
serpentinite 42
settling velocity 198
shale 44, 134 (weathering)
shear fractures 87
shear strength 15, 89, 107, 239, 366, (of discontinuities) 372–373
shear testing 86, 239 (lab)
shear zone 60
Shields parameter 198
siderite 33
silicate minerals 31, 152
sills 38
siltstone 44
sinkholes 319–323
site characterization model 220–222
slackwater flood deposits 142
slake durability testing 106
slaking 44
slate 49
slickensides 60, 102
slip (of a fault) 60
slip or failure surface of a landslide 133, 354
slip rate (of a fault) 340
slug tests 256
smear zone 287
smectite 168
softcopy photogrammetry 146
soft rocks 29
soil 127
soil behavior type index (for CPT) 228
soil-water characteristic curve (SWCC) 264, 369–370
solifluction 131
solubility (mineral) 155–156
soluble rock 316
solutes 154
sonic drilling 223
sorption, sorbent, sorbate 161–164
sorting 15, 43, 210, 232
specific discharge 16
specific retention 254
specific storage 252–253, 305–306
specific yield 254
spherical projection 72

standard penetration test 222
stereogram 72–73
stereographic projection 72
stick–slip process 330, 336
stiffness 13, 82
stochastic method (prediction of ground motion) 343–345
stockwork 136
storativity 254–258
storm surge 392–397
strain 83
strata (sing. stratum) 28, 53
strata-bound ore deposits 165
stream power 138
strength 11
strike 55
strike-slip fault 60, 77, 327
structural geology 5, 23
subduction 24
subglacial deposits 178, 184
subsidence 305 (land), 398–399 (coastal)
subsidence troughs (or bowls) 310, 314
subsurface storm flow 126
supergene ore deposits 165
supraglacial deposits 184
surface of rupture 133
surface waves (seismic) 331, 332
suspended load and suspended bed-material load 197
swash 394
syenite 40

taliks 191
tectonic forces 6
tectonics 23
tensile failure 84
tensile strength 88, 107, 116
terrain analysis 143–147
terrestrial laser scanning 116
terrigeneous sediments 231
thaw flows 193
thermokarst 190, 193
thrust fault 328, 337
tidal range and period 388
till 184–186
toe (of landslide) 133, 354
toppling failure of rock slopes 95, 356, 401–403
top-set beds 212
TDS (total dissolved solids) 155
transform fault 24
translational slide 356
transmissivity 115, 253–258
travertine 44

triaxial compression test 86
triggering mechanisms (landslides) 356–362
trough (of a syncline) 56
tsunami 397–398
tuff 40
tunnel support 110
turbidites 45
turbidity current 406–407

Udden–Wentworth scale 232–233
ultrasonic pulse velocities 106–107
unconfined compression test 86, 88
unconformity 65–66
underpressured systems 81
uniaxial compressive strength 88
Unified Engineering Geology Mapping System (UEGMS) 67
uniformitarian principle 67
unknown unknowns 98
Unified Soil Classification System (USCS) 226–227

vadose zone 261
varves 184
velocity (groundwater) 249
vermiculite 168
vesicle 102
volcanic rocks and hazards 24, 136
vugs 36

wall (of a fault) 60
wash material or load 139, 197
water–rock interaction 277
wave energy and wave power 387
wave runup and setup 394
weak layer (as cause of submarine landslide) 406
weak rocks 45, 106–107, 112–113, 401
weathering 6, 21, 127–136, 150–154, (chemical), 186
wedge failure of rock slopes 89, 95, 356, 403
well completion 80 (oilfield)
wellhead protection areas 297
wells (water) 255–256
wettability 283
Wisconsin glacial stage 175
whole rock analysis 37

yield strength (of debris flow) 358–360
yielding (stress-induced) 89
Young's modulus 13, 79

zeolites 46